IC

CROSS REFERENCE

BOOK

IC

CROSS REFERENCE

BOOK

FROM THE ENGINEERS OF
HOWARD W. SAMS & COMPANY

PROMPT® PUBLICATIONS

THIRD EDITION, 1998

PROMPT© Publications is an imprint of Howard W. Sams & Company, A Bell Atlantic Company, 2647 Waterfront Parkway, E. Dr., Indianapolis, IN 46214-2041.

International Standard Book Number: 0-7906-1141-4

Acquisitions Editor: Candace M. Hall
Editor: Loretta L. Yates
Assistant Editors: Pat Brady, Natalie Harris
Compilation: Barry Buchanan, William Skinner
Typesetting: Loretta Yates
Cover Design: Christy Peirce

PRINTED IN THE UNITED STATES OF AMERICA

9 8 7 6 5 4 3 2 1

TABLE OF CONTENTS

How to Use This Book

To find a replacement or substitution for an IC or module, follow these simple steps:

■Locate the part number stamped on the part you wish to replace, or find the part number in a parts list.

■Look up the part or type number in Section 1 of this guide. To the right of the number will be a replacement code/block number.

■Look up the replacement code/block number in Section 2 of this guide. The ICs listed are pin-for-pin compatible to each other. (Use insulating hardware and/or heat sinks if necessary. Observe all other antistatic and safety precautions.

Other Important Information

The engineering staff of Howard W. Sams & Company has assembled this cross reference guide to help you find replacements or substitutions for more than 35,000 ICs or modules.

It has been compiled from manufacturers' data and from the analysis of consumer electronics devices for PHOTOFACT® service data, which has been relied upon since 1946 by service technicians worldwide. Here are some important notes to help you use the two sections of the book effectively.

Section 1:
Original IC or Module Part or Type Numbers

This section lists IC or module part numbers in alphanumeric order by manufacturer's part number, type number, or other identification, including numbers from the U.S., Europe and the Far East. At the right of the part number is a replacement code/block number that you will use to look up comparable replacements in Section 2.

Section 2:
Replacements

This section provides substitutions and replacements for the ICs and modules listed in Section 1. This information was developed during PHOTOFACT® analysis and is the most complete replacement guide available.

SECTION 1

ORIGINAL IC

OR

MODULE PART OR TYPE NUMBERS

ORIGINAL DEVICE TYPES AND REPLACEMENT CODES

DEVICE TYPE	REPL CODE
00001201	BLOCK-4
000074010	BLOCK-7
000074020	BLOCK-43
000074030	BLOCK-138
0000LA3301	BLOCK-7
0000LD3000	BLOCK-15
000HA1306U	BLOCK-28
000LA1201B	BLOCK-4
000LA1306U	BLOCK-28
000LD1020A	BLOCK-12
001-0020-02	BLOCK-85
001-0036	BLOCK-1277
001-0091	BLOCK-136
0019-1313	BLOCK-398
002	BLOCK-1265
0031015012	BLOCK-1269
003501	BLOCK-4
003515	BLOCK-1825
003516	BLOCK-205
003519	BLOCK-223
003522	BLOCK-50
003526	BLOCK-159
003536	BLOCK-182
003542	BLOCK-1296
004	BLOCK-96
004(TOYOTA)	BLOCK-96
004795	BLOCK-1282
006-0000146	BLOCK-1293
006-0000147	BLOCK-1304
006-0000151	BLOCK-1409
006-0000162	BLOCK-1305
006-0004779	BLOCK-1956
006-0005545	BLOCK-1961
0061-232-70001	BLOCK-553
0061-238-00001	BLOCK-1093
0061-247-90001	BLOCK-2087
0061-257-40001	BLOCK-778
0061-258-60003	BLOCK-757
0061-258-60004	BLOCK-758
0061-262-50001	BLOCK-754
0061-920-01330	BLOCK-770
0061-920-60090	BLOCK-660
0061-969-90130	BLOCK-2144
007-009-00	BLOCK-1282
007-0150-00	BLOCK-1846
007-1671101	BLOCK-1963
007-1681301	BLOCK-1961
007-1695001	BLOCK-1293
007-1695101	BLOCK-1365
007-1695301	BLOCK-1297
007-1695701	BLOCK-1390
007-1695901	BLOCK-1304
007-1696001	BLOCK-1418
007-1696101	BLOCK-1320
007-1696201	BLOCK-1295
007-1696301	BLOCK-1384
007-1696701	BLOCK-1307
007-1696801	BLOCK-1325
007-1696901	BLOCK-1299
007-1697701	BLOCK-1352
007-1697801	BLOCK-1391
007-1698301	BLOCK-1357
007-1698401	BLOCK-1358
007-1698901	BLOCK-1401
007-1699201	BLOCK-1307
007-1699301	BLOCK-1301
007-1699401	BLOCK-1296
007-1699801	BLOCK-1411
0070-0120	BLOCK-265
0070-0210	BLOCK-603
0070-0500	BLOCK-732
0071-0030	BLOCK-198
0073-0020	BLOCK-1157
0073-0030	BLOCK-550
0073-0180	BLOCK-401
0079-0060	BLOCK-235
0079-0150	BLOCK-607
0079-0160	BLOCK-473
0079-0170	BLOCK-223
0079-0220	BLOCK-684
0079-0500	BLOCK-532
0079-0510	BLOCK-525
0079-0780	BLOCK-404
0079-1313	BLOCK-398
0098-0710	BLOCK-1002
00HA1306PU	BLOCK-28
00S08	BLOCK-221
01-119185-01	BLOCK-1303
01-119185-02	BLOCK-1303
01-119185-03	BLOCK-1303
01-121365	BLOCK-1269
013-005005	BLOCK-1791
013-005005-6	BLOCK-1791
013-005007	BLOCK-1791
013-005007-6	BLOCK-1791
013-005009-6	BLOCK-1794
015040/7	BLOCK-1297
02-004558	BLOCK-1801
02-010103	BLOCK-180
02-010241	BLOCK-151
02-010245	BLOCK-153
02-010331	BLOCK-157
02-010612	BLOCK-238
02-091128	BLOCK-1269
02-091366	BLOCK-248
02-103898	BLOCK-974
02-121201	BLOCK-4
02-121365	BLOCK-1269
02-124400	BLOCK-181
02-124422	BLOCK-145
02-151204	BLOCK-422
02-161204	BLOCK-422
02-161521	BLOCK-159
02-165143	BLOCK-1269
02-173710	BLOCK-182
02-173756	BLOCK-258
02-217660	BLOCK-1405
02-235104	BLOCK-244
02-252222	BLOCK-265
02-257130	BLOCK-223
02-257205	BLOCK-145
02-257222	BLOCK-265
02-257310	BLOCK-180
02-263001	BLOCK-180
02-280007	BLOCK-249
02-300023	BLOCK-44
02-300574	BLOCK-1157
02-300577	BLOCK-80
02-301181	BLOCK-272
02-301182	BLOCK-273
02-301380	BLOCK-184
02-310023	BLOCK-44
02-311380	BLOCK-184
02-341125	BLOCK-1269
02-341128	BLOCK-1269
02-341358	BLOCK-1269
02-343065	BLOCK-1269
02-360002	BLOCK-156
02-360003	BLOCK-259
02-373001	BLOCK-180
02-392816	BLOCK-515
02-403065	BLOCK-1269
02-435624	BLOCK-242
02-437205	BLOCK-145
02-437310	BLOCK-180
02-455804	BLOCK-1801
02-507120	BLOCK-523
02-537666	BLOCK-1269
02-540800	BLOCK-114
02-561171	BLOCK-1816
02-561201	BLOCK-1808
02-561251	BLOCK-1843
02-561410	BLOCK-1271
02-781050	BLOCK-2144
02-781060	BLOCK-2173
020-1114-006	BLOCK-1275
020-1114-007	BLOCK-1212
020-1114-008	BLOCK-1212
020-1114-009	BLOCK-1277
020-1114-016	BLOCK-1963
0201201	BLOCK-4
0207046	BLOCK-106
0207120	BLOCK-85
0207205	BLOCK-145
022-2844-001	BLOCK-158
022-2844-002	BLOCK-1212
022-2844-501	BLOCK-2173
022-2844-701	BLOCK-2173
022-2844-702	BLOCK-1859
022-2844-703	BLOCK-158
025-1	BLOCK-1286
025B1C	BLOCK-1286
036001	BLOCK-1212
03SI-MC1352P	BLOCK-1422
0404011-001	BLOCK-2203
0404012-001	BLOCK-2210
051-0010-00	BLOCK-22
051-0011-00	BLOCK-101
051-0011-00-04	BLOCK-101
051-0011-00-05	BLOCK-105
051-0011-04	BLOCK-550
051-0012-00	BLOCK-1277
051-0012-11	BLOCK-1277
051-0020-00	BLOCK-85
051-0020-02	BLOCK-85
051-0021-00	BLOCK-1405
051-0022-00	BLOCK-1264
051-0035-01	BLOCK-85
051-0035-02	BLOCK-85
051-0035-03	BLOCK-85
051-0035-04	BLOCK-85
051-0036-00	BLOCK-36
051-0036-01	BLOCK-36
051-0036-0102	BLOCK-36
051-0036-02	BLOCK-36
051-0036-03	BLOCK-36
051-0038-00	BLOCK-6
051-0039-00	BLOCK-138
051-0050-00	BLOCK-1825
051-0050-01	BLOCK-1825
051-0055-02	BLOCK-145
051-0055-0203	BLOCK-145
051-0055-03	BLOCK-145
051-0066-00	BLOCK-277
051-0068-02	BLOCK-4
051-0086-00	BLOCK-214
051-0087-00	BLOCK-243
051-0088-00	BLOCK-205
051-0088-00101	BLOCK-205
051-0100-00	BLOCK-156
058-001138	BLOCK-1048
05B2B	BLOCK-1269
05B2D	BLOCK-1269
05B2M	BLOCK-1269
06300001	BLOCK-162
06300003	BLOCK-153
06300004	BLOCK-184
06300005	BLOCK-44
06300009	BLOCK-1269
06300030	BLOCK-655
06300031	BLOCK-537
06300034	BLOCK-404
06300039	BLOCK-2087
06300049	BLOCK-532
06300050	BLOCK-519
06300051	BLOCK-519
06300052	BLOCK-484
06300056	BLOCK-740
06300058	BLOCK-535
06300066	BLOCK-510
06300069	BLOCK-1269
06300071	BLOCK-396
06300072	BLOCK-740
06300080	BLOCK-459
06300089	BLOCK-2087
06300090	BLOCK-459
06300095	BLOCK-695
06300105	BLOCK-2063
06300107	BLOCK-993
06300111	BLOCK-700
06300113	BLOCK-706
06300115	BLOCK-636
06300116	BLOCK-700
06300120	BLOCK-1269
06300121	BLOCK-644
06300144	BLOCK-2087
06300154	BLOCK-721
06300160	BLOCK-563
06300191	BLOCK-1088
06300192	BLOCK-1088
06300219	BLOCK-2097
06300248	BLOCK-1572
06300311	BLOCK-1859
06300315	BLOCK-1043
06300316	BLOCK-1041
06300317	BLOCK-1049
06300323	BLOCK-1182
06300334	BLOCK-773
06300336	BLOCK-764
06300342	BLOCK-2087
06300357	BLOCK-709
06300359	BLOCK-824
06300360	BLOCK-754
06300365	BLOCK-636
06300394	BLOCK-2063
06300403	BLOCK-818
06300456	BLOCK-839
06300457	BLOCK-840
06300488	BLOCK-1196
0630049	BLOCK-532
06300491	BLOCK-778
06300508	BLOCK-396
06300514	BLOCK-1188
06300519	BLOCK-817
06300546	BLOCK-459
06300554	BLOCK-1087
06300565	BLOCK-1993
06300574	BLOCK-709
06300616	BLOCK-824
06300621	BLOCK-1196
06300629	BLOCK-758
06300656	BLOCK-1629
06300688	BLOCK-723
06300738	BLOCK-1093
06300761	BLOCK-1912
06300763	BLOCK-1191
06300766	BLOCK-671
06300767	BLOCK-762
07-28777-40	BLOCK-121
07-28777-41	BLOCK-1269
07-28777-42	BLOCK-250
07-28777-45	BLOCK-253
07-28777-49	BLOCK-256
07-28777-56	BLOCK-395
07-28778-70	BLOCK-392
07-28778-72	BLOCK-413
07-28778-73	BLOCK-409
07-28778-74	BLOCK-400
07-28778-75	BLOCK-407
07-28778-76	BLOCK-609
07-28778-77	BLOCK-415
07-28778-78	BLOCK-2054
07-28815-100	BLOCK-389
07-28815-101	BLOCK-2085
07-28816-70	BLOCK-416
07-28816-71	BLOCK-387
07-28816-72	BLOCK-410
07-28816-73	BLOCK-211
07-28816-81	BLOCK-1157
07-28816-83	BLOCK-2055
07-28816-84	BLOCK-1048
07-28816-85	BLOCK-1085
07-28816-86	BLOCK-1096
075-045037	BLOCK-1411
075-046270	BLOCK-1409
0783Z	BLOCK-1264
07B27	BLOCK-1262
07B2B	BLOCK-1262
07B2Z	BLOCK-1813
07B3B	BLOCK-1262
07B3M	BLOCK-1262
07B3Z	BLOCK-1262
0801-000961	BLOCK-1088
08200040	BLOCK-422
08200043	BLOCK-2087
08200045	BLOCK-2074
08200050	BLOCK-882
08200064	BLOCK-882
08200069	BLOCK-2144
08200070	BLOCK-827
08200077	BLOCK-827
08200091	BLOCK-2063
08200093	BLOCK-906
08210001	BLOCK-1568
08210002	BLOCK-1572
08210004	BLOCK-1684
08210006	BLOCK-1299
08210008	BLOCK-2079
08210017	BLOCK-1573
08210022	BLOCK-1717
08210023	BLOCK-1593
08210029	BLOCK-1591
08210030	BLOCK-1600
08210032	BLOCK-1574
08210033	BLOCK-1706
08210038	BLOCK-1671
08210039	BLOCK-1602
08210040	BLOCK-1641
08210046	BLOCK-1592
08210047	BLOCK-1771
08210048	BLOCK-1772
08210050	BLOCK-1703
08210052	BLOCK-1672
08210053	BLOCK-1659
08210058	BLOCK-1579
08210059	BLOCK-1584
08210063	BLOCK-1628
08210064	BLOCK-1679
08210068	BLOCK-1666
08210069	BLOCK-1634
08210070	BLOCK-1713
08210071	BLOCK-1598
08210072	BLOCK-1570
08220014	BLOCK-1018
08220027	BLOCK-1004
08220030	BLOCK-1019
08220031	BLOCK-1004
08220033	BLOCK-1019
08220072	BLOCK-1570
08220090	BLOCK-1019
08220092	BLOCK-1873
08SI-UPD4556BC	BLOCK-1147
09-004	BLOCK-1275
09-011	BLOCK-1277
09-308002	BLOCK-1280
09-308003	BLOCK-1280
09-308004	BLOCK-1212
09-308007	BLOCK-1280
09-308008	BLOCK-4
09-308009	BLOCK-44
09-308010	BLOCK-1282
09-308011	BLOCK-52
09-308013	BLOCK-1212
09-308017	BLOCK-1223
09-308019	BLOCK-1212
09-308021	BLOCK-2208
09-308022	BLOCK-1293
09-308024	BLOCK-20
09-308025	BLOCK-74
09-308026	BLOCK-89
09-308027	BLOCK-45
09-308028	BLOCK-75
09-308029	BLOCK-84
09-308030	BLOCK-87
09-308031	BLOCK-46
09-308033	BLOCK-43
09-308034	BLOCK-7
09-308041	BLOCK-98
09-308043	BLOCK-101
09-308044	BLOCK-118
09-308045	BLOCK-1272
09-308046	BLOCK-1272
09-308047	BLOCK-1270
09-308048	BLOCK-1270
09-308049	BLOCK-19
09-308050	BLOCK-3
09-308051	BLOCK-111
09-308052	BLOCK-80
09-308053	BLOCK-73
09-308054	BLOCK-43
09-308059	BLOCK-98
09-308061	BLOCK-95
09-308062	BLOCK-4
09-308063	BLOCK-4
09-308064	BLOCK-7
09-308065	BLOCK-16
09-308066	BLOCK-104
09-308067	BLOCK-105
09-308069	BLOCK-45
09-308070	BLOCK-106
09-308071	BLOCK-138
09-308074	BLOCK-1431
09-308076	BLOCK-50
09-308077	BLOCK-12
09-308079	BLOCK-1271
09-308080	BLOCK-145
09-308083	BLOCK-47
09-308084	BLOCK-125
09-308086	BLOCK-143
09-308089	BLOCK-5
09-308090	BLOCK-1422
09-308094	BLOCK-232
09-308095	BLOCK-85
09-308096	BLOCK-90
09-308098	BLOCK-1269
09-308099	BLOCK-1816
09-308100	BLOCK-1269
09-308102	BLOCK-47
09-308103	BLOCK-444
0900296	BLOCK-1280
09391000A	BLOCK-1987
095C	BLOCK-295
09A02	BLOCK-1405
09A04	BLOCK-1621
09A05	BLOCK-1059
09A07	BLOCK-1812
09A08	BLOCK-228
09A10	BLOCK-158
0IGL301300A	BLOCK-758
0IGS406600A	BLOCK-1093
0IGS781200A	BLOCK-2119
0IGS781500A	BLOCK-2119
0IKE780120A	BLOCK-2097
0ISA426100A	BLOCK-671
0ISA701600A	BLOCK-762
0ISA783000A	BLOCK-754
0ISA783100A	BLOCK-778
0ISK500410A	BLOCK-873
0ISS330000A	BLOCK-1157
0IT0868000A	BLOCK-1191
0N049874	BLOCK-2187
0N088598	BLOCK-1941
0N101880	BLOCK-1940
0N101881	BLOCK-1943
0N188320-2	BLOCK-1029
0N198382-2	BLOCK-1048
1-000-099-00	BLOCK-1423
1-235-444-11	BLOCK-1389
1-235-971-12	BLOCK-1825
1-424-115-11	BLOCK-819
1-590-1261	BLOCK-1390
1-800-662-11	BLOCK-174
1-805-105-11	BLOCK-94
1-TR-016	BLOCK-973
10-001	BLOCK-243
10-004#	BLOCK-180
10-004(IC)	BLOCK-180
10-004IC	BLOCK-180
10-005	BLOCK-238
10-005(IC)	BLOCK-238
10-010#	BLOCK-2095
10-010(IC)	BLOCK-2095
10-010IC	BLOCK-2095
1000-23	BLOCK-1376
1000-25	BLOCK-1212
1000-6948	BLOCK-237
1000-6963	BLOCK-558
1000-6971	BLOCK-374
1000-7227	BLOCK-544
1000100	BLOCK-1425
1000100-000	BLOCK-1425
1000101	BLOCK-1426
1000101-000	BLOCK-1426
1001(EF-JOHNSON)	BLOCK-2193
1001-0036	BLOCK-1277
1001-5956	BLOCK-510
1001-5972	BLOCK-1269
1001-7150	BLOCK-175
1001-7168	BLOCK-176
1001-9677	BLOCK-175
1001-9685	BLOCK-176
1001-9693	BLOCK-1269
1001-9727	BLOCK-537
1001-9735	BLOCK-1157
1002-0006	BLOCK-4
1002-0345	BLOCK-453
1002-07	BLOCK-12
1003-02	BLOCK-43
1003-0567	BLOCK-184
1003-3116	BLOCK-453
1004-1994	BLOCK-237
1004-2042	BLOCK-547
1006-0887	BLOCK-237
1006-0903	BLOCK-454
1006-7445	BLOCK-363
1007-6537	BLOCK-975
1009-11	BLOCK-1304
1010-9932	BLOCK-1291
1010-9940	BLOCK-1269
1010-9965	BLOCK-1808
1010-9973	BLOCK-1270
10112563	BLOCK-2176
1014-25	BLOCK-1212
1016-80	BLOCK-1212
10176209	BLOCK-2189
1018-25	BLOCK-54
1019-25	BLOCK-20
1020-25	BLOCK-73
102005	BLOCK-1423
1021-25	BLOCK-98
1025-2712	BLOCK-1269
1025-4172	BLOCK-1269
1025-4712	BLOCK-1269
1027-4231	BLOCK-1993
1027-4264	BLOCK-674
1027-4389	BLOCK-760
1027-4462	BLOCK-792
1027-4850	BLOCK-793
1027-4959	BLOCK-794
1027-5006	BLOCK-808
1027-5113	BLOCK-809
1027-5535	BLOCK-795
1027-6244	BLOCK-768
102DE14020	BLOCK-774
102SD13780	BLOCK-662
103-237	BLOCK-1157
1030-25	BLOCK-1410
10302-01	BLOCK-1377
10302-02	BLOCK-1365
10302-03	BLOCK-1304
10302-04	BLOCK-1293
10302-05	BLOCK-1402
10302-06	BLOCK-1298
1031-25	BLOCK-1294
1032-25	BLOCK-136
1033-1247	BLOCK-692
1033-1262	BLOCK-693
1033-1338	BLOCK-685
1033-1874	BLOCK-615
1033-4340	BLOCK-1093
1035-7564	BLOCK-2087
103508-2	BLOCK-2189
103508-3	BLOCK-2190

ORIGINAL DEVICE TYPES AND REPLACEMENT CODES 3

DEVICE TYPE	REPL CODE
1037-7174	BLOCK-544
103705	BLOCK-2208
103717	BLOCK-2217
103729	BLOCK-2197
103731	BLOCK-2199
103743	BLOCK-2202
103755	BLOCK-2207
1038-1804	BLOCK-44
1039956300	BLOCK-1157
103B98M090	BLOCK-881
103B98N060	BLOCK-2093
103D079400	BLOCK-396
103DE75200	BLOCK-709
103SD78370	BLOCK-835
1041-68	BLOCK-13
1041-69	BLOCK-16
1042-11	BLOCK-1280
1042-7938	BLOCK-1422
1043-7275	BLOCK-563
1044-7035	BLOCK-2168
1044-7043	BLOCK-163
1044-7049	BLOCK-163
1045-2761	BLOCK-1269
104648D	BLOCK-1018
1047-25	BLOCK-50
104825	BLOCK-168
104830	BLOCK-2197
104833	BLOCK-2200
104836	BLOCK-2202
104844	BLOCK-2206
104846	BLOCK-2208
104862	BLOCK-2217
104925	BLOCK-169
1052-6390	BLOCK-150
1052-6408	BLOCK-153
1052-6416	BLOCK-162
105412(5	BLOCK-2198
10541284	BLOCK-2185
10541285	BLOCK-2198
105S925080	BLOCK-732
1061-5972	BLOCK-1269
1061-9153	BLOCK-126
1061-9161	BLOCK-131
1061-9526	BLOCK-188
1061-9666	BLOCK-127
1061-9856	BLOCK-130
1063-4939	BLOCK-164
1063-5019	BLOCK-1821
1065-2055	BLOCK-162
1065-2212	BLOCK-695
1065-4861	BLOCK-1293
10650A01	BLOCK-1801
106525	BLOCK-156
10655A01	BLOCK-1822
10655A03	BLOCK-1291
10655A05	BLOCK-1422
10655A13	BLOCK-1421
10655B01	BLOCK-1822
10655B02	BLOCK-1421
10655B03	BLOCK-1291
10655B13	BLOCK-1269
10655B14	BLOCK-1864
10655C05	BLOCK-1422
10658276	BLOCK-2197
10658278	BLOCK-2202
10658279	BLOCK-2206
10658280	BLOCK-2207
10658281	BLOCK-2208
10658282	BLOCK-2213
1066-9666	BLOCK-127
106625	BLOCK-155
106719	BLOCK-108
107225	BLOCK-180
107255	BLOCK-223
107625	BLOCK-223
1077	BLOCK-1270
1077-2382	BLOCK-4
1077-2390	BLOCK-7
1077-2408	BLOCK-50
1077-3844	BLOCK-206
1079	BLOCK-1808
10795-10	BLOCK-1953
10795-5	BLOCK-2207
10795-6	BLOCK-2208
10795-8	BLOCK-2217
10795-9	BLOCK-1950
1081-3541	BLOCK-164
1081-3558	BLOCK-2170
1081-4739	BLOCK-1943
1086-9931	BLOCK-973
1089-7619	BLOCK-50
1097	BLOCK-1808
1099-2949	BLOCK-214
1099-3616	BLOCK-2175

DEVICE TYPE	REPL CODE
1099-3624	BLOCK-1852
1099-6395	BLOCK-1894
1099-6544	BLOCK-947
11-102	BLOCK-245
11-103	BLOCK-1105
11-113	BLOCK-180
11-115	BLOCK-411
11-117	BLOCK-881
11-60	BLOCK-2173
1100-9750	BLOCK-453
110240-005	BLOCK-2209
110242-003	BLOCK-2217
110242-004	BLOCK-2217
1103-3669	BLOCK-1157
1103-7751	BLOCK-184
110472-003	BLOCK-2206
111-4-2020-04600	BLOCK-1269
111-4-2060-02300	BLOCK-5
111-4-2060-02400	BLOCK-5
111-4-2060-03900	BLOCK-78
111-4-2060-0400	BLOCK-1269
111-4-2060-04000	BLOCK-1269
111-4-2060-05100	BLOCK-147
111-4-2060-05200	BLOCK-148
11119016	BLOCK-1157
11119023	BLOCK-1269
11119024	BLOCK-602
11119025	BLOCK-607
11119026	BLOCK-473
11119027	BLOCK-223
11119030	BLOCK-453
11119031	BLOCK-821
11119032	BLOCK-396
11119033	BLOCK-740
11119034	BLOCK-695
11119112	BLOCK-754
11119113	BLOCK-360
11119114	BLOCK-761
11119118	BLOCK-756
11119139	BLOCK-840
11119140	BLOCK-839
11119225	BLOCK-843
11119226	BLOCK-835
11119399	BLOCK-1191
1111P	BLOCK-3
1114-4458	BLOCK-58
1116-6691	BLOCK-176
111825	BLOCK-2173
11200-1	BLOCK-1304
11202-1	BLOCK-1297
11203-1	BLOCK-1426
11204-1	BLOCK-1358
11205-1	BLOCK-1365
11206-1	BLOCK-1408
11207-1	BLOCK-1295
11208-1	BLOCK-1377
11209-1	BLOCK-1417
1121-1	BLOCK-1409
11211-1	BLOCK-1302
11213-1	BLOCK-1411
11214-1	BLOCK-1390
11216-1	BLOCK-1293
1122-8863	BLOCK-453
11233-2	BLOCK-1404
1124-0239	BLOCK-532
11253588	BLOCK-2187
1126-0239	BLOCK-532
11273-1	BLOCK-1311
11274-1	BLOCK-1296
11276-1	BLOCK-1429
1128-4239	BLOCK-602
1129	BLOCK-43
11292312	BLOCK-1471
11292313	BLOCK-1488
11292314	BLOCK-1490
11292315	BLOCK-1493
11369564	BLOCK-1474
1137	BLOCK-1270
1138-0789	BLOCK-1895
1138-0979	BLOCK-401
1138-9442	BLOCK-1422
1140-2302	BLOCK-175
1140-4118	BLOCK-425
1141-3119	BLOCK-219
1143-0584	BLOCK-1066
1145-2786	BLOCK-532
1145-2794	BLOCK-1887
1147-08	BLOCK-2087
1147-09	BLOCK-1039
1147-10	BLOCK-2191
1147-11	BLOCK-1068
1147-12	BLOCK-1122
1147-161	BLOCK-243
1147-8815	BLOCK-611

DEVICE TYPE	REPL CODE
1148-9499	BLOCK-605
1148-9523	BLOCK-597
1148-9556	BLOCK-223
1149-0729	BLOCK-1269
1149-6729	BLOCK-1269
1151	BLOCK-1422
1154	BLOCK-1808
1154-9938	BLOCK-1873
1155	BLOCK-973
1155-5570	BLOCK-214
1155-6578	BLOCK-973
1155-9259	BLOCK-550
1156(TC)	BLOCK-182
1157-4738	BLOCK-973
1157-8564	BLOCK-1039
1157-8572	BLOCK-562
1157-8689	BLOCK-1157
1159-2342	BLOCK-1893
116445	BLOCK-1262
1165(GE)	BLOCK-1269
1166-3829	BLOCK-1887
1166-3879	BLOCK-1887
1166-5197	BLOCK-638
1166-5205	BLOCK-553
116623	BLOCK-1963
1169-2076	BLOCK-1269
1170-0036	BLOCK-603
1172-0026	BLOCK-471
1172-0034	BLOCK-459
1172-0036	BLOCK-603
1172-0042	BLOCK-2097
1172-5157	BLOCK-223
1172-9878	BLOCK-453
1172-9886	BLOCK-547
1172-9894	BLOCK-974
1173-4639	BLOCK-643
1174	BLOCK-1422
1175-7879	BLOCK-675
1175-7903	BLOCK-617
1175-7911	BLOCK-1993
1175-7937	BLOCK-2063
1175-8000	BLOCK-1993
1175-8018	BLOCK-2063
1176-5005	BLOCK-616
1176-7266	BLOCK-534
1176-7274	BLOCK-225
1177-1144	BLOCK-428
1177-1151	BLOCK-398
1177-1169	BLOCK-286
1178	BLOCK-1269
1178-7454	BLOCK-532
1178-7462	BLOCK-534
1180-2055	BLOCK-560
1180-2071	BLOCK-727
1180-2089	BLOCK-563
1180-5991	BLOCK-636
1180-6445	BLOCK-695
1181-3201	BLOCK-214
1181-3565	BLOCK-548
1181-3763	BLOCK-534
1181-4910	BLOCK-1029
1181-4928	BLOCK-1048
1181-7129	BLOCK-428
1181-7137	BLOCK-398
1181-7145	BLOCK-1269
1181-7194	BLOCK-1157
1183-1633	BLOCK-164
118361	BLOCK-1223
1184-8108	BLOCK-404
1184-8728	BLOCK-537
1185-0161	BLOCK-691
1186-5672	BLOCK-603
1186-5888	BLOCK-471
1186-5912	BLOCK-650
119035	BLOCK-739
1191-3340	BLOCK-396
1191-3357	BLOCK-1041
1191-3365	BLOCK-1077
1191-4991	BLOCK-647
1191-8547	BLOCK-2063
1192(IC)	BLOCK-43
1192-7852	BLOCK-510
1193-0054	BLOCK-136
1195-6547	BLOCK-655
1195-6562	BLOCK-537
1195-6570	BLOCK-404
119609	BLOCK-1223
1197-6032	BLOCK-821
1197-6040	BLOCK-637
1199-7848	BLOCK-1269
1199-7855	BLOCK-459
11C44DC	BLOCK-2140
1200	BLOCK-109
1200-2200	BLOCK-740

DEVICE TYPE	REPL CODE
1200-2358	BLOCK-660
1200-2382	BLOCK-1157
1200-2416	BLOCK-1873
1200-2440	BLOCK-540
1200-3430	BLOCK-360
1200-3448	BLOCK-636
1200-3463	BLOCK-644
1200-3497	BLOCK-1087
1200-3505	BLOCK-1044
1200-3530	BLOCK-360
1204-000395	BLOCK-773
1204-000506	BLOCK-1253
1205-7832	BLOCK-532
1205-7907	BLOCK-540
1207-4707	BLOCK-1066
1207-6873	BLOCK-576
1207-6907	BLOCK-1066
1207-6931	BLOCK-1088
1214-4945	BLOCK-532
1215-8663	BLOCK-814
1215-8671	BLOCK-1182
1217-2284	BLOCK-722
1217-8935	BLOCK-786
1217-8943	BLOCK-788
1217-8950	BLOCK-789
1217-8968	BLOCK-790
1217-8976	BLOCK-790
1217-8992	BLOCK-791
1217-9008	BLOCK-793
1217-9016	BLOCK-794
1217-9024	BLOCK-1993
1217-9040	BLOCK-795
1217-9065	BLOCK-768
122199	BLOCK-1223
1223909	BLOCK-156
1223910	BLOCK-155
1224076	BLOCK-180
1224275	BLOCK-259
1225-4975	BLOCK-1093
1227-8610	BLOCK-660
1228-4410	BLOCK-534
1229-0862	BLOCK-787
1229-4625	BLOCK-516
1229-4633	BLOCK-1098
1229-4658	BLOCK-1098
1229-8725	BLOCK-1157
123-001	BLOCK-1280
1230	BLOCK-218
1231-0116	BLOCK-690
1231-0132	BLOCK-1801
1231-0140	BLOCK-1801
1231-0157	BLOCK-697
1231-0165	BLOCK-516
1231-0207	BLOCK-853
1231-0256	BLOCK-408
1231-8325	BLOCK-697
1231-8333	BLOCK-2063
1231-8341	BLOCK-396
1231-8358	BLOCK-1202
1231-8366	BLOCK-2063
1232-7292	BLOCK-761
1232-7300	BLOCK-360
1232-7573	BLOCK-664
1234-1095	BLOCK-720
1234-1674	BLOCK-663
1234-5922	BLOCK-756
1234-5948	BLOCK-663
1234-5955	BLOCK-1049
1234-5963	BLOCK-772
1237-7912	BLOCK-695
1239-6537	BLOCK-761
1239-6545	BLOCK-636
1239-7766	BLOCK-1093
1240-3432	BLOCK-1038
1240-3440	BLOCK-1135
1240-3457	BLOCK-1051
1250-5210	BLOCK-1087
1254-1694	BLOCK-2097
1254-2106	BLOCK-735
1254-3583	BLOCK-874
1254-5869	BLOCK-1188
1255-1063	BLOCK-786
1255-1081	BLOCK-788
1255-1123	BLOCK-791
1255-1131	BLOCK-813
1255-1198	BLOCK-768
1259-4917	BLOCK-563
1259-4958	BLOCK-695
1259-4966	BLOCK-853
1259-5005	BLOCK-1069
1259-6516	BLOCK-136
1259-6524	BLOCK-709
1259-6532	BLOCK-556
125C3	BLOCK-1809

DEVICE TYPE	REPL CODE
125C3RB	BLOCK-1809
126-1	BLOCK-401
126-40	BLOCK-1949
1264	BLOCK-1815
1266-2053	BLOCK-1157
1266-2425	BLOCK-396
1266-2433	BLOCK-740
1266-2441	BLOCK-700
1266-2458	BLOCK-706
1266-2466	BLOCK-636
1266-2953	BLOCK-1157
126604	BLOCK-1267
126871	BLOCK-1265
1270-1751	BLOCK-382
1270-1769	BLOCK-781
1270-1777	BLOCK-782
1270-4847	BLOCK-1147
1270-5299	BLOCK-721
1271-3111	BLOCK-818
127166	BLOCK-1810
1278	BLOCK-1422
1278(GE)	BLOCK-1422
1287-7908	BLOCK-1923
1287-7916	BLOCK-781
1287-7924	BLOCK-782
1287-7932	BLOCK-1084
1287-7940	BLOCK-563
1287-8989	BLOCK-1923
1288-0183	BLOCK-2083
128C212H01	BLOCK-2213
128C213H01	BLOCK-1955
128C830H05	BLOCK-1426
12951-1	BLOCK-1944
1296-1892	BLOCK-709
1296-1900	BLOCK-817
1296-1918	BLOCK-1093
1296-1926	BLOCK-876
1296-2718	BLOCK-1021
1298-0777	BLOCK-709
1298-0785	BLOCK-660
1298-0793	BLOCK-824
1298-3409	BLOCK-2097
1298-3888	BLOCK-2097
1298-3920	BLOCK-810
129821	BLOCK-1944
129871	BLOCK-1265
1299-7862	BLOCK-2097
13-0161	BLOCK-2160
13-1-6	BLOCK-1212
13-10-6	BLOCK-1212
13-100000	BLOCK-1828
13-1000000	BLOCK-1828
13-1000001	BLOCK-1828
13-11-6	BLOCK-1212
13-117-6	BLOCK-1616
13-26-6	BLOCK-1275
13-27-6	BLOCK-1267
13-28-6	BLOCK-1245
13-29-5	BLOCK-1269
13-29-6	BLOCK-1269
13-30-6	BLOCK-1805
13-3017119-2	BLOCK-1895
13-35059-1	BLOCK-1269
13-350591	BLOCK-1269
13-36-6	BLOCK-1275
13-40-6	BLOCK-1272
13-41-6	BLOCK-1815
13-42-6	BLOCK-1271
13-50-6	BLOCK-1794
13-56-6	BLOCK-1291
13-57-6	BLOCK-1292
13-59-6	BLOCK-1828
13-61-6	BLOCK-1828
13-64-6	BLOCK-1269
13-67-6	BLOCK-1292
13-73-6	BLOCK-1808
13-9-6	BLOCK-1212
130130	BLOCK-1805
130751	BLOCK-1269
1309-1004	BLOCK-1049
1310-8394	BLOCK-709
1310-8402	BLOCK-824
1310-8410	BLOCK-754
1311-8641	BLOCK-396
1313-5546	BLOCK-758
1313-5553	BLOCK-817
1313-5561	BLOCK-1568
1313-5793	BLOCK-1088
131300	BLOCK-2197
131301	BLOCK-2199
131303	BLOCK-2202
131304	BLOCK-2206
131305	BLOCK-2207
131306	BLOCK-2208

DEVICE TYPE	REPL CODE
131308	BLOCK-2217
131312	BLOCK-1950
1316-0951	BLOCK-563
1316-0963	BLOCK-1182
1316-0973	BLOCK-839
1316-0981	BLOCK-840
1316-1003	BLOCK-1196
1316-1005	BLOCK-1196
1316-1013	BLOCK-778
1316-1039	BLOCK-1188
1316-3217	BLOCK-1088
1316-3258	BLOCK-817
1316-7887	BLOCK-807
1316-7895	BLOCK-659
1316-7911	BLOCK-1157
1316-7929	BLOCK-1157
132313	BLOCK-1286
132314	BLOCK-1284
132315	BLOCK-1285
1325-1186	BLOCK-770
1327-1291	BLOCK-636
1327-1309	BLOCK-459
133-002	BLOCK-1265
133-0021-0	BLOCK-1944
1330-5313	BLOCK-836
1330-9190	BLOCK-841
1330-9208	BLOCK-1183
1330-9216	BLOCK-763
1330-9224	BLOCK-2097
1330-9232	BLOCK-723
1330-9240	BLOCK-1802
1330-9257	BLOCK-761
1330-9265	BLOCK-840
1330-9273	BLOCK-839
1330-9281	BLOCK-360
1330021-0	BLOCK-1944
1330021-1	BLOCK-1944
1331409	BLOCK-1015
133600	BLOCK-1286
133P100	BLOCK-1956
134195-001	BLOCK-2197
134196-001	BLOCK-2199
134197-001	BLOCK-2207
134198-001	BLOCK-2208
134254-001	BLOCK-2212
13428129	BLOCK-981
134340	BLOCK-1792
134509	BLOCK-1787
1346-6347	BLOCK-1194
1346-8905	BLOCK-644
1346-8921	BLOCK-843
1346-8939	BLOCK-835
1346-8947	BLOCK-757
1347	BLOCK-1292
1348	BLOCK-1292
1348-1593	BLOCK-735
1348-5024	BLOCK-758
1348-5032	BLOCK-1191
1348A12H01	BLOCK-1343
1348A14H01	BLOCK-1413
1348A30H01	BLOCK-2197
1348A32H01	BLOCK-2199
1348A36H01	BLOCK-2202
1348A44H01	BLOCK-2206
1348A45H01	BLOCK-2207
1348A46H01	BLOCK-2208
1348A62H01	BLOCK-2217
1351	BLOCK-1291
1352	BLOCK-404
1359-0161	BLOCK-2097
136145	BLOCK-1812
136146	BLOCK-1813
136147	BLOCK-1814
1365-5113	BLOCK-1088
1368-1846	BLOCK-671
1368-1853	BLOCK-762
1368-2638	BLOCK-1157
137161	BLOCK-1773
137245	BLOCK-1805
1374-6821	BLOCK-835
1374-6847	BLOCK-758
1377-2744	BLOCK-873
1377-2777	BLOCK-1157
1380-4687	BLOCK-814
1380-4729	BLOCK-881
1380-6070	BLOCK-723
138311	BLOCK-1293
138312	BLOCK-1294
138313	BLOCK-1295
138314	BLOCK-1297
138315	BLOCK-1301
138317	BLOCK-1307
138318	BLOCK-1365
138319	BLOCK-1366

4 ORIGINAL DEVICE TYPES AND REPLACEMENT CODES

DEVICE TYPE	REPL CODE
138320	BLOCK-1358
138380	BLOCK-1814
138380(IC)	BLOCK-1814
138381	BLOCK-1378
138403	BLOCK-1410
138681	BLOCK-2073
138699	BLOCK-1821
1393-9285	BLOCK-563
1393-9327	BLOCK-763
1393-9376	BLOCK-907
14-2007	BLOCK-1244
14-2007-00	BLOCK-1244
14-2007-00B	BLOCK-1256
14-2007-01	BLOCK-1256
14-2007-02	BLOCK-1244
14-2007-03	BLOCK-1256
14-2008-01	BLOCK-1262
14-2010-01	BLOCK-1270
14-2010-03	BLOCK-1815
14-2011-01	BLOCK-1804
14-201I-02	BLOCK-1804
141134	BLOCK-2168
141135	BLOCK-163
141259	BLOCK-1821
141270	BLOCK-164
141279	BLOCK-1821
141280	BLOCK-165
141290	BLOCK-166
142007-2	BLOCK-1256
14207	BLOCK-2069
142137	BLOCK-1157
142251(IC)	BLOCK-1797
142341	BLOCK-2169
142718	BLOCK-164
142719	BLOCK-2170
142903	BLOCK-2169
143033	BLOCK-164
143041(IC)	BLOCK-1967
14305	BLOCK-2087
143062	BLOCK-1956
14308	BLOCK-2095
143696	BLOCK-1894
143766	BLOCK-947
143808	BLOCK-2168
143821	BLOCK-163
143822	BLOCK-2170
144011	BLOCK-121
144012	BLOCK-250
144015	BLOCK-253
144018	BLOCK-256
144022	BLOCK-1978
144024	BLOCK-395
144026	BLOCK-1269
144072-2	BLOCK-2097
144612	BLOCK-415
144848	BLOCK-392
144849	BLOCK-416
144850	BLOCK-413
144851	BLOCK-409
144852	BLOCK-400
144853	BLOCK-407
144854	BLOCK-609
144855	BLOCK-2054
144920	BLOCK-389
144921	BLOCK-2085
144923	BLOCK-1157
144967	BLOCK-2055
144968	BLOCK-1039
144969	BLOCK-1048
144970	BLOCK-1085
144971	BLOCK-1096
14500001-001	BLOCK-2208
14500001-002	BLOCK-2210
14500001-003	BLOCK-2218
14500001-004	BLOCK-2216
14500004-001D	BLOCK-2202
14500004-001P	BLOCK-2202
14500004-002	BLOCK-2203
14500004-003	BLOCK-2201
14500005-001	BLOCK-2206
14511	BLOCK-1122
145113	BLOCK-1157
14529	BLOCK-1136
145586	BLOCK-1041
1458	BLOCK-1801
145803	BLOCK-1860
145818	BLOCK-1039
146052	BLOCK-164
146138	BLOCK-1157
146149	BLOCK-1885
146150	BLOCK-1884
146151	BLOCK-2169
146152	BLOCK-2170
146164	BLOCK-2175
1462	BLOCK-1313
1462434-1	BLOCK-1265
1462445-1	BLOCK-1267
1462506-1	BLOCK-1285
1462516	BLOCK-1269
1462516-001	BLOCK-1269
1462516-1	BLOCK-1269
1462554	BLOCK-1286
1462554-1	BLOCK-1286
1462554-2	BLOCK-1286
1462554-3	BLOCK-1286
1462554-4	BLOCK-1286
1462559-1	BLOCK-1284
1462560-001	BLOCK-1285
1462560-1	BLOCK-1285
1463672	BLOCK-1805
1463677-1	BLOCK-1805
1463677-2	BLOCK-1805
1463677-3	BLOCK-1805
1463681-1	BLOCK-1944
1463686-1	BLOCK-1269
1464295-1	BLOCK-1787
1464295-2	BLOCK-1792
1464437-2	BLOCK-1812
1464437-3	BLOCK-1812
1464438-2	BLOCK-1813
1464438-3	BLOCK-1813
1464439-2	BLOCK-1814
1464460-2	BLOCK-1814
1464460-3	BLOCK-1814
146460-2	BLOCK-1814
1464846-1	BLOCK-2073
1464846-2	BLOCK-2073
146514	BLOCK-407
1465158-1	BLOCK-1821
1465158-2	BLOCK-1821
1465188-1	BLOCK-1284
1465316-1	BLOCK-2168
1465345-1	BLOCK-163
1465615	BLOCK-165
1465615-1	BLOCK-165
1465615-2	BLOCK-165
1465617-2	BLOCK-164
146595	BLOCK-402
146641	BLOCK-2208
146642	BLOCK-2217
146643	BLOCK-2197
146644	BLOCK-2207
1466701-1	BLOCK-166
146732	BLOCK-1369
146742	BLOCK-1378
146857	BLOCK-532
146858	BLOCK-1887
146893	BLOCK-416
146894	BLOCK-387
146895	BLOCK-410
146896	BLOCK-1269
146911	BLOCK-1157
147-7040-01	BLOCK-112
147-7043-01	BLOCK-221
1471-4356	BLOCK-2202
1471-4364	BLOCK-2208
1471-4372	BLOCK-2217
1471-4380	BLOCK-2197
1471-4398	BLOCK-2206
1471-9860	BLOCK-2202
147232	BLOCK-646
147234	BLOCK-1039
147256	BLOCK-2206
147339	BLOCK-587
147340	BLOCK-541
1473528-1	BLOCK-1810
1473528-2	BLOCK-1810
1473677-3	BLOCK-1805
147437	BLOCK-457
147438	BLOCK-1269
147464	BLOCK-1157
147478	BLOCK-1093
1477-3436	BLOCK-1944
1478-6982	BLOCK-1961
147887	BLOCK-1041
1479-0224	BLOCK-2203
1479-0240	BLOCK-1471
1479-0257	BLOCK-1485
1479-0265	BLOCK-1484
1479-7971	BLOCK-1473
148132	BLOCK-646
148235	BLOCK-1801
148236	BLOCK-1084
1485-9896	BLOCK-1965
1486-8780	BLOCK-1476
148684	BLOCK-1648
148715	BLOCK-257
148716	BLOCK-1731
148717	BLOCK-612
148718	BLOCK-689
148719	BLOCK-608
148720	BLOCK-410
148721	BLOCK-680
148722	BLOCK-555
148723	BLOCK-681
148724	BLOCK-1801
148725	BLOCK-682
148726	BLOCK-683
148727	BLOCK-301
148728	BLOCK-1801
148729	BLOCK-600
148730	BLOCK-1039
148731	BLOCK-1040
148732	BLOCK-1064
148733	BLOCK-1084
148734	BLOCK-1093
148735	BLOCK-1104
148736	BLOCK-1035
148737	BLOCK-1135
148740	BLOCK-1873
148741	BLOCK-396
148742	BLOCK-1336
148744	BLOCK-428
148745	BLOCK-2087
148805	BLOCK-1157
148954	BLOCK-1014
148956	BLOCK-2093
148991	BLOCK-1961
149017	BLOCK-2087
149018	BLOCK-2171
149019	BLOCK-1264
149037	BLOCK-1889
149038	BLOCK-2119
149252	BLOCK-1269
149253	BLOCK-1923
149254	BLOCK-1887
149255	BLOCK-1984
149585	BLOCK-683
149586	BLOCK-699
149598	BLOCK-2093
1496	BLOCK-2139
149870	BLOCK-1893
149938	BLOCK-1059
14L0035	BLOCK-930
14LN033	BLOCK-58
14LN034	BLOCK-170
14LQ007	BLOCK-4
14LV233	BLOCK-773
15-14504-1	BLOCK-1819
15-26587-1	BLOCK-1212
15-3015129-1	BLOCK-1821
15-301513-1	BLOCK-221
15-3015130-1	BLOCK-1303
15-3015131-1	BLOCK-363
15-3015131-2	BLOCK-361
15-3015131-3	BLOCK-363
15-3015569-1	BLOCK-1029
15-3015570-1	BLOCK-1114
15-3015734-1	BLOCK-401
15-3015740-1	BLOCK-1836
15-3017045-1	BLOCK-220
15-3017119-1	BLOCK-1895
15-3017119-2	BLOCK-1895
15-31015-1	BLOCK-21
15-31015-7	BLOCK-1269
15-33201-1	BLOCK-1269
15-33201-2	BLOCK-1269
15-34005-1	BLOCK-1283
15-34048	BLOCK-1276
15-34048-1	BLOCK-1262
15-34202-1	BLOCK-1283
15-34379-1	BLOCK-1277
15-34401-1	BLOCK-1792
15-34408-1	BLOCK-1262
15-34452	BLOCK-1264
15-34452-1	BLOCK-1264
15-34502-1	BLOCK-1801
15-34502-2	BLOCK-1801
15-34502-3	BLOCK-1801
15-34906-1	BLOCK-1823
15-35059	BLOCK-1269
15-35059-1	BLOCK-1269
15-35059-2	BLOCK-1269
15-36446-1	BLOCK-1376
15-36647-1	BLOCK-1841
15-36994-1	BLOCK-1813
15-36995-1	BLOCK-1389
15-37534-1	BLOCK-1819
15-37534-2	BLOCK-1819
15-37700-1	BLOCK-1817
15-37701-1	BLOCK-1818
15-37701-2	BLOCK-1818
15-37702-1	BLOCK-1271
15-37703-1	BLOCK-1272
15-37704-1	BLOCK-1815
15-37833-2	BLOCK-1954
15-39060-1	BLOCK-1407
15-39061-1	BLOCK-1820
15-39075-1	BLOCK-1291
15-392007-1	BLOCK-1831
15-39207-1	BLOCK-1831
15-39208-1	BLOCK-1833
15-39209-1	BLOCK-1816
15-39600-1	BLOCK-1819
15-39600-2	BLOCK-1819
15-40140-2	BLOCK-2097
15-40183-1	BLOCK-1825
15-41545-1	BLOCK-1819
15-41627-1	BLOCK-1849
15-41627-2	BLOCK-1849
15-41627-3	BLOCK-1849
15-41764-1	BLOCK-1850
15-41764-2	BLOCK-1850
15-41856-1	BLOCK-1820
15-43098-1	BLOCK-1847
15-43251-1	BLOCK-1407
15-43251-2	BLOCK-1407
15-43312-1	BLOCK-1848
15-43312-2	BLOCK-1848
15-43636-1	BLOCK-219
15-43637-1	BLOCK-276
15-43638-1	BLOCK-1286
15-43703-1	BLOCK-1291
15-43704-1	BLOCK-1068
15-43705-1	BLOCK-1039
15-43706-1	BLOCK-1110
15-45141-1	BLOCK-1039
15-45184-1	BLOCK-1063
15-45185-1	BLOCK-1048
15-45186-1	BLOCK-2171
15-45300-1	BLOCK-1269
15-45693-1	BLOCK-1891
15-5016T/IC7313AP	BLOCK-453
15-53201-1	BLOCK-1276
15-53201-2	BLOCK-1276
1502MH833	BLOCK-638
150325	BLOCK-532
150340	BLOCK-545
150342	BLOCK-415
150365	BLOCK-1157
150368	BLOCK-638
150369	BLOCK-553
150371	BLOCK-1963
150401	BLOCK-623
150418	BLOCK-1873
150419	BLOCK-1146
150430	BLOCK-545
150440	BLOCK-1591
15105300	BLOCK-1022
151056	BLOCK-2063
15109200	BLOCK-1722
151333	BLOCK-396
151544	BLOCK-2199
151548	BLOCK-1953
152051	BLOCK-1887
152364	BLOCK-1873
1527-5282	BLOCK-2199
1527-5308	BLOCK-2207
152920	BLOCK-2055
152921	BLOCK-1993
152923	BLOCK-1039
152929	BLOCK-1135
152935	BLOCK-1096
153456	BLOCK-1873
153568	BLOCK-1993
153612	BLOCK-904
153684	BLOCK-164
153685	BLOCK-1887
153712	BLOCK-695
153875	BLOCK-1923
153937	BLOCK-1039
154027	BLOCK-1048
15405-1	BLOCK-2208
15405-2	BLOCK-2210
15405-4	BLOCK-2199
1542	BLOCK-1325
1542(GE)	BLOCK-1325
154222	BLOCK-1029
154515	BLOCK-1923
154572	BLOCK-692
154574	BLOCK-693
154576	BLOCK-675
154577	BLOCK-690
154581	BLOCK-1993
154648	BLOCK-1035
154649	BLOCK-1093
154652	BLOCK-615
154730	BLOCK-696
154753	BLOCK-685
154754	BLOCK-223
154767	BLOCK-1030
154768	BLOCK-1040
154770	BLOCK-1098
154771	BLOCK-1043
154772	BLOCK-1063
154773	BLOCK-1076
154774	BLOCK-1029
154782	BLOCK-2087
154822	BLOCK-695
155508	BLOCK-691
155509	BLOCK-691
155556	BLOCK-685
155632	BLOCK-1035
155708	BLOCK-1327
155709	BLOCK-1331
155710	BLOCK-1357
155711	BLOCK-988
155712	BLOCK-2079
155892	BLOCK-82
155911	BLOCK-226
156-0011-00	BLOCK-2189
156-0017	BLOCK-1943
156-0017-00	BLOCK-1943
156-0148-00	BLOCK-1297
156-0151-00	BLOCK-1967
156-0176-00	BLOCK-1022
156018	BLOCK-8
156025	BLOCK-112
156477	BLOCK-1269
156545	BLOCK-516
156546	BLOCK-872
156571	BLOCK-665
156580	BLOCK-686
156725	BLOCK-615
156730	BLOCK-644
156895	BLOCK-545
156912	BLOCK-672
156915	BLOCK-665
156918	BLOCK-686
156932	BLOCK-1802
156960	BLOCK-685
157564	BLOCK-1967
157575	BLOCK-1022
157653	BLOCK-537
157654	BLOCK-1041
157677	BLOCK-762
157806	BLOCK-1079
157971	BLOCK-699
157972	BLOCK-993
158040	BLOCK-2160
158061	BLOCK-1040
158063	BLOCK-1040
158066	BLOCK-1950
158238	BLOCK-1077
158240	BLOCK-1078
158249	BLOCK-1092
158578	BLOCK-1157
158624	BLOCK-1393
158625	BLOCK-1394
158626	BLOCK-1394
159249	BLOCK-847
159355	BLOCK-2144
15942-1	BLOCK-1791
159425	BLOCK-161
159433	BLOCK-720
159435	BLOCK-1105
159588	BLOCK-1823
159947	BLOCK-50
159970	BLOCK-396
159974	BLOCK-1157
15SI-CA3064E	BLOCK-1808
160175	BLOCK-545
160554	BLOCK-1140
160576	BLOCK-645
160741	BLOCK-645
160752	BLOCK-1330
160753	BLOCK-1327
160763	BLOCK-686
160764	BLOCK-1357
1607A80	BLOCK-1299
160832	BLOCK-1993
160844	BLOCK-906
160869	BLOCK-695
16088	BLOCK-1423
161-006-0001	BLOCK-2179
161-006-0002	BLOCK-2179
161-011-0001	BLOCK-2177
161-011-0002	BLOCK-2177
161-012-0002	BLOCK-2181
161-118-0001	BLOCK-2185
161-152-0101	BLOCK-2189
161-152-0102	BLOCK-2189
16103860	BLOCK-1851
161079	BLOCK-1087
161193	BLOCK-50
161325	BLOCK-776
161326	BLOCK-775
161392	BLOCK-1450
161573	BLOCK-871
161715	BLOCK-768
16173130	BLOCK-453
161732	BLOCK-1098
161743	BLOCK-1801
162-37	BLOCK-1021
162037	BLOCK-1801
16233010	BLOCK-7
162451	BLOCK-1792
162674	BLOCK-696
1627-1413	BLOCK-1959
1627-2064	BLOCK-1950
16271033	BLOCK-2202
16271041	BLOCK-2202
162922	BLOCK-525
163434	BLOCK-2171
163755	BLOCK-1093
163794	BLOCK-2144
163823	BLOCK-716
1643	BLOCK-2087
16453160	BLOCK-974
164602	BLOCK-2083
16483610	BLOCK-973
16483611	BLOCK-973
164981	BLOCK-2201
164982	BLOCK-2202
164984	BLOCK-1039
16605	BLOCK-1256
166907	BLOCK-3
167-001B	BLOCK-1269
167-006A	BLOCK-1157
167-006B	BLOCK-1157
167-013A	BLOCK-1422
167-014A	BLOCK-1407
167-018A	BLOCK-1269
167-019A	BLOCK-1269
16740135809	BLOCK-23
16778	BLOCK-1287
167900306081	BLOCK-4
169403	BLOCK-1411
17-12054-1	BLOCK-2197
17-12056-1	BLOCK-2199
17-12057-1	BLOCK-2200
17-12058-1	BLOCK-2208
17-12064-1	BLOCK-2217
17-12065-1	BLOCK-2207
17-12089-1	BLOCK-2213
1700-5182	BLOCK-1943
170954	BLOCK-41
170964	BLOCK-29
1710-1	BLOCK-1048
171179-027	BLOCK-43
171179-028	BLOCK-118
171179-036	BLOCK-4
171179-045	BLOCK-85
171179-051	BLOCK-7
171179-070	BLOCK-132
17184600	BLOCK-1298
171982	BLOCK-70
172252	BLOCK-4
172272	BLOCK-7
172318	BLOCK-1487
173116	BLOCK-871
1741-0051	BLOCK-1293
1741-0069	BLOCK-1471
1741-0085	BLOCK-1294
1741-0119	BLOCK-1295
1741-0143	BLOCK-1297
1741-0150	BLOCK-1473
1741-0176	BLOCK-1298
1741-0184	BLOCK-1474
1741-0200	BLOCK-1301
1741-0234	BLOCK-1304
1741-0242	BLOCK-1476
1741-0275	BLOCK-1482
1741-0291	BLOCK-1343
1741-0325	BLOCK-1365
1741-0333	BLOCK-1484
1741-0366	BLOCK-1486
1741-0416	BLOCK-1377
1741-0424	BLOCK-1487
1741-0440	BLOCK-1384
1741-0473	BLOCK-1390
1741-0481	BLOCK-1488
1741-0564	BLOCK-1402

DEVICE TYPE	REPL CODE
1741-0572	BLOCK-1490
1741-0598	BLOCK-1403
1741-0606	BLOCK-1492
1741-0622	BLOCK-1404
1741-0630	BLOCK-1493
1741-0663	BLOCK-1494
1741-0689	BLOCK-1406
1741-0697	BLOCK-1495
1741-0721	BLOCK-1501
1741-0747	BLOCK-1412
1741-0770	BLOCK-1417
1741-0804	BLOCK-1419
1741-0895	BLOCK-1426
1741-0952	BLOCK-1428
1741-1018	BLOCK-1305
1741-1042	BLOCK-1327
1741-1075	BLOCK-1328
1741-1133	BLOCK-1307
1741-1190	BLOCK-1391
1741-1224	BLOCK-1392
1741-1257	BLOCK-1311
1741-1299	BLOCK-1485
1741-1323	BLOCK-1483
1741-1349	BLOCK-1352
1741-1380	BLOCK-1491
1754-3	BLOCK-2171
1756-1	BLOCK-532
176222	BLOCK-1923
176223	BLOCK-781
176224	BLOCK-782
176225	BLOCK-1084
176226	BLOCK-563
176853	BLOCK-778
176854	BLOCK-771
177121	BLOCK-1157
177612	BLOCK-823
177619	BLOCK-861
177857	BLOCK-1923
177A07	BLOCK-1794
178476	BLOCK-762
178690	BLOCK-136
178692	BLOCK-662
178693	BLOCK-732
178694	BLOCK-771
178695	BLOCK-644
178697	BLOCK-1200
178737	BLOCK-1722
178738	BLOCK-1299
178740	BLOCK-1574
178741	BLOCK-1801
178742	BLOCK-814
178743	BLOCK-1590
178744	BLOCK-1629
178745	BLOCK-2144
178746	BLOCK-762
179-46444-01	BLOCK-2185
179-46444-02	BLOCK-2189
179-46444-05	BLOCK-2188
179-46445-01	BLOCK-2199
179-46445-02	BLOCK-2200
179-46445-03	BLOCK-2203
179-46445-05	BLOCK-2209
179-46445-06	BLOCK-2210
179-46445-08	BLOCK-2216
179-46445-09	BLOCK-2218
179-46445-10	BLOCK-1951
179-46447-21	BLOCK-1967
179726	BLOCK-829
1800	BLOCK-2193
180010-001	BLOCK-1963
1800DC	BLOCK-2150
1800PC	BLOCK-2150
1801DC	BLOCK-2151
1801PC	BLOCK-2151
180215	BLOCK-593
180217	BLOCK-1802
1802DC	BLOCK-2152
1802PC	BLOCK-2152
1803DC	BLOCK-2153
1803PC	BLOCK-2153
1804DC	BLOCK-2154
1804PC	BLOCK-2154
1805	BLOCK-1294
1805DC	BLOCK-2155
1805PC	BLOCK-2155
1806	BLOCK-1297
1806DC	BLOCK-2156
1806PC	BLOCK-2156
1807	BLOCK-1392
1807DC	BLOCK-2157
1807PC	BLOCK-2157
1808	BLOCK-1423
1808DC	BLOCK-2158
1808PC	BLOCK-2158

DEVICE TYPE	REPL CODE
1809DC	BLOCK-2159
1809PC	BLOCK-2159
181-000100	BLOCK-1790
181-000200	BLOCK-1275
1810DC	BLOCK-2161
181183	BLOCK-2063
1811DC	BLOCK-2162
1811PC	BLOCK-2162
1812DC	BLOCK-2163
1812PC	BLOCK-2163
181315	BLOCK-628
1813DC	BLOCK-2164
1813PC	BLOCK-2164
1814DC	BLOCK-2165
1814PC	BLOCK-2165
181860	BLOCK-1093
1820-0054	BLOCK-1293
1820-0055	BLOCK-1423
1820-0063	BLOCK-1402
1820-0068	BLOCK-1304
1820-0069	BLOCK-1365
1820-0070	BLOCK-1377
1820-0075	BLOCK-1410
1820-0077	BLOCK-1411
1820-0087	BLOCK-2212
1820-0095	BLOCK-2209
1820-0099	BLOCK-1426
1820-0122	BLOCK-1950
1820-0174	BLOCK-1297
1820-0214	BLOCK-1392
1820-0261	BLOCK-1311
1820-0301	BLOCK-1412
1820-0304	BLOCK-1409
1820-0328	BLOCK-1295
1820-0341	BLOCK-1951
1820-0352	BLOCK-1944
1820-0430	BLOCK-1022
1820-0476	BLOCK-1967
1820-0495	BLOCK-1307
1820-0512	BLOCK-1501
1820-0861	BLOCK-2208
1820-0862	BLOCK-2209
1820-0863	BLOCK-1969
1820-0864	BLOCK-1296
1820-0865	BLOCK-2206
1820-0869	BLOCK-2201
1820-0870	BLOCK-1301
1820-0894	BLOCK-1297
1820-1064	BLOCK-1341
1820-1068	BLOCK-1487
1820-1111	BLOCK-1390
1820-1172	BLOCK-1336
182503	BLOCK-1484
182510	BLOCK-1385
1826-1	BLOCK-1087
183013	BLOCK-1275
183044	BLOCK-1828
183405	BLOCK-1828
183532	BLOCK-2056
18410-148	BLOCK-1813
185233	BLOCK-1093
185895	BLOCK-807
185896	BLOCK-659
185897	BLOCK-647
185898	BLOCK-1157
185899	BLOCK-822
185900	BLOCK-821
185901	BLOCK-684
186-1	BLOCK-1048
18600-155	BLOCK-138
18600-156	BLOCK-56
186671	BLOCK-1157
186773	BLOCK-2097
188086	BLOCK-754
188242	BLOCK-841
188243	BLOCK-770
188244	BLOCK-728
188246	BLOCK-763
188660-01	BLOCK-2056
189152	BLOCK-762
1895989-1	BLOCK-1951
1895991-1	BLOCK-2208
1895992-1	BLOCK-2217
1895993-1	BLOCK-2197
1895994-1	BLOCK-2199
1895995-1	BLOCK-2200
18SI-SN76650N	BLOCK-1422
19-020-079	BLOCK-1212
19-05589	BLOCK-1408
19-076001	BLOCK-4
19-09973-0	BLOCK-1294
19-10476-0	BLOCK-1472
19-130-004	BLOCK-1325
19-130-005	BLOCK-1423

DEVICE TYPE	REPL CODE
1901-0557	BLOCK-1945
191296	BLOCK-2063
191297	BLOCK-2087
191299	BLOCK-2087
1914062-1	BLOCK-2202
19166123	BLOCK-2216
191862	BLOCK-1069
191868	BLOCK-1801
191869	BLOCK-2063
191985	BLOCK-1105
191986	BLOCK-1105
191989	BLOCK-1105
192062	BLOCK-1130
192276	BLOCK-853
192285	BLOCK-2097
192742	BLOCK-778
193022	BLOCK-1969
193082	BLOCK-771
193207	BLOCK-2056
193309	BLOCK-807
193333	BLOCK-697
194285	BLOCK-2097
194305	BLOCK-2087
195606	BLOCK-1993
195791	BLOCK-1088
196121	BLOCK-237
196122	BLOCK-773
196187	BLOCK-403
196188	BLOCK-853
196374	BLOCK-1332
196634	BLOCK-1476
196639	BLOCK-2185
197323	BLOCK-1598
197376	BLOCK-1646
197381	BLOCK-1649
197666	BLOCK-2207
198400	BLOCK-1949
198409-1	BLOCK-2210
198409-11	BLOCK-2208
198409-12	BLOCK-2217
198409-13	BLOCK-2197
198409-14	BLOCK-2202
198409-5	BLOCK-2206
198409-6	BLOCK-2199
198410-1	BLOCK-1949
198609	BLOCK-365
198801-1	BLOCK-1945
19A115913-1	BLOCK-2197
19A115913-10	BLOCK-1950
19A115913-19	BLOCK-2208
19A115913-2	BLOCK-2213
19A115913-20	BLOCK-2217
19A115913-3	BLOCK-2199
19A115913-4	BLOCK-2200
19A115913P14	BLOCK-2202
19A115913P2	BLOCK-2213
19A116180-11	BLOCK-1404
19A116180-18	BLOCK-1419
19A116180-22	BLOCK-1298
19A116180-24	BLOCK-1423
19A116180-25	BLOCK-1424
19A116180-27	BLOCK-1425
19A116180-29	BLOCK-1428
19A116180-7	BLOCK-1390
19A116180-8	BLOCK-1401
19A116180-P13	BLOCK-1408
19A116180P1	BLOCK-1293
19A116180P11	BLOCK-1404
19A116180P15	BLOCK-1410
19A116180P16	BLOCK-1411
19A116180P18	BLOCK-1419
19A116180P2	BLOCK-1294
19A116180P20	BLOCK-1297
19A116180P3	BLOCK-1295
19A116180P4	BLOCK-1304
19A116180P5	BLOCK-1365
19A116180P7	BLOCK-1390
19A116180P8	BLOCK-1401
19A11644P1	BLOCK-1262
19E15-1	BLOCK-2179
19E18-1	BLOCK-2178
19E19-2	BLOCK-2177
1A4102	BLOCK-216
1C-112	BLOCK-78
1C-126	BLOCK-103
1C-142	BLOCK-1269
1C-19	BLOCK-1804
1C-24	BLOCK-4
1C-289	BLOCK-1822
1C-30	BLOCK-1823
1C-502	BLOCK-1256
1C-507	BLOCK-4
1C-602	BLOCK-1828
1C12	BLOCK-2166

DEVICE TYPE	REPL CODE
1C13	BLOCK-1272
1C15	BLOCK-1269
1C17	BLOCK-1405
1C18	BLOCK-1407
1C20	BLOCK-1844
1C202	BLOCK-1407
1C204	BLOCK-1770
1C21	BLOCK-1807
1C26	BLOCK-1269
1C288	BLOCK-1804
1C311	BLOCK-1286
1C319	BLOCK-1811
1C504	BLOCK-1262
1C505	BLOCK-1264
1C507	BLOCK-43
1C508	BLOCK-1270
1C509	BLOCK-1271
1C510	BLOCK-1272
1C511	BLOCK-1275
1C6	BLOCK-1267
1C8	BLOCK-1421
1C91	BLOCK-1212
1CL-101-TY	BLOCK-160
1CL-101A-TY	BLOCK-160
1CL-201-TY	BLOCK-160
1CL-201A-TY	BLOCK-160
1CL-301A-PA	BLOCK-160
1CL-301A-TY	BLOCK-160
1CL-748-TY	BLOCK-160
1E703E	BLOCK-1212
1FP70020H	BLOCK-1245
1KK-HA11225	BLOCK-476
1KK-HA1196	BLOCK-472
1LA7837	BLOCK-1266
1LM340T12	BLOCK-2097
1M383T	BLOCK-221
1MC1358P	BLOCK-1269
1S-0142	BLOCK-1290
1S1699	BLOCK-1269
1S1700	BLOCK-1269
1TA7607AP	BLOCK-532
1TA7608CP5	BLOCK-519
1TA7609P-2	BLOCK-484
1UPC1363C	BLOCK-396
1UPD1937C	BLOCK-740
1UPD1986C	BLOCK-739
20-0352	BLOCK-1944
2000-001	BLOCK-186
2000-002	BLOCK-1039
2000-003	BLOCK-191
2000-004	BLOCK-2144
2000-005	BLOCK-182
2000-006	BLOCK-183
2000-007	BLOCK-80
2000-008	BLOCK-2144
2000-009	BLOCK-2139
2000-010	BLOCK-192
2000-011	BLOCK-192
2000-017	BLOCK-180
2000-029	BLOCK-249
2000-031	BLOCK-238
2000-032	BLOCK-265
2000-033	BLOCK-258
2000-036	BLOCK-515
2000-037	BLOCK-411
2000-052	BLOCK-242
200137	BLOCK-1191
2002010061	BLOCK-1426
2002010072	BLOCK-1410
2002020066	BLOCK-1088
2002020077	BLOCK-1093
2002100022	BLOCK-5
2002110206	BLOCK-1269
2002110269	BLOCK-1269
2002110332	BLOCK-654
2002110345	BLOCK-603
2002120012	BLOCK-1422
2002120022	BLOCK-5
2002120033	BLOCK-78
2002120046	BLOCK-510
2002120051	BLOCK-532
2002120067	BLOCK-401
2002300012	BLOCK-234
2002300023	BLOCK-459
2002300029	BLOCK-537
2002300033	BLOCK-459
2002300046	BLOCK-662
2002400617	BLOCK-1825
2002400638	BLOCK-563
2002400752	BLOCK-559
2002400779	BLOCK-1196
2002501006	BLOCK-1271
2002501300	BLOCK-1816
2002501708	BLOCK-184

DEVICE TYPE	REPL CODE
2002502202	BLOCK-283
2002502317	BLOCK-576
2002502338	BLOCK-774
2002502341	BLOCK-774
2002502385	BLOCK-774
2002600107	BLOCK-1157
2002600149	BLOCK-1157
2002600183	BLOCK-533
2002600198	BLOCK-1157
2002600213	BLOCK-723
2002610039	BLOCK-775
2002610042	BLOCK-776
2002610086	BLOCK-781
2002610091	BLOCK-782
2002610109	BLOCK-374
2002700101	BLOCK-727
2002700158	BLOCK-695
2002700175	BLOCK-663
2002800034	BLOCK-2058
2002800112	BLOCK-1801
2002800133	BLOCK-2063
2002800146	BLOCK-2063
2002800178	BLOCK-2063
2002900042	BLOCK-983
2002900395	BLOCK-556
2002900408	BLOCK-988
2003251	BLOCK-2087
2003321	BLOCK-1994
2004-70	BLOCK-1039
200420	BLOCK-1981
2004341	BLOCK-385
200X2100-022	BLOCK-5
200X2110-269	BLOCK-1269
200X2120-012	BLOCK-1422
200X2120-033	BLOCK-78
200X2300-033	BLOCK-459
200X2501-708	BLOCK-184
200X2600-183	BLOCK-533
20112	BLOCK-840
2016	BLOCK-1004
2016P	BLOCK-1004
2025-1	BLOCK-1286
2025-2	BLOCK-1286
2025-3	BLOCK-1286
202914-010	BLOCK-2197
202914-010/250-060	BLOCK-2207
202914-010/250-070	BLOCK-2201
202914-010/250-100	BLOCK-2202
202914-020	BLOCK-2208
202914-030	BLOCK-2217
202914-040	BLOCK-2199
202914-050	BLOCK-2206
202920-150	BLOCK-1424
202922-280	BLOCK-1273
2032-38	BLOCK-1039
203743	BLOCK-84
203751	BLOCK-94
203780	BLOCK-126
203781	BLOCK-127
2042-1	BLOCK-110
2042-2	BLOCK-109
2047-52	BLOCK-288
2049-03	BLOCK-2144
2056-04	BLOCK-80
2056-05	BLOCK-2138
2056-06	BLOCK-149
2056-105	BLOCK-186
2057A32-25	BLOCK-122
2057A32-26	BLOCK-138
2057A32-33	BLOCK-1825
205919	BLOCK-1912
206-5-0131-22210	BLOCK-215
206-5-0723-20110	BLOCK-168
206-5-0783-21011	BLOCK-454
206-5-0824-10110	BLOCK-169
206-5-0834-10214	BLOCK-216
206-5-2341-41610	BLOCK-500
2061-42	BLOCK-2144
2065-50	BLOCK-189
2065-51	BLOCK-1039
2065-52	BLOCK-2144
2065-53	BLOCK-186
2068-018-104	BLOCK-1825
2068-022-108	BLOCK-1813
2068510-0701	BLOCK-2202
2068510-0702	BLOCK-2208
2068510-0703	BLOCK-2217
2068510-0704	BLOCK-1950
2068510-0705	BLOCK-1953
207484	BLOCK-1872
2077-2	BLOCK-1813

DEVICE TYPE	REPL CODE
207827	BLOCK-2171
207890	BLOCK-1096
2091-02	BLOCK-149
2091-49	BLOCK-7
2091-50	BLOCK-6
2093/1097	BLOCK-95
20B1M	BLOCK-1212
20SI-TA7607AP	BLOCK-532
2102400203	BLOCK-1801
2102400429	BLOCK-1801
2102A	BLOCK-999
2102AN-4L	BLOCK-999
2102FDC	BLOCK-999
2102FPC	BLOCK-999
2104	BLOCK-1000
2107	BLOCK-1001
2109-101-5304	BLOCK-398
2109-102-5304	BLOCK-398
2109-102-5401	BLOCK-286
2109-103-2006	BLOCK-1269
2109-103-2200	BLOCK-532
2109-103-3001	BLOCK-534
2109-301-0707	BLOCK-1568
2109-301-6008	BLOCK-695
2109-301-6600	BLOCK-740
2109-301-6707	BLOCK-739
2111A	BLOCK-1262
2114	BLOCK-1002
2114-30L	BLOCK-1002
2114L	BLOCK-1002
2114L-3	BLOCK-1002
2114L-30	BLOCK-1002
2117	BLOCK-1003
2119-101-0500	BLOCK-1269
2119-101-1204	BLOCK-5
2119-101-1301	BLOCK-126
2119-101-1408	BLOCK-131
2119-101-1505	BLOCK-225
2119-101-1602	BLOCK-184
2119-101-2607	BLOCK-532
2119-101-3204	BLOCK-1873
2119-101-4306	BLOCK-1269
2119-101-5107	BLOCK-428
2119-101-5204	BLOCK-404
2119-101-9408	BLOCK-532
2119-101-9602	BLOCK-534
2119-101-9709	BLOCK-534
2119-102-0309	BLOCK-1269
2119-102-0503	BLOCK-709
2119-102-0600	BLOCK-824
2119-102-0907	BLOCK-709
2119-102-1003	BLOCK-824
2119-102-2600	BLOCK-532
2119-102-3003	BLOCK-660
2119-102-4309	BLOCK-664
2119-102-5100	BLOCK-428
2119-102-5401	BLOCK-286
2119-102-5702	BLOCK-660
2119-102-7304	BLOCK-761
2119-102-7401	BLOCK-360
2119-102-7605	BLOCK-2097
2119-103-0603	BLOCK-1157
2119-103-2302	BLOCK-1147
2119-201-1300	BLOCK-1873
2119-201-3009	BLOCK-1147
2119-202-0803	BLOCK-136
2119-601-2709	BLOCK-540
2119-601-5005	BLOCK-722
2119-601-5209	BLOCK-540
2119-901-0201	BLOCK-876
2119-901-1109	BLOCK-2097
21192010402	BLOCK-1813
21192021706	BLOCK-1825
21192050103	BLOCK-207
2135E	BLOCK-1288
2149-203-0607	BLOCK-540
215531	BLOCK-769
2165	BLOCK-1269
2169-103-2006	BLOCK-1269
2169-401-5808	BLOCK-1157
219-103-2006	BLOCK-1269
2199-101-9408	BLOCK-532
21A040-068	BLOCK-73
21A101-001	BLOCK-1245
21A101-001(IC)	BLOCK-1245
21A101-002	BLOCK-1267
21A101-004	BLOCK-45
21A101-005	BLOCK-44
21A101-006	BLOCK-87
21A101-007	BLOCK-46
21A101-008	BLOCK-75
21A101-009	BLOCK-74
21A101-010	BLOCK-89
21A101-011	BLOCK-82

6 ORIGINAL DEVICE TYPES AND REPLACEMENT CODES

DEVICE TYPE	REPL CODE
21A101-012	BLOCK-84
21A101-013	BLOCK-33
21A101-015	BLOCK-1407
21A101-016	BLOCK-1272
21A101-017	BLOCK-1271
21A101-018	BLOCK-48
21A119-045	BLOCK-1157
21A120-001	BLOCK-1422
21A120-002	BLOCK-1407
21A120-008	BLOCK-1269
21A120-009	BLOCK-247
21A120-016	BLOCK-1293
21A120-017	BLOCK-1301
21A120-044	BLOCK-184
21A120-045	BLOCK-246
21A120-049	BLOCK-404
21A120-050	BLOCK-393
21A120-051	BLOCK-532
21A120-052	BLOCK-405
21A120-071	BLOCK-130
21A120-074	BLOCK-425
21A120-099	BLOCK-532
21A120-178	BLOCK-399
21A120-185	BLOCK-398
21A120-204	BLOCK-537
21A120-208	BLOCK-235
21B1AH	BLOCK-1287
21B1M	BLOCK-1275
21B1Z	BLOCK-1275
21L02B	BLOCK-999
21M485	BLOCK-138
21M506	BLOCK-1823
21M532	BLOCK-7
21M582	BLOCK-122
221-0048	BLOCK-1269
221-101-01	BLOCK-2097
221-103	BLOCK-1892
221-103A	BLOCK-1892
221-104	BLOCK-1179
221-105	BLOCK-1190
221-105-01	BLOCK-1190
221-106	BLOCK-1201
221-106-01	BLOCK-1201
221-107	BLOCK-1389
221-108	BLOCK-1813
221-116	BLOCK-113
221-121	BLOCK-1873
221-129	BLOCK-2171
221-129A-01	BLOCK-2171
221-131	BLOCK-1824
221-132	BLOCK-1918
221-140	BLOCK-1888
221-141	BLOCK-1843
221-158	BLOCK-220
221-158-01	BLOCK-220
221-158-02	BLOCK-220
221-158-02001	BLOCK-220
221-158-03	BLOCK-220
221-166	BLOCK-2087
221-166-0	BLOCK-2087
221-166-00	BLOCK-2087
221-166-02	BLOCK-2095
221-166-04	BLOCK-2097
221-166-4	BLOCK-2097
221-167-01	BLOCK-2144
221-167-01A	BLOCK-2144
221-167-05	BLOCK-2074
221-171	BLOCK-2174
221-173	BLOCK-1093
221-175	BLOCK-1929
221-175-01	BLOCK-1929
221-177	BLOCK-363
221-178	BLOCK-620
221-179-01	BLOCK-1928
221-179-01-E-01	BLOCK-1928
221-179-02	BLOCK-1928
221-179-03	BLOCK-1928
221-179-1	BLOCK-1928
221-182-01	BLOCK-1063
221-187	BLOCK-668
221-187A	BLOCK-668
221-188	BLOCK-2171
221-190	BLOCK-1933
221-190-01	BLOCK-1930
221-190-01A	BLOCK-1930
221-190-E-01	BLOCK-1930
221-193	BLOCK-1892
221-193A	BLOCK-1892
221-202	BLOCK-281
221-213	BLOCK-2087
221-213-00	BLOCK-2087
221-213-02	BLOCK-2095
221-213-04	BLOCK-2097
221-213-4	BLOCK-2097

DEVICE TYPE	REPL CODE
221-217	BLOCK-2079
221-230	BLOCK-563
221-240	BLOCK-1801
221-261	BLOCK-1935
221-261-01	BLOCK-1935
221-261B03	BLOCK-1935
221-291	BLOCK-1182
221-296-04	BLOCK-2097
221-303	BLOCK-725
221-31	BLOCK-1212
221-316	BLOCK-367
221-32	BLOCK-1212
221-34	BLOCK-1262
221-346	BLOCK-1197
221-347	BLOCK-769
221-347-01	BLOCK-769
221-36	BLOCK-1244
221-364	BLOCK-1003
221-37	BLOCK-1256
221-383	BLOCK-773
221-39	BLOCK-1256
221-40	BLOCK-2056
221-41	BLOCK-1256
221-42	BLOCK-1271
221-422-01	BLOCK-361
221-43	BLOCK-1272
221-45	BLOCK-1287
221-45-01	BLOCK-1287
221-46	BLOCK-1270
221-467	BLOCK-771
221-468	BLOCK-763
221-475	BLOCK-839
221-4757D04	BLOCK-839
221-475C	BLOCK-839
221-476	BLOCK-840
221-4767D02	BLOCK-840
221-476D	BLOCK-840
221-48	BLOCK-1269
221-48-01	BLOCK-1269
221-516	BLOCK-1196
221-561	BLOCK-843
221-598	BLOCK-671
221-62	BLOCK-1815
221-637-01	BLOCK-1222
221-64	BLOCK-109
221-64(IC)	BLOCK-110
221-679	BLOCK-1238
221-69	BLOCK-1816
221-69-01	BLOCK-1816
221-76	BLOCK-1794
221-76-01	BLOCK-1794
221-77	BLOCK-1838
221-78	BLOCK-1816
221-796	BLOCK-754
221-80	BLOCK-110
221-82	BLOCK-1270
221-86	BLOCK-1876
221-87	BLOCK-1271
221-87-01	BLOCK-2166
221-89	BLOCK-1289
221-90	BLOCK-1279
221-91	BLOCK-1825
221-91-01	BLOCK-1825
221-93	BLOCK-2058
221-96	BLOCK-1846
221-96C	BLOCK-1846
221-97	BLOCK-1886
221-97-01	BLOCK-1886
221-98	BLOCK-219
221-98-05	BLOCK-219
221-98A05	BLOCK-219
221-992	BLOCK-838
221-Z9000	BLOCK-1223
221-Z9001	BLOCK-1245
221-Z9002	BLOCK-1264
221-Z9003	BLOCK-1265
221-Z9004	BLOCK-1267
221-Z9005	BLOCK-1275
221-Z9006	BLOCK-1277
221-Z9007	BLOCK-1284
221-Z9008	BLOCK-1285
221-Z9009	BLOCK-1286
221-Z9010	BLOCK-1291
221-Z9011	BLOCK-1376
221-Z9012	BLOCK-1421
221-Z9013	BLOCK-1422
221-Z9014	BLOCK-1805
221-Z9015	BLOCK-1814
221-Z9016	BLOCK-1828
221-Z9017	BLOCK-1949
221-Z9018	BLOCK-1963
221-Z9019	BLOCK-1983
221-Z9020	BLOCK-1984
221-Z9021	BLOCK-4

DEVICE TYPE	REPL CODE
221-Z9022	BLOCK-5
221-Z9023	BLOCK-98
221-Z9024	BLOCK-138
221-Z9025	BLOCK-1282
221-Z9026	BLOCK-7
221-Z9027	BLOCK-1825
221-Z9028	BLOCK-1405
221-Z9029	BLOCK-1407
221-Z9031	BLOCK-1819
221-Z9032	BLOCK-1292
221-Z9033	BLOCK-1364
221-Z9034	BLOCK-1801
221-Z9036	BLOCK-1808
221-Z9037	BLOCK-1813
221-Z9038	BLOCK-1817
221-Z9039	BLOCK-1818
221-Z9040	BLOCK-1820
221-Z9041	BLOCK-1821
221-Z9042	BLOCK-2079
221-Z9043	BLOCK-2087
221-Z9044	BLOCK-2144
221-Z9045	BLOCK-2160
221-Z9046	BLOCK-10
221-Z9047	BLOCK-44
221-Z9048	BLOCK-107
221-Z9049	BLOCK-115
221-Z9050	BLOCK-131
221-Z9051	BLOCK-145
221-Z9052	BLOCK-150
221-Z9053	BLOCK-153
221-Z9054	BLOCK-162
221-Z9055	BLOCK-163
221-Z9056	BLOCK-164
221-Z9057	BLOCK-165
221-Z9058	BLOCK-172
221-Z9059	BLOCK-175
221-Z9060	BLOCK-176
221-Z9061	BLOCK-180
221-Z9062	BLOCK-181
221-Z9063	BLOCK-182
221-Z9064	BLOCK-184
221-Z9065	BLOCK-222
221-Z9066	BLOCK-223
221-Z9067	BLOCK-229
221-Z9068	BLOCK-238
221-Z9069	BLOCK-239
221-Z9070	BLOCK-243
221-Z9071	BLOCK-248
221-Z9072	BLOCK-1064
221-Z9073	BLOCK-1068
221-Z9074	BLOCK-1084
221-Z9075	BLOCK-1293
221-Z9076	BLOCK-1297
221-Z9077	BLOCK-1365
221-Z9078	BLOCK-1426
221-Z9079	BLOCK-2197
221-Z9080	BLOCK-2199
221-Z9081	BLOCK-2202
221-Z9082	BLOCK-2205
221-Z9083	BLOCK-2207
221-Z9084	BLOCK-2208
221-Z9085	BLOCK-2217
221-Z9086	BLOCK-1313
221-Z9087	BLOCK-1279
221-Z9088	BLOCK-1400
221-Z9089	BLOCK-1770
221-Z9090	BLOCK-1786
221-Z9091	BLOCK-1790
221-Z9093	BLOCK-1823
221-Z9094	BLOCK-1826
221-Z9095	BLOCK-1827
221-Z9096	BLOCK-1843
221-Z9097	BLOCK-1844
221-Z9098	BLOCK-2138
221-Z9099	BLOCK-28
221-Z9100	BLOCK-29
221-Z9101	BLOCK-34
221-Z9102	BLOCK-35
221-Z9103	BLOCK-36
221-Z9104	BLOCK-38
221-Z9105	BLOCK-40
221-Z9106	BLOCK-41
221-Z9107	BLOCK-56
221-Z9108	BLOCK-59
221-Z9109	BLOCK-67
221-Z9110	BLOCK-69
221-Z9111	BLOCK-70
221-Z9112	BLOCK-73
221-Z9113	BLOCK-103
221-Z9114	BLOCK-108
221-Z9115	BLOCK-112
221-Z9116	BLOCK-125
221-Z9117	BLOCK-128
221-Z9118	BLOCK-129

DEVICE TYPE	REPL CODE
221-Z9119	BLOCK-143
221-Z9120	BLOCK-144
221-Z9121	BLOCK-147
221-Z9122	BLOCK-148
221-Z9123	BLOCK-114
221-Z9124	BLOCK-157
221-Z9125	BLOCK-158
221-Z9126	BLOCK-166
221-Z9127	BLOCK-178
221-Z9129	BLOCK-179
221-Z9130	BLOCK-212
221-Z9131	BLOCK-227
221-Z9132	BLOCK-237
221-Z9133	BLOCK-1039
221-Z9134	BLOCK-18
221-Z9135	BLOCK-136
221-Z9137	BLOCK-126
221-Z9138	BLOCK-122
221-Z9139	BLOCK-1809
221-Z9140	BLOCK-1830
221-Z9141	BLOCK-1944
221-Z9142	BLOCK-1948
221-Z9143	BLOCK-1955
221-Z9144	BLOCK-2056
221-Z9145	BLOCK-2057
221-Z9146	BLOCK-114
221-Z9147	BLOCK-2069
221-Z9148	BLOCK-2070
221-Z9149	BLOCK-76
221-Z9150	BLOCK-2093
221-Z9151	BLOCK-58
221-Z9152	BLOCK-2145
221-Z9153	BLOCK-1847
221-Z9154	BLOCK-1849
221-Z9155	BLOCK-1850
221-Z9156	BLOCK-46
221-Z9157	BLOCK-87
221-Z9158	BLOCK-121
221-Z9159	BLOCK-157
221-Z9160	BLOCK-174
221-Z9161	BLOCK-211
221-Z9162	BLOCK-250
221-Z9163	BLOCK-1085
221-Z9164	BLOCK-1093
221-Z9166	BLOCK-204
221-Z9167	BLOCK-1848
221-Z9168	BLOCK-1860
221-Z9169	BLOCK-1875
221-Z9170	BLOCK-1885
221-Z9171	BLOCK-1887
221-Z9172	BLOCK-1895
221-Z9173	BLOCK-2071
221-Z9174	BLOCK-2168
221-Z9175	BLOCK-2169
221-Z9176	BLOCK-2170
221-Z9177	BLOCK-130
221-Z9178	BLOCK-172
221-Z9179	BLOCK-235
221-Z9180	BLOCK-253
221-Z9181	BLOCK-256
221-Z9182	BLOCK-257
22114594	BLOCK-1801
22114626	BLOCK-474
22114634	BLOCK-498
221G3114	BLOCK-1943
221G3114-1	BLOCK-1943
223-18	BLOCK-535
223-18-01	BLOCK-757
223-28	BLOCK-819
223-40	BLOCK-758
223-40-01	BLOCK-758
223-Z9000	BLOCK-81
223-Z9003	BLOCK-88
223-Z9004	BLOCK-99
223-Z9005	BLOCK-100
223-Z9006	BLOCK-101
223-Z9007	BLOCK-102
223-Z9008	BLOCK-139
2234494	BLOCK-2176
2234496	BLOCK-2189
2234497	BLOCK-2194
223883	BLOCK-822
225A6946-P000	BLOCK-1293
225A6946-P003	BLOCK-1296
225A6946-P004	BLOCK-1297
225A6946-P010	BLOCK-1304
225A6946-P020	BLOCK-1365
225A6946-P050	BLOCK-1401
225A6946-P093	BLOCK-1426
225A6976-P095	BLOCK-1428
226P10601	BLOCK-418
2275-1	BLOCK-1859
229-1301-39	BLOCK-1819
229-1301-40	BLOCK-1364

DEVICE TYPE	REPL CODE
229-1301-41	BLOCK-1815
229-1301-42	BLOCK-1407
229-1301-43	BLOCK-1820
229-1301-44	BLOCK-1291
2295361	BLOCK-2217
2304-1	BLOCK-433
23114420	BLOCK-759
23115822	BLOCK-1157
23115878	BLOCK-1157
23115922	BLOCK-1157
23119003	BLOCK-1267
23119004	BLOCK-1265
23119005	BLOCK-1267
23119007	BLOCK-1265
23119011	BLOCK-99
23119012	BLOCK-5
23119013	BLOCK-107
23119014	BLOCK-1407
23119016	BLOCK-103
23119017	BLOCK-60
23119019	BLOCK-1422
23119022	BLOCK-99
23119023	BLOCK-107
23119025	BLOCK-1421
23119032	BLOCK-107
23119033	BLOCK-107
23119142	BLOCK-763
23119228	BLOCK-773
23119355	BLOCK-771
23119489	BLOCK-725
23119548	BLOCK-670
23119566	BLOCK-772
23119710	BLOCK-761
23119723	BLOCK-814
23119742	BLOCK-1644
23119763	BLOCK-360
23119796	BLOCK-534
23119901	BLOCK-532
23119950	BLOCK-1269
23119960	BLOCK-1426
23119961	BLOCK-132
23119966	BLOCK-184
23119971	BLOCK-188
23119978	BLOCK-225
23119981	BLOCK-126
23119988	BLOCK-130
23119989	BLOCK-131
23119990	BLOCK-126
23119993	BLOCK-127
23119994	BLOCK-129
23119995	BLOCK-128
23119999	BLOCK-1421
2319990	BLOCK-126
2327302	BLOCK-1212
2327311	BLOCK-38
2327312	BLOCK-38
2327411	BLOCK-40
2327421	BLOCK-1277
2327422	BLOCK-1277
23314068	BLOCK-820
23314158	BLOCK-873
23316411	BLOCK-1157
23318086	BLOCK-826
23318409	BLOCK-814
23319802	BLOCK-1993
2335991	BLOCK-1157
236-0002	BLOCK-2199
236-0003	BLOCK-2208
236-0005	BLOCK-1293
236-0006	BLOCK-1296
236-0007	BLOCK-1297
236-0008	BLOCK-1298
236-0009	BLOCK-1410
236-0012	BLOCK-2157
236-0017	BLOCK-2217
2360021	BLOCK-59
2360042	BLOCK-1269
2360141	BLOCK-5
2360151	BLOCK-201
2360171	BLOCK-5
2360201	BLOCK-1269
2360221	BLOCK-92
2360231	BLOCK-119
2360271	BLOCK-247
2360281	BLOCK-240
2360291	BLOCK-246
2360331	BLOCK-1269
2360361	BLOCK-5
2360391	BLOCK-1269
2360401	BLOCK-5
2360431	BLOCK-184
2360441	BLOCK-184
2360501	BLOCK-636
2360501(SIF)	BLOCK-1269

DEVICE TYPE	REPL CODE
2360511	BLOCK-5
2360561	BLOCK-1305
2360611	BLOCK-562
2360631	BLOCK-201
2360691	BLOCK-537
2360741	BLOCK-1293
2360751	BLOCK-1744
2360782	BLOCK-457
2360851	BLOCK-1039
2360951	BLOCK-1572
2361251	BLOCK-1299
2361281	BLOCK-404
2361301	BLOCK-740
2361461	BLOCK-2097
2361571	BLOCK-461
2361572	BLOCK-461
2362051	BLOCK-1574
2362301	BLOCK-1659
2362541	BLOCK-580
2362601	BLOCK-1801
2362605	BLOCK-1801
2362991	BLOCK-501
2363191	BLOCK-1093
2363261	BLOCK-461
2364131	BLOCK-533
2364172	BLOCK-510
2364181	BLOCK-459
2365021	BLOCK-654
2365061	BLOCK-636
2365062	BLOCK-636
2365411	BLOCK-1873
2365452	BLOCK-2063
2366151	BLOCK-639
2366201	BLOCK-724
2366392	BLOCK-761
2366591	BLOCK-221
2366621	BLOCK-644
2366631	BLOCK-668
2366721	BLOCK-664
2368211	BLOCK-710
2368282	BLOCK-642
2368501	BLOCK-662
23685015	BLOCK-662
2369151	BLOCK-576
23691515	BLOCK-576
2370873	BLOCK-723
2380426	BLOCK-2097
23904303	BLOCK-1993
23904391	BLOCK-1993
2420SLM339N	BLOCK-1873
2420SN07406	BLOCK-1299
2420SN07407	BLOCK-1300
2420SN75188	BLOCK-1771
2420SN75189	BLOCK-1772
2420SNLS151	BLOCK-1598
2420SNNE555	BLOCK-2079
2420T78005P	BLOCK-2087
2420TMM16P2	BLOCK-1004
2432	BLOCK-1265
2434	BLOCK-1265
2470-1724	BLOCK-1472
2470-1732	BLOCK-1474
2473-2109	BLOCK-1301
24MW1028	BLOCK-4
24MW1110	BLOCK-31
24MW996	BLOCK-20
24MW997	BLOCK-23
24S10	BLOCK-1749
2500389-432	BLOCK-2199
2500389-444	BLOCK-2206
2500390-435	BLOCK-2201
2500390-436	BLOCK-2202
2500390-437	BLOCK-2203
2500390-446	BLOCK-2208
2500390-449	BLOCK-2210
2500390-461	BLOCK-2216
2500390-462	BLOCK-2217
2500391-445	BLOCK-2207
2500747	BLOCK-2199
2501337-433	BLOCK-2200
2501557-430	BLOCK-2197
2501557-436	BLOCK-2202
2501557-437	BLOCK-2203
2501557-446	BLOCK-2208
2501557-449	BLOCK-2210
2501557-461	BLOCK-2216
2501557-462	BLOCK-2217
2516	BLOCK-1269
2516-1	BLOCK-1269
2516JL-45	BLOCK-1019
2554-1(RCA)	BLOCK-1286
2554-2(RCA)	BLOCK-1286
2554-3	BLOCK-1286
2554-3(RCA)	BLOCK-1286

ORIGINAL DEVICE TYPES AND REPLACEMENT CODES 7

DEVICE TYPE	REPL CODE
2554-4	BLOCK-1286
2554-4(RCA)	BLOCK-1286
2559	BLOCK-1284
2559-1	BLOCK-1284
2560	BLOCK-1285
2560(J.C.PENNEY)	BLOCK-1285
2560-1	BLOCK-1285
25810-107	BLOCK-79
25810-164	BLOCK-98
25810-165	BLOCK-138
25810-167	BLOCK-79
25840-165	BLOCK-138
25840-167	BLOCK-79
25B1C	BLOCK-1286
25B1T	BLOCK-1286
25B2C	BLOCK-1286
25B2T	BLOCK-1286
260-10-036	BLOCK-98
2610783	BLOCK-1295
2610784	BLOCK-1429
2610786	BLOCK-1293
2610788	BLOCK-1311
2626-1	BLOCK-1963
263P05209	BLOCK-1087
263P06602	BLOCK-1093
263P17402	BLOCK-1049
263P701010	BLOCK-1545
264P24401	BLOCK-1157
264P244010	BLOCK-1157
264P244020	BLOCK-1157
2652613	BLOCK-2206
2652614	BLOCK-2207
2656211-1	BLOCK-2197
2656212	BLOCK-2208
2656212-1	BLOCK-2208
2656213	BLOCK-2206
2656213-1	BLOCK-2206
2656214	BLOCK-2207
2656214-1	BLOCK-2207
2656747	BLOCK-2206
2656748	BLOCK-2208
26630109	BLOCK-1269
2666293	BLOCK-1416
2666294-1	BLOCK-1417
266P00101	BLOCK-1223
266P00102	BLOCK-1223
266P00301	BLOCK-94
266P00602	BLOCK-97
266P00801	BLOCK-94
266P01002	BLOCK-1157
266P016010	BLOCK-762
266P0501	BLOCK-390
266P10101	BLOCK-1422
266P10103	BLOCK-1422
266P10201	BLOCK-1407
266P10202	BLOCK-1407
266P10601	BLOCK-418
266P11301	BLOCK-532
266P11601	BLOCK-401
266P12101	BLOCK-709
266P15401	BLOCK-2063
266P154010	BLOCK-2063
266P19201	BLOCK-644
266P20402	BLOCK-285
266P28101	BLOCK-725
266P30102	BLOCK-1269
266P30103	BLOCK-1269
266P30109	BLOCK-1269
266P30201	BLOCK-73
266P30601	BLOCK-28
266P30706	BLOCK-101
266P32301	BLOCK-1269
266P32302	BLOCK-1269
266P32402	BLOCK-248
266P32501	BLOCK-180
266P32801	BLOCK-1813
266P35001	BLOCK-205
266P35801	BLOCK-207
266P36402	BLOCK-206
266P38802	BLOCK-761
266P50101	BLOCK-390
266P57001	BLOCK-719
266P60304	BLOCK-172
266P60502	BLOCK-1291
266P61401	BLOCK-284
266P62103	BLOCK-2058
266P63001	BLOCK-534
266P71301	BLOCK-1281
266P75201	BLOCK-1971
266P922010	BLOCK-2087
266P922020	BLOCK-2087
266P923020	BLOCK-881
266P92304	BLOCK-2144
266P931010	BLOCK-881
266P93402	BLOCK-2097
266P934020	BLOCK-2097
266P934040	BLOCK-2097
266P934060	BLOCK-2087
267P90202	BLOCK-535
267P90205	BLOCK-533
267P91002	BLOCK-758
267P91003	BLOCK-757
267P91006	BLOCK-758
26810-156	BLOCK-223
26810-157	BLOCK-1825
26810-158	BLOCK-50
26810-159	BLOCK-145
2709759	BLOCK-1022
2711904	BLOCK-1265
2716	BLOCK-1017
2716-SA2-A0	BLOCK-1017
2716-SC2-A1	BLOCK-1017
272P01401	BLOCK-782
272P026010	BLOCK-826
272P14001	BLOCK-779
272P140010	BLOCK-779
272P15901	BLOCK-670
272P238010	BLOCK-843
272P238020	BLOCK-843
272P239010	BLOCK-835
272P461010	BLOCK-1191
2732	BLOCK-1018
2732/2332	BLOCK-1018
2732A	BLOCK-1018
2732A-2	BLOCK-1018
2732A-30	BLOCK-1018
2732D	BLOCK-1018
274-69BIC	BLOCK-1272
276-007	BLOCK-2058
276-009	BLOCK-1983
276-010	BLOCK-2056
276-038	BLOCK-1801
276-1251	BLOCK-1019
276-1705	BLOCK-1931
276-1707	BLOCK-496
276-1708	BLOCK-497
276-1709	BLOCK-536
276-1711	BLOCK-2171
276-1712	BLOCK-1873
276-1713	BLOCK-2191
276-1714	BLOCK-1912
276-1715	BLOCK-1910
276-1718	BLOCK-2080
276-1720	BLOCK-2174
276-1721	BLOCK-1868
276-1723	BLOCK-2079
276-1725	BLOCK-1303
276-1728	BLOCK-2145
276-1731	BLOCK-1851
276-1737	BLOCK-1852
276-1740	BLOCK-1984
276-1743	BLOCK-1872
276-1757	BLOCK-1277
276-1758	BLOCK-1405
276-1759	BLOCK-1269
276-1762	BLOCK-879
276-1770	BLOCK-2087
276-1771	BLOCK-2097
276-1772	BLOCK-2119
276-1773	BLOCK-2091
276-1774	BLOCK-2108
276-1777	BLOCK-2135
276-1778	BLOCK-2083
276-1792	BLOCK-968
276-1801	BLOCK-1293
276-1802	BLOCK-1297
276-1803	BLOCK-1410
276-1805	BLOCK-1397
276-1808	BLOCK-1423
276-1811	BLOCK-1295
276-1813	BLOCK-1413
276-1818	BLOCK-1411
276-1822	BLOCK-1301
276-1828	BLOCK-1325
276-1835	BLOCK-1668
276-1915	BLOCK-1659
276-2334	BLOCK-1919
276-2401	BLOCK-1029
276-2411	BLOCK-1039
276-2413	BLOCK-1041
276-2417	BLOCK-1051
276-2423	BLOCK-1062
276-2449	BLOCK-1084
276-2466	BLOCK-1093
276-2481	BLOCK-1105
276-2503	BLOCK-1003
276-2505	BLOCK-1003
276-2520	BLOCK-1771
276-2521	BLOCK-1772
276-703	BLOCK-221
276-705	BLOCK-145
276-706	BLOCK-1303
2764	BLOCK-1019
2764-2	BLOCK-1019
2764-25	BLOCK-1019
2764-250	BLOCK-1019
2764-30	BLOCK-1019
2764-450	BLOCK-1019
2764-FA4-A2	BLOCK-1019
2764-FA5-A3	BLOCK-1019
2764-FC5-A3	BLOCK-1019
27840-164	BLOCK-79
2799-1	BLOCK-1814
2799-2	BLOCK-1814
27A10446-101-11	BLOCK-1953
280-0001	BLOCK-98
280-0002	BLOCK-1825
2808577	BLOCK-2199
2868536-1	BLOCK-1411
28810-175	BLOCK-80
28810-176	BLOCK-138
28810-177	BLOCK-50
28810-178	BLOCK-79
2898431-2	BLOCK-2208
2899000-00	BLOCK-2208
2899001-00	BLOCK-2217
2899002-00	BLOCK-2197
2899003-00	BLOCK-2202
2899004-00	BLOCK-2207
2901	BLOCK-1874
2901(SM)	BLOCK-1874
2902798-2	BLOCK-1967
2903D	BLOCK-2063
2904D	BLOCK-1993
2906-004	BLOCK-244
2906-005	BLOCK-1405
2910011	BLOCK-824
2910012	BLOCK-752
2910021	BLOCK-754
2911481	BLOCK-725
2911501	BLOCK-205
2912175	BLOCK-759
2912176	BLOCK-759
2912177	BLOCK-758
2912178	BLOCK-759
2912561	BLOCK-779
2913981	BLOCK-763
2914931	BLOCK-840
2914941	BLOCK-839
2914961	BLOCK-1187
2915192	BLOCK-1188
2916681	BLOCK-748
2917361	BLOCK-748
2917601	BLOCK-780
2917611	BLOCK-762
293118	BLOCK-1342
295-95-1	BLOCK-1873
29611ADC	BLOCK-1763
29611DC	BLOCK-1763
29621ADC	BLOCK-1755
29621DC	BLOCK-1755
29631ADC	BLOCK-1759
29631DC	BLOCK-1759
29651ADC	BLOCK-1754
29651DC	BLOCK-1754
29810-179	BLOCK-1813
29810-180	BLOCK-1825
29B1B	BLOCK-1815
29B1Z	BLOCK-1815
2N5466	BLOCK-1808
300003220005	BLOCK-621
300007313167	BLOCK-453
300012026007	BLOCK-237
300012413001	BLOCK-544
3001-201	BLOCK-2184
3007359-00	BLOCK-2203
3007359-01	BLOCK-2210
3007359-02	BLOCK-2218
3007359-03	BLOCK-2199
3007472-00	BLOCK-2207
3007473-00	BLOCK-2199
3007473-01	BLOCK-2199
3007474-00	BLOCK-2197
3007474-01	BLOCK-2197
3007475-00	BLOCK-2217
3007475-01	BLOCK-2217
3007476-00	BLOCK-2208
3007476-01	BLOCK-2208
3007477-00	BLOCK-2202
3007477-01	BLOCK-2202
3007572-00	BLOCK-2200
3007573-00	BLOCK-1950
3007604-00	BLOCK-1965
3008321-00	BLOCK-1022
301-576-14	BLOCK-1303
301-576-2	BLOCK-1969
301-576-3	BLOCK-1262
301-576-4	BLOCK-1293
301-679-1	BLOCK-1845
301-680-1	BLOCK-1774
301003450375	BLOCK-975
301AL	BLOCK-1998
301CL	BLOCK-1998
302AL	BLOCK-1999
302CL	BLOCK-1999
303-1(IC)	BLOCK-404
3034725-1	BLOCK-1943
303AL	BLOCK-2000
303CL	BLOCK-2000
304AL	BLOCK-2001
304CL	BLOCK-2001
305-1	BLOCK-612
3052P	BLOCK-905
3052V	BLOCK-906
3064TC	BLOCK-1805
3066DC	BLOCK-1284
3067DC	BLOCK-1285
306AL	BLOCK-2002
306CL	BLOCK-2002
307-005-9-001	BLOCK-4
307-007-9-001	BLOCK-98
307-007-9-002	BLOCK-100
307-008-9-001	BLOCK-1280
307-029-1-001	BLOCK-98
307-047-9-001	BLOCK-2138
307-095-9-002	BLOCK-1039
307-095-9-003	BLOCK-2144
307-095-9-004	BLOCK-186
307-107-9-001	BLOCK-224
307-107-9-003	BLOCK-145
307-107-9-004	BLOCK-196
307-107-9-005	BLOCK-185
307-107-9-006	BLOCK-195
307-108-9-001	BLOCK-190
307-112-9-001	BLOCK-182
307-112-9-002	BLOCK-2144
307-112-9-003	BLOCK-2144
307-112-9-005	BLOCK-2138
307-112-9-006	BLOCK-183
307-112-9-007	BLOCK-80
307-112-9-009	BLOCK-186
307-113-9-001	BLOCK-1039
307-113-9-002	BLOCK-191
307-113-9-003	BLOCK-2144
307-113-9-004	BLOCK-183
307-115-9-001	BLOCK-80
307-120-9-001	BLOCK-229
307-131-9-001	BLOCK-191
307-131-9-002	BLOCK-248
307-131-9-006	BLOCK-214
307-131-9001	BLOCK-191
307-133-9-004	BLOCK-180
307-133-9-005	BLOCK-258
307-133-9001	BLOCK-258
307-143-9-001	BLOCK-249
307-143-9-002	BLOCK-238
307-143-9-003	BLOCK-265
307-145-9-001	BLOCK-515
307-152-9-004	BLOCK-1105
307-152-9-005	BLOCK-1624
307-152-9-007	BLOCK-1568
307-152-9-009	BLOCK-1717
307-152-9-012	BLOCK-1039
30710	BLOCK-1269
3071339004	BLOCK-180
3071459001	BLOCK-515
3071519001	BLOCK-411
3075DC	BLOCK-1279
307AL	BLOCK-2003
307CL	BLOCK-2003
309-033-0	BLOCK-2173
30900361	BLOCK-1801
30900460	BLOCK-202
30900530	BLOCK-503
30900540	BLOCK-477
30900720	BLOCK-476
30900740	BLOCK-1801
30900741	BLOCK-1801
30901050	BLOCK-2144
30901080	BLOCK-1684
30901120	BLOCK-476
30901150	BLOCK-929
30901350	BLOCK-685
3100001	BLOCK-1984
3100002	BLOCK-2191
3100004	BLOCK-2087
3102006	BLOCK-1568
3102007	BLOCK-1570
3102008	BLOCK-1572
3102009	BLOCK-1573
3102010	BLOCK-1579
3102011	BLOCK-1626
3102013	BLOCK-1658
3102014	BLOCK-1659
3102015	BLOCK-1706
3102016	BLOCK-1716
3102017	BLOCK-1717
3102018	BLOCK-1588
3102019	BLOCK-1599
3102020	BLOCK-1602
3102021	BLOCK-1610
3102022	BLOCK-1615
3102023	BLOCK-1616
3102024	BLOCK-1668
3102025	BLOCK-1669
3102026	BLOCK-1431
3102027	BLOCK-1433
3102028	BLOCK-1601
310254	BLOCK-1301
3106002	BLOCK-1777
3108002	BLOCK-999
3108003	BLOCK-1003
3108009	BLOCK-1003
3110001	BLOCK-1025
3119-901-110	BLOCK-2097
311AL	BLOCK-2005
311CL	BLOCK-2005
3122P	BLOCK-907
312AJ	BLOCK-2006
312AL	BLOCK-2006
312CJ	BLOCK-2006
312CL	BLOCK-2006
3130-0000-014	BLOCK-2160
3130-3167-909	BLOCK-1993
3130-3193-512	BLOCK-1812
3130-3248-801	BLOCK-112
3135/709	BLOCK-1262
3153GM1	BLOCK-2168
317-2627-1	BLOCK-2208
3172626	BLOCK-2200
3172627-2	BLOCK-2210
3172629-1	BLOCK-2197
3172629-2	BLOCK-2216
3172630	BLOCK-2206
3172631	BLOCK-2209
3172632-1	BLOCK-2217
3172638	BLOCK-2212
3176135-1	BLOCK-2202
3176135-2	BLOCK-2203
3177200	BLOCK-2199
317K	BLOCK-2135
317T	BLOCK-2083
31SI-UPC1360C	BLOCK-508
32-23555-1	BLOCK-1223
32-23555-2	BLOCK-1223
32-23555-3	BLOCK-1223
32-23555-4	BLOCK-1223
32109-301-070	BLOCK-1568
32109-301-610	BLOCK-814
32109-301-890	BLOCK-1088
32109-410-270	BLOCK-1093
32109-410-280	BLOCK-1093
32117-901-110	BLOCK-2097
32119-101-940	BLOCK-532
32119-102-010	BLOCK-817
32119-102-020	BLOCK-836
32119-102-040	BLOCK-1188
32119-102-050	BLOCK-709
32119-102-060	BLOCK-824
32119-102-090	BLOCK-709
32119-102-100	BLOCK-824
32119-102-260	BLOCK-532
32119-102-300	BLOCK-660
32119-102-570	BLOCK-660
32119-102-730	BLOCK-761
32119-102-760	BLOCK-2097
32119-103-110	BLOCK-810
32119-103-420	BLOCK-735
32119-110-016	BLOCK-735
32119-110-050	BLOCK-1222
32119-201-310	BLOCK-1088
32119-201-360	BLOCK-2056
32119-401-010	BLOCK-1182
32119-801-130	BLOCK-758
32119-901-020	BLOCK-876
32119-901-110	BLOCK-2097
32119-901-130	BLOCK-758
32199-901-200	BLOCK-874
321AJ	BLOCK-2008
321AL	BLOCK-2008
321CJ	BLOCK-2008
321CL	BLOCK-2008
322AJ	BLOCK-2009
322AL	BLOCK-2009
322CJ	BLOCK-2009
322CL	BLOCK-2009
32309-024-010	BLOCK-1021
323AJ	BLOCK-2010
323AL	BLOCK-2010
323CJ	BLOCK-2010
323CL	BLOCK-2010
324(IC)	BLOCK-2171
324AL	BLOCK-2011
324CL	BLOCK-2011
325502-01	BLOCK-1004
325502-03	BLOCK-1004
325AJ	BLOCK-2012
325AL	BLOCK-2012
325CJ	BLOCK-2012
325CL	BLOCK-2012
3267390-01	BLOCK-1488
326830	BLOCK-2197
326832	BLOCK-2199
326833	BLOCK-2200
326836	BLOCK-2202
326844	BLOCK-2206
326845	BLOCK-2207
326846	BLOCK-2208
326852	BLOCK-1953
326853	BLOCK-1950
326862	BLOCK-2217
326AJ	BLOCK-2013
326AL	BLOCK-2013
326CJ	BLOCK-2013
326CL	BLOCK-2013
327-1	BLOCK-553
3283-00160-000	BLOCK-874
329-1	BLOCK-554
32980	BLOCK-1264
329814	BLOCK-2207
331378	BLOCK-2199
331CL	BLOCK-2015
332AJ	BLOCK-2016
332AL	BLOCK-2016
333AJ	BLOCK-2017
333AL	BLOCK-2017
333CJ	BLOCK-2017
333CL	BLOCK-2017
334AL	BLOCK-2018
334CL	BLOCK-2018
335AJ	BLOCK-2019
335AL	BLOCK-2019
335CJ	BLOCK-2019
335CL	BLOCK-2019
336637-20	BLOCK-2176
339	BLOCK-1873
339002	BLOCK-1471
339003	BLOCK-1476
339009	BLOCK-1488
339300	BLOCK-1293
339300-2	BLOCK-1293
339486	BLOCK-1419
33A14668	BLOCK-1191
33A17056	BLOCK-770
33A17057	BLOCK-670
33A17059	BLOCK-2097
33A17062	BLOCK-757
3400-2412-3	BLOCK-1943
34001PC	BLOCK-1029
34002412	BLOCK-1943
34002412-3	BLOCK-1943
34002PC	BLOCK-1030
340085PC	BLOCK-1033
340097PC	BLOCK-1035
340098PC	BLOCK-1036
34011PC	BLOCK-1039
34012PC	BLOCK-1040
34013PC	BLOCK-1041
34014DC	BLOCK-1042
34014PC	BLOCK-1042
34015DC	BLOCK-1043
34015DM	BLOCK-1043
34015PC	BLOCK-1043
340160PC	BLOCK-1044
340161PC	BLOCK-1045
340162PC	BLOCK-1046
340163PC	BLOCK-1047
34016PC	BLOCK-1048
340174DC	BLOCK-1049
340174DM	BLOCK-1049
340174PC	BLOCK-1049
340175DC	BLOCK-1050
340175DM	BLOCK-1050

8 ORIGINAL DEVICE TYPES AND REPLACEMENT CODES

DEVICE TYPE	REPL CODE
340175PC	BLOCK-1050
340017PC	BLOCK-1051
340192PC	BLOCK-1054
340193PC	BLOCK-1055
340194DC	BLOCK-1056
340194DM	BLOCK-1056
340194PC	BLOCK-1056
340195DC	BLOCK-1057
340195DM	BLOCK-1057
340195PC	BLOCK-1057
34019PC	BLOCK-1058
34020PC	BLOCK-1059
34021DC	BLOCK-1060
34021PC	BLOCK-1060
34023PC	BLOCK-1062
34025PC	BLOCK-1064
34027PC	BLOCK-1066
34028PC	BLOCK-1067
34029PC	BLOCK-1068
3403	BLOCK-2144
34030PC	BLOCK-1069
34040PC	BLOCK-1076
34042PC	BLOCK-1077
34049PC	BLOCK-1084
34050PC	BLOCK-1085
34051PC	BLOCK-1086
34052PC	BLOCK-1087
34055PC	BLOCK-1089
34066PC	BLOCK-1093
34068PC	BLOCK-1095
34069PC	BLOCK-1096
34070PC	BLOCK-1097
34071PC	BLOCK-1098
34077PC	BLOCK-1103
34078PC	BLOCK-1104
34081PC	BLOCK-1105
34085PC	BLOCK-1107
34086PC	BLOCK-1108
34099PC	BLOCK-1115
3412951	BLOCK-1944
3412986-1	BLOCK-1943
342AJ	BLOCK-2021
342AL	BLOCK-2021
342CJ	BLOCK-2021
342CL	BLOCK-2021
343AJ	BLOCK-2022
343AL	BLOCK-2022
343CJ	BLOCK-2022
343CL	BLOCK-2022
34512PC	BLOCK-1123
34518PC	BLOCK-1129
34520PC	BLOCK-1130
34527PC	BLOCK-1134
34539PC	BLOCK-1141
34555PC	BLOCK-1146
34556PC	BLOCK-1147
347AL	BLOCK-2023
349-113-011	BLOCK-2197
349-113-012	BLOCK-2199
349-113-013	BLOCK-2207
349-113-014	BLOCK-2208
349-113-015	BLOCK-2213
349-113-022	BLOCK-2207
349-113-024	BLOCK-2212
349-113-025	BLOCK-2205
34SI-UPC1373HA	BLOCK-695
35059-2	BLOCK-1269
351-029-020	BLOCK-2056
351-3006	BLOCK-2189
351-7008-010	BLOCK-2185
351-7008-020	BLOCK-2188
351-7011-020	BLOCK-2176
351-7011-030	BLOCK-2177
351-7011-040	BLOCK-2177
351-7011-050	BLOCK-2179
351-7011-060	BLOCK-2179
351-7011-070	BLOCK-2185
351-7011-080	BLOCK-2185
351-7011-090	BLOCK-2186
351-7011-100	BLOCK-2189
351-7011-110	BLOCK-2189
351-7011-120	BLOCK-2189
351-7011-150	BLOCK-2192
351-7011-160	BLOCK-2192
351-7015-010	BLOCK-2182
351-7015-020	BLOCK-2183
351-7015-030	BLOCK-2185
351-7015-040	BLOCK-2186
351-7015-050	BLOCK-2187
351-7015-060	BLOCK-2188
351-7015-070	BLOCK-2192
351-7025-010	BLOCK-2181
351-7025-020	BLOCK-2189
351-7026-010	BLOCK-2182

DEVICE TYPE	REPL CODE
351-7026-030	BLOCK-2185
351-7026-060	BLOCK-2188
351-7026-070	BLOCK-2192
351-7121-020	BLOCK-2189
351-7206-080	BLOCK-2182
351-7206-090	BLOCK-2183
351-7206-100	BLOCK-2185
351-7206-110	BLOCK-2186
351-7206-140	BLOCK-2192
351-7206-150	BLOCK-2182
351-7206-160	BLOCK-2183
351-7206-170	BLOCK-2185
351-7206-180	BLOCK-2186
351-7206-190	BLOCK-2187
351-7206-200	BLOCK-2188
351-7206-210	BLOCK-2192
351-7577-010	BLOCK-1956
352-0023-001	BLOCK-2199
3520025-001	BLOCK-2202
3520041-001	BLOCK-1293
3520042-001	BLOCK-1304
3520043-001	BLOCK-1410
3520044-001	BLOCK-1390
3520045-001	BLOCK-1402
3520046-001	BLOCK-1411
3520047-001	BLOCK-1377
3520048-001	BLOCK-1297
3520050-001	BLOCK-1409
353(IC)	BLOCK-1910
3531-021-000	BLOCK-1317
3531-022-000	BLOCK-1355
3531-023-000	BLOCK-1321
3531-033-000	BLOCK-109
35392DC	BLOCK-1023
3539DC	BLOCK-1023
354(CHRYSLER)	BLOCK-1279
35682-104-620	BLOCK-874
35683-112-220	BLOCK-735
35684-107-534	BLOCK-2097
3596353	BLOCK-1289
3596354	BLOCK-1279
3596808	BLOCK-1282
3596809	BLOCK-1814
3596810	BLOCK-1813
3597049	BLOCK-112
3597280	BLOCK-112
3598173	BLOCK-112
35SI-UPC574J	BLOCK-1157
35SI-UPC580C	BLOCK-184
35SZ-UPC574J	BLOCK-1157
36-0041	BLOCK-138
36-0083	BLOCK-136
36003049	BLOCK-1157
3610003	BLOCK-2189
3610005	BLOCK-2176
36186200	BLOCK-1968
36188000	BLOCK-1412
36188700	BLOCK-1471
3633-1	BLOCK-1265
3633-1(IC)	BLOCK-1265
363AL	BLOCK-2028
363CL	BLOCK-2028
36446-2	BLOCK-1376
3681-1	BLOCK-1944
3686-1	BLOCK-1269
36SI-CA3065E	BLOCK-1269
370-2	BLOCK-567
370010011	BLOCK-1269
37001003	BLOCK-45
37001034	BLOCK-603
37001039	BLOCK-563
37001040	BLOCK-136
37001047	BLOCK-360
37001059	BLOCK-761
37001064	BLOCK-382
37001065	BLOCK-781
37001066	BLOCK-782
37003001(IC)	BLOCK-87
37003012	BLOCK-404
37003020	BLOCK-398
37003023	BLOCK-576
37003027	BLOCK-636
37003029	BLOCK-636
37007014	BLOCK-471
37007021	BLOCK-709
37009010	BLOCK-537
37009016	BLOCK-650
37011020	BLOCK-695
37011039	BLOCK-644
37011041	BLOCK-853
37011063	BLOCK-556
37051036	BLOCK-1093
37051100	BLOCK-1069
37051246	BLOCK-1066

DEVICE TYPE	REPL CODE
37051310	BLOCK-1039
37051378	BLOCK-1088
37053002	BLOCK-1087
37053003	BLOCK-1044
370AJ	BLOCK-2032
370AL	BLOCK-2032
370CJ	BLOCK-2032
370CL	BLOCK-2032
371AJ	BLOCK-2033
371AL	BLOCK-2033
371CJ	BLOCK-2033
371CL	BLOCK-2033
372AL	BLOCK-2034
372CL	BLOCK-2034
373401-1	BLOCK-1293
373404-1	BLOCK-1297
373405-1	BLOCK-1304
373406-1	BLOCK-1365
373407-1	BLOCK-1377
373408-1	BLOCK-1390
373409-1	BLOCK-1411
373410-1	BLOCK-1419
373411-1	BLOCK-1408
373412-1	BLOCK-1417
373413-1	BLOCK-1342
373414-1	BLOCK-1413
373423-1	BLOCK-1302
373424-1	BLOCK-1409
373427-1	BLOCK-1423
373428-1	BLOCK-1429
373429-1	BLOCK-1299
373708-1	BLOCK-1351
373712-1	BLOCK-1425
373713-1	BLOCK-1412
373714-1	BLOCK-1403
373714-2	BLOCK-1404
373715-1	BLOCK-1402
373716-1	BLOCK-1416
373718-1	BLOCK-1426
373721-1	BLOCK-1300
374109-1	BLOCK-1301
374110-1	BLOCK-1406
37510-164	BLOCK-98
37510-165	BLOCK-50
37510-166	BLOCK-50
37510-167	BLOCK-79
376-0062	BLOCK-2150
380(IC)	BLOCK-1303
380049	BLOCK-1969
380A/C	BLOCK-1303
380B/M	BLOCK-1303
383(IC)	BLOCK-221
38510-169	BLOCK-1813
38510-170	BLOCK-1825
38510-171	BLOCK-85
386	BLOCK-1851
38927-00000	BLOCK-1945
38SI-UPC1373H	BLOCK-695
39-033-0	BLOCK-2173
39-033-2	BLOCK-2173
39-047-1	BLOCK-1859
39-054-1	BLOCK-881
39-059-0	BLOCK-202
39-060-1	BLOCK-248
39-061-0	BLOCK-180
39-077-0	BLOCK-1873
3900	BLOCK-2191
3909	BLOCK-1931
39209-1	BLOCK-1816
394-1	BLOCK-401
394-2	BLOCK-401
398-13223-1	BLOCK-1293
398-13224-1	BLOCK-1297
398-13225-1	BLOCK-1298
398-13226-1	BLOCK-1294
398-13632-1	BLOCK-1297
398-8418-1	BLOCK-2178
398-8972-1	BLOCK-1406
3H82-00020-000	BLOCK-2097
3H83-00050-000	BLOCK-735
3H83-00160-000	BLOCK-874
3L4-9002-1	BLOCK-1811
3L4-9004-1	BLOCK-1376
3L4-9004-3	BLOCK-1376
3L4-9004-4	BLOCK-1376
3L4-9004-51	BLOCK-1376
3L4-9004-6	BLOCK-1376
3L4-9006-1	BLOCK-1282
3L4-9006-51	BLOCK-1282
3L4-9007-0	BLOCK-1279
3L4-9007-1	BLOCK-1290
3L4-9007-51	BLOCK-1290
3L4-9008-01	BLOCK-1814
3L4-9008-51	BLOCK-1814

DEVICE TYPE	REPL CODE
3L4-9013-01	BLOCK-2208
3L4-9015-1	BLOCK-1852
3L4-9020-1	BLOCK-221
4-009	BLOCK-1277
4-08018-667	BLOCK-2185
4-2060-02300	BLOCK-5
4-2060-02400	BLOCK-5
4-2060-02600	BLOCK-78
4-2060-02900	BLOCK-78
4-2060-03900	BLOCK-78
4-2060-04000	BLOCK-1269
4-2060-04200	BLOCK-92
4-2060-04300	BLOCK-119
4-2060-04600	BLOCK-1269
4-2060-04800	BLOCK-120
4-2060-04900	BLOCK-120
4-2060-05200	BLOCK-148
4-2060-07200	BLOCK-92
4-2060-07300	BLOCK-119
4-2060-07500	BLOCK-147
4-2060-09100	BLOCK-167
4-2060-09200	BLOCK-1291
4-2061-05170	BLOCK-1422
4-2061-05370	BLOCK-107
4-2069-70232	BLOCK-550
4-2069-71660	BLOCK-643
4-2069-71710	BLOCK-505
4-2069-71730	BLOCK-821
4-3036	BLOCK-28
40-035-0	BLOCK-1039
40-065-19-001	BLOCK-2216
40-065-19-002	BLOCK-2218
40-065-19-003	BLOCK-2210
40-065-19-004	BLOCK-2203
40-065-19-005	BLOCK-2206
40-065-19-006	BLOCK-2199
40-065-19-007	BLOCK-2209
40-065-19-008	BLOCK-2200
40-065-19-012	BLOCK-1488
40-065-19-013	BLOCK-2208
40-065-19-014	BLOCK-1952
40-065-19-016	BLOCK-1390
40-065-19-027	BLOCK-1412
40-065-19-029	BLOCK-1419
40-065-19-030	BLOCK-1428
40-13184-3	BLOCK-175
400-1735	BLOCK-2190
400-1736	BLOCK-2194
4000	BLOCK-1028
4000(IC)	BLOCK-1028
4001(IC)	BLOCK-1029
4001(MODULE)	BLOCK-2
4001BDC	BLOCK-1029
4001BDM	BLOCK-1029
4001BFC	BLOCK-1029
4001BFM	BLOCK-1029
4001BPC	BLOCK-1029
4002(IC)	BLOCK-1030
4002A	BLOCK-1030
4002BDC	BLOCK-1030
4002BDM	BLOCK-1030
4002BFC	BLOCK-1030
4002BFM	BLOCK-1030
4002BPC	BLOCK-1030
4006BDC	BLOCK-1031
4006BPC	BLOCK-1031
4007(IC)	BLOCK-1032
4007UBDC	BLOCK-1032
4007UBPC	BLOCK-1032
40085BDC	BLOCK-1033
40085BDM	BLOCK-1033
40085BFC	BLOCK-1033
40085BFM	BLOCK-1033
40085BPC	BLOCK-1033
4008BDC	BLOCK-1034
4008BDM	BLOCK-1034
4008BFC	BLOCK-1034
4008BFM	BLOCK-1034
4008BPC	BLOCK-1034
4009(IC)	BLOCK-1084
400931	BLOCK-2201
40097BDC	BLOCK-1035
40097BPC	BLOCK-1035
40098BDC	BLOCK-1036
40098BPC	BLOCK-1036
4011(IC)	BLOCK-1039
4011-PC	BLOCK-1039
401113-1	BLOCK-2208
401113-2	BLOCK-2197
401113-3	BLOCK-2199
401113-4	BLOCK-2207
401113-5	BLOCK-2217
401182	BLOCK-2213
4011A	BLOCK-1039

DEVICE TYPE	REPL CODE
4011BDC	BLOCK-1039
4011BDM	BLOCK-1039
4011BFC	BLOCK-1039
4011BFM	BLOCK-1039
4011BPC	BLOCK-1039
4011PC	BLOCK-1039
4012(IC)	BLOCK-1040
4012A	BLOCK-1040
4012BDC	BLOCK-1040
4012BDM	BLOCK-1040
4012BFC	BLOCK-1040
4012BFM	BLOCK-1040
4012BPC	BLOCK-1040
4013(IC)	BLOCK-1041
4013373-0701	BLOCK-2220
4013B	BLOCK-1041
4013BDC	BLOCK-1041
4013BDM	BLOCK-1041
4013BFC	BLOCK-1041
4013BFM	BLOCK-1041
4013BPC	BLOCK-1041
4014BDC	BLOCK-1042
4014BPC	BLOCK-1042
4015	BLOCK-1043
4015(IC)	BLOCK-1043
4015BDC	BLOCK-1043
4015BPC	BLOCK-1043
4016	BLOCK-1048
4016(IC)	BLOCK-1048
4016(RAM)	BLOCK-1004
40160BDC	BLOCK-1044
40160BDM	BLOCK-1044
40160BFC	BLOCK-1044
40160BFM	BLOCK-1044
40160BPC	BLOCK-1044
40161BDC	BLOCK-1045
40161BDM	BLOCK-1045
40161BFC	BLOCK-1045
40161BFM	BLOCK-1045
40161BPC	BLOCK-1045
40162BDC	BLOCK-1046
40162BDM	BLOCK-1046
40162BFC	BLOCK-1046
40162BFM	BLOCK-1046
40162BPC	BLOCK-1046
40163BDC	BLOCK-1047
40163BDM	BLOCK-1047
40163BFC	BLOCK-1047
40163BFM	BLOCK-1047
40163BPC	BLOCK-1047
4016BDC	BLOCK-1048
4016BPC	BLOCK-1048
4016CX	BLOCK-1004
4017	BLOCK-1051
4017(IC)	BLOCK-1051
4017BDC	BLOCK-1051
4017BDM	BLOCK-1051
4017BFC	BLOCK-1051
4017BFM	BLOCK-1051
4017BPC	BLOCK-1051
4018BDC	BLOCK-1053
4018BDM	BLOCK-1053
4018BFC	BLOCK-1053
4018BFM	BLOCK-1053
4018BPC	BLOCK-1053
4019	BLOCK-1058
4019(IC)	BLOCK-1058
40192BDC	BLOCK-1054
40192BDM	BLOCK-1054
40192BFC	BLOCK-1054
40192BFM	BLOCK-1054
40192BPC	BLOCK-1054
40193BDC	BLOCK-1055
40193BDM	BLOCK-1055
40193BFC	BLOCK-1055
40193BFM	BLOCK-1055
40193BPC	BLOCK-1055
40194BDC	BLOCK-1056
40194BPC	BLOCK-1056
40195BDC	BLOCK-1057
40195BPC	BLOCK-1057
4019BDC	BLOCK-1058
4019BPC	BLOCK-1058
4020	BLOCK-1810
4020(CMOS)	BLOCK-1059
4020(IC)	BLOCK-1059
4020BDC	BLOCK-1059
4020BDM	BLOCK-1059
4020BFC	BLOCK-1059
4020BFM	BLOCK-1059
4020BPC	BLOCK-1059
4021(IC)	BLOCK-1060
4021BDC	BLOCK-1060
4021BPC	BLOCK-1060

DEVICE TYPE	REPL CODE
4022BDC	BLOCK-1061
4022BDM	BLOCK-1061
4022BFC	BLOCK-1061
4022BFM	BLOCK-1061
4022BPC	BLOCK-1061
4023	BLOCK-1062
4023(IC)	BLOCK-1062
4023A	BLOCK-1062
4023BDC	BLOCK-1062
4023BDM	BLOCK-1062
4023BFC	BLOCK-1062
4023BFM	BLOCK-1062
4023BPC	BLOCK-1062
4024	BLOCK-1063
4024(IC)	BLOCK-1063
4024BDC	BLOCK-1063
4024BDM	BLOCK-1063
4024BFC	BLOCK-1063
4024BFM	BLOCK-1063
4024BPC	BLOCK-1063
4025	BLOCK-1064
4025(IC)	BLOCK-1064
4025A	BLOCK-1064
4025BDC	BLOCK-1064
4025BDM	BLOCK-1064
4025BFC	BLOCK-1064
4025BFM	BLOCK-1064
4025BPC	BLOCK-1064
4027	BLOCK-1066
4027(IC)	BLOCK-1066
4027B	BLOCK-1066
4027BDC	BLOCK-1066
4027BDM	BLOCK-1066
4027BFC	BLOCK-1066
4027BFM	BLOCK-1066
4027BPC	BLOCK-1066
4028BDC	BLOCK-1067
4028BDM	BLOCK-1067
4028BFC	BLOCK-1067
4028BFM	BLOCK-1067
4028BPC	BLOCK-1067
4029	BLOCK-1068
4029BDC	BLOCK-1068
4029BDM	BLOCK-1068
4029BFC	BLOCK-1068
4029BFM	BLOCK-1068
4029BPC	BLOCK-1068
4030	BLOCK-1069
4030(IC)	BLOCK-1069
40306312	BLOCK-1421
40306400	BLOCK-1422
40306604	BLOCK-107
4030BDC	BLOCK-1069
4030BDM	BLOCK-1069
4030BFC	BLOCK-1069
4030BFM	BLOCK-1069
4030BPC	BLOCK-1069
4031BDC	BLOCK-1070
4031BPC	BLOCK-1070
4034BDC	BLOCK-1073
4034BPC	BLOCK-1073
4035BDC	BLOCK-1074
4035BPC	BLOCK-1074
4040	BLOCK-1076
4040(IC)	BLOCK-1076
4040BDC	BLOCK-1076
4040BDM	BLOCK-1076
4040BFC	BLOCK-1076
4040BFM	BLOCK-1076
4040BPC	BLOCK-1076
4042	BLOCK-1077
4042(IC)	BLOCK-1077
4042BDC	BLOCK-1077
4042BDM	BLOCK-1077
4042BFC	BLOCK-1077
4042BFM	BLOCK-1077
4042BPC	BLOCK-1077
4043BDC	BLOCK-1078
4043BDM	BLOCK-1078
4043BFC	BLOCK-1078
4043BFM	BLOCK-1078
4043BPC	BLOCK-1078
4044BDC	BLOCK-1079
4044BDM	BLOCK-1079
4044BFC	BLOCK-1079
4044BFM	BLOCK-1079
4044BPC	BLOCK-1079
4047BDC	BLOCK-1082
4047BDM	BLOCK-1082
4047BFC	BLOCK-1082
4047BFM	BLOCK-1082
4047BPC	BLOCK-1082
4049	BLOCK-1084
4049(IC)	BLOCK-1084

ORIGINAL DEVICE TYPES AND REPLACEMENT CODES 9

DEVICE TYPE	REPL CODE
4049B	BLOCK-1084
4049BDC	BLOCK-1084
4049BDM	BLOCK-1084
4049BPC	BLOCK-1084
4050	BLOCK-1085
4050(IC)	BLOCK-1085
4050BDC	BLOCK-1085
4050BDM	BLOCK-1085
4050BPC	BLOCK-1085
4051	BLOCK-1086
4051(IC)	BLOCK-1086
4051BDC	BLOCK-1086
4051BDM	BLOCK-1086
4051BPC	BLOCK-1086
4052	BLOCK-1087
4052(IC)	BLOCK-1087
4052BDC	BLOCK-1087
4052BDM	BLOCK-1087
4052BPC	BLOCK-1087
4053	BLOCK-1088
4053BDC	BLOCK-1088
4053BDM	BLOCK-1088
4053BPC	BLOCK-1088
4055	BLOCK-1089
4055(IC)	BLOCK-1089
4066(IC)	BLOCK-1093
4066BDC	BLOCK-1093
4066BDM	BLOCK-1093
4066BPC	BLOCK-1093
4067BDC	BLOCK-1094
4067BDM	BLOCK-1094
4067BPC	BLOCK-1094
4068BDC	BLOCK-1095
4068BDM	BLOCK-1095
4068BFC	BLOCK-1095
4068BFM	BLOCK-1095
4068BPC	BLOCK-1095
4069UBDC	BLOCK-1096
4069UBDM	BLOCK-1096
4069UBPC	BLOCK-1096
4070BDC	BLOCK-1097
4070BDM	BLOCK-1097
4070BFC	BLOCK-1097
4070BFM	BLOCK-1097
4070BPC	BLOCK-1097
4071BDC	BLOCK-1098
4071BDM	BLOCK-1098
4071BFC	BLOCK-1098
4071BFM	BLOCK-1098
4071BPC	BLOCK-1098
4072BDC	BLOCK-1099
4072BDM	BLOCK-1099
4072BFC	BLOCK-1099
4072BFM	BLOCK-1099
4072BPC	BLOCK-1099
4073BDC	BLOCK-1100
4073BDM	BLOCK-1100
4073BFC	BLOCK-1100
4073BFM	BLOCK-1100
4073BPC	BLOCK-1100
4075BDC	BLOCK-1101
4075BDM	BLOCK-1101
4075BFC	BLOCK-1101
4075BFM	BLOCK-1101
4075BPC	BLOCK-1101
4076BDC	BLOCK-1102
4076BDM	BLOCK-1102
4076BPC	BLOCK-1102
4077BDC	BLOCK-1103
4077BDM	BLOCK-1103
4077BFC	BLOCK-1103
4077BFM	BLOCK-1103
4077BPC	BLOCK-1103
4078BDC	BLOCK-1104
4078BDM	BLOCK-1104
4078BFC	BLOCK-1104
4078BFM	BLOCK-1104
4078BPC	BLOCK-1104
408 000 0301	BLOCK-1021
408-000-0301	BLOCK-1021
4081	BLOCK-1105
4081(IC)	BLOCK-1105
4081BDC	BLOCK-1105
4081BDM	BLOCK-1105
4081BFC	BLOCK-1105
4081BFM	BLOCK-1105
4081BPC	BLOCK-1105
4082626-0001	BLOCK-1963
4082664-0001	BLOCK-1282
4082665	BLOCK-1279
4082665-0001	BLOCK-1279
4082665-0002	BLOCK-1279
4082665-0003	BLOCK-1279
4082799-0001	BLOCK-1814

DEVICE TYPE	REPL CODE
4082799-0002	BLOCK-1814
4082799-1	BLOCK-1814
4082799-2	BLOCK-1814
4082802-0001	BLOCK-1806
4082802-0002	BLOCK-1806
4082BDC	BLOCK-1106
4082BDM	BLOCK-1106
4082BFC	BLOCK-1106
4082BFM	BLOCK-1106
4082BPC	BLOCK-1106
4084117-0001	BLOCK-2073
4084117-0002	BLOCK-2073
4085BDC	BLOCK-1107
4085BDM	BLOCK-1107
4085BFC	BLOCK-1107
4085BFM	BLOCK-1107
4085BPC	BLOCK-1107
4086BDC	BLOCK-1108
4086BDM	BLOCK-1108
4086BFC	BLOCK-1108
4086BFM	BLOCK-1108
4086BPC	BLOCK-1108
409 017 7208	BLOCK-671
409 017 7505	BLOCK-671
409 051 3006	BLOCK-1088
409 172 1509	BLOCK-2087
409 241 5407	BLOCK-2087
409 243 0806	BLOCK-758
409 320 5700	BLOCK-2087
409-001-0604	BLOCK-761
409-004-5705	BLOCK-839
409-004-8409	BLOCK-840
409-006-0203	BLOCK-1601
409-009-2709	BLOCK-1088
409-017-7208	BLOCK-671
409-017-7505	BLOCK-779
409-019-3109	BLOCK-709
409-019-4205	BLOCK-817
409-019-4700	BLOCK-525
409-019-5608	BLOCK-754
409-019-5707	BLOCK-778
409-019-6209	BLOCK-644
409-019-6605	BLOCK-815
409-036-0105	BLOCK-826
409-042-9000	BLOCK-1601
409-047-5601	BLOCK-802
409-047-8602	BLOCK-759
409-047-9104	BLOCK-723
409-047-9203	BLOCK-724
409-051-2801	BLOCK-1087
409-051-3006	BLOCK-1088
409-053-7309	BLOCK-663
409-083-0103	BLOCK-1184
409-117-9607	BLOCK-1189
409-132-7909	BLOCK-1088
409-135-5506	BLOCK-1200
409-146-7001	BLOCK-843
409-156-6605	BLOCK-843
409-243-0806	BLOCK-758
4090605-1	BLOCK-1828
4093BDC	BLOCK-1110
4093BDM	BLOCK-1110
40SI-MC1364P	BLOCK-1808
410-051-5808	BLOCK-1601
410-067-3300	BLOCK-1601
4102	BLOCK-999
4116	BLOCK-1003
412-2	BLOCK-784
4150002031	BLOCK-52
4150003770	BLOCK-1828
415100574J	BLOCK-1157
4151005753	BLOCK-136
415101379C	BLOCK-647
4151014908	BLOCK-1194
415200050	BLOCK-26
4152032010	BLOCK-168
4152041000	BLOCK-169
4152056300	BLOCK-1157
415207530	BLOCK-807
4152075300	BLOCK-807
415207530N	BLOCK-807
415207601W	BLOCK-659
4152076200	BLOCK-824
4152076550	BLOCK-843
4152078300	BLOCK-754
4152078350	BLOCK-835
4152079100	BLOCK-644
4152085600	BLOCK-975
415307313P	BLOCK-453
4154011370	BLOCK-1813
4154011510	BLOCK-226
4154011560	BLOCK-1825
415401160	BLOCK-1825
4157013102	BLOCK-1825

DEVICE TYPE	REPL CODE
4159301230	BLOCK-757
416	BLOCK-1003
4164	BLOCK-1006
4164-2	BLOCK-1006
4164C	BLOCK-1006
417-119	BLOCK-1223
41C-402	BLOCK-93
41C-406	BLOCK-95
41C-407	BLOCK-4
41C-409	BLOCK-108
41SI-HA1128	BLOCK-1269
41SI-LM3065N	BLOCK-1269
42-1	BLOCK-110
42-2	BLOCK-109
42-27377	BLOCK-4
4206002400	BLOCK-5
4206002600	BLOCK-78
4206003900	BLOCK-78
4206004000	BLOCK-1269
4206004400	BLOCK-5
4206004600	BLOCK-1269
4206005200	BLOCK-148
4206007500	BLOCK-147
4206009100	BLOCK-172
4206009200	BLOCK-1291
4206009700	BLOCK-1291
4206009800	BLOCK-172
4206012500	BLOCK-204
4206042400	BLOCK-5
4206104970	BLOCK-1269
4206105170	BLOCK-1422
4206105370	BLOCK-107
4206105470	BLOCK-1269
4206105670	BLOCK-9
4206105770	BLOCK-1269
4206105870	BLOCK-1269
4206970020	BLOCK-930
423-800-235	BLOCK-2200
423-800175	BLOCK-2199
423-800176	BLOCK-2217
423-800177	BLOCK-2199
423-800178	BLOCK-2208
423-800194	BLOCK-2210
423-800202	BLOCK-2202
423-800223	BLOCK-1952
423-800234	BLOCK-2197
423-800236	BLOCK-1950
4295-1	BLOCK-1787
4295-1(RCA)	BLOCK-1787
4295-2	BLOCK-1787
4295-2(RCA)	BLOCK-1792
43122874	BLOCK-1282
43200	BLOCK-1293
43201	BLOCK-1297
43202	BLOCK-1301
43205	BLOCK-1410
435-21026-0A	BLOCK-1293
435-21027-0A	BLOCK-1295
435-21028-0A	BLOCK-1297
435-21029-0A	BLOCK-1301
435-21030-0A	BLOCK-1304
435-21033-0A	BLOCK-1365
435-21034-0A	BLOCK-1402
435-21035-0A	BLOCK-1419
435-23006-0A	BLOCK-1410
435-23007-0A	BLOCK-1411
436-10010-0A	BLOCK-1425
436-10011-0A	BLOCK-1429
4381P1	BLOCK-2189
43A168135-1	BLOCK-2199
43A168135-10	BLOCK-2217
43A168135-2	BLOCK-2197
43A168135-3	BLOCK-2200
43A168135-4	BLOCK-2208
43A168135-6	BLOCK-2202
43A168135-7	BLOCK-2206
43A168135-8	BLOCK-2207
43A168135-9	BLOCK-1950
43A223006P1	BLOCK-1471
43A223007	BLOCK-1294
43A223008	BLOCK-1472
43A223009	BLOCK-1295
43A223012	BLOCK-1476
43A223015	BLOCK-1377
43A223017	BLOCK-1488
43A223018	BLOCK-1488
43A223025	BLOCK-1410
43A223026P1	BLOCK-1411
43A223028	BLOCK-1413
43A223029P1	BLOCK-1392
43A223030	BLOCK-1428
43A223031	BLOCK-1429
43A223033P1	BLOCK-1417
43A223034P1	BLOCK-1426

DEVICE TYPE	REPL CODE
43C216408P1	BLOCK-1471
43C216409P1	BLOCK-1472
43C216410P1	BLOCK-1473
43C216411P1	BLOCK-1484
43C216414P1	BLOCK-1490
43C216447	BLOCK-1358
43C216447P1	BLOCK-1358
4400Y	BLOCK-199
442-10	BLOCK-1244
442-18	BLOCK-1405
442-2	BLOCK-1954
442-22	BLOCK-2058
442-24	BLOCK-903
442-25	BLOCK-1985
442-28	BLOCK-1262
442-30	BLOCK-1022
442-30-2897	BLOCK-1022
442-33	BLOCK-1270
442-39	BLOCK-2141
442-5	BLOCK-1805
442-54	BLOCK-2087
442-57	BLOCK-1271
442-58	BLOCK-1272
442-60	BLOCK-2137
442-602	BLOCK-2171
442-604	BLOCK-930
442-612	BLOCK-1851
442-613	BLOCK-2128
442-615	BLOCK-1281
442-623	BLOCK-1997
442-63	BLOCK-2119
442-630	BLOCK-2091
442-635	BLOCK-1881
442-636	BLOCK-472
442-640	BLOCK-1942
442-644	BLOCK-2074
442-647	BLOCK-1081
442-648	BLOCK-1986
442-651	BLOCK-2007
442-654	BLOCK-2174
442-659	BLOCK-945
442-66	BLOCK-2042
442-663	BLOCK-2097
442-664	BLOCK-2108
442-665	BLOCK-893
442-672	BLOCK-933
442-673	BLOCK-470
442-674	BLOCK-2097
442-675	BLOCK-2108
442-677	BLOCK-367
442-682	BLOCK-944
442-686	BLOCK-1303
442-688	BLOCK-1868
442-691	BLOCK-2095
442-702	BLOCK-2004
442-704	BLOCK-2083
442-705	BLOCK-2084
442-71	BLOCK-2191
442-713	BLOCK-221
442-715	BLOCK-1997
442-728	BLOCK-1993
442-74	BLOCK-1804
442-75	BLOCK-1981
442-82	BLOCK-1270
442-9	BLOCK-1277
442-96	BLOCK-2138
442-99	BLOCK-1048
443-1	BLOCK-1293
443-12	BLOCK-1304
443-13	BLOCK-1412
443-15	BLOCK-1401
443-16	BLOCK-1413
443-162	BLOCK-1358
443-17	BLOCK-1321
443-18	BLOCK-1297
443-2	BLOCK-1365
443-22	BLOCK-1311
443-23	BLOCK-1312
443-25	BLOCK-1328
443-26	BLOCK-1719
443-27	BLOCK-1950
443-3	BLOCK-1377
443-32	BLOCK-1311
443-34	BLOCK-1425
443-35	BLOCK-1391
443-36	BLOCK-1397
443-4	BLOCK-1409
443-42	BLOCK-1728
443-43	BLOCK-1478
443-44	BLOCK-1317
443-44-2854	BLOCK-1317
443-45	BLOCK-1301
443-46	BLOCK-1295
443-5	BLOCK-1410

DEVICE TYPE	REPL CODE
443-53	BLOCK-1392
443-54	BLOCK-1296
443-58	BLOCK-1482
443-59	BLOCK-1504
443-6	BLOCK-1411
443-60	BLOCK-1040
443-603	BLOCK-1039
443-604	BLOCK-1032
443-606	BLOCK-1066
443-607	BLOCK-1041
443-61	BLOCK-1048
443-612	BLOCK-1358
443-62	BLOCK-2140
443-622	BLOCK-1326
443-623	BLOCK-1307
443-623-51644	BLOCK-1307
443-625	BLOCK-1318
443-628	BLOCK-1360
443-629	BLOCK-1423
443-640	BLOCK-1426
443-642	BLOCK-1298
443-65	BLOCK-1372
443-66	BLOCK-1357
443-67	BLOCK-2089
443-68	BLOCK-1476
443-680	BLOCK-1428
443-698	BLOCK-1419
443-7	BLOCK-1423
443-7-16088	BLOCK-1423
443-70	BLOCK-1479
443-703	BLOCK-1029
443-704	BLOCK-1030
443-706	BLOCK-1098
443-707	BLOCK-1126
443-708	BLOCK-1127
443-709	BLOCK-1469
443-71	BLOCK-1471
443-711	BLOCK-1636
443-712	BLOCK-1064
443-713	BLOCK-1067
443-717	BLOCK-1315
443-718	BLOCK-1623
443-719	BLOCK-1646
443-72	BLOCK-1343
443-720	BLOCK-1837
443-722	BLOCK-2089
443-727	BLOCK-2132
443-728	BLOCK-1568
443-729	BLOCK-1587
443-73	BLOCK-1336
443-730	BLOCK-1706
443-731	BLOCK-1653
443-732	BLOCK-1658
443-733	BLOCK-1654
443-737	BLOCK-1129
443-738	BLOCK-1147
443-74	BLOCK-1777
443-745	BLOCK-1571
443-751	BLOCK-1105
443-752	BLOCK-1616
443-754	BLOCK-1630
443-755	BLOCK-1572
443-757	BLOCK-1605
443-760	BLOCK-1076
443-762	BLOCK-1832
443-764	BLOCK-1002
443-77	BLOCK-1385
443-777	BLOCK-1114
443-778	BLOCK-1110
443-779	BLOCK-1570
443-780	BLOCK-1574
443-781	BLOCK-1707
443-782	BLOCK-1600
443-783	BLOCK-1061
443-784	BLOCK-1069
443-785	BLOCK-1449
443-792	BLOCK-1588
443-794	BLOCK-1771
443-795	BLOCK-1772
443-797	BLOCK-1576
443-798	BLOCK-1626
443-799	BLOCK-1602
443-800	BLOCK-1647
443-801	BLOCK-1624
443-802	BLOCK-1641
443-804	BLOCK-1643
443-805	BLOCK-1648
443-807	BLOCK-1684
443-811	BLOCK-1585
443-813	BLOCK-1714
443-815	BLOCK-1621
443-816	BLOCK-1575
443-817	BLOCK-1620
443-818	BLOCK-1573

DEVICE TYPE	REPL CODE
443-822	BLOCK-1592
443-824	BLOCK-1631
443-827	BLOCK-1164
443-828	BLOCK-1705
443-829	BLOCK-1708
443-836	BLOCK-1049
443-837	BLOCK-1671
443-839	BLOCK-1633
443-840	BLOCK-1172
443-843	BLOCK-1169
443-854	BLOCK-1649
443-855	BLOCK-1652
443-857	BLOCK-1668
443-858	BLOCK-1320
443-863	BLOCK-1672
443-864	BLOCK-1579
443-87	BLOCK-1325
443-871	BLOCK-1125
443-872	BLOCK-1593
443-875	BLOCK-1659
443-877	BLOCK-1591
443-879	BLOCK-1615
443-881	BLOCK-1025
443-884	BLOCK-1632
443-885	BLOCK-1635
443-886	BLOCK-1040
443-887	BLOCK-1062
443-888	BLOCK-1119
443-889	BLOCK-1594
443-89	BLOCK-1302
443-891	BLOCK-1713
443-892	BLOCK-1610
443-893	BLOCK-1678
443-896	BLOCK-1720
443-897	BLOCK-1722
443-899	BLOCK-1767
443-90	BLOCK-1313
443-900	BLOCK-1768
443-904	BLOCK-1003
443-909	BLOCK-1821
443-912	BLOCK-1596
443-915	BLOCK-1769
443-916	BLOCK-1114
443-919	BLOCK-1586
443-920	BLOCK-1712
443-921	BLOCK-1678
443-928	BLOCK-1034
443-929	BLOCK-1051
443-930	BLOCK-1082
443-934	BLOCK-1607
443-935	BLOCK-1737
443-942	BLOCK-1584
443-948	BLOCK-1580
443-951	BLOCK-1689
443-955	BLOCK-1599
443-958	BLOCK-1091
443-961	BLOCK-1627
443-964	BLOCK-1738
443-973	BLOCK-1679
443-976	BLOCK-1724
443-981	BLOCK-1733
443-982	BLOCK-1753
443-983	BLOCK-1741
443-999	BLOCK-1694
4437-3	BLOCK-1812
4438-3	BLOCK-1813
4439-1	BLOCK-1814
4439-1(RCA)	BLOCK-1814
4439-2	BLOCK-1814
4439-2(RCA)	BLOCK-1814
444-1(IC)	BLOCK-735
4460-2	BLOCK-1814
4460-3	BLOCK-1814
449-1	BLOCK-733
44A417779-001	BLOCK-2058
44B1	BLOCK-1291
44B1Z	BLOCK-1291
44T-100-120	BLOCK-145
44T-300-100	BLOCK-2160
44T-300-101	BLOCK-1859
44T-300-102	BLOCK-145
44T-300-359	BLOCK-503
44T-300-99	BLOCK-102
4506	BLOCK-1119
4510BDC	BLOCK-1121
4510BDM	BLOCK-1121
4510BFC	BLOCK-1121
4510BFM	BLOCK-1121
4510BPC	BLOCK-1121
4511	BLOCK-1122
4511424	BLOCK-1378
4511BDC	BLOCK-1122
4511BDM	BLOCK-1122
4511BPC	BLOCK-1122

10 ORIGINAL DEVICE TYPES AND REPLACEMENT CODES

DEVICE TYPE	REPL CODE
4512BDC	BLOCK-1123
4512BDM	BLOCK-1123
4512BPC	BLOCK-1123
4514BDC	BLOCK-1125
4514BDM	BLOCK-1125
4514BPC	BLOCK-1125
4515BDC	BLOCK-1126
4515BDM	BLOCK-1126
4515BPC	BLOCK-1126
4516BDC	BLOCK-1127
4516BDM	BLOCK-1127
4516BFC	BLOCK-1127
4516BFM	BLOCK-1127
4516BPC	BLOCK-1127
4518	BLOCK-1129
4518(IC)	BLOCK-1129
4518BDC	BLOCK-1129
4518BDM	BLOCK-1129
4518BFC	BLOCK-1129
4518BFM	BLOCK-1129
4518BPC	BLOCK-1129
4520BDC	BLOCK-1130
4520BDM	BLOCK-1130
4520BFC	BLOCK-1130
4520BFM	BLOCK-1130
4520BPC	BLOCK-1130
4528BDC	BLOCK-1129
4528BDM	BLOCK-1135
4528BPC	BLOCK-1135
45299	BLOCK-1945
4531BDC	BLOCK-1137
4531BDM	BLOCK-1137
4531BPC	BLOCK-1137
4532BDC	BLOCK-1138
4532BDM	BLOCK-1138
4532BPC	BLOCK-1138
45380	BLOCK-1275
45390	BLOCK-1376
45393	BLOCK-1828
45394	BLOCK-1303
45395	BLOCK-1828
4539BDC	BLOCK-1141
4539BDM	BLOCK-1141
4539BPC	BLOCK-1141
4543BDC	BLOCK-1143
4543BDM	BLOCK-1143
4543BPC	BLOCK-1143
4555BDC	BLOCK-1146
4555BDM	BLOCK-1146
4555BPC	BLOCK-1146
4556BDC	BLOCK-1147
4556BDM	BLOCK-1147
4556BPC	BLOCK-1147
4558C	BLOCK-1801
4558D	BLOCK-1801
4558DD	BLOCK-1801
4558DV	BLOCK-1801
455947	BLOCK-1093
45810-167	BLOCK-1813
45810-168	BLOCK-1825
45810-169	BLOCK-85
45810-170	BLOCK-145
4582BDC	BLOCK-1052
4582BDM	BLOCK-1052
4582BPC	BLOCK-1052
4583BDC	BLOCK-1153
4583BDM	BLOCK-1153
4583BPC	BLOCK-1153
46-13101-3	BLOCK-78
46-131014-3	BLOCK-556
46-13103-3	BLOCK-92
46-131030-3	BLOCK-872
46-13104-3	BLOCK-119
46-13105-3	BLOCK-120
46-131060-3	BLOCK-516
46-131073-3	BLOCK-1801
46-131098-3	BLOCK-866
46-131099-3	BLOCK-753
46-131100-3	BLOCK-763
46-131126-3	BLOCK-1713
46-131131-3	BLOCK-814
46-131172-3	BLOCK-1087
46-131188-3	BLOCK-396
46-13119-3	BLOCK-148
46-131203-3	BLOCK-663
46-131206-3	BLOCK-709
46-131207-3	BLOCK-824
46-131208-3	BLOCK-754
46-13121-3	BLOCK-225
46-131230-3	BLOCK-215
46-131237-3	BLOCK-725
46-131238-3	BLOCK-205
46-13124-3	BLOCK-1291
46-131245-3	BLOCK-670

DEVICE TYPE	REPL CODE
46-13125-3	BLOCK-172
46-131262-3	BLOCK-2097
46-131265-3	BLOCK-764
46-131266-3	BLOCK-772
46-131274-3	BLOCK-768
46-13131-3	BLOCK-9
46-131315-3	BLOCK-752
46-131341-3	BLOCK-778
46-131342-3	BLOCK-780
46-131343-3	BLOCK-779
46-131351-3	BLOCK-724
46-131357-3	BLOCK-206
46-131396-3	BLOCK-1021
46-131401-3	BLOCK-815
46-13143-3	BLOCK-184
46-131432-3	BLOCK-771
46-131433-3	BLOCK-759
46-131438-9	BLOCK-759
46-13145-3	BLOCK-1269
46-131507-3	BLOCK-752
46-131514-3	BLOCK-840
46-131515-3	BLOCK-839
46-131516-3	BLOCK-881
46-13152-3	BLOCK-458
46-13153-3	BLOCK-509
46-13156-3	BLOCK-817
46-131577-3	BLOCK-1193
46-131585-3	BLOCK-723
46-13162-3	BLOCK-585
46-131629-3	BLOCK-826
46-13164-3	BLOCK-575
46-13165-3	BLOCK-588
46-131656-3	BLOCK-817
46-13166-3	BLOCK-586
46-13167-3	BLOCK-581
46-131678-3	BLOCK-1088
46-13168-3	BLOCK-579
46-131683-3	BLOCK-861
46-13169-3	BLOCK-571
46-13171-3	BLOCK-583
46-13172-3	BLOCK-250
46-13173-3	BLOCK-85
46-13174-3	BLOCK-573
46-13175-3	BLOCK-592
46-13176-3	BLOCK-442
46-13177-3	BLOCK-973
46-131772-3	BLOCK-1184
46-13178-3	BLOCK-2058
46-131787-3	BLOCK-771
46-131793-3	BLOCK-2087
46-13180-3	BLOCK-204
46-131801-3	BLOCK-2087
46-13184-3	BLOCK-175
46-13185-3	BLOCK-176
46-13188-3	BLOCK-532
46-13190-3	BLOCK-484
46-131905-3	BLOCK-396
46-131969-3	BLOCK-813
46-132070-3	BLOCK-863
46-13208-3	BLOCK-418
46-13216-3	BLOCK-519
46-13223-3	BLOCK-234
46-13225-3	BLOCK-543
46-132286-3	BLOCK-740
46-132287-3	BLOCK-818
46-132303-3	BLOCK-1189
46-132305-3	BLOCK-773
46-13233-3	BLOCK-577
46-13237-3	BLOCK-467
46-13238-3	BLOCK-519
46-132394-3	BLOCK-1093
46-132398-3	BLOCK-840
46-132422-3	BLOCK-1088
46-1325-3	BLOCK-172
46-13257-3	BLOCK-525
46-13258-3	BLOCK-1039
46-132637-3	BLOCK-706
46-132638-3	BLOCK-636
46-132647-3	BLOCK-843
46-132648-3	BLOCK-835
46-132649-3	BLOCK-1200
46-13267-3	BLOCK-418
46-13272-3	BLOCK-1029
46-13279-3	BLOCK-1993
46-132850-3	BLOCK-2087
46-132884-3	BLOCK-1601
46-132886-3	BLOCK-770
46-132931-3	BLOCK-1191
46-13296-3	BLOCK-1093
46-13297-3	BLOCK-1993
46-13299-3	BLOCK-569
46-13301-3	BLOCK-456
46-13302-3	BLOCK-471
46-13307-3	BLOCK-986

DEVICE TYPE	REPL CODE
46-13309-3	BLOCK-993
46-13310-3	BLOCK-1801
46-13311-3	BLOCK-2119
46-13312-3	BLOCK-402
46-13313-3	BLOCK-1041
46-13314-3	BLOCK-1076
46-13315-3	BLOCK-1041
46-13316-3	BLOCK-1106
46-13317-3	BLOCK-1029
46-13319-3	BLOCK-1096
46-13322-3	BLOCK-454
46-13323-3	BLOCK-471
46-13327-3	BLOCK-433
46-133275-3	BLOCK-732
46-133293-3	BLOCK-1039
46-133325-3	BLOCK-1801
46-133344-3	BLOCK-1191
46-13338-3	BLOCK-604
46-133389-3	BLOCK-1021
46-13339-3	BLOCK-543
46-13340-3	BLOCK-1269
46-13341-3	BLOCK-442
46-13345-3	BLOCK-1801
46-13356-3	BLOCK-534
46-133692-3	BLOCK-1021
46-13379-3	BLOCK-471
46-133831-3	BLOCK-1021
46-13392-3	BLOCK-550
46-1340-3	BLOCK-1267
46-13404-3	BLOCK-1801
46-13411	BLOCK-467
46-13411-3	BLOCK-467
46-13428-3	BLOCK-467
46-1343-3	BLOCK-4
46-13448-3	BLOCK-654
46-13455-3	BLOCK-1584
46-1346-3	BLOCK-1267
46-1347-3	BLOCK-1265
46-1348-3	BLOCK-1265
46-13484-3	BLOCK-545
46-13487-3	BLOCK-1067
46-13491-3	BLOCK-1130
46-13493-3	BLOCK-770
46-13495-3	BLOCK-2063
46-13502-3	BLOCK-985
46-1352-3	BLOCK-1256
46-13527-3	BLOCK-525
46-13528-3	BLOCK-602
46-13538-3	BLOCK-459
46-13546-3	BLOCK-540
46-13547-3	BLOCK-642
46-13548-3	BLOCK-659
46-13550-3	BLOCK-428
46-13551-3	BLOCK-398
46-13552-3	BLOCK-286
46-1356-3	BLOCK-1265
46-13561-3	BLOCK-225
46-1357-3	BLOCK-5
46-13581-3	BLOCK-644
46-13588-3	BLOCK-535
46-13588-9	BLOCK-535
46-13590-3	BLOCK-563
46-13593-3	BLOCK-1706
46-13604	BLOCK-659
46-13604-3	BLOCK-659
46-1361-3	BLOCK-1269
46-1362-3	BLOCK-99
46-1363-3	BLOCK-103
46-1364-3	BLOCK-60
46-13645-3	BLOCK-643
46-13649-3	BLOCK-727
46-1365-3	BLOCK-1422
46-1366-3	BLOCK-1421
46-13662-3	BLOCK-1088
46-13686-3	BLOCK-762
46-13689-3	BLOCK-697
46-1369-3	BLOCK-5
46-1370-3	BLOCK-107
46-13704-3	BLOCK-525
46-13730-3	BLOCK-1087
46-13815-3	BLOCK-695
46-13816-3	BLOCK-1098
46-13829-3	BLOCK-756
46-1383-3	BLOCK-18
46-13830-3	BLOCK-560
46-13839-3	BLOCK-529
46-1391-3	BLOCK-218
46-13911-3	BLOCK-722
46-13912-3	BLOCK-642
46-1392-3	BLOCK-10
46-13921-3	BLOCK-761
46-13922-3	BLOCK-360
46-13924-3	BLOCK-663
46-1393-3	BLOCK-127

DEVICE TYPE	REPL CODE
46-13938-3	BLOCK-659
46-1394-3	BLOCK-128
46-1395-3	BLOCK-131
46-1396-3	BLOCK-126
46-13963-3	BLOCK-659
46-13969-3	BLOCK-671
46-1397-3	BLOCK-129
46-1398-3	BLOCK-130
46-14367-3	BLOCK-525
46-29-6	BLOCK-1269
46-5002-1	BLOCK-1212
46-5002-11	BLOCK-1815
46-5002-12	BLOCK-1271
46-5002-13	BLOCK-1272
46-5002-15	BLOCK-1269
46-5002-16	BLOCK-1805
46-5002-19	BLOCK-1804
46-5002-21	BLOCK-1807
46-5002-23	BLOCK-1303
46-5002-26	BLOCK-1269
46-5002-27	BLOCK-1804
46-5002-28	BLOCK-1422
46-5002-3	BLOCK-1274
46-5002-31	BLOCK-60
46-5002-4	BLOCK-1212
46-5002-5	BLOCK-1807
46-5002-6	BLOCK-1267
46-5002-7	BLOCK-1212
46-5002-8	BLOCK-1421
46-5290-0042	BLOCK-453
46-5290-0178	BLOCK-454
46-5291-0164	BLOCK-973
46-722066-3	BLOCK-306
46-861739-3	BLOCK-820
46-86465-3	BLOCK-1157
46-86474-3	BLOCK-1157
4613443	BLOCK-93
463984-1	BLOCK-1390
4663001A905	BLOCK-1411
4663001A909	BLOCK-1297
4663001A911	BLOCK-1390
4663001A912	BLOCK-1304
4663001A915	BLOCK-1377
4663001D907	BLOCK-1293
46SI-SAF1039P	BLOCK-741
477-0375-001	BLOCK-2199
477-0377-001	BLOCK-2197
477-0379-001	BLOCK-2217
477-0380-001	BLOCK-2207
477-0381-001	BLOCK-2208
477-0412-004	BLOCK-1410
477-0415-002	BLOCK-1401
477-0417-002	BLOCK-1403
48-1050-SL12318	BLOCK-2197
48-1050-SL12319	BLOCK-2199
48-1050-SL12320	BLOCK-2200
48-1050-SL12321	BLOCK-2202
48-1050-SL12322	BLOCK-2206
48-1050-SL12323	BLOCK-2207
48-1050-SL12324	BLOCK-2208
48-1050-SL12325	BLOCK-2217
48-1050-SL12326	BLOCK-1950
48-1050-SL12327	BLOCK-1953
48-90445A97	BLOCK-1157
4800155144	BLOCK-1157
4822 209 60955	BLOCK-554
4835 209 17033	BLOCK-1088
4835 209 87067	BLOCK-2087
4835 209 87085	BLOCK-779
4835 209 87089	BLOCK-1252
4835 209 87486	BLOCK-2087
4835 209 88003	BLOCK-1222
4835-130-47042	BLOCK-1021
4835-130-97006	BLOCK-1021
4835-130-97042	BLOCK-1021
4835-209-17004	BLOCK-1550
4835-209-17005	BLOCK-1545
4835-209-17032	BLOCK-1912
4835-209-17033	BLOCK-1088
4835-209-17148	BLOCK-784
4835-209-47005	BLOCK-1235
4835-209-47056	BLOCK-758
4835-209-87065	BLOCK-567
4835-209-87066	BLOCK-831
4835-209-87067	BLOCK-2087
4835-209-87069	BLOCK-754
4835-209-87071	BLOCK-746
4835-209-87073	BLOCK-778
4835-209-87074	BLOCK-762
4835-209-87079	BLOCK-1993
4835-209-87081	BLOCK-2063
4835-209-87082	BLOCK-207
4835-209-87085	BLOCK-779
4835-209-87089	BLOCK-1252

DEVICE TYPE	REPL CODE
4835-209-87091	BLOCK-838
4835-209-87094	BLOCK-824
4835-209-87105	BLOCK-1093
4835-209-87252	BLOCK-2144
4835-209-87259	BLOCK-2087
4835-209-87261	BLOCK-1873
4835-209-87262	BLOCK-1993
4835-209-87263	BLOCK-1873
4835-209-87264	BLOCK-1157
4835-209-87268	BLOCK-747
4835-209-87274	BLOCK-2087
4835-209-87275	BLOCK-2085
4835-209-87277	BLOCK-1087
4835-209-87283	BLOCK-2097
4835-209-87406	BLOCK-770
4835-209-87552	BLOCK-884
4835-209-87612	BLOCK-660
4835-209-87803	BLOCK-710
4835-209-87821	BLOCK-2144
4835-209-87827	BLOCK-2097
4835-209-87834	BLOCK-746
4835-209-87835	BLOCK-831
4835-209-88003	BLOCK-1222
483520917279	BLOCK-1096
483520947083	BLOCK-1088
483520987112	BLOCK-1995
483520987113	BLOCK-1995
489751-175	BLOCK-44
48S00155144	BLOCK-1157
48S155144	BLOCK-1157
48X90445A97	BLOCK-1157
48X90456A34	BLOCK-1157
4914296	BLOCK-2180
4915702	BLOCK-1336
4915705	BLOCK-1307
49A0000	BLOCK-1412
49A0002-000	BLOCK-1410
49A0005-000	BLOCK-1304
49A0006-000	BLOCK-1365
49A0010	BLOCK-2206
49A0012-000	BLOCK-1411
49A0510	BLOCK-2206
4H-209-80751	BLOCK-383
4H-209-80821	BLOCK-237
4H-209-80967	BLOCK-550
4H-209-81252	BLOCK-558
4H-209-81473	BLOCK-275
4H2098-0357	BLOCK-279
4H20980587	BLOCK-2171
4H20981194	BLOCK-220
4H20981787	BLOCK-532
4IC-402	BLOCK-93
4IC-407	BLOCK-4
4IC-409	BLOCK-108
5-113641	BLOCK-2208
5000	BLOCK-1791
5002-030	BLOCK-156
5002-031	BLOCK-145
5005	BLOCK-1791
5005(THOMAS)	BLOCK-1791
5007	BLOCK-1791
5007(THOMAS)	BLOCK-1791
5009	BLOCK-1794
5009(THOMAS)	BLOCK-1794
50210-2	BLOCK-2199
50210-7	BLOCK-1951
50210-8	BLOCK-2217
50254200	BLOCK-1299
50254400	BLOCK-1967
50254600	BLOCK-1719
50254700	BLOCK-1726
50254900	BLOCK-1745
50255000	BLOCK-1729
505254	BLOCK-2205
5076204	BLOCK-1719
5076205BZ	BLOCK-1745
5082-4350	BLOCK-1021
508590	BLOCK-1297
50SI-HA11446	BLOCK-636
51-03009A02	BLOCK-1405
51-10276A01	BLOCK-1223
51-10302A01	BLOCK-1212
51-10302A1	BLOCK-1212
51-10422801	BLOCK-1275
51-10422A01	BLOCK-1275
51-10422A02	BLOCK-1275
51-10437A01	BLOCK-1275
51-10534A01	BLOCK-1288
51-10534A03	BLOCK-1786
51-10534A04	BLOCK-1288
51-10542A01	BLOCK-1794
51-10594A01	BLOCK-1279
51-1059A01	BLOCK-1279
51-10600A	BLOCK-1774

DEVICE TYPE	REPL CODE
51-10600A01	BLOCK-1774
51-10611A09	BLOCK-1790
51-10611A11	BLOCK-1293
51-10611A12	BLOCK-1297
51-10611A15	BLOCK-1401
51-10611A16	BLOCK-1412
51-10619A01	BLOCK-1279
51-10631A01	BLOCK-1262
51-10636A	BLOCK-1797
51-10636A01	BLOCK-1797
51-10637A01	BLOCK-1806
51-10638A01	BLOCK-1814
51-10650A01	BLOCK-1801
51-10655A03	BLOCK-1291
51-10655A17	BLOCK-1029
51-10655A19	BLOCK-1041
51-10655A21	BLOCK-1048
51-10655B02	BLOCK-1421
51-10658A01	BLOCK-1813
51-10658A02	BLOCK-1813
51-10658A03	BLOCK-1813
51-10672A01	BLOCK-1289
51-10678A01	BLOCK-1814
51-10679A13	BLOCK-1421
51-10711A01	BLOCK-1376
51-10711A02	BLOCK-1376
51-13753-A19	BLOCK-1821
51-13753A01	BLOCK-184
51-13753A04	BLOCK-151
51-13753A10	BLOCK-153
51-13753A11	BLOCK-1269
51-13753A18	BLOCK-162
51-13753A19	BLOCK-1821
51-13753A20	BLOCK-165
51-13753A22	BLOCK-398
51-13753A23	BLOCK-403
51-13753A24	BLOCK-286
51-13753A25	BLOCK-473
51-13753A26	BLOCK-428
51-13753A27	BLOCK-425
51-13753A28	BLOCK-398
51-13753A29	BLOCK-393
51-13753A35	BLOCK-428
51-13753A39	BLOCK-428
51-13753A40	BLOCK-219
51-13753A46	BLOCK-163
51-13753A47	BLOCK-2168
51-13753A48	BLOCK-388
51-13753A49	BLOCK-390
51-13753A50	BLOCK-2170
51-13753A52	BLOCK-535
51-13753A53	BLOCK-655
51-13753A54	BLOCK-398
51-13753A55	BLOCK-537
51-13763A27	BLOCK-398
51-40000S08	BLOCK-221
51-40464P01	BLOCK-1825
51-41850J01	BLOCK-7
51-42211P01	BLOCK-159
51-42908J01	BLOCK-1825
51-43639802	BLOCK-1786
51-43639B02	BLOCK-1786
51-43684B01	BLOCK-1288
51-43684B02	BLOCK-1786
51-43684B0P	BLOCK-1786
51-44789J01	BLOCK-223
51-44789J02	BLOCK-223
51-44837J04	BLOCK-228
51-70177A07	BLOCK-1794
51-84320A09	BLOCK-1797
51-903	BLOCK-1048
51-90305A04	BLOCK-175
51-90305A05	BLOCK-176
51-90305A20	BLOCK-121
51-90305A21	BLOCK-250
51-90305A24	BLOCK-253
51-90305A27	BLOCK-256
51-90305A30	BLOCK-2052
51-90305A31	BLOCK-1978
51-90305A37	BLOCK-415
51-90305A39	BLOCK-392
51-90305A40	BLOCK-416
51-90305A41	BLOCK-413
51-90305A43	BLOCK-400
51-90305A44	BLOCK-407
51-90305A45	BLOCK-609
51-90305A46	BLOCK-2054
51-90305A50	BLOCK-1422
51-90305A59	BLOCK-1157
51-90305A6	BLOCK-1269
51-90305A60	BLOCK-58
51-90305A61	BLOCK-1269
51-90305A62	BLOCK-395
51-90305A72	BLOCK-973

DEVICE TYPE	REPL CODE
51-90305A73	BLOCK-409
51-90305A77	BLOCK-370
51-90305A78	BLOCK-237
51-90305A87	BLOCK-1269
51-90305A88	BLOCK-602
51-90305A89	BLOCK-537
51-90305A90	BLOCK-1422
51-90305A91	BLOCK-1269
51-90305A92	BLOCK-393
51-90433A07	BLOCK-389
51-90433A08	BLOCK-2085
51-90433A10	BLOCK-1096
51-90433A11	BLOCK-1085
51-90433A12	BLOCK-2055
51-90433A13	BLOCK-1048
51-90433A27	BLOCK-2085
51-90433A30	BLOCK-2085
51-90433A32	BLOCK-416
51-90433A33	BLOCK-387
51-90433A34	BLOCK-410
51-90433A35	BLOCK-407
51-90433A52	BLOCK-607
51-90433A53	BLOCK-473
51-90433A54	BLOCK-223
51-90433A60	BLOCK-151
51-90433A63	BLOCK-973
51-90433A66	BLOCK-176
5113753A37	BLOCK-668
5113753A67	BLOCK-360
5113753A68	BLOCK-649
51310000	BLOCK-2199
51310001	BLOCK-2200
51310002	BLOCK-2202
51310003	BLOCK-2206
51310004	BLOCK-2208
51310005	BLOCK-2209
51310006	BLOCK-2217
51310007	BLOCK-1951
51320000	BLOCK-1293
51320001	BLOCK-1295
51320002	BLOCK-1297
51320003	BLOCK-1304
51320004	BLOCK-1365
51320005	BLOCK-1390
51320006	BLOCK-1344
51320008	BLOCK-1409
51320009	BLOCK-1296
51320011	BLOCK-1377
51320012	BLOCK-1293
51320016	BLOCK-1402
51320017	BLOCK-1298
51320018	BLOCK-1419
51330005	BLOCK-1301
514-048	BLOCK-1791
5147013102	BLOCK-1825
515104221401	BLOCK-1275
51510422A01	BLOCK-1275
51513753A09	BLOCK-1269
51544789J01	BLOCK-223
51544789J02	BLOCK-223
51577500	BLOCK-2197
51577600	BLOCK-2199
51577700	BLOCK-2200
51577800	BLOCK-2206
51577900	BLOCK-2208
51578000	BLOCK-2209
51578100	BLOCK-2217
5158	BLOCK-1821
5158-1	BLOCK-1821
51624200	BLOCK-2206
5165440	BLOCK-2202
51717000	BLOCK-1951
51717100	BLOCK-1952
51750	BLOCK-2056
5175460	BLOCK-1301
518022S	BLOCK-138
5190266A68	BLOCK-760
5190458A07	BLOCK-1873
5190480A92	BLOCK-685
5190480A96	BLOCK-660
5190495A88	BLOCK-728
5190495A90	BLOCK-669
5190507A20	BLOCK-223
5190507A39	BLOCK-1993
5190507A40	BLOCK-1993
5190507A46	BLOCK-675
5190507A60	BLOCK-617
5190507A78	BLOCK-692
5190507A87	BLOCK-615
5190518A19	BLOCK-691
5190528A01	BLOCK-1993
5190528A27	BLOCK-728
5190528A28	BLOCK-669
5190528A29	BLOCK-2097
5190528A41	BLOCK-723
5190528A43	BLOCK-764
5190528A53	BLOCK-522
5190528A55	BLOCK-792
5190528A56	BLOCK-793
5190528A57	BLOCK-794
5190528A58	BLOCK-808
5190528A59	BLOCK-809
5190528A61	BLOCK-795
5190528A64	BLOCK-768
5190538A18	BLOCK-1066
5190538A20	BLOCK-760
5190538A22	BLOCK-2097
5190538A98	BLOCK-770
5190555A50	BLOCK-761
5190555A53	BLOCK-669
5190555A60	BLOCK-1088
51C43684B02	BLOCK-1786
51C436884B01	BLOCK-1786
51D70177A01	BLOCK-1822
51D70177A02	BLOCK-1421
51D70177A02(C)	BLOCK-43
51D70177A07	BLOCK-1269
51D70177B02	BLOCK-1421
51G10679A03	BLOCK-1291
51G10679A13	BLOCK-1421
51M33801A01	BLOCK-151
51M33801A04	BLOCK-184
51M33801A05	BLOCK-162
51M33801A06	BLOCK-150
51M33801A07	BLOCK-153
51M70177A01	BLOCK-1822
51M70177A03	BLOCK-1291
51M70177A05	BLOCK-1422
51M70177A07	BLOCK-1794
51M70177B02	BLOCK-1421
51R84320A09	BLOCK-1797
51R8432A09	BLOCK-1797
51S10276A01	BLOCK-1223
51S10302A01	BLOCK-1212
51S10382A	BLOCK-1275
51S10408A01	BLOCK-1265
51S10422A	BLOCK-1275
51S10422A01	BLOCK-1275
51S10432A01	BLOCK-1274
51S10437A01	BLOCK-1275
51S1048A01	BLOCK-1269
51S10534A01	BLOCK-1786
51S10534A03	BLOCK-1786
51S1059401	BLOCK-1816
51S10594A01	BLOCK-1279
51S10600A	BLOCK-1287
51S10600A01	BLOCK-1774
51S10611A09	BLOCK-1790
51S10611A11	BLOCK-1293
51S10611A12	BLOCK-1297
51S10611A15	BLOCK-1401
51S10611A16	BLOCK-1412
51S10636A	BLOCK-1797
51S10636A01	BLOCK-1797
51S10638A01	BLOCK-1814
51S10655A01	BLOCK-1822
51S10655A17	BLOCK-1029
51S10655A18	BLOCK-1039
51S10655A19	BLOCK-1041
51S10655B01	BLOCK-1822
51S10655B03	BLOCK-1291
51S10655B13	BLOCK-1269
51S10655C05	BLOCK-1422
51S1373A40	BLOCK-219
51S13749A46	BLOCK-163
51S13752A47	BLOCK-2168
51S13753A01	BLOCK-184
51S13753A02	BLOCK-153
51S13753A03	BLOCK-150
51S13753A06	BLOCK-162
51S13753A07	BLOCK-1269
51S13753A08	BLOCK-157
51S13753A09	BLOCK-1269
51S13753A10	BLOCK-153
51S13753A11	BLOCK-1269
51S13753A23	BLOCK-403
51S13753A28	BLOCK-398
51S13753A37	BLOCK-668
51S13753A39	BLOCK-428
51S13753A40	BLOCK-219
51S13753A46	BLOCK-163
51S13753A47	BLOCK-2168
51S13753A48	BLOCK-388
51S13753A49	BLOCK-390
51S13753A50	BLOCK-2170
51S13753A52	BLOCK-535
51S13753A53	BLOCK-655
51S13753A55	BLOCK-537
51S13753A56	BLOCK-398
51S13753A60	BLOCK-388
51S13753A64	BLOCK-2170
51S13753A65	BLOCK-636
51S13753A67	BLOCK-360
51S13753A68	BLOCK-649
51S137A48	BLOCK-388
51S137A50	BLOCK-2170
51S23753A47	BLOCK-2168
51S33753A47	BLOCK-2168
51S3753A48	BLOCK-388
51S44789J01	BLOCK-223
51S44789J02	BLOCK-223
51T40113T01	BLOCK-932
51X13753A24	BLOCK-286
51X90266A15	BLOCK-2063
51X90266A45	BLOCK-1993
51X90305A21	BLOCK-250
51X90305A41	BLOCK-413
51X90305A45	BLOCK-609
51X90305A46	BLOCK-2054
51X90305A59	BLOCK-1157
51X90305A61	BLOCK-1269
51X90305A88	BLOCK-602
51X904080A96	BLOCK-660
51X90433A12	BLOCK-2055
51X90433A28	BLOCK-402
51X90433A33	BLOCK-387
51X90433A35	BLOCK-407
51X90433A52	BLOCK-607
51X90433A53	BLOCK-473
51X90433A54	BLOCK-223
51X90433A99	BLOCK-608
51X90458A01	BLOCK-410
51X90458A02	BLOCK-681
51X90458A03	BLOCK-682
51X90458A05	BLOCK-301
51X90458A06	BLOCK-600
51X90458A07	BLOCK-1873
51X90458A15	BLOCK-1035
51X90458A30	BLOCK-428
51X90458A50	BLOCK-428
51X90458A51	BLOCK-1269
51X90458A52	BLOCK-282
51X90458A55	BLOCK-533
51X90458A56	BLOCK-415
51X90458A71	BLOCK-605
51X90458A72	BLOCK-597
51X90458A73	BLOCK-223
51X90458A75	BLOCK-616
51X90458A78	BLOCK-257
51X90458A80	BLOCK-612
51X90458A81	BLOCK-680
51X90458A82	BLOCK-555
51X90458A83	BLOCK-1801
51X90458A84	BLOCK-1801
51X90458A88	BLOCK-1093
51X90458A89	BLOCK-1104
51X90458A90	BLOCK-1135
51X90458A93	BLOCK-1873
51X90458A94	BLOCK-1873
51X90458A95	BLOCK-1336
51X90458A96	BLOCK-1299
51X90458A98	BLOCK-2093
51X90480A07	BLOCK-2087
51X90480A08	BLOCK-2087
51X90480A09	BLOCK-2097
51X90480A16	BLOCK-1093
51X90480A17	BLOCK-1104
51X90480A18	BLOCK-1035
51X90480A20	BLOCK-396
51X90480A32	BLOCK-699
51X90480A41	BLOCK-683
51X90480A46	BLOCK-2093
51X90480A57	BLOCK-398
51X90480A64	BLOCK-538
51X90480A65	BLOCK-708
51X90480A66	BLOCK-2097
51X90480A92	BLOCK-685
51X90480A93	BLOCK-616
51X90480A94	BLOCK-219
51X90480A95	BLOCK-533
51X90480A96	BLOCK-660
51X90495A14	BLOCK-533
51X90507A20	BLOCK-223
51X90507A24	BLOCK-2093
51X90507A28	BLOCK-696
51X90507A29	BLOCK-695
51X90507A39	BLOCK-1993
51X90507A40	BLOCK-1993
51X90507A46	BLOCK-675
51X90507A77	BLOCK-690
51X90507A78	BLOCK-692
51X90507A80	BLOCK-693
51X90507A83	BLOCK-696
51X90507A84	BLOCK-1029
51X90507A85	BLOCK-1098
51X90507A87	BLOCK-615
51X90507A92	BLOCK-1043
51X90507A93	BLOCK-1063
51X90507A97	BLOCK-1030
51X90507A98	BLOCK-1040
51X90507A99	BLOCK-1096
51X90518A07	BLOCK-2087
51X90518A08	BLOCK-1076
51X90518A09	BLOCK-684
51X90518A19	BLOCK-691
51X90518A28	BLOCK-1135
51X90518A34	BLOCK-1035
51X90518A35	BLOCK-2055
51X90518A36	BLOCK-1039
51X90518A69	BLOCK-2079
51X90518A71	BLOCK-988
51X90518A97	BLOCK-598
52-040-009-0	BLOCK-4
52-040-010-0	BLOCK-7
52-045-001-0	BLOCK-1
52000-030	BLOCK-2197
52000-032	BLOCK-2199
52000-033	BLOCK-2200
52000-036	BLOCK-2202
52000-037	BLOCK-2203
52000-044	BLOCK-2206
52000-046	BLOCK-2208
52000-048	BLOCK-2209
52000-049	BLOCK-2210
52000-052	BLOCK-1953
52000-055	BLOCK-1952
52000-057	BLOCK-1968
52000-058	BLOCK-1969
52000-061	BLOCK-2216
52000-062	BLOCK-2217
52000-063	BLOCK-2218
5233-7100	BLOCK-1307
52335500	BLOCK-1428
52335600	BLOCK-1473
52335600HL	BLOCK-1473
52335700	BLOCK-1484
52335800	BLOCK-1489
52335800HL	BLOCK-1489
52335900	BLOCK-1501
52335900HL	BLOCK-1501
52TA7611AP-S	BLOCK-401
53-1487	BLOCK-111
530176-1	BLOCK-1157
5302390086	BLOCK-1157
5303340001	BLOCK-1021
5310	BLOCK-398
5320200050	BLOCK-1269
5320200150	BLOCK-532
5320200160	BLOCK-519
5320200170	BLOCK-484
5320500100	BLOCK-2168
5320500120	BLOCK-163
5320600070	BLOCK-1821
5329500060	BLOCK-1269
5350121	BLOCK-28
5350132	BLOCK-29
5350136	BLOCK-29
5350141	BLOCK-30
5350151	BLOCK-31
5350152	BLOCK-31
5350161	BLOCK-32
5350182	BLOCK-73
5350211	BLOCK-42
5350231	BLOCK-56
5350251	BLOCK-646
5350321	BLOCK-158
5350491	BLOCK-158
5350604	BLOCK-516
5350611	BLOCK-1157
5350702	BLOCK-451
5350961	BLOCK-646
5351021	BLOCK-38
5351031	BLOCK-41
5351051	BLOCK-40
5351061	BLOCK-70
5351062	BLOCK-70
5351351	BLOCK-1269
5351361	BLOCK-5
5351411	BLOCK-478
5351501	BLOCK-572
5351521	BLOCK-582
5351551	BLOCK-584
5352631	BLOCK-545
5353591	BLOCK-716
5355262	BLOCK-695
5355611	BLOCK-644
5359031	BLOCK-1293
5359251	BLOCK-244
5359261	BLOCK-243
5359262	BLOCK-243
5359271	BLOCK-1426
5359281	BLOCK-2160
5359522	BLOCK-980
5359701	BLOCK-973
5359841	BLOCK-578
5362741	BLOCK-1039
5364091	BLOCK-871
5364201	BLOCK-768
5364601	BLOCK-2144
5369181	BLOCK-1802
5369431	BLOCK-762
5391032	BLOCK-685
544-0003-043	BLOCK-1878
544-2002-004	BLOCK-1282
544-2002-008	BLOCK-1963
544-2003-002	BLOCK-1421
544-2003-005	BLOCK-2119
544-2006-001	BLOCK-112
544-2006-011	BLOCK-221
544-2006-11	BLOCK-263
544-2009-555	BLOCK-2079
544-2020-002	BLOCK-2171
544-3001-001	BLOCK-2193
544-3001-103	BLOCK-1029
544-3001-117	BLOCK-1051
544-3001-140	BLOCK-1084
544-3001-143	BLOCK-1122
544-3001-201	BLOCK-2184
54LS02	BLOCK-1570
54LS02DM	BLOCK-1570
54SI-TA7075P	BLOCK-78
55001	BLOCK-1293
55002	BLOCK-1295
55003	BLOCK-1297
55004	BLOCK-1298
55005	BLOCK-1304
55006	BLOCK-1365
55007	BLOCK-1377
55008	BLOCK-1392
55009	BLOCK-1397
55010	BLOCK-1401
55011	BLOCK-1411
55012	BLOCK-1413
55017	BLOCK-1483
55021	BLOCK-1302
55022	BLOCK-1310
55023	BLOCK-1367
55024	BLOCK-1336
55027	BLOCK-1317
55029	BLOCK-1404
55032	BLOCK-1294
55034	BLOCK-1390
55035	BLOCK-1300
55036	BLOCK-1299
551-010-00	BLOCK-2208
551-013-00	BLOCK-2207
551A	BLOCK-1791
552	BLOCK-934
555	BLOCK-2079
555D	BLOCK-2079
556	BLOCK-2145
556D	BLOCK-2145
55810-167	BLOCK-1212
55810-168	BLOCK-43
55810-169	BLOCK-1
55810-170	BLOCK-118
558875	BLOCK-1293
558876	BLOCK-1295
558877	BLOCK-1304
558878	BLOCK-1365
558879	BLOCK-1377
558880	BLOCK-1390
558881	BLOCK-1410
558882	BLOCK-1411
558883	BLOCK-1423
558885	BLOCK-1426
558CP	BLOCK-1988
559-1492-001	BLOCK-2186
559-1493-001	BLOCK-2192
559-1494-001	BLOCK-2188
559-1495-001	BLOCK-2183
559-1496-001	BLOCK-2185
559507	BLOCK-1294
559509	BLOCK-1401
559510	BLOCK-1406
559613	BLOCK-1403
55975-1	BLOCK-2207
55976-1	BLOCK-2208
55977	BLOCK-2197
55978-1	BLOCK-2206
55979-1	BLOCK-2213
55980	BLOCK-2200
55982-1	BLOCK-2199
55987-1	BLOCK-2217
55SI-MC14011BCP	BLOCK-1039
56-4833	BLOCK-71
56-4834	BLOCK-1825
561-0884-006	BLOCK-1957
561-0884-041	BLOCK-2057
561-0884-043	BLOCK-2058
561-0884-047	BLOCK-2070
561103	BLOCK-1815
5632-AN217(BB)	BLOCK-58
5635-HZT33	BLOCK-1157
565	BLOCK-2174
56519	BLOCK-2208
5652-AN217	BLOCK-58
5652-AN217(BB)	BLOCK-58
5652-AN217BB	BLOCK-58
5652-AN5316	BLOCK-707
5652-AN5436	BLOCK-708
5652-AN5512	BLOCK-660
5652-AN7218	BLOCK-550
5652-AN7410	BLOCK-547
5652-AN7410NS	BLOCK-547
5652-BA5402	BLOCK-606
5652-BA6124	BLOCK-548
5652-HA1156	BLOCK-1825
5652-HA1325	BLOCK-227
5652-LA7520	BLOCK-709
5652-M51513L	BLOCK-245
5652-TA7230P	BLOCK-374
5652-TA7658P	BLOCK-558
5652-TA78L005	BLOCK-2144
5652-UPC1018C	BLOCK-550
5652-UPC1373H	BLOCK-695
5653-AN5265	BLOCK-770
5653-STR30130	BLOCK-758
5653-TA7777N	BLOCK-771
5653-TA78012A	BLOCK-2097
5653-TA7812S	BLOCK-921
5654-HD38991A	BLOCK-974
56552	BLOCK-1950
56553	BLOCK-2197
56557	BLOCK-2202
56558	BLOCK-2207
56571	BLOCK-2217
5662-HA1325	BLOCK-227
567	BLOCK-1868
569-0320-809	BLOCK-2056
569-0320-813	BLOCK-1281
569-0320-814	BLOCK-1950
569-0320-815	BLOCK-1961
569-0320-818	BLOCK-2202
569-0320-820	BLOCK-2208
569-0320-821	BLOCK-2217
569-0320-822	BLOCK-2208
569-0320-826	BLOCK-2202
569-0320-831	BLOCK-1984
569-0540-355	BLOCK-2079
569-0540-358	BLOCK-1801
569-0540-623	BLOCK-1984
569-0541-000	BLOCK-2098
569-0541-006	BLOCK-2101
569-0541-014	BLOCK-2105
569-0541-018	BLOCK-2106
569-0541-024	BLOCK-2111
569-0541-030	BLOCK-2114
569-0541-031	BLOCK-2115
569-0541-036	BLOCK-2118
569-0541-037	BLOCK-2120
569-0541-038	BLOCK-2121
569-0541-041	BLOCK-2124
569-0541-042	BLOCK-2125
569-0541-043	BLOCK-2126
569-0542-744	BLOCK-2140
569-0544-300	BLOCK-2197
569-0544-304	BLOCK-2199
569-0544-305	BLOCK-2200
569-0544-308	BLOCK-2202
569-0544-312	BLOCK-2201
569-0544-318	BLOCK-2206
569-0544-322	BLOCK-2208
569-0544-328	BLOCK-1953
569-0544-348	BLOCK-2217
569-0690-300	BLOCK-1940
569-0690-301	BLOCK-1963
569-0690-500	BLOCK-1029
569-0690-501	BLOCK-1039
569-0690-502	BLOCK-1041
569-0690-550	BLOCK-1085
569-0690-681	BLOCK-1970
569-0773-600	BLOCK-2079

DEVICE TYPE	REPL CODE
569-0773-610	BLOCK-1801
569-0773-800	BLOCK-1293
569-0773-811	BLOCK-1307
569-0880-430	BLOCK-2197
569-0880-444	BLOCK-2206
569-0880-445	BLOCK-2207
569-0880-446	BLOCK-2208
569-0880-450	BLOCK-2212
569-0880-458	BLOCK-1969
569-0880-462	BLOCK-2217
569-0881-801	BLOCK-1472
569-0881-804	BLOCK-1473
569-0881-810	BLOCK-1476
569-0881-811	BLOCK-1482
569-0881-820	BLOCK-1484
569-0881-821	BLOCK-1485
569-0881-822	BLOCK-1486
569-0881-830	BLOCK-1487
569-0881-840	BLOCK-1488
569-0881-850	BLOCK-1489
569-0881-851	BLOCK-1490
569-0881-852	BLOCK-1491
569-0881-853	BLOCK-1492
569-0881-855	BLOCK-1494
569-0881-860	BLOCK-1495
569-0881-874	BLOCK-1501
569-0881-908	BLOCK-1481
569-0882-600	BLOCK-1293
569-0882-601	BLOCK-1294
569-0882-602	BLOCK-1295
569-0882-603	BLOCK-1296
569-0882-604	BLOCK-1297
569-0882-605	BLOCK-1298
569-0882-606	BLOCK-1299
569-0882-607	BLOCK-1300
569-0882-608	BLOCK-1301
569-0882-609	BLOCK-1302
569-0882-610	BLOCK-1304
569-0882-612	BLOCK-1310
569-0882-613	BLOCK-1317
569-0882-614	BLOCK-1320
569-0882-616	BLOCK-1336
569-0882-617	BLOCK-1343
569-0882-620	BLOCK-1365
569-0882-627	BLOCK-1372
569-0882-630	BLOCK-1377
569-0882-632	BLOCK-1378
569-0882-633	BLOCK-1379
569-0882-637	BLOCK-1384
569-0882-638	BLOCK-1385
569-0882-640	BLOCK-1390
569-0882-641	BLOCK-1391
569-0882-642	BLOCK-1392
569-0882-647	BLOCK-1397
569-0882-650	BLOCK-1401
569-0882-651	BLOCK-1402
569-0882-653	BLOCK-1403
569-0882-654	BLOCK-1404
569-0882-660	BLOCK-1406
569-0882-672	BLOCK-1409
569-0882-673	BLOCK-1410
569-0882-674	BLOCK-1411
569-0882-675	BLOCK-1412
569-0882-676	BLOCK-1413
569-0882-680	BLOCK-1414
569-0882-682	BLOCK-1416
569-0882-685	BLOCK-1418
569-0882-686	BLOCK-1419
569-0882-689	BLOCK-1420
569-0882-690	BLOCK-1423
569-0882-692	BLOCK-1425
569-0882-693	BLOCK-1426
569-0882-696	BLOCK-1429
569-0882-707	BLOCK-1305
569-0882-709	BLOCK-1306
569-0882-721	BLOCK-1311
569-0882-722	BLOCK-1312
569-0882-723	BLOCK-1313
569-0882-726	BLOCK-1315
569-0882-732	BLOCK-1318
569-0882-741	BLOCK-1321
569-0882-745	BLOCK-1325
569-0882-750	BLOCK-1327
569-0882-751	BLOCK-1328
569-0882-753	BLOCK-1330
569-0882-754	BLOCK-1331
569-0882-755	BLOCK-1332
569-0882-760	BLOCK-1337
569-0882-761	BLOCK-1338
569-0882-764	BLOCK-1341
569-0882-765	BLOCK-1342
569-0882-774	BLOCK-1346
569-0882-775	BLOCK-1347
569-0882-776	BLOCK-1348
569-0882-777	BLOCK-1349
569-0882-779	BLOCK-1351
569-0882-780	BLOCK-1352
569-0882-781	BLOCK-1353
569-0882-782	BLOCK-1354
569-0882-790	BLOCK-1355
569-0882-791	BLOCK-1356
569-0882-792	BLOCK-1357
569-0882-793	BLOCK-1358
569-0882-796	BLOCK-1360
569-0882-797	BLOCK-1361
569-0883-200	BLOCK-1719
569-0883-203	BLOCK-1721
569-0883-204	BLOCK-1722
569-0883-205	BLOCK-1723
569-0883-210	BLOCK-1726
569-0883-211	BLOCK-1727
569-0883-215	BLOCK-1735
569-0883-220	BLOCK-1745
569-0883-222	BLOCK-1746
569-0883-240	BLOCK-1753
569-0883-264	BLOCK-1766
569-0883-265	BLOCK-1767
569-0883-274	BLOCK-1768
569-0883-312	BLOCK-1728
569-0883-338	BLOCK-1733
569-0883-340	BLOCK-1734
569-0883-353	BLOCK-1737
569-0883-357	BLOCK-1738
569-0883-358	BLOCK-1739
569-0883-374	BLOCK-1740
569-0885-850	BLOCK-1775
569-0885-853	BLOCK-1778
569-0885-854	BLOCK-1777
569-0886-031	BLOCK-1276
569-0886-800	BLOCK-1706
569-0964-001	BLOCK-1954
569-0964-002	BLOCK-2141
569-0965-006	BLOCK-2163
56A1	BLOCK-1212
56A1-1	BLOCK-1212
56A101-1	BLOCK-279
56A102-1	BLOCK-484
56A107-1	BLOCK-401
56A11-1	BLOCK-1365
56A113-1	BLOCK-1157
56A120-1	BLOCK-510
56A121-1	BLOCK-234
56A122-1	BLOCK-220
56A13-1	BLOCK-1395
56A137-1	BLOCK-534
56A138-1	BLOCK-559
56A14-1	BLOCK-1409
56A141-1	BLOCK-724
56A141-4	BLOCK-818
56A15-1	BLOCK-1423
56A16-1	BLOCK-951
56A17-1	BLOCK-1816
56A19-1	BLOCK-176
56A20-1	BLOCK-1808
56A21-1	BLOCK-2137
56A223-2	BLOCK-1194
56A23-1	BLOCK-1843
56A24-1	BLOCK-1818
56A25-1	BLOCK-1843
56A265-1	BLOCK-1093
56A3-1	BLOCK-1269
56A34-1	BLOCK-561
56A355-1	BLOCK-644
56A359-1	BLOCK-843
56A360-1	BLOCK-835
56A4-1	BLOCK-1271
56A42-1	BLOCK-114
56A46-1	BLOCK-1287
56A49-1	BLOCK-1898
56A49-2	BLOCK-1898
56A5-1	BLOCK-1272
56A55-1	BLOCK-1860
56A551	BLOCK-1860
56A6-1	BLOCK-1815
56A60-1	BLOCK-764
56A65-1	BLOCK-219
56A74-1	BLOCK-428
56A75-1	BLOCK-404
56A75-1A	BLOCK-404
56A9-1	BLOCK-1421
56B74-1	BLOCK-428
56B75-1	BLOCK-398
56C1	BLOCK-1212
56C1-1	BLOCK-1212
56C17-1	BLOCK-1816
56C49-1	BLOCK-1898
56D17-1	BLOCK-1816
56D20-1	BLOCK-1808
56D3-1	BLOCK-1269
56D4-1	BLOCK-1271
56D49-1	BLOCK-1898
56D49-2	BLOCK-1898
56D5-1	BLOCK-1272
56D6-1	BLOCK-1815
56D65-1	BLOCK-219
56D75-1	BLOCK-398
56D75-1A	BLOCK-404
56D9-1	BLOCK-1421
56SI-TA7644BP	BLOCK-534
5757	BLOCK-1806
576-0004-035(IC)	BLOCK-115
576-14	BLOCK-1303
57A29-2	BLOCK-1245
57A32-1	BLOCK-1280
57A32-11	BLOCK-73
57A32-17	BLOCK-29
57A32-19	BLOCK-101
57A32-2	BLOCK-138
57A32-21	BLOCK-138
57A32-22	BLOCK-1277
57A32-25	BLOCK-122
57A32-27	BLOCK-122
57A32-28	BLOCK-138
57A32-29	BLOCK-138
57A32-3	BLOCK-38
57A32-32	BLOCK-1825
57A32-6	BLOCK-40
57C28	BLOCK-1223
57C29-1	BLOCK-1245
57C29-2	BLOCK-1245
57D17-1	BLOCK-1816
57SI-HA11423	BLOCK-459
586-151	BLOCK-2197
586-152	BLOCK-2202
586-153	BLOCK-2208
586-154	BLOCK-1953
586-155	BLOCK-2217
586-187	BLOCK-2207
586-303	BLOCK-2199
586-308	BLOCK-2203
586-331	BLOCK-2156
586-412	BLOCK-2163
586-415	BLOCK-1968
586-425	BLOCK-1950
586-442	BLOCK-1969
586-517	BLOCK-2154
586-528	BLOCK-2152
586-546	BLOCK-2150
586-547	BLOCK-2158
586-780	BLOCK-2161
587-033	BLOCK-1344
588-40-202	BLOCK-138
58810-170	BLOCK-98
58810-171	BLOCK-138
58840-201	BLOCK-98
58840-203	BLOCK-85
58SI-LM393N	BLOCK-2063
58SI-LM393P	BLOCK-2063
590	BLOCK-1844
592-027	BLOCK-2185
592-028	BLOCK-2186
592-029-0	BLOCK-2189
592-081-0	BLOCK-2189
5932	BLOCK-1814
59B402787	BLOCK-2189
59B402788	BLOCK-2220
59SI-LM393N	BLOCK-2063
59SI-UPA53C	BLOCK-993
5D20-1	BLOCK-1808
5K4164ANP-15	BLOCK-1006
6001	BLOCK-112
601-0100865	BLOCK-1317
60213(1C)	BLOCK-1282
610002-011	BLOCK-1579
610002-014	BLOCK-1593
610002-026	BLOCK-1644
610002-038	BLOCK-1385
610002-074	BLOCK-1706
610002-086	BLOCK-1713
610002-123	BLOCK-1584
610010-413	BLOCK-936
610020-917	BLOCK-2211
6105080001	BLOCK-870
611064	BLOCK-1298
611065	BLOCK-1412
611066	BLOCK-1419
611071	BLOCK-1392
611563	BLOCK-1293
611564	BLOCK-1295
611565	BLOCK-1297
611566	BLOCK-1304
611567	BLOCK-1365
611568	BLOCK-1390
611569	BLOCK-1401
611570	BLOCK-1403
611571	BLOCK-1411
611572	BLOCK-1423
611573	BLOCK-1425
611730	BLOCK-1358
611731	BLOCK-1357
611844	BLOCK-1419
611845	BLOCK-1298
611870	BLOCK-1413
611872	BLOCK-1305
611878	BLOCK-1494
611900	BLOCK-1305
611901	BLOCK-1352
612005-1	BLOCK-1269
612005-2	BLOCK-1269
612007	BLOCK-1262
612007-1	BLOCK-1262
612007-2	BLOCK-1262
612007-3	BLOCK-1262
612008-2	BLOCK-1281
612020-1	BLOCK-1212
612021-1	BLOCK-1275
612024-1	BLOCK-1287
612024-3	BLOCK-1287
612025-1	BLOCK-1286
612025-2	BLOCK-1286
612025-3	BLOCK-1286
612029-1	BLOCK-1815
612029-3	BLOCK-1815
612042-1	BLOCK-110
612042-2	BLOCK-109
612044-1	BLOCK-1291
612045-1	BLOCK-1828
612048-4	BLOCK-2143
6120480004	BLOCK-2058
612054-2	BLOCK-1971
612061-1	BLOCK-1808
612067	BLOCK-1272
612069-1	BLOCK-1272
612070-1	BLOCK-1271
612072-1	BLOCK-1292
612074-1	BLOCK-1041
612075-1	BLOCK-1825
612075-3	BLOCK-1825
6120750003	BLOCK-1825
612076-1	BLOCK-1364
612076-2	BLOCK-1364
612076-4	BLOCK-1364
612077-2	BLOCK-1813
612080-1	BLOCK-2191
612082-1	BLOCK-1843
612082-2	BLOCK-1843
612082-3	BLOCK-1843
612091-1	BLOCK-1310
612092-1	BLOCK-1411
612100-1	BLOCK-1397
612103-3	BLOCK-2087
612103-4	BLOCK-2137
612107-2	BLOCK-2137
612107-4	BLOCK-2087
612113-1	BLOCK-1801
612120-1	BLOCK-1875
612126	BLOCK-532
612126-1	BLOCK-532
6121260001	BLOCK-401
612144-1	BLOCK-1828
612159-1	BLOCK-1299
612160030	BLOCK-791
6121630001	BLOCK-2171
612184-1	BLOCK-1020
612185-1	BLOCK-1298
612186-1	BLOCK-1048
6121860001	BLOCK-1048
612188-1	BLOCK-1395
612189-1	BLOCK-1908
612189-2	BLOCK-1908
612192-1	BLOCK-1168
612194-1	BLOCK-1659
612195-1	BLOCK-1596
612197-1	BLOCK-1666
612199-1	BLOCK-1568
612200-1	BLOCK-1706
612230-1	BLOCK-1601
612247-2	BLOCK-741
6122470002	BLOCK-741
6122500001	BLOCK-1096
612261-1	BLOCK-1568
612272-1	BLOCK-163
612273-1	BLOCK-2168
612275-1	BLOCK-1859
6122750001	BLOCK-1859
6122850001	BLOCK-1993
6122850003	BLOCK-1993
612291-1	BLOCK-1717
612294-1	BLOCK-1894
612303-1	BLOCK-404
6123030001	BLOCK-404
612304-1	BLOCK-433
6123040001	BLOCK-433
612305	BLOCK-220
612305-1	BLOCK-220
612305-2	BLOCK-220
6123050001	BLOCK-612
612308-1	BLOCK-1705
6123180001	BLOCK-1010
6123270001	BLOCK-553
6123280001	BLOCK-619
6123290001	BLOCK-554
6123300023	BLOCK-657
6123300024	BLOCK-603
6123300025	BLOCK-662
6123300027	BLOCK-401
6123300101	BLOCK-559
6123300102	BLOCK-774
6123300140	BLOCK-647
6123300215	BLOCK-768
6123300334	BLOCK-659
6123300418	BLOCK-807
612331-2	BLOCK-1848
6123310002	BLOCK-1848
612332-1	BLOCK-361
612332-6	BLOCK-363
6123320001	BLOCK-361
6123320003	BLOCK-361
6123320006	BLOCK-363
612334-1	BLOCK-1821
6123340001	BLOCK-1821
6123360001	BLOCK-1920
6123370003	BLOCK-1921
6123370005	BLOCK-1921
612338-3	BLOCK-1850
6123380002	BLOCK-1850
6123380003	BLOCK-1850
6123390001	BLOCK-401
6123420001	BLOCK-1007
612347-1	BLOCK-1849
6123470002	BLOCK-1849
6123500001	BLOCK-1039
612351-1	BLOCK-1269
6123510001	BLOCK-1269
612352-1	BLOCK-1891
6123520001	BLOCK-1891
6123540001	BLOCK-1303
6123590001	BLOCK-1908
6123600001	BLOCK-2087
6123620001	BLOCK-220
6123620002	BLOCK-220
6123630001	BLOCK-1895
6123630005	BLOCK-1895
6123700001	BLOCK-567
6123700002	BLOCK-567
6123800001	BLOCK-1093
6123940001	BLOCK-401
6123940002	BLOCK-401
6123999001	BLOCK-2144
6124120002	BLOCK-784
6124140001	BLOCK-306
6124160003	BLOCK-1009
6124240001	BLOCK-767
6124440001	BLOCK-735
6124450001	BLOCK-619
612448	BLOCK-2087
6124480001	BLOCK-2087
6124490001	BLOCK-733
6124500001	BLOCK-743
6124500002	BLOCK-742
6124670001	BLOCK-831
6124740001	BLOCK-1087
612479-1	BLOCK-2087
612479-5	BLOCK-2119
6124790001	BLOCK-2087
6124790003	BLOCK-2095
6124790004	BLOCK-2097
6124790005	BLOCK-2119
6124800001	BLOCK-1912
6124890001	BLOCK-725
61249-1	BLOCK-1801
6124920001	BLOCK-936
612493-1	BLOCK-1088
6124930001	BLOCK-1088
612494-1	BLOCK-1801
6124940001	BLOCK-1801
612496-1	BLOCK-641
6124960001	BLOCK-641
6125040001	BLOCK-207
61250700001	BLOCK-1199
6125070001	BLOCK-1199
6125080001	BLOCK-870
6125390001	BLOCK-1912
6125450001	BLOCK-767
6125630001	BLOCK-781
6125640001	BLOCK-782
6125660001	BLOCK-1030
6125710001	BLOCK-1190
6125740001	BLOCK-778
6125750001	BLOCK-825
6125860001	BLOCK-757
6125860002	BLOCK-758
6125960001	BLOCK-722
6126060001	BLOCK-773
6126120001	BLOCK-1199
6126150001	BLOCK-870
6126150K01	BLOCK-870
6126250001	BLOCK-754
6126280001	BLOCK-779
6126640001	BLOCK-746
6126710001	BLOCK-2087
6126760001	BLOCK-1588
6126810001	BLOCK-762
6126880001	BLOCK-837
6190000200	BLOCK-2058
6190000210	BLOCK-205
6192000040	BLOCK-768
6192000990	BLOCK-773
6192001330	BLOCK-770
6192020160	BLOCK-822
6192020190	BLOCK-821
6192020400	BLOCK-709
6192040031	BLOCK-794
6192040040	BLOCK-795
6192040260	BLOCK-824
6192060090	BLOCK-660
6192060140	BLOCK-754
6192080080	BLOCK-786
6192080250	BLOCK-662
6192100150	BLOCK-516
6192120110	BLOCK-816
6192140020	BLOCK-1157
6192140080	BLOCK-2087
6192140081	BLOCK-2144
6192140170	BLOCK-723
6192140210	BLOCK-664
6192140400	BLOCK-851
6192140620	BLOCK-1157
6192149480	BLOCK-2087
6192160030	BLOCK-791
6192180010	BLOCK-788
6192180061	BLOCK-690
6192180080	BLOCK-768
6192180110	BLOCK-813
6192180160	BLOCK-791
6193220010	BLOCK-808
619550-1	BLOCK-1859
61A0001-10	BLOCK-80
61A0001-11	BLOCK-138
61A0001-12	BLOCK-50
61A001-10	BLOCK-80
61A001-11	BLOCK-138
61A001-12	BLOCK-50
61A023-2	BLOCK-43
61A030-6	BLOCK-1269
61A030-9	BLOCK-81
61A0306	BLOCK-1269
61B042-9	BLOCK-4
61B1C	BLOCK-1808
61C001-11	BLOCK-256
61C001-12	BLOCK-211
61C001-2	BLOCK-121
61C001-30	BLOCK-392
61C001-32	BLOCK-416
61C001-33	BLOCK-413
61C001-34	BLOCK-409
61C001-35	BLOCK-400
61C001-36	BLOCK-407
61C001-38	BLOCK-395
61C001-4	BLOCK-250
61C001-7	BLOCK-253
61C002-2	BLOCK-415
61C002-3	BLOCK-2054
61K001-10	BLOCK-1825
61K001-12	BLOCK-11
61K001-13	BLOCK-174
61K001-9	BLOCK-11
61L001-3	BLOCK-201
61L001-5	BLOCK-246
61L001-6	BLOCK-247
61L001-8	BLOCK-240
61SI-SN74145N	BLOCK-1325
6207159	BLOCK-232
62737654	BLOCK-1421

ORIGINAL DEVICE TYPES AND REPLACEMENT CODES 13

DEVICE TYPE	REPL CODE
6300-1N	BLOCK-1752
640000002	BLOCK-180
640000003	BLOCK-221
6458D	BLOCK-1801
649-3	BLOCK-638
65-11137-23	BLOCK-1813
65-20820-00	BLOCK-115
65-33361-00	BLOCK-237
65-45402-00	BLOCK-973
6502	BLOCK-1158
6502A	BLOCK-1158
6502AD	BLOCK-1158
6502B	BLOCK-1158
6502C	BLOCK-1158
6507	BLOCK-1159
6532	BLOCK-1162
6552	BLOCK-1801
65600500	BLOCK-2197
65600600	BLOCK-2199
65600700	BLOCK-2208
65600800	BLOCK-1950
65600900	BLOCK-1953
65611000	BLOCK-1951
65611100	BLOCK-1952
6562	BLOCK-1993
6570	BLOCK-2058
657161	BLOCK-1773
6572	BLOCK-1801
65SI-HA11580	BLOCK-184
6604	BLOCK-1000
6605	BLOCK-1001
6640000410	BLOCK-2087
6644000100	BLOCK-1269
6644001100	BLOCK-126
6644001400	BLOCK-131
6644001500	BLOCK-130
6644001700	BLOCK-128
6644001800	BLOCK-129
6644001900	BLOCK-127
6644004202	BLOCK-1269
6644004301	BLOCK-184
6644004400	BLOCK-5
6644012500	BLOCK-532
6644012602	BLOCK-519
6644012700	BLOCK-484
6664	BLOCK-1163
668-009A	BLOCK-1813
668-028B	BLOCK-206
668-033A	BLOCK-1825
668-045A	BLOCK-1873
668-614A	BLOCK-202
668-616A	BLOCK-207
669A464H01	BLOCK-2199
669A471H01	BLOCK-1409
669A492H01	BLOCK-2217
66F125-1	BLOCK-1817
66F1271	BLOCK-1815
66F136-1	BLOCK-1818
66F1551	BLOCK-1291
66F175-1	BLOCK-1831
66F1751	BLOCK-1831
66F176-1	BLOCK-1833
66F1761	BLOCK-1833
67-01603-01	BLOCK-719
67-10430-01	BLOCK-1157
67-32009-01	BLOCK-534
67-32109-02	BLOCK-559
67-32720-01	BLOCK-1028
67-32720-02	BLOCK-1066
67-32808-02	BLOCK-663
67-33215-01	BLOCK-1193
67-51703-02	BLOCK-389
67-53315-01	BLOCK-761
67-90430-01	BLOCK-1157
67-93401-01	BLOCK-781
67-93501-01	BLOCK-782
671A290H01	BLOCK-2197
671A291H01	BLOCK-2206
671A292H01	BLOCK-2207
671A293H01	BLOCK-2208
679-1	BLOCK-1798
680-1(IC)	BLOCK-1774
6800	BLOCK-1164
6802	BLOCK-1165
6821	BLOCK-1169
6838	BLOCK-685
6850	BLOCK-1170
6889	BLOCK-1178
68995-1	BLOCK-1956
68A7349-D32	BLOCK-2199
68A7349-D46	BLOCK-2208
68A7349-D62	BLOCK-2217
68A7349-PD32	BLOCK-2199
68A7349PD30	BLOCK-2197
68A7349PD36	BLOCK-2202
68A7349PD45	BLOCK-2207
68A7349PD46	BLOCK-2208
68A7349PD62	BLOCK-2217
68A7652P1	BLOCK-1956
68A7652P2	BLOCK-1959
68A9025	BLOCK-1293
68A9026	BLOCK-1471
68A9027	BLOCK-1295
68A9028	BLOCK-1297
68A9030	BLOCK-1304
68A9031	BLOCK-1476
68A9032	BLOCK-1299
68A9033	BLOCK-1365
68A9034	BLOCK-1369
68A9035	BLOCK-1377
68A9036	BLOCK-1384
68A9037	BLOCK-1385
68A9038	BLOCK-1401
68A9040	BLOCK-1406
68A9041	BLOCK-1412
68A9042	BLOCK-1413
68A9047	BLOCK-1480
68A9048	BLOCK-1328
68A9049	BLOCK-1330
69-3116	BLOCK-1828
6900K91-002	BLOCK-2206
6900K91-003	BLOCK-2208
6900K91-004	BLOCK-2217
6900K91-006	BLOCK-2202
6900K91-007	BLOCK-2199
6900K91-009	BLOCK-1968
6900K93-001	BLOCK-2213
6900K93-002	BLOCK-2205
6900K95-001	BLOCK-1953
6900K95-002	BLOCK-1950
6914	BLOCK-2063
692-0016-00	BLOCK-2193
692-0020-00	BLOCK-2224
69B1M	BLOCK-1272
6H740AC	BLOCK-2051
6N136	BLOCK-1021
6N136-020	BLOCK-1021
6N136-HP	BLOCK-1021
7-759-651-35	BLOCK-5
7011200-02	BLOCK-2200
7011201-02	BLOCK-2210
7011203-02	BLOCK-1412
7011203-03	BLOCK-1412
70119023	BLOCK-250
70119034	BLOCK-588
70119036	BLOCK-583
70119037	BLOCK-581
70119041	BLOCK-255
70119048	BLOCK-585
70119049	BLOCK-573
70119062	BLOCK-1039
70119063	BLOCK-1067
70119064	BLOCK-1085
70119526	BLOCK-840
70119527	BLOCK-839
7012128	BLOCK-2202
7012128-02	BLOCK-2202
7012130	BLOCK-1956
7012131	BLOCK-1961
7012132	BLOCK-1416
7012133-02	BLOCK-1417
7012142-03	BLOCK-1358
7012166	BLOCK-1295
7012167-02	BLOCK-1423
70177A01	BLOCK-1822
70177A02	BLOCK-1421
70177A07	BLOCK-1794
70177B02	BLOCK-1421
70270740	BLOCK-93
703639-1	BLOCK-1364
703639-2	BLOCK-1364
7044260991	BLOCK-743
70622(RCA)	BLOCK-1277
709CE	BLOCK-1949
70B1C	BLOCK-1271
70B1Z	BLOCK-1271
710CE	BLOCK-1955
7130A	BLOCK-43
7149A	BLOCK-1269
715HC	BLOCK-1967
7161	BLOCK-1773
7161(WURLITZER)	BLOCK-1773
717126-505	BLOCK-1473
717136-1	BLOCK-1719
717136-15	BLOCK-1745
717399-22	BLOCK-1956
717399-73	BLOCK-1961
72045600	BLOCK-1959
7204A	BLOCK-1828
720DC	BLOCK-1389
72132-1	BLOCK-1859
72133-1	BLOCK-1794
72133-2	BLOCK-1794
72181	BLOCK-1311
72185	BLOCK-1413
722114299	BLOCK-930
7221A	BLOCK-1269
723	BLOCK-1984
723BE	BLOCK-1983
723C	BLOCK-1984
723CE	BLOCK-1983
723CJ	BLOCK-1984
72B1Z	BLOCK-1292
72SI-TA7074P	BLOCK-1422
7311325	BLOCK-1264
7313	BLOCK-1270
733W00024	BLOCK-2197
733W00025	BLOCK-2199
733W00026	BLOCK-2200
733W00027	BLOCK-2202
733W00028	BLOCK-2208
733W00029	BLOCK-2217
733W00030	BLOCK-1950
733W00039	BLOCK-1423
733W00042	BLOCK-1293
733W00043	BLOCK-1365
733W00048	BLOCK-2203
733W00049	BLOCK-2210
733W00050	BLOCK-2218
733W00064	BLOCK-1297
733W00065	BLOCK-1304
733W00067	BLOCK-1411
733W00068	BLOCK-1419
733W00069	BLOCK-1413
733W00076	BLOCK-1358
733W00126	BLOCK-1378
733W00133	BLOCK-1426
733W00183	BLOCK-1327
733W00211	BLOCK-1039
739	BLOCK-1281
739DC	BLOCK-1281
73C180475	BLOCK-1269
73C180476-5	BLOCK-1269
73C180837-1	BLOCK-1270
73C180837-2	BLOCK-1270
73C180837-3	BLOCK-1270
73C180843	BLOCK-1808
73C180843-1	BLOCK-1808
73C180843-2	BLOCK-1808
73C180843-3	BLOCK-1808
73C180843-4	BLOCK-1808
73C182186	BLOCK-1269
73C182763-1	BLOCK-1292
73C182763-2	BLOCK-1292
73C182764-1	BLOCK-1291
73C182764-2	BLOCK-1291
73C182764-4	BLOCK-1291
73SI-MC1353P	BLOCK-78
73SI-MC1358P	BLOCK-1269
73W00124	BLOCK-1301
740-1003-150	BLOCK-20
740-2001-110	BLOCK-14
740-2001-306	BLOCK-28
740-2002-111	BLOCK-1262
740-2003-150	BLOCK-20
740-2007-120	BLOCK-85
740-5903-301	BLOCK-7
740-8120-160	BLOCK-6
740-8160-190	BLOCK-4
740-9000-554	BLOCK-138
740-9000-566	BLOCK-50
740-9000-754	BLOCK-138
740-9003-301	BLOCK-7
740-9003-350	BLOCK-205
740-9007-046	BLOCK-106
740-9007-205	BLOCK-145
740-9011-322	BLOCK-36
740-9016-105	BLOCK-1277
740-9017-092	BLOCK-22
740-9037-120	BLOCK-85
740-9037-204	BLOCK-143
740-937-120	BLOCK-85
740-9607-205	BLOCK-145
7400	BLOCK-1293
7400-6A	BLOCK-1293
7400-9A	BLOCK-1293
7400/9N00	BLOCK-1293
7400A	BLOCK-1293
7400DC	BLOCK-1293
7400PC	BLOCK-1293
7401-6A	BLOCK-1294
7401-9A	BLOCK-1294
7401DC	BLOCK-1294
7401PC	BLOCK-1294
7402	BLOCK-1295
7402-6A	BLOCK-1295
7402-9A	BLOCK-1295
74021-1	BLOCK-1487
7402DC	BLOCK-1295
7402N	BLOCK-1295
7402PC	BLOCK-1295
7403-6A	BLOCK-1296
7403-9A	BLOCK-1296
7403DC	BLOCK-1296
7403N	BLOCK-1296
7403PC	BLOCK-1296
7404	BLOCK-1297
7404-6A	BLOCK-1297
7404-9A	BLOCK-1297
7404A	BLOCK-1297
7404DC	BLOCK-1297
7404N	BLOCK-1297
7404PC	BLOCK-1297
7405	BLOCK-1298
7405-6A	BLOCK-1298
7405-9A	BLOCK-1298
740502	BLOCK-105
740538	BLOCK-101
740543	BLOCK-105
740583	BLOCK-101
7405DC	BLOCK-1298
7405PC	BLOCK-1298
7406	BLOCK-1299
740622	BLOCK-1277
7406DC	BLOCK-1298
7406N	BLOCK-1299
7406PC	BLOCK-1299
7407	BLOCK-1300
740781	BLOCK-36
740782	BLOCK-85
7407A	BLOCK-1300
7407DC	BLOCK-1300
7407N	BLOCK-1300
7407PC	BLOCK-1300
7408	BLOCK-1301
7408-6A	BLOCK-1301
7408-9A	BLOCK-1301
7408A	BLOCK-1301
7408DC	BLOCK-1301
7408N	BLOCK-1301
7408PC	BLOCK-1301
7409DC	BLOCK-1302
7409PC	BLOCK-1302
741	BLOCK-2058
7410	BLOCK-1304
7410-6A	BLOCK-1304
7410-9A	BLOCK-1304
74107DC	BLOCK-1305
74107PC	BLOCK-1305
741098	BLOCK-7
7410DC	BLOCK-1304
7410N	BLOCK-1304
7410PC	BLOCK-1304
7411-6A	BLOCK-1307
7411-9A	BLOCK-1307
7411DC	BLOCK-1307
7411PC	BLOCK-1307
74121DC	BLOCK-1311
74121N	BLOCK-1311
74121PC	BLOCK-1311
74122DC	BLOCK-1312
74122PC	BLOCK-1312
74123	BLOCK-1313
74123DC	BLOCK-1313
74123PC	BLOCK-1313
74125DC	BLOCK-1314
74125N	BLOCK-1314
74125PC	BLOCK-1314
74126DC	BLOCK-1315
74126PC	BLOCK-1315
7412DC	BLOCK-1310
7412PC	BLOCK-1310
7413(IC)	BLOCK-1317
74132DC	BLOCK-1318
74132PC	BLOCK-1318
7413DC	BLOCK-1317
7413PC	BLOCK-1317
7414(IC)	BLOCK-1320
74141DC	BLOCK-1321
74141PC	BLOCK-1321
74145	BLOCK-1325
74145DC	BLOCK-1325
74145N	BLOCK-1325
74145PC	BLOCK-1325
741473	BLOCK-159
7414DC	BLOCK-1320
7414PC	BLOCK-1320
74150	BLOCK-1327
74150DC	BLOCK-1327
74150PC	BLOCK-1327
741518	BLOCK-80
741519	BLOCK-1825
74151ADC	BLOCK-1328
74151APC	BLOCK-1328
74151DC	BLOCK-1328
74151PC	BLOCK-1328
74152DC	BLOCK-1329
74152PC	BLOCK-1329
74153DC	BLOCK-1330
74153PC	BLOCK-1330
74154	BLOCK-1331
74154DC	BLOCK-1331
74154PC	BLOCK-1331
74155DC	BLOCK-1332
74155PC	BLOCK-1332
74156DC	BLOCK-1333
74156PC	BLOCK-1333
74157DC	BLOCK-1334
74157PC	BLOCK-1334
74158	BLOCK-1335
7416	BLOCK-1336
74160DC	BLOCK-1337
74160PC	BLOCK-1337
74161DC	BLOCK-1338
74161PC	BLOCK-1338
74162PC	BLOCK-1339
74163DC	BLOCK-1340
74163N	BLOCK-1340
74163PC	BLOCK-1340
74164DC	BLOCK-1341
74164PC	BLOCK-1341
74165	BLOCK-1342
74165DC	BLOCK-1342
74165PC	BLOCK-1342
741673	BLOCK-149
741686	BLOCK-156
741687	BLOCK-145
7416DC	BLOCK-1336
7416N	BLOCK-1336
7416NA	BLOCK-1584
7416PC	BLOCK-1336
7417	BLOCK-1343
74170DC	BLOCK-1344
74170PC	BLOCK-1344
74174	BLOCK-1346
74174DC	BLOCK-1346
74174PC	BLOCK-1346
74175	BLOCK-1347
74175DC	BLOCK-1347
74175PC	BLOCK-1347
74176DC	BLOCK-1348
74176PC	BLOCK-1348
74177	BLOCK-1349
74177DC	BLOCK-1349
74177PC	BLOCK-1349
74178DC	BLOCK-1350
74178PC	BLOCK-1350
74179DC	BLOCK-1351
74179PC	BLOCK-1351
7417DC	BLOCK-1343
7417N	BLOCK-1343
7417NA	BLOCK-1343
7417PC	BLOCK-1343
74180DC	BLOCK-1352
74180PC	BLOCK-1352
74181DC	BLOCK-1353
74181PC	BLOCK-1353
74182DC	BLOCK-1354
74182PC	BLOCK-1354
741852	BLOCK-80
741853	BLOCK-54
741854	BLOCK-56
7419	BLOCK-1262
74190DC	BLOCK-1355
74190PC	BLOCK-1355
74191DC	BLOCK-1356
74191PC	BLOCK-1356
74192	BLOCK-1357
74192DC	BLOCK-1357
74192PC	BLOCK-1357
74193	BLOCK-1358
74193DC	BLOCK-1358
74193PC	BLOCK-1358
74195DC	BLOCK-1359
74195N	BLOCK-1359
74195PC	BLOCK-1359
74196	BLOCK-1360
74196DC	BLOCK-1360
74196PC	BLOCK-1360
74197DC	BLOCK-1361
74197PC	BLOCK-1361
74198DC	BLOCK-1362
74198PC	BLOCK-1362
74199DC	BLOCK-1363
74199PC	BLOCK-1363
741C	BLOCK-2058
7420(IC)	BLOCK-1365
7420-6A	BLOCK-1365
7420-9A	BLOCK-1365
7420DC	BLOCK-1365
7420PC	BLOCK-1365
7421	BLOCK-1366
7421DC	BLOCK-1366
7421PC	BLOCK-1366
74221	BLOCK-1629
74221N	BLOCK-1368
7422DC	BLOCK-1367
7422PC	BLOCK-1367
742362	BLOCK-214
742363	BLOCK-85
742364	BLOCK-145
7423DC	BLOCK-1369
7423PC	BLOCK-1369
742510	BLOCK-180
7427(IC)	BLOCK-1372
742723	BLOCK-237
742725	BLOCK-1411
742726	BLOCK-2144
742727	BLOCK-2191
7427DC	BLOCK-1372
7427PC	BLOCK-1372
74290PC	BLOCK-1374
74293PC	BLOCK-1375
7430-6A	BLOCK-1377
7430-9A	BLOCK-1377
7430DC	BLOCK-1377
7430PC	BLOCK-1377
7432(IC)	BLOCK-1378
7432DC	BLOCK-1378
7432PC	BLOCK-1378
7437(IC)	BLOCK-1384
7437DC	BLOCK-1384
7437PC	BLOCK-1384
7438	BLOCK-1385
7438DC	BLOCK-1385
7438N	BLOCK-1385
7438PC	BLOCK-1385
74390	BLOCK-1387
7439DC	BLOCK-1386
7439PC	BLOCK-1386
7440-6A	BLOCK-1390
7440-9A	BLOCK-1390
7440PC	BLOCK-1390
7441	BLOCK-1391
7441-6A	BLOCK-1391
7441-9A	BLOCK-1391
7441DC	BLOCK-1391
7441PC	BLOCK-1391
7442ADC	BLOCK-1392
7442APC	BLOCK-1392
7442DC	BLOCK-1392
7442PC	BLOCK-1392
7443	BLOCK-1393
7443ADC	BLOCK-1393
7443APC	BLOCK-1393
7443DC	BLOCK-1393
7443PC	BLOCK-1393
7444	BLOCK-1394
7444ADC	BLOCK-1394
7444APC	BLOCK-1394
7444DC	BLOCK-1394
7444PC	BLOCK-1394
7445	BLOCK-1395
7445DC	BLOCK-1395
7445PC	BLOCK-1395
7446	BLOCK-1396
7446ADC	BLOCK-1396
7446APC	BLOCK-1396
7446PC	BLOCK-1396
7447	BLOCK-1397
7447ADC	BLOCK-1397
7447APC	BLOCK-1397
7447BDC	BLOCK-1397
7447BPC	BLOCK-1397
7447DC	BLOCK-1397
7447PC	BLOCK-1397
7448	BLOCK-1398
7448DC	BLOCK-1398
7448PC	BLOCK-1398
7450-6A	BLOCK-1401
7450-9A	BLOCK-1401
7450DC	BLOCK-1401
7450PC	BLOCK-1401

DEVICE TYPE	REPL CODE
7451	BLOCK-1402
7451-6A	BLOCK-1402
7451-9A	BLOCK-1402
7451DC	BLOCK-1402
7451PC	BLOCK-1402
7453-6A	BLOCK-1403
7453-9A	BLOCK-1403
7453DC	BLOCK-1403
7453PC	BLOCK-1403
7454-6A	BLOCK-1404
7454-9A	BLOCK-1404
7454DC	BLOCK-1404
7454PC	BLOCK-1404
7460	BLOCK-1406
7460-6A	BLOCK-1406
7460-9A	BLOCK-1406
7460DC	BLOCK-1406
7460PC	BLOCK-1406
746HC	BLOCK-1256
7470-6A	BLOCK-1408
7470-9A	BLOCK-1408
7470DC	BLOCK-1408
7470PC	BLOCK-1408
7472	BLOCK-1409
7472-6A	BLOCK-1409
7472-9A	BLOCK-1409
7472DC	BLOCK-1409
7472PC	BLOCK-1409
7473	BLOCK-1410
7473-6A	BLOCK-1410
7473-9A	BLOCK-1410
7473DC	BLOCK-1410
7473F	BLOCK-1410
7473PC	BLOCK-1410
7474	BLOCK-1411
7474-6A	BLOCK-1411
7474-9A	BLOCK-1411
7474/9N74	BLOCK-1411
7474DC	BLOCK-1411
7474F	BLOCK-1411
7474N	BLOCK-1411
7474PC	BLOCK-1411
7475	BLOCK-1412
7475-6A	BLOCK-1412
7475-9A	BLOCK-1412
7475DC	BLOCK-1412
7475PC	BLOCK-1412
7476	BLOCK-1413
7476-6A	BLOCK-1413
7476-9A	BLOCK-1413
7476DC	BLOCK-1413
7476N	BLOCK-1413
7476PC	BLOCK-1413
7480DC	BLOCK-1414
7480PC	BLOCK-1414
7482-6A	BLOCK-1416
7482-9A	BLOCK-1416
7482DC	BLOCK-1416
7482PC	BLOCK-1416
7483ADC	BLOCK-1415
7483APC	BLOCK-1415
7483DC	BLOCK-1417
7483PC	BLOCK-1417
7485	BLOCK-1418
7485DC	BLOCK-1418
7485PC	BLOCK-1418
7486	BLOCK-1419
7486DC	BLOCK-1419
7486PC	BLOCK-1419
7489DC	BLOCK-1420
7489PC	BLOCK-1420
749-8160-190	BLOCK-4
7490	BLOCK-1423
7490-6A	BLOCK-1423
7490-9A	BLOCK-1423
7490ADC	BLOCK-1423
7490APC	BLOCK-1423
7490DC	BLOCK-1423
7490PC	BLOCK-1423
7491ADC	BLOCK-1424
7491APC	BLOCK-1424
7491DC	BLOCK-1424
7491PC	BLOCK-1424
7492	BLOCK-1425
7492-6A	BLOCK-1425
7492-9A	BLOCK-1425
7492ADC	BLOCK-1425
7492APC	BLOCK-1425
7492DC	BLOCK-1425
7492PC	BLOCK-1425
7493	BLOCK-1426
7493-6A	BLOCK-1426
7493-9A	BLOCK-1426
7493/9393	BLOCK-1426
7493ADC	BLOCK-1426
7493APC	BLOCK-1426
7493DC	BLOCK-1426
7493PC	BLOCK-1426
7494DC	BLOCK-1427
7494PC	BLOCK-1427
7495ADC	BLOCK-1428
7495APC	BLOCK-1428
7495DC	BLOCK-1428
7495PC	BLOCK-1428
7496DC	BLOCK-1429
7496PC	BLOCK-1429
749DHC	BLOCK-2073
74C04	BLOCK-1433
74H00DC	BLOCK-1471
74H00PC	BLOCK-1471
74H01DC	BLOCK-1472
74H01PC	BLOCK-1472
74H04DC	BLOCK-1473
74H04PC	BLOCK-1473
74H05DC	BLOCK-1474
74H05PC	BLOCK-1474
74H08PC	BLOCK-1475
74H101DC	BLOCK-1477
74H101PC	BLOCK-1477
74H102DC	BLOCK-1478
74H102PC	BLOCK-1478
74H103DC	BLOCK-1479
74H103PC	BLOCK-1479
74H106DC	BLOCK-1480
74H106PC	BLOCK-1480
74H108DC	BLOCK-1481
74H108PC	BLOCK-1481
74H10DC	BLOCK-1476
74H10PC	BLOCK-1476
74H11DC	BLOCK-1482
74H11PC	BLOCK-1482
74H183DC	BLOCK-1483
74H183PC	BLOCK-1483
74H20DC	BLOCK-1484
74H20PC	BLOCK-1484
74H21DC	BLOCK-1485
74H21PC	BLOCK-1485
74H22	BLOCK-1486
74H22/94H22	BLOCK-1486
74H22/9H22	BLOCK-1486
74H22DC	BLOCK-1486
74H22PC	BLOCK-1486
74H30DC	BLOCK-1487
74H30PC	BLOCK-1487
74H40DC	BLOCK-1488
74H40PC	BLOCK-1488
74H50DC	BLOCK-1489
74H50FC	BLOCK-1489
74H50PC	BLOCK-1489
74H51DC	BLOCK-1490
74H51PC	BLOCK-1490
74H52DC	BLOCK-1491
74H52PC	BLOCK-1491
74H53DC	BLOCK-1492
74H53PC	BLOCK-1492
74H54DC	BLOCK-1493
74H54PC	BLOCK-1493
74H55DC	BLOCK-1494
74H55PC	BLOCK-1494
74H60DC	BLOCK-1495
74H60PC	BLOCK-1495
74H61DC	BLOCK-1496
74H61PC	BLOCK-1496
74H62DC	BLOCK-1497
74H62PC	BLOCK-1497
74H71DC	BLOCK-1498
74H71PC	BLOCK-1498
74H72DC	BLOCK-1499
74H72PC	BLOCK-1499
74H73DC	BLOCK-1500
74H73PC	BLOCK-1500
74H74DC	BLOCK-1501
74H74PC	BLOCK-1501
74H76DC	BLOCK-1502
74H76PC	BLOCK-1502
74H78DC	BLOCK-1503
74H78PC	BLOCK-1503
74H87DC	BLOCK-1505
74H87PC	BLOCK-1505
74HC02N	BLOCK-1507
74HC04AP	BLOCK-1508
74HC240N	BLOCK-1530
74HC373N	BLOCK-1454
74HCT04	BLOCK-1552
74HCT04N	BLOCK-1552
74HCT08	BLOCK-1553
74HCT08N	BLOCK-1553
74HCT14	BLOCK-1555
74HCT14N	BLOCK-1555
74HCT158N	BLOCK-1554
74HCT244	BLOCK-1560
74HCT244N	BLOCK-1560
74HCT32	BLOCK-1562
74HCT32N	BLOCK-1562
74HCT373N	BLOCK-1563
74L03	BLOCK-1571
74LS00	BLOCK-1568
74LS00CH	BLOCK-1568
74LS00DC	BLOCK-1568
74LS00J	BLOCK-1568
74LS00N	BLOCK-1568
74LS00NA	BLOCK-1568
74LS00PC	BLOCK-1568
74LS00W	BLOCK-1568
74LS02	BLOCK-1570
74LS02CH	BLOCK-1570
74LS02DC	BLOCK-1570
74LS02J	BLOCK-1570
74LS02N	BLOCK-1570
74LS02PC	BLOCK-1570
74LS02W	BLOCK-1570
74LS03	BLOCK-1571
74LS03CH	BLOCK-1571
74LS03DC	BLOCK-1571
74LS03J	BLOCK-1571
74LS03P	BLOCK-1571
74LS03PC	BLOCK-1571
74LS03W	BLOCK-1571
74LS04	BLOCK-1572
74LS04DC	BLOCK-1572
74LS04NA	BLOCK-1572
74LS04P	BLOCK-1572
74LS04PC	BLOCK-1572
74LS05	BLOCK-1573
74LS05DC	BLOCK-1573
74LS05PC	BLOCK-1573
74LS08	BLOCK-1574
74LS08CH	BLOCK-1574
74LS08DC	BLOCK-1574
74LS08J	BLOCK-1574
74LS08N	BLOCK-1574
74LS08NA	BLOCK-1574
74LS08PC	BLOCK-1574
74LS08W	BLOCK-1574
74LS09	BLOCK-1575
74LS09PC	BLOCK-1575
74LS10	BLOCK-1576
74LS107CH	BLOCK-1577
74LS107DC	BLOCK-1577
74LS107J	BLOCK-1577
74LS107PC	BLOCK-1577
74LS107W	BLOCK-1577
74LS109	BLOCK-1578
74LS109CH	BLOCK-1578
74LS109DC	BLOCK-1578
74LS109J	BLOCK-1578
74LS109PC	BLOCK-1578
74LS109W	BLOCK-1578
74LS10CH	BLOCK-1576
74LS10DC	BLOCK-1576
74LS10J	BLOCK-1576
74LS10N	BLOCK-1576
74LS10PC	BLOCK-1576
74LS10W	BLOCK-1576
74LS11	BLOCK-1579
74LS112	BLOCK-1580
74LS112DC	BLOCK-1580
74LS112PC	BLOCK-1580
74LS114PC	BLOCK-1581
74LS11CH	BLOCK-1579
74LS11DC	BLOCK-1579
74LS11J	BLOCK-1579
74LS11N	BLOCK-1579
74LS11PC	BLOCK-1579
74LS11W	BLOCK-1579
74LS12	BLOCK-1582
74LS123	BLOCK-1584
74LS123CH	BLOCK-1584
74LS123J	BLOCK-1584
74LS123N	BLOCK-1584
74LS123W	BLOCK-1584
74LS125	BLOCK-1585
74LS125A	BLOCK-1585
74LS125ADC	BLOCK-1585
74LS125AN	BLOCK-1585
74LS125APC	BLOCK-1585
74LS126	BLOCK-1586
74LS126DC	BLOCK-1586
74LS126PC	BLOCK-1586
74LS132	BLOCK-1588
74LS132DC	BLOCK-1588
74LS132N	BLOCK-1588
74LS132PC	BLOCK-1588
74LS133DC	BLOCK-1589
74LS133PC	BLOCK-1589
74LS136CH	BLOCK-1590
74LS136DC	BLOCK-1590
74LS136J	BLOCK-1590
74LS136PC	BLOCK-1590
74LS136W	BLOCK-1590
74LS138	BLOCK-1591
74LS138DC	BLOCK-1591
74LS138N	BLOCK-1591
74LS138P	BLOCK-1591
74LS138PC	BLOCK-1591
74LS139	BLOCK-1592
74LS139DC	BLOCK-1592
74LS139N	BLOCK-1592
74LS139PC	BLOCK-1592
74LS13DC	BLOCK-1587
74LS13N	BLOCK-1587
74LS13PC	BLOCK-1587
74LS14	BLOCK-1593
74LS145	BLOCK-1594
74LS148	BLOCK-1596
74LS14DC	BLOCK-1593
74LS14N	BLOCK-1593
74LS14PC	BLOCK-1593
74LS14PF	BLOCK-1593
74LS15	BLOCK-1597
74LS151	BLOCK-1598
74LS151DC	BLOCK-1598
74LS151PC	BLOCK-1598
74LS153	BLOCK-1599
74LS153DC	BLOCK-1599
74LS153N	BLOCK-1599
74LS153PC	BLOCK-1599
74LS155	BLOCK-1600
74LS155DC	BLOCK-1600
74LS155PC	BLOCK-1600
74LS156	BLOCK-1601
74LS156DC	BLOCK-1601
74LS156N	BLOCK-1601
74LS156PC	BLOCK-1601
74LS157	BLOCK-1602
74LS157DC	BLOCK-1602
74LS157N	BLOCK-1602
74LS157PC	BLOCK-1602
74LS158	BLOCK-1603
74LS158DC	BLOCK-1603
74LS158PC	BLOCK-1603
74LS15PC	BLOCK-1597
74LS160PC	BLOCK-1604
74LS161	BLOCK-1605
74LS161AN	BLOCK-1605
74LS161CH	BLOCK-1605
74LS161DC	BLOCK-1605
74LS161J	BLOCK-1605
74LS161PC	BLOCK-1605
74LS161W	BLOCK-1605
74LS162PC	BLOCK-1606
74LS163	BLOCK-1607
74LS163CH	BLOCK-1607
74LS163DC	BLOCK-1607
74LS163J	BLOCK-1607
74LS163PC	BLOCK-1607
74LS163W	BLOCK-1607
74LS164	BLOCK-1608
74LS164DC	BLOCK-1608
74LS164FC	BLOCK-1608
74LS164N	BLOCK-1608
74LS164PC	BLOCK-1608
74LS165DC	BLOCK-1609
74LS165FC	BLOCK-1609
74LS165PC	BLOCK-1609
74LS166	BLOCK-1610
74LS168DC	BLOCK-1611
74LS168PC	BLOCK-1611
74LS169DC	BLOCK-1612
74LS169PC	BLOCK-1612
74LS170DC	BLOCK-1613
74LS170FC	BLOCK-1613
74LS170PC	BLOCK-1613
74LS173A	BLOCK-1614
74LS174	BLOCK-1615
74LS174N	BLOCK-1615
74LS174PC	BLOCK-1615
74LS175	BLOCK-1616
74LS175DC	BLOCK-1616
74LS175N	BLOCK-1616
74LS175PC	BLOCK-1616
74LS181PC	BLOCK-1617
74LS190PC	BLOCK-1618
74LS191CH	BLOCK-1619
74LS191DC	BLOCK-1619
74LS191J	BLOCK-1619
74LS191PC	BLOCK-1619
74LS191W	BLOCK-1619
74LS192	BLOCK-1620
74LS192DC	BLOCK-1620
74LS192PC	BLOCK-1620
74LS193	BLOCK-1621
74LS193CH	BLOCK-1621
74LS193DC	BLOCK-1621
74LS193J	BLOCK-1621
74LS193N	BLOCK-1621
74LS193PC	BLOCK-1621
74LS193W	BLOCK-1621
74LS194	BLOCK-1622
74LS194ADC	BLOCK-1622
74LS194AFC	BLOCK-1622
74LS194APC	BLOCK-1622
74LS195ADC	BLOCK-1623
74LS195AFC	BLOCK-1623
74LS195APC	BLOCK-1623
74LS196	BLOCK-1624
74LS196PC	BLOCK-1624
74LS197	BLOCK-1625
74LS197DC	BLOCK-1625
74LS197PC	BLOCK-1625
74LS20	BLOCK-1626
74LS20CH	BLOCK-1626
74LS20DC	BLOCK-1626
74LS20J	BLOCK-1626
74LS20N	BLOCK-1626
74LS20PC	BLOCK-1626
74LS20W	BLOCK-1626
74LS21	BLOCK-1627
74LS21DC	BLOCK-1627
74LS21PC	BLOCK-1627
74LS22	BLOCK-1628
74LS221	BLOCK-1629
74LS221CH	BLOCK-1629
74LS221J	BLOCK-1629
74LS221W	BLOCK-1629
74LS22PC	BLOCK-1628
74LS240	BLOCK-1630
74LS240DC	BLOCK-1630
74LS240N	BLOCK-1630
74LS240PC	BLOCK-1630
74LS241	BLOCK-1631
74LS241DC	BLOCK-1631
74LS241N	BLOCK-1631
74LS241PC	BLOCK-1631
74LS242	BLOCK-1632
74LS242DC	BLOCK-1632
74LS242PC	BLOCK-1632
74LS243DC	BLOCK-1633
74LS243N	BLOCK-1633
74LS243PC	BLOCK-1633
74LS244	BLOCK-1634
74LS244DC	BLOCK-1634
74LS244N	BLOCK-1634
74LS244PC	BLOCK-1634
74LS245	BLOCK-1635
74LS245DC	BLOCK-1635
74LS245N	BLOCK-1635
74LS245NA	BLOCK-1635
74LS245PC	BLOCK-1635
74LS247DC	BLOCK-1636
74LS247PC	BLOCK-1636
74LS248DC	BLOCK-1636
74LS248PC	BLOCK-1637
74LS249DC	BLOCK-1638
74LS249PC	BLOCK-1638
74LS251DC	BLOCK-1639
74LS251PC	BLOCK-1639
74LS253DC	BLOCK-1640
74LS253PC	BLOCK-1640
74LS257	BLOCK-1641
74LS257A	BLOCK-1641
74LS257AN	BLOCK-1641
74LS257APC	BLOCK-1641
74LS257DC	BLOCK-1641
74LS257N	BLOCK-1641
74LS257PC	BLOCK-1641
74LS258	BLOCK-1642
74LS258APC	BLOCK-1642
74LS258DC	BLOCK-1642
74LS258N	BLOCK-1642
74LS258PC	BLOCK-1642
74LS259	BLOCK-1643
74LS259DC	BLOCK-1643
74LS259PC	BLOCK-1643
74LS26	BLOCK-1644
74LS260DC	BLOCK-1645
74LS260PC	BLOCK-1645
74LS266	BLOCK-1646
74LS266CH	BLOCK-1646
74LS266DC	BLOCK-1646
74LS266J	BLOCK-1646
74LS266PC	BLOCK-1646
74LS266W	BLOCK-1646
74LS26DC	BLOCK-1644
74LS26PC	BLOCK-1644
74LS27	BLOCK-1647
74LS273	BLOCK-1648
74LS273DC	BLOCK-1648
74LS273N	BLOCK-1648
74LS273PC	BLOCK-1648
74LS279DC	BLOCK-1649
74LS279PC	BLOCK-1649
74LS27CH	BLOCK-1647
74LS27DC	BLOCK-1647
74LS27J	BLOCK-1647
74LS27PC	BLOCK-1647
74LS27W	BLOCK-1647
74LS280	BLOCK-1651
74LS283	BLOCK-1652
74LS283DC	BLOCK-1652
74LS283PC	BLOCK-1652
74LS28DC	BLOCK-1650
74LS28PC	BLOCK-1650
74LS290	BLOCK-1653
74LS290DC	BLOCK-1653
74LS290PC	BLOCK-1653
74LS293	BLOCK-1654
74LS293DC	BLOCK-1654
74LS293PC	BLOCK-1654
74LS295ADC	BLOCK-1655
74LS295AFC	BLOCK-1655
74LS295APC	BLOCK-1655
74LS298DC	BLOCK-1656
74LS298PC	BLOCK-1656
74LS299DC	BLOCK-1657
74LS299FC	BLOCK-1657
74LS30	BLOCK-1658
74LS30CH	BLOCK-1658
74LS30DC	BLOCK-1658
74LS30J	BLOCK-1658
74LS30PC	BLOCK-1658
74LS30W	BLOCK-1658
74LS32	BLOCK-1659
74LS32CH	BLOCK-1659
74LS32DC	BLOCK-1659
74LS32J	BLOCK-1659
74LS32L	BLOCK-1659
74LS32N	BLOCK-1659
74LS32P	BLOCK-1659
74LS32PC	BLOCK-1659
74LS32W	BLOCK-1659
74LS33	BLOCK-1660
74LS33PC	BLOCK-1660
74LS352	BLOCK-1662
74LS352DC	BLOCK-1662
74LS352PC	BLOCK-1662
74LS353DC	BLOCK-1663
74LS353PC	BLOCK-1663
74LS365	BLOCK-1666
74LS365A	BLOCK-1666
74LS365ADC	BLOCK-1666
74LS365APC	BLOCK-1666
74LS366ADC	BLOCK-1667
74LS366APC	BLOCK-1667
74LS367	BLOCK-1668
74LS367A	BLOCK-1668
74LS367ADC	BLOCK-1668
74LS367APC	BLOCK-1668
74LS368	BLOCK-1669
74LS368A	BLOCK-1669
74LS368ADC	BLOCK-1669
74LS368AN	BLOCK-1669
74LS368APC	BLOCK-1669
74LS37	BLOCK-1670
74LS373	BLOCK-1671
74LS373DC	BLOCK-1671
74LS373N	BLOCK-1671
74LS373PC	BLOCK-1671
74LS374	BLOCK-1672
74LS374DC	BLOCK-1672
74LS374N	BLOCK-1672
74LS374P	BLOCK-1672
74LS374PC	BLOCK-1672
74LS377DC	BLOCK-1673
74LS377PC	BLOCK-1673
74LS378	BLOCK-1674
74LS378DC	BLOCK-1674
74LS378FC	BLOCK-1674
74LS378PC	BLOCK-1674
74LS379DC	BLOCK-1675
74LS379FC	BLOCK-1675
74LS379PC	BLOCK-1675
74LS37DC	BLOCK-1670
74LS37FC	BLOCK-1670

ORIGINAL DEVICE TYPES AND REPLACEMENT CODES 15

DEVICE TYPE	REPL CODE
74LS37PC	BLOCK-1670
74LS38	BLOCK-1676
74LS38-1	BLOCK-1676
74LS38CH	BLOCK-1676
74LS38DC	BLOCK-1676
74LS38FC	BLOCK-1676
74LS38J	BLOCK-1676
74LS38N	BLOCK-1676
74LS38PC	BLOCK-1676
74LS38W	BLOCK-1676
74LS390	BLOCK-1678
74LS390DC	BLOCK-1678
74LS390PC	BLOCK-1678
74LS393	BLOCK-1679
74LS393DC	BLOCK-1679
74LS393N	BLOCK-1679
74LS393NA	BLOCK-1679
74LS393PC	BLOCK-1679
74LS395A	BLOCK-1680
74LS395DC	BLOCK-1680
74LS395FC	BLOCK-1680
74LS395PC	BLOCK-1680
74LS40DC	BLOCK-1683
74LS40PC	BLOCK-1683
74LS42	BLOCK-1684
74LS42CH	BLOCK-1684
74LS42DC	BLOCK-1684
74LS42J	BLOCK-1684
74LS42PC	BLOCK-1684
74LS42W	BLOCK-1684
74LS47DC	BLOCK-1685
74LS47PC	BLOCK-1685
74LS48DC	BLOCK-1686
74LS48PC	BLOCK-1686
74LS490	BLOCK-1688
74LS490PC	BLOCK-1688
74LS49DC	BLOCK-1687
74LS49PC	BLOCK-1687
74LS51	BLOCK-1689
74LS51CH	BLOCK-1689
74LS51DC	BLOCK-1689
74LS51J	BLOCK-1689
74LS51PC	BLOCK-1689
74LS51W	BLOCK-1689
74LS540DC	BLOCK-1691
74LS540PC	BLOCK-1691
74LS541	BLOCK-1692
74LS541DC	BLOCK-1692
74LS541N	BLOCK-1692
74LS541PC	BLOCK-1692
74LS54DC	BLOCK-1690
74LS54PC	BLOCK-1690
74LS55PC	BLOCK-1693
74LS624	BLOCK-1694
74LS629N	BLOCK-1698
74LS645	BLOCK-1703
74LS670	BLOCK-1704
74LS670DC	BLOCK-1704
74LS670FC	BLOCK-1704
74LS670PC	BLOCK-1704
74LS73	BLOCK-1705
74LS73CH	BLOCK-1705
74LS73J	BLOCK-1705
74LS73W	BLOCK-1705
74LS74	BLOCK-1706
74LS74A	BLOCK-1706
74LS74AN	BLOCK-1706
74LS74APC	BLOCK-1706
74LS74CH	BLOCK-1706
74LS74J	BLOCK-1706
74LS74PC	BLOCK-1706
74LS74W	BLOCK-1706
74LS75	BLOCK-1707
74LS75CH	BLOCK-1707
74LS75J	BLOCK-1707
74LS75W	BLOCK-1707
74LS76	BLOCK-1708
74LS76DC	BLOCK-1708
74LS76PC	BLOCK-1708
74LS77	BLOCK-1709
74LS83ACH	BLOCK-1711
74LS83ADC	BLOCK-1711
74LS83AJ	BLOCK-1711
74LS83APC	BLOCK-1711
74LS83AW	BLOCK-1711
74LS85	BLOCK-1712
74LS85CH	BLOCK-1712
74LS85DC	BLOCK-1712
74LS85J	BLOCK-1712
74LS85PC	BLOCK-1712
74LS85W	BLOCK-1712
74LS86	BLOCK-1713
74LS86CH	BLOCK-1713
74LS86DC	BLOCK-1713
74LS86J	BLOCK-1713
74LS86N	BLOCK-1713
74LS86PC	BLOCK-1713
74LS86W	BLOCK-1713
74LS90	BLOCK-1714
74LS90DC	BLOCK-1714
74LS90PC	BLOCK-1714
74LS91	BLOCK-1715
74LS92	BLOCK-1716
74LS92PC	BLOCK-1716
74LS93	BLOCK-1717
74LS93CH	BLOCK-1717
74LS93DC	BLOCK-1717
74LS93J	BLOCK-1717
74LS93N	BLOCK-1717
74LS93PC	BLOCK-1717
74LS93W	BLOCK-1717
74LSP14APC	BLOCK-1622
74S00	BLOCK-1719
74S00DC	BLOCK-1719
74S00N	BLOCK-1719
74S00PC	BLOCK-1719
74S02	BLOCK-1720
74S02PC	BLOCK-1720
74S03DC	BLOCK-1721
74S03PC	BLOCK-1721
74S04	BLOCK-1722
74S04ADC	BLOCK-1722
74S04APC	BLOCK-1722
74S04DC	BLOCK-1722
74S04PC	BLOCK-1722
74S05	BLOCK-1723
74S05ADC	BLOCK-1723
74S05APC	BLOCK-1723
74S05DC	BLOCK-1723
74S05PC	BLOCK-1723
74S08	BLOCK-1724
74S08DC	BLOCK-1724
74S08PC	BLOCK-1724
74S09DC	BLOCK-1725
74S09PC	BLOCK-1725
74S10	BLOCK-1726
74S10DC	BLOCK-1726
74S10PC	BLOCK-1726
74S11	BLOCK-1727
74S112	BLOCK-1728
74S112A	BLOCK-1728
74S112DC	BLOCK-1728
74S112N	BLOCK-1728
74S112PC	BLOCK-1728
74S113DC	BLOCK-1729
74S113PC	BLOCK-1729
74S11DC	BLOCK-1727
74S11PC	BLOCK-1727
74S133DC	BLOCK-1731
74S133PC	BLOCK-1731
74S134DC	BLOCK-1732
74S134PC	BLOCK-1732
74S138	BLOCK-1733
74S138DC	BLOCK-1733
74S138PC	BLOCK-1733
74S140DC	BLOCK-1734
74S140PC	BLOCK-1734
74S151	BLOCK-1736
74S151DC	BLOCK-1736
74S151PC	BLOCK-1736
74S153	BLOCK-1737
74S153DC	BLOCK-1737
74S153PC	BLOCK-1737
74S157	BLOCK-1738
74S157DC	BLOCK-1738
74S157PC	BLOCK-1738
74S158DC	BLOCK-1739
74S158PC	BLOCK-1739
74S15DC	BLOCK-1735
74S15PC	BLOCK-1735
74S163	BLOCK-1607
74S174	BLOCK-1740
74S174N	BLOCK-1740
74S174NA	BLOCK-1740
74S174PC	BLOCK-1740
74S175	BLOCK-1741
74S175DC	BLOCK-1741
74S175N	BLOCK-1741
74S175PC	BLOCK-1741
74S181DC	BLOCK-1742
74S181PC	BLOCK-1742
74S194DC	BLOCK-1744
74S194FC	BLOCK-1744
74S194PC	BLOCK-1744
74S20DC	BLOCK-1745
74S20PC	BLOCK-1745
74S22DC	BLOCK-1746
74S22PC	BLOCK-1746
74S240DC	BLOCK-1753
74S251DC	BLOCK-1747
74S258DC	BLOCK-1748
74S258PC	BLOCK-1748
74S288	BLOCK-1750
74S30DC	BLOCK-1751
74S30PC	BLOCK-1751
74S40	BLOCK-1753
74S40PC	BLOCK-1753
74S51	BLOCK-1761
74S51DC	BLOCK-1761
74S51NA	BLOCK-1761
74S51PC	BLOCK-1761
74S64	BLOCK-1766
74S64DC	BLOCK-1766
74S64PC	BLOCK-1766
74S65	BLOCK-1767
74S65DC	BLOCK-1767
74S65PC	BLOCK-1767
74S74	BLOCK-1768
74S74DC	BLOCK-1768
74S74PC	BLOCK-1768
74S86	BLOCK-1769
74S86DC	BLOCK-1769
74S86PC	BLOCK-1769
75188	BLOCK-1771
75188N	BLOCK-1771
75189	BLOCK-1772
75189A	BLOCK-1772
75189AN	BLOCK-1772
7528046-P4	BLOCK-2199
7528048-P4	BLOCK-2208
7528153-P4	BLOCK-2209
7528156-P3	BLOCK-2185
7528158-P4	BLOCK-2185
7528159-P4	BLOCK-2206
7528160-P4	BLOCK-2213
7528374P3	BLOCK-2213
75450APC	BLOCK-1775
75450BDC	BLOCK-1775
75450BN	BLOCK-1775
75450BPC	BLOCK-1775
75451ATC	BLOCK-1776
75451BN	BLOCK-1776
75451BRC	BLOCK-1776
75451BT	BLOCK-1776
75451BTC	BLOCK-1776
75452	BLOCK-1777
75452ATC	BLOCK-1777
75452BN	BLOCK-1777
75452BRC	BLOCK-1777
75452BT	BLOCK-1777
75452BTC	BLOCK-1777
75453ATC	BLOCK-1778
75453BN	BLOCK-1778
75453BRC	BLOCK-1778
75453BT	BLOCK-1778
75453BTC	BLOCK-1778
75454BN	BLOCK-1779
75454BT	BLOCK-1779
75468	BLOCK-936
75491PC	BLOCK-1780
75492PC	BLOCK-1781
7555	BLOCK-1872
76-005-003	BLOCK-1573
76-005-005	BLOCK-1644
76-005-007	BLOCK-1706
76-005-013	BLOCK-1668
76-100-001	BLOCK-1990
76-120-001	BLOCK-2079
76-140-001	BLOCK-2211
76-500-001	BLOCK-936
760011	BLOCK-1293
760013	BLOCK-1423
760015	BLOCK-1413
760048	BLOCK-1272
760105	BLOCK-1859
760106	BLOCK-1122
760522-0008	BLOCK-1376
760522-0012	BLOCK-115
762200-14	BLOCK-1950
7625	BLOCK-94
7636	BLOCK-2184
76600P	BLOCK-1405
7666	BLOCK-1269
76SI-HD14011BP	BLOCK-1039
77027-1	BLOCK-1950
77027-2	BLOCK-1950
770339	BLOCK-1790
7753	BLOCK-1289
77C800-005	BLOCK-2197
77C800-007	BLOCK-2199
77C800-008	BLOCK-2208
77C813-002	BLOCK-2207
78005AP	BLOCK-2144
78010AP	BLOCK-904
78012AP	BLOCK-2097
7805	BLOCK-2087
7805A	BLOCK-2087
7805C	BLOCK-2087
7805CDA	BLOCK-1022
7805CT	BLOCK-2087
7805H	BLOCK-2087
7806CU	BLOCK-2093
7808C	BLOCK-2095
7808CU	BLOCK-2095
781-1	BLOCK-720
78106C	BLOCK-2173
7812	BLOCK-2097
7812(IC)	BLOCK-2097
7812A	BLOCK-2097
7812CDA	BLOCK-2014
7812CT	BLOCK-2097
7812CU	BLOCK-2097
7812P	BLOCK-2097
7815	BLOCK-2119
7815(IC)	BLOCK-2119
7815CDA	BLOCK-892
7815CU	BLOCK-2119
7818CU	BLOCK-2085
7824CU	BLOCK-2137
78606C	BLOCK-2173
7874	BLOCK-1791
7874(WURLITZER)	BLOCK-1773
78A200010P4	BLOCK-1293
78H05-KC	BLOCK-2007
78H05KC	BLOCK-2007
78HO5KC	BLOCK-2007
78HV05CDA	BLOCK-2004
78HV05CU	BLOCK-2087
78HV12CDA	BLOCK-2014
78HV12CU	BLOCK-2097
78L05	BLOCK-2144
78L05-AN	BLOCK-2144
78L05-AV	BLOCK-2144
78L05A	BLOCK-2144
78L05ACP	BLOCK-2144
78L05AV	BLOCK-2144
78L05AVP	BLOCK-2144
78L05J	BLOCK-2144
78L05V	BLOCK-2144
78L06	BLOCK-2173
78L06A	BLOCK-2173
78L06AC	BLOCK-2173
78L06C	BLOCK-2173
78L08	BLOCK-2160
78L08A	BLOCK-2160
78L08AC	BLOCK-2160
78L08AWC	BLOCK-2160
78L09	BLOCK-881
78L09A	BLOCK-881
78L09CLP	BLOCK-881
78L12	BLOCK-2074
78L12A	BLOCK-2074
78L12ACZ	BLOCK-2074
78L62	BLOCK-2173
78L62AC	BLOCK-2173
78L62AWC	BLOCK-2173
78L62WV	BLOCK-2173
78L82AC	BLOCK-2160
78M05	BLOCK-2087
78M05A	BLOCK-2087
78M05C	BLOCK-2087
78M05HF	BLOCK-2087
78M08	BLOCK-2095
78M09	BLOCK-881
78M09A	BLOCK-881
78M12	BLOCK-2097
78M12KC	BLOCK-2097
78M18A	BLOCK-2085
7905	BLOCK-2091
7905CDA	BLOCK-889
7905CU	BLOCK-2091
7905DA	BLOCK-889
7905UC	BLOCK-2091
7906C	BLOCK-2094
7906CU	BLOCK-2094
7908CU	BLOCK-2096
7912	BLOCK-2108
7912C	BLOCK-2108
7915C	BLOCK-2128
7932367	BLOCK-1276
7932980	BLOCK-1264
79L05	BLOCK-893
79L05AC	BLOCK-893
79L12	BLOCK-882
79M05	BLOCK-2091
79M05C	BLOCK-2091
79M05CT	BLOCK-2091
79M12	BLOCK-2108
79M12C	BLOCK-2108
79M12CKC	BLOCK-2108
79SI-UPC393C	BLOCK-2063
8-719-113-63	BLOCK-396
8-719-800-43	BLOCK-1021
8-729-100-09	BLOCK-539
8-749-901-35	BLOCK-759
8-749-930-35	BLOCK-724
8-749-956-30	BLOCK-713
8-749-958-30	BLOCK-713
8-750-105-11	BLOCK-94
8-751-001-00	BLOCK-592
8-751-300-00	BLOCK-590
8-751-310-00	BLOCK-588
8-751-3300-00	BLOCK-586
8-751-340-00	BLOCK-585
8-751-350-00	BLOCK-583
8-751-360-00	BLOCK-581
8-751-370-00	BLOCK-579
8-751-380-00	BLOCK-577
8-751-390-00	BLOCK-575
8-751-410-00	BLOCK-573
8-751-430-00	BLOCK-569
8-751-450-00	BLOCK-571
8-751-700-10	BLOCK-289
8-751-771-00	BLOCK-293
8-751-771-10	BLOCK-293
8-752-011-20	BLOCK-840
8-752-030-26	BLOCK-839
8-758-480-00	BLOCK-651
8-759-000-49	BLOCK-1093
8-759-001-20	BLOCK-567
8-759-013-06	BLOCK-2087
8-759-013-09	BLOCK-2097
8-759-040-46	BLOCK-1081
8-759-040-53	BLOCK-1088
8-759-045-38	BLOCK-1140
8-759-10-31	BLOCK-234
8-759-100-07	BLOCK-539
8-759-100-09	BLOCK-539
8-759-100-60	BLOCK-650
8-759-100-75	BLOCK-664
8-759-101-60	BLOCK-43
8-759-101-80	BLOCK-280
8-759-103-00	BLOCK-408
8-759-103-68	BLOCK-414
8-759-103-93	BLOCK-2063
8-759-105-56	BLOCK-398
8-759-105-57	BLOCK-405
8-759-105-82	BLOCK-662
8-759-106-61	BLOCK-823
8-759-108-05	BLOCK-2144
8-759-110-17	BLOCK-550
8-759-110-31	BLOCK-234
8-759-111-85	BLOCK-280
8-759-111-88	BLOCK-705
8-759-112-38	BLOCK-221
8-759-112-77	BLOCK-613
8-759-113-53	BLOCK-235
8-759-113-58	BLOCK-408
8-759-113-63	BLOCK-396
8-759-113-68	BLOCK-414
8-759-113-73	BLOCK-695
8-759-113-78	BLOCK-662
8-759-115-56	BLOCK-398
8-759-125-56	BLOCK-404
8-759-131-11	BLOCK-1981
8-759-132-40	BLOCK-2171
8-759-133-90	BLOCK-1873
8-759-135-80	BLOCK-1993
8-759-140-01	BLOCK-1029
8-759-140-11	BLOCK-1039
8-759-140-29	BLOCK-1068
8-759-140-53	BLOCK-1088
8-759-140-66	BLOCK-1093
8-759-145-27	BLOCK-773
8-759-145-58	BLOCK-1801
8-759-150-61	BLOCK-2144
8-759-157-40	BLOCK-1157
8-759-157-41	BLOCK-1157
8-759-157-52	BLOCK-136
8-759-157-60	BLOCK-174
8-759-170-08	BLOCK-2095
8-759-171-05	BLOCK-2087
8-759-178-05	BLOCK-2144
8-759-178-12	BLOCK-2074
8-759-206-28	BLOCK-1513
8-759-231-53	BLOCK-912
8-759-240-01	BLOCK-1029
8-759-240-11	BLOCK-1039
8-759-240-12	BLOCK-1040
8-759-240-13	BLOCK-1041
8-759-240-16	BLOCK-1093
8-759-240-27	BLOCK-1066
8-759-240-29	BLOCK-1068
8-759-240-30	BLOCK-1069
8-759-240-49	BLOCK-1084
8-759-240-51	BLOCK-1086
8-759-240-52	BLOCK-1087
8-759-240-53	BLOCK-1088
8-759-240-66	BLOCK-1093
8-759-240-69	BLOCK-1096
8-759-240-71	BLOCK-1098
8-759-240-81	BLOCK-525
8-759-245-20	BLOCK-1130
8-759-245-28	BLOCK-1135
8-759-271-20	BLOCK-85
8-759-276-07	BLOCK-532
8-759-276-14	BLOCK-640
8-759-276-30	BLOCK-563
8-759-276-58	BLOCK-558
8-759-312-21	BLOCK-480
8-759-312-44	BLOCK-531
8-759-400-01	BLOCK-391
8-759-400-88	BLOCK-725
8-759-402-35	BLOCK-763
8-759-424	BLOCK-151
8-759-424-00	BLOCK-1269
8-759-425	BLOCK-151
8-759-425-00	BLOCK-1269
8-759-600-05	BLOCK-399
8-759-600-43	BLOCK-406
8-759-600-95	BLOCK-295
8-759-601-95	BLOCK-295
8-759-604-29	BLOCK-2087
8-759-605-55	BLOCK-399
8-759-608-43	BLOCK-406
8-759-610-95	BLOCK-295
8-759-618-48	BLOCK-2081
8-759-619-03	BLOCK-498
8-759-645-19	BLOCK-988
8-759-651-34	BLOCK-94
8-759-651-35	BLOCK-5
8-759-651-42	BLOCK-94
8-759-684-78	BLOCK-568
8-759-701-01	BLOCK-1995
8-759-701-56	BLOCK-2087
8-759-701-79	BLOCK-921
8-759-800-12	BLOCK-688
8-759-800-15	BLOCK-364
8-759-800-65	BLOCK-644
8-759-800-81	BLOCK-762
8-759-801-25	BLOCK-369
8-759-801-98	BLOCK-754
8-759-802-10	BLOCK-454
8-759-803-29	BLOCK-779
8-759-812-01	BLOCK-4
8-759-814-05	BLOCK-503
8-759-820-92	BLOCK-835
8-759-820-93	BLOCK-843
8-759-822-02	BLOCK-835
8-759-826-01	BLOCK-427
8-759-832-10	BLOCK-454
8-759-833-01	BLOCK-7
8-759-833-61	BLOCK-237
8-759-841-25	BLOCK-369
8-759-841-40	BLOCK-453
8-759-878-02	BLOCK-526
8-759-878-03	BLOCK-526
8-759-879-03	BLOCK-526
8-759-900-04	BLOCK-1572
8-759-900-09	BLOCK-1575
8-759-900-10	BLOCK-1576
8-759-900-74	BLOCK-1706
8-759-900-86	BLOCK-1713
8-759-900-93	BLOCK-1717
8-759-901-23	BLOCK-1584
8-759-901-38	BLOCK-1591
8-759-903-86	BLOCK-1851
8-759-904-69	BLOCK-1096
8-759-904-94	BLOCK-770
8-759-924-12	BLOCK-2087
8-759-929-62	BLOCK-2097
8-759-932-33	BLOCK-1093
8-759-932-80	BLOCK-645
8-759-937-59	BLOCK-710
8-759-945-58	BLOCK-1801
8-759-951-02	BLOCK-545
8-759-953-87	BLOCK-973
8-759-980-58	BLOCK-769
8-759-981-69	BLOCK-1995
8-759-982-13	BLOCK-2097
8-759-982-21	BLOCK-2144
8-759-982-26	BLOCK-2074
8-759-982-31	BLOCK-2087

16 ORIGINAL DEVICE TYPES AND REPLACEMENT CODES

DEVICE TYPE	REPL CODE
8-759-985-20	BLOCK-1130
8-759-987-16	BLOCK-2063
8-759-994-51	BLOCK-660
8-795-145-58	BLOCK-1801
8-795-270-70	BLOCK-5
8-795-600-95	BLOCK-295
8-851-340-00	BLOCK-585
800-0016	BLOCK-1336
800-619	BLOCK-1838
8000-00004-305	BLOCK-1212
8000-00004-306	BLOCK-43
8000-00004-307	BLOCK-73
8000-00004-308	BLOCK-20
8000-00006-011	BLOCK-1
8000-00012-041	BLOCK-1212
8000-00016-127	BLOCK-98
8000-00028-042	BLOCK-1297
8000-00028-043	BLOCK-1426
8000-00028-044	BLOCK-1325
8000-00032-030	BLOCK-56
8000-00032-031	BLOCK-1280
8000-00038-002	BLOCK-2140
8000-00038-003	BLOCK-1338
8000-00038-004	BLOCK-1293
8000-00038-005	BLOCK-1486
8000-00038-006	BLOCK-1426
8000-00038-007	BLOCK-1411
8000-0004-P090	BLOCK-1280
8000-00042-008	BLOCK-1293
8000-00042-009	BLOCK-1304
8000-00042-011	BLOCK-1358
8000-00045-004	BLOCK-1338
8000-00047-002	BLOCK-1151
8000-00047-003	BLOCK-1133
8000-00047-006	BLOCK-2173
8000-00049-025	BLOCK-243
8000-00053-004	BLOCK-1405
8000-00057-009	BLOCK-1122
8000-00058-006	BLOCK-1405
8000-00058-009	BLOCK-180
8000-0016-127	BLOCK-98
8000-0016-129	BLOCK-98
8000-0028-042	BLOCK-1297
8000-0058-006	BLOCK-1405
8000016	BLOCK-1336
8000017	BLOCK-1343
800020-001	BLOCK-1365
800021-001	BLOCK-1377
800022-001	BLOCK-1390
800023-001	BLOCK-1304
800024-001	BLOCK-1293
800025-001	BLOCK-1403
800026-001	BLOCK-1401
800080-001	BLOCK-1295
800382-001	BLOCK-1412
800383-001	BLOCK-1417
800385-001	BLOCK-1392
800386-001	BLOCK-1358
800387-001	BLOCK-1297
800400-001	BLOCK-1411
800491-001	BLOCK-1311
80053	BLOCK-1223
800651-001	BLOCK-1299
80070	BLOCK-1223
80071	BLOCK-1223
80073	BLOCK-1223
80074	BLOCK-1223
800806-001	BLOCK-1300
80081	BLOCK-1282
80083	BLOCK-1223
80090	BLOCK-1223
80094	BLOCK-1223
8010-171	BLOCK-43
8010-172	BLOCK-118
80114	BLOCK-1245
801800	BLOCK-2193
801805	BLOCK-1294
801806	BLOCK-1297
801807	BLOCK-1392
801808	BLOCK-1423
802-0027	BLOCK-1647
802-0038	BLOCK-1676
802-0245	BLOCK-1635
802-0260	BLOCK-1645
802-0266	BLOCK-1646
8020244	BLOCK-1634
8020245	BLOCK-1635
8020373	BLOCK-1671
8026004	BLOCK-1552
8026008	BLOCK-1553
8026014	BLOCK-1555
8026032	BLOCK-1562
8026244	BLOCK-1560
8026373	BLOCK-1563
80287	BLOCK-1245
803-0073	BLOCK-1100
8030529	BLOCK-1136
8041016	BLOCK-1003
8041016A	BLOCK-1003
8050339	BLOCK-1873
8050358	BLOCK-1993
8050386	BLOCK-1851
8052805	BLOCK-2144
80710	BLOCK-1269
8080A	BLOCK-1832
80829	BLOCK-1814
80910-147	BLOCK-98
8098	BLOCK-1422
80C97	BLOCK-1837
80SI-M51358B	BLOCK-532
80SI-M51358P	BLOCK-532
81-46128001-8	BLOCK-95
81-46128002-6	BLOCK-108
810	BLOCK-1813
810002-269	BLOCK-2197
8114	BLOCK-1002
8116	BLOCK-1003
811790	BLOCK-2197
811791	BLOCK-2208
811793	BLOCK-2217
811794	BLOCK-2199
8128	BLOCK-1004
8175-7937	BLOCK-2063
8190005	BLOCK-2091
8255A	BLOCK-1855
8255A-5	BLOCK-1855
8255AC-5	BLOCK-1855
8280	BLOCK-1348
829704-6	BLOCK-1485
829704-7	BLOCK-1485
82M432B2	BLOCK-1967
82S123	BLOCK-1750
83-11	BLOCK-1303
84011	BLOCK-1039
84011U	BLOCK-1039
8409-08004	BLOCK-1572
8410	BLOCK-1304
8411	BLOCK-1307
8412	BLOCK-1310
8413	BLOCK-1317
8414	BLOCK-1320
8416	BLOCK-1336
8417	BLOCK-1343
84323	BLOCK-2198
84324	BLOCK-2217
84353-2	BLOCK-2195
84626	BLOCK-2189
84626-1	BLOCK-2189
84626-2	BLOCK-2189
84626-3	BLOCK-2189
84626-4	BLOCK-2189
84628	BLOCK-2192
84628-1	BLOCK-2192
84628-2	BLOCK-2192
84628-3	BLOCK-2192
84628-4	BLOCK-2192
84630	BLOCK-2176
84630-1	BLOCK-2176
84630-2	BLOCK-2176
84630-3	BLOCK-2176
84631	BLOCK-2220
84631-1	BLOCK-2220
84631-2	BLOCK-2220
84631-3	BLOCK-2220
84631-4	BLOCK-2220
8474	BLOCK-1411
8475	BLOCK-1412
8476	BLOCK-1413
848	BLOCK-651
84S157PC	BLOCK-1738
8505870-1	BLOCK-2185
8508331RB	BLOCK-1943
8520	BLOCK-221
8530	BLOCK-1423
86-5009-2	BLOCK-1794
860003-101	BLOCK-2197
860003-111	BLOCK-2198
860003-121	BLOCK-2199
860003-141	BLOCK-2207
860003-151	BLOCK-2207
860003-161	BLOCK-2202
860003-99	BLOCK-2206
862200-16	BLOCK-1950
862209-16	BLOCK-2217
86SI-74145N	BLOCK-1325
86X0024-001	BLOCK-1223
86X0027-001	BLOCK-1223
86X0053-001	BLOCK-1269
86X0055-001	BLOCK-1284
86X0056-001	BLOCK-1285
86X0056-002	BLOCK-1285
86X0064-001	BLOCK-1823
86X0082-001	BLOCK-1821
86X0084-001	BLOCK-1843
86X0086-001	BLOCK-164
86X53-1	BLOCK-1269
86X56-2	BLOCK-1285
86X82-1	BLOCK-1821
86X86-1	BLOCK-164
87-0004	BLOCK-7
87-0022	BLOCK-1
87-0216	BLOCK-138
87-0217	BLOCK-174
87-0233	BLOCK-231
87-0234	BLOCK-1825
87-0246	BLOCK-211
870233	BLOCK-231
870246	BLOCK-211
87322-04	BLOCK-109
875910582	BLOCK-662
87SI-TA7680AP	BLOCK-559
88-18920	BLOCK-56
88-20372	BLOCK-1813
88-20404	BLOCK-1825
88-9302	BLOCK-1389
88-9302RS	BLOCK-1389
88-9302S	BLOCK-1389
88-9574	BLOCK-1290
88-9779	BLOCK-1289
88-9779F	BLOCK-1289
88-9841R	BLOCK-1814
88-9842F	BLOCK-1279
88-9842R	BLOCK-1279
88-9842RS	BLOCK-1279
88-9842S	BLOCK-1279
88-B-7-258	BLOCK-1264
88-B-7258	BLOCK-1264
880-101-00	BLOCK-1212
880-102-00	BLOCK-1282
880-103-00	BLOCK-2189
88060-147	BLOCK-98
881916	BLOCK-1411
8840-171	BLOCK-138
88510-177	BLOCK-98
88510-178	BLOCK-138
885540026-3	BLOCK-1365
885540031-2	BLOCK-1485
88901010	BLOCK-1019
88902010	BLOCK-1019
88921010	BLOCK-1019
88922010	BLOCK-1018
889302	BLOCK-1389
889304	BLOCK-1828
88940010	BLOCK-1018
88941010	BLOCK-1018
88942010	BLOCK-1018
88962010	BLOCK-1019
88A752271	BLOCK-2073
88B7258	BLOCK-1264
88X0053-001	BLOCK-43
89028-2	BLOCK-2207
89028-3	BLOCK-2206
89028-4	BLOCK-2208
89028-4-1-4	BLOCK-2208
89028-6	BLOCK-2200
8910-147	BLOCK-98
8910-148	BLOCK-138
8T28	BLOCK-1178
8T95	BLOCK-1174
9-901-371-01	BLOCK-1157
9-901-504-01	BLOCK-817
9-901-505-01	BLOCK-762
9-901-507-01	BLOCK-835
90-35	BLOCK-100
90-36	BLOCK-98
90-37	BLOCK-102
90-39	BLOCK-1411
90-67	BLOCK-1411
90-68	BLOCK-1294
90-72	BLOCK-100
90-73	BLOCK-98
90-74	BLOCK-102
9001345-02	BLOCK-1426
9001346	BLOCK-1307
9001349-02	BLOCK-1419
9001349-03	BLOCK-1419
9001549-02	BLOCK-1390
9001551-02	BLOCK-1297
9001551-03	BLOCK-1297
9001567	BLOCK-1968
9001567-02	BLOCK-1968
9001570-02	BLOCK-1296
9002097-03	BLOCK-1305
9003091-02	BLOCK-1304
9003091-03	BLOCK-1304
9003096-02	BLOCK-1401
9003096-03	BLOCK-1401
9003097-02	BLOCK-1305
9003148	BLOCK-2197
9003148-01	BLOCK-2197
9003148-02	BLOCK-2197
9003149	BLOCK-2208
9003149-02	BLOCK-2208
9003150	BLOCK-2217
9003150-02	BLOCK-2217
9003150-03	BLOCK-2217
9003151	BLOCK-1293
9003151-03	BLOCK-1293
9003152	BLOCK-1411
9003152-01	BLOCK-1411
9003234-04	BLOCK-1385
9003398-03	BLOCK-1301
9003398-04	BLOCK-1301
9003420-03	BLOCK-1429
9003445-03	BLOCK-1425
9003642-03	BLOCK-1377
9003911	BLOCK-2199
9003911-01	BLOCK-2199
9003911-03	BLOCK-2199
9004075-03	BLOCK-1294
9004076	BLOCK-1365
9004076-03	BLOCK-1365
9004076-04	BLOCK-1365
9004093-03	BLOCK-1410
9004300-03	BLOCK-1413
9004360-03	BLOCK-1404
9004896-04	BLOCK-1403
9004898-04	BLOCK-1307
9005	BLOCK-1212
9006	BLOCK-1212
9008	BLOCK-1790
900HC	BLOCK-2176
9010	BLOCK-1405
901251-30	BLOCK-1593
901435-01	BLOCK-1158
901453-01	BLOCK-1002
901458-01	BLOCK-1162
901502-01	BLOCK-1093
901505-01	BLOCK-1006
901510-01	BLOCK-2090
901521-01	BLOCK-1568
901521-02	BLOCK-1572
901521-03	BLOCK-1574
901521-06	BLOCK-1706
901521-17	BLOCK-1684
901521-18	BLOCK-1592
901521-20	BLOCK-1585
901521-21	BLOCK-1570
901521-22	BLOCK-1647
901521-26	BLOCK-1621
901521-29	BLOCK-1671
901521-30	BLOCK-1593
901521-32	BLOCK-1713
901521-34	BLOCK-1616
901521-54	BLOCK-1625
901521-57	BLOCK-1641
901521-58	BLOCK-1642
901521-68	BLOCK-1698
901522-01	BLOCK-1343
901522-03	BLOCK-1349
901522-04	BLOCK-1293
901522-06	BLOCK-1299
901522-30	BLOCK-1300
901523-01	BLOCK-2079
901523-03	BLOCK-2145
901523-04	BLOCK-1980
901523-08	BLOCK-1990
901527-01	BLOCK-2097
901527-02	BLOCK-2087
901528-03	BLOCK-1022
901528-04	BLOCK-890
90200090	BLOCK-7
90200100	BLOCK-136
903-10056	BLOCK-1157
903-483	BLOCK-1157
903-885	BLOCK-1157
903HC	BLOCK-2177
904HC	BLOCK-2178
905-00066	BLOCK-973
905-00160	BLOCK-1376
905-00161	BLOCK-973
905-10085	BLOCK-758
905-101	BLOCK-569
905-102	BLOCK-1423
905-103	BLOCK-573
905-104	BLOCK-592
905-105	BLOCK-5
905-106	BLOCK-295
905-107	BLOCK-588
905-108	BLOCK-586
905-1085	BLOCK-2063
905-109	BLOCK-583
905-110	BLOCK-581
905-1106	BLOCK-1029
905-111	BLOCK-579
905-112	BLOCK-571
905-1126	BLOCK-1093
905-114	BLOCK-585
905-115	BLOCK-577
905-116	BLOCK-575
905-118	BLOCK-973
905-118-01	BLOCK-973
905-120	BLOCK-1813
905-121	BLOCK-205
905-122	BLOCK-206
905-1224	BLOCK-804
905-1227	BLOCK-2171
905-123	BLOCK-42
905-1238	BLOCK-1993
905-1239	BLOCK-1041
905-1240	BLOCK-1041
905-125	BLOCK-1029
905-126	BLOCK-1039
905-1289	BLOCK-2172
905-13	BLOCK-52
905-131	BLOCK-933
905-137	BLOCK-183
905-1372	BLOCK-1874
905-1373	BLOCK-1041
905-138	BLOCK-206
905-139	BLOCK-207
905-13B	BLOCK-52
905-141	BLOCK-1825
905-144	BLOCK-207
905-147	BLOCK-1043
905-152	BLOCK-9
905-155	BLOCK-202
905-165	BLOCK-234
905-170	BLOCK-183
905-171	BLOCK-207
905-172	BLOCK-931
905-173	BLOCK-159
905-177	BLOCK-85
905-178	BLOCK-202
905-186	BLOCK-1041
905-189	BLOCK-293
905-190	BLOCK-1157
905-191	BLOCK-5
905-198	BLOCK-568
905-200	BLOCK-1066
905-201	BLOCK-85
905-202	BLOCK-1981
905-204	BLOCK-974
905-209	BLOCK-1897
905-210	BLOCK-973
905-214	BLOCK-169
905-217	BLOCK-396
905-218	BLOCK-1068
905-219	BLOCK-2144
905-220	BLOCK-525
905-221	BLOCK-988
905-223	BLOCK-295
905-226	BLOCK-115
905-229	BLOCK-932
905-233	BLOCK-1433
905-234	BLOCK-295
905-235	BLOCK-454
905-238	BLOCK-1996
905-239	BLOCK-2036
905-241	BLOCK-1135
905-242	BLOCK-2097
905-243	BLOCK-1897
905-244	BLOCK-112
905-253	BLOCK-237
905-254	BLOCK-216
905-255	BLOCK-1801
905-256	BLOCK-2081
905-260	BLOCK-1096
905-262	BLOCK-1130
905-265	BLOCK-1130
905-267	BLOCK-2144
905-268	BLOCK-1105
905-27	BLOCK-1828
905-28	BLOCK-1828
905-284	BLOCK-453
905-287	BLOCK-531
905-288	BLOCK-1873
905-289	BLOCK-454
905-291	BLOCK-216
905-298	BLOCK-115
905-301	BLOCK-974
905-325	BLOCK-214
905-326	BLOCK-504
905-329	BLOCK-503
905-344	BLOCK-1706
905-347	BLOCK-545
905-354	BLOCK-1088
905-368	BLOCK-459
905-369	BLOCK-1093
905-380	BLOCK-1086
905-39B	BLOCK-73
905-465	BLOCK-500
905-5	BLOCK-23
905-55	BLOCK-1276
905-59	BLOCK-112
905-658	BLOCK-1993
905-659	BLOCK-2063
905-66	BLOCK-973
905-662	BLOCK-1041
905-69	BLOCK-973
905-717	BLOCK-1041
905-9-B	BLOCK-26
905-922	BLOCK-2063
905-940	BLOCK-1196
90500160	BLOCK-1376
905HC	BLOCK-2179
906128-01	BLOCK-2140
906HC	BLOCK-2180
907HC	BLOCK-2181
908HC	BLOCK-2182
9093-1-6A	BLOCK-1950
9093-9-6A	BLOCK-1950
9093-9-7A	BLOCK-1950
9093DC	BLOCK-1950
9093PC	BLOCK-1950
9094-1-6A	BLOCK-1951
9094-9-6A	BLOCK-1951
9094-9-7A	BLOCK-1951
9094DC	BLOCK-1951
9094PC	BLOCK-1951
9097-1-6A	BLOCK-1952
9097-9-6A	BLOCK-1952
9097-9-7A	BLOCK-1952
9097DC	BLOCK-1952
9097PC	BLOCK-1952
9099-1-6A	BLOCK-1953
9099-9-6A	BLOCK-1953
9099-9-7A	BLOCK-1953
9099DC	BLOCK-1953
9099PC	BLOCK-1953
909HC	BLOCK-2183
9101600	BLOCK-1967
9109DC	BLOCK-1956
910HC	BLOCK-2185
9110DC	BLOCK-1959
9111DC	BLOCK-1960
9112DC	BLOCK-1961
911HC	BLOCK-2186
912HC	BLOCK-2187
9135PC	BLOCK-1965
9137-C-1004	BLOCK-2207
9137-C-1020	BLOCK-2217
913HC	BLOCK-2188
914HC	BLOCK-2189
9157PC	BLOCK-1968
9158PC	BLOCK-1969
915HC	BLOCK-2190
916002	BLOCK-174
916061	BLOCK-28
916063	BLOCK-90
916064	BLOCK-43
916067	BLOCK-101
916070	BLOCK-138
916072	BLOCK-7
916081	BLOCK-10
916083	BLOCK-79
916084	BLOCK-50
916092	BLOCK-90
916098	BLOCK-79
916101	BLOCK-125
916102	BLOCK-174
916105	BLOCK-102
916106	BLOCK-159
916109	BLOCK-143
916110	BLOCK-145
916111	BLOCK-1405
916112	BLOCK-1264
916113	BLOCK-1825
916121	BLOCK-223
916125	BLOCK-182
916138	BLOCK-1338
916150	BLOCK-2144
916155	BLOCK-214
916157	BLOCK-424

ORIGINAL DEVICE TYPES AND REPLACEMENT CODES 17

DEVICE TYPE	REPL CODE
917-1201-0	BLOCK-4
917-12010	BLOCK-4
921-1188	BLOCK-1801
921HC	BLOCK-2192
926HC	BLOCK-2194
927HC	BLOCK-2195
928510-1	BLOCK-2213
928510-101	BLOCK-2213
928512-101	BLOCK-2197
928514-1	BLOCK-2208
928514-101	BLOCK-2208
928515-101	BLOCK-2207
928517-101	BLOCK-2200
928533-101	BLOCK-2217
928560-1	BLOCK-2206
928560-101	BLOCK-2206
928571-1	BLOCK-2206
92BIC	BLOCK-1279
92BLC	BLOCK-1279
9301DC	BLOCK-1861
9301PC	BLOCK-1861
930347-1	BLOCK-1365
930347-10	BLOCK-1365
930347-11	BLOCK-1295
930347-12	BLOCK-1402
930347-13	BLOCK-1297
930347-15	BLOCK-1298
930347-2	BLOCK-1365
930347-3	BLOCK-1293
930347-4	BLOCK-1409
930347-5	BLOCK-1390
930347-6	BLOCK-1404
930347-7	BLOCK-1410
930347-9	BLOCK-1424
9307PC	BLOCK-1398
9308DC	BLOCK-1862
9308PC	BLOCK-1862
9309DC	BLOCK-1863
9309PC	BLOCK-1863
930DC	BLOCK-2197
930PC	BLOCK-2197
9310PC	BLOCK-1337
9311PC	BLOCK-1331
93141DC	BLOCK-1321
93141PC	BLOCK-1321
93145DC	BLOCK-1325
93145PC	BLOCK-1325
9314DC	BLOCK-1865
9314PC	BLOCK-1865
93150DC	BLOCK-1327
93150PC	BLOCK-1327
93151DC	BLOCK-1328
93151PC	BLOCK-1328
93152DC	BLOCK-1329
93152PC	BLOCK-1329
93153DC	BLOCK-1330
93153PC	BLOCK-1330
93154DC	BLOCK-1331
93154PC	BLOCK-1331
93155DC	BLOCK-1332
93155PC	BLOCK-1332
93156DC	BLOCK-1333
93156PC	BLOCK-1333
9315DC	BLOCK-1391
9315PC	BLOCK-1391
9316/74161	BLOCK-1338
93164DC	BLOCK-1341
93164PC	BLOCK-1341
93165DC	BLOCK-1342
93165PC	BLOCK-1342
9316DC	BLOCK-1866
9316DM	BLOCK-1866
9316PC	BLOCK-1866
93170DC	BLOCK-1344
93170PC	BLOCK-1344
93174DC	BLOCK-1346
93174PC	BLOCK-1346
93175DC	BLOCK-1347
93175PC	BLOCK-1347
93176DC	BLOCK-1348
93176PC	BLOCK-1348
93177DC	BLOCK-1349
93177PC	BLOCK-1349
93178DC	BLOCK-1350
93178PC	BLOCK-1350
93179DC	BLOCK-1351
93179PC	BLOCK-1351
9317BDC	BLOCK-1396
9317BDM	BLOCK-1396
9317BPC	BLOCK-1396
9317CDC	BLOCK-1396
9317CDM	BLOCK-1396
93180DC	BLOCK-1352
93180PC	BLOCK-1352

DEVICE TYPE	REPL CODE
9318DC	BLOCK-1867
9318PC	BLOCK-1867
93190DC	BLOCK-1355
93190PC	BLOCK-1355
93191DC	BLOCK-1356
93191PC	BLOCK-1356
93196DC	BLOCK-1360
93196PC	BLOCK-1360
93197DC	BLOCK-1361
93197PC	BLOCK-1361
93198DC	BLOCK-1362
93198PC	BLOCK-1362
93199DC	BLOCK-1363
93199PC	BLOCK-1363
931DC	BLOCK-2198
931PC	BLOCK-2198
9321DC	BLOCK-1869
9321PC	BLOCK-1869
932292-1C	BLOCK-2197
9322DM	BLOCK-1334
9322PC	BLOCK-1334
9328DC	BLOCK-1870
9328PC	BLOCK-1870
932DC	BLOCK-2199
932PC	BLOCK-2199
933044-2D	BLOCK-2199
9334DC	BLOCK-1641
9334DM	BLOCK-1641
9334PC	BLOCK-1643
933DC	BLOCK-2200
933PC	BLOCK-2200
9341DC	BLOCK-1353
9341PC	BLOCK-1353
9342DC	BLOCK-1354
93438DM	BLOCK-1758
93438PC	BLOCK-1758
93448DC	BLOCK-1757
93448DM	BLOCK-1757
93448PC	BLOCK-1757
93450DC	BLOCK-1760
93450DM	BLOCK-1760
93450PC	BLOCK-1760
93451DC	BLOCK-1759
93451DM	BLOCK-1759
93451PC	BLOCK-1759
93452DC	BLOCK-1764
93452DM	BLOCK-1764
93452PC	BLOCK-1764
93453DM	BLOCK-1764
93453PC	BLOCK-1765
9345DC	BLOCK-1395
9345PC	BLOCK-1395
9352DC	BLOCK-1392
9352PC	BLOCK-1392
9353DC	BLOCK-1393
9353PC	BLOCK-1393
9354DC	BLOCK-1394
9354PC	BLOCK-1394
9357A	BLOCK-1396
9357ADC	BLOCK-1396
9357APC	BLOCK-1396
9357B	BLOCK-1397
9357BDC	BLOCK-1397
9357BPC	BLOCK-1397
9358	BLOCK-1398
9358DC	BLOCK-1398
9358PC	BLOCK-1398
935DC	BLOCK-2201
935PC	BLOCK-2201
9360DC	BLOCK-1357
9360PC	BLOCK-1357
9366DC	BLOCK-1358
9366PC	BLOCK-1358
9368DC	BLOCK-1877
9368PC	BLOCK-1877
936DC	BLOCK-2202
936PC	BLOCK-2202
9370DC	BLOCK-1879
9370PC	BLOCK-1879
9374DC	BLOCK-1880
9374PC	BLOCK-1880
9375DC	BLOCK-1412
9375PC	BLOCK-1412
9377	BLOCK-1430
937DC	BLOCK-2203
937PC	BLOCK-2203
9380DC	BLOCK-1414
9380PC	BLOCK-1414
9382DC	BLOCK-1416
9382PC	BLOCK-1416
9383DC	BLOCK-1417
9383PC	BLOCK-1417
9385DC	BLOCK-1418
9385PC	BLOCK-1418

DEVICE TYPE	REPL CODE
9386DC	BLOCK-1646
9386DM	BLOCK-1646
9386PC	BLOCK-1646
9390DC	BLOCK-1423
9390PC	BLOCK-1423
9391DC	BLOCK-1424
9391PC	BLOCK-1424
9392DC	BLOCK-1425
9392PC	BLOCK-1425
9393	BLOCK-1426
9393DC	BLOCK-1426
9393PC	BLOCK-1426
9394DC	BLOCK-1427
9394PC	BLOCK-1427
9395DC	BLOCK-1428
9395PC	BLOCK-1428
9396DC	BLOCK-1429
9396PC	BLOCK-1429
93BLC	BLOCK-1812
93C	BLOCK-2198
93ERH-1X1128	BLOCK-1269
93H183DC	BLOCK-1483
93H193PC	BLOCK-1483
93H87DC	BLOCK-1505
93H87PC	BLOCK-1505
93L00DC	BLOCK-1623
93L00DM	BLOCK-1623
93L08DC	BLOCK-2049
93L08DM	BLOCK-2049
93L08PC	BLOCK-2049
93L16DC	BLOCK-2050
93L16PC	BLOCK-2050
93S153DC	BLOCK-1737
93S157DC	BLOCK-1738
93S157PC	BLOCK-1738
93S158DC	BLOCK-1739
93S158PC	BLOCK-1739
93S174DC	BLOCK-1740
93S174PC	BLOCK-1740
93S175DC	BLOCK-1741
93S175PC	BLOCK-1741
93S194DC	BLOCK-1744
93S194PC	BLOCK-1744
93T4150133000	BLOCK-114
93T4150182072	BLOCK-454
94152	BLOCK-1301
941DC	BLOCK-2205
941PC	BLOCK-2205
94325	BLOCK-2203
94327	BLOCK-2206
94331	BLOCK-2200
94333	BLOCK-2199
944DC	BLOCK-2206
944PC	BLOCK-2206
945DC	BLOCK-2207
945PC	BLOCK-2207
946DC	BLOCK-2208
946PC	BLOCK-2208
94825000-05	BLOCK-1950
94825000-11	BLOCK-2202
948DC	BLOCK-2209
948PC	BLOCK-2209
949DC	BLOCK-2210
949PC	BLOCK-2210
94BLC	BLOCK-1280
950DC	BLOCK-2212
950PC	BLOCK-2212
951DC	BLOCK-2213
951PC	BLOCK-2213
95286	BLOCK-1376
95287	BLOCK-1813
9550-1	BLOCK-1859
95KUCB0027AZ	BLOCK-2087
96-5238-01	BLOCK-1212
96-5238-02	BLOCK-1212
9600DC	BLOCK-2088
9600DM	BLOCK-2088
9600PC	BLOCK-2088
9601DC	BLOCK-2089
9601PC	BLOCK-2089
9602	BLOCK-2090
9602DC	BLOCK-2090
9602DM	BLOCK-2090
9602PC	BLOCK-2090
9615DC	BLOCK-2092
9615DM	BLOCK-2092
9615PC	BLOCK-2092
961DC	BLOCK-2216
961PC	BLOCK-2216
962DC	BLOCK-2217
962PC	BLOCK-2217
963DC	BLOCK-2218
963PC	BLOCK-2218
9664PC	BLOCK-1781

DEVICE TYPE	REPL CODE
9665PC	BLOCK-934
9666DC	BLOCK-935
9666PC	BLOCK-935
9667DC	BLOCK-936
9667PC	BLOCK-936
9668DC	BLOCK-937
9668PC	BLOCK-937
96L02	BLOCK-2132
96L02DC	BLOCK-2132
96L02DM	BLOCK-2132
96L02PC	BLOCK-2132
96LS02DC	BLOCK-2133
96LS02DM	BLOCK-2133
96LS02PC	BLOCK-2133
96S02DC	BLOCK-2134
96S02PC	BLOCK-2134
974HC	BLOCK-2220
991-019542-1	BLOCK-1791
991-025515	BLOCK-1039
991-025515-001	BLOCK-1039
991-025515-002	BLOCK-1039
991-025517	BLOCK-1105
991-026005	BLOCK-1069
991-026022	BLOCK-1066
991-026022-00	BLOCK-1066
991-026025	BLOCK-1029
991-026605	BLOCK-1062
991-026609	BLOCK-1043
991-026615	BLOCK-1051
991-027358	BLOCK-1041
991-027358-001	BLOCK-1041
991-028650-001	BLOCK-1085
991-028679-001	BLOCK-1048
991-028680-001	BLOCK-1058
991-028681-001	BLOCK-1063
991-028682-001	BLOCK-1064
991-029866	BLOCK-1060
99195-1	BLOCK-1087
99195-3	BLOCK-1087
9942PC	BLOCK-1354
995022	BLOCK-1212
995042	BLOCK-1265
995081-1	BLOCK-1270
998280-930	BLOCK-2197
998280-945	BLOCK-2207
998280-962	BLOCK-2217
99E16-1	BLOCK-2176
99S022	BLOCK-1212
99S042	BLOCK-1265
99S049	BLOCK-1282
99S082-1	BLOCK-1269
99S094-1	BLOCK-1271
99S095-1	BLOCK-1272
99S096-1	BLOCK-1270
99S097-1	BLOCK-1808
99SO22	BLOCK-1212
9H00DC	BLOCK-1471
9H00PC	BLOCK-1471
9H01DC	BLOCK-1472
9H01PC	BLOCK-1472
9H04DC	BLOCK-1473
9H04PC	BLOCK-1473
9H05DC	BLOCK-1474
9H05PC	BLOCK-1474
9H08DC	BLOCK-1475
9H08PC	BLOCK-1475
9H101DC	BLOCK-1477
9H101PC	BLOCK-1477
9H102DC	BLOCK-1478
9H102PC	BLOCK-1478
9H103DC	BLOCK-1479
9H103PC	BLOCK-1479
9H106DC	BLOCK-1480
9H106PC	BLOCK-1480
9H108DC	BLOCK-1481
9H108PC	BLOCK-1481
9H10DC	BLOCK-1476
9H10PC	BLOCK-1476
9H11DC	BLOCK-1482
9H11PC	BLOCK-1482
9H20DC	BLOCK-1484
9H20PC	BLOCK-1484
9H21DC	BLOCK-1485
9H21PC	BLOCK-1485
9H22	BLOCK-1486
9H22DC	BLOCK-1486
9H22PC	BLOCK-1486
9H30DC	BLOCK-1487
9H30PC	BLOCK-1487
9H40DC	BLOCK-1488
9H40PC	BLOCK-1488
9H50DC	BLOCK-1489
9H50PC	BLOCK-1489
9H51DC	BLOCK-1490

DEVICE TYPE	REPL CODE
9H51PC	BLOCK-1490
9H52DC	BLOCK-1491
9H52PC	BLOCK-1491
9H53DC	BLOCK-1492
9H53PC	BLOCK-1492
9H54DC	BLOCK-1493
9H54PC	BLOCK-1493
9H55DC	BLOCK-1494
9H55PC	BLOCK-1494
9H60DC	BLOCK-1495
9H60PC	BLOCK-1495
9H61DC	BLOCK-1496
9H61PC	BLOCK-1496
9H62DC	BLOCK-1497
9H62PC	BLOCK-1497
9H71DC	BLOCK-1498
9H71PC	BLOCK-1498
9H72DC	BLOCK-1499
9H72PC	BLOCK-1499
9H73DC	BLOCK-1500
9H73PC	BLOCK-1500
9H74DC	BLOCK-1501
9H74PC	BLOCK-1501
9H76DC	BLOCK-1502
9H76PC	BLOCK-1502
9H78DC	BLOCK-1503
9H78PC	BLOCK-1503
9LS164PC	BLOCK-1608
9LS174DC	BLOCK-1615
9LS174DM	BLOCK-1615
9LS174PC	BLOCK-1615
9LS175DC	BLOCK-1616
9LS175PC	BLOCK-1616
9LS194DC	BLOCK-1622
9LS194DM	BLOCK-1622
9LS194PC	BLOCK-1622
9LS195PC	BLOCK-1623
9N00	BLOCK-1293
9N00DC	BLOCK-1293
9N00PC	BLOCK-1293
9N01DC	BLOCK-1294
9N01PC	BLOCK-1294
9N02	BLOCK-1295
9N02DC	BLOCK-1295
9N02PC	BLOCK-1295
9N03DC	BLOCK-1296
9N03PC	BLOCK-1296
9N04	BLOCK-1297
9N04DC	BLOCK-1297
9N04PC	BLOCK-1297
9N05DC	BLOCK-1298
9N05PC	BLOCK-1298
9N06	BLOCK-1299
9N06DC	BLOCK-1299
9N06PC	BLOCK-1299
9N07DC	BLOCK-1300
9N07PC	BLOCK-1300
9N08	BLOCK-1301
9N08DC	BLOCK-1301
9N08PC	BLOCK-1301
9N09DC	BLOCK-1302
9N09PC	BLOCK-1302
9N10	BLOCK-1304
9N107DC	BLOCK-1305
9N107PC	BLOCK-1305
9N10DC	BLOCK-1304
9N10PC	BLOCK-1304
9N11DC	BLOCK-1307
9N11PC	BLOCK-1307
9N122DC	BLOCK-1312
9N122PC	BLOCK-1312
9N123	BLOCK-1313
9N123DC	BLOCK-1313
9N123PC	BLOCK-1313
9N12DC	BLOCK-1310
9N12PC	BLOCK-1310
9N13	BLOCK-1317
9N132DC	BLOCK-1318
9N132PC	BLOCK-1318
9N13DC	BLOCK-1317
9N13PC	BLOCK-1317
9N14DC	BLOCK-1320
9N14PC	BLOCK-1320
9N16DC	BLOCK-1336
9N16PC	BLOCK-1336
9N17DC	BLOCK-1343
9N17PC	BLOCK-1343
9N20	BLOCK-1365
9N20DC	BLOCK-1365
9N20PC	BLOCK-1365
9N21DC	BLOCK-1366
9N21PC	BLOCK-1366
9N23DC	BLOCK-1369
9N23PC	BLOCK-1369

DEVICE TYPE	REPL CODE
9N27	BLOCK-1372
9N27DC	BLOCK-1372
9N27PC	BLOCK-1372
9N30DC	BLOCK-1377
9N30PC	BLOCK-1377
9N32	BLOCK-1378
9N32DC	BLOCK-1378
9N32PC	BLOCK-1378
9N37DC	BLOCK-1384
9N37PC	BLOCK-1384
9N38DC	BLOCK-1385
9N39DC	BLOCK-1386
9N39PC	BLOCK-1386
9N40DC	BLOCK-1390
9N40PC	BLOCK-1390
9N50DC	BLOCK-1401
9N50PC	BLOCK-1401
9N51	BLOCK-1402
9N51DC	BLOCK-1402
9N51PC	BLOCK-1402
9N53DC	BLOCK-1403
9N53PC	BLOCK-1403
9N54DC	BLOCK-1404
9N54PC	BLOCK-1404
9N60DC	BLOCK-1406
9N60PC	BLOCK-1406
9N70DC	BLOCK-1408
9N70PC	BLOCK-1408
9N72DC	BLOCK-1409
9N72PC	BLOCK-1409
9N73	BLOCK-1410
9N73DC	BLOCK-1410
9N73PC	BLOCK-1410
9N74	BLOCK-1411
9N74DC	BLOCK-1411
9N74PC	BLOCK-1411
9N76	BLOCK-1413
9N76DC	BLOCK-1413
9N76PC	BLOCK-1413
9N86	BLOCK-1419
9N86DC	BLOCK-1419
9N86PC	BLOCK-1419
9S00DC	BLOCK-1719
9S00PC	BLOCK-1719
9S03DC	BLOCK-1721
9S03PC	BLOCK-1721
9S04DC	BLOCK-1722
9S04PC	BLOCK-1722
9S05DC	BLOCK-1723
9S05PC	BLOCK-1723
9S10DC	BLOCK-1726
9S10PC	BLOCK-1726
9S112DC	BLOCK-1728
9S112PC	BLOCK-1728
9S113DC	BLOCK-1729
9S113PC	BLOCK-1729
9S11DC	BLOCK-1727
9S11PC	BLOCK-1727
9S133DC	BLOCK-1731
9S133PC	BLOCK-1731
9S134DC	BLOCK-1732
9S134PC	BLOCK-1732
9S140DC	BLOCK-1734
9S140PC	BLOCK-1734
9S15DC	BLOCK-1735
9S15PC	BLOCK-1735
9S20DC	BLOCK-1745
9S20PC	BLOCK-1745
9S22DC	BLOCK-1746
9S22PC	BLOCK-1746
9S40DC	BLOCK-1753
9S40PC	BLOCK-1753
9S64DC	BLOCK-1766
9S64PC	BLOCK-1766
9S65DC	BLOCK-1767
9S65PC	BLOCK-1767
9S74DC	BLOCK-1768
9S74PC	BLOCK-1768
A-1201B	BLOCK-4
A-13284	BLOCK-535
A-9982	BLOCK-220
A-9982-02A	BLOCK-620
A-9982-07	BLOCK-220
A-9982A	BLOCK-220
A-U2569	BLOCK-1276
A00	BLOCK-1293
A02(I.C.)	BLOCK-1296
A03	BLOCK-1297
A0311400	BLOCK-126
A0354805	BLOCK-519
A04(I.C.)	BLOCK-1298
A05(I.C.)	BLOCK-1304
A054-165	BLOCK-1212
A06(I.C.)	BLOCK-1365

18 ORIGINAL DEVICE TYPES AND REPLACEMENT CODES

DEVICE TYPE	REPL CODE
A066-125	BLOCK-1944
A072133-001	BLOCK-1794
A08(I.C.)	BLOCK-1377
A09	BLOCK-1390
A10	BLOCK-1488
A1011(IC)	BLOCK-839
A12(I.C.)	BLOCK-1402
A1201	BLOCK-4
A1201-H-75	BLOCK-1269
A1201-H45	BLOCK-43
A1201B	BLOCK-4
A1357N	BLOCK-471
A1364N	BLOCK-5
A1365	BLOCK-1269
A13658E3	BLOCK-1269
A1368	BLOCK-1291
A1369	BLOCK-167
A15	BLOCK-1411
A17(I.C.)	BLOCK-1423
A18(I.C.)	BLOCK-1426
A19(I.C.)	BLOCK-1428
A23-1101-01A	BLOCK-771
A3300	BLOCK-6
A3301	BLOCK-7
A3350	BLOCK-205
A4030	BLOCK-8
A455	BLOCK-1264
A6-5002-9	BLOCK-1275
A6358S	BLOCK-1994
A6458S	BLOCK-516
A7016	BLOCK-762
A7205	BLOCK-1287
A7520(IC)	BLOCK-709
A7530N	BLOCK-807
A780	BLOCK-1262
A781	BLOCK-1272
A7M42394B39	BLOCK-1786
A7M42894B39	BLOCK-1786
A8641301	BLOCK-1021
A8641302	BLOCK-1021
A8641303	BLOCK-1021
A9982-02	BLOCK-620
AB4164ANP-12	BLOCK-1006
AB4164ANP-15	BLOCK-1006
AC4045NL	BLOCK-1002
AD101AH	BLOCK-160
AD1408-7D	BLOCK-971
AD201AH	BLOCK-160
AD201H	BLOCK-160
AD301AH	BLOCK-160
AD301AN	BLOCK-2143
AD741	BLOCK-2056
AD741C	BLOCK-2056
AD741CH	BLOCK-2056
AD741CN	BLOCK-2057
AD741H	BLOCK-2056
AD741KH	BLOCK-2056
AD741KN	BLOCK-2058
AD7516JN	BLOCK-1048
ADC0801	BLOCK-968
ADC0802	BLOCK-968
ADC0802LCD	BLOCK-968
ADC0802LCN	BLOCK-968
ADC0803LCD	BLOCK-968
ADC0803LCN	BLOCK-968
ADC0804LCD	BLOCK-968
ADC0804LCN	BLOCK-968
ADD3501	BLOCK-967
AE-904-03	BLOCK-1376
AE-907	BLOCK-1290
AE304	BLOCK-1376
AE900	BLOCK-1811
AE904	BLOCK-1376
AE904-04	BLOCK-1376
AE904-4	BLOCK-1376
AE904-51	BLOCK-1376
AE904-6	BLOCK-1376
AE906	BLOCK-1282
AE907	BLOCK-1290
AE907-51	BLOCK-1290
AE907@	BLOCK-1279
AE908	BLOCK-1814
AE908-51	BLOCK-1814
AE915	BLOCK-1852
AE920	BLOCK-221
AGH-001	BLOCK-200
AM167001A	BLOCK-1269
AM167013A	BLOCK-1422
AM167014A	BLOCK-1407
AM224D	BLOCK-2171
AM25LS151PC	BLOCK-1598
AM26L02PC	BLOCK-2132
AM2708DC	BLOCK-1016
AM2716DC	BLOCK-1017
AM2764ADC	BLOCK-1019
AM27S12DC	BLOCK-1762
AM27S13DC	BLOCK-1763
AM27S180DC	BLOCK-1760
AM27S181DC	BLOCK-1759
AM27S185DC	BLOCK-1754
AM27S18DC	BLOCK-1743
AM27S19DC	BLOCK-1750
AM27S20DC	BLOCK-1752
AM27S21DC	BLOCK-1749
AM27S28DC	BLOCK-1756
AM27S29DC	BLOCK-1755
AM27S30DC	BLOCK-1758
AM27S31DC	BLOCK-1757
AM27S32DC	BLOCK-1764
AM27S33DC	BLOCK-1765
AM324D	BLOCK-2171
AM324N	BLOCK-2171
AM723HM	BLOCK-1983
AM741HC	BLOCK-2056
AM747HC	BLOCK-2069
AM747HM	BLOCK-2069
AM748DC	BLOCK-2141
AM748DM	BLOCK-2141
AM748HC	BLOCK-160
AM748HM	BLOCK-160
AM74LS373J	BLOCK-1671
AM74LS373N	BLOCK-1671
AM74LS373X	BLOCK-1671
AM74S194N	BLOCK-1744
AM8255A-5PC	BLOCK-1855
AM9016CDC	BLOCK-1003
AM9016CPC	BLOCK-1003
AM9016DDC	BLOCK-1003
AM9016DPC	BLOCK-1003
AM9016EDC	BLOCK-1003
AM9016EPC	BLOCK-1003
AM9060CDC	BLOCK-1001
AM9060CPC	BLOCK-1001
AM9060DDC	BLOCK-1001
AM9060DPC	BLOCK-1001
AM9060EDC	BLOCK-1001
AM9060EPC	BLOCK-1001
AM9102ADC	BLOCK-999
AM9102APC	BLOCK-999
AM9102BDC	BLOCK-999
AM9102BPC	BLOCK-999
AM9102CDC	BLOCK-999
AM9102CPC	BLOCK-999
AM9102DC	BLOCK-999
AM9102PC	BLOCK-999
AM9114BDC	BLOCK-1002
AM9114BPC	BLOCK-1002
AM9114CDC	BLOCK-1002
AM9114CPC	BLOCK-1002
AMD-746	BLOCK-1270
AMD-780	BLOCK-1271
AMD-781	BLOCK-1272
AMD746	BLOCK-1815
AMD780	BLOCK-1271
AMD781	BLOCK-1272
AMLM101AH	BLOCK-160
AMLM101H	BLOCK-160
AMLM111D	BLOCK-1981
AMLM201	BLOCK-160
AMLM201A	BLOCK-160
AMLM201AH	BLOCK-160
AMLM201H	BLOCK-160
AMLM207D	BLOCK-2143
AMLM301	BLOCK-160
AMLM301A	BLOCK-160
AMLM301AH	BLOCK-160
AMLM308A	BLOCK-2037
AMLM311D	BLOCK-1981
AMLM319N	BLOCK-1976
AMLM348N	BLOCK-2071
AMU5R7723312	BLOCK-1983
AMX-3551	BLOCK-1570
AMX-3675	BLOCK-1299
AMX-3683	BLOCK-1385
AMX-3716	BLOCK-1593
AMX-4591	BLOCK-1299
AMX-4945	BLOCK-1722
AMX3548	BLOCK-1984
AMX3550	BLOCK-1568
AMX3551	BLOCK-1570
AMX3552	BLOCK-1572
AMX3553	BLOCK-1573
AMX3554	BLOCK-1579
AMX3555	BLOCK-1626
AMX3556	BLOCK-1658
AMX3557	BLOCK-1659
AMX3558	BLOCK-1706
AMX3560	BLOCK-1717
AMX3561	BLOCK-1588
AMX3562	BLOCK-1599
AMX3563	BLOCK-1602
AMX3564	BLOCK-1610
AMX3565	BLOCK-1615
AMX3566	BLOCK-1616
AMX3567	BLOCK-1668
AMX3568	BLOCK-1669
AMX3573	BLOCK-1777
AMX3586	BLOCK-1025
AMX3655	BLOCK-1297
AMX3675	BLOCK-1299
AMX3683	BLOCK-1385
AMX3684	BLOCK-1300
AMX3688	BLOCK-1990
AMX3698	BLOCK-1574
AMX3701	BLOCK-1713
AMX3706	BLOCK-1679
AMX3716	BLOCK-1593
AMX3800	BLOCK-1592
AMX3803	BLOCK-1584
AMX3804	BLOCK-1714
AMX3810	BLOCK-1629
AMX3864	BLOCK-1634
AMX3867	BLOCK-1771
AMX3868	BLOCK-1772
AMX3898	BLOCK-1576
AMX3955	BLOCK-1313
AMX4181	BLOCK-2211
AMX4188	BLOCK-2108
AMX4200	BLOCK-1873
AMX4225	BLOCK-1630
AMX4227	BLOCK-1648
AMX4258	BLOCK-2058
AMX4260	BLOCK-2091
AMX4261	BLOCK-1173
AMX4321	BLOCK-1777
AMX4326	BLOCK-1990
AMX4327	BLOCK-1981
AMX4328	BLOCK-1676
AMX4470	BLOCK-1635
AMX4574	BLOCK-1901
AMX4577	BLOCK-2097
AMX4578	BLOCK-1169
AMX4583	BLOCK-1591
AMX4584	BLOCK-1085
AMX4585	BLOCK-1136
AMX4658	BLOCK-1647
AMX4659	BLOCK-1594
AMX4660	BLOCK-1646
AMX4661	BLOCK-1801
AMX4663	BLOCK-1698
AMX4666	BLOCK-1076
AN-203	BLOCK-52
AN-206	BLOCK-55
AN-210	BLOCK-52
AN-211	BLOCK-54
AN-214	BLOCK-56
AN-214-G	BLOCK-56
AN-214-P	BLOCK-56
AN-214P	BLOCK-56
AN-214Q	BLOCK-56
AN-246	BLOCK-153
AN-253BB	BLOCK-70
AN-315	BLOCK-229
AN103	BLOCK-180
AN103-0	BLOCK-180
AN103G	BLOCK-180
AN115	BLOCK-214
AN1339	BLOCK-1873
AN1358	BLOCK-1993
AN1358S	BLOCK-1995
AN136	BLOCK-51
AN1364	BLOCK-219
AN1393	BLOCK-2063
AN1431	BLOCK-2227
AN1458	BLOCK-1801
AN158	BLOCK-1422
AN1741	BLOCK-2058
AN203	BLOCK-52
AN203(C)	BLOCK-52
AN203AA	BLOCK-52
AN203BA	BLOCK-52
AN203BB	BLOCK-52
AN206	BLOCK-55
AN2065	BLOCK-55
AN206AB	BLOCK-55
AN206B	BLOCK-55
AN206S	BLOCK-55
AN208	BLOCK-133
AN210	BLOCK-53
AN210B	BLOCK-53
AN210C	BLOCK-53
AN211	BLOCK-54
AN211A	BLOCK-54
AN211AB	BLOCK-54
AN211B	BLOCK-54
AN213	BLOCK-57
AN213-AB	BLOCK-57
AN213AB	BLOCK-57
AN214	BLOCK-56
AN214-QR	BLOCK-56
AN214D	BLOCK-56
AN214P	BLOCK-56
AN214PQR	BLOCK-56
AN214Q	BLOCK-56
AN214QR	BLOCK-56
AN214R	BLOCK-56
AN216	BLOCK-123
AN217	BLOCK-58
AN217(BB)	BLOCK-58
AN217BB	BLOCK-58
AN217CB	BLOCK-58
AN217P	BLOCK-58
AN217PBB	BLOCK-58
AN220	BLOCK-59
AN221	BLOCK-1269
AN222	BLOCK-5
AN225	BLOCK-60
AN227	BLOCK-61
AN228V	BLOCK-63
AN228W	BLOCK-63
AN229	BLOCK-62
AN230	BLOCK-63
AN231	BLOCK-64
AN234	BLOCK-65
AN236	BLOCK-121
AN238S	BLOCK-1422
AN239	BLOCK-150
AN239Q	BLOCK-150
AN239QA	BLOCK-150
AN239QB	BLOCK-150
AN240	BLOCK-1269
AN240D	BLOCK-1269
AN240P	BLOCK-1269
AN240PD	BLOCK-1269
AN240PN	BLOCK-1269
AN240S	BLOCK-1269
AN241	BLOCK-151
AN241A	BLOCK-151
AN241B	BLOCK-151
AN241C	BLOCK-151
AN241D	BLOCK-151
AN241P	BLOCK-151
AN241PD	BLOCK-151
AN242	BLOCK-66
AN245	BLOCK-153
AN246	BLOCK-153
AN247	BLOCK-162
AN247P	BLOCK-162
AN252	BLOCK-70
AN253	BLOCK-70
AN253AB	BLOCK-70
AN253BB	BLOCK-70
AN253P	BLOCK-70
AN259	BLOCK-475
AN260	BLOCK-72
AN262	BLOCK-250
AN271	BLOCK-170
AN271B	BLOCK-170
AN271FB	BLOCK-170
AN272	BLOCK-438
AN274	BLOCK-134
AN277	BLOCK-71
AN277AB	BLOCK-71
AN277B	BLOCK-71
AN277BA	BLOCK-71
AN278	BLOCK-436
AN287	BLOCK-556
AN287NT	BLOCK-556
AN288	BLOCK-67
AN289	BLOCK-171
AN295	BLOCK-393
AN295N	BLOCK-393
AN301	BLOCK-434
AN302	BLOCK-251
AN303	BLOCK-252
AN304	BLOCK-387
AN305	BLOCK-253
AN313	BLOCK-432
AN315	BLOCK-229
AN316	BLOCK-254
AN318	BLOCK-255
AN3210K	BLOCK-786
AN3211K	BLOCK-786
AN3211NK	BLOCK-786
AN326	BLOCK-239
AN328	BLOCK-68
AN331	BLOCK-157
AN3310K	BLOCK-787
AN3312	BLOCK-788
AN3313	BLOCK-788
AN3320K	BLOCK-789
AN337	BLOCK-256
AN340	BLOCK-442
AN340P	BLOCK-442
AN342	BLOCK-69
AN343	BLOCK-69
AN352	BLOCK-237
AN360	BLOCK-211
AN362	BLOCK-237
AN362L	BLOCK-237
AN363N	BLOCK-205
AN366	BLOCK-231
AN374	BLOCK-212
AN374P	BLOCK-440
AN380	BLOCK-179
AN3821K	BLOCK-790
AN3822K	BLOCK-790
AN4102	BLOCK-368
AN5010	BLOCK-389
AN5020	BLOCK-668
AN5070	BLOCK-674
AN5111	BLOCK-428
AN5125	BLOCK-760
AN5125M	BLOCK-760
AN5132	BLOCK-538
AN5135K	BLOCK-791
AN5135NK	BLOCK-791
AN5136K	BLOCK-841
AN5136K-R	BLOCK-841
AN5136KR	BLOCK-841
AN5153NK	BLOCK-791
AN5160NK	BLOCK-1246
AN5160NK-N	BLOCK-1246
AN5215	BLOCK-223
AN5250	BLOCK-391
AN5255	BLOCK-649
AN5256	BLOCK-649
AN5262	BLOCK-764
AN5265	BLOCK-770
AN5301K	BLOCK-1183
AN5301NK	BLOCK-1183
AN5302K	BLOCK-1268
AN5310	BLOCK-398
AN5310K	BLOCK-398
AN5310KL	BLOCK-398
AN5310L	BLOCK-398
AN5310M	BLOCK-398
AN5310N	BLOCK-398
AN5310U	BLOCK-398
AN5311	BLOCK-398
AN5311KL	BLOCK-398
AN5315	BLOCK-707
AN5316	BLOCK-707
AN5318	BLOCK-728
AN5318A	BLOCK-728
AN5318N	BLOCK-728
AN5320	BLOCK-403
AN5322NK	BLOCK-1189
AN5330	BLOCK-425
AN5340	BLOCK-388
AN5352	BLOCK-814
AN5352N	BLOCK-814
AN5410	BLOCK-405
AN5411	BLOCK-286
AN5416	BLOCK-669
AN5435	BLOCK-282
AN5436	BLOCK-708
AN5436N	BLOCK-708
AN5440	BLOCK-390
AN5510	BLOCK-393
AN5512	BLOCK-660
AN5515	BLOCK-670
AN5515X	BLOCK-670
AN5520	BLOCK-473
AN5521	BLOCK-763
AN5700	BLOCK-602
AN5703	BLOCK-612
AN5710	BLOCK-607
AN5712	BLOCK-605
AN5720	BLOCK-473
AN5722	BLOCK-597
AN5730	BLOCK-223
AN5732	BLOCK-223
AN5740	BLOCK-599
AN5752	BLOCK-616
AN5753	BLOCK-616
AN5836	BLOCK-761
AN5900	BLOCK-1228
AN6041	BLOCK-689
AN612	BLOCK-238
AN6209	BLOCK-690
AN6209K	BLOCK-690
AN6291	BLOCK-725
AN6300	BLOCK-400
AN6306	BLOCK-692
AN6307	BLOCK-691
AN6310	BLOCK-608
AN6320	BLOCK-407
AN6320N	BLOCK-407
AN6321	BLOCK-609
AN6326	BLOCK-675
AN6326N	BLOCK-675
AN6328	BLOCK-693
AN6331	BLOCK-410
AN6332	BLOCK-410
AN6340	BLOCK-416
AN6341	BLOCK-413
AN6341N	BLOCK-413
AN6342	BLOCK-415
AN6342N	BLOCK-415
AN6343	BLOCK-680
AN6344	BLOCK-416
AN6345	BLOCK-555
AN6350	BLOCK-681
AN6352	BLOCK-1801
AN6356N	BLOCK-792
AN6359	BLOCK-793
AN6359N	BLOCK-793
AN6360	BLOCK-682
AN6361	BLOCK-683
AN6361N	BLOCK-683
AN6362	BLOCK-301
AN6366	BLOCK-794
AN6366K	BLOCK-794
AN6366NK	BLOCK-794
AN6387	BLOCK-617
AN6390	BLOCK-610
AN640G	BLOCK-257
AN6551	BLOCK-516
AN6552	BLOCK-1801
AN6552(F)	BLOCK-1801
AN6553	BLOCK-1801
AN6558	BLOCK-1801
AN6561	BLOCK-1994
AN6562	BLOCK-1993
AN6564	BLOCK-2171
AN6570	BLOCK-2058
AN6571	BLOCK-556
AN6572	BLOCK-1801
AN6677	BLOCK-600
AN6811	BLOCK-409
AN6821	BLOCK-487
AN6873	BLOCK-696
AN6873N	BLOCK-696
AN6875	BLOCK-491
AN6884	BLOCK-548
AN6912	BLOCK-1873
AN6912N	BLOCK-1873
AN6913	BLOCK-699
AN6914	BLOCK-2063
AN7062	BLOCK-845
AN7074P	BLOCK-1422
AN7110	BLOCK-684
AN7110E	BLOCK-684
AN7115	BLOCK-368
AN7115F	BLOCK-368
AN7120	BLOCK-451
AN7130	BLOCK-350
AN7131	BLOCK-352
AN7140	BLOCK-352
AN7145	BLOCK-370
AN7145H	BLOCK-370
AN7145L	BLOCK-370
AN7145M	BLOCK-370
AN7146H	BLOCK-354
AN7146M	BLOCK-370
AN7154	BLOCK-356
AN7156N	BLOCK-358
AN7158	BLOCK-360
AN7158N	BLOCK-360
AN7161N	BLOCK-1192
AN7166	BLOCK-360
AN7168	BLOCK-382
AN7168N	BLOCK-382
AN7218	BLOCK-550
AN7218H	BLOCK-550
AN7273	BLOCK-869
AN7311	BLOCK-423
AN7410	BLOCK-547
AN7410NS	BLOCK-547
AN74LS107AN	BLOCK-1577
AN7805	BLOCK-2087
AN7805F	BLOCK-2087
AN7805LB	BLOCK-2087

ORIGINAL DEVICE TYPES AND REPLACEMENT CODES 19

DEVICE TYPE	REPL CODE
AN7806	BLOCK-2093
AN7812	BLOCK-2097
AN7812F	BLOCK-2097
AN7815	BLOCK-2119
AN7818F	BLOCK-2085
AN78L05	BLOCK-2144
AN78L05-Y	BLOCK-2144
AN78L06	BLOCK-2173
AN78L08	BLOCK-2160
AN78L09	BLOCK-881
AN78L09-Y	BLOCK-881
AN78L12	BLOCK-2074
AN78L12-Y	BLOCK-2074
AN78L15	BLOCK-2075
AN78L18	BLOCK-884
AN78L24	BLOCK-886
AN78M05	BLOCK-2087
AN78M05F	BLOCK-2087
AN78M05LB	BLOCK-2087
AN78M08	BLOCK-2095
AN78M09-(LC)	BLOCK-881
AN78M09F	BLOCK-881
AN78M10	BLOCK-904
AN78M12	BLOCK-2097
AN78M12(LB)	BLOCK-2097
AN78M12-(LB)	BLOCK-2097
AN78M12-LB	BLOCK-2097
AN78M12LB	BLOCK-2097
AN78M15	BLOCK-2119
AN78M15(LB)	BLOCK-2119
AN78M15LB	BLOCK-2119
AN78N05	BLOCK-2144
AN78N05LB	BLOCK-2144
AN7905T	BLOCK-2091
AN79L05	BLOCK-893
AN79L15	BLOCK-883
AN79L18	BLOCK-885
AN79M05	BLOCK-2091
AN79M06	BLOCK-2094
AN829	BLOCK-427
AN829S	BLOCK-427
AN90C21	BLOCK-808
AN90C22	BLOCK-809
AXX3021	BLOCK-1003
AXX3038	BLOCK-1002
AXX3044	BLOCK-1000
AXX3051	BLOCK-1167
AXX3055	BLOCK-1003
AY1-0212	BLOCK-960
AZQQ001GEA	BLOCK-154
B00	BLOCK-1392
B0075660	BLOCK-1410
B01	BLOCK-1412
B02	BLOCK-1413
B0272050	BLOCK-727
B0272054	BLOCK-727
B0272120	BLOCK-663
B0272490	BLOCK-663
B03	BLOCK-1502
B0305401	BLOCK-224
B0306000	BLOCK-5
B0306004	BLOCK-5
B0311000	BLOCK-85
B0311006	BLOCK-85
B0311007	BLOCK-85
B0311008	BLOCK-85
B0311400	BLOCK-126
B0311402	BLOCK-126
B0311405	BLOCK-126
B0313300	BLOCK-131
B0313400	BLOCK-130
B0313500	BLOCK-131
B0313600	BLOCK-128
B0313700	BLOCK-129
B0313800	BLOCK-127
B0315500	BLOCK-188
B0316403	BLOCK-225
B0316451	BLOCK-1269
B0318920	BLOCK-184
B0319200	BLOCK-145
B0325350	BLOCK-643
B0325355	BLOCK-643
B0345450	BLOCK-2058
B0345710	BLOCK-532
B0347500	BLOCK-1993
B0350000	BLOCK-1801
B0350500	BLOCK-1801
B0350602	BLOCK-1801
B0351500	BLOCK-1265
B03544710	BLOCK-532
B0354700	BLOCK-532
B0354710	BLOCK-532
B0354804	BLOCK-519
B0354805	BLOCK-519

DEVICE TYPE	REPL CODE
B0354806	BLOCK-519
B0354830	BLOCK-519
B0354832	BLOCK-519
B0354833	BLOCK-519
B0354901	BLOCK-484
B0355431	BLOCK-640
B0356190	BLOCK-563
B0356385	BLOCK-821
B0356446	BLOCK-534
B0356448	BLOCK-534
B0356602	BLOCK-756
B0356640	BLOCK-560
B0356641	BLOCK-560
B0356690	BLOCK-637
B0356711	BLOCK-225
B0358265	BLOCK-840
B0358268	BLOCK-771
B0358272	BLOCK-771
B0364710	BLOCK-532
B0372540	BLOCK-2087
B0373230	BLOCK-2097
B0376795	BLOCK-1254
B0379400	BLOCK-771
B0383100	BLOCK-1191
B0383111	BLOCK-1191
B0470116	BLOCK-1039
B0470130	BLOCK-1041
B0470494	BLOCK-1084
B0470522	BLOCK-1087
B0470532	BLOCK-1088
B0470662	BLOCK-1093
B0470932	BLOCK-1110
B0474370	BLOCK-1049
B77T0049	BLOCK-2189
BA10324	BLOCK-2171
BA10339	BLOCK-1873
BA10358	BLOCK-1993
BA12003	BLOCK-936
BA1310	BLOCK-214
BA1310F	BLOCK-214
BA1320	BLOCK-237
BA1330	BLOCK-205
BA1350	BLOCK-618
BA15218N	BLOCK-1802
BA178M05T	BLOCK-2087
BA222	BLOCK-596
BA301	BLOCK-132
BA301B	BLOCK-132
BA308	BLOCK-50
BA311	BLOCK-83
BA313	BLOCK-198
BA318	BLOCK-230
BA328	BLOCK-645
BA328LN	BLOCK-645
BA328MR	BLOCK-159
BA333	BLOCK-454
BA340	BLOCK-646
BA3704	BLOCK-548
BA401	BLOCK-102
BA402	BLOCK-98
BA403	BLOCK-528
BA404	BLOCK-512
BA4110	BLOCK-591
BA4220	BLOCK-544
BA4558	BLOCK-1801
BA4558DX	BLOCK-1801
BA4558F	BLOCK-1803
BA501	BLOCK-233
BA501A	BLOCK-233
BA5102	BLOCK-545
BA5102A	BLOCK-545
BA511	BLOCK-154
BA511A	BLOCK-154
BA521	BLOCK-155
BA521A	BLOCK-155
BA521AX	BLOCK-155
BA526	BLOCK-599
BA532	BLOCK-362
BA532S	BLOCK-362
BA536	BLOCK-606
BA5402A	BLOCK-606
BA5406	BLOCK-606
BA546	BLOCK-614
BA6122A	BLOCK-698
BA6124	BLOCK-548
BA6144	BLOCK-846
BA6209	BLOCK-697
BA6209U	BLOCK-697
BA6209U1	BLOCK-697
BA6209UI	BLOCK-697
BA6219	BLOCK-747
BA6219B	BLOCK-747
BA6238A	BLOCK-813
BA6238AU	BLOCK-813

DEVICE TYPE	REPL CODE
BA6238AU4	BLOCK-813
BA6304	BLOCK-665
BA6304A	BLOCK-665
BA6304AL	BLOCK-665
BA6993	BLOCK-2063
BA715	BLOCK-516
BA718	BLOCK-516
BA728	BLOCK-1993
BA847	BLOCK-623
BB-42-1B	BLOCK-109
BFW70	BLOCK-1276
BN-5416	BLOCK-669
BN5111	BLOCK-428
BN5111A	BLOCK-428
BN5111B	BLOCK-428
BN5416	BLOCK-669
BN5436N	BLOCK-708
BO316415	BLOCK-1269
BO319200	BLOCK-145
BO351500	BLOCK-1265
BO354710	BLOCK-532
BO354830	BLOCK-519
BO356446	BLOCK-534
BO356620	BLOCK-519
BO470016	BLOCK-1029
BO470116	BLOCK-1039
BS74154	BLOCK-1331
BU4011B	BLOCK-1039
BU4013B	BLOCK-1041
BU4052B	BLOCK-1087
BU4053B	BLOCK-1088
BU4066B	BLOCK-1093
BU4066BL	BLOCK-1093
BU4066BP	BLOCK-1093
BZX74C10	BLOCK-1435
C-555A	BLOCK-1212
C0-10750	BLOCK-1162
C0-10816	BLOCK-1085
C010174-02	BLOCK-1963
C010745-03	BLOCK-1159
C010745-12	BLOCK-1159
C010750-03	BLOCK-1162
C010816	BLOCK-1085
C010816-01	BLOCK-1085
C014331-09	BLOCK-1003
C014336	BLOCK-1086
C014345	BLOCK-1603
C014348	BLOCK-2087
C014377	BLOCK-1158
C014795	BLOCK-1158
C014806	BLOCK-1158
C014806-03	BLOCK-1158
C017097	BLOCK-1574
C060474	BLOCK-1689
C060612	BLOCK-1006
C061428	BLOCK-1591
C061702	BLOCK-1993
C061850	BLOCK-1593
C0900P	BLOCK-2100
C0901P	BLOCK-2101
C0902P	BLOCK-2106
C0903P	BLOCK-2107
C0904P	BLOCK-2109
C0905P	BLOCK-2099
C0906P	BLOCK-2102
C0907P	BLOCK-2122
C0908P	BLOCK-2121
C0909P	BLOCK-2111
C0910P	BLOCK-2103
C0911P	BLOCK-2105
C0912P	BLOCK-2101
C1032P	BLOCK-2199
C1033P	BLOCK-2200
C1035P	BLOCK-2201
C1036P	BLOCK-2202
C1044P	BLOCK-2206
C1045P	BLOCK-2207
C1046P	BLOCK-2208
C1046P(HEP)	BLOCK-2208
C1052P	BLOCK-1953
C1053P	BLOCK-1950
C1057P	BLOCK-1969
C1058P	BLOCK-1969
C1062P	BLOCK-2217
C1156H	BLOCK-182
C1252H2	BLOCK-775
C1253H2	BLOCK-776
C1330P	BLOCK-1407
C1353H2	BLOCK-776
C1356C2	BLOCK-471
C1363CA	BLOCK-396
C1366C2	BLOCK-510
C1373HA	BLOCK-695
C1406HA	BLOCK-773

DEVICE TYPE	REPL CODE
C1474HA	BLOCK-772
C1480CA	BLOCK-781
C1481CA	BLOCK-782
C1490HA	BLOCK-1194
C1513HA	BLOCK-768
C2001P(HEP)	BLOCK-2193
C2005P	BLOCK-2224
C2007G	BLOCK-2186
C2102A	BLOCK-999
C2708	BLOCK-1016
C3000P	BLOCK-1293
C3001	BLOCK-180
C3001-0	BLOCK-180
C3001-O	BLOCK-180
C3001A	BLOCK-180
C3001M	BLOCK-180
C3001P	BLOCK-1294
C3001T	BLOCK-180
C3002P	BLOCK-1295
C3004P	BLOCK-1297
C3010P	BLOCK-1304
C3020P	BLOCK-1365
C3030P	BLOCK-1377
C3040P	BLOCK-1390
C3041P	BLOCK-1391
C3050P	BLOCK-1401
C305C(IC)	BLOCK-903
C3073P	BLOCK-1410
C3075P	BLOCK-1412
C31C	BLOCK-46
C358C	BLOCK-1993
C3800P	BLOCK-1423
C3801P	BLOCK-1425
C3806P	BLOCK-161
C393C	BLOCK-2063
C393C(IC)	BLOCK-2063
C4059P	BLOCK-970
C4558C	BLOCK-1801
C4LS374AP	BLOCK-1672
C554C	BLOCK-138
C555A	BLOCK-1212
C575C2	BLOCK-136
C577H	BLOCK-80
C596(IC)	BLOCK-176
C596C(IC)	BLOCK-176
C6001	BLOCK-1790
C6003	BLOCK-1773
C6009P	BLOCK-1859
C6010	BLOCK-1774
C6010P	BLOCK-1881
C6011	BLOCK-1791
C6013	BLOCK-1786
C6014	BLOCK-1787
C6016	BLOCK-1797
C6017	BLOCK-1798
C6050G	BLOCK-2138
C6052M	BLOCK-2056
C6052P	BLOCK-2058
C6052P(HEP)	BLOCK-2058
C6055L(HEP)	BLOCK-1281
C6057P	BLOCK-1815
C6057P(HEP)	BLOCK-1815
C6059P	BLOCK-1405
C6059P(HEP)	BLOCK-1405
C6060P	BLOCK-1421
C6060P(HEP)	BLOCK-1421
C6061P	BLOCK-1770
C6062P	BLOCK-1262
C6062P(HEP)	BLOCK-1262
C6063P	BLOCK-1269
C6063P(HEP)	BLOCK-1269
C6065P	BLOCK-1276
C6065P(HEP)	BLOCK-1276
C6066P	BLOCK-1822
C6067G	BLOCK-1256
C6068P	BLOCK-1277
C6068P(HEP)	BLOCK-1277
C6070P	BLOCK-1271
C6070P(HEP)	BLOCK-1271
C6071P	BLOCK-1272
C6071P(HEP)	BLOCK-1272
C6072P	BLOCK-1292
C6072P(HEP)	BLOCK-1292
C6073P	BLOCK-1842
C6074P	BLOCK-1275
C6074P(HEP)	BLOCK-1275
C6075P	BLOCK-1291
C6076P	BLOCK-1422
C6076P(HEP)	BLOCK-1422
C6079P	BLOCK-1407
C6082P	BLOCK-1262
C6082P(HEP)	BLOCK-1262
C6083P	BLOCK-1269
C6083P(HEP)	BLOCK-1269

DEVICE TYPE	REPL CODE
C6085P(HEP)	BLOCK-1287
C6089	BLOCK-1244
C6089(HEP)	BLOCK-1256
C6091	BLOCK-1844
C6091G	BLOCK-1844
C6095P	BLOCK-1277
C6096P	BLOCK-1825
C6096P(HEP)	BLOCK-1825
C6099P	BLOCK-1820
C6100P	BLOCK-1808
C6100P(HEP)	BLOCK-1808
C6101P	BLOCK-1279
C6101P(HEP)	BLOCK-1279
C6102P	BLOCK-1801
C6102P(HEP)	BLOCK-1801
C6103P	BLOCK-1949
C6105	BLOCK-2137
C6107P	BLOCK-2141
C6111P	BLOCK-2093
C6112P	BLOCK-2095
C6113P	BLOCK-2097
C6114P	BLOCK-2119
C6116P	BLOCK-2137
C6118P	BLOCK-2091
C6120P	BLOCK-2094
C6122P	BLOCK-2108
C6123P	BLOCK-2128
C6130P	BLOCK-2079
C6131P	BLOCK-2079
C6132P	BLOCK-2144
C7400P	BLOCK-1293
C7400P(HEP)	BLOCK-1293
C7401P	BLOCK-1294
C7401P(HEP)	BLOCK-1294
C7402P	BLOCK-1295
C7402P(HEP)	BLOCK-1295
C7403P	BLOCK-1296
C7404P	BLOCK-1297
C7405P	BLOCK-1298
C7406P	BLOCK-1299
C7407P	BLOCK-1300
C7408P	BLOCK-1301
C7408P(HEP)	BLOCK-1301
C7409P	BLOCK-1302
C74107P	BLOCK-1490
C74109P	BLOCK-1306
C7410P	BLOCK-1304
C7411P	BLOCK-1307
C74121P	BLOCK-1311
C74123P	BLOCK-1313
C74125P	BLOCK-1314
C74126P	BLOCK-1315
C74132P	BLOCK-1318
C7413P	BLOCK-1317
C74141P	BLOCK-1321
C74145P	BLOCK-1325
C74147P	BLOCK-1326
C7414P	BLOCK-1320
C74150P	BLOCK-1327
C74151AP	BLOCK-1328
C74153P	BLOCK-1330
C74154P	BLOCK-1331
C74155P	BLOCK-1332
C74156P	BLOCK-1333
C74157P	BLOCK-1334
C74160AP	BLOCK-1337
C74162AP	BLOCK-1339
C74163AP	BLOCK-1340
C74164P	BLOCK-1341
C74165P	BLOCK-1342
C7416P	BLOCK-1336
C74170P	BLOCK-1344
C74173P	BLOCK-1345
C74174P	BLOCK-1346
C74175P	BLOCK-1347
C74176P	BLOCK-1348
C74177P	BLOCK-1349
C7417P	BLOCK-1343
C74180P	BLOCK-1352
C74181P	BLOCK-1353
C74182P	BLOCK-1354
C74190P	BLOCK-1355
C74191P	BLOCK-1356
C74192P	BLOCK-1357
C74193P	BLOCK-1358
C74195P	BLOCK-1359
C74196P	BLOCK-1360
C74197P	BLOCK-1361
C74198P	BLOCK-1362
C74199P	BLOCK-1363
C7420P	BLOCK-1365
C7423P	BLOCK-1369
C74251P	BLOCK-1371
C7427P	BLOCK-1372

DEVICE TYPE	REPL CODE
C7430P	BLOCK-1377
C7432P	BLOCK-1378
C74365P	BLOCK-1380
C74366P	BLOCK-1381
C74367P	BLOCK-1382
C74368P	BLOCK-1383
C7437P	BLOCK-1384
C7440P	BLOCK-1390
C7441AP	BLOCK-1391
C7445P	BLOCK-1395
C7446AP	BLOCK-1396
C7447AP	BLOCK-1397
C7448P	BLOCK-1398
C7450P	BLOCK-1401
C7451P	BLOCK-1402
C7453P	BLOCK-1403
C7454P	BLOCK-1404
C7460P	BLOCK-1406
C7470P	BLOCK-1408
C7472P	BLOCK-1409
C7473P	BLOCK-1410
C7473P(HEP)	BLOCK-1410
C7474P	BLOCK-1411
C7474P(HEP)	BLOCK-1411
C7475P	BLOCK-1412
C7476P	BLOCK-1413
C7476P(HEP)	BLOCK-1413
C7483P	BLOCK-1417
C7485P	BLOCK-1418
C7486P	BLOCK-1419
C7490AP	BLOCK-1423
C7490P(HEP)	BLOCK-1423
C7491AP	BLOCK-1424
C7492AP	BLOCK-1425
C7493AP	BLOCK-1426
C7495P	BLOCK-1428
C7496P	BLOCK-1429
C74LS00P	BLOCK-1568
C74LS02P	BLOCK-1570
C74LS03P	BLOCK-1571
C74LS04P	BLOCK-1572
C74LS05P	BLOCK-1573
C74LS08P	BLOCK-1574
C74LS109P	BLOCK-1578
C74LS10P	BLOCK-1576
C74LS11P	BLOCK-1579
C74LS136P	BLOCK-1590
C74LS138P	BLOCK-1591
C74LS13P	BLOCK-1587
C74LS14P	BLOCK-1593
C74LS151P	BLOCK-1598
C74LS153P	BLOCK-1599
C74LS155P	BLOCK-1600
C74LS157P	BLOCK-1602
C74LS158P	BLOCK-1603
C74LS170P	BLOCK-1613
C74LS174P	BLOCK-1615
C74LS175P	BLOCK-1616
C74LS181P	BLOCK-1617
C74LS190P	BLOCK-1618
C74LS191P	BLOCK-1619
C74LS192P	BLOCK-1620
C74LS193P	BLOCK-1621
C74LS196P	BLOCK-1624
C74LS197P	BLOCK-1625
C74LS20P	BLOCK-1626
C74LS251P	BLOCK-1639
C74LS253P	BLOCK-1640
C74LS257P	BLOCK-1641
C74LS258P	BLOCK-1642
C74LS259P	BLOCK-1643
C74LS279P	BLOCK-1649
C74LS27P	BLOCK-1647
C74LS298P	BLOCK-1656
C74LS30P	BLOCK-1658
C74LS32P	BLOCK-1659
C74LS367P	BLOCK-1668
C74LS368P	BLOCK-1669
C74LS42P	BLOCK-1684
C74LS51P	BLOCK-1689
C74LS54P	BLOCK-1690
C74LS55P	BLOCK-1693
C74LS670P	BLOCK-1704
C74LS74P	BLOCK-1768
C74LS86P	BLOCK-1713
C74LS90P	BLOCK-1714
C74LS93P	BLOCK-1717
C8080A	BLOCK-1832
C92-0025-R0	BLOCK-1280
CA-3088E	BLOCK-1812
CA-3089E	BLOCK-1813
CA0324E	BLOCK-2171
CA0345	BLOCK-1963
CA101AT	BLOCK-160

DEVICE TYPE	REPL CODE
CA101T	BLOCK-160
CA1190G	BLOCK-219
CA1190GM	BLOCK-219
CA1190GQ	BLOCK-219
CA1190GQ/M	BLOCK-219
CA124E	BLOCK-2171
CA1310	BLOCK-1825
CA1310E	BLOCK-1825
CA1352	BLOCK-1422
CA1352E	BLOCK-1422
CA1365	BLOCK-1269
CA1391E	BLOCK-1843
CA1394E	BLOCK-1876
CA1394G	BLOCK-1876
CA1398E	BLOCK-1291
CA139AE	BLOCK-1873
CA139F	BLOCK-1873
CA139G	BLOCK-1873
CA1458	BLOCK-1801
CA1458E	BLOCK-1801
CA1458F	BLOCK-1801
CA1458G	BLOCK-1801
CA1458S	BLOCK-1801
CA1558E	BLOCK-1801
CA158AS	BLOCK-1992
CA158AT	BLOCK-1992
CA158S	BLOCK-1992
CA158T	BLOCK-1992
CA2002	BLOCK-221
CA2002M	BLOCK-221
CA2002V	BLOCK-221
CA2004	BLOCK-221
CA201T	BLOCK-160
CA2065	BLOCK-1269
CA211	BLOCK-1980
CA2111	BLOCK-1262
CA2111A	BLOCK-1262
CA2111AE	BLOCK-1262
CA2111AQ	BLOCK-1262
CA211AE	BLOCK-1262
CA211AQ	BLOCK-1262
CA211E	BLOCK-1981
CA211G	BLOCK-1981
CA2136A	BLOCK-1290
CA2136AE	BLOCK-1290
CA224E	BLOCK-2171
CA224G	BLOCK-2171
CA239	BLOCK-1873
CA239A	BLOCK-1873
CA239AE	BLOCK-1873
CA239AG	BLOCK-1873
CA239E	BLOCK-1873
CA239G	BLOCK-1873
CA258S	BLOCK-1992
CA258T	BLOCK-1992
CA2904E	BLOCK-1993
CA3000	BLOCK-1940
CA3000T	BLOCK-1940
CA3001	BLOCK-1941
CA3001T	BLOCK-1941
CA3010	BLOCK-1943
CA3010A	BLOCK-1943
CA3010AT	BLOCK-1943
CA3010T	BLOCK-1943
CA3011	BLOCK-1282
CA3011T	BLOCK-1282
CA3012	BLOCK-1282
CA3012T	BLOCK-1282
CA3013	BLOCK-1223
CA3013S	BLOCK-1223
CA3013T	BLOCK-1223
CA3014	BLOCK-1223
CA3014S	BLOCK-1223
CA3014T	BLOCK-1223
CA3015	BLOCK-1943
CA3015A	BLOCK-1943
CA3015AT	BLOCK-1943
CA3015T	BLOCK-1943
CA3017	BLOCK-1272
CA3018	BLOCK-1944
CA3018A	BLOCK-1944
CA3018T	BLOCK-1944
CA3019	BLOCK-1945
CA3019T	BLOCK-1945
CA301AE	BLOCK-2141
CA301AG	BLOCK-2141
CA301AS	BLOCK-160
CA301AT	BLOCK-160
CA301E	BLOCK-2141
CA301G	BLOCK-2141
CA301T	BLOCK-160
CA3020	BLOCK-1809
CA3020A	BLOCK-1809
CA3020T	BLOCK-1809
CA3021E	BLOCK-1816
CA3026	BLOCK-1946
CA3026T	BLOCK-1946
CA3028	BLOCK-1280
CA3028A	BLOCK-1280
CA3028AF	BLOCK-1280
CA3028AS	BLOCK-1280
CA3028AT	BLOCK-1280
CA3028B	BLOCK-1280
CA3028BF	BLOCK-1280
CA3028BS	BLOCK-1280
CA3028T	BLOCK-1280
CA3029	BLOCK-1948
CA3029A	BLOCK-1948
CA3029E	BLOCK-1948
CA3030	BLOCK-1948
CA3030A	BLOCK-1948
CA3030AE	BLOCK-1948
CA3030E	BLOCK-1948
CA3035	BLOCK-1810
CA3035T	BLOCK-1810
CA3035V	BLOCK-1810
CA3035V1	BLOCK-1810
CA3037	BLOCK-1948
CA3037A	BLOCK-1948
CA3037AE	BLOCK-1948
CA3037E	BLOCK-1948
CA3038	BLOCK-1948
CA3038A	BLOCK-1948
CA3038AE	BLOCK-1948
CA3038E	BLOCK-1948
CA3039	BLOCK-1947
CA3039T	BLOCK-1947
CA3041	BLOCK-1245
CA3041P	BLOCK-1245
CA3041T	BLOCK-1245
CA3042	BLOCK-1265
CA3042E	BLOCK-1265
CA3043	BLOCK-1811
CA3043T	BLOCK-1811
CA3044	BLOCK-1267
CA3044T	BLOCK-1267
CA3044V	BLOCK-1267
CA3044V1	BLOCK-1267
CA3045	BLOCK-1963
CA3045E	BLOCK-1963
CA3045F	BLOCK-1963
CA3045L	BLOCK-1963
CA3046	BLOCK-1963
CA3046E	BLOCK-1963
CA3047E	BLOCK-1964
CA3048	BLOCK-1283
CA3049T	BLOCK-1946
CA3052	BLOCK-1283
CA3052E	BLOCK-1283
CA3053	BLOCK-1280
CA3053T	BLOCK-1280
CA3054	BLOCK-1971
CA3054E	BLOCK-1971
CA3056	BLOCK-2056
CA3056AE	BLOCK-1966
CA3058	BLOCK-1966
CA3058E	BLOCK-1966
CA3059	BLOCK-1966
CA3059E	BLOCK-1966
CA3064	BLOCK-1805
CA3064E	BLOCK-1808
CA3065	BLOCK-1269
CA3065,RCA	BLOCK-1269
CA3065D	BLOCK-1269
CA3065E	BLOCK-1269
CA3065F	BLOCK-1269
CA3065F.C.	BLOCK-1269
CA3065FC	BLOCK-1269
CA3065N	BLOCK-1269
CA3065PC	BLOCK-1269
CA3065RCA	BLOCK-1269
CA3066	BLOCK-1284
CA3066AE	BLOCK-1284
CA3066E	BLOCK-1284
CA3067	BLOCK-1285
CA3067AE	BLOCK-1285
CA3067E	BLOCK-1285
CA3068	BLOCK-1286
CA307	BLOCK-2143
CA3070	BLOCK-1271
CA3070E	BLOCK-1271
CA3070G	BLOCK-1271
CA3071	BLOCK-1272
CA3071E	BLOCK-1272
CA3072	BLOCK-1270
CA3072E	BLOCK-1270
CA3075	BLOCK-1279
CA3075E	BLOCK-1279
CA3076	BLOCK-1806
CA3076T	BLOCK-1806
CA3079	BLOCK-1966
CA3079E	BLOCK-1966
CA307E	BLOCK-2143
CA307G	BLOCK-2143
CA3080	BLOCK-1942
CA3080A	BLOCK-1942
CA3080E	BLOCK-2215
CA3080S	BLOCK-1942
CA3081	BLOCK-1970
CA3081E	BLOCK-1970
CA3082	BLOCK-946
CA3083	BLOCK-1996
CA3085A	BLOCK-1983
CA3086	BLOCK-1963
CA3086E	BLOCK-1963
CA3088E	BLOCK-1812
CA3089E	BLOCK-1813
CA3089F	BLOCK-1813
CA3090	BLOCK-1813
CA30900	BLOCK-1814
CA3090AB	BLOCK-1814
CA3090AQ	BLOCK-1814
CA3090E	BLOCK-1814
CA3090Q	BLOCK-1814
CA3100S	BLOCK-1926
CA3100T	BLOCK-1926
CA3100TY	BLOCK-1926
CA311	BLOCK-1980
CA311E	BLOCK-1981
CA311G	BLOCK-1981
CA311S	BLOCK-1980
CA311T	BLOCK-1980
CA3120E	BLOCK-1287
CA3121	BLOCK-1816
CA3121E	BLOCK-1816
CA3121E-G	BLOCK-1816
CA3121EG	BLOCK-1816
CA3121G	BLOCK-1816
CA3123	BLOCK-1389
CA3123E	BLOCK-1389
CA3125	BLOCK-1822
CA3125E	BLOCK-1822
CA3126	BLOCK-1821
CA3126EM	BLOCK-1821
CA3126EM1	BLOCK-1821
CA3126EMI	BLOCK-1821
CA3126FM1	BLOCK-1821
CA3126Q	BLOCK-1821
CA3130	BLOCK-1997
CA3130A	BLOCK-1997
CA3130AE	BLOCK-1997
CA3130AS	BLOCK-1997
CA3130AT	BLOCK-1997
CA3130B	BLOCK-1997
CA3130S	BLOCK-1997
CA3130T	BLOCK-1997
CA3134	BLOCK-164
CA3134EM	BLOCK-164
CA3134GQM	BLOCK-164
CA3134Q	BLOCK-164
CA3134QM	BLOCK-164
CA3135E	BLOCK-1846
CA3135G	BLOCK-1846
CA3136E	BLOCK-1886
CA3136G	BLOCK-1886
CA3137	BLOCK-165
CA3137E	BLOCK-165
CA3137EM	BLOCK-165
CA3137EM1	BLOCK-165
CA3137EMI	BLOCK-165
CA3139	BLOCK-163
CA3139E	BLOCK-163
CA3139G	BLOCK-163
CA3139GM1	BLOCK-163
CA3139GMI	BLOCK-163
CA3139GQ	BLOCK-163
CA3139Q	BLOCK-163
CA3142	BLOCK-1287
CA3142E	BLOCK-1287
CA3143	BLOCK-166
CA3143E	BLOCK-166
CA3143Q	BLOCK-166
CA3144E	BLOCK-2169
CA3144G	BLOCK-2169
CA3144Q	BLOCK-2169
CA3145	BLOCK-1201
CA314E	BLOCK-166
CA3151G	BLOCK-2170
CA3151GM1	BLOCK-2170
CA3151GMI	BLOCK-2170
CA3152E	BLOCK-1422
CA3153G	BLOCK-2168
CA3153GM1	BLOCK-2168
CA3153GMI	BLOCK-2168
CA3154G	BLOCK-1190
CA3156	BLOCK-1850
CA3156E	BLOCK-1850
CA3156G	BLOCK-1850
CA3157G	BLOCK-1892
CA3158G	BLOCK-1179
CA3159	BLOCK-1848
CA3159G	BLOCK-1848
CA3161E	BLOCK-955
CA3162E	BLOCK-1821
CA3163G	BLOCK-1894
CA3168	BLOCK-947
CA3168E	BLOCK-947
CA3170	BLOCK-2166
CA3170E	BLOCK-2166
CA3172	BLOCK-1849
CA3172E	BLOCK-1849
CA3172G	BLOCK-1849
CA3177G	BLOCK-1918
CA3189E	BLOCK-1197
CA3191E	BLOCK-1885
CA3192E	BLOCK-1884
CA3195	BLOCK-641
CA3195E	BLOCK-641
CA3218E	BLOCK-720
CA3221	BLOCK-1847
CA3221G	BLOCK-1847
CA324E	BLOCK-2171
CA324G	BLOCK-2171
CA3289E	BLOCK-1813
CA3290AE	BLOCK-2063
CA3290E	BLOCK-2063
CA339	BLOCK-1873
CA339A	BLOCK-1873
CA339AE	BLOCK-1873
CA339AG	BLOCK-1873
CA339E	BLOCK-1873
CA339G	BLOCK-1873
CA3401E	BLOCK-2191
CA3524E	BLOCK-701
CA3539	BLOCK-1948
CA3545	BLOCK-1947
CA358AS	BLOCK-1992
CA358AT	BLOCK-1992
CA358S	BLOCK-1992
CA358T	BLOCK-1992
CA3741CE	BLOCK-2058
CA3741E	BLOCK-2058
CA3748CT	BLOCK-160
CA555CE	BLOCK-2079
CA555CG	BLOCK-2079
CA555E	BLOCK-2079
CA555G	BLOCK-2079
CA723	BLOCK-1983
CA723CE	BLOCK-1984
CA723CT	BLOCK-1983
CA723E	BLOCK-1984
CA723T	BLOCK-1983
CA741	BLOCK-2058
CA741C	BLOCK-2058
CA741CE	BLOCK-2058
CA741CG	BLOCK-2058
CA741CS	BLOCK-2058
CA741CT	BLOCK-2056
CA741E	BLOCK-2058
CA741G	BLOCK-2058
CA741S	BLOCK-1949
CA741T	BLOCK-2056
CA747CE	BLOCK-2070
CA747CF	BLOCK-2070
CA747CG	BLOCK-2070
CA747CT	BLOCK-2069
CA747E	BLOCK-2070
CA747F	BLOCK-2070
CA747G	BLOCK-2070
CA747T	BLOCK-2069
CA748CE	BLOCK-2141
CA748CG	BLOCK-2141
CA748CJ	BLOCK-2141
CA748CN	BLOCK-2141
CA748CT	BLOCK-160
CA748E	BLOCK-2141
CA748G	BLOCK-2141
CA748J	BLOCK-2141
CA748N	BLOCK-2141
CA748T	BLOCK-160
CA7520	BLOCK-709
CA758E	BLOCK-1376
CA810Q	BLOCK-113
CD-4066B	BLOCK-1093
CD2300E	BLOCK-2197
CD2300E/830	BLOCK-2197
CD2301E	BLOCK-2216
CD2301E/861	BLOCK-2216
CD2302E	BLOCK-2208
CD2302E/846	BLOCK-2208
CD2303E	BLOCK-2210
CD2303E/849	BLOCK-2210
CD2304E	BLOCK-2207
CD2304E/845	BLOCK-2207
CD2305E	BLOCK-2209
CD2305E/848	BLOCK-2209
CD2306E	BLOCK-2199
CD2306E/832	BLOCK-2199
CD2307E	BLOCK-2206
CD2307E/844	BLOCK-2206
CD2308E	BLOCK-2217
CD2308E/862	BLOCK-2217
CD2309E	BLOCK-2218
CD2309E/863	BLOCK-2218
CD2310E	BLOCK-2202
CD2310E/836	BLOCK-2202
CD2311E/837	BLOCK-2203
CD2312E	BLOCK-2201
CD2314E	BLOCK-2200
CD2314E/833	BLOCK-2200
CD2315E	BLOCK-1950
CD2318E	BLOCK-1953
CD4000AD	BLOCK-1028
CD4000AE	BLOCK-1028
CD4000AF	BLOCK-1028
CD4000AZ	BLOCK-1028
CD4000BD	BLOCK-1028
CD4000BE	BLOCK-1028
CD4000BF	BLOCK-1028
CD4000CJ	BLOCK-1028
CD4000CN	BLOCK-1028
CD4000MJ	BLOCK-1028
CD4000MW	BLOCK-1028
CD4000UBD	BLOCK-1028
CD4000UBE	BLOCK-1028
CD4000UBF	BLOCK-1028
CD4001	BLOCK-1029
CD4001AD	BLOCK-1029
CD4001AE	BLOCK-1029
CD4001AF	BLOCK-1029
CD4001BCJ	BLOCK-1029
CD4001BCN	BLOCK-1029
CD4001BD	BLOCK-1029
CD4001BE	BLOCK-1029
CD4001BF	BLOCK-1029
CD4001BMJ	BLOCK-1029
CD4001BMW	BLOCK-1029
CD4001CJ	BLOCK-1029
CD4001CN	BLOCK-1029
CD4001MJ	BLOCK-1029
CD4001MW	BLOCK-1029
CD4001UBD	BLOCK-1029
CD4001UBE	BLOCK-1029
CD4001UBF	BLOCK-1029
CD4002	BLOCK-1030
CD4002AD	BLOCK-1030
CD4002AE	BLOCK-1030
CD4002AF	BLOCK-1030
CD4002BD	BLOCK-1030
CD4002BE	BLOCK-1030
CD4002BF	BLOCK-1030
CD4002CJ	BLOCK-1030
CD4002CN	BLOCK-1030
CD4002MJ	BLOCK-1030
CD4002MW	BLOCK-1030
CD4002UBD	BLOCK-1030
CD4002UBE	BLOCK-1030
CD4002UBF	BLOCK-1030
CD4006	BLOCK-1031
CD4006AE	BLOCK-1031
CD4006BE	BLOCK-1031
CD4007AE	BLOCK-1032
CD4007AF	BLOCK-1032
CD4007CJ	BLOCK-1032
CD4007CN	BLOCK-1032
CD4007UBE	BLOCK-1032
CD4008AD	BLOCK-1034
CD4008AE	BLOCK-1034
CD4008AF	BLOCK-1034
CD4008B	BLOCK-1034
CD4008BCJ	BLOCK-1034
CD4008BCN	BLOCK-1034
CD4008BD	BLOCK-1034
CD4008BE	BLOCK-1034
CD4008BF	BLOCK-1034
CD4008BJ	BLOCK-1034
CD4008BMJ	BLOCK-1034
CD4008BMW	BLOCK-1034
CD4009AD	BLOCK-1084
CD4009AE	BLOCK-1084
CD4009AF	BLOCK-1084
CD4009CJ	BLOCK-1084
CD4009CN	BLOCK-1084
CD4009UBD	BLOCK-1084
CD4009UBE	BLOCK-1084
CD4009UBF	BLOCK-1084
CD40100BD	BLOCK-1037
CD40100BE	BLOCK-1037
CD40100BF	BLOCK-1037
CD40106BCJ	BLOCK-1038
CD40106BCN	BLOCK-1038
CD40106BD	BLOCK-1038
CD40106BE	BLOCK-1038
CD40106BF	BLOCK-1038
CD40106BMJ	BLOCK-1038
CD4010AE	BLOCK-1085
CD4010BE	BLOCK-1085
CD4010CJ	BLOCK-1085
CD4010CN	BLOCK-1085
CD4011	BLOCK-1039
CD4011AD	BLOCK-1039
CD4011AE	BLOCK-1039
CD4011AF	BLOCK-1039
CD4011BCJ	BLOCK-1039
CD4011BCN	BLOCK-1039
CD4011BD	BLOCK-1039
CD4011BE	BLOCK-1039
CD4011BF	BLOCK-1039
CD4011BMJ	BLOCK-1039
CD4011BMW	BLOCK-1039
CD4011CJ	BLOCK-1039
CD4011CN	BLOCK-1039
CD4011MJ	BLOCK-1039
CD4011MW	BLOCK-1039
CD4011UBD	BLOCK-1039
CD4011UBE	BLOCK-1039
CD4011UBF	BLOCK-1039
CD4012	BLOCK-1040
CD4012AD	BLOCK-1040
CD4012AE	BLOCK-1040
CD4012AF	BLOCK-1040
CD4012BD	BLOCK-1040
CD4012BE	BLOCK-1040
CD4012BF	BLOCK-1040
CD4012CJ	BLOCK-1040
CD4012CN	BLOCK-1040
CD4012MJ	BLOCK-1040
CD4012MW	BLOCK-1040
CD4012UBD	BLOCK-1040
CD4012UBE	BLOCK-1040
CD4012UBF	BLOCK-1040
CD4013	BLOCK-1041
CD4013AD	BLOCK-1041
CD4013AE	BLOCK-1041
CD4013AF	BLOCK-1041
CD4013B	BLOCK-1041
CD4013BCJ	BLOCK-1041
CD4013BCN	BLOCK-1041
CD4013BD	BLOCK-1041
CD4013BE	BLOCK-1041
CD4013BF	BLOCK-1041
CD4013BMJ	BLOCK-1041
CD4013BMW	BLOCK-1041
CD4013BN	BLOCK-1041
CD4014AD	BLOCK-1042
CD4014AE	BLOCK-1042
CD4014AF	BLOCK-1042
CD4014BD	BLOCK-1042
CD4014BE	BLOCK-1042
CD4014BF	BLOCK-1042
CD4015	BLOCK-1043
CD4015AD	BLOCK-1043
CD4015AE	BLOCK-1043
CD4015AF	BLOCK-1043
CD4015BD	BLOCK-1043
CD4015BE	BLOCK-1043
CD4015BF	BLOCK-1043
CD4016	BLOCK-1048
CD40160BCJ	BLOCK-1044
CD40160BCN	BLOCK-1044
CD40160BD	BLOCK-1044
CD40160BE	BLOCK-1044
CD40160BF	BLOCK-1044
CD40160BMJ	BLOCK-1044
CD40160BMW	BLOCK-1044
CD40161BCJ	BLOCK-1045
CD40161BCN	BLOCK-1045
CD40161BD	BLOCK-1045
CD40161BE	BLOCK-1045
CD40161BF	BLOCK-1045
CD40161BMJ	BLOCK-1045
CD40161BMW	BLOCK-1045
CD40162BCJ	BLOCK-1046
CD40162BCN	BLOCK-1046

ORIGINAL DEVICE TYPES AND REPLACEMENT CODES 21

DEVICE TYPE	REPL CODE
CD40162BD	BLOCK-1046
CD40162BE	BLOCK-1046
CD40162BF	BLOCK-1046
CD40162BMJ	BLOCK-1046
CD40162BMW	BLOCK-1046
CD40163BCJ	BLOCK-1047
CD40163BCN	BLOCK-1047
CD40163BD	BLOCK-1047
CD40163BE	BLOCK-1047
CD40163BF	BLOCK-1047
CD40163BMJ	BLOCK-1047
CD40163BMW	BLOCK-1047
CD4016AD	BLOCK-1048
CD4016AE	BLOCK-1048
CD4016AF	BLOCK-1048
CD4016B	BLOCK-1048
CD4016BD	BLOCK-1048
CD4016BE	BLOCK-1048
CD4016BF	BLOCK-1048
CD4016CJ	BLOCK-1048
CD4016CN	BLOCK-1048
CD4016MJ	BLOCK-1048
CD4017	BLOCK-1051
CD40174BC	BLOCK-1049
CD40174BD	BLOCK-1049
CD40174BE	BLOCK-1049
CD40174BF	BLOCK-1049
CD40175BC	BLOCK-1050
CD4017AD	BLOCK-1051
CD4017AE	BLOCK-1051
CD4017AF	BLOCK-1051
CD4017BCJ	BLOCK-1051
CD4017BCN	BLOCK-1051
CD4017BD	BLOCK-1051
CD4017BE	BLOCK-1051
CD4017BF	BLOCK-1051
CD4017BMJ	BLOCK-1051
CD4017BMW	BLOCK-1051
CD4018	BLOCK-1053
CD40182BD	BLOCK-1052
CD40182BE	BLOCK-1052
CD40182BF	BLOCK-1052
CD4018AD	BLOCK-1053
CD4018AE	BLOCK-1053
CD4018AF	BLOCK-1053
CD4018BCJ	BLOCK-1053
CD4018BCN	BLOCK-1053
CD4018BD	BLOCK-1053
CD4018BE	BLOCK-1053
CD4018BF	BLOCK-1053
CD4018BMJ	BLOCK-1053
CD4018BMW	BLOCK-1053
CD4019	BLOCK-1058
CD40192BCJ	BLOCK-1054
CD40192BCN	BLOCK-1054
CD40192BD	BLOCK-1054
CD40192BE	BLOCK-1054
CD40192BF	BLOCK-1054
CD40192BMJ	BLOCK-1054
CD40192BMW	BLOCK-1054
CD40192CJ	BLOCK-1054
CD40192CN	BLOCK-1054
CD40193BCJ	BLOCK-1055
CD40193BCN	BLOCK-1055
CD40193BD	BLOCK-1055
CD40193BE	BLOCK-1055
CD40193BF	BLOCK-1055
CD40193BMJ	BLOCK-1055
CD40193BMW	BLOCK-1055
CD40194BD	BLOCK-1056
CD40194BE	BLOCK-1056
CD40194BF	BLOCK-1056
CD4019AD	BLOCK-1058
CD4019AE	BLOCK-1058
CD4019AF	BLOCK-1058
CD4019BC	BLOCK-1058
CD4019BCJ	BLOCK-1058
CD4019BCN	BLOCK-1058
CD4019BD	BLOCK-1058
CD4019BE	BLOCK-1058
CD4019BF	BLOCK-1058
CD4020	BLOCK-1059
CD4020AD	BLOCK-1059
CD4020AE	BLOCK-1059
CD4020AF	BLOCK-1059
CD4020BCJ	BLOCK-1059
CD4020BCN	BLOCK-1059
CD4020BD	BLOCK-1059
CD4020BE	BLOCK-1059
CD4020BF	BLOCK-1059
CD4020BMJ	BLOCK-1059
CD4020BMW	BLOCK-1059
CD4021	BLOCK-1060
CD4021AD	BLOCK-1060
CD4021AE	BLOCK-1060
CD4021AF	BLOCK-1060
CD4021BD	BLOCK-1060
CD4021BE	BLOCK-1060
CD4021BF	BLOCK-1060
CD4022	BLOCK-1061
CD4022AD	BLOCK-1061
CD4022AE	BLOCK-1061
CD4022AF	BLOCK-1061
CD4022BCJ	BLOCK-1061
CD4022BCN	BLOCK-1061
CD4022BD	BLOCK-1061
CD4022BE	BLOCK-1061
CD4022BF	BLOCK-1061
CD4022BMJ	BLOCK-1061
CD4022BMW	BLOCK-1061
CD4023	BLOCK-1062
CD4023AD	BLOCK-1062
CD4023AE	BLOCK-1062
CD4023AF	BLOCK-1062
CD4023BCJ	BLOCK-1062
CD4023BCN	BLOCK-1062
CD4023BD	BLOCK-1062
CD4023BE	BLOCK-1062
CD4023BF	BLOCK-1062
CD4023BMJ	BLOCK-1062
CD4023BMW	BLOCK-1062
CD4023C	BLOCK-1062
CD4023CJ	BLOCK-1062
CD4023CN	BLOCK-1062
CD4023MJ	BLOCK-1062
CD4023MW	BLOCK-1062
CD4023UBD	BLOCK-1062
CD4023UBE	BLOCK-1062
CD4023UBF	BLOCK-1062
CD4024	BLOCK-1063
CD4024AD	BLOCK-1063
CD4024AE	BLOCK-1063
CD4024AF	BLOCK-1063
CD4024BCJ	BLOCK-1063
CD4024BCN	BLOCK-1063
CD4024BD	BLOCK-1063
CD4024BE	BLOCK-1063
CD4024BF	BLOCK-1063
CD4024BMJ	BLOCK-1063
CD4024BMW	BLOCK-1063
CD4025	BLOCK-1064
CD4025AD	BLOCK-1064
CD4025AE	BLOCK-1064
CD4025AF	BLOCK-1064
CD4025BCJ	BLOCK-1064
CD4025BCN	BLOCK-1064
CD4025BD	BLOCK-1064
CD4025BE	BLOCK-1064
CD4025BF	BLOCK-1064
CD4025BMJ	BLOCK-1064
CD4025BMW	BLOCK-1064
CD4025CJ	BLOCK-1064
CD4025CN	BLOCK-1064
CD4025MJ	BLOCK-1064
CD4025MW	BLOCK-1064
CD4025UBD	BLOCK-1064
CD4025UBE	BLOCK-1064
CD4025UBF	BLOCK-1064
CD4026AD	BLOCK-1065
CD4026AE	BLOCK-1065
CD4026AF	BLOCK-1065
CD4026BD	BLOCK-1065
CD4026BE	BLOCK-1065
CD4026BF	BLOCK-1065
CD4027	BLOCK-1066
CD4027AD	BLOCK-1066
CD4027AE	BLOCK-1066
CD4027AF	BLOCK-1066
CD4027BCJ	BLOCK-1066
CD4027BCN	BLOCK-1066
CD4027BD	BLOCK-1066
CD4027BE	BLOCK-1066
CD4027BF	BLOCK-1066
CD4027BMJ	BLOCK-1066
CD4027BMW	BLOCK-1066
CD4028AD	BLOCK-1067
CD4028AE	BLOCK-1067
CD4028AF	BLOCK-1067
CD4028BCJ	BLOCK-1067
CD4028BCN	BLOCK-1067
CD4028BD	BLOCK-1067
CD4028BE	BLOCK-1067
CD4028BF	BLOCK-1067
CD4028BMJ	BLOCK-1067
CD4028BMW	BLOCK-1067
CD4028F	BLOCK-1067
CD4029AD	BLOCK-1068
CD4029AE	BLOCK-1068
CD4029AF	BLOCK-1068
CD4029BCJ	BLOCK-1068
CD4029BCN	BLOCK-1068
CD4029BD	BLOCK-1068
CD4029BE	BLOCK-1068
CD4029BF	BLOCK-1068
CD4029BMJ	BLOCK-1068
CD4029BMW	BLOCK-1068
CD4030	BLOCK-1069
CD4030AD	BLOCK-1069
CD4030AE	BLOCK-1069
CD4030AF	BLOCK-1069
CD4030BD	BLOCK-1069
CD4030BE	BLOCK-1069
CD4030BF	BLOCK-1069
CD4030BMW	BLOCK-1069
CD4030CJ	BLOCK-1069
CD4030CN	BLOCK-1069
CD4030MJ	BLOCK-1069
CD4030MW	BLOCK-1069
CD4031AD	BLOCK-1070
CD4031AE	BLOCK-1070
CD4031AF	BLOCK-1070
CD4031BD	BLOCK-1070
CD4031BE	BLOCK-1070
CD4031BF	BLOCK-1070
CD4032AD	BLOCK-1071
CD4032AE	BLOCK-1071
CD4032AF	BLOCK-1071
CD4032BD	BLOCK-1071
CD4032BE	BLOCK-1071
CD4032BF	BLOCK-1071
CD4033	BLOCK-1072
CD4033AD	BLOCK-1072
CD4033AE	BLOCK-1072
CD4033AF	BLOCK-1072
CD4033BD	BLOCK-1072
CD4033BE	BLOCK-1072
CD4033BF	BLOCK-1072
CD4034AD	BLOCK-1073
CD4034AE	BLOCK-1073
CD4034AF	BLOCK-1073
CD4034AY	BLOCK-1073
CD4034BD	BLOCK-1073
CD4034BE	BLOCK-1073
CD4034BF	BLOCK-1073
CD4035AD	BLOCK-1074
CD4035AE	BLOCK-1074
CD4035AF	BLOCK-1074
CD4035AY	BLOCK-1074
CD4035BD	BLOCK-1074
CD4035BE	BLOCK-1074
CD4035BF	BLOCK-1074
CD4038AD	BLOCK-1075
CD4038AE	BLOCK-1075
CD4038AF	BLOCK-1075
CD4038BD	BLOCK-1075
CD4038BE	BLOCK-1075
CD4038BF	BLOCK-1075
CD4040	BLOCK-1076
CD4040AD	BLOCK-1076
CD4040AE	BLOCK-1076
CD4040AF	BLOCK-1076
CD4040BCJ	BLOCK-1076
CD4040BCN	BLOCK-1076
CD4040BD	BLOCK-1076
CD4040BE	BLOCK-1076
CD4040BF	BLOCK-1076
CD4040BMJ	BLOCK-1076
CD4040BMW	BLOCK-1076
CD4042	BLOCK-1077
CD4042AD	BLOCK-1077
CD4042AE	BLOCK-1077
CD4042AF	BLOCK-1077
CD4042BCJ	BLOCK-1077
CD4042BCN	BLOCK-1077
CD4042BD	BLOCK-1077
CD4042BE	BLOCK-1077
CD4042BF	BLOCK-1077
CD4042BMJ	BLOCK-1077
CD4042BMW	BLOCK-1077
CD4043AD	BLOCK-1078
CD4043AE	BLOCK-1078
CD4043AF	BLOCK-1078
CD4043BD	BLOCK-1078
CD4043BE	BLOCK-1078
CD4043BF	BLOCK-1078
CD4043CJ	BLOCK-1078
CD4043CN	BLOCK-1078
CD4043MJ	BLOCK-1078
CD4043MW	BLOCK-1078
CD4044AD	BLOCK-1079
CD4044AE	BLOCK-1079
CD4044AF	BLOCK-1079
CD4044BD	BLOCK-1079
CD4044BE	BLOCK-1079
CD4044BF	BLOCK-1079
CD4044CJ	BLOCK-1079
CD4044CN	BLOCK-1079
CD4044MJ	BLOCK-1079
CD4044MW	BLOCK-1079
CD4045AE	BLOCK-1080
CD4045AF	BLOCK-1080
CD4045BD	BLOCK-1080
CD4045BE	BLOCK-1080
CD4045BF	BLOCK-1080
CD4046A	BLOCK-1081
CD4046AD	BLOCK-1081
CD4046AE	BLOCK-2149
CD4046AF	BLOCK-1081
CD4046BD	BLOCK-1081
CD4046BE	BLOCK-1081
CD4046BF	BLOCK-1081
CD4047AD	BLOCK-1082
CD4047AE	BLOCK-1082
CD4047AF	BLOCK-1082
CD4047BCJ	BLOCK-1082
CD4047BCN	BLOCK-1082
CD4047BD	BLOCK-1082
CD4047BE	BLOCK-1082
CD4047BF	BLOCK-1082
CD4047BMJ	BLOCK-1082
CD4047BMW	BLOCK-1082
CD4048AD	BLOCK-1083
CD4048AE	BLOCK-1083
CD4048AF	BLOCK-1083
CD4048BCJ	BLOCK-1083
CD4048BCN	BLOCK-1083
CD4048BD	BLOCK-1083
CD4048BE	BLOCK-1083
CD4048BF	BLOCK-1083
CD4048BMJ	BLOCK-1083
CD4049AD	BLOCK-1084
CD4049AE	BLOCK-1084
CD4049AF	BLOCK-1084
CD4049CJ	BLOCK-1084
CD4049CN	BLOCK-1084
CD4049MJ	BLOCK-1084
CD4049UBD	BLOCK-1084
CD4049UBE	BLOCK-1084
CD4049UBF	BLOCK-1084
CD4050	BLOCK-1085
CD4050AD	BLOCK-1085
CD4050AE	BLOCK-1085
CD4050AF	BLOCK-1085
CD4050B	BLOCK-1085
CD4050BCJ	BLOCK-1085
CD4050BCN	BLOCK-1085
CD4050BD	BLOCK-1085
CD4050BE	BLOCK-1085
CD4050BF	BLOCK-1085
CD4051	BLOCK-1086
CD4051AE	BLOCK-1086
CD4051B	BLOCK-1086
CD4051BCJ	BLOCK-1086
CD4051BCN	BLOCK-1086
CD4051BD	BLOCK-1086
CD4051BE	BLOCK-1086
CD4051BF	BLOCK-1086
CD4051BMJ	BLOCK-1086
CD4052	BLOCK-1087
CD4052AE	BLOCK-1086
CD4052BCJ	BLOCK-1087
CD4052BCN	BLOCK-1087
CD4052BD	BLOCK-1087
CD4052BE	BLOCK-1087
CD4052BF	BLOCK-1087
CD4052BMJ	BLOCK-1087
CD4053BCJ	BLOCK-1088
CD4053BCN	BLOCK-1088
CD4053BD	BLOCK-1088
CD4053BE	BLOCK-1088
CD4053BF	BLOCK-1088
CD4053BP	BLOCK-1088
CD4055	BLOCK-1089
CD4055AE	BLOCK-1089
CD4055BD	BLOCK-1089
CD4055BE	BLOCK-1089
CD4055BF	BLOCK-1089
CD4055BY	BLOCK-1089
CD4056BD	BLOCK-1090
CD4056BE	BLOCK-1090
CD4056BY	BLOCK-1090
CD4060	BLOCK-1091
CD4060AD	BLOCK-1091
CD4060AE	BLOCK-1091
CD4060AF	BLOCK-1091
CD4060BCJ	BLOCK-1091
CD4060BCN	BLOCK-1091
CD4060BD	BLOCK-1091
CD4060BE	BLOCK-1091
CD4060BF	BLOCK-1091
CD4060BH	BLOCK-1091
CD4060BMJ	BLOCK-1091
CD4060BMW	BLOCK-1091
CD4063BD	BLOCK-1092
CD4063BE	BLOCK-1091
CD4063BF	BLOCK-1092
CD4066	BLOCK-1093
CD4066AD	BLOCK-1093
CD4066AE	BLOCK-1093
CD4066AF	BLOCK-1093
CD4066B	BLOCK-1093
CD4066BCJ	BLOCK-1093
CD4066BCN	BLOCK-1093
CD4066BD	BLOCK-1093
CD4066BE	BLOCK-1093
CD4066BF	BLOCK-1093
CD4066BMJ	BLOCK-1093
CD4067BD	BLOCK-1094
CD4067BE	BLOCK-1094
CD4067BF	BLOCK-1094
CD4068BD	BLOCK-1095
CD4068BE	BLOCK-1095
CD4068BF	BLOCK-1095
CD4069BE	BLOCK-1096
CD4069C	BLOCK-1096
CD4069CJ	BLOCK-1096
CD4069CN	BLOCK-1096
CD4069M	BLOCK-1096
CD4069MJ	BLOCK-1096
CD4069N	BLOCK-1096
CD4069UBD	BLOCK-1096
CD4069UBE	BLOCK-1096
CD4069UBF	BLOCK-1096
CD4070	BLOCK-1069
CD4070BCJ	BLOCK-1097
CD4070BCN	BLOCK-1097
CD4070BD	BLOCK-1097
CD4070BE	BLOCK-1097
CD4070BF	BLOCK-1097
CD4070BMJ	BLOCK-1097
CD4070BMW	BLOCK-1097
CD4071BCJ	BLOCK-1098
CD4071BCN	BLOCK-1098
CD4071BD	BLOCK-1098
CD4071BE	BLOCK-1098
CD4071BF	BLOCK-1098
CD4071BMJ	BLOCK-1098
CD4071BMW	BLOCK-1098
CD4072BD	BLOCK-1099
CD4072BE	BLOCK-1099
CD4072BF	BLOCK-1099
CD4073BCJ	BLOCK-1100
CD4073BCN	BLOCK-1100
CD4073BD	BLOCK-1100
CD4073BE	BLOCK-1100
CD4073BF	BLOCK-1100
CD4073BMJ	BLOCK-1100
CD4073BMW	BLOCK-1100
CD4075BCJ	BLOCK-1101
CD4075BCN	BLOCK-1101
CD4075BD	BLOCK-1101
CD4075BE	BLOCK-1101
CD4075BF	BLOCK-1101
CD4075BMJ	BLOCK-1101
CD4075BMN	BLOCK-1101
CD4075BMW	BLOCK-1101
CD4075BT	BLOCK-1101
CD4076BC	BLOCK-1102
CD4076BD	BLOCK-1102
CD4076BE	BLOCK-1102
CD4076BF	BLOCK-1102
CD4076BM	BLOCK-1102
CD4077BD	BLOCK-1103
CD4077BE	BLOCK-1103
CD4077BF	BLOCK-1103
CD4078BD	BLOCK-1104
CD4078BE	BLOCK-1104
CD4078BF	BLOCK-1104
CD4081	BLOCK-1105
CD4081BCJ	BLOCK-1105
CD4081BCN	BLOCK-1105
CD4081BD	BLOCK-1105
CD4081BE	BLOCK-1105
CD4081BF	BLOCK-1105
CD4081BMJ	BLOCK-1105
CD4081BMW	BLOCK-1105
CD4082	BLOCK-1106
CD4082B	BLOCK-1106
CD4082BD	BLOCK-1106
CD4082BE	BLOCK-1106
CD4082BF	BLOCK-1106
CD4085BD	BLOCK-1107
CD4085BE	BLOCK-1107
CD4085BF	BLOCK-1107
CD4086BD	BLOCK-1108
CD4086BE	BLOCK-1108
CD4086BF	BLOCK-1108
CD4089BCJ	BLOCK-1109
CD4089BCN	BLOCK-1109
CD4089BD	BLOCK-1109
CD4089BE	BLOCK-1109
CD4089BF	BLOCK-1109
CD4089BMJ	BLOCK-1109
CD4093BCJ	BLOCK-1110
CD4093BCN	BLOCK-1110
CD4093BD	BLOCK-1110
CD4093BE	BLOCK-1110
CD4093BF	BLOCK-1110
CD4093BMJ	BLOCK-1110
CD4095BD	BLOCK-1111
CD4095BE	BLOCK-1111
CD4095BF	BLOCK-1111
CD4096BD	BLOCK-1112
CD4096BE	BLOCK-1112
CD4096BF	BLOCK-1112
CD4097BD	BLOCK-1113
CD4097BE	BLOCK-1113
CD4097BF	BLOCK-1113
CD4098BD	BLOCK-1114
CD4098BE	BLOCK-1114
CD4098BF	BLOCK-1114
CD4099	BLOCK-1115
CD4099BCJ	BLOCK-1115
CD4099BCN	BLOCK-1115
CD4099BD	BLOCK-1115
CD4099BE	BLOCK-1115
CD4099BF	BLOCK-1115
CD4099BMJ	BLOCK-1115
CD4099BMW	BLOCK-1115
CD4502BD	BLOCK-1118
CD4502BE	BLOCK-1118
CD4502BF	BLOCK-1118
CD4503B	BLOCK-1035
CD4503BD	BLOCK-1035
CD4503BE	BLOCK-1035
CD4503BF	BLOCK-1035
CD4508BE	BLOCK-1120
CD4508BF	BLOCK-1120
CD4510BCJ	BLOCK-1121
CD4510BCN	BLOCK-1121
CD4510BD	BLOCK-1121
CD4510BE	BLOCK-1121
CD4510BF	BLOCK-1121
CD4510BMJ	BLOCK-1121
CD4510BMW	BLOCK-1121
CD4511BCJ	BLOCK-1122
CD4511BCN	BLOCK-1122
CD4511BD	BLOCK-1122
CD4511BE	BLOCK-1122
CD4511BF	BLOCK-1122
CD4512BD	BLOCK-1123
CD4512BE	BLOCK-1123
CD4512BF	BLOCK-1123
CD4514BC	BLOCK-1125
CD4514BD	BLOCK-1125
CD4514BE	BLOCK-1125
CD4514BF	BLOCK-1125
CD4514BH	BLOCK-1125
CD4514BK	BLOCK-1125
CD4514BM	BLOCK-1125
CD4515BC	BLOCK-1126
CD4515BD	BLOCK-1126
CD4515BE	BLOCK-1126
CD4515BF	BLOCK-1126
CD4515BH	BLOCK-1126
CD4515BK	BLOCK-1126
CD4515BM	BLOCK-1126
CD4516BCJ	BLOCK-1127
CD4516BCN	BLOCK-1127
CD4516BD	BLOCK-1127
CD4516BE	BLOCK-1127
CD4516BF	BLOCK-1127
CD4516BMJ	BLOCK-1127
CD4516BMW	BLOCK-1127
CD4517BE	BLOCK-1128
CD4517BF	BLOCK-1128
CD4518B	BLOCK-1129
CD4518BCJ	BLOCK-1129
CD4518BCN	BLOCK-1129
CD4518BD	BLOCK-1129
CD4518BE	BLOCK-1129
CD4518BF	BLOCK-1129
CD4518BMJ	BLOCK-1129
CD4518BMW	BLOCK-1129

22 ORIGINAL DEVICE TYPES AND REPLACEMENT CODES

DEVICE TYPE	REPL CODE
CD4520B	BLOCK-1130
CD4520BCJ	BLOCK-1130
CD4520BCN	BLOCK-1130
CD4520BD	BLOCK-1130
CD4520BE	BLOCK-1130
CD4520BEX	BLOCK-1130
CD4520BF	BLOCK-1130
CD4520BMJ	BLOCK-1130
CD4520BMW	BLOCK-1130
CD4521BE	BLOCK-1131
CD4522BC	BLOCK-1132
CD4522BE	BLOCK-1132
CD4522BM	BLOCK-1132
CD4526BC	BLOCK-1133
CD4526BCN	BLOCK-1133
CD4526BM	BLOCK-1133
CD4527BCJ	BLOCK-1134
CD4527BCN	BLOCK-1134
CD4527BD	BLOCK-1134
CD4527BE	BLOCK-1134
CD4527BF	BLOCK-1134
CD4527BMJ	BLOCK-1134
CD4527BMW	BLOCK-1134
CD4529BCN	BLOCK-1136
CD4532BD	BLOCK-1138
CD4532BE	BLOCK-1138
CD4532BF	BLOCK-1138
CD4536BD	BLOCK-1139
CD4536BE	BLOCK-1139
CD4536BF	BLOCK-1139
CD4538BD	BLOCK-1140
CD4538BE	BLOCK-1140
CD4538BF	BLOCK-1140
CD4543BD	BLOCK-1143
CD4543BE	BLOCK-1143
CD4543BF	BLOCK-1143
CD4555BD	BLOCK-1146
CD4555BE	BLOCK-1146
CD4555BF	BLOCK-1146
CD4556BD	BLOCK-1147
CD4556BE	BLOCK-1147
CD4556BF	BLOCK-1147
CD4566	BLOCK-1150
CD4566A	BLOCK-1150
CD4585BD	BLOCK-1154
CD4585BE	BLOCK-1154
CD4585BF	BLOCK-1154
CD5328	BLOCK-2185
CD6039	BLOCK-2189
CD73/187/72	BLOCK-1426
CD73/187/73	BLOCK-1487
CD74HCT04E	BLOCK-1552
CD74HCT08E	BLOCK-1553
CD74HCT138E	BLOCK-1554
CD74HCT14E	BLOCK-1555
CD74HCT244E	BLOCK-1560
CD74HCT32E	BLOCK-1562
CD74HCT373E	BLOCK-1563
CG24015A	BLOCK-2206
CH3065	BLOCK-1269
CI-1002	BLOCK-1280
CI-1003	BLOCK-1279
CI-1004	BLOCK-1275
CJSE067	BLOCK-2014
CJSE071	BLOCK-2004
CK-134A	BLOCK-585
CLEAK-05710	BLOCK-1018
CLEAK-05903	BLOCK-1018
CLEAK-06002	BLOCK-1018
CLEAK-06102	BLOCK-1019
CLEAK-06201	BLOCK-1019
CLEAK-12004	BLOCK-1019
CM0SMM74C00N	BLOCK-1431
CM301AN	BLOCK-2141
CM3900	BLOCK-2191
CM4000	BLOCK-1028
CM4000AD	BLOCK-1028
CM4000AE	BLOCK-1028
CM4001	BLOCK-1029
CM4001AD	BLOCK-1029
CM4001AE	BLOCK-1029
CM4002	BLOCK-1030
CM4002AD	BLOCK-1030
CM4002AE	BLOCK-1030
CM4006AD	BLOCK-1031
CM4006AE	BLOCK-1031
CM4007	BLOCK-1032
CM4007AD	BLOCK-1032
CM4007AE	BLOCK-1032
CM4008AD	BLOCK-1034
CM4008AE	BLOCK-1034
CM4009	BLOCK-1084
CM4009AD	BLOCK-1084
CM4009AE	BLOCK-1084

DEVICE TYPE	REPL CODE
CM4010AE	BLOCK-1085
CM4011	BLOCK-1039
CM4011AD	BLOCK-1039
CM4011AE	BLOCK-1039
CM4012	BLOCK-1040
CM4012AD	BLOCK-1040
CM4012AE	BLOCK-1040
CM4013	BLOCK-1041
CM4013AD	BLOCK-1041
CM4013AE	BLOCK-1041
CM4014AD	BLOCK-1042
CM4014AE	BLOCK-1042
CM4015	BLOCK-1043
CM4015AD	BLOCK-1043
CM4015AE	BLOCK-1043
CM4016	BLOCK-1048
CM4016AD	BLOCK-1048
CM4016AE	BLOCK-1048
CM4017	BLOCK-1051
CM4017AD	BLOCK-1051
CM4017AE	BLOCK-1051
CM4018AD	BLOCK-1053
CM4018AE	BLOCK-1053
CM4019	BLOCK-1058
CM4019AD	BLOCK-1058
CM4019AE	BLOCK-1058
CM4020	BLOCK-1059
CM4020AD	BLOCK-1059
CM4020AE	BLOCK-1059
CM4021	BLOCK-1060
CM4021AD	BLOCK-1060
CM4021AE	BLOCK-1060
CM4022AD	BLOCK-1061
CM4022AE	BLOCK-1061
CM4023	BLOCK-1062
CM4023AD	BLOCK-1062
CM4023AE	BLOCK-1062
CM4024	BLOCK-1063
CM4024AD	BLOCK-1063
CM4024AE	BLOCK-1063
CM4025	BLOCK-1064
CM4025AD	BLOCK-1064
CM4025AE	BLOCK-1064
CM4026AD	BLOCK-1065
CM4026AE	BLOCK-1065
CM4027	BLOCK-1066
CM4027AD	BLOCK-1066
CM4027AE	BLOCK-1066
CM4028AD	BLOCK-1067
CM4028AE	BLOCK-1067
CM4029AD	BLOCK-1068
CM4029AE	BLOCK-1068
CM4030	BLOCK-1069
CM4030AD	BLOCK-1069
CM4030AE	BLOCK-1069
CM4032AD	BLOCK-1071
CM4032AE	BLOCK-1071
CM4033AD	BLOCK-1072
CM4033AE	BLOCK-1072
CM4035AD	BLOCK-1074
CM4035AE	BLOCK-1074
CM4038AD	BLOCK-1075
CM4038AE	BLOCK-1075
CM4040	BLOCK-1076
CM4040AD	BLOCK-1076
CM4040AE	BLOCK-1076
CM4042	BLOCK-1077
CM4042AD	BLOCK-1077
CM4042AE	BLOCK-1077
CM4043AD	BLOCK-1078
CM4043AE	BLOCK-1078
CM4044AD	BLOCK-1079
CM4044AE	BLOCK-1079
CM4047AD	BLOCK-1082
CM4047AE	BLOCK-1082
CM4047AF	BLOCK-1082
CM4048AD	BLOCK-1083
CM4048AE	BLOCK-1083
CM4049	BLOCK-1084
CM4049AD	BLOCK-1084
CM4049AE	BLOCK-1084
CM4050	BLOCK-1085
CM4050AD	BLOCK-1085
CM4050AE	BLOCK-1085
CM4051	BLOCK-1086
CM4052	BLOCK-1087
CM4066AD	BLOCK-1093
CM4066AE	BLOCK-1093
CM4069B	BLOCK-1096
CM4081	BLOCK-1105
CM4116AD	BLOCK-1003
CM4116AE	BLOCK-1003
CM4518	BLOCK-1129
CM4520	BLOCK-1130

DEVICE TYPE	REPL CODE
CN5310	BLOCK-398
CN5310(K)	BLOCK-398
CN5310(L)	BLOCK-398
CN5310(M)	BLOCK-398
CN5310K	BLOCK-398
CN5310L	BLOCK-398
CN5310M	BLOCK-398
CN5311	BLOCK-398
CN5311CL	BLOCK-398
CN5311K	BLOCK-398
CN5311KL	BLOCK-398
CN5311LM	BLOCK-398
CN5411	BLOCK-286
CN5510	BLOCK-393
CN5520	BLOCK-473
CO-10174-02	BLOCK-1963
CO-10745-03	BLOCK-1159
CO-10750-03	BLOCK-1162
CO-10816	BLOCK-1085
CO-14348	BLOCK-2087
CO10816-XX	BLOCK-1085
CO14313	BLOCK-1634
CO1433-09	BLOCK-1003
CO14336	BLOCK-1086
CO14341	BLOCK-1576
CO14344	BLOCK-1591
CO14345	BLOCK-1603
CO14377	BLOCK-1158
CO14806-12	BLOCK-1158
CP3120E	BLOCK-1287
CPS-16676-1	BLOCK-2213
CPS15553-101	BLOCK-2199
CQEAK-01103	BLOCK-1019
CS2917	BLOCK-2211
CS5995	BLOCK-1212
CX-064-2	BLOCK-298
CX-075	BLOCK-288
CX-075B	BLOCK-288
CX-093	BLOCK-1269
CX-095C	BLOCK-295
CX-095D	BLOCK-295
CX-095E	BLOCK-295
CX-1000	BLOCK-592
CX-100D	BLOCK-592
CX-130	BLOCK-590
CX-131A	BLOCK-588
CX-133A	BLOCK-586
CX-134A	BLOCK-585
CX-135	BLOCK-583
CX-136A	BLOCK-581
CX-137A	BLOCK-579
CX-138	BLOCK-577
CX-138A	BLOCK-577
CX-139A	BLOCK-575
CX-141	BLOCK-573
CX-143	BLOCK-569
CX-143A	BLOCK-569
CX-145	BLOCK-571
CX-177B	BLOCK-293
CX-555A	BLOCK-399
CX-556	BLOCK-398
CX-557	BLOCK-539
CX-557A	BLOCK-405
CX-557S	BLOCK-405
CX-843	BLOCK-406
CX-843A	BLOCK-406
CX-848	BLOCK-651
CX064	BLOCK-298
CX065	BLOCK-287
CX065B	BLOCK-287
CX075B	BLOCK-288
CX089	BLOCK-94
CX089D	BLOCK-94
CX093D	BLOCK-1269
CX095	BLOCK-295
CX095A	BLOCK-295
CX095C	BLOCK-295
CX095D	BLOCK-295
CX095E	BLOCK-295
CX100B	BLOCK-592
CX100D	BLOCK-592
CX130	BLOCK-590
CX131A	BLOCK-588
CX133A	BLOCK-586
CX134A	BLOCK-585
CX135	BLOCK-583
CX136	BLOCK-581
CX136A	BLOCK-581
CX137A	BLOCK-579
CX138	BLOCK-577
CX139A	BLOCK-575
CX141	BLOCK-573
CX141A	BLOCK-573
CX143	BLOCK-569

DEVICE TYPE	REPL CODE
CX143A	BLOCK-569
CX145	BLOCK-571
CX157	BLOCK-299
CX158	BLOCK-300
CX162	BLOCK-297
CX170	BLOCK-289
CX172	BLOCK-291
CX173	BLOCK-292
CX177	BLOCK-293
CX177B	BLOCK-293
CX181	BLOCK-294
CX20112	BLOCK-840
CX555	BLOCK-399
CX555A	BLOCK-399
CX556	BLOCK-398
CX556A	BLOCK-398
CX556B	BLOCK-404
CX557	BLOCK-405
CX557A	BLOCK-405
CX557B	BLOCK-539
CX557S	BLOCK-539
CX842B	BLOCK-295
CX843	BLOCK-406
CX843A	BLOCK-406
CX848	BLOCK-651
CXA1011P	BLOCK-839
CXL1025	BLOCK-24
CXL1027	BLOCK-26
CXL1062	BLOCK-60
CXL1063	BLOCK-61
CXL1090	BLOCK-88
CXL1139	BLOCK-135
CXL309K	BLOCK-1022
CXL370	BLOCK-1024
CXL703A	BLOCK-1212
CXL704	BLOCK-1223
CXL705A	BLOCK-1256
CXL706	BLOCK-4
CXL708	BLOCK-1262
CXL709	BLOCK-1264
CXL710	BLOCK-1265
CXL711	BLOCK-1267
CXL713	BLOCK-1270
CXL714	BLOCK-1271
CXL715	BLOCK-1272
CXL718	BLOCK-1275
CXL725	BLOCK-1281
CXL726	BLOCK-1282
CXL727	BLOCK-1283
CXL728	BLOCK-1284
CXL729	BLOCK-1285
CXL730	BLOCK-1286
CXL731	BLOCK-1287
CXL734	BLOCK-1289
CXL738	BLOCK-1291
CXL747	BLOCK-1407
CXL749	BLOCK-1422
CXL760	BLOCK-1790
CXL780	BLOCK-1805
CXL781	BLOCK-1806
CXL783	BLOCK-1808
CXL786	BLOCK-1811
CXL787	BLOCK-1812
CXL788	BLOCK-1813
CXL790	BLOCK-1815
CXL949	BLOCK-2073
D-TEX-8	BLOCK-1944
D2114	BLOCK-1002
D2114-2	BLOCK-1002
D2114A4	BLOCK-1002
D2114A5	BLOCK-1002
D2114AL3	BLOCK-1002
D2114AL4	BLOCK-1002
D2114L	BLOCK-1002
D2114L2	BLOCK-1002
D2114L3	BLOCK-1002
D2716D	BLOCK-1017
D2716D-Q1P-4	BLOCK-1017
D2716D-Q1X-0	BLOCK-1017
D2732A	BLOCK-1018
D2732D	BLOCK-1018
D2764	BLOCK-1019
D2764A-2	BLOCK-1019
D2764D	BLOCK-1019
D34002410-001	BLOCK-2189
D34013890-002	BLOCK-1473
D3628A	BLOCK-1759
D3628A-1	BLOCK-1759
D369	BLOCK-79
D4002BC	BLOCK-1030
D4011BC	BLOCK-1039
D4012BC	BLOCK-1040
D4013BC	BLOCK-1041
D4016CX-15	BLOCK-1004

DEVICE TYPE	REPL CODE
D4017BC	BLOCK-1051
D4027BC	BLOCK-1066
D4030BC	BLOCK-1069
D4040BC	BLOCK-1076
D4049UBC	BLOCK-1084
D4052BC	BLOCK-1087
D4053BC	BLOCK-1088
D4066BC	BLOCK-1093
D4069C	BLOCK-1096
D4069UBC	BLOCK-1096
D4081	BLOCK-1105
D4081BC	BLOCK-1105
D416	BLOCK-1003
D4160BC	BLOCK-1044
D4164C-15	BLOCK-1006
D4164C-2	BLOCK-1006
D4164C-3	BLOCK-1006
D7486J	BLOCK-1419
D780C-1	BLOCK-1025
D8255A-5	BLOCK-1855
D8255AC-2	BLOCK-1855
D8255AC-5	BLOCK-1855
D858(IC)	BLOCK-186
D919695	BLOCK-2176
D919698	BLOCK-2177
D919699	BLOCK-2178
D919700	BLOCK-2179
DAC0806LCN	BLOCK-971
DAC0807LCN	BLOCK-971
DAP601	BLOCK-666
DBXAN6291	BLOCK-725
DDEY001001	BLOCK-2138
DDEY002001	BLOCK-1280
DDEY004001	BLOCK-98
DDEY015001	BLOCK-1339
DDEY017001	BLOCK-161
DDEY019001	BLOCK-2140
DDEY020001	BLOCK-2139
DDEY026001	BLOCK-2141
DDEY027001	BLOCK-79
DDEY028001	BLOCK-2095
DDEY029001	BLOCK-1423
DDEY030001	BLOCK-1293
DDEY032001	BLOCK-161
DDEY037001	BLOCK-187
DDEY042001	BLOCK-2144
DDEY046001	BLOCK-2144
DDEY055001	BLOCK-186
DDEY058001	BLOCK-1039
DDEY061001	BLOCK-182
DDEY064001	BLOCK-80
DDEY082001	BLOCK-192
DDEY084001	BLOCK-1039
DDEY087001	BLOCK-191
DDEY088001	BLOCK-2144
DDEY089001	BLOCK-1039
DDEY091001	BLOCK-182
DDEY093001	BLOCK-2144
DDEY097001	BLOCK-183
DDEY109001	BLOCK-180
DDEY123001	BLOCK-159
DDEY130001	BLOCK-238
DDEY131001	BLOCK-258
DDEY133001	BLOCK-249
DDEY146001	BLOCK-265
DDEY147001	BLOCK-515
DDEY149001	BLOCK-273
DDEY157001	BLOCK-411
DDEY171001	BLOCK-411
DDEY172001	BLOCK-412
DDEY179001	BLOCK-214
DDF4091001	BLOCK-182
DDFY055001	BLOCK-186
DDFY091001	BLOCK-182
DE-087	BLOCK-191
DEEY004001	BLOCK-98
DGL5630	BLOCK-1157
DHHZT33	BLOCK-1157
DM-103	BLOCK-1105
DM-104	BLOCK-2144
DM-106	BLOCK-2160
DM-11	BLOCK-1264
DM-110	BLOCK-1128
DM-11A	BLOCK-1264
DM-14	BLOCK-1275
DM-20	BLOCK-1389
DM-24	BLOCK-1276
DM-31	BLOCK-1264
DM-32	BLOCK-1389
DM-33	BLOCK-2095
DM-35	BLOCK-1829
DM-36	BLOCK-1825
DM-37	BLOCK-565
DM-40	BLOCK-1801

DEVICE TYPE	REPL CODE
DM-41	BLOCK-1290
DM-44	BLOCK-1376
DM-50	BLOCK-2167
DM-51	BLOCK-1813
DM-54	BLOCK-1276
DM-59	BLOCK-264
DM-65	BLOCK-934
DM-77	BLOCK-1096
DM-84	BLOCK-264
DM-86	BLOCK-1048
DM-87	BLOCK-1873
DM-88	BLOCK-1812
DM-9	BLOCK-1289
DM-90	BLOCK-1997
DM-91	BLOCK-2095
DM-92	BLOCK-2095
DM-94	BLOCK-2087
DM-98	BLOCK-264
DM106	BLOCK-2160
DM11	BLOCK-1264
DM14	BLOCK-1275
DM24	BLOCK-1276
DM31	BLOCK-1264
DM32	BLOCK-1389
DM44	BLOCK-1376
DM54	BLOCK-1276
DM7373N	BLOCK-1410
DM7400	BLOCK-1293
DM7400J	BLOCK-1293
DM7400N	BLOCK-1293
DM7401J	BLOCK-1294
DM7401N	BLOCK-1294
DM7402J	BLOCK-1295
DM7402N	BLOCK-1295
DM7403J	BLOCK-1296
DM7403N	BLOCK-1296
DM7404N	BLOCK-1297
DM7405N	BLOCK-1298
DM7406N	BLOCK-1299
DM7407N	BLOCK-1300
DM7408J	BLOCK-1301
DM7408N	BLOCK-1301
DM7409J	BLOCK-1302
DM7409N	BLOCK-1302
DM74107J	BLOCK-1305
DM74107N	BLOCK-1305
DM7410J	BLOCK-1304
DM7410N	BLOCK-1304
DM7411J	BLOCK-1307
DM7411N	BLOCK-1307
DM74121J	BLOCK-1311
DM74121N	BLOCK-1311
DM74123	BLOCK-1313
DM74123N	BLOCK-1313
DM74125J	BLOCK-1314
DM74125N	BLOCK-1314
DM74126N	BLOCK-1315
DM74132J	BLOCK-1318
DM74132N	BLOCK-1318
DM7413J	BLOCK-1317
DM7413N	BLOCK-1317
DM74141J	BLOCK-1321
DM74141N	BLOCK-1321
DM74145	BLOCK-1325
DM74145N	BLOCK-1325
DM74147J	BLOCK-1326
DM74147N	BLOCK-1326
DM7414J	BLOCK-1320
DM7414N	BLOCK-1320
DM74150	BLOCK-1327
DM74150N	BLOCK-1327
DM74151	BLOCK-1328
DM74151N	BLOCK-1328
DM74153	BLOCK-1330
DM74153N	BLOCK-1330
DM74154	BLOCK-1331
DM74154J	BLOCK-1331
DM74154N	BLOCK-1331
DM74155N	BLOCK-1332
DM74156N	BLOCK-1333
DM74157N	BLOCK-1334
DM7415O	BLOCK-1327
DM74160AN	BLOCK-1337
DM74160N	BLOCK-1337
DM74161AN	BLOCK-1338
DM74161N	BLOCK-1338
DM74162	BLOCK-1339
DM74162AN	BLOCK-1339
DM74162N	BLOCK-1339
DM74163AN	BLOCK-1340
DM74163N	BLOCK-1340
DM74164N	BLOCK-1341
DM74165N	BLOCK-1342
DM7416N	BLOCK-1336

ORIGINAL DEVICE TYPES AND REPLACEMENT CODES 23

DEVICE TYPE	REPL CODE
DM74170N	BLOCK-1344
DM74173N	BLOCK-1345
DM74174	BLOCK-1346
DM74174N	BLOCK-1346
DM74175N	BLOCK-1347
DM74176J	BLOCK-1348
DM74176N	BLOCK-1348
DM74177J	BLOCK-1349
DM74177N	BLOCK-1349
DM7417N	BLOCK-1343
DM74180J	BLOCK-1352
DM74180N	BLOCK-1352
DM74181J	BLOCK-1353
DM74181N	BLOCK-1353
DM74182N	BLOCK-1354
DM74190	BLOCK-1355
DM74190J	BLOCK-1355
DM74190N	BLOCK-1355
DM74191J	BLOCK-1356
DM74191N	BLOCK-1356
DM74192J	BLOCK-1357
DM74192N	BLOCK-1357
DM74193	BLOCK-1358
DM74193J	BLOCK-1358
DM74193N	BLOCK-1358
DM74195N	BLOCK-1359
DM74196N	BLOCK-1360
DM74197N	BLOCK-1361
DM74198N	BLOCK-1362
DM74199N	BLOCK-1363
DM7420J	BLOCK-1365
DM7420N	BLOCK-1365
DM7423J	BLOCK-1369
DM7423N	BLOCK-1369
DM7423W	BLOCK-1369
DM74251N	BLOCK-1371
DM7427J	BLOCK-1372
DM7427N	BLOCK-1372
DM7427W	BLOCK-1372
DM7430J	BLOCK-1377
DM7430N	BLOCK-1377
DM7432J	BLOCK-1378
DM7432N	BLOCK-1378
DM74365J	BLOCK-1380
DM74365N	BLOCK-1380
DM74366J	BLOCK-1381
DM74366N	BLOCK-1381
DM74367J	BLOCK-1382
DM74367N	BLOCK-1382
DM74368J	BLOCK-1383
DM74368N	BLOCK-1383
DM7437J	BLOCK-1384
DM7437N	BLOCK-1384
DM7437W	BLOCK-1384
DM7438J	BLOCK-1385
DM7438N	BLOCK-1385
DM7438W	BLOCK-1385
DM7440J	BLOCK-1390
DM7440N	BLOCK-1390
DM7441AJ	BLOCK-1391
DM7441AN	BLOCK-1391
DM7441N	BLOCK-1391
DM7442J	BLOCK-1392
DM7442N	BLOCK-1392
DM7445J	BLOCK-1395
DM7445N	BLOCK-1395
DM7445W	BLOCK-1395
DM7446AJ	BLOCK-1396
DM7446AN	BLOCK-1396
DM7447AJ	BLOCK-1397
DM7447AN	BLOCK-1397
DM7447N	BLOCK-1397
DM7448J	BLOCK-1398
DM7448N	BLOCK-1398
DM7448W	BLOCK-1398
DM7450J	BLOCK-1401
DM7450N	BLOCK-1401
DM7451J	BLOCK-1402
DM7451N	BLOCK-1402
DM7453J	BLOCK-1403
DM7453N	BLOCK-1403
DM7454J	BLOCK-1404
DM7454N	BLOCK-1404
DM7460J	BLOCK-1406
DM7460N	BLOCK-1406
DM7470J	BLOCK-1408
DM7470N	BLOCK-1408
DM7472J	BLOCK-1409
DM7472N	BLOCK-1409
DM7473J	BLOCK-1410
DM7473N	BLOCK-1410
DM7474J	BLOCK-1411
DM7474M	BLOCK-1411
DM7474N	BLOCK-1411

DEVICE TYPE	REPL CODE
DM7475J	BLOCK-1412
DM7475N	BLOCK-1412
DM7476J	BLOCK-1413
DM7476N	BLOCK-1413
DM7483J	BLOCK-1415
DM7483N	BLOCK-1417
DM7485N	BLOCK-1412
DM7486J	BLOCK-1419
DM7486N	BLOCK-1419
DM7489N	BLOCK-1420
DM7490	BLOCK-1423
DM7490AN	BLOCK-1423
DM7490J	BLOCK-1423
DM7490N	BLOCK-1423
DM7491AN	BLOCK-1424
DM7492AN	BLOCK-1425
DM7492J	BLOCK-1425
DM7492N	BLOCK-1425
DM7493AN	BLOCK-1426
DM7493J	BLOCK-1426
DM7493N	BLOCK-1426
DM7495N	BLOCK-1428
DM7496J	BLOCK-1429
DM7496N	BLOCK-1429
DM7497J	BLOCK-1430
DM74H00J	BLOCK-1471
DM74H00N	BLOCK-1471
DM74H01J	BLOCK-1472
DM74H01N	BLOCK-1472
DM74H04J	BLOCK-1473
DM74H04N	BLOCK-1473
DM74H05	BLOCK-1474
DM74H05J	BLOCK-1474
DM74H05L	BLOCK-1474
DM74H05N	BLOCK-1474
DM74H08J	BLOCK-1475
DM74H08N	BLOCK-1475
DM74H10J	BLOCK-1476
DM74H10N	BLOCK-1476
DM74H11J	BLOCK-1482
DM74H11N	BLOCK-1482
DM74H20J	BLOCK-1484
DM74H20N	BLOCK-1484
DM74H21J	BLOCK-1485
DM74H21N	BLOCK-1485
DM74H22J	BLOCK-1486
DM74H22N	BLOCK-1486
DM74H30J	BLOCK-1487
DM74H30N	BLOCK-1487
DM74H40J	BLOCK-1488
DM74H40N	BLOCK-1488
DM74H50J	BLOCK-1489
DM74H50N	BLOCK-1489
DM74H51J	BLOCK-1490
DM74H51N	BLOCK-1490
DM74H52J	BLOCK-1491
DM74H52N	BLOCK-1491
DM74H53J	BLOCK-1492
DM74H53N	BLOCK-1492
DM74H54J	BLOCK-1493
DM74H54N	BLOCK-1493
DM74H55J	BLOCK-1494
DM74H55N	BLOCK-1494
DM74H60J	BLOCK-1495
DM74H60N	BLOCK-1495
DM74H61J	BLOCK-1496
DM74H61N	BLOCK-1496
DM74H62J	BLOCK-1497
DM74H62N	BLOCK-1497
DM74H71J	BLOCK-1498
DM74H71N	BLOCK-1498
DM74H72J	BLOCK-1499
DM74H72N	BLOCK-1499
DM74H73J	BLOCK-1500
DM74H73N	BLOCK-1500
DM74H74J	BLOCK-1501
DM74H74N	BLOCK-1501
DM74H76J	BLOCK-1502
DM74H76N	BLOCK-1502
DM74H78J	BLOCK-1503
DM74H78N	BLOCK-1503
DM74L93F	BLOCK-1567
DM74L93J	BLOCK-1567
DM74L93N	BLOCK-1567
DM74LS00N	BLOCK-1568
DM74LS01N	BLOCK-1569
DM74LS02N	BLOCK-1570
DM74LS03N	BLOCK-1571
DM74LS04N	BLOCK-1572
DM74LS05N	BLOCK-1573
DM74LS08N	BLOCK-1574
DM74LS09N	BLOCK-1575
DM74LS109AN	BLOCK-1578
DM74LS10N	BLOCK-1576

DEVICE TYPE	REPL CODE
DM74LS112AN	BLOCK-1580
DM74LS11N	BLOCK-1579
DM74LS125AN	BLOCK-1585
DM74LS126AJ	BLOCK-1586
DM74LS126AN	BLOCK-1586
DM74LS126AW	BLOCK-1586
DM74LS12N	BLOCK-1582
DM74LS132J	BLOCK-1588
DM74LS132N	BLOCK-1588
DM74LS132W	BLOCK-1588
DM74LS138N	BLOCK-1591
DM74LS139N	BLOCK-1592
DM74LS13J	BLOCK-1587
DM74LS13N	BLOCK-1587
DM74LS13W	BLOCK-1587
DM74LS14J	BLOCK-1593
DM74LS14N	BLOCK-1593
DM74LS14W	BLOCK-1593
DM74LS151N	BLOCK-1598
DM74LS153N	BLOCK-1599
DM74LS155N	BLOCK-1600
DM74LS156N	BLOCK-1601
DM74LS160AN	BLOCK-1604
DM74LS161AN	BLOCK-1605
DM74LS162AN	BLOCK-1606
DM74LS163AN	BLOCK-1607
DM74LS163N	BLOCK-1607
DM74LS164N	BLOCK-1608
DM74LS165N	BLOCK-1609
DM74LS166N	BLOCK-1610
DM74LS170N	BLOCK-1613
DM74LS173N	BLOCK-1614
DM74LS174N	BLOCK-1615
DM74LS175N	BLOCK-1616
DM74LS190N	BLOCK-1618
DM74LS191N	BLOCK-1619
DM74LS192N	BLOCK-1620
DM74LS193N	BLOCK-1621
DM74LS194AN	BLOCK-1622
DM74LS195AN	BLOCK-1623
DM74LS196	BLOCK-1624
DM74LS196N	BLOCK-1624
DM74LS197N	BLOCK-1625
DM74LS20N	BLOCK-1626
DM74LS240N	BLOCK-1630
DM74LS241N	BLOCK-1631
DM74LS242N	BLOCK-1632
DM74LS243N	BLOCK-1633
DM74LS244N	BLOCK-1634
DM74LS245N	BLOCK-1635
DM74LS247N	BLOCK-1636
DM74LS251N	BLOCK-1639
DM74LS253N	BLOCK-1640
DM74LS257BN	BLOCK-1641
DM74LS258BN	BLOCK-1642
DM74LS259N	BLOCK-1643
DM74LS26N	BLOCK-1644
DM74LS279N	BLOCK-1649
DM74LS27N	BLOCK-1647
DM74LS283N	BLOCK-1652
DM74LS290N	BLOCK-1653
DM74LS293N	BLOCK-1654
DM74LS32N	BLOCK-1659
DM74LS352J	BLOCK-1663
DM74LS352N	BLOCK-1662
DM74LS352W	BLOCK-1662
DM74LS353J	BLOCK-1663
DM74LS353N	BLOCK-1663
DM74LS353W	BLOCK-1663
DM74LS365AJ	BLOCK-1666
DM74LS365AN	BLOCK-1666
DM74LS365AW	BLOCK-1666
DM74LS366AJ	BLOCK-1667
DM74LS366AN	BLOCK-1667
DM74LS366AW	BLOCK-1667
DM74LS367AJ	BLOCK-1668
DM74LS367AN	BLOCK-1668
DM74LS367AW	BLOCK-1668
DM74LS367N	BLOCK-1668
DM74LS368AJ	BLOCK-1669
DM74LS368AN	BLOCK-1669
DM74LS368AW	BLOCK-1669
DM74LS373N	BLOCK-1671
DM74LS374N	BLOCK-1672
DM74LS38N	BLOCK-1676
DM74LS390N	BLOCK-1678
DM74LS393N	BLOCK-1679
DM74LS40N	BLOCK-1683
DM74LS42N	BLOCK-1684
DM74LS47N	BLOCK-1685
DM74LS48N	BLOCK-1686
DM74LS51	BLOCK-1689
DM74LS51N	BLOCK-1689
DM74LS55N	BLOCK-1693

DEVICE TYPE	REPL CODE
DM74LS670J	BLOCK-1704
DM74LS670N	BLOCK-1704
DM74LS670W	BLOCK-1704
DM74LS74AN	BLOCK-1706
DM74LS75N	BLOCK-1707
DM74LS83AN	BLOCK-1711
DM74LS85N	BLOCK-1712
DM74LS86N	BLOCK-1713
DM74LS92N	BLOCK-1716
DM74LS93N	BLOCK-1717
DM74S00J	BLOCK-1719
DM74S00N	BLOCK-1719
DM74S03J	BLOCK-1721
DM74S03N	BLOCK-1721
DM74S04J	BLOCK-1722
DM74S04N	BLOCK-1722
DM74S05J	BLOCK-1723
DM74S05N	BLOCK-1723
DM74S10J	BLOCK-1726
DM74S10N	BLOCK-1726
DM74S112N	BLOCK-1728
DM74S113N	BLOCK-1729
DM74S11J	BLOCK-1727
DM74S11N	BLOCK-1727
DM74S140N	BLOCK-1734
DM74S151N	BLOCK-1736
DM74S153N	BLOCK-1737
DM74S157N	BLOCK-1738
DM74S15N	BLOCK-1735
DM74S163N	BLOCK-1607
DM74S174N	BLOCK-1740
DM74S175N	BLOCK-1741
DM74S181N	BLOCK-1742
DM74S188J	BLOCK-1743
DM74S188N	BLOCK-1743
DM74S194N	BLOCK-1744
DM74S20J	BLOCK-1745
DM74S20N	BLOCK-1745
DM74S22J	BLOCK-1746
DM74S22N	BLOCK-1746
DM74S251N	BLOCK-1747
DM74S258N	BLOCK-1748
DM74S287J	BLOCK-1749
DM74S287N	BLOCK-1749
DM74S288J	BLOCK-1750
DM74S288N	BLOCK-1750
DM74S387J	BLOCK-1752
DM74S387N	BLOCK-1752
DM74S40N	BLOCK-1753
DM74S472AJ	BLOCK-1755
DM74S472AN	BLOCK-1755
DM74S472N	BLOCK-1755
DM74S473AJ	BLOCK-1756
DM74S473AN	BLOCK-1756
DM74S473J	BLOCK-1756
DM74S473N	BLOCK-1756
DM74S474AJ	BLOCK-1757
DM74S474AN	BLOCK-1757
DM74S474J	BLOCK-1757
DM74S474N	BLOCK-1757
DM74S475AJ	BLOCK-1758
DM74S475AN	BLOCK-1758
DM74S475J	BLOCK-1758
DM74S475N	BLOCK-1758
DM74S51N	BLOCK-1761
DM74S570J	BLOCK-1762
DM74S570N	BLOCK-1762
DM74S571J	BLOCK-1763
DM74S571N	BLOCK-1763
DM74S572AJ	BLOCK-1764
DM74S572AN	BLOCK-1764
DM74S572J	BLOCK-1764
DM74S572N	BLOCK-1764
DM74S573AJ	BLOCK-1765
DM74S573AN	BLOCK-1765
DM74S573J	BLOCK-1765
DM74S573N	BLOCK-1765
DM74S64N	BLOCK-1766
DM74S65N	BLOCK-1767
DM74S74N	BLOCK-1768
DM74S86N	BLOCK-1769
DM8121N	BLOCK-1371
DM8123N	BLOCK-1840
DM8214N	BLOCK-1278
DM8280J	BLOCK-1348
DM8280N	BLOCK-1348
DM8281J	BLOCK-1349
DM8281N	BLOCK-1349
DM8301J	BLOCK-1861
DM8301N	BLOCK-1861
DM8316N	BLOCK-1866
DM8318N	BLOCK-1867
DM8542N	BLOCK-1899
DM8546N	BLOCK-1900

DEVICE TYPE	REPL CODE
DM8551N	BLOCK-1358
DM8553N	BLOCK-1903
DM8554N	BLOCK-1904
DM8555N	BLOCK-1905
DM8556N	BLOCK-1906
DM8560	BLOCK-1357
DM8560J	BLOCK-1357
DM8560N	BLOCK-1357
DM8563J	BLOCK-1358
DM8563N	BLOCK-1358
DM8570N	BLOCK-1341
DM8590N	BLOCK-1342
DM8601J	BLOCK-2089
DM8601N	BLOCK-2089
DM8602N	BLOCK-2090
DM8613N	BLOCK-1916
DM87	BLOCK-1873
DM87S185J	BLOCK-1754
DM87S185N	BLOCK-1754
DM88	BLOCK-1812
DM8853N	BLOCK-1936
DM8890N	BLOCK-1818
DM90	BLOCK-1997
DM9093N	BLOCK-1950
DM9094N	BLOCK-1951
DM9097N	BLOCK-1952
DM9099N	BLOCK-1953
DM91	BLOCK-2095
DM9301J	BLOCK-1861
DM9301N	BLOCK-1861
DM930N	BLOCK-2197
DM9318J	BLOCK-1867
DM932N	BLOCK-2199
DM933N	BLOCK-2200
DM935N	BLOCK-2201
DM936N	BLOCK-2202
DM937N	BLOCK-2203
DM944N	BLOCK-2206
DM945N	BLOCK-2207
DM946N	BLOCK-2208
DM948N	BLOCK-2209
DM949N	BLOCK-2210
DM958N	BLOCK-1969
DM9601J	BLOCK-2089
DM9601N	BLOCK-2089
DM961N	BLOCK-2216
DM962N	BLOCK-2217
DM963N	BLOCK-2218
DN1946	BLOCK-2208
DN6338-A	BLOCK-685
DN6811	BLOCK-409
DN6838	BLOCK-685
DN6838A	BLOCK-685
DN6851H1	BLOCK-1828
DN6851HI	BLOCK-1828
DN74LS02	BLOCK-1570
DN74LS08	BLOCK-1574
DN74LS09	BLOCK-1575
DN74LS125P	BLOCK-1585
DN74LS132	BLOCK-1588
DN74LS14P	BLOCK-1593
DN74LS156	BLOCK-1601
DN74LS174P	BLOCK-1615
DN74LS240P	BLOCK-1630
DN74LS244	BLOCK-1634
DN74LS245P	BLOCK-1635
DN74LS32	BLOCK-1659
DN74LS373	BLOCK-1671
DN811	BLOCK-2052
DN819	BLOCK-1978
DN838	BLOCK-2054
DN850	BLOCK-2053
DN852	BLOCK-2055
DN852P	BLOCK-2055
DNMPC574J	BLOCK-1157
DNUPC574J	BLOCK-1157
DS1488J	BLOCK-1771
DS1488N	BLOCK-1771
DS1489AJ	BLOCK-1772
DS1489AN	BLOCK-1772
DS1489J	BLOCK-1772
DS1489N	BLOCK-1772
DS75450N	BLOCK-1775
DS75451N	BLOCK-1776
DS75452N	BLOCK-1777
DS75453N	BLOCK-1778
DS75454N	BLOCK-1779
DS75491N	BLOCK-1780
DS75492N	BLOCK-1781
DS75493N	BLOCK-1782
DS75494N	BLOCK-1783
DS8870	BLOCK-950
DS8870J	BLOCK-950
DS8870N	BLOCK-950

DEVICE TYPE	REPL CODE
DS8880F	BLOCK-951
DS8880J	BLOCK-951
DS8880N	BLOCK-951
DS8889J	BLOCK-953
DS8889N	BLOCK-953
DS8T26AJ	BLOCK-1173
DS8T26AN	BLOCK-1173
DS8T28J	BLOCK-1178
DS8T28MJ	BLOCK-1178
DS8T28N	BLOCK-1178
DSD84	BLOCK-532
DTML9093	BLOCK-1950
DTML9099	BLOCK-1953
DTML9930	BLOCK-2197
DTML9932	BLOCK-2199
DTML9933	BLOCK-2200
DTML9935	BLOCK-2201
DTML9936	BLOCK-2202
DTML9944	BLOCK-2206
DTML9945	BLOCK-2207
DTML9946	BLOCK-2208
DTML9948	BLOCK-2209
DTML9949	BLOCK-2210
DTML9961	BLOCK-2216
DTML9962	BLOCK-2217
DTML9963	BLOCK-2218
E-AN7110E	BLOCK-684
E-HA12413	BLOCK-544
E-LA4192	BLOCK-653
E-TA7343P	BLOCK-643
E-TA7343PS	BLOCK-643
E17336	BLOCK-45
E1CM-0060	BLOCK-145
E23036	BLOCK-75
E2495	BLOCK-1283
E25056	BLOCK-75
EA-7316B	BLOCK-973
EA000-13000	BLOCK-1039
EA000-17200	BLOCK-1593
EA000-22600	BLOCK-1671
EA000-44600	BLOCK-1855
EA001-02200	BLOCK-1855
EA001-02700	BLOCK-1004
EA001-02703	BLOCK-1004
EA16X35	BLOCK-4
EA1760	BLOCK-1274
EA3282	BLOCK-4
EA33C8390	BLOCK-454
EA33X8333	BLOCK-288
EA33X8351	BLOCK-26
EA33X8352	BLOCK-19
EA33X8356	BLOCK-6
EA33X8363	BLOCK-169
EA33X8364	BLOCK-1823
EA33X8367	BLOCK-4
EA33X8368	BLOCK-35
EA33X8371	BLOCK-122
EA33X8372	BLOCK-138
EA33X8373	BLOCK-1814
EA33X8374	BLOCK-168
EA33X8375	BLOCK-1825
EA33X8383	BLOCK-288
EA33X8384	BLOCK-181
EA33X8385	BLOCK-1410
EA33X8386	BLOCK-1410
EA33X8388	BLOCK-156
EA33X8389	BLOCK-145
EA33X8392	BLOCK-180
EA33X8394	BLOCK-180
EA33X8395	BLOCK-238
EA33X8396	BLOCK-145
EA33X8399	BLOCK-2208
EA33X8450	BLOCK-973
EA33X8465	BLOCK-974
EA33X8468	BLOCK-2079
EA33X8469	BLOCK-1984
EA33X8479	BLOCK-2058
EA33X8500	BLOCK-98
EA33X8501	BLOCK-259
EA33X8508	BLOCK-238
EA33X8509	BLOCK-180
EA33X8511	BLOCK-1825
EA33X8521	BLOCK-503
EA33X8522	BLOCK-427
EA33X8532	BLOCK-611
EA33X8535	BLOCK-214
EA33X8537	BLOCK-611
EA33X8545	BLOCK-273
EA33X8548	BLOCK-1851
EA33X8559	BLOCK-453
EA33X8564	BLOCK-611
EA33X8569	BLOCK-640
EA33X8572	BLOCK-673
EA33X8583	BLOCK-550

DEVICE TYPE	REPL CODE
EA33X8584	BLOCK-974
EA33X8585	BLOCK-50
EA33X8600	BLOCK-974
EA33X8605	BLOCK-1568
EA33X8607	BLOCK-237
EA33X8705	BLOCK-237
EA33X8706	BLOCK-621
EA33X8707	BLOCK-366
EA33X8717	BLOCK-544
EA6801	BLOCK-1
EA7316A-1	BLOCK-973
EA7316B	BLOCK-973
EA7317B	BLOCK-973
EAO01-00600	BLOCK-1004
EAQ00-02000	BLOCK-1297
EAQ00-05000	BLOCK-1777
EAQ00-07500	BLOCK-1299
EAQ00-07514	BLOCK-1299
EAQ00-11100	BLOCK-2079
EAQ00-11300	BLOCK-1384
EAQ00-1200	BLOCK-1572
EAQ00-12100	BLOCK-1568
EAQ00-12200	BLOCK-1572
EAQ00-12300	BLOCK-1574
EAQ00-12400	BLOCK-1576
EAQ00-12700	BLOCK-1706
EAQ00-12900	BLOCK-1584
EAQ00-12914	BLOCK-1584
EAQ00-13000	BLOCK-1600
EAQ00-13100	BLOCK-1615
EAQ00-15200	BLOCK-1300
EAQ00-17200	BLOCK-1593
EAQ00-17600	BLOCK-1659
EAQ00-18700	BLOCK-1591
EAQ00-19000	BLOCK-1666
EAQ00-19300	BLOCK-1672
EAQ00-19400	BLOCK-1634
EAQ00-20300	BLOCK-1607
EAQ00-21500	BLOCK-1592
EAQ00-22100	BLOCK-1635
EAQ00-22600	BLOCK-1671
EAQ00-22614	BLOCK-1671
EAQ00-22900	BLOCK-1600
EAQ00-23600	BLOCK-1641
EAQ00-24000	BLOCK-1660
EAQ00-24700	BLOCK-1601
EAQ00-30000	BLOCK-1771
EAQ00-30200	BLOCK-1772
EAQ00-30300	BLOCK-1674
EAS00-00700	BLOCK-2087
EAS00-05800	BLOCK-1873
EAS00-06100	BLOCK-1801
EAS00-09000	BLOCK-2058
EAS00-09700	BLOCK-2108
EAS00-12900	BLOCK-2063
EAS00-13500	BLOCK-2097
EAT00-09100	BLOCK-718
EC-A063	BLOCK-718
ECC-01262	BLOCK-2199
ECC-01263	BLOCK-2207
ECC-01264	BLOCK-2208
ECC-01265	BLOCK-2217
ECC-01266	BLOCK-2213
ECC-01267	BLOCK-2197
ECG1000	BLOCK-1
ECG1001	BLOCK-2
ECG1002	BLOCK-3
ECG1003	BLOCK-4
ECG1004	BLOCK-5
ECG1005	BLOCK-6
ECG1006	BLOCK-7
ECG1008	BLOCK-8
ECG1009	BLOCK-9
ECG1010	BLOCK-10
ECG1011	BLOCK-11
ECG1012	BLOCK-12
ECG1013	BLOCK-13
ECG1014	BLOCK-14
ECG1015	BLOCK-15
ECG1016	BLOCK-16
ECG1017	BLOCK-17
ECG1018	BLOCK-18
ECG1019	BLOCK-19
ECG1021	BLOCK-20
ECG1022	BLOCK-21
ECG1023	BLOCK-22
ECG1024	BLOCK-23
ECG1025	BLOCK-24
ECG1026	BLOCK-25
ECG1027	BLOCK-26
ECG1028	BLOCK-27
ECG1029	BLOCK-28
ECG1030	BLOCK-29
ECG1031	BLOCK-30
ECG1032	BLOCK-31
ECG1033	BLOCK-32
ECG1034	BLOCK-33
ECG1035	BLOCK-34
ECG1036	BLOCK-35
ECG1037	BLOCK-36
ECG1038	BLOCK-37
ECG1039	BLOCK-38
ECG1040	BLOCK-39
ECG1041	BLOCK-40
ECG1042	BLOCK-41
ECG1043	BLOCK-42
ECG1045	BLOCK-43
ECG1046	BLOCK-44
ECG1047	BLOCK-45
ECG1048	BLOCK-46
ECG1049	BLOCK-47
ECG1050	BLOCK-48
ECG1051	BLOCK-49
ECG1052	BLOCK-50
ECG1053	BLOCK-51
ECG1054	BLOCK-52
ECG1055	BLOCK-53
ECG1056	BLOCK-54
ECG1057	BLOCK-55
ECG1058	BLOCK-56
ECG1059	BLOCK-57
ECG1060	BLOCK-58
ECG1061	BLOCK-59
ECG1062	BLOCK-60
ECG1063	BLOCK-61
ECG1064	BLOCK-62
ECG1065	BLOCK-63
ECG1066	BLOCK-64
ECG1067	BLOCK-65
ECG1068	BLOCK-66
ECG1069	BLOCK-67
ECG1070	BLOCK-68
ECG1071	BLOCK-69
ECG1072	BLOCK-70
ECG1073	BLOCK-71
ECG1074	BLOCK-72
ECG1075	BLOCK-73
ECG1075A	BLOCK-73
ECG1076	BLOCK-74
ECG1077	BLOCK-75
ECG1078	BLOCK-76
ECG1079	BLOCK-77
ECG1080	BLOCK-78
ECG1081	BLOCK-79
ECG1081A	BLOCK-79
ECG1082	BLOCK-80
ECG1083	BLOCK-81
ECG1084	BLOCK-82
ECG1085	BLOCK-83
ECG1086	BLOCK-84
ECG1087	BLOCK-85
ECG1088	BLOCK-86
ECG1089	BLOCK-87
ECG1090	BLOCK-88
ECG1091	BLOCK-89
ECG1092	BLOCK-90
ECG1093	BLOCK-91
ECG1094	BLOCK-92
ECG1095	BLOCK-93
ECG1096	BLOCK-94
ECG1097	BLOCK-95
ECG1098	BLOCK-96
ECG1099	BLOCK-97
ECG1100	BLOCK-98
ECG1101	BLOCK-99
ECG1102	BLOCK-100
ECG1103	BLOCK-101
ECG1104	BLOCK-102
ECG1105	BLOCK-103
ECG1106	BLOCK-104
ECG1107	BLOCK-105
ECG1108	BLOCK-106
ECG1109	BLOCK-107
ECG1110	BLOCK-108
ECG1111	BLOCK-109
ECG1112	BLOCK-110
ECG1113	BLOCK-111
ECG1115	BLOCK-112
ECG1115A	BLOCK-113
ECG1116	BLOCK-114
ECG1117	BLOCK-115
ECG1118	BLOCK-116
ECG1119	BLOCK-117
ECG1120	BLOCK-118
ECG1121	BLOCK-119
ECG1122	BLOCK-120
ECG1123	BLOCK-121
ECG1124	BLOCK-122
ECG1125	BLOCK-123
ECG1126	BLOCK-124
ECG1127	BLOCK-125
ECG1128	BLOCK-126
ECG1130	BLOCK-127
ECG1131	BLOCK-128
ECG1132	BLOCK-129
ECG1133	BLOCK-130
ECG1134	BLOCK-131
ECG1135	BLOCK-132
ECG1136	BLOCK-133
ECG1137	BLOCK-134
ECG1139	BLOCK-135
ECG1140	BLOCK-136
ECG1141	BLOCK-137
ECG1142	BLOCK-138
ECG1148	BLOCK-139
ECG1149	BLOCK-140
ECG1150	BLOCK-141
ECG1152	BLOCK-142
ECG1153	BLOCK-143
ECG1154	BLOCK-144
ECG1155	BLOCK-145
ECG1156	BLOCK-146
ECG1158	BLOCK-147
ECG1159	BLOCK-148
ECG1160	BLOCK-149
ECG1161	BLOCK-150
ECG1162	BLOCK-151
ECG1163	BLOCK-152
ECG1164	BLOCK-153
ECG1165	BLOCK-154
ECG1166	BLOCK-155
ECG1167	BLOCK-156
ECG1168	BLOCK-157
ECG1169	BLOCK-158
ECG1170	BLOCK-159
ECG1171	BLOCK-160
ECG1172	BLOCK-161
ECG1173	BLOCK-162
ECG1174	BLOCK-163
ECG1175	BLOCK-164
ECG1176	BLOCK-165
ECG1177	BLOCK-166
ECG1178	BLOCK-167
ECG1179	BLOCK-168
ECG1180	BLOCK-169
ECG1181	BLOCK-170
ECG1182	BLOCK-171
ECG1183	BLOCK-172
ECG1184	BLOCK-173
ECG1185	BLOCK-174
ECG1186	BLOCK-175
ECG1187	BLOCK-176
ECG1188	BLOCK-177
ECG1189	BLOCK-178
ECG1191	BLOCK-179
ECG1192	BLOCK-180
ECG1193	BLOCK-181
ECG1194	BLOCK-182
ECG1195	BLOCK-183
ECG1196	BLOCK-184
ECG1197	BLOCK-185
ECG1198	BLOCK-186
ECG1199	BLOCK-187
ECG1200	BLOCK-188
ECG1201	BLOCK-189
ECG1202	BLOCK-190
ECG1203	BLOCK-191
ECG1204	BLOCK-192
ECG1205	BLOCK-193
ECG1206	BLOCK-194
ECG1207	BLOCK-195
ECG1208	BLOCK-196
ECG1209	BLOCK-197
ECG1210	BLOCK-198
ECG1211	BLOCK-199
ECG1212	BLOCK-200
ECG1213	BLOCK-201
ECG1214	BLOCK-202
ECG1215	BLOCK-203
ECG1216	BLOCK-204
ECG1217	BLOCK-205
ECG1218	BLOCK-206
ECG1219	BLOCK-207
ECG1220	BLOCK-208
ECG1221	BLOCK-209
ECG1222	BLOCK-210
ECG1223	BLOCK-211
ECG1224	BLOCK-212
ECG1225	BLOCK-213
ECG1226	BLOCK-214
ECG1227	BLOCK-215
ECG1228	BLOCK-216
ECG1229	BLOCK-217
ECG1230	BLOCK-218
ECG1231	BLOCK-219
ECG1231A	BLOCK-220
ECG1232	BLOCK-221
ECG1233	BLOCK-222
ECG1234	BLOCK-223
ECG1235	BLOCK-224
ECG1236	BLOCK-225
ECG1237	BLOCK-226
ECG1238	BLOCK-227
ECG1239	BLOCK-228
ECG1240	BLOCK-229
ECG1241	BLOCK-230
ECG1242	BLOCK-231
ECG1243	BLOCK-232
ECG1244	BLOCK-233
ECG1245	BLOCK-234
ECG1246	BLOCK-235
ECG1247	BLOCK-236
ECG1248	BLOCK-237
ECG1249	BLOCK-238
ECG1250	BLOCK-239
ECG1251	BLOCK-240
ECG1252	BLOCK-241
ECG1253	BLOCK-242
ECG1254	BLOCK-243
ECG1255	BLOCK-244
ECG1256	BLOCK-245
ECG1257	BLOCK-246
ECG1258	BLOCK-247
ECG1261	BLOCK-248
ECG1262	BLOCK-249
ECG1263	BLOCK-250
ECG1264	BLOCK-251
ECG1265	BLOCK-252
ECG1266	BLOCK-253
ECG1267	BLOCK-254
ECG1268	BLOCK-255
ECG1269	BLOCK-256
ECG1270	BLOCK-257
ECG1271	BLOCK-258
ECG1272	BLOCK-259
ECG1273	BLOCK-260
ECG1274	BLOCK-261
ECG1275	BLOCK-262
ECG1276	BLOCK-263
ECG1277	BLOCK-264
ECG1278	BLOCK-265
ECG1279	BLOCK-266
ECG1280	BLOCK-267
ECG1281	BLOCK-268
ECG1282	BLOCK-269
ECG1283	BLOCK-270
ECG1284	BLOCK-271
ECG1285	BLOCK-272
ECG1286	BLOCK-273
ECG1287	BLOCK-274
ECG1288	BLOCK-275
ECG1289	BLOCK-276
ECG1290	BLOCK-277
ECG1291	BLOCK-278
ECG1292	BLOCK-279
ECG1293	BLOCK-280
ECG1294	BLOCK-281
ECG1295	BLOCK-282
ECG1296	BLOCK-283
ECG1297	BLOCK-284
ECG1298	BLOCK-285
ECG1299	BLOCK-286
ECG1300	BLOCK-287
ECG1301	BLOCK-288
ECG1302	BLOCK-289
ECG1303	BLOCK-290
ECG1304	BLOCK-291
ECG1305	BLOCK-292
ECG1306	BLOCK-293
ECG1307	BLOCK-294
ECG1308	BLOCK-295
ECG1309	BLOCK-296
ECG1310	BLOCK-297
ECG1311	BLOCK-298
ECG1312	BLOCK-299
ECG1313	BLOCK-300
ECG1314	BLOCK-301
ECG1315	BLOCK-302
ECG1316	BLOCK-303
ECG1317	BLOCK-304
ECG1318	BLOCK-305
ECG1319	BLOCK-306
ECG1320	BLOCK-307
ECG1321	BLOCK-308
ECG1322	BLOCK-309
ECG1323	BLOCK-310
ECG1324	BLOCK-311
ECG1325	BLOCK-312
ECG1326	BLOCK-313
ECG1327	BLOCK-314
ECG1328	BLOCK-315
ECG1329	BLOCK-316
ECG1330	BLOCK-317
ECG1331	BLOCK-318
ECG1332	BLOCK-319
ECG1333	BLOCK-320
ECG1334	BLOCK-321
ECG1335	BLOCK-322
ECG1336	BLOCK-323
ECG1337	BLOCK-324
ECG1338	BLOCK-325
ECG1339	BLOCK-326
ECG1340	BLOCK-327
ECG1341	BLOCK-328
ECG1342	BLOCK-329
ECG1343	BLOCK-330
ECG1344	BLOCK-331
ECG1345	BLOCK-332
ECG1346	BLOCK-333
ECG1347	BLOCK-334
ECG1348	BLOCK-335
ECG1349	BLOCK-336
ECG1350	BLOCK-337
ECG1351	BLOCK-338
ECG1352	BLOCK-339
ECG1353	BLOCK-340
ECG1354	BLOCK-341
ECG1355	BLOCK-342
ECG1356	BLOCK-343
ECG1357	BLOCK-344
ECG1358	BLOCK-345
ECG1359	BLOCK-346
ECG1360	BLOCK-347
ECG1361	BLOCK-348
ECG1362	BLOCK-349
ECG1363	BLOCK-350
ECG1364	BLOCK-351
ECG1365	BLOCK-352
ECG1366	BLOCK-353
ECG1367	BLOCK-354
ECG1368	BLOCK-355
ECG1369	BLOCK-356
ECG1370	BLOCK-357
ECG1371	BLOCK-358
ECG1372	BLOCK-359
ECG1373	BLOCK-360
ECG1374	BLOCK-361
ECG1375	BLOCK-362
ECG1376	BLOCK-363
ECG1377	BLOCK-364
ECG1378	BLOCK-365
ECG1379	BLOCK-366
ECG1380	BLOCK-367
ECG1381	BLOCK-368
ECG1382	BLOCK-369
ECG1383	BLOCK-370
ECG1384	BLOCK-371
ECG1385	BLOCK-372
ECG1386	BLOCK-373
ECG1387	BLOCK-374
ECG1388	BLOCK-375
ECG1389	BLOCK-376
ECG1390	BLOCK-377
ECG1391	BLOCK-378
ECG1392	BLOCK-379
ECG1393	BLOCK-380
ECG1394	BLOCK-381
ECG1395	BLOCK-382
ECG1396	BLOCK-383
ECG1397	BLOCK-384
ECG1398	BLOCK-385
ECG1399	BLOCK-386
ECG1400	BLOCK-387
ECG1401	BLOCK-388
ECG1402	BLOCK-389
ECG1403	BLOCK-390
ECG1404	BLOCK-391
ECG1405	BLOCK-392
ECG1406	BLOCK-393
ECG1407	BLOCK-394
ECG1408	BLOCK-395
ECG1409	BLOCK-396
ECG1409C	BLOCK-397
ECG1410	BLOCK-398
ECG1411	BLOCK-399
ECG1412	BLOCK-400
ECG1413	BLOCK-401
ECG1414	BLOCK-402
ECG1415	BLOCK-403
ECG1416	BLOCK-404
ECG1417	BLOCK-405
ECG1418	BLOCK-406
ECG1419	BLOCK-407
ECG1420	BLOCK-408
ECG1421	BLOCK-409
ECG1422	BLOCK-410
ECG1423	BLOCK-411
ECG1424	BLOCK-412
ECG1425	BLOCK-413
ECG1426	BLOCK-414
ECG1427	BLOCK-415
ECG1428	BLOCK-416
ECG1429	BLOCK-417
ECG1430	BLOCK-418
ECG1431	BLOCK-419
ECG1432	BLOCK-420
ECG1433	BLOCK-421
ECG1434	BLOCK-422
ECG1435	BLOCK-423
ECG1436	BLOCK-424
ECG1437	BLOCK-425
ECG1438	BLOCK-426
ECG1439	BLOCK-427
ECG1440	BLOCK-428
ECG1441	BLOCK-429
ECG1442	BLOCK-430
ECG1443	BLOCK-431
ECG1444	BLOCK-432
ECG1445	BLOCK-433
ECG1446	BLOCK-434
ECG1447	BLOCK-435
ECG1448	BLOCK-436
ECG1449	BLOCK-437
ECG1450	BLOCK-438
ECG1451	BLOCK-439
ECG1452	BLOCK-440
ECG1453	BLOCK-441
ECG1454	BLOCK-442
ECG1455	BLOCK-443
ECG1456	BLOCK-444
ECG1457	BLOCK-445
ECG1458	BLOCK-446
ECG1459	BLOCK-447
ECG1460	BLOCK-448
ECG1461	BLOCK-449
ECG1462	BLOCK-450
ECG1463	BLOCK-451
ECG1464	BLOCK-452
ECG1465	BLOCK-453
ECG1466	BLOCK-454
ECG1467	BLOCK-455
ECG1468	BLOCK-456
ECG1469	BLOCK-457
ECG1470	BLOCK-458
ECG1471	BLOCK-459
ECG1472	BLOCK-460
ECG1473	BLOCK-461
ECG1474	BLOCK-462
ECG1475	BLOCK-463
ECG1476	BLOCK-464
ECG1477	BLOCK-465
ECG1478	BLOCK-466
ECG1479	BLOCK-467
ECG1480	BLOCK-468
ECG1481	BLOCK-469
ECG1482	BLOCK-470
ECG1483	BLOCK-471
ECG1484	BLOCK-472
ECG1485	BLOCK-473
ECG1486	BLOCK-474
ECG1487	BLOCK-475
ECG1488	BLOCK-476
ECG1489	BLOCK-477
ECG1490	BLOCK-478
ECG1491	BLOCK-479
ECG1492	BLOCK-480
ECG1493	BLOCK-481
ECG1494	BLOCK-482
ECG1495	BLOCK-483
ECG1496	BLOCK-484
ECG1497	BLOCK-485
ECG1498	BLOCK-486
ECG1499	BLOCK-487
ECG1500	BLOCK-488
ECG1501	BLOCK-489
ECG1502	BLOCK-490
ECG1503	BLOCK-491
ECG1504	BLOCK-492
ECG1505	BLOCK-493
ECG1506	BLOCK-494
ECG1507	BLOCK-495
ECG1508	BLOCK-496
ECG1509	BLOCK-497
ECG1510	BLOCK-498

ORIGINAL DEVICE TYPES AND REPLACEMENT CODES 25

DEVICE TYPE	REPL CODE
ECG1511	BLOCK-499
ECG1512	BLOCK-500
ECG1513	BLOCK-501
ECG1514	BLOCK-502
ECG1515	BLOCK-503
ECG1516	BLOCK-504
ECG1517	BLOCK-505
ECG1518	BLOCK-506
ECG1519	BLOCK-507
ECG1520	BLOCK-508
ECG1521	BLOCK-509
ECG1522	BLOCK-510
ECG1523	BLOCK-511
ECG1525	BLOCK-512
ECG1526	BLOCK-513
ECG1527	BLOCK-514
ECG1528	BLOCK-515
ECG1529	BLOCK-516
ECG1530	BLOCK-517
ECG1531	BLOCK-518
ECG1532	BLOCK-519
ECG1533	BLOCK-520
ECG1534	BLOCK-521
ECG1535	BLOCK-522
ECG1536	BLOCK-523
ECG1537	BLOCK-524
ECG1538	BLOCK-525
ECG1539	BLOCK-526
ECG1540	BLOCK-527
ECG1541	BLOCK-528
ECG1542	BLOCK-529
ECG1543	BLOCK-530
ECG1544	BLOCK-531
ECG1545	BLOCK-532
ECG1546	BLOCK-533
ECG1547	BLOCK-534
ECG1548	BLOCK-535
ECG1549	BLOCK-536
ECG1550	BLOCK-537
ECG1551	BLOCK-538
ECG1552	BLOCK-539
ECG1553	BLOCK-540
ECG1554	BLOCK-541
ECG1555	BLOCK-542
ECG1556	BLOCK-543
ECG1557	BLOCK-544
ECG1558	BLOCK-545
ECG1559	BLOCK-546
ECG1560	BLOCK-547
ECG1561	BLOCK-548
ECG1562	BLOCK-549
ECG1563	BLOCK-550
ECG1564	BLOCK-551
ECG1565	BLOCK-552
ECG1566	BLOCK-553
ECG1567	BLOCK-554
ECG1568	BLOCK-555
ECG1569	BLOCK-556
ECG1570	BLOCK-557
ECG1571	BLOCK-558
ECG1572	BLOCK-559
ECG1573	BLOCK-560
ECG1574	BLOCK-561
ECG1575	BLOCK-562
ECG1576	BLOCK-563
ECG1577	BLOCK-564
ECG1578	BLOCK-565
ECG1579	BLOCK-566
ECG1580	BLOCK-567
ECG1581	BLOCK-568
ECG1582	BLOCK-569
ECG1583	BLOCK-570
ECG1584	BLOCK-571
ECG1585	BLOCK-572
ECG1586	BLOCK-573
ECG1587	BLOCK-574
ECG1588	BLOCK-575
ECG1589	BLOCK-576
ECG1590	BLOCK-577
ECG1591	BLOCK-578
ECG1592	BLOCK-579
ECG1593	BLOCK-580
ECG1594	BLOCK-581
ECG1595	BLOCK-582
ECG1596	BLOCK-583
ECG1597	BLOCK-584
ECG1598	BLOCK-585
ECG1599	BLOCK-586
ECG1600	BLOCK-587
ECG1601	BLOCK-588
ECG1602	BLOCK-589
ECG1603	BLOCK-590
ECG1604	BLOCK-591
ECG1605	BLOCK-592
ECG1606	BLOCK-593
ECG1607	BLOCK-594
ECG1608	BLOCK-595
ECG1609	BLOCK-596
ECG1610	BLOCK-597
ECG1611	BLOCK-598
ECG1612	BLOCK-599
ECG1613	BLOCK-600
ECG1614	BLOCK-601
ECG1615	BLOCK-602
ECG1616	BLOCK-603
ECG1617	BLOCK-604
ECG1618	BLOCK-605
ECG1619	BLOCK-606
ECG1620	BLOCK-607
ECG1621	BLOCK-608
ECG1622	BLOCK-609
ECG1623	BLOCK-610
ECG1624	BLOCK-611
ECG1625	BLOCK-612
ECG1626	BLOCK-613
ECG1627	BLOCK-614
ECG1628	BLOCK-615
ECG1629	BLOCK-616
ECG1630	BLOCK-617
ECG1631	BLOCK-618
ECG1632	BLOCK-619
ECG1633	BLOCK-620
ECG1634	BLOCK-621
ECG1635	BLOCK-622
ECG1636	BLOCK-623
ECG1637	BLOCK-624
ECG1638	BLOCK-625
ECG1639	BLOCK-626
ECG1640	BLOCK-627
ECG1641	BLOCK-628
ECG1642	BLOCK-629
ECG1644	BLOCK-630
ECG1645	BLOCK-631
ECG1646	BLOCK-632
ECG1647	BLOCK-633
ECG1648	BLOCK-634
ECG1649	BLOCK-635
ECG1650	BLOCK-636
ECG1651	BLOCK-637
ECG1652	BLOCK-638
ECG1653	BLOCK-639
ECG1654	BLOCK-640
ECG1655	BLOCK-641
ECG1656	BLOCK-642
ECG1657	BLOCK-643
ECG1658	BLOCK-644
ECG1659	BLOCK-645
ECG1660	BLOCK-646
ECG1661	BLOCK-647
ECG1662	BLOCK-648
ECG1663	BLOCK-649
ECG1664	BLOCK-650
ECG1665	BLOCK-651
ECG1666	BLOCK-652
ECG1667	BLOCK-653
ECG1668	BLOCK-654
ECG1669	BLOCK-655
ECG1670	BLOCK-656
ECG1671	BLOCK-657
ECG1672	BLOCK-658
ECG1673	BLOCK-659
ECG1674	BLOCK-660
ECG1675	BLOCK-661
ECG1676	BLOCK-662
ECG1677	BLOCK-663
ECG1678	BLOCK-664
ECG1679	BLOCK-665
ECG1680	BLOCK-666
ECG1681	BLOCK-667
ECG1682	BLOCK-668
ECG1683	BLOCK-669
ECG1684	BLOCK-670
ECG1685	BLOCK-671
ECG1686	BLOCK-672
ECG1687	BLOCK-673
ECG1688	BLOCK-674
ECG1689	BLOCK-675
ECG1690	BLOCK-676
ECG1691	BLOCK-677
ECG1692	BLOCK-678
ECG1693	BLOCK-679
ECG1700	BLOCK-680
ECG1701	BLOCK-681
ECG1702	BLOCK-682
ECG1703	BLOCK-683
ECG1704	BLOCK-684
ECG1705	BLOCK-685
ECG1706	BLOCK-686
ECG1707	BLOCK-687
ECG1708	BLOCK-688
ECG1709	BLOCK-689
ECG1710	BLOCK-690
ECG1711	BLOCK-691
ECG1712	BLOCK-692
ECG1713	BLOCK-693
ECG1714M	BLOCK-694
ECG1714S	BLOCK-695
ECG1715	BLOCK-696
ECG1716	BLOCK-697
ECG1717	BLOCK-698
ECG1718	BLOCK-699
ECG1719	BLOCK-700
ECG1720	BLOCK-701
ECG1721	BLOCK-702
ECG1722	BLOCK-703
ECG1723	BLOCK-704
ECG1724	BLOCK-705
ECG1725	BLOCK-706
ECG1726	BLOCK-707
ECG1727	BLOCK-708
ECG1728	BLOCK-709
ECG1729	BLOCK-710
ECG1730	BLOCK-711
ECG1731	BLOCK-712
ECG1732	BLOCK-713
ECG1733	BLOCK-714
ECG1734	BLOCK-715
ECG1735	BLOCK-716
ECG1736	BLOCK-717
ECG1737	BLOCK-718
ECG1738	BLOCK-719
ECG1739	BLOCK-720
ECG1740	BLOCK-721
ECG1741	BLOCK-722
ECG1742	BLOCK-723
ECG1743	BLOCK-724
ECG1744	BLOCK-725
ECG1745	BLOCK-726
ECG1746	BLOCK-727
ECG1747	BLOCK-728
ECG1748	BLOCK-729
ECG1749	BLOCK-730
ECG1750	BLOCK-731
ECG1751	BLOCK-732
ECG1752	BLOCK-733
ECG1753	BLOCK-734
ECG1754	BLOCK-735
ECG1755	BLOCK-736
ECG1756	BLOCK-737
ECG1757	BLOCK-738
ECG1758	BLOCK-739
ECG1759	BLOCK-740
ECG1760	BLOCK-741
ECG1761	BLOCK-742
ECG1762	BLOCK-743
ECG1763	BLOCK-744
ECG1764	BLOCK-745
ECG1765	BLOCK-746
ECG1766	BLOCK-747
ECG1767	BLOCK-748
ECG1768	BLOCK-749
ECG1769	BLOCK-750
ECG1770	BLOCK-751
ECG1771	BLOCK-752
ECG1772	BLOCK-753
ECG1773	BLOCK-754
ECG1774	BLOCK-755
ECG1775	BLOCK-756
ECG1776	BLOCK-757
ECG1777	BLOCK-758
ECG1778	BLOCK-759
ECG1779	BLOCK-760
ECG1780	BLOCK-761
ECG1781	BLOCK-762
ECG1782	BLOCK-763
ECG1783	BLOCK-764
ECG1784	BLOCK-765
ECG1785	BLOCK-766
ECG1786	BLOCK-767
ECG1787	BLOCK-768
ECG1788	BLOCK-769
ECG1789	BLOCK-770
ECG1790	BLOCK-771
ECG1791	BLOCK-772
ECG1792	BLOCK-773
ECG1793	BLOCK-774
ECG1794	BLOCK-775
ECG1795	BLOCK-776
ECG1796	BLOCK-777
ECG1797	BLOCK-778
ECG1798	BLOCK-779
ECG1799	BLOCK-780
ECG1800	BLOCK-781
ECG1801	BLOCK-782
ECG1802	BLOCK-783
ECG1803	BLOCK-784
ECG1804	BLOCK-785
ECG1805	BLOCK-786
ECG1806	BLOCK-787
ECG1807	BLOCK-788
ECG1808	BLOCK-789
ECG1809	BLOCK-790
ECG1810	BLOCK-791
ECG1811	BLOCK-792
ECG1812	BLOCK-793
ECG1813	BLOCK-794
ECG1814	BLOCK-795
ECG1815	BLOCK-796
ECG1816	BLOCK-797
ECG1817	BLOCK-798
ECG1818	BLOCK-799
ECG1819	BLOCK-800
ECG1820	BLOCK-801
ECG1821	BLOCK-802
ECG1822	BLOCK-803
ECG1823	BLOCK-804
ECG1824	BLOCK-805
ECG1825	BLOCK-806
ECG1827	BLOCK-807
ECG1828	BLOCK-808
ECG1829	BLOCK-809
ECG1830	BLOCK-810
ECG1831	BLOCK-811
ECG1832	BLOCK-812
ECG1834	BLOCK-813
ECG1835	BLOCK-814
ECG1836	BLOCK-815
ECG1837	BLOCK-816
ECG1838	BLOCK-817
ECG1839	BLOCK-818
ECG1840	BLOCK-819
ECG1841	BLOCK-820
ECG1842	BLOCK-821
ECG1843	BLOCK-822
ECG1844	BLOCK-823
ECG1845	BLOCK-824
ECG1846	BLOCK-825
ECG1847	BLOCK-826
ECG1848	BLOCK-827
ECG1849	BLOCK-828
ECG1850	BLOCK-829
ECG1851	BLOCK-830
ECG1852	BLOCK-831
ECG1853	BLOCK-832
ECG1854D	BLOCK-833
ECG1854M	BLOCK-834
ECG1855	BLOCK-835
ECG1856	BLOCK-836
ECG1857	BLOCK-837
ECG1858	BLOCK-838
ECG1859	BLOCK-839
ECG1860	BLOCK-840
ECG1861	BLOCK-841
ECG1862	BLOCK-842
ECG1863	BLOCK-843
ECG1864	BLOCK-844
ECG1865	BLOCK-845
ECG1866	BLOCK-846
ECG1867	BLOCK-847
ECG1868	BLOCK-848
ECG1869	BLOCK-849
ECG1869SM	BLOCK-850
ECG1870	BLOCK-851
ECG1872	BLOCK-852
ECG1873	BLOCK-853
ECG1874	BLOCK-854
ECG1875	BLOCK-855
ECG1876	BLOCK-856
ECG1877	BLOCK-857
ECG1878	BLOCK-858
ECG1879	BLOCK-859
ECG1880	BLOCK-860
ECG1881	BLOCK-861
ECG1882	BLOCK-862
ECG1883	BLOCK-863
ECG1884	BLOCK-864
ECG1885	BLOCK-865
ECG1886	BLOCK-866
ECG1887	BLOCK-867
ECG1888	BLOCK-868
ECG1889	BLOCK-869
ECG1890	BLOCK-870
ECG1892	BLOCK-871
ECG1893	BLOCK-872
ECG1894	BLOCK-873
ECG1895	BLOCK-874
ECG1896	BLOCK-875
ECG1897	BLOCK-876
ECG1898	BLOCK-877
ECG1899	BLOCK-878
ECG1900	BLOCK-879
ECG1901	BLOCK-880
ECG1902	BLOCK-881
ECG1903	BLOCK-882
ECG1905	BLOCK-883
ECG1906	BLOCK-884
ECG1907	BLOCK-885
ECG1908	BLOCK-886
ECG1911	BLOCK-887
ECG1912	BLOCK-888
ECG1913	BLOCK-889
ECG1914	BLOCK-890
ECG1915	BLOCK-891
ECG1916	BLOCK-892
ECG1917	BLOCK-893
ECG1918	BLOCK-894
ECG1919	BLOCK-895
ECG1920	BLOCK-896
ECG1923	BLOCK-897
ECG1924	BLOCK-898
ECG1925	BLOCK-899
ECG1926	BLOCK-900
ECG1927	BLOCK-901
ECG1928	BLOCK-902
ECG1930	BLOCK-903
ECG1932	BLOCK-904
ECG1934	BLOCK-905
ECG1934X	BLOCK-906
ECG1936	BLOCK-907
ECG1938	BLOCK-908
ECG1940	BLOCK-909
ECG1941	BLOCK-910
ECG1942	BLOCK-911
ECG1960	BLOCK-912
ECG1961	BLOCK-913
ECG1962	BLOCK-914
ECG1963	BLOCK-915
ECG1964	BLOCK-916
ECG1965	BLOCK-917
ECG1966	BLOCK-918
ECG1967	BLOCK-919
ECG1968	BLOCK-920
ECG1970	BLOCK-921
ECG1971	BLOCK-922
ECG1972	BLOCK-923
ECG1973	BLOCK-924
ECG1974	BLOCK-925
ECG1975	BLOCK-926
ECG1976	BLOCK-927
ECG1977	BLOCK-928
ECG2000	BLOCK-929
ECG2001	BLOCK-930
ECG2002	BLOCK-931
ECG2003	BLOCK-932
ECG2004	BLOCK-933
ECG2011	BLOCK-934
ECG2012	BLOCK-935
ECG2013	BLOCK-936
ECG2014	BLOCK-937
ECG2015	BLOCK-938
ECG2016	BLOCK-939
ECG2017	BLOCK-940
ECG2018	BLOCK-941
ECG2019	BLOCK-942
ECG2020	BLOCK-943
ECG2021	BLOCK-944
ECG2022	BLOCK-945
ECG2023	BLOCK-946
ECG2024	BLOCK-947
ECG2025	BLOCK-948
ECG2026	BLOCK-949
ECG2027	BLOCK-950
ECG2028	BLOCK-951
ECG2029	BLOCK-952
ECG2030	BLOCK-953
ECG2031	BLOCK-954
ECG2032	BLOCK-955
ECG2033	BLOCK-956
ECG2040	BLOCK-957
ECG2041	BLOCK-958
ECG2042	BLOCK-959
ECG2043	BLOCK-960
ECG2045	BLOCK-961
ECG2046	BLOCK-962
ECG2047	BLOCK-963
ECG2049	BLOCK-964
ECG2050	BLOCK-965
ECG2051	BLOCK-966
ECG2052	BLOCK-967
ECG2053	BLOCK-968
ECG2054	BLOCK-969
ECG2055	BLOCK-970
ECG2056	BLOCK-971
ECG2057	BLOCK-972
ECG2060	BLOCK-973
ECG2061	BLOCK-974
ECG2062	BLOCK-975
ECG2063	BLOCK-976
ECG2064	BLOCK-977
ECG2065	BLOCK-978
ECG2070	BLOCK-979
ECG2071	BLOCK-980
ECG2072	BLOCK-981
ECG2073	BLOCK-982
ECG2074	BLOCK-983
ECG2075	BLOCK-984
ECG2076	BLOCK-985
ECG2077	BLOCK-986
ECG2078	BLOCK-987
ECG2079	BLOCK-988
ECG2080	BLOCK-989
ECG2081	BLOCK-990
ECG2082	BLOCK-991
ECG2083	BLOCK-992
ECG2084	BLOCK-993
ECG2085	BLOCK-994
ECG2086	BLOCK-995
ECG2087	BLOCK-996
ECG2088	BLOCK-997
ECG2090	BLOCK-998
ECG2102	BLOCK-999
ECG2104	BLOCK-1000
ECG2107	BLOCK-1001
ECG2114	BLOCK-1002
ECG2117	BLOCK-1003
ECG2128	BLOCK-1004
ECG2147	BLOCK-1005
ECG2164	BLOCK-1006
ECG2200	BLOCK-1007
ECG2201	BLOCK-1008
ECG2202	BLOCK-1009
ECG2203	BLOCK-1010
ECG2204	BLOCK-1011
ECG2205	BLOCK-1012
ECG2206	BLOCK-1013
ECG2207	BLOCK-1014
ECG2532	BLOCK-1015
ECG2708	BLOCK-1016
ECG2716	BLOCK-1017
ECG2732	BLOCK-1018
ECG2764	BLOCK-1019
ECG2800	BLOCK-1020
ECG3092	BLOCK-1021
ECG309K	BLOCK-1022
ECG3539	BLOCK-1023
ECG370	BLOCK-1024
ECG370A	BLOCK-1024
ECG3880	BLOCK-1025
ECG3881	BLOCK-1026
ECG3882	BLOCK-1027
ECG4000	BLOCK-1028
ECG4001	BLOCK-1029
ECG4001B	BLOCK-1029
ECG4002	BLOCK-1030
ECG4002B	BLOCK-1030
ECG4006B	BLOCK-1031
ECG4007	BLOCK-1032
ECG4007B	BLOCK-1032
ECG40085	BLOCK-1033
ECG40085B	BLOCK-1033
ECG4008B	BLOCK-1034
ECG40097	BLOCK-1035
ECG40097B	BLOCK-1035
ECG40098	BLOCK-1036
ECG40098B	BLOCK-1036
ECG40100B	BLOCK-1037
ECG40106B	BLOCK-1038
ECG4011	BLOCK-1039
ECG4011B	BLOCK-1039
ECG4012	BLOCK-1040
ECG4012B	BLOCK-1040
ECG4013	BLOCK-1041
ECG4013B	BLOCK-1041
ECG4014	BLOCK-1042
ECG4014B	BLOCK-1042
ECG4015	BLOCK-1043
ECG4015B	BLOCK-1043
ECG4016	BLOCK-1048
ECG40160	BLOCK-1044
ECG40160B	BLOCK-1044
ECG40161	BLOCK-1045
ECG40161B	BLOCK-1045
ECG40162	BLOCK-1046
ECG40162B	BLOCK-1046

26 ORIGINAL DEVICE TYPES AND REPLACEMENT CODES

DEVICE TYPE	REPL CODE
ECG40163	BLOCK-1047
ECG40163B	BLOCK-1047
ECG4016B	BLOCK-1048
ECG4017	BLOCK-1051
ECG40174	BLOCK-1049
ECG40174B	BLOCK-1049
ECG40175B	BLOCK-1050
ECG4017B	BLOCK-1051
ECG40182B	BLOCK-1052
ECG4018B	BLOCK-1053
ECG4019	BLOCK-1058
ECG40192	BLOCK-1054
ECG40192B	BLOCK-1054
ECG40193	BLOCK-1055
ECG40193B	BLOCK-1055
ECG40194	BLOCK-1056
ECG40194B	BLOCK-1056
ECG40195	BLOCK-1057
ECG40195B	BLOCK-1057
ECG4019B	BLOCK-1058
ECG4020	BLOCK-1059
ECG4020B	BLOCK-1059
ECG4021	BLOCK-1060
ECG4021B	BLOCK-1060
ECG4022B	BLOCK-1061
ECG4023	BLOCK-1062
ECG4023B	BLOCK-1062
ECG4024	BLOCK-1063
ECG4024B	BLOCK-1063
ECG4025	BLOCK-1064
ECG4025B	BLOCK-1064
ECG4026B	BLOCK-1065
ECG4027	BLOCK-1066
ECG4027B	BLOCK-1066
ECG4028	BLOCK-1067
ECG4028B	BLOCK-1067
ECG4029	BLOCK-1068
ECG4029B	BLOCK-1068
ECG4030	BLOCK-1069
ECG4030B	BLOCK-1069
ECG4031B	BLOCK-1070
ECG4032B	BLOCK-1071
ECG4033B	BLOCK-1072
ECG4034B	BLOCK-1073
ECG4035B	BLOCK-1074
ECG4038B	BLOCK-1075
ECG4040	BLOCK-1076
ECG4040B	BLOCK-1076
ECG4042	BLOCK-1077
ECG4042B	BLOCK-1077
ECG4043B	BLOCK-1078
ECG4044B	BLOCK-1079
ECG4045B	BLOCK-1080
ECG4046B	BLOCK-1081
ECG4047B	BLOCK-1082
ECG4048B	BLOCK-1083
ECG4049	BLOCK-1084
ECG4049B	BLOCK-1084
ECG4050	BLOCK-1085
ECG4050B	BLOCK-1085
ECG4051	BLOCK-1086
ECG4051B	BLOCK-1086
ECG4052	BLOCK-1087
ECG4052B	BLOCK-1087
ECG4053B	BLOCK-1088
ECG4055	BLOCK-1089
ECG4055B	BLOCK-1089
ECG4056B	BLOCK-1090
ECG4060B	BLOCK-1091
ECG4063B	BLOCK-1092
ECG4066B	BLOCK-1093
ECG4067B	BLOCK-1094
ECG4068	BLOCK-1095
ECG4068B	BLOCK-1095
ECG4069	BLOCK-1096
ECG4070B	BLOCK-1097
ECG4071	BLOCK-1098
ECG4071B	BLOCK-1098
ECG4072B	BLOCK-1099
ECG4073B	BLOCK-1100
ECG4075B	BLOCK-1101
ECG4076B	BLOCK-1102
ECG4077	BLOCK-1103
ECG4077B	BLOCK-1103
ECG4078	BLOCK-1104
ECG4078B	BLOCK-1104
ECG4081	BLOCK-1105
ECG4081B	BLOCK-1105
ECG4082B	BLOCK-1106
ECG4085	BLOCK-1107
ECG4085B	BLOCK-1107
ECG4086	BLOCK-1108
ECG4086B	BLOCK-1108
ECG4089B	BLOCK-1109

DEVICE TYPE	REPL CODE
ECG4093B	BLOCK-1110
ECG4095B	BLOCK-1111
ECG4096B	BLOCK-1112
ECG4097B	BLOCK-1113
ECG4098B	BLOCK-1114
ECG4099	BLOCK-1115
ECG4099B	BLOCK-1115
ECG4256	BLOCK-1116
ECG4501	BLOCK-1117
ECG4502	BLOCK-1118
ECG4502B	BLOCK-1118
ECG4506B	BLOCK-1119
ECG4508B	BLOCK-1120
ECG4510B	BLOCK-1121
ECG4511B	BLOCK-1122
ECG4512	BLOCK-1123
ECG4512B	BLOCK-1123
ECG4513B	BLOCK-1124
ECG4514B	BLOCK-1125
ECG4515B	BLOCK-1126
ECG4516B	BLOCK-1127
ECG4517B	BLOCK-1128
ECG4518	BLOCK-1129
ECG4518B	BLOCK-1129
ECG4520	BLOCK-1130
ECG4520B	BLOCK-1130
ECG4521B	BLOCK-1131
ECG4522	BLOCK-1132
ECG4522B	BLOCK-1132
ECG4526B	BLOCK-1133
ECG4527	BLOCK-1134
ECG4527B	BLOCK-1134
ECG4528B	BLOCK-1135
ECG4529B	BLOCK-1136
ECG4531B	BLOCK-1137
ECG4532B	BLOCK-1138
ECG4536B	BLOCK-1139
ECG4538B	BLOCK-1140
ECG4539	BLOCK-1141
ECG4539B	BLOCK-1141
ECG4541B	BLOCK-1142
ECG4543B	BLOCK-1143
ECG4547B	BLOCK-1144
ECG4553B	BLOCK-1145
ECG4555	BLOCK-1146
ECG4555B	BLOCK-1146
ECG4556	BLOCK-1147
ECG4556B	BLOCK-1147
ECG4558B	BLOCK-1148
ECG4562B	BLOCK-1149
ECG4566B	BLOCK-1150
ECG4568B	BLOCK-1151
ECG4569B	BLOCK-1152
ECG4582B	BLOCK-1052
ECG4583B	BLOCK-1153
ECG4585B	BLOCK-1154
ECG4597B	BLOCK-1155
ECG4598B	BLOCK-1156
ECG615	BLOCK-1157
ECG615A	BLOCK-1157
ECG6502	BLOCK-1158
ECG6507	BLOCK-1159
ECG6508	BLOCK-1160
ECG65101	BLOCK-1161
ECG6532	BLOCK-1162
ECG6664	BLOCK-1163
ECG6800	BLOCK-1164
ECG6802	BLOCK-1165
ECG6809	BLOCK-1166
ECG6809E	BLOCK-1167
ECG6810	BLOCK-1168
ECG6821	BLOCK-1169
ECG6850	BLOCK-1170
ECG6860	BLOCK-1171
ECG6875	BLOCK-1172
ECG6880	BLOCK-1173
ECG6885	BLOCK-1174
ECG6886	BLOCK-1175
ECG6887	BLOCK-1176
ECG6888	BLOCK-1177
ECG6889	BLOCK-1178
ECG700	BLOCK-1179
ECG7000	BLOCK-1180
ECG7001	BLOCK-1181
ECG7002	BLOCK-1182
ECG7003	BLOCK-1183
ECG7004	BLOCK-1184
ECG7005	BLOCK-1185
ECG7006	BLOCK-1186
ECG7007	BLOCK-1187
ECG7008	BLOCK-1188
ECG7009	BLOCK-1189
ECG701	BLOCK-1190
ECG7010	BLOCK-1191

DEVICE TYPE	REPL CODE
ECG7011	BLOCK-1192
ECG7012	BLOCK-1193
ECG7013	BLOCK-1194
ECG7014	BLOCK-1195
ECG7015	BLOCK-1196
ECG7016	BLOCK-1197
ECG7017	BLOCK-1198
ECG7018	BLOCK-1199
ECG7019	BLOCK-1200
ECG702	BLOCK-1201
ECG7020	BLOCK-1202
ECG7021	BLOCK-1203
ECG7022	BLOCK-1204
ECG7023	BLOCK-1205
ECG7024	BLOCK-1206
ECG7025	BLOCK-1207
ECG7026	BLOCK-1208
ECG7027	BLOCK-1209
ECG7028	BLOCK-1210
ECG7029	BLOCK-1211
ECG703	BLOCK-1212
ECG7030	BLOCK-1213
ECG7031	BLOCK-1214
ECG7032	BLOCK-1215
ECG7033	BLOCK-1216
ECG7034	BLOCK-1217
ECG7035	BLOCK-1218
ECG7036	BLOCK-1219
ECG7037	BLOCK-1220
ECG7038	BLOCK-1221
ECG7039	BLOCK-1222
ECG703A	BLOCK-1212
ECG704	BLOCK-1223
ECG7040	BLOCK-1224
ECG7041	BLOCK-1225
ECG7042	BLOCK-1226
ECG7043	BLOCK-1227
ECG7044	BLOCK-1228
ECG7045	BLOCK-1229
ECG7046	BLOCK-1230
ECG7047	BLOCK-1231
ECG7048	BLOCK-1232
ECG7049	BLOCK-1233
ECG705	BLOCK-1244
ECG7050	BLOCK-1234
ECG7051	BLOCK-1235
ECG7052	BLOCK-1236
ECG7053	BLOCK-1237
ECG7054	BLOCK-1238
ECG7055	BLOCK-1239
ECG7056	BLOCK-1240
ECG7057	BLOCK-1241
ECG7058	BLOCK-1242
ECG7059	BLOCK-1243
ECG705A	BLOCK-1244
ECG706	BLOCK-1245
ECG7060	BLOCK-1246
ECG7061	BLOCK-1247
ECG7062	BLOCK-1248
ECG7063	BLOCK-1249
ECG7064	BLOCK-1250
ECG7065	BLOCK-1251
ECG7066	BLOCK-1252
ECG7067	BLOCK-1253
ECG7068	BLOCK-1254
ECG7069	BLOCK-1255
ECG707	BLOCK-1256
ECG7070	BLOCK-1257
ECG7071	BLOCK-1258
ECG7072	BLOCK-1259
ECG7073	BLOCK-1260
ECG7074	BLOCK-1261
ECG708	BLOCK-1262
ECG7085	BLOCK-1263
ECG709	BLOCK-1264
ECG710	BLOCK-1265
ECG7104	BLOCK-1266
ECG711	BLOCK-1267
ECG7111	BLOCK-1268
ECG712	BLOCK-1269
ECG713	BLOCK-1270
ECG714	BLOCK-1271
ECG715	BLOCK-1272
ECG716	BLOCK-1273
ECG717	BLOCK-1274
ECG718	BLOCK-1275
ECG719	BLOCK-1276
ECG720	BLOCK-1277
ECG7214	BLOCK-1278
ECG723	BLOCK-1279
ECG724	BLOCK-1280
ECG725	BLOCK-1281
ECG726	BLOCK-1282
ECG727	BLOCK-1283

DEVICE TYPE	REPL CODE
ECG728	BLOCK-1284
ECG729	BLOCK-1285
ECG730	BLOCK-1286
ECG731	BLOCK-1287
ECG732	BLOCK-1828
ECG733	BLOCK-1828
ECG734	BLOCK-1289
ECG735	BLOCK-1288
ECG736	BLOCK-1289
ECG737	BLOCK-1290
ECG738	BLOCK-1291
ECG739	BLOCK-1292
ECG740	BLOCK-1303
ECG7400	BLOCK-1293
ECG7401	BLOCK-1294
ECG7402	BLOCK-1295
ECG7403	BLOCK-1296
ECG7404	BLOCK-1297
ECG7405	BLOCK-1298
ECG7406	BLOCK-1299
ECG7407	BLOCK-1300
ECG7408	BLOCK-1301
ECG7409	BLOCK-1302
ECG740A	BLOCK-1303
ECG741	BLOCK-1917
ECG7410	BLOCK-1304
ECG74107	BLOCK-1305
ECG74109	BLOCK-1306
ECG7411	BLOCK-1307
ECG74110	BLOCK-1308
ECG74111	BLOCK-1309
ECG7412	BLOCK-1310
ECG74121	BLOCK-1311
ECG74122	BLOCK-1312
ECG74123	BLOCK-1313
ECG74125	BLOCK-1314
ECG74126	BLOCK-1315
ECG74128	BLOCK-1316
ECG7413	BLOCK-1317
ECG74132	BLOCK-1318
ECG74136	BLOCK-1319
ECG7414	BLOCK-1320
ECG74141	BLOCK-1321
ECG74142	BLOCK-1322
ECG74143	BLOCK-1323
ECG74144	BLOCK-1324
ECG74145	BLOCK-1325
ECG74147	BLOCK-1326
ECG74150	BLOCK-1327
ECG74151	BLOCK-1328
ECG74152	BLOCK-1329
ECG74153	BLOCK-1330
ECG74154	BLOCK-1331
ECG74155	BLOCK-1332
ECG74156	BLOCK-1333
ECG74157	BLOCK-1334
ECG74158	BLOCK-1335
ECG7416	BLOCK-1336
ECG74160	BLOCK-1337
ECG74161	BLOCK-1338
ECG74162	BLOCK-1339
ECG74163	BLOCK-1340
ECG74164	BLOCK-1341
ECG74165	BLOCK-1342
ECG7417	BLOCK-1343
ECG74170	BLOCK-1344
ECG74173	BLOCK-1345
ECG74174	BLOCK-1346
ECG74175	BLOCK-1347
ECG74176	BLOCK-1348
ECG74177	BLOCK-1349
ECG74178	BLOCK-1350
ECG74179	BLOCK-1351
ECG74180	BLOCK-1352
ECG74181	BLOCK-1353
ECG74182	BLOCK-1354
ECG74190	BLOCK-1355
ECG74191	BLOCK-1356
ECG74192	BLOCK-1357
ECG74193	BLOCK-1358
ECG74195	BLOCK-1359
ECG74196	BLOCK-1360
ECG74197	BLOCK-1361
ECG74198	BLOCK-1362
ECG74199	BLOCK-1363
ECG742	BLOCK-1364
ECG7420	BLOCK-1365
ECG7421	BLOCK-1366
ECG7422	BLOCK-1367
ECG74221	BLOCK-1368
ECG7423	BLOCK-1369
ECG74249	BLOCK-1370
ECG74251	BLOCK-1371
ECG7427	BLOCK-1372

DEVICE TYPE	REPL CODE
ECG7428	BLOCK-1373
ECG74290	BLOCK-1374
ECG74293	BLOCK-1375
ECG743	BLOCK-1376
ECG7430	BLOCK-1377
ECG7432	BLOCK-1378
ECG7433	BLOCK-1379
ECG74365	BLOCK-1380
ECG74366	BLOCK-1381
ECG74367	BLOCK-1382
ECG74368	BLOCK-1383
ECG7437	BLOCK-1384
ECG7438	BLOCK-1385
ECG7439	BLOCK-1386
ECG74390	BLOCK-1387
ECG74393	BLOCK-1388
ECG744	BLOCK-1389
ECG7440	BLOCK-1390
ECG7441	BLOCK-1391
ECG7442	BLOCK-1392
ECG7443	BLOCK-1393
ECG7444	BLOCK-1394
ECG7445	BLOCK-1395
ECG7446	BLOCK-1396
ECG7447	BLOCK-1397
ECG7448	BLOCK-1398
ECG74490	BLOCK-1399
ECG745	BLOCK-1400
ECG7450	BLOCK-1401
ECG7451	BLOCK-1402
ECG7453	BLOCK-1403
ECG7454	BLOCK-1404
ECG746	BLOCK-1405
ECG7460	BLOCK-1406
ECG747	BLOCK-1407
ECG7470	BLOCK-1408
ECG7472	BLOCK-1409
ECG7473	BLOCK-1410
ECG7474	BLOCK-1411
ECG7475	BLOCK-1412
ECG7476	BLOCK-1413
ECG748	BLOCK-1421
ECG7480	BLOCK-1414
ECG7481	BLOCK-1415
ECG7482	BLOCK-1416
ECG7483	BLOCK-1417
ECG7485	BLOCK-1418
ECG7486	BLOCK-1419
ECG7489	BLOCK-1420
ECG748A	BLOCK-1421
ECG749	BLOCK-1422
ECG7490	BLOCK-1423
ECG7491	BLOCK-1424
ECG7492	BLOCK-1425
ECG7493A	BLOCK-1426
ECG7494	BLOCK-1427
ECG7495	BLOCK-1428
ECG7496	BLOCK-1429
ECG7497	BLOCK-1430
ECG74C00	BLOCK-1431
ECG74C02	BLOCK-1432
ECG74C04	BLOCK-1433
ECG74C08	BLOCK-1434
ECG74C10	BLOCK-1435
ECG74C107	BLOCK-1436
ECG74C14	BLOCK-1437
ECG74C151	BLOCK-1438
ECG74C154	BLOCK-1439
ECG74C157	BLOCK-1440
ECG74C161	BLOCK-1441
ECG74C164	BLOCK-1442
ECG74C173	BLOCK-1443
ECG74C174	BLOCK-1444
ECG74C175	BLOCK-1445
ECG74C192	BLOCK-1446
ECG74C193	BLOCK-1447
ECG74C20	BLOCK-1448
ECG74C221	BLOCK-1449
ECG74C240	BLOCK-1450
ECG74C244	BLOCK-1451
ECG74C30	BLOCK-1452
ECG74C32	BLOCK-1453
ECG74C373	BLOCK-1454
ECG74C374	BLOCK-1455
ECG74C48	BLOCK-1456
ECG74C73	BLOCK-1457
ECG74C74	BLOCK-1458
ECG74C76	BLOCK-1459
ECG74C85	BLOCK-1460
ECG74C90	BLOCK-1461
ECG74C901	BLOCK-1462
ECG74C902	BLOCK-1463
ECG74C903	BLOCK-1464
ECG74C904	BLOCK-1465

DEVICE TYPE	REPL CODE
ECG74C922	BLOCK-1466
ECG74C923	BLOCK-1467
ECG74C925	BLOCK-1468
ECG74C93	BLOCK-1469
ECG74C95	BLOCK-1470
ECG74H00	BLOCK-1471
ECG74H01	BLOCK-1472
ECG74H04	BLOCK-1473
ECG74H05	BLOCK-1474
ECG74H08	BLOCK-1475
ECG74H10	BLOCK-1476
ECG74H101	BLOCK-1477
ECG74H102	BLOCK-1478
ECG74H103	BLOCK-1479
ECG74H106	BLOCK-1480
ECG74H108	BLOCK-1481
ECG74H11	BLOCK-1482
ECG74H183	BLOCK-1483
ECG74H20	BLOCK-1484
ECG74H21	BLOCK-1485
ECG74H22	BLOCK-1486
ECG74H30	BLOCK-1487
ECG74H40	BLOCK-1488
ECG74H50	BLOCK-1489
ECG74H51	BLOCK-1490
ECG74H52	BLOCK-1491
ECG74H53	BLOCK-1492
ECG74H54	BLOCK-1493
ECG74H55	BLOCK-1494
ECG74H60	BLOCK-1495
ECG74H61	BLOCK-1496
ECG74H62	BLOCK-1497
ECG74H71	BLOCK-1498
ECG74H72	BLOCK-1499
ECG74H73	BLOCK-1500
ECG74H74	BLOCK-1501
ECG74H76	BLOCK-1502
ECG74H78	BLOCK-1503
ECG74H86	BLOCK-1504
ECG74H87	BLOCK-1505
ECG74HC00	BLOCK-1506
ECG74HC02	BLOCK-1507
ECG74HC04	BLOCK-1508
ECG74HC08	BLOCK-1509
ECG74HC10	BLOCK-1510
ECG74HC109	BLOCK-1511
ECG74HC11	BLOCK-1512
ECG74HC123	BLOCK-1513
ECG74HC125	BLOCK-1514
ECG74HC126	BLOCK-1515
ECG74HC132	BLOCK-1516
ECG74HC138	BLOCK-1517
ECG74HC139	BLOCK-1518
ECG74HC14	BLOCK-1519
ECG74HC151	BLOCK-1520
ECG74HC153	BLOCK-1521
ECG74HC154	BLOCK-1522
ECG74HC161	BLOCK-1523
ECG74HC163	BLOCK-1524
ECG74HC164	BLOCK-1525
ECG74HC165	BLOCK-1526
ECG74HC173	BLOCK-1527
ECG74HC174	BLOCK-1528
ECG74HC175	BLOCK-1529
ECG74HC240	BLOCK-1530
ECG74HC244	BLOCK-1531
ECG74HC257	BLOCK-1532
ECG74HC259	BLOCK-1533
ECG74HC273	BLOCK-1534
ECG74HC299	BLOCK-1535
ECG74HC32	BLOCK-1536
ECG74HC373	BLOCK-1537
ECG74HC374	BLOCK-1538
ECG74HC377	BLOCK-1539
ECG74HC390	BLOCK-1540
ECG74HC393	BLOCK-1541
ECG74HC40105	BLOCK-1542
ECG74HC4020	BLOCK-1543
ECG74HC4040	BLOCK-1544
ECG74HC4053	BLOCK-1545
ECG74HC4060	BLOCK-1546
ECG74HC4067	BLOCK-1547
ECG74HC573	BLOCK-1548
ECG74HC574	BLOCK-1549
ECG74HC86	BLOCK-1550
ECG74HCT00	BLOCK-1551
ECG74HCT04	BLOCK-1552
ECG74HCT08	BLOCK-1553
ECG74HCT138	BLOCK-1554
ECG74HCT14	BLOCK-1555
ECG74HCT161	BLOCK-1556
ECG74HCT163	BLOCK-1557
ECG74HCT174	BLOCK-1558
ECG74HCT240	BLOCK-1559

ORIGINAL DEVICE TYPES AND REPLACEMENT CODES 27

DEVICE TYPE	REPL CODE
ECG74HCT244	BLOCK-1560
ECG74HCT273	BLOCK-1561
ECG74HCT32	BLOCK-1562
ECG74HCT373	BLOCK-1563
ECG74HCT374	BLOCK-1564
ECG74HCT573	BLOCK-1565
ECG74HCT574	BLOCK-1566
ECG74L93	BLOCK-1567
ECG74LS00	BLOCK-1568
ECG74LS01	BLOCK-1569
ECG74LS02	BLOCK-1570
ECG74LS03	BLOCK-1571
ECG74LS04	BLOCK-1572
ECG74LS05	BLOCK-1573
ECG74LS08	BLOCK-1574
ECG74LS09	BLOCK-1575
ECG74LS10	BLOCK-1576
ECG74LS107	BLOCK-1577
ECG74LS109A	BLOCK-1578
ECG74LS11	BLOCK-1579
ECG74LS112A	BLOCK-1580
ECG74LS114	BLOCK-1581
ECG74LS12	BLOCK-1582
ECG74LS122	BLOCK-1583
ECG74LS123	BLOCK-1584
ECG74LS125A	BLOCK-1585
ECG74LS126	BLOCK-1586
ECG74LS13	BLOCK-1587
ECG74LS132	BLOCK-1588
ECG74LS133	BLOCK-1589
ECG74LS136	BLOCK-1590
ECG74LS138	BLOCK-1591
ECG74LS139	BLOCK-1592
ECG74LS14	BLOCK-1593
ECG74LS145	BLOCK-1594
ECG74LS147	BLOCK-1595
ECG74LS148	BLOCK-1596
ECG74LS15	BLOCK-1597
ECG74LS151	BLOCK-1598
ECG74LS153	BLOCK-1599
ECG74LS155	BLOCK-1600
ECG74LS156	BLOCK-1601
ECG74LS157	BLOCK-1602
ECG74LS158	BLOCK-1603
ECG74LS160A	BLOCK-1604
ECG74LS161A	BLOCK-1605
ECG74LS162A	BLOCK-1606
ECG74LS163A	BLOCK-1607
ECG74LS164	BLOCK-1608
ECG74LS165	BLOCK-1609
ECG74LS166	BLOCK-1610
ECG74LS168A	BLOCK-1611
ECG74LS169A	BLOCK-1612
ECG74LS170	BLOCK-1613
ECG74LS173	BLOCK-1614
ECG74LS174	BLOCK-1615
ECG74LS175	BLOCK-1616
ECG74LS181	BLOCK-1617
ECG74LS190	BLOCK-1618
ECG74LS191	BLOCK-1619
ECG74LS192	BLOCK-1620
ECG74LS193	BLOCK-1621
ECG74LS194A	BLOCK-1622
ECG74LS195A	BLOCK-1623
ECG74LS196	BLOCK-1624
ECG74LS197	BLOCK-1625
ECG74LS20	BLOCK-1626
ECG74LS21	BLOCK-1627
ECG74LS22	BLOCK-1628
ECG74LS221	BLOCK-1629
ECG74LS240	BLOCK-1630
ECG74LS241	BLOCK-1631
ECG74LS242	BLOCK-1632
ECG74LS243	BLOCK-1633
ECG74LS244	BLOCK-1634
ECG74LS245	BLOCK-1635
ECG74LS247	BLOCK-1636
ECG74LS248	BLOCK-1637
ECG74LS249	BLOCK-1638
ECG74LS251	BLOCK-1639
ECG74LS253	BLOCK-1640
ECG74LS257	BLOCK-1641
ECG74LS258	BLOCK-1642
ECG74LS259	BLOCK-1643
ECG74LS26	BLOCK-1644
ECG74LS260	BLOCK-1645
ECG74LS266	BLOCK-1646
ECG74LS27	BLOCK-1647
ECG74LS273	BLOCK-1648
ECG74LS279	BLOCK-1649
ECG74LS28	BLOCK-1650
ECG74LS280	BLOCK-1651
ECG74LS283	BLOCK-1652
ECG74LS290	BLOCK-1653
ECG74LS293	BLOCK-1654
ECG74LS295A	BLOCK-1655
ECG74LS298	BLOCK-1656
ECG74LS299	BLOCK-1657
ECG74LS30	BLOCK-1658
ECG74LS32	BLOCK-1659
ECG74LS33	BLOCK-1660
ECG74LS348	BLOCK-1661
ECG74LS352	BLOCK-1662
ECG74LS353	BLOCK-1663
ECG74LS363	BLOCK-1664
ECG74LS364	BLOCK-1665
ECG74LS365A	BLOCK-1666
ECG74LS366A	BLOCK-1667
ECG74LS367	BLOCK-1668
ECG74LS368	BLOCK-1669
ECG74LS37	BLOCK-1670
ECG74LS373	BLOCK-1671
ECG74LS374	BLOCK-1672
ECG74LS377	BLOCK-1673
ECG74LS378	BLOCK-1674
ECG74LS379	BLOCK-1675
ECG74LS38	BLOCK-1676
ECG74LS386	BLOCK-1677
ECG74LS390	BLOCK-1678
ECG74LS393	BLOCK-1679
ECG74LS395A	BLOCK-1680
ECG74LS398	BLOCK-1681
ECG74LS399	BLOCK-1682
ECG74LS40	BLOCK-1683
ECG74LS42	BLOCK-1684
ECG74LS47	BLOCK-1685
ECG74LS48	BLOCK-1686
ECG74LS49	BLOCK-1687
ECG74LS490	BLOCK-1688
ECG74LS51	BLOCK-1689
ECG74LS54	BLOCK-1690
ECG74LS540	BLOCK-1691
ECG74LS541	BLOCK-1692
ECG74LS55	BLOCK-1693
ECG74LS624	BLOCK-1694
ECG74LS625	BLOCK-1695
ECG74LS626	BLOCK-1696
ECG74LS627	BLOCK-1697
ECG74LS629	BLOCK-1698
ECG74LS640	BLOCK-1699
ECG74LS641	BLOCK-1700
ECG74LS642	BLOCK-1701
ECG74LS643	BLOCK-1702
ECG74LS645	BLOCK-1703
ECG74LS670	BLOCK-1704
ECG74LS73	BLOCK-1705
ECG74LS74A	BLOCK-1706
ECG74LS75	BLOCK-1707
ECG74LS76A	BLOCK-1708
ECG74LS77	BLOCK-1709
ECG74LS78	BLOCK-1710
ECG74LS83A	BLOCK-1711
ECG74LS85	BLOCK-1712
ECG74LS86	BLOCK-1713
ECG74LS90	BLOCK-1714
ECG74LS91	BLOCK-1715
ECG74LS92	BLOCK-1716
ECG74LS93	BLOCK-1717
ECG74LS95	BLOCK-1718
ECG74S00	BLOCK-1719
ECG74S02	BLOCK-1720
ECG74S03	BLOCK-1721
ECG74S04	BLOCK-1722
ECG74S05	BLOCK-1723
ECG74S08	BLOCK-1724
ECG74S09	BLOCK-1725
ECG74S10	BLOCK-1726
ECG74S11	BLOCK-1727
ECG74S112	BLOCK-1728
ECG74S113	BLOCK-1729
ECG74S124	BLOCK-1730
ECG74S133	BLOCK-1731
ECG74S134	BLOCK-1732
ECG74S138	BLOCK-1733
ECG74S140	BLOCK-1734
ECG74S15	BLOCK-1735
ECG74S151	BLOCK-1736
ECG74S153	BLOCK-1737
ECG74S157	BLOCK-1738
ECG74S158	BLOCK-1739
ECG74S174	BLOCK-1740
ECG74S175	BLOCK-1741
ECG74S181	BLOCK-1742
ECG74S188	BLOCK-1743
ECG74S194	BLOCK-1744
ECG74S20	BLOCK-1745
ECG74S22	BLOCK-1746
ECG74S251	BLOCK-1747
ECG74S258	BLOCK-1748
ECG74S287	BLOCK-1749
ECG74S288	BLOCK-1750
ECG74S30	BLOCK-1751
ECG74S387	BLOCK-1752
ECG74S40	BLOCK-1753
ECG74S454	BLOCK-1754
ECG74S472	BLOCK-1755
ECG74S473	BLOCK-1756
ECG74S474	BLOCK-1757
ECG74S475	BLOCK-1758
ECG74S478	BLOCK-1759
ECG74S479	BLOCK-1760
ECG74S51	BLOCK-1761
ECG74S570	BLOCK-1762
ECG74S571	BLOCK-1763
ECG74S572	BLOCK-1764
ECG74S573	BLOCK-1765
ECG74S64	BLOCK-1766
ECG74S65	BLOCK-1767
ECG74S74	BLOCK-1768
ECG74S86	BLOCK-1769
ECG750	BLOCK-1770
ECG75188	BLOCK-1771
ECG75189	BLOCK-1772
ECG752	BLOCK-1773
ECG753	BLOCK-1774
ECG75450B	BLOCK-1775
ECG75451B	BLOCK-1776
ECG75452B	BLOCK-1777
ECG75453B	BLOCK-1778
ECG75454B	BLOCK-1779
ECG75491B	BLOCK-1780
ECG75492B	BLOCK-1781
ECG75493	BLOCK-1782
ECG75494	BLOCK-1783
ECG75497	BLOCK-1784
ECG75498	BLOCK-1785
ECG755	BLOCK-1786
ECG756	BLOCK-1787
ECG757	BLOCK-1787
ECG758	BLOCK-1788
ECG759	BLOCK-1789
ECG760	BLOCK-1790
ECG761	BLOCK-1791
ECG763	BLOCK-1792
ECG764	BLOCK-1793
ECG765	BLOCK-1793
ECG766	BLOCK-1859
ECG766A	BLOCK-1859
ECG767	BLOCK-1794
ECG768	BLOCK-1795
ECG770	BLOCK-1796
ECG772	BLOCK-1797
ECG772A	BLOCK-1797
ECG773	BLOCK-1798
ECG775	BLOCK-1799
ECG776	BLOCK-1800
ECG778	BLOCK-1801
ECG778A	BLOCK-1801
ECG778S	BLOCK-1802
ECG778SM	BLOCK-1803
ECG779	BLOCK-1804
ECG779-1	BLOCK-1804
ECG779A	BLOCK-1804
ECG780	BLOCK-1805
ECG781	BLOCK-1806
ECG782	BLOCK-1807
ECG783	BLOCK-1808
ECG784	BLOCK-1809
ECG785	BLOCK-1810
ECG786	BLOCK-1811
ECG787	BLOCK-1812
ECG788	BLOCK-1813
ECG789	BLOCK-1814
ECG790	BLOCK-1815
ECG791	BLOCK-1816
ECG792	BLOCK-1817
ECG793	BLOCK-1818
ECG794	BLOCK-1819
ECG795	BLOCK-1820
ECG796	BLOCK-1797
ECG797	BLOCK-1821
ECG798	BLOCK-1822
ECG799	BLOCK-1823
ECG800	BLOCK-1824
ECG801	BLOCK-1825
ECG802	BLOCK-1826
ECG803	BLOCK-1827
ECG804	BLOCK-1828
ECG806	BLOCK-1829
ECG807	BLOCK-1830
ECG808	BLOCK-1831
ECG8080A	BLOCK-1832
ECG809	BLOCK-1833
ECG8092	BLOCK-1834
ECG80C95	BLOCK-1835
ECG80C96	BLOCK-1836
ECG80C97	BLOCK-1837
ECG810	BLOCK-1838
ECG810A	BLOCK-1838
ECG812	BLOCK-1839
ECG8123	BLOCK-1840
ECG813	BLOCK-1841
ECG814	BLOCK-1842
ECG814A	BLOCK-1842
ECG815	BLOCK-1843
ECG816	BLOCK-1844
ECG817	BLOCK-1845
ECG818	BLOCK-1846
ECG819	BLOCK-1847
ECG820	BLOCK-1848
ECG821	BLOCK-1849
ECG822	BLOCK-1850
ECG823	BLOCK-1851
ECG824	BLOCK-1852
ECG825	BLOCK-1853
ECG8250	BLOCK-1854
ECG8255	BLOCK-1855
ECG826	BLOCK-1856
ECG8266	BLOCK-1857
ECG828	BLOCK-1858
ECG829	BLOCK-1859
ECG830	BLOCK-1860
ECG8301	BLOCK-1861
ECG8308	BLOCK-1862
ECG8309	BLOCK-1863
ECG831	BLOCK-1864
ECG8314	BLOCK-1865
ECG8316	BLOCK-1866
ECG8318	BLOCK-1867
ECG832	BLOCK-1868
ECG8321	BLOCK-1869
ECG8328	BLOCK-1870
ECG832SM	BLOCK-1871
ECG833	BLOCK-1872
ECG834	BLOCK-1873
ECG834SM	BLOCK-1874
ECG835	BLOCK-1875
ECG836	BLOCK-1876
ECG8368	BLOCK-1877
ECG837	BLOCK-1878
ECG8370	BLOCK-1879
ECG8374	BLOCK-1880
ECG838	BLOCK-1881
ECG839	BLOCK-1882
ECG840	BLOCK-1883
ECG841	BLOCK-1884
ECG842	BLOCK-1885
ECG843	BLOCK-1886
ECG844	BLOCK-1887
ECG845	BLOCK-1888
ECG846	BLOCK-1889
ECG847	BLOCK-1890
ECG848	BLOCK-1891
ECG849	BLOCK-1892
ECG850	BLOCK-1893
ECG851	BLOCK-1894
ECG852	BLOCK-1895
ECG8520	BLOCK-1896
ECG853	BLOCK-1897
ECG854	BLOCK-1898
ECG8542	BLOCK-1899
ECG8546	BLOCK-1900
ECG855	BLOCK-1901
ECG8552	BLOCK-1902
ECG8553	BLOCK-1903
ECG8554	BLOCK-1904
ECG8555	BLOCK-1905
ECG8556	BLOCK-1906
ECG856	BLOCK-1907
ECG857M	BLOCK-1908
ECG857SM	BLOCK-1909
ECG858M	BLOCK-1910
ECG858SM	BLOCK-1911
ECG859	BLOCK-1912
ECG859SM	BLOCK-1913
ECG860	BLOCK-1914
ECG861	BLOCK-1915
ECG8613	BLOCK-1916
ECG862	BLOCK-1917
ECG863	BLOCK-1918
ECG864	BLOCK-1919
ECG865	BLOCK-1920
ECG866	BLOCK-1921
ECG867	BLOCK-1922
ECG868	BLOCK-1923
ECG869	BLOCK-1924
ECG870	BLOCK-1925
ECG871	BLOCK-1926
ECG872	BLOCK-1927
ECG873	BLOCK-1928
ECG874	BLOCK-1929
ECG875	BLOCK-1930
ECG876	BLOCK-1931
ECG877	BLOCK-1932
ECG878	BLOCK-1933
ECG879	BLOCK-1934
ECG880	BLOCK-1935
ECG8853	BLOCK-1936
ECG887M	BLOCK-1937
ECG888M	BLOCK-1938
ECG889M	BLOCK-1939
ECG900	BLOCK-1940
ECG901	BLOCK-1941
ECG902	BLOCK-1942
ECG903	BLOCK-1943
ECG904	BLOCK-1944
ECG905	BLOCK-1945
ECG906	BLOCK-1946
ECG907	BLOCK-1947
ECG908	BLOCK-1948
ECG909	BLOCK-1949
ECG9093	BLOCK-1950
ECG9094	BLOCK-1951
ECG9097	BLOCK-1952
ECG9099	BLOCK-1953
ECG909D	BLOCK-1954
ECG910	BLOCK-1955
ECG9109	BLOCK-1956
ECG910D	BLOCK-1957
ECG911	BLOCK-1958
ECG9110	BLOCK-1959
ECG9111	BLOCK-1960
ECG9112	BLOCK-1961
ECG911D	BLOCK-1962
ECG912	BLOCK-1963
ECG913	BLOCK-1964
ECG9135	BLOCK-1965
ECG914	BLOCK-1966
ECG915	BLOCK-1967
ECG9157	BLOCK-1968
ECG9158	BLOCK-1969
ECG916	BLOCK-1970
ECG917	BLOCK-1971
ECG918	BLOCK-1972
ECG918M	BLOCK-1973
ECG918SM	BLOCK-1974
ECG919	BLOCK-1975
ECG919D	BLOCK-1976
ECG920	BLOCK-1977
ECG9200	BLOCK-1978
ECG921	BLOCK-1979
ECG922	BLOCK-1980
ECG922M	BLOCK-1981
ECG922SM	BLOCK-1982
ECG923	BLOCK-1983
ECG923D	BLOCK-1984
ECG924	BLOCK-1985
ECG924M	BLOCK-1986
ECG925	BLOCK-1987
ECG926	BLOCK-1988
ECG927	BLOCK-1989
ECG927D	BLOCK-1990
ECG927SM	BLOCK-1991
ECG928	BLOCK-1992
ECG928M	BLOCK-1993
ECG928S	BLOCK-1994
ECG928SM	BLOCK-1995
ECG929	BLOCK-1996
ECG930	BLOCK-1997
ECG9301	BLOCK-1998
ECG9302	BLOCK-1999
ECG9303	BLOCK-2000
ECG9304	BLOCK-2001
ECG9306	BLOCK-2002
ECG9307	BLOCK-2003
ECG931	BLOCK-2004
ECG9311	BLOCK-2005
ECG9312	BLOCK-2006
ECG932	BLOCK-2007
ECG9321	BLOCK-2008
ECG9322	BLOCK-2009
ECG9323	BLOCK-2010
ECG9324	BLOCK-2011
ECG9325	BLOCK-2012
ECG9326	BLOCK-2013
ECG933	BLOCK-2014
ECG9331	BLOCK-2015
ECG9332	BLOCK-2016
ECG9333	BLOCK-2017
ECG9334	BLOCK-2018
ECG9335	BLOCK-2019
ECG934	BLOCK-2020
ECG9342	BLOCK-2021
ECG9343	BLOCK-2022
ECG9347	BLOCK-2023
ECG935	BLOCK-2024
ECG936	BLOCK-2025
ECG9361	BLOCK-2026
ECG9362	BLOCK-2027
ECG9363	BLOCK-2028
ECG9367	BLOCK-2029
ECG9368	BLOCK-2030
ECG937	BLOCK-2031
ECG9370	BLOCK-2032
ECG9371	BLOCK-2033
ECG9372	BLOCK-2034
ECG9375	BLOCK-2035
ECG937M	BLOCK-2036
ECG938	BLOCK-2037
ECG9380	BLOCK-2038
ECG9381	BLOCK-2039
ECG9382	BLOCK-2040
ECG9383	BLOCK-2041
ECG938M	BLOCK-2042
ECG939	BLOCK-2043
ECG9390	BLOCK-2044
ECG9391	BLOCK-2045
ECG9392	BLOCK-2046
ECG9393	BLOCK-2047
ECG9394	BLOCK-2048
ECG93L08	BLOCK-2049
ECG93L16	BLOCK-2050
ECG940	BLOCK-2051
ECG9400	BLOCK-2052
ECG9401	BLOCK-2053
ECG9402	BLOCK-2054
ECG9403	BLOCK-2055
ECG941	BLOCK-2056
ECG941D	BLOCK-2057
ECG941M	BLOCK-2058
ECG941S	BLOCK-2059
ECG941SM	BLOCK-2060
ECG942	BLOCK-2061
ECG943	BLOCK-2062
ECG943M	BLOCK-2063
ECG943SM	BLOCK-2064
ECG944	BLOCK-2065
ECG944M	BLOCK-2066
ECG945	BLOCK-2067
ECG946	BLOCK-2068
ECG947	BLOCK-2069
ECG947D	BLOCK-2070
ECG948	BLOCK-2071
ECG948SM	BLOCK-2072
ECG949	BLOCK-2073
ECG950	BLOCK-2074
ECG951	BLOCK-2075
ECG952	BLOCK-2076
ECG953	BLOCK-2077
ECG954	BLOCK-2078
ECG955M	BLOCK-2079
ECG955MC	BLOCK-2080
ECG955S	BLOCK-2081
ECG955SM	BLOCK-2082
ECG956	BLOCK-2083
ECG957	BLOCK-2084
ECG958	BLOCK-2085
ECG959	BLOCK-2086
ECG960	BLOCK-2087
ECG9600	BLOCK-2088
ECG9601	BLOCK-2089
ECG9602	BLOCK-2090
ECG961	BLOCK-2091
ECG9615	BLOCK-2092
ECG962	BLOCK-2093
ECG963	BLOCK-2094
ECG964	BLOCK-2095
ECG965	BLOCK-2096
ECG966	BLOCK-2097
ECG9660	BLOCK-2098
ECG9661	BLOCK-2099
ECG9662	BLOCK-2100
ECG9663	BLOCK-2101
ECG9664	BLOCK-2102
ECG9665	BLOCK-2103
ECG9666	BLOCK-2104
ECG9667	BLOCK-2105
ECG9668	BLOCK-2106
ECG9669	BLOCK-2107
ECG967	BLOCK-2108
ECG9670	BLOCK-2109
ECG9671	BLOCK-2110
ECG9672	BLOCK-2111
ECG9673	BLOCK-2112

28 ORIGINAL DEVICE TYPES AND REPLACEMENT CODES

DEVICE TYPE	REPL CODE
ECG9674	BLOCK-2113
ECG9675	BLOCK-2114
ECG9676	BLOCK-2115
ECG9677	BLOCK-2116
ECG9678	BLOCK-2117
ECG9679	BLOCK-2118
ECG968	BLOCK-2119
ECG9680	BLOCK-2120
ECG9681	BLOCK-2121
ECG9682	BLOCK-2122
ECG9683	BLOCK-2123
ECG9684	BLOCK-2124
ECG9685	BLOCK-2125
ECG9686	BLOCK-2126
ECG9689	BLOCK-2127
ECG969	BLOCK-2128
ECG9690	BLOCK-2129
ECG9691	BLOCK-2130
ECG9696	BLOCK-2131
ECG96L02	BLOCK-2132
ECG96LS02	BLOCK-2133
ECG96S02	BLOCK-2134
ECG970	BLOCK-2135
ECG971	BLOCK-2136
ECG972	BLOCK-2137
ECG973	BLOCK-2138
ECG973D	BLOCK-2139
ECG974	BLOCK-2140
ECG975	BLOCK-2141
ECG975SM	BLOCK-2142
ECG976	BLOCK-2143
ECG977	BLOCK-2144
ECG978	BLOCK-2145
ECG978C	BLOCK-2146
ECG978SM	BLOCK-2147
ECG979	BLOCK-2148
ECG980	BLOCK-2149
ECG9800	BLOCK-2150
ECG9801	BLOCK-2151
ECG9802	BLOCK-2152
ECG9803	BLOCK-2153
ECG9804	BLOCK-2154
ECG9805	BLOCK-2155
ECG9806	BLOCK-2156
ECG9807	BLOCK-2157
ECG9808	BLOCK-2158
ECG9809	BLOCK-2159
ECG981	BLOCK-2160
ECG9810	BLOCK-2161
ECG9811	BLOCK-2162
ECG9812	BLOCK-2163
ECG9813	BLOCK-2164
ECG9814	BLOCK-2165
ECG982	BLOCK-2166
ECG983	BLOCK-2167
ECG984	BLOCK-2168
ECG985	BLOCK-2169
ECG986	BLOCK-2170
ECG987	BLOCK-2171
ECG987SM	BLOCK-2172
ECG988	BLOCK-2173
ECG989	BLOCK-2174
ECG990	BLOCK-2175
ECG9900	BLOCK-2176
ECG9903	BLOCK-2177
ECG9904	BLOCK-2178
ECG9905	BLOCK-2179
ECG9906	BLOCK-2180
ECG9907	BLOCK-2181
ECG9908	BLOCK-2182
ECG9909	BLOCK-2183
ECG991	BLOCK-2184
ECG9910	BLOCK-2185
ECG9911	BLOCK-2186
ECG9912	BLOCK-2187
ECG9913	BLOCK-2188
ECG9914	BLOCK-2189
ECG9915	BLOCK-2190
ECG992	BLOCK-2191
ECG9921	BLOCK-2192
ECG9924	BLOCK-2193
ECG9926	BLOCK-2194
ECG9927	BLOCK-2195
ECG993	BLOCK-2196
ECG9930	BLOCK-2197
ECG9931	BLOCK-2198
ECG9932	BLOCK-2199
ECG9933	BLOCK-2200
ECG9935	BLOCK-2201
ECG9936	BLOCK-2202
ECG9937	BLOCK-2203
ECG994	BLOCK-2204
ECG9941	BLOCK-2205
ECG9944	BLOCK-2206

DEVICE TYPE	REPL CODE
ECG9945	BLOCK-2207
ECG9946	BLOCK-2208
ECG9948	BLOCK-2209
ECG9949	BLOCK-2210
ECG995	BLOCK-2211
ECG9950	BLOCK-2212
ECG9951	BLOCK-2213
ECG995M	BLOCK-2214
ECG996	BLOCK-2215
ECG9961	BLOCK-2216
ECG9962	BLOCK-2217
ECG9963	BLOCK-2218
ECG997	BLOCK-2219
ECG9974	BLOCK-2220
ECG9976	BLOCK-2221
ECG998	BLOCK-2222
ECG9982	BLOCK-2223
ECG9989	BLOCK-2224
ECG999	BLOCK-2225
ECG9990	BLOCK-2226
ECG999M	BLOCK-2227
ECG999SM	BLOCK-2228
EF6810C	BLOCK-1168
EF6810P	BLOCK-1168
EI307200	BLOCK-477
EI703843	BLOCK-476
EI703907	BLOCK-500
EICM-0060	BLOCK-145
EICM-14	BLOCK-101
EICM-19	BLOCK-177
EN11235	BLOCK-537
EN11238	BLOCK-655
EN11301	BLOCK-219
EN11401	BLOCK-388
EN11414	BLOCK-390
EN11436	BLOCK-636
EN11438	BLOCK-2170
EN1364	BLOCK-219
EN1364/1231	BLOCK-219
EP15X90(IC)	BLOCK-2087
EP16X54	BLOCK-1157
EP84X1	BLOCK-175
EP84X10	BLOCK-175
EP84X100	BLOCK-578
EP84X109	BLOCK-363
EP84X11	BLOCK-1365
EP84X119	BLOCK-2171
EP84X12	BLOCK-1291
EP84X13	BLOCK-1293
EP84X154	BLOCK-471
EP84X18	BLOCK-1426
EP84X19	BLOCK-1313
EP84X2	BLOCK-1269
EP84X20	BLOCK-2191
EP84X213	BLOCK-471
EP84X214	BLOCK-636
EP84X215	BLOCK-459
EP84X216	BLOCK-603
EP84X217	BLOCK-2097
EP84X221	BLOCK-636
EP84X227	BLOCK-695
EP84X229	BLOCK-563
EP84X240	BLOCK-1038
EP84X241	BLOCK-1135
EP84X242	BLOCK-1051
EP84X25	BLOCK-561
EP84X250	BLOCK-853
EP84X3	BLOCK-1270
EP84X30	BLOCK-1048
EP84X34	BLOCK-401
EP84X35	BLOCK-184
EP84X36	BLOCK-1364
EP84X37	BLOCK-1843
EP84X4	BLOCK-1808
EP84X5	BLOCK-1808
EP84X6	BLOCK-1269
EP84X60	BLOCK-425
EP84X62	BLOCK-572
EP84X63	BLOCK-582
EP84X67	BLOCK-584
EP84X69	BLOCK-1801
EP84X7	BLOCK-1284
EP84X71	BLOCK-646
EP84X72	BLOCK-42
EP84X73	BLOCK-1269
EP84X74	BLOCK-5
EP84X78	BLOCK-508
EP84X8	BLOCK-1422
EP84X80	BLOCK-418
EP84X81	BLOCK-219
EP84X82	BLOCK-1895
EP84X9	BLOCK-1270
ER1400	BLOCK-1020
ES75X1	BLOCK-1269

DEVICE TYPE	REPL CODE
ES84X1	BLOCK-8
ES84X2	BLOCK-18
ES84X3	BLOCK-1269
ES84X6	BLOCK-1269
ESI-LA1111P	BLOCK-3
ESI-UPC1020H	BLOCK-149
ESI-UPC30C	BLOCK-47
ETI-13	BLOCK-444
ETI-21	BLOCK-3
ETI-22	BLOCK-1825
ETI-23	BLOCK-145
EW15X570	BLOCK-722
EW16X381	BLOCK-1157
EW84X0425	BLOCK-707
EW84X1007	BLOCK-1571
EW84X1008	BLOCK-770
EW84X1009	BLOCK-708
EW84X1010	BLOCK-758
EW84X114	BLOCK-1269
EW84X115	BLOCK-510
EW84X12	BLOCK-1157
EW84X15	BLOCK-643
EW84X156	BLOCK-1873
EW84X172	BLOCK-1035
EW84X18	BLOCK-503
EW84X192	BLOCK-1801
EW84X196	BLOCK-2063
EW84X217	BLOCK-223
EW84X224	BLOCK-1994
EW84X232	BLOCK-1993
EW84X233	BLOCK-1993
EW84X239	BLOCK-675
EW84X252	BLOCK-617
EW84X255	BLOCK-853
EW84X267	BLOCK-756
EW84X268	BLOCK-1269
EW84X269	BLOCK-526
EW84X270	BLOCK-404
EW84X272	BLOCK-1020
EW84X273	BLOCK-1157
EW84X296	BLOCK-615
EW84X30	BLOCK-1051
EW84X310	BLOCK-459
EW84X312	BLOCK-692
EW84X314	BLOCK-691
EW84X315	BLOCK-674
EW84X317	BLOCK-793
EW84X319	BLOCK-808
EW84X320	BLOCK-792
EW84X322	BLOCK-809
EW84X323	BLOCK-768
EW84X327	BLOCK-760
EW84X352	BLOCK-768
EW84X370	BLOCK-710
EW84X372	BLOCK-1873
EW84X375	BLOCK-693
EW84X377	BLOCK-711
EW84X381	BLOCK-790
EW84X385	BLOCK-696
EW84X394	BLOCK-403
EW84X416	BLOCK-396
EW84X422	BLOCK-538
EW84X423	BLOCK-649
EW84X424	BLOCK-282
EW84X425	BLOCK-707
EW84X426	BLOCK-660
EW84X427	BLOCK-533
EW84X431	BLOCK-1157
EW84X440	BLOCK-786
EW84X441	BLOCK-788
EW84X442	BLOCK-789
EW84X443	BLOCK-790
EW84X444	BLOCK-791
EW84X445	BLOCK-794
EW84X446	BLOCK-795
EW84X451	BLOCK-760
EW84X452	BLOCK-728
EW84X453	BLOCK-2087
EW84X464	BLOCK-1066
EW84X470	BLOCK-1087
EW84X479	BLOCK-690
EW84X480	BLOCK-725
EW84X481	BLOCK-675
EW84X483	BLOCK-1801
EW84X484	BLOCK-516
EW84X486	BLOCK-596
EW84X487	BLOCK-132
EW84X488	BLOCK-1801
EW84X490	BLOCK-697
EW84X496	BLOCK-1093
EW84X499	BLOCK-516
EW84X526	BLOCK-2119
EW84X536	BLOCK-789
EW84X589	BLOCK-787

DEVICE TYPE	REPL CODE
EW84X595	BLOCK-853
EW84X60	BLOCK-695
EW84X624	BLOCK-768
EW84X625	BLOCK-1202
EW84X738	BLOCK-684
EW84X744	BLOCK-788
EW84X747	BLOCK-2097
EW84X756	BLOCK-1096
EW84X761	BLOCK-786
EW84X763	BLOCK-813
EW84X767	BLOCK-791
EW84X774	BLOCK-786
EW84X782	BLOCK-636
EW84X783	BLOCK-721
EW84X784	BLOCK-706
EW84X785	BLOCK-2058
EW84X788	BLOCK-1801
EW84X793	BLOCK-205
EW84X810	BLOCK-816
EW84X863	BLOCK-1041
EW84X865	BLOCK-1801
EW84X888	BLOCK-644
EW84X894	BLOCK-821
EW84X895	BLOCK-822
EW84X897	BLOCK-647
EW84X899	BLOCK-532
EW84X900	BLOCK-534
EW84X901	BLOCK-1269
EW84X905	BLOCK-762
EW84X906	BLOCK-709
EW84X909	BLOCK-757
EW84X910	BLOCK-659
EW84X911	BLOCK-807
EW84X913	BLOCK-649
EW84X914	BLOCK-669
EW84X976	BLOCK-761
EW84X977	BLOCK-1195
EW84X978	BLOCK-360
EW84X979	BLOCK-1296
EW84Z316	BLOCK-795
EX33X8392	BLOCK-180
EX39	BLOCK-1244
EX39-X	BLOCK-1244
EX42	BLOCK-2166
EX42-X	BLOCK-1271
EX42X	BLOCK-1271
EX46-X	BLOCK-1270
EX48	BLOCK-1269
EX48-X	BLOCK-1272
EX62-X	BLOCK-1815
EXT95247C	BLOCK-540
EXT954-20C	BLOCK-700
EXT954-25C	BLOCK-540
EXT95423C	BLOCK-535
EXT95427C	BLOCK-540
EZT308	BLOCK-1944
F-51056	BLOCK-838
F-7402PC	BLOCK-1295
F-7404PC	BLOCK-1297
F-7417PC	BLOCK-1343
F-74LS04PC	BLOCK-1572
F-74LS390PC	BLOCK-1678
F-74LS86PC	BLOCK-1713
F-74S138PC	BLOCK-1733
F16K3DC	BLOCK-1003
F16K4DC	BLOCK-1003
F16K5DC	BLOCK-1003
F274	BLOCK-1272
F4-005	BLOCK-1281
F4001	BLOCK-1029
F4002	BLOCK-1030
F4007	BLOCK-1032
F4011	BLOCK-1039
F4011PC	BLOCK-1039
F4012	BLOCK-1040
F4013	BLOCK-1041
F4015	BLOCK-1043
F4016	BLOCK-1048
F4017	BLOCK-1051
F4019	BLOCK-1058
F4020	BLOCK-1059
F4021	BLOCK-1060
F4023	BLOCK-1062
F4024	BLOCK-1063
F4024B	BLOCK-1063
F4025	BLOCK-1064
F4025B	BLOCK-1064
F4026B	BLOCK-1065
F4027	BLOCK-1066
F4029	BLOCK-1068
F4029PC	BLOCK-1068
F4030	BLOCK-1069
F4032B	BLOCK-1071
F4040	BLOCK-1076

DEVICE TYPE	REPL CODE
F4042	BLOCK-1077
F4049	BLOCK-1084
F4049PC	BLOCK-1084
F4050	BLOCK-1085
F4051	BLOCK-1086
F4052	BLOCK-1087
F4075	BLOCK-1101
F4076	BLOCK-1102
F4077BPC	BLOCK-1103
F4081	BLOCK-1105
F4511	BLOCK-1122
F4511PC	BLOCK-1122
F4518	BLOCK-1129
F4528B	BLOCK-1135
F4846-1	BLOCK-2073
F5404DM	BLOCK-1297
F6810CP	BLOCK-1168
F6810CS	BLOCK-1168
F6810DM	BLOCK-1168
F6810P	BLOCK-1168
F6810S	BLOCK-1168
F6821CP	BLOCK-1169
F6821DM	BLOCK-1169
F6821P	BLOCK-1169
F6821S	BLOCK-1169
F6850CP	BLOCK-1170
F6850CS	BLOCK-1170
F6850DL	BLOCK-1170
F6850DM	BLOCK-1170
F6850P	BLOCK-1170
F6850S	BLOCK-1170
F7417PC	BLOCK-1343
F7430PC	BLOCK-1377
F7474PC	BLOCK-1411
F7476PC	BLOCK-1413
F74S138	BLOCK-1733
F78105AV	BLOCK-2144
F78L05AC	BLOCK-2144
F78L05AV	BLOCK-2144
F78L05AWC	BLOCK-2144
F78L06	BLOCK-2173
F78L062AC	BLOCK-2173
F78L06AC	BLOCK-2173
F78L06C	BLOCK-2173
F78L62AC	BLOCK-2173
F78L62AWC	BLOCK-2173
F78L62WV	BLOCK-2173
F9316PC	BLOCK-1338
FA-6008T	BLOCK-40
FA6001D	BLOCK-1212
FB274	BLOCK-43
FC74LS196N	BLOCK-1624
FC74LS90	BLOCK-1714
FC74LS93	BLOCK-1717
FD-1073-BF	BLOCK-1293
FD-1073-BG	BLOCK-1295
FD-1073-BH	BLOCK-1296
FD-1073-BJ	BLOCK-1297
FD-1073-BM	BLOCK-1301
FD-1073-BN	BLOCK-1304
FD-1073-BR	BLOCK-1365
FD-1073-BS	BLOCK-1377
FD-1073-BU	BLOCK-1390
FD-1073-BW	BLOCK-1402
FD-1073-CA	BLOCK-1419
FF274	BLOCK-1272
FH2000571-4	BLOCK-843
FH70401E	BLOCK-1245
FJH101	BLOCK-1377
FJH101-7430	BLOCK-1377
FJH111	BLOCK-1365
FJH111-7420	BLOCK-1365
FJH121	BLOCK-1304
FJH121-7410	BLOCK-1304
FJH131	BLOCK-1293
FJH131-7400	BLOCK-1293
FJH141	BLOCK-1390
FJH141-7440	BLOCK-1390
FJH151	BLOCK-1401
FJH151-7450	BLOCK-1401
FJH161	BLOCK-1402
FJH161-7451	BLOCK-1402
FJH171	BLOCK-1403
FJH171-7453	BLOCK-1403
FJH181	BLOCK-1404
FJH181-7454	BLOCK-1404
FJH191	BLOCK-1414
FJH201	BLOCK-1416
FJH211	BLOCK-1417
FJH221	BLOCK-1295
FJH221-7402	BLOCK-1295
FJH231	BLOCK-1294
FJH231-7401	BLOCK-1294
FJH241	BLOCK-1297

DEVICE TYPE	REPL CODE
FJH251	BLOCK-1298
FJH261	BLOCK-1392
FJH261-7442	BLOCK-1392
FJH291	BLOCK-1296
FJH291-7403	BLOCK-1296
FJH311	BLOCK-1294
FJH311-7401	BLOCK-1294
FJH321	BLOCK-1298
FJJ101	BLOCK-1408
FJJ111	BLOCK-1409
FJJ111-7472	BLOCK-1409
FJJ121	BLOCK-1410
FJJ121-7473	BLOCK-1410
FJJ131	BLOCK-1411
FJJ131-7474	BLOCK-1411
FJJ141	BLOCK-1423
FJJ141-7490	BLOCK-1423
FJJ152	BLOCK-1425
FJJ181	BLOCK-1412
FJJ181-7475	BLOCK-1412
FJJ191	BLOCK-1413
FJJ191-7476	BLOCK-1413
FJJ211	BLOCK-1426
FJJ211-7493	BLOCK-1426
FJJ251-7492	BLOCK-1425
FJJ261	BLOCK-1305
FJJ261-74107	BLOCK-1305
FJK101	BLOCK-1311
FJK101-74121	BLOCK-1311
FJL101	BLOCK-1391
FJY101	BLOCK-1406
FJY101-7460	BLOCK-1406
FL274	BLOCK-1269
FLH101	BLOCK-1293
FLH111	BLOCK-1304
FLH121	BLOCK-1365
FLH131	BLOCK-1377
FLH141	BLOCK-1390
FLH151	BLOCK-1401
FLH161	BLOCK-1402
FLH171	BLOCK-1403
FLH181	BLOCK-1404
FLH191	BLOCK-1295
FLH201	BLOCK-1294
FLH211	BLOCK-1297
FLH221	BLOCK-1414
FLH231	BLOCK-1416
FLH241	BLOCK-1417
FLH271	BLOCK-1298
FLH281	BLOCK-1392
FLH291	BLOCK-1296
FLH341	BLOCK-1419
FLH351	BLOCK-1317
FLH361	BLOCK-1393
FLH371	BLOCK-1394
FLH381	BLOCK-1301
FLH391	BLOCK-1302
FLH401	BLOCK-1353
FLH411	BLOCK-1354
FLH421	BLOCK-1352
FLH431	BLOCK-1418
FLH441	BLOCK-1505
FLH451	BLOCK-1483
FLH481	BLOCK-1299
FLH481T	BLOCK-1336
FLH491	BLOCK-1300
FLH491T	BLOCK-1343
FLH501	BLOCK-1310
FLH511	BLOCK-1369
FLH531	BLOCK-1384
FLH541	BLOCK-1385
FLH551	BLOCK-1398
FLH581	BLOCK-1307
FLH601	BLOCK-1318
FLH611	BLOCK-1367
FLH621	BLOCK-1372
FLH631	BLOCK-1378
FLJ101	BLOCK-1408
FLJ111	BLOCK-1409
FLJ121	BLOCK-1410
FLJ131	BLOCK-1413
FLJ141	BLOCK-1411
FLJ151	BLOCK-1412
FLJ161	BLOCK-1423
FLJ171	BLOCK-1425
FLJ181	BLOCK-1426
FLJ191	BLOCK-1428
FLJ201	BLOCK-1355
FLJ211	BLOCK-1356
FLJ221	BLOCK-1424
FLJ231	BLOCK-1427
FLJ241	BLOCK-1357
FLJ251	BLOCK-1358
FLJ261	BLOCK-1429

DEVICE TYPE	REPL CODE
FLJ271	BLOCK-1305
FLJ311	BLOCK-1362
FLJ321	BLOCK-1363
FLJ331	BLOCK-1430
FLJ381	BLOCK-1360
FLJ391	BLOCK-1361
FLJ401	BLOCK-1337
FLJ411	BLOCK-1338
FLJ421	BLOCK-1339
FLJ431	BLOCK-1340
FLJ441	BLOCK-1341
FLJ451	BLOCK-1342
FLJ531	BLOCK-1346
FLJ541	BLOCK-1347
FLJ561	BLOCK-1359
FLK101	BLOCK-1311
FLK111	BLOCK-1312
FLK121	BLOCK-1313
FLL101	BLOCK-1321
FLL111	BLOCK-1395
FLL111T	BLOCK-1325
FLL121	BLOCK-1396
FLL121T	BLOCK-1397
FLL121U	BLOCK-1396
FLL121V	BLOCK-1397
FLQ101	BLOCK-1420
FLQ111	BLOCK-1415
FLQ131	BLOCK-1344
FS-7812	BLOCK-2097
FS7812	BLOCK-2097
FU5F7715393	BLOCK-1967
FU6A7709393	BLOCK-1954
FUL914-28	BLOCK-2189
FUN14LH026	BLOCK-1277
FV5D770339	BLOCK-1212
FX0014CE	BLOCK-1021
FX274	BLOCK-1271
FZL111	BLOCK-2041
G09-006-A	BLOCK-28
G09-006-B	BLOCK-28
G09-006-C	BLOCK-28
G09-007-A	BLOCK-101
G09-007-B	BLOCK-101
G09-008-B	BLOCK-85
G09-008-C	BLOCK-85
G09-008-D	BLOCK-85
G09-008-E	BLOCK-85
G09-009-A	BLOCK-7
G09-010-A	BLOCK-28
G09-011-A	BLOCK-37
G09-012-A	BLOCK-37
G09-012-B	BLOCK-37
G09-012-C	BLOCK-37
G09-013-A	BLOCK-56
G09-013-A(1)	BLOCK-56
G09-015-B	BLOCK-4
G09-017-A	BLOCK-646
G09-017-B	BLOCK-646
G09-017-C	BLOCK-646
G09-017-D	BLOCK-646
G09-018-A	BLOCK-1813
G09-028-A	BLOCK-1825
G09-029-B	BLOCK-50
G09-029-C	BLOCK-101
G09-029-D	BLOCK-101
G09-029B	BLOCK-101
G09-034-A	BLOCK-1825
G09007	BLOCK-101
G09009	BLOCK-7
G09015	BLOCK-4
G390503S1	BLOCK-1967
G612994	BLOCK-2177
GD4052B	BLOCK-1087
GD4066	BLOCK-1093
GD4066B	BLOCK-1093
GD4066BD	BLOCK-1093
GD4066BP	BLOCK-1093
GD74LS00	BLOCK-1568
GD74LS04	BLOCK-1572
GD74LS125A	BLOCK-1585
GD74LS138	BLOCK-1591
GD74LS157	BLOCK-1602
GD74LS158	BLOCK-1603
GD74LS175	BLOCK-1616
GD74LS21	BLOCK-1627
GD74LS240	BLOCK-1630
GD74LS244	BLOCK-1634
GD74LS245	BLOCK-1635
GD74LS32	BLOCK-1659
GD74LS373	BLOCK-1671
GD74LS74A	BLOCK-1706
GD74S04	BLOCK-1722
GD74S08	BLOCK-1724
GD74S74	BLOCK-1768

DEVICE TYPE	REPL CODE
GD75188	BLOCK-1771
GD75189A	BLOCK-1772
GE-1005	BLOCK-6
GE-1021	BLOCK-20
GE-1038	BLOCK-37
GE-1110	BLOCK-108
GE-1127	BLOCK-125
GE-1162	BLOCK-151
GE-1164	BLOCK-153
GE-1169	BLOCK-158
GE-117	BLOCK-73
GE-1175	BLOCK-164
GE-1178	BLOCK-167
GE-1179	BLOCK-168
GE-1183	BLOCK-172
GE-1192	BLOCK-180
GE-1193	BLOCK-181
GE-1194	BLOCK-182
GE-1196	BLOCK-184
GE-1198	BLOCK-186
GE-1211	BLOCK-199
GE-1218	BLOCK-206
GE-1219	BLOCK-207
GE-1228	BLOCK-216
GE-1231	BLOCK-219
GE-1232	BLOCK-221
GE-1239	BLOCK-228
GE-1257	BLOCK-246
GE-1258	BLOCK-247
GE-205	BLOCK-1223
GE-309K	BLOCK-1022
GE-4000	BLOCK-1028
GE-4001	BLOCK-1029
GE-4002	BLOCK-1030
GE-4007	BLOCK-1032
GE-4009	BLOCK-1084
GE-4011	BLOCK-1039
GE-4012	BLOCK-1040
GE-4013	BLOCK-1041
GE-4015	BLOCK-1043
GE-4016	BLOCK-1048
GE-4017	BLOCK-1051
GE-4019	BLOCK-1058
GE-4020	BLOCK-1059
GE-4021	BLOCK-1060
GE-4023	BLOCK-1062
GE-4024	BLOCK-1063
GE-4025	BLOCK-1064
GE-4027	BLOCK-1066
GE-4030	BLOCK-1069
GE-4040	BLOCK-1076
GE-4042	BLOCK-1077
GE-4049	BLOCK-1084
GE-4050	BLOCK-1085
GE-4051	BLOCK-1086
GE-4052	BLOCK-1087
GE-4055	BLOCK-1089
GE-4081	BLOCK-1105
GE-4518	BLOCK-1129
GE-706	BLOCK-1245
GE-720	BLOCK-23
GE-721	BLOCK-26
GE-722	BLOCK-24
GE-723	BLOCK-88
GE-724	BLOCK-27
GE-725	BLOCK-200
GE-730	BLOCK-1286
GE-7400	BLOCK-1293
GE-7402	BLOCK-1295
GE-7404	BLOCK-1297
GE-7406	BLOCK-1299
GE-7408	BLOCK-1301
GE-7410	BLOCK-1304
GE-74123	BLOCK-1313
GE-7413	BLOCK-1317
GE-74145	BLOCK-1325
GE-74150	BLOCK-1327
GE-74154	BLOCK-1331
GE-7416	BLOCK-1336
GE-7417	BLOCK-1343
GE-74192	BLOCK-1357
GE-74193	BLOCK-1358
GE-74196	BLOCK-1360
GE-7420	BLOCK-1365
GE-7427	BLOCK-1372
GE-7430	BLOCK-1377
GE-7432	BLOCK-1378
GE-7441	BLOCK-1391
GE-7447	BLOCK-1397
GE-7448	BLOCK-1398
GE-7451	BLOCK-1402
GE-7473	BLOCK-1410
GE-7474	BLOCK-1411
GE-7475	BLOCK-1412

DEVICE TYPE	REPL CODE
GE-7476	BLOCK-1413
GE-7485	BLOCK-1418
GE-7486	BLOCK-1419
GE-7490	BLOCK-1423
GE-7492	BLOCK-1425
GE-761	BLOCK-1791
GE-792	BLOCK-1817
GE-793	BLOCK-1818
GE-794	BLOCK-1819
GE-812	BLOCK-1839
GE-818	BLOCK-1846
GE-822	BLOCK-1850
GE-961	BLOCK-2091
GE-962	BLOCK-2093
GE-972	BLOCK-2137
GE-973	BLOCK-2138
GE-984	BLOCK-2168
GE-IC11	BLOCK-1264
GE-IC2	BLOCK-1269
GE-IC6	BLOCK-1272
GE-IC7	BLOCK-1277
GE-IC8	BLOCK-1275
GE1C-166	BLOCK-33
GEIC-10	BLOCK-1262
GEIC-101	BLOCK-103
GEIC-102	BLOCK-21
GEIC-103	BLOCK-85
GEIC-104	BLOCK-83
GEIC-105	BLOCK-126
GEIC-106	BLOCK-131
GEIC-107	BLOCK-130
GEIC-109	BLOCK-128
GEIC-11	BLOCK-1264
GEIC-110	BLOCK-129
GEIC-111	BLOCK-127
GEIC-113	BLOCK-232
GEIC-114	BLOCK-225
GEIC-115	BLOCK-1421
GEIC-116	BLOCK-45
GEIC-117	BLOCK-73
GEIC-118	BLOCK-44
GEIC-119	BLOCK-122
GEIC-12	BLOCK-1212
GEIC-120	BLOCK-87
GEIC-121	BLOCK-47
GEIC-122	BLOCK-46
GEIC-123	BLOCK-48
GEIC-124	BLOCK-75
GEIC-125	BLOCK-74
GEIC-126	BLOCK-89
GEIC-127	BLOCK-82
GEIC-128	BLOCK-138
GEIC-13	BLOCK-1287
GEIC-130	BLOCK-90
GEIC-131	BLOCK-49
GEIC-132	BLOCK-77
GEIC-133	BLOCK-48
GEIC-135	BLOCK-50
GEIC-136	BLOCK-76
GEIC-137	BLOCK-173
GEIC-138	BLOCK-136
GEIC-139	BLOCK-174
GEIC-140	BLOCK-80
GEIC-142	BLOCK-84
GEIC-143	BLOCK-59
GEIC-147	BLOCK-1269
GEIC-148	BLOCK-1269
GEIC-149	BLOCK-5
GEIC-15	BLOCK-1279
GEIC-150	BLOCK-201
GEIC-153	BLOCK-217
GEIC-154	BLOCK-226
GEIC-155	BLOCK-1825
GEIC-156	BLOCK-120
GEIC-157	BLOCK-92
GEIC-158	BLOCK-119
GEIC-159	BLOCK-38
GEIC-16	BLOCK-1290
GEIC-160	BLOCK-40
GEIC-161	BLOCK-39
GEIC-162	BLOCK-28
GEIC-163	BLOCK-29
GEIC-164	BLOCK-31
GEIC-165	BLOCK-32
GEIC-166	BLOCK-33
GEIC-167	BLOCK-34
GEIC-168	BLOCK-35
GEIC-169	BLOCK-42
GEIC-17	BLOCK-1289
GEIC-170	BLOCK-36
GEIC-171	BLOCK-646
GEIC-172	BLOCK-1963
GEIC-173	BLOCK-160
GEIC-175	BLOCK-175

DEVICE TYPE	REPL CODE
GEIC-176	BLOCK-176
GEIC-178	BLOCK-193
GEIC-179	BLOCK-145
GEIC-18	BLOCK-1815
GEIC-180	BLOCK-1941
GEIC-181	BLOCK-223
GEIC-182	BLOCK-143
GEIC-183	BLOCK-1421
GEIC-184	BLOCK-60
GEIC-185	BLOCK-1949
GEIC-19	BLOCK-1281
GEIC-190	BLOCK-2087
GEIC-191	BLOCK-2095
GEIC-193	BLOCK-1411
GEIC-194	BLOCK-1294
GEIC-196	BLOCK-149
GEIC-197	BLOCK-79
GEIC-199	BLOCK-161
GEIC-2	BLOCK-1269
GEIC-20	BLOCK-1805
GEIC-201	BLOCK-91
GEIC-202	BLOCK-209
GEIC-203	BLOCK-218
GEIC-204	BLOCK-213
GEIC-205	BLOCK-1223
GEIC-207	BLOCK-1267
GEIC-208	BLOCK-1273
GEIC-209	BLOCK-1274
GEIC-21	BLOCK-1808
GEIC-210	BLOCK-1283
GEIC-212	BLOCK-1288
GEIC-213	BLOCK-1364
GEIC-214	BLOCK-1376
GEIC-215	BLOCK-1389
GEIC-216	BLOCK-1400
GEIC-217	BLOCK-1405
GEIC-218	BLOCK-1407
GEIC-219	BLOCK-1770
GEIC-22	BLOCK-1284
GEIC-220	BLOCK-1801
GEIC-221	BLOCK-1804
GEIC-222	BLOCK-1805
GEIC-223	BLOCK-1806
GEIC-224	BLOCK-1807
GEIC-225	BLOCK-1808
GEIC-226	BLOCK-1810
GEIC-227	BLOCK-1811
GEIC-228	BLOCK-1812
GEIC-229	BLOCK-1813
GEIC-23	BLOCK-1285
GEIC-230	BLOCK-1815
GEIC-231	BLOCK-1816
GEIC-232	BLOCK-1820
GEIC-233	BLOCK-1821
GEIC-234	BLOCK-1822
GEIC-235	BLOCK-1826
GEIC-236	BLOCK-1809
GEIC-237	BLOCK-1827
GEIC-239	BLOCK-1829
GEIC-24	BLOCK-1389
GEIC-240	BLOCK-1830
GEIC-241	BLOCK-1838
GEIC-242	BLOCK-1839
GEIC-243	BLOCK-1842
GEIC-244	BLOCK-1843
GEIC-245	BLOCK-1940
GEIC-246	BLOCK-1946
GEIC-247	BLOCK-1947
GEIC-248	BLOCK-1948
GEIC-249	BLOCK-1949
GEIC-25	BLOCK-2073
GEIC-250	BLOCK-1954
GEIC-251	BLOCK-1955
GEIC-252	BLOCK-1957
GEIC-253	BLOCK-1958
GEIC-254	BLOCK-1962
GEIC-255	BLOCK-1964
GEIC-256	BLOCK-1966
GEIC-257	BLOCK-1970
GEIC-258	BLOCK-1971
GEIC-259	BLOCK-1983
GEIC-26	BLOCK-1421
GEIC-260	BLOCK-1984
GEIC-261	BLOCK-1987
GEIC-262	BLOCK-2051
GEIC-263	BLOCK-2056
GEIC-264	BLOCK-2057
GEIC-265	BLOCK-2058
GEIC-266	BLOCK-2068
GEIC-267	BLOCK-2069
GEIC-268	BLOCK-2069
GEIC-269	BLOCK-2079
GEIC-27	BLOCK-1828
GEIC-271	BLOCK-8

DEVICE TYPE	REPL CODE
GEIC-272	BLOCK-22
GEIC-273	BLOCK-30
GEIC-274	BLOCK-37
GEIC-275	BLOCK-81
GEIC-276	BLOCK-105
GEIC-278	BLOCK-112
GEIC-279	BLOCK-114
GEIC-28	BLOCK-1276
GEIC-280	BLOCK-115
GEIC-281	BLOCK-121
GEIC-282	BLOCK-123
GEIC-284	BLOCK-133
GEIC-285	BLOCK-140
GEIC-286	BLOCK-141
GEIC-287	BLOCK-144
GEIC-288	BLOCK-1943
GEIC-289	BLOCK-1944
GEIC-29	BLOCK-1291
GEIC-290	BLOCK-1945
GEIC-291	BLOCK-231
GEIC-292	BLOCK-237
GEIC-293	BLOCK-250
GEIC-295	BLOCK-211
GEIC-296	BLOCK-212
GEIC-298	BLOCK-432
GEIC-299	BLOCK-438
GEIC-3	BLOCK-1244
GEIC-30	BLOCK-1292
GEIC-300	BLOCK-150
GEIC-301	BLOCK-153
GEIC-302	BLOCK-162
GEIC-303	BLOCK-171
GEIC-304	BLOCK-2141
GEIC-31	BLOCK-1303
GEIC-311	BLOCK-96
GEIC-312	BLOCK-94
GEIC-313	BLOCK-124
GEIC-314	BLOCK-132
GEIC-315	BLOCK-154
GEIC-316	BLOCK-155
GEIC-317	BLOCK-156
GEIC-318	BLOCK-132
GEIC-319	BLOCK-108
GEIC-32	BLOCK-1376
GEIC-320	BLOCK-151
GEIC-321	BLOCK-153
GEIC-322	BLOCK-158
GEIC-323	BLOCK-167
GEIC-324	BLOCK-168
GEIC-325	BLOCK-172
GEIC-326	BLOCK-180
GEIC-327	BLOCK-182
GEIC-328	BLOCK-184
GEIC-329	BLOCK-199
GEIC-33	BLOCK-1816
GEIC-330	BLOCK-206
GEIC-331	BLOCK-207
GEIC-332	BLOCK-216
GEIC-333	BLOCK-219
GEIC-334	BLOCK-221
GEIC-335	BLOCK-228
GEIC-34	BLOCK-1823
GEIC-35	BLOCK-1825
GEIC-36	BLOCK-5
GEIC-37	BLOCK-10
GEIC-38	BLOCK-7
GEIC-39	BLOCK-119
GEIC-4	BLOCK-1271
GEIC-40	BLOCK-120
GEIC-41	BLOCK-78
GEIC-42	BLOCK-6
GEIC-43	BLOCK-1245
GEIC-44	BLOCK-51
GEIC-45	BLOCK-52
GEIC-46	BLOCK-55
GEIC-47	BLOCK-53
GEIC-48	BLOCK-54
GEIC-49	BLOCK-56
GEIC-5	BLOCK-1270
GEIC-50	BLOCK-58
GEIC-51	BLOCK-59
GEIC-52	BLOCK-60
GEIC-53	BLOCK-61
GEIC-54	BLOCK-62
GEIC-55	BLOCK-63
GEIC-56	BLOCK-64
GEIC-57	BLOCK-65
GEIC-58	BLOCK-66
GEIC-59	BLOCK-70
GEIC-6	BLOCK-1272
GEIC-60	BLOCK-72
GEIC-61	BLOCK-170
GEIC-62	BLOCK-134
GEIC-63	BLOCK-71

DEVICE TYPE	REPL CODE
GEIC-64	BLOCK-67
GEIC-65	BLOCK-68
GEIC-66	BLOCK-157
GEIC-67	BLOCK-69
GEIC-68	BLOCK-152
GEIC-69	BLOCK-3
GEIC-7	BLOCK-1277
GEIC-70	BLOCK-215
GEIC-71	BLOCK-147
GEIC-72	BLOCK-148
GEIC-73	BLOCK-92
GEIC-74	BLOCK-203
GEIC-75	BLOCK-168
GEIC-76	BLOCK-205
GEIC-77	BLOCK-9
GEIC-79	BLOCK-11
GEIC-8	BLOCK-1275
GEIC-80	BLOCK-169
GEIC-81	BLOCK-1282
GEIC-82	BLOCK-177
GEIC-83	BLOCK-1282
GEIC-84	BLOCK-1223
GEIC-86	BLOCK-1280
GEIC-87	BLOCK-106
GEIC-88	BLOCK-59
GEIC-89	BLOCK-1265
GEIC-90	BLOCK-104
GEIC-91	BLOCK-102
GEIC-92	BLOCK-98
GEIC-93	BLOCK-100
GEIC-94	BLOCK-101
GEIC-95	BLOCK-99
GEIC-96	BLOCK-5
GEIC-97	BLOCK-1422
GEIC-98	BLOCK-78
GEIC-99	BLOCK-107
GEIC191	BLOCK-2095
GEIC197	BLOCK-149
GEL2111	BLOCK-1262
GEL2111AL1	BLOCK-1262
GEL2111F1	BLOCK-1262
GEL2113	BLOCK-1264
GEL2113AL1	BLOCK-1264
GEL2113F1	BLOCK-1264
GEL2114	BLOCK-1270
GEL211AL1	BLOCK-1262
GEL211F1	BLOCK-1262
GEL3072F1	BLOCK-1815
GEVR-100	BLOCK-2144
GEVR-101	BLOCK-2087
GEVR-102	BLOCK-2087
GEVR-104	BLOCK-2091
GEVR-105	BLOCK-2091
GEVR-106	BLOCK-2160
GEVR-107	BLOCK-2095
GEVR-108	BLOCK-2095
GEVR-109	BLOCK-2074
GEVR-110	BLOCK-2097
GEVR-111	BLOCK-2097
GEVR-112	BLOCK-882
GEVR-113	BLOCK-2108
GEVR-114	BLOCK-2108
GEVR-115	BLOCK-1983
GEVR-116	BLOCK-1983
GL-3201	BLOCK-1269
GL1010	BLOCK-383
GL1130	BLOCK-754
GL3101A	BLOCK-706
GL3101B	BLOCK-706
GL3120	BLOCK-709
GL3201	BLOCK-1269
GL3301	BLOCK-636
GL3301B	BLOCK-636
GL3320	BLOCK-824
GL358	BLOCK-1993
GL4558	BLOCK-1801
GL7805	BLOCK-2087
GL7806	BLOCK-2093
GL7812	BLOCK-2097
GL7815	BLOCK-2119
GM320MP-5.2	BLOCK-2091
GN13G	BLOCK-1021
H-IX0065CE	BLOCK-537
H102D1	BLOCK-2111
H102D2	BLOCK-2111
H102D6	BLOCK-2111
H103D1	BLOCK-2110
H103D2	BLOCK-2110
H103D6	BLOCK-2110
H104D1	BLOCK-2098
H104D2	BLOCK-2098
H104D6	BLOCK-2098
H105D1	BLOCK-2112
H105D2	BLOCK-2112

DEVICE TYPE	REPL CODE
H105D6	BLOCK-2112
H112D1	BLOCK-2121
H112D6	BLOCK-2121
H115D1	BLOCK-2117
H115D6	BLOCK-2117
H118D1	BLOCK-2120
H118D6	BLOCK-2120
H119D1	BLOCK-2116
H119D6	BLOCK-2116
H122D1	BLOCK-2106
H122D2	BLOCK-2106
H122D6	BLOCK-2106
H124D1	BLOCK-2099
H124D2	BLOCK-2099
H124D6	BLOCK-2099
H202B1	BLOCK-2111
H203B1	BLOCK-2111
H204B1	BLOCK-2098
H205B1	BLOCK-2112
H215	BLOCK-1286
H221	BLOCK-1808
H222B1	BLOCK-2106
H224B1	BLOCK-2099
H31100127	BLOCK-1043
H31100128	BLOCK-1105
H31100129	BLOCK-1084
H31100130	BLOCK-1768
HA-1202	BLOCK-40
HA-1306P	BLOCK-28
HA-1339	BLOCK-158
HA1-4741-5	BLOCK-2071
HA1108	BLOCK-59
HA11103	BLOCK-1269
HA11107	BLOCK-1269
HA11112	BLOCK-178
HA11115W	BLOCK-1277
HA1111SW	BLOCK-1277
HA1112	BLOCK-178
HA1115	BLOCK-1277
HA1115(W)	BLOCK-1277
HA1115W	BLOCK-1277
HA1117	BLOCK-120
HA1118	BLOCK-92
HA1119	BLOCK-119
HA11211	BLOCK-474
HA11215A	BLOCK-457
HA11219	BLOCK-530
HA11220	BLOCK-392
HA11221	BLOCK-480
HA11223W	BLOCK-470
HA11225	BLOCK-476
HA11226	BLOCK-929
HA11227	BLOCK-552
HA11228	BLOCK-201
HA11229	BLOCK-562
HA11231	BLOCK-1179
HA11235	BLOCK-537
HA11238	BLOCK-655
HA11238N	BLOCK-655
HA1124	BLOCK-1269
HA11244	BLOCK-531
HA11247	BLOCK-283
HA1124D	BLOCK-1269
HA1124DS	BLOCK-1269
HA1124S	BLOCK-1269
HA1124Z	BLOCK-1269
HA1125	BLOCK-1269
HA11251	BLOCK-478
HA11251A	BLOCK-457
HA1125D	BLOCK-1269
HA1125S	BLOCK-1269
HA1126	BLOCK-5
HA1126AW	BLOCK-5
HA1126D	BLOCK-5
HA1126DW	BLOCK-5
HA1128	BLOCK-1269
HA1128E	BLOCK-1269
HA1133	BLOCK-1286
HA1137	BLOCK-1813
HA1137(W)	BLOCK-1813
HA1137P	BLOCK-1813
HA1137W	BLOCK-1813
HA1138	BLOCK-482
HA1139A	BLOCK-158
HA11401	BLOCK-388
HA11409	BLOCK-580
HA1141	BLOCK-1269
HA11412	BLOCK-461
HA11412A	BLOCK-461
HA11413	BLOCK-544
HA11414	BLOCK-390
HA11417	BLOCK-587
HA11423	BLOCK-459
HA11431	BLOCK-576
HA11431NT	BLOCK-576
HA114323	BLOCK-459
HA11434	BLOCK-461
HA11436	BLOCK-636
HA11436A	BLOCK-636
HA1144	BLOCK-201
HA11440	BLOCK-706
HA11440A	BLOCK-706
HA11446	BLOCK-636
HA1146	BLOCK-5
HA1148	BLOCK-217
HA11480	BLOCK-636
HA11480A	BLOCK-636
HA1150	BLOCK-1825
HA1151	BLOCK-226
HA11510NT	BLOCK-1187
HA11516C4	BLOCK-226
HA11516E1	BLOCK-226
HA11516J	BLOCK-226
HA1152	BLOCK-78
HA1154	BLOCK-1269
HA1156	BLOCK-1825
HA1156(W)	BLOCK-1825
HA1156-6C	BLOCK-1825
HA1156H	BLOCK-1825
HA1156W	BLOCK-1825
HA1156WP	BLOCK-1825
HA1157	BLOCK-120
HA1158	BLOCK-92
HA1158-O	BLOCK-92
HA11580	BLOCK-184
HA1159	BLOCK-119
HA11701	BLOCK-572
HA11702	BLOCK-570
HA11703	BLOCK-582
HA11706	BLOCK-584
HA11711	BLOCK-578
HA11714	BLOCK-587
HA11715	BLOCK-541
HA11747-A	BLOCK-686
HA11747A	BLOCK-686
HA11747ANT	BLOCK-686
HA11747BNT	BLOCK-686
HA11749	BLOCK-672
HA1177	BLOCK-184
HA1196	BLOCK-472
HA1197	BLOCK-202
HA1197W	BLOCK-202
HA1199	BLOCK-277
HA1199P	BLOCK-277
HA12002	BLOCK-622
HA12002-W	BLOCK-622
HA12003	BLOCK-214
HA1201	BLOCK-38
HA12013	BLOCK-453
HA1202	BLOCK-40
HA12026	BLOCK-237
HA1203	BLOCK-41
HA1211	BLOCK-39
HA12411	BLOCK-566
HA12412	BLOCK-546
HA12413	BLOCK-544
HA12413-03	BLOCK-544
HA1301	BLOCK-1943
HA1304	BLOCK-2056
HA1306	BLOCK-28
HA1306P	BLOCK-28
HA1306PU	BLOCK-28
HA1306W	BLOCK-28
HA1306WU	BLOCK-28
HA1308	BLOCK-29
HA1310	BLOCK-30
HA1311	BLOCK-31
HA1311W	BLOCK-31
HA1312	BLOCK-32
HA1313	BLOCK-33
HA1316	BLOCK-35
HA1318PU	BLOCK-37
HA1318PU-1	BLOCK-37
HA1318PU-2	BLOCK-37
HA1318PU-3	BLOCK-37
HA1319	BLOCK-42
HA1322	BLOCK-36
HA1322C	BLOCK-36
HA1325	BLOCK-227
HA1338	BLOCK-29
HA1339	BLOCK-158
HA1339A	BLOCK-158
HA13403	BLOCK-872
HA1342	BLOCK-228
HA13421	BLOCK-1185
HA13421A	BLOCK-1185
HA1342A	BLOCK-228
HA1353	BLOCK-78
HA1364	BLOCK-219
HA1366R	BLOCK-248
HA1366WR	BLOCK-248
HA1374	BLOCK-384
HA1374A	BLOCK-384
HA1377	BLOCK-385
HA1377A	BLOCK-385
HA1388	BLOCK-386
HA1389R	BLOCK-380
HA1392	BLOCK-593
HA1393	BLOCK-589
HA1406	BLOCK-646
HA1406-2	BLOCK-646
HA1406-3	BLOCK-646
HA1406-4	BLOCK-646
HA1452	BLOCK-468
HA1452W	BLOCK-468
HA17094PS	BLOCK-1993
HA17301G	BLOCK-2191
HA17301P	BLOCK-2191
HA17339	BLOCK-1873
HA17358	BLOCK-1993
HA17393	BLOCK-2063
HA17408P	BLOCK-971
HA17458GS	BLOCK-1801
HA17458PS	BLOCK-1801
HA17524P	BLOCK-701
HA17555	BLOCK-2079
HA17555PS	BLOCK-2079
HA17558	BLOCK-1801
HA17723G	BLOCK-1984
HA17741	BLOCK-2058
HA17741G	BLOCK-2058
HA17741GS	BLOCK-2058
HA17741PS	BLOCK-2058
HA17747G	BLOCK-2070
HA17747P	BLOCK-2070
HA17805	BLOCK-2087
HA17805H	BLOCK-2087
HA17805P	BLOCK-2087
HA17806P	BLOCK-2093
HA17808P	BLOCK-2095
HA17812	BLOCK-2097
HA17812P	BLOCK-2097
HA17815P	BLOCK-2119
HA17818P	BLOCK-2085
HA17824P	BLOCK-2137
HA178M05	BLOCK-2087
HA178M05P	BLOCK-2087
HA178M06P	BLOCK-2093
HA178M08	BLOCK-2095
HA178M08P	BLOCK-2095
HA178M12P	BLOCK-2097
HA178M15P	BLOCK-2119
HA178M18P	BLOCK-2085
HA178M24P	BLOCK-2137
HA17901G	BLOCK-1873
HA17901P	BLOCK-1873
HA17902G	BLOCK-2171
HA17902P	BLOCK-2171
HA17904	BLOCK-1993
HA17904PS	BLOCK-1993
HA3-4741-5	BLOCK-2071
HA4741-5	BLOCK-2071
HA5062-5	BLOCK-1910
HA5084-5	BLOCK-1912
HBC4000AD	BLOCK-1028
HBC4000AF	BLOCK-1028
HBC4000AK	BLOCK-1028
HBC4001AD	BLOCK-1029
HBC4001AF	BLOCK-1029
HBC4001AK	BLOCK-1029
HBC4002AD	BLOCK-1030
HBC4002AF	BLOCK-1030
HBC4002AK	BLOCK-1030
HBC4008AD	BLOCK-1034
HBC4008AF	BLOCK-1034
HBC4008AK	BLOCK-1034
HBC4011AD	BLOCK-1039
HBC4011AF	BLOCK-1039
HBC4011AK	BLOCK-1039
HBC4012AD	BLOCK-1040
HBC4012AF	BLOCK-1040
HBC4012AK	BLOCK-1040
HBC4013AD	BLOCK-1041
HBC4013AF	BLOCK-1041
HBC4013AK	BLOCK-1041
HBC4017AD	BLOCK-1051
HBC4017AF	BLOCK-1051
HBC4017AK	BLOCK-1051
HBC4018AD	BLOCK-1053
HBC4018AF	BLOCK-1053
HBC4018AK	BLOCK-1053
HBC4020AD	BLOCK-1059
HBC4020AF	BLOCK-1059
HBC4020AK	BLOCK-1059
HBC4022AD	BLOCK-1061
HBC4022AE	BLOCK-1061
HBC4022AF	BLOCK-1061
HBC4022AK	BLOCK-1061
HBC4023AD	BLOCK-1062
HBC4023AF	BLOCK-1062
HBC4023AK	BLOCK-1062
HBC4024AD	BLOCK-1063
HBC4024AF	BLOCK-1063
HBC4024AK	BLOCK-1063
HBC4025AD	BLOCK-1064
HBC4025AF	BLOCK-1064
HBC4025AK	BLOCK-1064
HBC4027AD	BLOCK-1066
HBC4027AF	BLOCK-1066
HBC4027AK	BLOCK-1066
HBC4028AD	BLOCK-1067
HBC4028AF	BLOCK-1067
HBC4028AK	BLOCK-1067
HBC4029AD	BLOCK-1068
HBC4029AF	BLOCK-1068
HBC4029AK	BLOCK-1068
HBC4030AD	BLOCK-1069
HBC4030AF	BLOCK-1069
HBC4030AK	BLOCK-1069
HBC4040AD	BLOCK-1076
HBC4040AF	BLOCK-1076
HBC4040AK	BLOCK-1076
HBC4042AD	BLOCK-1077
HBC4042AF	BLOCK-1077
HBC4042AK	BLOCK-1077
HBC4043AD	BLOCK-1078
HBC4043AF	BLOCK-1078
HBC4043AK	BLOCK-1078
HBC4044AD	BLOCK-1079
HBC4044AF	BLOCK-1079
HBC4044AK	BLOCK-1079
HBC4047AD	BLOCK-1082
HBC4047AK	BLOCK-1082
HBC4055AD	BLOCK-1089
HBC4055AF	BLOCK-1089
HBC4055AK	BLOCK-1089
HBC4060AD	BLOCK-1091
HBC4060AF	BLOCK-1091
HBC4060AK	BLOCK-1091
HBF4000AE	BLOCK-1028
HBF4000AF	BLOCK-1028
HBF4001	BLOCK-1029
HBF4001A	BLOCK-1029
HBF4001AE	BLOCK-1029
HBF4001AF	BLOCK-1029
HBF4002	BLOCK-1030
HBF4002A	BLOCK-1030
HBF4002AE	BLOCK-1030
HBF4002AF	BLOCK-1030
HBF4006AE	BLOCK-1031
HBF4006AF	BLOCK-1031
HBF4007	BLOCK-1032
HBF4008AE	BLOCK-1034
HBF4008AF	BLOCK-1034
HBF4009	BLOCK-1084
HBF4009A	BLOCK-1084
HBF4011	BLOCK-1039
HBF4011A	BLOCK-1039
HBF4011AE	BLOCK-1039
HBF4011AF	BLOCK-1039
HBF4012	BLOCK-1040
HBF4012A	BLOCK-1040
HBF4012AE	BLOCK-1040
HBF4012AF	BLOCK-1040
HBF4013	BLOCK-1041
HBF4013A	BLOCK-1041
HBF4013AE	BLOCK-1041
HBF4013AF	BLOCK-1041
HBF4014A	BLOCK-1042
HBF4014AE	BLOCK-1042
HBF4015A	BLOCK-1043
HBF4016	BLOCK-1048
HBF4016A	BLOCK-1048
HBF4017	BLOCK-1051
HBF4017A	BLOCK-1051
HBF4017AE	BLOCK-1051
HBF4017AF	BLOCK-1051
HBF4018AE	BLOCK-1053
HBF4018AF	BLOCK-1053
HBF4019A	BLOCK-1058
HBF4020	BLOCK-1059
HBF4020A	BLOCK-1059
HBF4020AE	BLOCK-1059
HBF4020AF	BLOCK-1059
HBF4021	BLOCK-1060
HBF4021A	BLOCK-1060
HBF4022AE	BLOCK-1061
HBF4022AF	BLOCK-1061
HBF4023	BLOCK-1062
HBF4023A	BLOCK-1062
HBF4023AE	BLOCK-1062
HBF4023AF	BLOCK-1062
HBF4024	BLOCK-1063
HBF4024A	BLOCK-1063
HBF4024AE	BLOCK-1063
HBF4024AF	BLOCK-1063
HBF4025	BLOCK-1064
HBF4025A	BLOCK-1064
HBF4025AE	BLOCK-1064
HBF4025AF	BLOCK-1064
HBF4027	BLOCK-1066
HBF4027A	BLOCK-1066
HBF4027AE	BLOCK-1066
HBF4027AF	BLOCK-1066
HBF4028A	BLOCK-1067
HBF4028AE	BLOCK-1067
HBF4028AF	BLOCK-1067
HBF4029A	BLOCK-1068
HBF4029AE	BLOCK-1068
HBF4029AF	BLOCK-1068
HBF4030A	BLOCK-1069
HBF4030AE	BLOCK-1069
HBF4030AF	BLOCK-1069
HBF4032AE	BLOCK-1071
HBF4032AF	BLOCK-1071
HBF4033AE	BLOCK-1072
HBF4033AF	BLOCK-1072
HBF4034AE	BLOCK-1073
HBF4038AE	BLOCK-1075
HBF4038AF	BLOCK-1075
HBF4040A	BLOCK-1076
HBF4040AE	BLOCK-1076
HBF4040AF	BLOCK-1076
HBF4042A	BLOCK-1077
HBF4042AE	BLOCK-1077
HBF4042AF	BLOCK-1077
HBF4043AE	BLOCK-1078
HBF4043AF	BLOCK-1078
HBF4044AE	BLOCK-1079
HBF4044AF	BLOCK-1079
HBF4047AE	BLOCK-1082
HBF4048AE	BLOCK-1083
HBF4048AF	BLOCK-1083
HBF4049A	BLOCK-1084
HBF4050A	BLOCK-1085
HBF4051A	BLOCK-1086
HBF4052A	BLOCK-1087
HBF4055AE	BLOCK-1089
HBF4055AF	BLOCK-1089
HBF4060AE	BLOCK-1091
HBF4060AF	BLOCK-1091
HC1000109	BLOCK-1212
HC10001090	BLOCK-1212
HC1000111-0	BLOCK-1212
HC1000114-0	BLOCK-1282
HC1000117-0	BLOCK-1275
HC10001200	BLOCK-159
HC10002050	BLOCK-98
HC1000217	BLOCK-2056
HC1000217-0	BLOCK-2057
HC10002170	BLOCK-2056
HC10004010	BLOCK-1825
HC1000403	BLOCK-6
HC10004110	BLOCK-1293
HC1000417	BLOCK-1823
HC1000503	BLOCK-19
HC1000505	BLOCK-83
HC1000603	BLOCK-19
HC10006170	BLOCK-1825
HC1000703	BLOCK-26
HC10008060	BLOCK-243
HC1001001	BLOCK-42
HC10019210	BLOCK-230
HC10022010	BLOCK-223
HC10023010	BLOCK-228
HC10034050	BLOCK-223
HC10036010	BLOCK-544
HC10044030	BLOCK-369
HC100603	BLOCK-19
HCC4000BD	BLOCK-1028
HCC4000BF	BLOCK-1028
HCC4000BK	BLOCK-1028
HCC4001BD	BLOCK-1029
HCC4001BF	BLOCK-1029
HCC4001BK	BLOCK-1029
HCC4002BD	BLOCK-1030
HCC4002BF	BLOCK-1030
HCC4002BK	BLOCK-1030
HCC4006BF	BLOCK-1031
HCC4008BD	BLOCK-1034
HCC4008BF	BLOCK-1034
HCC4008BK	BLOCK-1034
HCC40100BD	BLOCK-1037
HCC40100BF	BLOCK-1037
HCC40106BF	BLOCK-1038
HCC4011BD	BLOCK-1039
HCC4011BF	BLOCK-1039
HCC4011BK	BLOCK-1039
HCC4012BD	BLOCK-1040
HCC4012BF	BLOCK-1040
HCC4012BK	BLOCK-1040
HCC4013BD	BLOCK-1041
HCC4013BF	BLOCK-1041
HCC4013BK	BLOCK-1041
HCC40160BD	BLOCK-1044
HCC40160BF	BLOCK-1044
HCC40160BK	BLOCK-1044
HCC40161BD	BLOCK-1045
HCC40161BF	BLOCK-1045
HCC40161BK	BLOCK-1045
HCC40162BF	BLOCK-1046
HCC40162BK	BLOCK-1046
HCC40163BD	BLOCK-1047
HCC40163BF	BLOCK-1047
HCC40163BK	BLOCK-1047
HCC4016BF	BLOCK-1048
HCC40174BD	BLOCK-1049
HCC40174BF	BLOCK-1049
HCC40174BK	BLOCK-1049
HCC4017BD	BLOCK-1051
HCC4017BF	BLOCK-1051
HCC4017BK	BLOCK-1051
HCC40182BF	BLOCK-1052
HCC4018BD	BLOCK-1053
HCC4018BF	BLOCK-1053
HCC4018BK	BLOCK-1053
HCC4019BF	BLOCK-1058
HCC4020BD	BLOCK-1059
HCC4020BF	BLOCK-1059
HCC4020BK	BLOCK-1059
HCC4021BF	BLOCK-1060
HCC4022BD	BLOCK-1061
HCC4022BE	BLOCK-1061
HCC4022BF	BLOCK-1061
HCC4022BK	BLOCK-1061
HCC4023BD	BLOCK-1062
HCC4023BF	BLOCK-1062
HCC4023BK	BLOCK-1062
HCC4024BD	BLOCK-1063
HCC4024BF	BLOCK-1063
HCC4024BK	BLOCK-1063
HCC4025BD	BLOCK-1064
HCC4025BF	BLOCK-1064
HCC4025BK	BLOCK-1064
HCC4026BF	BLOCK-1065
HCC4027BD	BLOCK-1066
HCC4027BF	BLOCK-1066
HCC4027BK	BLOCK-1066
HCC4028BD	BLOCK-1067
HCC4028BF	BLOCK-1067
HCC4028BK	BLOCK-1067
HCC4029BD	BLOCK-1068
HCC4029BF	BLOCK-1068
HCC4029BK	BLOCK-1068
HCC4030BF	BLOCK-1069
HCC4031BF	BLOCK-1070
HCC4032BF	BLOCK-1071
HCC4033BF	BLOCK-1072
HCC4035BF	BLOCK-1074
HCC4038BF	BLOCK-1075
HCC4040BD	BLOCK-1076
HCC4040BF	BLOCK-1076
HCC4040BK	BLOCK-1076
HCC4042BD	BLOCK-1077
HCC4042BF	BLOCK-1077
HCC4042BK	BLOCK-1077
HCC4043BD	BLOCK-1078
HCC4043BF	BLOCK-1078
HCC4043BK	BLOCK-1078
HCC4044BD	BLOCK-1079
HCC4044BF	BLOCK-1079
HCC4044BK	BLOCK-1079
HCC4045BF	BLOCK-1080
HCC4047BF	BLOCK-1082
HCC4048BF	BLOCK-1083
HCC4051BF	BLOCK-1086
HCC4052BF	BLOCK-1087
HCC4053BF	BLOCK-1088
HCC4055BF	BLOCK-1089
HCC4056BD	BLOCK-1090
HCC4056BF	BLOCK-1090
HCC4056BK	BLOCK-1090
HCC4060BD	BLOCK-1091
HCC4060BF	BLOCK-1091

DEVICE TYPE	REPL CODE
HCC4060BK	BLOCK-1091
HCC4063BD	BLOCK-1092
HCC4063BF	BLOCK-1092
HCC4063BK	BLOCK-1092
HCC4066BF	BLOCK-1093
HCC4068BD	BLOCK-1095
HCC4068BF	BLOCK-1095
HCC4068BK	BLOCK-1095
HCC4070BD	BLOCK-1097
HCC4070BF	BLOCK-1097
HCC4070BK	BLOCK-1097
HCC4071BD	BLOCK-1098
HCC4071BF	BLOCK-1098
HCC4071BK	BLOCK-1098
HCC4072BD	BLOCK-1099
HCC4072BF	BLOCK-1099
HCC4072BK	BLOCK-1099
HCC4073BD	BLOCK-1100
HCC4073BF	BLOCK-1100
HCC4073BK	BLOCK-1100
HCC4075BD	BLOCK-1101
HCC4075BF	BLOCK-1101
HCC4075BK	BLOCK-1101
HCC4076BF	BLOCK-1102
HCC4077BD	BLOCK-1103
HCC4077BF	BLOCK-1103
HCC4077BK	BLOCK-1103
HCC4078BD	BLOCK-1104
HCC4078BF	BLOCK-1104
HCC4078BK	BLOCK-1104
HCC4081BD	BLOCK-1105
HCC4081BF	BLOCK-1105
HCC4081BK	BLOCK-1105
HCC4082BD	BLOCK-1106
HCC4082BF	BLOCK-1106
HCC4082BK	BLOCK-1106
HCC4085BD	BLOCK-1107
HCC4085BF	BLOCK-1107
HCC4085BK	BLOCK-1107
HCC4086BD	BLOCK-1108
HCC4086BF	BLOCK-1108
HCC4086BK	BLOCK-1108
HCC4089BF	BLOCK-1109
HCC4095BF	BLOCK-1111
HCC4096BF	BLOCK-1112
HCC4098BD	BLOCK-1114
HCC4098BF	BLOCK-1114
HCC4098BK	BLOCK-1114
HCC4099BD	BLOCK-1115
HCC4099BF	BLOCK-1115
HCC4099BK	BLOCK-1115
HCC4510BD	BLOCK-1121
HCC4510BE	BLOCK-1121
HCC4510BF	BLOCK-1121
HCC4510BK	BLOCK-1121
HCC4511BD	BLOCK-1122
HCC4511BF	BLOCK-1122
HCC4511BK	BLOCK-1122
HCC4514BD	BLOCK-1125
HCC4515BD	BLOCK-1126
HCC4516BD	BLOCK-1127
HCC4516BF	BLOCK-1127
HCC4516BK	BLOCK-1127
HCC4518BD	BLOCK-1129
HCC4518BF	BLOCK-1129
HCC4518BK	BLOCK-1129
HCC4520BD	BLOCK-1130
HCC4520BF	BLOCK-1130
HCC4520BK	BLOCK-1130
HCC4527BD	BLOCK-1134
HCC4527BF	BLOCK-1134
HCC4527BK	BLOCK-1134
HCC4532BF	BLOCK-1138
HCC4555BD	BLOCK-1146
HCC4555BF	BLOCK-1146
HCC4555BK	BLOCK-1146
HCC4556BD	BLOCK-1147
HCC4556BF	BLOCK-1147
HCC4556BK	BLOCK-1147
HCF4000BE	BLOCK-1028
HCF4000BF	BLOCK-1028
HCF4001BE	BLOCK-1029
HCF4001BF	BLOCK-1029
HCF4002BE	BLOCK-1030
HCF4002BF	BLOCK-1030
HCF4006BE	BLOCK-1031
HCF4006BF	BLOCK-1031
HCF4008BE	BLOCK-1034
HCF4008BF	BLOCK-1034
HCF40100BE	BLOCK-1037
HCF40100BF	BLOCK-1037
HCF40106BE	BLOCK-1038
HCF40106BF	BLOCK-1038
HCF4011BE	BLOCK-1039

DEVICE TYPE	REPL CODE
HCF4011BF	BLOCK-1039
HCF4012BE	BLOCK-1040
HCF4012BF	BLOCK-1040
HCF4013BE	BLOCK-1041
HCF4013BF	BLOCK-1041
HCF4014BE	BLOCK-1042
HCF4014BF	BLOCK-1042
HCF4015BE	BLOCK-1043
HCF4015BF	BLOCK-1043
HCF40160BE	BLOCK-1044
HCF40160BF	BLOCK-1044
HCF40161BE	BLOCK-1045
HCF40161BF	BLOCK-1045
HCF40162BE	BLOCK-1046
HCF40162BF	BLOCK-1046
HCF40163BE	BLOCK-1047
HCF40163BF	BLOCK-1047
HCF4016BE	BLOCK-1048
HCF4016BF	BLOCK-1048
HCF40174BE	BLOCK-1049
HCF40174BF	BLOCK-1049
HCF4017BE	BLOCK-1051
HCF4017BF	BLOCK-1051
HCF40182BE	BLOCK-1052
HCF40182BF	BLOCK-1052
HCF4018BE	BLOCK-1053
HCF4018BF	BLOCK-1053
HCF4019BE	BLOCK-1058
HCF4019BF	BLOCK-1058
HCF4020BE	BLOCK-1059
HCF4020BF	BLOCK-1059
HCF4021BE	BLOCK-1060
HCF4021BF	BLOCK-1060
HCF4022BE	BLOCK-1061
HCF4022BF	BLOCK-1061
HCF4023BE	BLOCK-1062
HCF4023BF	BLOCK-1062
HCF4024BE	BLOCK-1063
HCF4024BF	BLOCK-1063
HCF4025BE	BLOCK-1064
HCF4025BF	BLOCK-1064
HCF4026BE	BLOCK-1065
HCF4026BF	BLOCK-1065
HCF4027BE	BLOCK-1066
HCF4027BF	BLOCK-1066
HCF4028BE	BLOCK-1067
HCF4028BF	BLOCK-1067
HCF4029BE	BLOCK-1068
HCF4029BF	BLOCK-1068
HCF4030BE	BLOCK-1069
HCF4030BF	BLOCK-1069
HCF4031BE	BLOCK-1070
HCF4031BF	BLOCK-1070
HCF4032BE	BLOCK-1071
HCF4032BF	BLOCK-1071
HCF4033BE	BLOCK-1072
HCF4033BF	BLOCK-1072
HCF4034BE	BLOCK-1073
HCF4035BE	BLOCK-1074
HCF4035BF	BLOCK-1074
HCF4038BF	BLOCK-1075
HCF4040BE	BLOCK-1076
HCF4040BF	BLOCK-1076
HCF4042BE	BLOCK-1077
HCF4042BF	BLOCK-1077
HCF4043BE	BLOCK-1078
HCF4043BF	BLOCK-1078
HCF4044BE	BLOCK-1079
HCF4044BF	BLOCK-1079
HCF4045BE	BLOCK-1080
HCF4045BF	BLOCK-1080
HCF4047BE	BLOCK-1082
HCF4047BF	BLOCK-1082
HCF4048BE	BLOCK-1083
HCF4048BF	BLOCK-1083
HCF4049BE	BLOCK-1084
HCF4049BF	BLOCK-1084
HCF4050BE	BLOCK-1085
HCF4050BF	BLOCK-1085
HCF4051BE	BLOCK-1086
HCF4051BF	BLOCK-1086
HCF4052BE	BLOCK-1087
HCF4052BF	BLOCK-1087
HCF4053BE	BLOCK-1088
HCF4053BF	BLOCK-1088
HCF4055BE	BLOCK-1089
HCF4055BF	BLOCK-1089
HCF4056BE	BLOCK-1090
HCF4056BF	BLOCK-1090
HCF4060BE	BLOCK-1091
HCF4060BF	BLOCK-1091
HCF4063BE	BLOCK-1092
HCF4063BF	BLOCK-1092
HCF4066BE	BLOCK-1093

DEVICE TYPE	REPL CODE
HCF4066BF	BLOCK-1093
HCF4068BE	BLOCK-1095
HCF4068BF	BLOCK-1095
HCF4070BE	BLOCK-1097
HCF4070BF	BLOCK-1097
HCF4071BE	BLOCK-1098
HCF4071BF	BLOCK-1098
HCF4072BE	BLOCK-1099
HCF4072BF	BLOCK-1099
HCF4073BE	BLOCK-1100
HCF4073BF	BLOCK-1100
HCF4075BE	BLOCK-1101
HCF4075BF	BLOCK-1101
HCF4076BE	BLOCK-1102
HCF4076BF	BLOCK-1102
HCF4077BE	BLOCK-1103
HCF4077BF	BLOCK-1103
HCF4078BE	BLOCK-1104
HCF4078BF	BLOCK-1104
HCF4081BE	BLOCK-1105
HCF4081BF	BLOCK-1105
HCF4082BE	BLOCK-1106
HCF4082BF	BLOCK-1106
HCF4085BE	BLOCK-1107
HCF4085BF	BLOCK-1107
HCF4086BE	BLOCK-1108
HCF4086BF	BLOCK-1108
HCF4086BK	BLOCK-1108
HCF4089BE	BLOCK-1109
HCF4089BF	BLOCK-1109
HCF4095BE	BLOCK-1111
HCF4095BF	BLOCK-1111
HCF4096BE	BLOCK-1112
HCF4096BF	BLOCK-1112
HCF4098BE	BLOCK-1114
HCF4098BF	BLOCK-1114
HCF4099BE	BLOCK-1115
HCF4099BF	BLOCK-1115
HCF4502BE	BLOCK-1118
HCF4510BE	BLOCK-1121
HCF4510BF	BLOCK-1121
HCF4511BE	BLOCK-1122
HCF4511BF	BLOCK-1122
HCF4514BD	BLOCK-1125
HCF4514BE	BLOCK-1125
HCF4515BD	BLOCK-1126
HCF4515BE	BLOCK-1126
HCF4516BE	BLOCK-1127
HCF4516BF	BLOCK-1127
HCF4518BE	BLOCK-1129
HCF4518BF	BLOCK-1129
HCF4520BE	BLOCK-1130
HCF4520BF	BLOCK-1130
HCF4527BE	BLOCK-1134
HCF4527BF	BLOCK-1134
HCF4532BE	BLOCK-1138
HCF4532BF	BLOCK-1138
HCF4555BE	BLOCK-1146
HCF4555BF	BLOCK-1146
HCF4556BE	BLOCK-1147
HCF4556BF	BLOCK-1147
HCT373	BLOCK-1563
HD1-74C00	BLOCK-1431
HD1-74C02	BLOCK-1432
HD1-74C08	BLOCK-1434
HD1-74C10	BLOCK-1435
HD1-74C154	BLOCK-1439
HD1-74C161	BLOCK-1441
HD1-74C192	BLOCK-1446
HD1-74C193	BLOCK-1447
HD1-74C20	BLOCK-1448
HD1-74C48	BLOCK-1456
HD1-74C73	BLOCK-1457
HD1-74C74	BLOCK-1458
HD1-74C76	BLOCK-1459
HD1-74C85	BLOCK-1460
HD1-74C90	BLOCK-1461
HD1-74C93	BLOCK-1469
HD14000UB	BLOCK-1028
HD14001B	BLOCK-1029
HD14002B	BLOCK-1030
HD14006B	BLOCK-1031
HD14007UB	BLOCK-1032
HD14008B	BLOCK-1034
HD14011B	BLOCK-1039
HD14011BP	BLOCK-1039
HD14012B	BLOCK-1040
HD14013B	BLOCK-1041
HD14014B	BLOCK-1042
HD14015B	BLOCK-1043
HD14016B	BLOCK-1048
HD14017B	BLOCK-1051
HD14018B	BLOCK-1053
HD14020B	BLOCK-1059

DEVICE TYPE	REPL CODE
HD14021B	BLOCK-1060
HD14022B	BLOCK-1061
HD14023B	BLOCK-1062
HD14024B	BLOCK-1063
HD14025B	BLOCK-1064
HD14027B	BLOCK-1066
HD14028B	BLOCK-1067
HD14032B	BLOCK-1071
HD14034B	BLOCK-1073
HD14035B	BLOCK-1074
HD14038B	BLOCK-1075
HD14040B	BLOCK-1076
HD14042B	BLOCK-1077
HD14042BP	BLOCK-1077
HD14042P	BLOCK-1077
HD14043B	BLOCK-1078
HD14044B	BLOCK-1079
HD14044BP	BLOCK-1079
HD14049UB	BLOCK-1084
HD14050B	BLOCK-1085
HD14050BP	BLOCK-1085
HD14051B	BLOCK-1086
HD14051BP	BLOCK-1086
HD14052B	BLOCK-1087
HD14052BP	BLOCK-1087
HD14053B	BLOCK-1088
HD14053BP	BLOCK-1088
HD14066B	BLOCK-1093
HD14066BCP	BLOCK-1093
HD14066BP	BLOCK-1093
HD14068B	BLOCK-1095
HD14069UB	BLOCK-1096
HD14070B	BLOCK-1097
HD14071B	BLOCK-1098
HD14072B	BLOCK-1099
HD14073B	BLOCK-1100
HD14075B	BLOCK-1101
HD14076B	BLOCK-1102
HD14077B	BLOCK-1103
HD14078B	BLOCK-1104
HD14081B	BLOCK-1105
HD14081BP	BLOCK-1105
HD14082B	BLOCK-1106
HD14093B	BLOCK-1110
HD14174B	BLOCK-1049
HD14175B	BLOCK-1050
HD14194B	BLOCK-1056
HD14501UB	BLOCK-1117
HD14502B	BLOCK-1118
HD14503B	BLOCK-1035
HD14503P	BLOCK-1035
HD14506B	BLOCK-1119
HD14508B	BLOCK-1120
HD14510B	BLOCK-1121
HD14511B	BLOCK-1122
HD14512B	BLOCK-1123
HD14514B	BLOCK-1125
HD14515B	BLOCK-1125
HD14516B	BLOCK-1127
HD14517B	BLOCK-1128
HD14518B	BLOCK-1129
HD14520B	BLOCK-1130
HD14526B	BLOCK-1133
HD14527B	BLOCK-1134
HD14532B	BLOCK-1138
HD14536B	BLOCK-1139
HD14538B	BLOCK-1140
HD14538BP	BLOCK-1140
HD14541B	BLOCK-1142
HD14543B	BLOCK-1143
HD14555B	BLOCK-1146
HD14556B	BLOCK-1147
HD14558B	BLOCK-1148
HD14568B	BLOCK-1151
HD14569B	BLOCK-1151
HD14584BP	BLOCK-1038
HD2201	BLOCK-2199
HD2201P	BLOCK-2199
HD2202	BLOCK-2200
HD2202P	BLOCK-2200
HD2203	BLOCK-2208
HD2203P	BLOCK-2208
HD2204	BLOCK-2197
HD2204P	BLOCK-2197
HD2205	BLOCK-2207
HD2205P	BLOCK-2207
HD2206	BLOCK-2202
HD2206P	BLOCK-2202
HD2207	BLOCK-2217
HD2207P	BLOCK-2217
HD2208	BLOCK-2201
HD2208P	BLOCK-2201
HD2209	BLOCK-2206
HD2209P	BLOCK-2206

DEVICE TYPE	REPL CODE
HD2210	BLOCK-1953
HD2210P	BLOCK-1953
HD2211	BLOCK-1950
HD2211P	BLOCK-1950
HD2213	BLOCK-1968
HD2213&	BLOCK-1968
HD2214	BLOCK-1969
HD2214P	BLOCK-1969
HD2215	BLOCK-2217
HD2215P	BLOCK-2217
HD2216	BLOCK-2203
HD2216P	BLOCK-2203
HD2501	BLOCK-1390
HD2501P	BLOCK-1390
HD2502	BLOCK-1406
HD2502P	BLOCK-1406
HD2503	BLOCK-1293
HD2503P	BLOCK-1293
HD2504	BLOCK-1365
HD2504P	BLOCK-1365
HD2505	BLOCK-1402
HD2505P	BLOCK-1402
HD2506	BLOCK-1401
HD2506P	BLOCK-1401
HD2507	BLOCK-1304
HD2507P	BLOCK-1304
HD2508	BLOCK-1377
HD2508P	BLOCK-1377
HD2509	BLOCK-1294
HD2509P	BLOCK-1294
HD2510	BLOCK-1411
HD2510P	BLOCK-1411
HD2511	BLOCK-1295
HD2511P	BLOCK-1295
HD2512	BLOCK-1403
HD2512P	BLOCK-1403
HD2513	BLOCK-1416
HD2513P	BLOCK-1416
HD2514	BLOCK-1404
HD2514P	BLOCK-1404
HD2515	BLOCK-1410
HD2515P	BLOCK-1410
HD2516	BLOCK-1413
HD2516P	BLOCK-1413
HD2517	BLOCK-1412
HD2517P	BLOCK-1412
HD2519	BLOCK-1423
HD2519P	BLOCK-1423
HD2520	BLOCK-1426
HD2520P	BLOCK-1426
HD2521	BLOCK-1425
HD2521P	BLOCK-1425
HD2522	BLOCK-1297
HD2522P	BLOCK-1297
HD2523	BLOCK-1298
HD2523P	BLOCK-1298
HD2524	BLOCK-1424
HD2524P	BLOCK-1424
HD252538	BLOCK-1394
HD2526	BLOCK-1419
HD2526P	BLOCK-1419
HD2528	BLOCK-1296
HD2528P	BLOCK-1296
HD2529	BLOCK-1409
HD2529P	BLOCK-1409
HD2530	BLOCK-1305
HD2530P	BLOCK-1305
HD2531	BLOCK-1395
HD2531P	BLOCK-1395
HD2532	BLOCK-1397
HD2532P	BLOCK-1397
HD2533	BLOCK-1427
HD2533P	BLOCK-1427
HD2534	BLOCK-1428
HD2534P	BLOCK-1428
HD2535	BLOCK-1417
HD2535P	BLOCK-1417
HD2536	BLOCK-1392
HD2536P	BLOCK-1392
HD2537	BLOCK-1393
HD2537P	BLOCK-1393
HD2538P	BLOCK-1394
HD2539	BLOCK-1408
HD2539P	BLOCK-1408
HD2540	BLOCK-1344
HD2540P	BLOCK-1344
HD2541	BLOCK-1357
HD2541P	BLOCK-1357
HD2542	BLOCK-1358
HD2542P	BLOCK-1358
HD2543	BLOCK-1311
HD2543P	BLOCK-1311
HD2544	BLOCK-1385
HD2544P	BLOCK-1385

DEVICE TYPE	REPL CODE
HD2545	BLOCK-1317
HD2545P	BLOCK-1317
HD2546	BLOCK-1429
HD2546P	BLOCK-1429
HD2547	BLOCK-1353
HD2547P	BLOCK-1353
HD2548	BLOCK-1327
HD2548P	BLOCK-1327
HD2549	BLOCK-1328
HD2549P	BLOCK-1328
HD2550	BLOCK-1301
HD2550P	BLOCK-1301
HD2551	BLOCK-1302
HD2551P	BLOCK-1302
HD2552	BLOCK-1384
HD2552P	BLOCK-1384
HD2555	BLOCK-1325
HD2555P	BLOCK-1325
HD2558	BLOCK-1321
HD2558P	BLOCK-1321
HD2561	BLOCK-1313
HD2561P	BLOCK-1313
HD2562	BLOCK-1354
HD2562P	BLOCK-1354
HD2563	BLOCK-1483
HD2563P	BLOCK-1483
HD2564	BLOCK-1330
HD2564P	BLOCK-1330
HD2572	BLOCK-1360
HD2572P	BLOCK-1360
HD2573	BLOCK-1361
HD2573P	BLOCK-1361
HD2580	BLOCK-1331
HD2580P	BLOCK-1331
HD38980A	BLOCK-974
HD38980C	BLOCK-974
HD38991	BLOCK-973
HD38991A	BLOCK-974
HD4000	BLOCK-1028
HD4001	BLOCK-1029
HD4002	BLOCK-1030
HD4007	BLOCK-1032
HD4009	BLOCK-1084
HD4011	BLOCK-1039
HD4012	BLOCK-1040
HD4013	BLOCK-1041
HD4015	BLOCK-1043
HD4016	BLOCK-1048
HD4017	BLOCK-1051
HD4019	BLOCK-1058
HD4020	BLOCK-1059
HD4021	BLOCK-1060
HD4023	BLOCK-1062
HD4024	BLOCK-1063
HD4025	BLOCK-1064
HD4027	BLOCK-1066
HD4030	BLOCK-1069
HD4040	BLOCK-1076
HD4049	BLOCK-1084
HD4050	BLOCK-1085
HD42851	BLOCK-243
HD42851A2	BLOCK-243
HD42853	BLOCK-242
HD4518	BLOCK-1129
HD6809	BLOCK-1166
HD6809EP	BLOCK-1167
HD6821P	BLOCK-1169
HD68A09EP	BLOCK-1167
HD7400	BLOCK-1293
HD7400P	BLOCK-1293
HD7401P	BLOCK-1294
HD7402	BLOCK-1295
HD7402P	BLOCK-1295
HD7403	BLOCK-1296
HD7403P	BLOCK-1296
HD7404	BLOCK-1297
HD7404P	BLOCK-1297
HD7405	BLOCK-1298
HD7405P	BLOCK-1298
HD7406	BLOCK-1299
HD7406N	BLOCK-1299
HD7406P	BLOCK-1299
HD7407	BLOCK-1300
HD7407P	BLOCK-1300
HD740LP	BLOCK-1294
HD7410	BLOCK-1304
HD74107	BLOCK-1305
HD74107P	BLOCK-1305
HD7410P	BLOCK-1304
HD7412	BLOCK-1310
HD74121	BLOCK-1311
HD74121P	BLOCK-1311
HD74123P	BLOCK-1313
HD74125	BLOCK-1314

32 ORIGINAL DEVICE TYPES AND REPLACEMENT CODES

DEVICE TYPE	REPL CODE
HD74125P	BLOCK-1314
HD74126	BLOCK-1315
HD74126P	BLOCK-1315
HD7412P	BLOCK-1310
HD74132	BLOCK-1318
HD74132P	BLOCK-1318
HD7414	BLOCK-1320
HD7414P	BLOCK-1320
HD74150	BLOCK-1327
HD74150P	BLOCK-1327
HD74151A	BLOCK-1328
HD74151AP	BLOCK-1328
HD74155	BLOCK-1332
HD74155P	BLOCK-1332
HD74156	BLOCK-1333
HD74156P	BLOCK-1333
HD7416	BLOCK-1336
HD74160	BLOCK-1337
HD74160P	BLOCK-1337
HD74161	BLOCK-1338
HD74161P	BLOCK-1338
HD74162	BLOCK-1339
HD74162P	BLOCK-1339
HD74163	BLOCK-1340
HD74163P	BLOCK-1340
HD74164	BLOCK-1341
HD74164P	BLOCK-1341
HD7416P	BLOCK-1336
HD7417	BLOCK-1343
HD74174	BLOCK-1346
HD74174P	BLOCK-1346
HD74175	BLOCK-1347
HD74175P	BLOCK-1347
HD74177P	BLOCK-1349
HD7417P	BLOCK-1343
HD74180	BLOCK-1352
HD74180P	BLOCK-1352
HD74190	BLOCK-1355
HD74190P	BLOCK-1355
HD74191	BLOCK-1356
HD74191P	BLOCK-1356
HD74194	BLOCK-1744
HD74194P	BLOCK-1744
HD74198	BLOCK-1362
HD74198P	BLOCK-1362
HD74199	BLOCK-1363
HD74199P	BLOCK-1363
HD7420	BLOCK-1365
HD7420P	BLOCK-1365
HD7427	BLOCK-1372
HD7427P	BLOCK-1372
HD7430	BLOCK-1377
HD7430P	BLOCK-1377
HD7432	BLOCK-1378
HD7432P	BLOCK-1378
HD7438P	BLOCK-1385
HD7440	BLOCK-1390
HD7440P	BLOCK-1390
HD7442A	BLOCK-1392
HD7442AP	BLOCK-1392
HD7443A	BLOCK-1393
HD7443AP	BLOCK-1393
HD7444A	BLOCK-1394
HD7444AP	BLOCK-1394
HD7450	BLOCK-1401
HD7450P	BLOCK-1401
HD7451	BLOCK-1402
HD7451P	BLOCK-1402
HD7453	BLOCK-1403
HD7453P	BLOCK-1403
HD7454	BLOCK-1404
HD7454P	BLOCK-1404
HD7460	BLOCK-1406
HD7460P	BLOCK-1406
HD7471P	BLOCK-1343
HD7472	BLOCK-1409
HD7472P	BLOCK-1409
HD7473AP	BLOCK-1410
HD7473P	BLOCK-1410
HD7474	BLOCK-1411
HD7474P	BLOCK-1411
HD7475	BLOCK-1412
HD7475P	BLOCK-1412
HD7485	BLOCK-1418
HD7485P	BLOCK-1418
HD7486	BLOCK-1419
HD7486P	BLOCK-1419
HD7490A	BLOCK-1423
HD7490AP	BLOCK-1423
HD7492A	BLOCK-1425
HD7492AP	BLOCK-1425
HD7493A	BLOCK-1426
HD7493AD	BLOCK-1426
HD7493AP	BLOCK-1426
HD7493P	BLOCK-1426
HD7496	BLOCK-1429
HD7496P	BLOCK-1429
HD74LS00	BLOCK-1568
HD74LS00P	BLOCK-1568
HD74LS01	BLOCK-1569
HD74LS01P	BLOCK-1569
HD74LS02	BLOCK-1570
HD74LS02P	BLOCK-1570
HD74LS03	BLOCK-1571
HD74LS03P	BLOCK-1571
HD74LS04	BLOCK-1572
HD74LS04P	BLOCK-1572
HD74LS05	BLOCK-1573
HD74LS05P	BLOCK-1573
HD74LS08	BLOCK-1574
HD74LS08P	BLOCK-1574
HD74LS09	BLOCK-1575
HD74LS09P	BLOCK-1575
HD74LS10	BLOCK-1576
HD74LS107	BLOCK-1577
HD74LS107P	BLOCK-1577
HD74LS109A	BLOCK-1578
HD74LS109AP	BLOCK-1578
HD74LS10P	BLOCK-1576
HD74LS11	BLOCK-1579
HD74LS112	BLOCK-1580
HD74LS112P	BLOCK-1580
HD74LS114	BLOCK-1581
HD74LS114P	BLOCK-1581
HD74LS11P	BLOCK-1579
HD74LS12	BLOCK-1582
HD74LS122	BLOCK-1583
HD74LS122P	BLOCK-1583
HD74LS123	BLOCK-1584
HD74LS123P	BLOCK-1584
HD74LS125AP	BLOCK-1585
HD74LS12P	BLOCK-1582
HD74LS13	BLOCK-1587
HD74LS136	BLOCK-1590
HD74LS136P	BLOCK-1590
HD74LS138	BLOCK-1591
HD74LS138P	BLOCK-1591
HD74LS139	BLOCK-1592
HD74LS139P	BLOCK-1592
HD74LS13P	BLOCK-1587
HD74LS14	BLOCK-1593
HD74LS145	BLOCK-1594
HD74LS145P	BLOCK-1594
HD74LS148	BLOCK-1596
HD74LS148P	BLOCK-1596
HD74LS14P	BLOCK-1593
HD74LS15	BLOCK-1597
HD74LS151	BLOCK-1598
HD74LS151P	BLOCK-1598
HD74LS153	BLOCK-1599
HD74LS153P	BLOCK-1599
HD74LS155	BLOCK-1600
HD74LS155P	BLOCK-1600
HD74LS156	BLOCK-1601
HD74LS156P	BLOCK-1601
HD74LS157	BLOCK-1602
HD74LS157P	BLOCK-1602
HD74LS158	BLOCK-1603
HD74LS158P	BLOCK-1603
HD74LS15P	BLOCK-1597
HD74LS160	BLOCK-1604
HD74LS161	BLOCK-1605
HD74LS161P	BLOCK-1605
HD74LS162	BLOCK-1606
HD74LS162P	BLOCK-1606
HD74LS163	BLOCK-1607
HD74LS163P	BLOCK-1607
HD74LS164	BLOCK-1608
HD74LS164P	BLOCK-1608
HD74LS166	BLOCK-1610
HD74LS166P	BLOCK-1610
HD74LS174	BLOCK-1615
HD74LS174P	BLOCK-1615
HD74LS175	BLOCK-1616
HD74LS175P	BLOCK-1616
HD74LS181	BLOCK-1617
HD74LS181P	BLOCK-1617
HD74LS190	BLOCK-1618
HD74LS190P	BLOCK-1618
HD74LS191	BLOCK-1619
HD74LS191P	BLOCK-1619
HD74LS192	BLOCK-1620
HD74LS192P	BLOCK-1620
HD74LS193	BLOCK-1621
HD74LS193P	BLOCK-1621
HD74LS194A	BLOCK-1622
HD74LS194AP	BLOCK-1622
HD74LS195A	BLOCK-1623
HD74LS195AP	BLOCK-1623
HD74LS20	BLOCK-1626
HD74LS20P	BLOCK-1626
HD74LS21	BLOCK-1627
HD74LS21P	BLOCK-1627
HD74LS22	BLOCK-1628
HD74LS221	BLOCK-1629
HD74LS221P	BLOCK-1629
HD74LS22P	BLOCK-1628
HD74LS240	BLOCK-1630
HD74LS240P	BLOCK-1630
HD74LS241	BLOCK-1631
HD74LS241P	BLOCK-1631
HD74LS244	BLOCK-1634
HD74LS244P	BLOCK-1634
HD74LS245WP	BLOCK-1635
HD74LS247	BLOCK-1636
HD74LS247P	BLOCK-1636
HD74LS248	BLOCK-1637
HD74LS248P	BLOCK-1637
HD74LS249	BLOCK-1638
HD74LS249P	BLOCK-1638
HD74LS251	BLOCK-1639
HD74LS251P	BLOCK-1639
HD74LS253	BLOCK-1640
HD74LS253P	BLOCK-1640
HD74LS257	BLOCK-1641
HD74LS257P	BLOCK-1641
HD74LS258	BLOCK-1642
HD74LS258P	BLOCK-1642
HD74LS259	BLOCK-1643
HD74LS259P	BLOCK-1643
HD74LS26	BLOCK-1644
HD74LS266	BLOCK-1646
HD74LS266P	BLOCK-1646
HD74LS26P	BLOCK-1644
HD74LS27	BLOCK-1647
HD74LS273P	BLOCK-1648
HD74LS279	BLOCK-1649
HD74LS279P	BLOCK-1649
HD74LS27P	BLOCK-1647
HD74LS280	BLOCK-1651
HD74LS280P	BLOCK-1651
HD74LS283	BLOCK-1652
HD74LS283P	BLOCK-1652
HD74LS290	BLOCK-1653
HD74LS290P	BLOCK-1653
HD74LS293	BLOCK-1654
HD74LS293P	BLOCK-1654
HD74LS295B	BLOCK-1655
HD74LS295BP	BLOCK-1655
HD74LS298	BLOCK-1656
HD74LS298P	BLOCK-1656
HD74LS299	BLOCK-1657
HD74LS299P	BLOCK-1657
HD74LS30	BLOCK-1658
HD74LS30P	BLOCK-1658
HD74LS32	BLOCK-1659
HD74LS32P	BLOCK-1659
HD74LS365AP	BLOCK-1666
HD74LS367AP	BLOCK-1668
HD74LS37	BLOCK-1670
HD74LS373P	BLOCK-1671
HD74LS374P	BLOCK-1672
HD74LS37P	BLOCK-1670
HD74LS38	BLOCK-1676
HD74LS386	BLOCK-1677
HD74LS386P	BLOCK-1677
HD74LS38P	BLOCK-1676
HD74LS390	BLOCK-1678
HD74LS390P	BLOCK-1678
HD74LS393	BLOCK-1679
HD74LS393P	BLOCK-1679
HD74LS40	BLOCK-1683
HD74LS40P	BLOCK-1683
HD74LS42	BLOCK-1684
HD74LS42P	BLOCK-1684
HD74LS47	BLOCK-1685
HD74LS47P	BLOCK-1685
HD74LS48	BLOCK-1686
HD74LS48P	BLOCK-1686
HD74LS49	BLOCK-1687
HD74LS490	BLOCK-1688
HD74LS490P	BLOCK-1688
HD74LS49P	BLOCK-1687
HD74LS51	BLOCK-1689
HD74LS51P	BLOCK-1689
HD74LS54	BLOCK-1690
HD74LS54P	BLOCK-1690
HD74LS55	BLOCK-1693
HD74LS55P	BLOCK-1693
HD74LS73	BLOCK-1705
HD74LS73P	BLOCK-1705
HD74LS74	BLOCK-1706
HD74LS74A	BLOCK-1706
HD74LS74AP	BLOCK-1706
HD74LS75	BLOCK-1707
HD74LS75P	BLOCK-1707
HD74LS76	BLOCK-1708
HD74LS76P	BLOCK-1708
HD74LS77	BLOCK-1709
HD74LS77P	BLOCK-1709
HD74LS78	BLOCK-1710
HD74LS78P	BLOCK-1710
HD74LS83A	BLOCK-1711
HD74LS83AP	BLOCK-1711
HD74LS85	BLOCK-1712
HD74LS85P	BLOCK-1712
HD74LS86	BLOCK-1713
HD74LS86P	BLOCK-1713
HD74LS90	BLOCK-1714
HD74LS90P	BLOCK-1714
HD74LS91	BLOCK-1715
HD74LS91P	BLOCK-1715
HD74LS92	BLOCK-1716
HD74LS92P	BLOCK-1716
HD74LS93	BLOCK-1717
HD74LS93P	BLOCK-1717
HD74LSS76P	BLOCK-1708
HD74S00	BLOCK-1719
HD74S00P	BLOCK-1719
HD74S03	BLOCK-1721
HD74S03P	BLOCK-1721
HD74S04	BLOCK-1722
HD74S04P	BLOCK-1722
HD74S05	BLOCK-1723
HD74S05P	BLOCK-1723
HD74S10	BLOCK-1726
HD74S10P	BLOCK-1726
HD74S11	BLOCK-1727
HD74S112	BLOCK-1728
HD74S112P	BLOCK-1728
HD74S113	BLOCK-1729
HD74S113P	BLOCK-1729
HD74S11P	BLOCK-1727
HD74S11S	BLOCK-1727
HD74S133	BLOCK-1731
HD74S133P	BLOCK-1731
HD74S134	BLOCK-1732
HD74S134P	BLOCK-1732
HD74S140	BLOCK-1734
HD74S140P	BLOCK-1734
HD74S15	BLOCK-1735
HD74S151	BLOCK-1736
HD74S151P	BLOCK-1736
HD74S15P	BLOCK-1735
HD74S174	BLOCK-1740
HD74S174P	BLOCK-1740
HD74S175P	BLOCK-1741
HD74S181	BLOCK-1742
HD74S181P	BLOCK-1742
HD74S20	BLOCK-1745
HD74S20P	BLOCK-1745
HD74S22	BLOCK-1746
HD74S22P	BLOCK-1746
HD74S251	BLOCK-1747
HD74S251P	BLOCK-1747
HD74S40	BLOCK-1753
HD74S40P	BLOCK-1753
HD74S64	BLOCK-1766
HD74S64P	BLOCK-1766
HD74S65	BLOCK-1767
HD74S65P	BLOCK-1767
HD74S74	BLOCK-1768
HD74S74P	BLOCK-1768
HD74S86	BLOCK-1769
HD74S86P	BLOCK-1769
HD75188	BLOCK-1771
HD75188P	BLOCK-1771
HD75189	BLOCK-1772
HD75189P	BLOCK-1772
HD75450A	BLOCK-1775
HD75451A	BLOCK-1776
HD75451AP	BLOCK-1776
HD75452	BLOCK-1777
HD75452P	BLOCK-1777
HD75453	BLOCK-1778
HD75453P	BLOCK-1778
HD75454	BLOCK-1779
HD75454P	BLOCK-1779
HD9-74C00	BLOCK-1431
HD9-74C02	BLOCK-1432
HD9-74C08	BLOCK-1434
HD9-74C10	BLOCK-1435
HD9-74C154	BLOCK-1439
HD9-74C161	BLOCK-1441
HD9-74C192	BLOCK-1446
HD9-74C193	BLOCK-1447
HD9-74C20	BLOCK-1448
HD9-74C221	BLOCK-1449
HD9-74C48	BLOCK-1456
HD9-74C73	BLOCK-1457
HD9-74C74	BLOCK-1458
HD9-74C76	BLOCK-1459
HD9-74C85	BLOCK-1460
HD9-74C90	BLOCK-1461
HD9-74C93	BLOCK-1469
HE-442-21	BLOCK-1801
HE-442-22	BLOCK-2058
HE-442-24	BLOCK-903
HE-442-25	BLOCK-1985
HE-442-30	BLOCK-1022
HE-442-39	BLOCK-2141
HE-442-4	BLOCK-1810
HE-442-54	BLOCK-2087
HE-442-60	BLOCK-2137
HE-442-602	BLOCK-2171
HE-442-604	BLOCK-930
HE-442-613	BLOCK-2128
HE-442-615	BLOCK-1281
HE-442-623	BLOCK-1997
HE-442-627	BLOCK-2144
HE-442-63	BLOCK-2119
HE-442-630	BLOCK-2091
HE-442-635	BLOCK-1881
HE-442-636	BLOCK-472
HE-442-640	BLOCK-1942
HE-442-644	BLOCK-2074
HE-442-647	BLOCK-1081
HE-442-648	BLOCK-1986
HE-442-651	BLOCK-2007
HE-442-654	BLOCK-2174
HE-442-659	BLOCK-945
HE-442-66	BLOCK-2042
HE-442-663	BLOCK-2097
HE-442-664	BLOCK-2108
HE-442-665	BLOCK-893
HE-442-672	BLOCK-933
HE-442-673	BLOCK-470
HE-442-674	BLOCK-2097
HE-442-675	BLOCK-2108
HE-442-677	BLOCK-367
HE-442-682	BLOCK-944
HE-442-686	BLOCK-1303
HE-442-688	BLOCK-1868
HE-442-691	BLOCK-2095
HE-442-702	BLOCK-2004
HE-442-71	BLOCK-2191
HE-442-715	BLOCK-1997
HE-442-74	BLOCK-1804
HE-442-75	BLOCK-1981
HE-442-99	BLOCK-1048
HE-443-1001	BLOCK-1651
HE-443-1020	BLOCK-1300
HE-443-1021	BLOCK-1855
HE-443-1024	BLOCK-1669
HE-443-1034	BLOCK-1676
HE-443-1036	BLOCK-1601
HE-443-1037	BLOCK-1641
HE-443-1040	BLOCK-2133
HE-443-1053	BLOCK-1740
HE-443-1058	BLOCK-1692
HE-443-1112	BLOCK-2090
HE-443-1173	BLOCK-1704
HE-443-12	BLOCK-1304
HE-443-13	BLOCK-1412
HE-443-15	BLOCK-1401
HE-443-16	BLOCK-1413
HE-443-17	BLOCK-1321
HE-443-22	BLOCK-1311
HE-443-23	BLOCK-1312
HE-443-25	BLOCK-1328
HE-443-26	BLOCK-1719
HE-443-3	BLOCK-1377
HE-443-34	BLOCK-1425
HE-443-36	BLOCK-1397
HE-443-4	BLOCK-1409
HE-443-44	BLOCK-1317
HE-443-45	BLOCK-1301
HE-443-46	BLOCK-1295
HE-443-5	BLOCK-1410
HE-443-53	BLOCK-1392
HE-443-54	BLOCK-1296
HE-443-6	BLOCK-1411
HE-443-603	BLOCK-1039
HE-443-604	BLOCK-1032
HE-443-606	BLOCK-1066
HE-443-607	BLOCK-1041
HE-443-612	BLOCK-1358
HE-443-622	BLOCK-1326
HE-443-623	BLOCK-1331
HE-443-625	BLOCK-1318
HE-443-628	BLOCK-1360
HE-443-629	BLOCK-1423
HE-443-640	BLOCK-1426
HE-443-642	BLOCK-1298
HE-443-66	BLOCK-1357
HE-443-68	BLOCK-1476
HE-443-680	BLOCK-1428
HE-443-695	BLOCK-1029
HE-443-7	BLOCK-1423
HE-443-70	BLOCK-1479
HE-443-703	BLOCK-1029
HE-443-704	BLOCK-1030
HE-443-706	BLOCK-1098
HE-443-707	BLOCK-1126
HE-443-708	BLOCK-1127
HE-443-709	BLOCK-1469
HE-443-711	BLOCK-1636
HE-443-712	BLOCK-1064
HE-443-713	BLOCK-1067
HE-443-717	BLOCK-1315
HE-443-718	BLOCK-1623
HE-443-719	BLOCK-1646
HE-443-72	BLOCK-1343
HE-443-720	BLOCK-1837
HE-443-722	BLOCK-2089
HE-443-727	BLOCK-2132
HE-443-728	BLOCK-1568
HE-443-729	BLOCK-1587
HE-443-73	BLOCK-1336
HE-443-730	BLOCK-1706
HE-443-731	BLOCK-1653
HE-443-732	BLOCK-1658
HE-443-733	BLOCK-1654
HE-443-737	BLOCK-1129
HE-443-738	BLOCK-1147
HE-443-74	BLOCK-1777
HE-443-745	BLOCK-1571
HE-443-751	BLOCK-1105
HE-443-752	BLOCK-1616
HE-443-754	BLOCK-1630
HE-443-755	BLOCK-1572
HE-443-757	BLOCK-1605
HE-443-762	BLOCK-1832
HE-443-764	BLOCK-1002
HE-443-769	BLOCK-1608
HE-443-77	BLOCK-1385
HE-443-777	BLOCK-1114
HE-443-778	BLOCK-1110
HE-443-779	BLOCK-1570
HE-443-780	BLOCK-1574
HE-443-781	BLOCK-1707
HE-443-782	BLOCK-1600
HE-443-783	BLOCK-1061
HE-443-784	BLOCK-1069
HE-443-785	BLOCK-1449
HE-443-791	BLOCK-1634
HE-443-792	BLOCK-1588
HE-443-794	BLOCK-1771
HE-443-795	BLOCK-1772
HE-443-797	BLOCK-1576
HE-443-798	BLOCK-1626
HE-443-799	BLOCK-1602
HE-443-800	BLOCK-1647
HE-443-801	BLOCK-1624
HE-443-802	BLOCK-1641
HE-443-804	BLOCK-1643
HE-443-805	BLOCK-1648
HE-443-807	BLOCK-1684
HE-443-811	BLOCK-1585
HE-443-813	BLOCK-1714
HE-443-815	BLOCK-1621
HE-443-816	BLOCK-1575
HE-443-817	BLOCK-1620
HE-443-818	BLOCK-1573
HE-443-822	BLOCK-1592
HE-443-824	BLOCK-1631
HE-443-827	BLOCK-1164
HE-443-828	BLOCK-1705
HE-443-829	BLOCK-1708
HE-443-836	BLOCK-1049
HE-443-837	BLOCK-1671
HE-443-838	BLOCK-1717
HE-443-839	BLOCK-1633
HE-443-840	BLOCK-1172
HE-443-843	BLOCK-1169
HE-443-854	BLOCK-1649
HE-443-855	BLOCK-1652
HE-443-857	BLOCK-1668
HE-443-858	BLOCK-1320
HE-443-863	BLOCK-1672
HE-443-864	BLOCK-1579
HE-443-867	BLOCK-1671
HE-443-87	BLOCK-1325
HE-443-871	BLOCK-1125

ORIGINAL DEVICE TYPES AND REPLACEMENT CODES 33

DEVICE TYPE	REPL CODE
HE-443-872	BLOCK-1593
HE-443-875	BLOCK-1659
HE-443-877	BLOCK-1591
HE-443-879	BLOCK-1615
HE-443-881	BLOCK-1025
HE-443-884	BLOCK-1632
HE-443-885	BLOCK-1635
HE-443-886	BLOCK-1040
HE-443-887	BLOCK-1062
HE-443-888	BLOCK-1119
HE-443-889	BLOCK-1594
HE-443-89	BLOCK-1302
HE-443-891	BLOCK-1713
HE-443-892	BLOCK-1610
HE-443-896	BLOCK-1720
HE-443-897	BLOCK-1722
HE-443-899	BLOCK-1767
HE-443-90	BLOCK-1313
HE-443-900	BLOCK-1768
HE-443-904	BLOCK-1003
HE-443-909	BLOCK-1821
HE-443-912	BLOCK-1596
HE-443-915	BLOCK-1769
HE-443-916	BLOCK-1114
HE-443-919	BLOCK-1586
HE-443-920	BLOCK-1712
HE-443-921	BLOCK-1678
HE-443-928	BLOCK-1034
HE-443-929	BLOCK-1051
HE-443-930	BLOCK-1082
HE-443-934	BLOCK-1607
HE-443-935	BLOCK-1737
HE-443-942	BLOCK-1584
HE-443-958	BLOCK-1091
HE-443-967	BLOCK-1299
HE-443-970	BLOCK-1006
HE-443-973	BLOCK-1679
HE-443-983	BLOCK-1741
HE-443-985	BLOCK-1741
HE-443-991	BLOCK-1085
HE-443-999	BLOCK-1694
HE-444-228	BLOCK-1749
HEF4000B	BLOCK-1028
HEF4000BD	BLOCK-1028
HEF4000BP	BLOCK-1028
HEF4000P	BLOCK-1028
HEF4001BD	BLOCK-1029
HEF4001BP	BLOCK-1029
HEF4001P	BLOCK-1029
HEF4002BD	BLOCK-1030
HEF4002BP	BLOCK-1030
HEF4002P	BLOCK-1030
HEF4006B	BLOCK-1031
HEF4006BD	BLOCK-1031
HEF4006BP	BLOCK-1031
HEF4006P	BLOCK-1031
HEF4008B	BLOCK-1034
HEF4008BD	BLOCK-1034
HEF4008BP	BLOCK-1034
HEF4008P	BLOCK-1034
HEF40097BD	BLOCK-1035
HEF40097BP	BLOCK-1035
HEF40098BD	BLOCK-1036
HEF40098BN	BLOCK-1036
HEF40098BP	BLOCK-1036
HEF40098P	BLOCK-1036
HEF4011BD	BLOCK-1039
HEF4011BP	BLOCK-1039
HEF4011P	BLOCK-1039
HEF4012	BLOCK-1040
HEF4012BD	BLOCK-1040
HEF4012BP	BLOCK-1040
HEF4012P	BLOCK-1040
HEF4013BD	BLOCK-1041
HEF4013BP	BLOCK-1041
HEF4013P	BLOCK-1041
HEF4014B	BLOCK-1042
HEF4014BD	BLOCK-1042
HEF4014BP	BLOCK-1042
HEF4014P	BLOCK-1042
HEF4015B	BLOCK-1043
HEF4015BD	BLOCK-1125
HEF4015BP	BLOCK-1125
HEF4015P	BLOCK-1043
HEF40160B	BLOCK-1044
HEF40160BD	BLOCK-1044
HEF40160BP	BLOCK-1044
HEF40161B	BLOCK-1045
HEF40161BD	BLOCK-1045
HEF40161BP	BLOCK-1045
HEF40162B	BLOCK-1046
HEF40162BD	BLOCK-1046
HEF40162BP	BLOCK-1046
HEF40163B	BLOCK-1047
HEF40163BD	BLOCK-1047
HEF40163BP	BLOCK-1047
HEF4016BD	BLOCK-1048
HEF4016BP	BLOCK-1048
HEF40174B	BLOCK-1049
HEF40174BD	BLOCK-1049
HEF40174BP	BLOCK-1049
HEF40174P	BLOCK-1049
HEF40175B	BLOCK-1050
HEF40175BD	BLOCK-1050
HEF40175BP	BLOCK-1050
HEF4017BD	BLOCK-1051
HEF4017BP	BLOCK-1051
HEF4017P	BLOCK-1051
HEF4018BD	BLOCK-1053
HEF4018BP	BLOCK-1053
HEF4018P	BLOCK-1053
HEF40192BD	BLOCK-1054
HEF40192BP	BLOCK-1054
HEF40192P	BLOCK-1054
HEF40193BD	BLOCK-1055
HEF40193BP	BLOCK-1055
HEF40193P	BLOCK-1055
HEF40194B	BLOCK-1056
HEF40194BD	BLOCK-1056
HEF40194BP	BLOCK-1056
HEF40194P	BLOCK-1056
HEF40195B	BLOCK-1057
HEF40195BD	BLOCK-1057
HEF40195BP	BLOCK-1057
HEF4019BD	BLOCK-1058
HEF4019BP	BLOCK-1058
HEF4020BD	BLOCK-1059
HEF4020BP	BLOCK-1059
HEF4020P	BLOCK-1059
HEF4021B	BLOCK-1060
HEF4021BD	BLOCK-1060
HEF4021BP	BLOCK-1060
HEF4021P	BLOCK-1060
HEF4022BD	BLOCK-1061
HEF4022BP	BLOCK-1061
HEF4022P	BLOCK-1061
HEF4023BD	BLOCK-1062
HEF4023BP	BLOCK-1062
HEF4023P	BLOCK-1062
HEF4024BD	BLOCK-1063
HEF4024BP	BLOCK-1063
HEF4024P	BLOCK-1063
HEF4025BD	BLOCK-1064
HEF4025BP	BLOCK-1064
HEF4025P	BLOCK-1064
HEF4027BD	BLOCK-1066
HEF4027BP	BLOCK-1066
HEF4027P	BLOCK-1066
HEF4028B	BLOCK-1067
HEF4028BD	BLOCK-1067
HEF4028BP	BLOCK-1067
HEF4028P	BLOCK-1067
HEF4029B	BLOCK-1068
HEF4029BD	BLOCK-1068
HEF4029BP	BLOCK-1068
HEF4029P	BLOCK-1068
HEF4030BD	BLOCK-1069
HEF4030BP	BLOCK-1069
HEF4030P	BLOCK-1069
HEF4031B	BLOCK-1070
HEF4031BD	BLOCK-1070
HEF4031BP	BLOCK-1070
HEF4031P	BLOCK-1070
HEF4035B	BLOCK-1074
HEF4035BD	BLOCK-1074
HEF4035BP	BLOCK-1074
HEF4035P	BLOCK-1074
HEF4040BD	BLOCK-1076
HEF4040BP	BLOCK-1076
HEF4040P	BLOCK-1076
HEF4042BD	BLOCK-1077
HEF4042BP	BLOCK-1077
HEF4042P	BLOCK-1077
HEF4043BD	BLOCK-1078
HEF4043BP	BLOCK-1078
HEF4043P	BLOCK-1078
HEF4044B	BLOCK-1079
HEF4044BD	BLOCK-1079
HEF4044BP	BLOCK-1079
HEF4044P	BLOCK-1079
HEF4047B	BLOCK-1082
HEF4047BD	BLOCK-1082
HEF4047BP	BLOCK-1082
HEF4049BD	BLOCK-1084
HEF4049BP	BLOCK-1084
HEF4050BD	BLOCK-1085
HEF4050BP	BLOCK-1085
HEF4051BD	BLOCK-1086
HEF4051BP	BLOCK-1086
HEF4052BD	BLOCK-1087
HEF4052BP	BLOCK-1087
HEF4053BD	BLOCK-1088
HEF4053BP	BLOCK-1088
HEF4066BD	BLOCK-1093
HEF4066BP	BLOCK-1093
HEF4068BD	BLOCK-1095
HEF4068BP	BLOCK-1095
HEF4068P	BLOCK-1095
HEF4069P	BLOCK-1096
HEF4069UBD	BLOCK-1096
HEF4069UBP	BLOCK-1096
HEF4070BD	BLOCK-1097
HEF4070BP	BLOCK-1097
HEF4070P	BLOCK-1097
HEF4071BD	BLOCK-1098
HEF4071BP	BLOCK-1098
HEF4071P	BLOCK-1098
HEF4072BD	BLOCK-1099
HEF4072BP	BLOCK-1099
HEF4072P	BLOCK-1099
HEF4073B	BLOCK-1100
HEF4073BD	BLOCK-1100
HEF4073BP	BLOCK-1100
HEF4073P	BLOCK-1100
HEF4075B	BLOCK-1101
HEF4075BD	BLOCK-1101
HEF4075BP	BLOCK-1101
HEF4076B	BLOCK-1102
HEF4076BD	BLOCK-1102
HEF4076BP	BLOCK-1102
HEF4076P	BLOCK-1102
HEF4077B	BLOCK-1103
HEF4077BD	BLOCK-1103
HEF4077BP	BLOCK-1103
HEF4077P	BLOCK-1103
HEF4078BD	BLOCK-1104
HEF4078BP	BLOCK-1104
HEF4078P	BLOCK-1104
HEF4081BD	BLOCK-1105
HEF4081BP	BLOCK-1105
HEF4081P	BLOCK-1105
HEF4082BD	BLOCK-1106
HEF4082BP	BLOCK-1106
HEF4082P	BLOCK-1106
HEF4085BD	BLOCK-1107
HEF4085BP	BLOCK-1107
HEF4085P	BLOCK-1107
HEF4086B	BLOCK-1108
HEF4086BD	BLOCK-1108
HEF4086BP	BLOCK-1108
HEF4086P	BLOCK-1108
HEF4093BD	BLOCK-1110
HEF4093BP	BLOCK-1110
HEF4502BD	BLOCK-1118
HEF4502BP	BLOCK-1118
HEF4508BD	BLOCK-1120
HEF4508BP	BLOCK-1120
HEF4510B	BLOCK-1121
HEF4510BD	BLOCK-1121
HEF4510BP	BLOCK-1121
HEF4511B	BLOCK-1122
HEF4511BD	BLOCK-1122
HEF4511BP	BLOCK-1122
HEF4512BD	BLOCK-1123
HEF4512BP	BLOCK-1123
HEF4514BD	BLOCK-1126
HEF4514BP	BLOCK-1125
HEF4515BD	BLOCK-1126
HEF4515BP	BLOCK-1126
HEF4516B	BLOCK-1127
HEF4516BD	BLOCK-1127
HEF4516BP	BLOCK-1127
HEF4518BD	BLOCK-1129
HEF4518BP	BLOCK-1129
HEF4518P	BLOCK-1129
HEF4520BD	BLOCK-1130
HEF4520BP	BLOCK-1130
HEF4520P	BLOCK-1130
HEF4522BD	BLOCK-1132
HEF4522BP	BLOCK-1132
HEF4528B	BLOCK-1135
HEF4528BD	BLOCK-1135
HEF4528BP	BLOCK-1135
HEF4531B	BLOCK-1137
HEF4531BD	BLOCK-1137
HEF4531BP	BLOCK-1137
HEF4538BP	BLOCK-1140
HEF4539BD	BLOCK-1141
HEF4539BP	BLOCK-1141
HEF4555BD	BLOCK-1146
HEF4555BP	BLOCK-1146
HEF4556BD	BLOCK-1147
HEF4556BP	BLOCK-1147
HEF4585BD	BLOCK-1154
HEF4585BP	BLOCK-1154
HEP-594	BLOCK-1275
HEP-595	BLOCK-1277
HEP-C6055L	BLOCK-1281
HEP-C6057P	BLOCK-1815
HEP-C6059P	BLOCK-1405
HEP-C6060P	BLOCK-1421
HEP-C6061P	BLOCK-1770
HEP-C6062P	BLOCK-1262
HEP-C6063P	BLOCK-1269
HEP-C6065P	BLOCK-1276
HEP-C6066P	BLOCK-1822
HEP-C6068P	BLOCK-1277
HEP-C6069G	BLOCK-1805
HEP-C6070P	BLOCK-1271
HEP-C6071P	BLOCK-1272
HEP-C6072P	BLOCK-1292
HEP-C6074P	BLOCK-1275
HEP-C6075P	BLOCK-1291
HEP-C6076P	BLOCK-1422
HEP-C6079P	BLOCK-1407
HEP-C6082P	BLOCK-1262
HEP-C6083P	BLOCK-1269
HEP-C6085P	BLOCK-1287
HEP-C6089	BLOCK-1244
HEP-C6094P	BLOCK-1275
HEP-C6095P	BLOCK-1277
HEP-C6096P	BLOCK-1825
HEP-C6099P	BLOCK-1820
HEP-C6100P	BLOCK-1808
HEP-C6101P	BLOCK-1279
HEP-C6102P	BLOCK-1801
HEP-C6103P	BLOCK-1949
HEP-C6132P	BLOCK-2144
HEP-C6133P	BLOCK-2074
HEP-C6137P	BLOCK-882
HEP3806P	BLOCK-161
HEP572	BLOCK-2221
HEP573	BLOCK-2224
HEP581	BLOCK-2186
HEP584	BLOCK-2189
HEP590	BLOCK-1844
HEP591	BLOCK-1223
HEP594	BLOCK-1275
HEP595	BLOCK-1277
HEP6060P	BLOCK-1421
HEP6061P	BLOCK-1770
HEPC0900P	BLOCK-2100
HEPC0901P	BLOCK-2101
HEPC0902P	BLOCK-2106
HEPC0903P	BLOCK-2107
HEPC0904P	BLOCK-2109
HEPC0905P	BLOCK-2099
HEPC0906P	BLOCK-2102
HEPC0907P	BLOCK-2122
HEPC0908P	BLOCK-2121
HEPC0909P	BLOCK-2111
HEPC0910P	BLOCK-2103
HEPC0911P	BLOCK-2105
HEPC0912P	BLOCK-2101
HEPC1030P	BLOCK-2197
HEPC1032P	BLOCK-2199
HEPC1033P	BLOCK-2200
HEPC1035P	BLOCK-2201
HEPC1036P	BLOCK-2202
HEPC1044P	BLOCK-2206
HEPC1045P	BLOCK-2207
HEPC1046P	BLOCK-2208
HEPC1052P	BLOCK-1953
HEPC1053P	BLOCK-1950
HEPC1056P	BLOCK-2208
HEPC1057P	BLOCK-1968
HEPC1058P	BLOCK-1969
HEPC1062P	BLOCK-2217
HEPC2001P	BLOCK-2193
HEPC2004P	BLOCK-2221
HEPC2005P	BLOCK-2224
HEPC2006G	BLOCK-2189
HEPC2007G	BLOCK-2186
HEPC2010G	BLOCK-2189
HEPC2012G	BLOCK-2189
HEPC2502P	BLOCK-2193
HEPC3000L	BLOCK-1293
HEPC3000P	BLOCK-1293
HEPC3001L	BLOCK-1294
HEPC3001P	BLOCK-1294
HEPC3002P	BLOCK-1295
HEPC3004L	BLOCK-1297
HEPC3004P	BLOCK-1297
HEPC3010L	BLOCK-1304
HEPC3010P	BLOCK-1304
HEPC3020L	BLOCK-1365
HEPC3020P	BLOCK-1365
HEPC3030L	BLOCK-1377
HEPC3030P	BLOCK-1377
HEPC3040L	BLOCK-1390
HEPC3040P	BLOCK-1390
HEPC3041L	BLOCK-1391
HEPC3050L	BLOCK-1401
HEPC3050P	BLOCK-1401
HEPC3073L	BLOCK-1410
HEPC3073P	BLOCK-1410
HEPC3075L	BLOCK-1412
HEPC3075P	BLOCK-1412
HEPC3401L	BLOCK-1391
HEPC3800L	BLOCK-1423
HEPC3800P	BLOCK-1423
HEPC3801L	BLOCK-1425
HEPC3801P	BLOCK-1425
HEPC3803P	BLOCK-2089
HEPC3804P	BLOCK-1048
HEPC3806P	BLOCK-161
HEPC3807P	BLOCK-2090
HEPC4000P	BLOCK-1028
HEPC4001P	BLOCK-1029
HEPC4002P	BLOCK-1030
HEPC4003P	BLOCK-1064
HEPC4004P	BLOCK-1097
HEPC4005P	BLOCK-1032
HEPC4007P	BLOCK-1032
HEPC4008P	BLOCK-1079
HEPC4009P	BLOCK-1084
HEPC4020P	BLOCK-1059
HEPC4021P	BLOCK-1060
HEPC4030P	BLOCK-1069
HEPC4031P	BLOCK-1121
HEPC4032P	BLOCK-1132
HEPC4033P	BLOCK-1129
HEPC4040P	BLOCK-1076
HEPC4041P	BLOCK-1122
HEPC4042P	BLOCK-1077
HEPC4050P	BLOCK-1085
HEPC4051P	BLOCK-1086
HEPC4052P	BLOCK-1087
HEPC4053P	BLOCK-1081
HEPC4054P	BLOCK-1125
HEPC4055P	BLOCK-1089
HEPC4058P	BLOCK-1139
HEPC4059P	BLOCK-970
HEPC6001	BLOCK-1212
HEPC6003	BLOCK-1773
HEPC6003P	BLOCK-1773
HEPC6009	BLOCK-1859
HEPC6009P	BLOCK-1859
HEPC6010	BLOCK-1774
HEPC6010P	BLOCK-1774
HEPC6011	BLOCK-1791
HEPC6013	BLOCK-1786
HEPC6014	BLOCK-1787
HEPC6015	BLOCK-1794
HEPC6016	BLOCK-1797
HEPC6016C	BLOCK-1797
HEPC6016P	BLOCK-1882
HEPC6017	BLOCK-1798
HEPC6049R	BLOCK-2068
HEPC6050G	BLOCK-2138
HEPC6052G	BLOCK-2056
HEPC6052P	BLOCK-2058
HEPC6055L	BLOCK-1281
HEPC6057P	BLOCK-1270
HEPC6058P	BLOCK-1272
HEPC6059P	BLOCK-1405
HEPC6060P	BLOCK-7
HEPC6061P	BLOCK-1770
HEPC6062P	BLOCK-1262
HEPC6063P	BLOCK-43
HEPC6065P	BLOCK-1276
HEPC6066P	BLOCK-1822
HEPC6067G	BLOCK-1256
HEPC6067P	BLOCK-1244
HEPC6069G	BLOCK-1805
HEPC6070P	BLOCK-1271
HEPC6071P	BLOCK-1272
HEPC6072P	BLOCK-1822
HEPC6074P	BLOCK-1275
HEPC6075P	BLOCK-1291
HEPC6076P	BLOCK-1422
HEPC6079P	BLOCK-138
HEPC6082P	BLOCK-1264
HEPC6083P	BLOCK-43
HEPC6085P	BLOCK-1287
HEPC6089	BLOCK-1256
HEPC6090	BLOCK-1828
HEPC6091	BLOCK-1844
HEPC6091G	BLOCK-1844
HEPC6094P	BLOCK-1275
HEPC6095P	BLOCK-1277
HEPC6096P	BLOCK-1825
HEPC6099P	BLOCK-1820
HEPC6100P	BLOCK-1808
HEPC6101P	BLOCK-1279
HEPC6102P	BLOCK-1801
HEPC6103P	BLOCK-1949
HEPC6104L	BLOCK-1983
HEPC6105P	BLOCK-2137
HEPC6107P	BLOCK-2141
HEPC6110P	BLOCK-2087
HEPC6111P	BLOCK-2093
HEPC6112P	BLOCK-2095
HEPC6113P	BLOCK-2097
HEPC6114P	BLOCK-2119
HEPC6115P	BLOCK-2085
HEPC6116P	BLOCK-2137
HEPC6118P	BLOCK-2091
HEPC6120P	BLOCK-2094
HEPC6121P	BLOCK-2096
HEPC6122P	BLOCK-2108
HEPC6123P	BLOCK-2128
HEPC6124P	BLOCK-2086
HEPC6125P	BLOCK-2136
HEPC6126P	BLOCK-1823
HEPC6127P	BLOCK-1826
HEPC6128P	BLOCK-1827
HEPC6129P	BLOCK-2071
HEPC6130P	BLOCK-2079
HEPC6131P	BLOCK-2079
HEPC6132P	BLOCK-2144
HEPC6133P	BLOCK-2074
HEPC6134P	BLOCK-2075
HEPC6135P	BLOCK-886
HEPC6136P	BLOCK-2091
HEPC6137P	BLOCK-882
HEPC6138P	BLOCK-883
HEPC6142P	BLOCK-2144
HEPC6144P	BLOCK-2160
HEPC6800P	BLOCK-1164
HEPC6802P	BLOCK-1165
HEPC6810P	BLOCK-1168
HEPC6821P	BLOCK-1169
HEPC6850P	BLOCK-1170
HEPC6860P	BLOCK-1171
HEPC7400P	BLOCK-1293
HEPC7401P	BLOCK-1294
HEPC7402P	BLOCK-1295
HEPC7403P	BLOCK-1296
HEPC7404P	BLOCK-1297
HEPC7405P	BLOCK-1298
HEPC7406P	BLOCK-1299
HEPC7407P	BLOCK-1300
HEPC7408P	BLOCK-1301
HEPC7409P	BLOCK-1302
HEPC74107P	BLOCK-1305
HEPC74109P	BLOCK-1306
HEPC7410P	BLOCK-1304
HEPC7411P	BLOCK-1307
HEPC74121P	BLOCK-1311
HEPC74123P	BLOCK-1313
HEPC74125P	BLOCK-1314
HEPC74126P	BLOCK-1315
HEPC74132P	BLOCK-1318
HEPC7413P	BLOCK-1317
HEPC74141P	BLOCK-1321
HEPC74145P	BLOCK-1325
HEPC74147P	BLOCK-1326
HEPC7414P	BLOCK-1320
HEPC74150P	BLOCK-1327
HEPC74151AP	BLOCK-1328
HEPC74153P	BLOCK-1330
HEPC74154P	BLOCK-1331
HEPC74155P	BLOCK-1332
HEPC74156P	BLOCK-1333
HEPC74157P	BLOCK-1334
HEPC74160AP	BLOCK-1337
HEPC74161AP	BLOCK-1338
HEPC74162AP	BLOCK-1339
HEPC74163AP	BLOCK-1340
HEPC74164P	BLOCK-1341
HEPC74165P	BLOCK-1342
HEPC7416P	BLOCK-1336
HEPC74170P	BLOCK-1344
HEPC74173P	BLOCK-1345
HEPC74174P	BLOCK-1346
HEPC74175P	BLOCK-1347
HEPC74176P	BLOCK-1348
HEPC74177P	BLOCK-1349
HEPC7417P	BLOCK-1343
HEPC74180P	BLOCK-1352
HEPC74181P	BLOCK-1353
HEPC74182P	BLOCK-1354
HEPC74190P	BLOCK-1355

34 ORIGINAL DEVICE TYPES AND REPLACEMENT CODES

DEVICE TYPE	REPL CODE
HEPC74191P	BLOCK-1356
HEPC74192P	BLOCK-1357
HEPC74193P	BLOCK-1358
HEPC74195P	BLOCK-1359
HEPC74196P	BLOCK-1360
HEPC74197P	BLOCK-1361
HEPC74198P	BLOCK-1362
HEPC74199P	BLOCK-1363
HEPC7420P	BLOCK-1365
HEPC7422P	BLOCK-1367
HEPC7423P	BLOCK-1369
HEPC74251P	BLOCK-1371
HEPC7427P	BLOCK-1372
HEPC7430P	BLOCK-1377
HEPC7432P	BLOCK-1378
HEPC74365P	BLOCK-1380
HEPC74366P	BLOCK-1381
HEPC74367P	BLOCK-1382
HEPC74368P	BLOCK-1383
HEPC7437P	BLOCK-1384
HEPC7438P	BLOCK-1385
HEPC7440P	BLOCK-1390
HEPC7441AP	BLOCK-1391
HEPC7441P	BLOCK-1391
HEPC7442P	BLOCK-1392
HEPC7445P	BLOCK-1395
HEPC7446AP	BLOCK-1396
HEPC7447AP	BLOCK-1397
HEPC7447P	BLOCK-1397
HEPC7448P	BLOCK-1398
HEPC7450P	BLOCK-1401
HEPC7451P	BLOCK-1402
HEPC7453P	BLOCK-1403
HEPC7454P	BLOCK-1404
HEPC7460P	BLOCK-1406
HEPC7470P	BLOCK-1408
HEPC7472P	BLOCK-1409
HEPC7473P	BLOCK-1410
HEPC7474P	BLOCK-1411
HEPC7475P	BLOCK-1412
HEPC7476P	BLOCK-1413
HEPC7483P	BLOCK-1417
HEPC7485P	BLOCK-1418
HEPC7486P	BLOCK-1419
HEPC7490AP	BLOCK-1423
HEPC7490P	BLOCK-1423
HEPC7491AP	BLOCK-1424
HEPC7492AP	BLOCK-1425
HEPC7492P	BLOCK-1425
HEPC7493AP	BLOCK-1426
HEPC7495P	BLOCK-1428
HEPC7496P	BLOCK-1429
HEPC74LS00P	BLOCK-1568
HEPC74LS02P	BLOCK-1570
HEPC74LS03P	BLOCK-1571
HEPC74LS04P	BLOCK-1572
HEPC74LS05P	BLOCK-1573
HEPC74LS08P	BLOCK-1574
HEPC74LS109P	BLOCK-1578
HEPC74LS10P	BLOCK-1576
HEPC74LS11P	BLOCK-1579
HEPC74LS136P	BLOCK-1590
HEPC74LS138P	BLOCK-1591
HEPC74LS13P	BLOCK-1587
HEPC74LS14P	BLOCK-1593
HEPC74LS151P	BLOCK-1598
HEPC74LS153P	BLOCK-1599
HEPC74LS155P	BLOCK-1600
HEPC74LS157P	BLOCK-1602
HEPC74LS158P	BLOCK-1603
HEPC74LS170P	BLOCK-1613
HEPC74LS174P	BLOCK-1615
HEPC74LS175P	BLOCK-1616
HEPC74LS181P	BLOCK-1617
HEPC74LS190P	BLOCK-1618
HEPC74LS191P	BLOCK-1619
HEPC74LS192P	BLOCK-1620
HEPC74LS193P	BLOCK-1621
HEPC74LS196P	BLOCK-1624
HEPC74LS197P	BLOCK-1625
HEPC74LS20P	BLOCK-1626
HEPC74LS251P	BLOCK-1639
HEPC74LS253P	BLOCK-1640
HEPC74LS257P	BLOCK-1641
HEPC74LS258P	BLOCK-1642
HEPC74LS259P	BLOCK-1643
HEPC74LS279P	BLOCK-1649
HEPC74LS27P	BLOCK-1647
HEPC74LS298P	BLOCK-1656
HEPC74LS30P	BLOCK-1658
HEPC74LS32P	BLOCK-1659
HEPC74LS367P	BLOCK-1668
HEPC74LS368P	BLOCK-1669
HEPC74LS42P	BLOCK-1684
HEPC74LS51P	BLOCK-1689
HEPC74LS54P	BLOCK-1690
HEPC74LS55P	BLOCK-1693
HEPC74LS670P	BLOCK-1704
HEPC74LS74P	BLOCK-1706
HEPC74LS86P	BLOCK-1713
HEPC74LS90P	BLOCK-1714
HEPC74LS93P	BLOCK-1717
HL-55661	BLOCK-1719
HL18998	BLOCK-1293
HL18999	BLOCK-1411
HL19000	BLOCK-1297
HL19001	BLOCK-1304
HL19002	BLOCK-1410
HL19003	BLOCK-1365
HL19004	BLOCK-1295
HL19005	BLOCK-1296
HL19006	BLOCK-1426
HL19008	BLOCK-1311
HL19009	BLOCK-1392
HL19010	BLOCK-1413
HL19011	BLOCK-1390
HL19012	BLOCK-1412
HL19013	BLOCK-1377
HL19014	BLOCK-1419
HL19015	BLOCK-1423
HL53424	BLOCK-2197
HL53426	BLOCK-2201
HL53428	BLOCK-2206
HL53429	BLOCK-2208
HL55660	BLOCK-1307
HL55663	BLOCK-1733
HL55723	BLOCK-1737
HL55763	BLOCK-1301
HL55764	BLOCK-1330
HL55861	BLOCK-1327
HL55862	BLOCK-1297
HL56320	BLOCK-1726
HL56420	BLOCK-1293
HL56421	BLOCK-1297
HL56422	BLOCK-1365
HL56423	BLOCK-1377
HL56424	BLOCK-1385
HL56425	BLOCK-1411
HL56426	BLOCK-1418
HL56427	BLOCK-1428
HL56429	BLOCK-1357
HL56430	BLOCK-1358
HL56431	BLOCK-1480
HL56842	BLOCK-1417
HL56899	BLOCK-1304
HM1-6508-5	BLOCK-1160
HM1-7602-5	BLOCK-1743
HM1-7603-5	BLOCK-1750
HM1-7610-5	BLOCK-1752
HM1-7611-5	BLOCK-1749
HM1-7620-5	BLOCK-1762
HM1-7621-5	BLOCK-1763
HM1-7640-2	BLOCK-1758
HM1-7640-5	BLOCK-1758
HM1-7641-5	BLOCK-1757
HM1-7642-5	BLOCK-1764
HM1-7642A-5	BLOCK-1764
HM1-7643-5	BLOCK-1765
HM1-7643A-5	BLOCK-1765
HM1-7648-5	BLOCK-1756
HM1-7649-5	BLOCK-1755
HM1-7680-5	BLOCK-1760
HM1-7681-5	BLOCK-1759
HM1-7685-5	BLOCK-1754
HM3-6508-5	BLOCK-1160
HM3-7602-5	BLOCK-1743
HM3-7603-5	BLOCK-1750
HM3-7610-5	BLOCK-1752
HM3-7610A5	BLOCK-1752
HM3-7611-5	BLOCK-1749
HM3-7611A5	BLOCK-1749
HM3-7620-5	BLOCK-1762
HM3-7620A5	BLOCK-1762
HM3-7621-5	BLOCK-1763
HM3-7621A5	BLOCK-1763
HM3-7640-5	BLOCK-1758
HM3-7640A5	BLOCK-1758
HM3-7641-5	BLOCK-1757
HM3-7641A5	BLOCK-1757
HM3-7642-5	BLOCK-1764
HM3-7643-5	BLOCK-1765
HM3-7643A-5	BLOCK-1765
HM3-7648-5	BLOCK-1756
HM3-7649-5	BLOCK-1755
HM3-7680-5	BLOCK-1760
HM3-7681-5	BLOCK-1759
HM3-7685-5	BLOCK-1754
HM472114-3	BLOCK-1002
HM472114-4	BLOCK-1002
HM472114P-3	BLOCK-1002
HM472114P-4	BLOCK-1002
HM4864-2	BLOCK-1006
HM4864-3	BLOCK-1006
HM4864P-2	BLOCK-1006
HM50256-15	BLOCK-1116
HM50256-20	BLOCK-1116
HM50256P-12	BLOCK-1116
HM50256P-15	BLOCK-1116
HM50256P-20	BLOCK-1116
HM6116	BLOCK-1004
HM6116-3	BLOCK-1004
HM6116-4	BLOCK-1004
HM6116FP3	BLOCK-1004
HM6116FP4	BLOCK-1004
HM6116LP-4	BLOCK-1004
HM6116P-3	BLOCK-1004
HM6116P-4	BLOCK-1004
HM6116P4	BLOCK-1004
HN25044	BLOCK-1764
HN25045	BLOCK-1765
HN25085	BLOCK-1754
HN25085S	BLOCK-1754
HN25088	BLOCK-1760
HN25088S	BLOCK-1760
HN25089	BLOCK-1759
HN25089S	BLOCK-1759
HN462532	BLOCK-1015
HN462532G	BLOCK-1015
HN462532G2	BLOCK-1015
HN462716G	BLOCK-1017
HN462732	BLOCK-1018
HN462732G	BLOCK-1018
HN482764G	BLOCK-1019
HN482764G-3	BLOCK-1019
HP4510	BLOCK-1021
HP74LS240P	BLOCK-1630
HYB4116-P2	BLOCK-1003
HYB4116-P3	BLOCK-1003
HYB4164-P2	BLOCK-1006
HZT-33	BLOCK-1157
HZT-3301	BLOCK-1157
HZT33	BLOCK-1157
HZT33-01	BLOCK-1157
HZT33-02	BLOCK-1157
HZT33-02-TE	BLOCK-1157
HZT33-02T	BLOCK-1157
HZT33-04	BLOCK-1157
HZT33-04T	BLOCK-1157
HZT33-05	BLOCK-1157
HZT33-10	BLOCK-1157
HZT33-12	BLOCK-1157
HZT33S1	BLOCK-1157
I00DF01120	BLOCK-840
I00DF10110	BLOCK-839
I01B98M090	BLOCK-881
I01S058360	BLOCK-761
I01SD57530	BLOCK-616
I02190574J	BLOCK-1157
I02990574J	BLOCK-1157
I02A978120	BLOCK-2097
I02A98L050	BLOCK-2144
I02DB13820	BLOCK-603
I02DC13710	BLOCK-398
I02DE14020	BLOCK-774
I02DF1480C	BLOCK-781
I02DF1481C	BLOCK-782
I02DF148C	BLOCK-782
I02I90574J	BLOCK-1157
I02J98L050	BLOCK-2144
I02S010310	BLOCK-234
I02SD13780	BLOCK-662
I02SD14980	BLOCK-748
I02SF12520	BLOCK-775
I02SF12530	BLOCK-776
I02SP12770	BLOCK-613
I031956310	BLOCK-1157
I039956300	BLOCK-1157
I03A98M050	BLOCK-2087
I03A98M060	BLOCK-2093
I03A98M090	BLOCK-881
I03A98M120	BLOCK-2097
I03B98M050	BLOCK-2087
I03B98M060	BLOCK-2093
I03B98M090	BLOCK-881
I03B98M120	BLOCK-2097
I03D040660	BLOCK-1093
I03D063240	BLOCK-2171
I03D072200	BLOCK-780
I03D079100	BLOCK-644
I03D079130	BLOCK-816
I03D079400	BLOCK-396
I03DA7530N	BLOCK-807
I03DD78000	BLOCK-525
I03DE75200	BLOCK-709
I03S063580	BLOCK-1994
I03S063930	BLOCK-2063
I03S064580	BLOCK-516
I03S070160	BLOCK-762
I03S073080	BLOCK-851
I03S079100	BLOCK-644
I03SD78350	BLOCK-835
I03SD78370	BLOCK-1266
I03SP41400	BLOCK-453
I03SP44450	BLOCK-687
I05D076300	BLOCK-563
I05DA7607A	BLOCK-532
I05DE76440	BLOCK-534
I05DE76810	BLOCK-557
I05DE7777P	BLOCK-771
I05DE86770	BLOCK-771
I05S925080	BLOCK-732
I05SP72220	BLOCK-265
I06DA365SP	BLOCK-1196
I07D069930	BLOCK-2063
I0QK98M090	BLOCK-918
I0QT98L050	BLOCK-2144
I0XDP12130	BLOCK-504
I1KA978090	BLOCK-918
I23S953320	BLOCK-860
I2764	BLOCK-1019
I2764-2	BLOCK-1019
I2B4900410	BLOCK-873
I2B4910410	BLOCK-874
I41D41256C	BLOCK-1116
I52D003930	BLOCK-2063
I52D040530	BLOCK-1088
I53D040010	BLOCK-1029
I55D040010	BLOCK-1029
I55D04052B	BLOCK-1087
I55D04066B	BLOCK-1093
I55D04538B	BLOCK-1140
I55D0HC020	BLOCK-1507
I55D4HC040	BLOCK-1508
I5PD14053B	BLOCK-1088
I61D4LS140	BLOCK-1593
I61DLS1250	BLOCK-1585
I61DLS1740	BLOCK-1615
I61DLS2400	BLOCK-1630
I61DLS2440	BLOCK-1634
I61DLS2450	BLOCK-1635
I61DLS3730	BLOCK-1671
I64D7406P0	BLOCK-1299
I6PDC14880	BLOCK-1771
I6PDC14890	BLOCK-1772
I8255	BLOCK-1855
IC-100(ELCOM)	BLOCK-1425
IC-101(ELCOM)	BLOCK-1397
IC-101-004	BLOCK-96
IC-102(ELCOM)	BLOCK-1301
IC-103(ELCOM)	BLOCK-1317
IC-104(ELCOM)	BLOCK-1299
IC-105(ELCOM)	BLOCK-1336
IC-106(ELCOM)	BLOCK-1311
IC-112	BLOCK-78
IC-12	BLOCK-1262
IC-12(PHILCO)	BLOCK-1271
IC-13	BLOCK-1272
IC-13(PHILCO)	BLOCK-1272
IC-14	BLOCK-1292
IC-14(PHILCO)	BLOCK-1292
IC-1489	BLOCK-1772
IC-15	BLOCK-1269
IC-15(PHILCO)	BLOCK-1269
IC-16	BLOCK-1805
IC-16(PHILCO)	BLOCK-1805
IC-17	BLOCK-1405
IC-17(ELCOM)	BLOCK-1275
IC-17(PHILCO)	BLOCK-1405
IC-18	BLOCK-1277
IC-18(PHILCO)	BLOCK-1407
IC-19	BLOCK-1804
IC-19#	BLOCK-1244
IC-19(PHILCO)	BLOCK-1804
IC-2(RCA)	BLOCK-1265
IC-20	BLOCK-1984
IC-20(PHILCO)	BLOCK-1984
IC-21(ELCOM)	BLOCK-1773
IC-211(ELCOM)	BLOCK-1943
IC-2114-L3	BLOCK-1002
IC-213(ELCOM)	BLOCK-1881
IC-217(ELCOM)	BLOCK-1798
IC-220(ELCOM)	BLOCK-2100
IC-221(ELCOM)	BLOCK-2103
IC-222(ELCOM)	BLOCK-2104
IC-223(ELCOM)	BLOCK-2105
IC-224(ELCOM)	BLOCK-2111
IC-225(ELCOM)	BLOCK-2101
IC-23	BLOCK-1303
IC-23(PHILCO)	BLOCK-1303
IC-234(ELCOM)	BLOCK-1953
IC-237(ELCOM)	BLOCK-2157
IC-243(ELCOM)	BLOCK-1859
IC-244(ELCOM)	BLOCK-1794
IC-25	BLOCK-1244
IC-25(ELCOM)	BLOCK-1256
IC-26	BLOCK-1421
IC-26(ELCOM)	BLOCK-1421
IC-27(PHILCO)	BLOCK-1804
IC-276(ELCOM)	BLOCK-2206
IC-279(ELCOM)	BLOCK-2209
IC-28	BLOCK-1422
IC-280(ELCOM)	BLOCK-2199
IC-281(ELCOM)	BLOCK-2200
IC-286(ELCOM)	BLOCK-1791
IC-288(ELCOM)	BLOCK-1804
IC-290(ELCOM)	BLOCK-1786
IC-298(ELCOM)	BLOCK-1790
IC-312	BLOCK-2166
IC-3130	BLOCK-1997
IC-32	BLOCK-1804
IC-34	BLOCK-1405
IC-4	BLOCK-59
IC-409(ELCOM)	BLOCK-2101
IC-410(ELCOM)	BLOCK-2106
IC-412(ELCOM)	BLOCK-2109
IC-413(ELCOM)	BLOCK-2099
IC-414(ELCOM)	BLOCK-2102
IC-415(ELCOM)	BLOCK-2122
IC-419(ELCOM)	BLOCK-1968
IC-420(ELCOM)	BLOCK-1969
IC-433(ELCOM)	BLOCK-1787
IC-435(ELCOM)	BLOCK-1786
IC-441(ELCOM)	BLOCK-2107
IC-442(ELCOM)	BLOCK-1786
IC-443(ELCOM)	BLOCK-1787
IC-444(ELCOM)	BLOCK-1788
IC-445(ELCOM)	BLOCK-1789
IC-446(ELCOM)	BLOCK-1792
IC-447(ELCOM)	BLOCK-1793
IC-448(ELCOM)	BLOCK-1793
IC-449(ELCOM)	BLOCK-1795
IC-450(ELCOM)	BLOCK-1796
IC-451(ELCOM)	BLOCK-1797
IC-453(ELCOM)	BLOCK-1800
IC-455(ELCOM)	BLOCK-1950
IC-5	BLOCK-1807
IC-5(ELCOM)	BLOCK-1212
IC-5(PHILCO)	BLOCK-1807
IC-500(ELCOM)	BLOCK-1416
IC-502	BLOCK-1244
IC-503	BLOCK-1244
IC-504	BLOCK-1262
IC-505	BLOCK-1264
IC-507	BLOCK-1269
IC-508	BLOCK-1270
IC-508(ELCOM)	BLOCK-73
IC-509	BLOCK-1271
IC-510	BLOCK-1272
IC-511	BLOCK-1275
IC-512(ELCOM)	BLOCK-38
IC-517(ELCOM)	BLOCK-41
IC-520(ELCOM)	BLOCK-2079
IC-524(ELCOM)	BLOCK-1303
IC-526(ELCOM)	BLOCK-1024
IC-547(ELCOM)	BLOCK-13
IC-548(ELCOM)	BLOCK-15
IC-549(ELCOM)	BLOCK-12
IC-55(ELCOM)	BLOCK-2193
IC-550(ELCOM)	BLOCK-142
IC-551(ELCOM)	BLOCK-18
IC-555	BLOCK-2079
IC-56(ELCOM)	BLOCK-2176
IC-58(ELCOM)	BLOCK-2185
IC-59(ELCOM)	BLOCK-2186
IC-6	BLOCK-1267
IC-6(PHILCO)	BLOCK-1267
IC-603(ELCOM)	BLOCK-1828
IC-605(ELCOM)	BLOCK-1303
IC-61(ELCOM)	BLOCK-2189
IC-62(ELCOM)	BLOCK-2189
IC-65(ELCOM)	BLOCK-2224
IC-6502	BLOCK-1158
IC-6850	BLOCK-1170
IC-7	BLOCK-1212
IC-71(ELCOM)	BLOCK-1377
IC-74(ELCOM)	BLOCK-1294
IC-7400	BLOCK-1293
IC-7402	BLOCK-1295
IC-7403	BLOCK-1296
IC-7404	BLOCK-1297
IC-7408	BLOCK-1301
IC-7408N	BLOCK-1301
IC-7410	BLOCK-1304
IC-74123	BLOCK-1313
IC-74151	BLOCK-1328
IC-74165	BLOCK-1342
IC-7417	BLOCK-1343
IC-74174	BLOCK-1346
IC-74175	BLOCK-1347
IC-7420	BLOCK-1365
IC-7430	BLOCK-1377
IC-7438	BLOCK-1385
IC-74390	BLOCK-1387
IC-7474	BLOCK-1411
IC-7475	BLOCK-1412
IC-7476	BLOCK-1413
IC-7492	BLOCK-1425
IC-74L03	BLOCK-1571
IC-74LS02	BLOCK-1570
IC-74LS04	BLOCK-1572
IC-74LS12	BLOCK-1582
IC-74LS138	BLOCK-1591
IC-74LS139	BLOCK-1592
IC-74LS157	BLOCK-1602
IC-74LS163	BLOCK-1607
IC-74LS165	BLOCK-1609
IC-74LS193	BLOCK-1621
IC-74LS20	BLOCK-1626
IC-74LS390	BLOCK-1678
IC-74LS75	BLOCK-1707
IC-74LS86	BLOCK-1713
IC-74S138	BLOCK-1733
IC-75(ELCOM)	BLOCK-1401
IC-79(ELCOM)	BLOCK-1428
IC-8	BLOCK-1421
IC-80(ELCOM)	BLOCK-1293
IC-81(ELCOM)	BLOCK-1426
IC-82(ELCOM)	BLOCK-1295
IC-83(ELCOM)	BLOCK-1296
IC-84(ELCOM)	BLOCK-1297
IC-85(ELCOM)	BLOCK-1298
IC-86(ELCOM)	BLOCK-1304
IC-87(ELCOM)	BLOCK-1365
IC-88(ELCOM)	BLOCK-1390
IC-8B	BLOCK-1223
IC-8B#	BLOCK-1421
IC-8T26	BLOCK-1173
IC-8T28	BLOCK-1178
IC-8T95	BLOCK-1174
IC-90(ELCOM)	BLOCK-1392
IC-903	BLOCK-1256
IC-91	BLOCK-1212
IC-91(ELCOM)	BLOCK-1402
IC-92(ELCOM)	BLOCK-1403
IC-93(ELCOM)	BLOCK-1404
IC-94(ELCOM)	BLOCK-1409
IC-95(ELCOM)	BLOCK-1410
IC-96(ELCOM)	BLOCK-1412
IC-97(ELCOM)	BLOCK-1411
IC-98(ELCOM)	BLOCK-1423
IC-99(ELCOM)	BLOCK-1413
IC-L2114-550	BLOCK-1002
IC112	BLOCK-78
IC12	BLOCK-1256
IC126	BLOCK-103
IC13	BLOCK-1274
IC132	BLOCK-72
IC136	BLOCK-49
IC14	BLOCK-1292
IC142	BLOCK-48
IC146	BLOCK-5
IC15	BLOCK-1269
IC17	BLOCK-1275
IC18	BLOCK-1277
IC19	BLOCK-1256
IC2(ECS)	BLOCK-1944
IC21	BLOCK-1807
IC21(ELCOM)	BLOCK-1773
IC213(ELCOM)	BLOCK-1881
IC217(ELOCM)	BLOCK-1798
IC229(ELCOM)	BLOCK-2207
IC23	BLOCK-1303
IC23(PHILCO)	BLOCK-1303
IC25(ELCOM)	BLOCK-1256
IC26	BLOCK-43
IC286(ELCOM)	BLOCK-1791
IC288(ELCOM)	BLOCK-1804
IC29(ELCOM)	BLOCK-1944
IC290(ELCOM)	BLOCK-1786
IC298(ELCOM)	BLOCK-1790
IC3	BLOCK-1274
IC3(PHILCO)@	BLOCK-1274
IC30(ELCOM)	BLOCK-2197
IC31(ELCOM)	BLOCK-2202

ORIGINAL DEVICE TYPES AND REPLACEMENT CODES 35

DEVICE TYPE	REPL CODE
IC311(ELCOM)	BLOCK-1286
IC32(ELCOM)	BLOCK-2208
IC320(ELCOM)	BLOCK-1814
IC33(ELCOM)	BLOCK-1950
IC34(ELCOM)	BLOCK-2217
IC35(ELCOM)	BLOCK-1967
IC4001BP	BLOCK-1029
IC41(ELCOM)	BLOCK-1943
IC5	BLOCK-1212
IC500	BLOCK-1212
IC500(IR)	BLOCK-1212
IC501	BLOCK-1223
IC501(ELCOM)	BLOCK-1790
IC502	BLOCK-1244
IC503	BLOCK-1256
IC503(IR)	BLOCK-1256
IC504	BLOCK-1262
IC505	BLOCK-1264
IC505(IR)	BLOCK-1264
IC506	BLOCK-1265
IC507	BLOCK-1269
IC507(IR)	BLOCK-1269
IC508	BLOCK-1270
IC508(IR)	BLOCK-1270
IC509	BLOCK-1271
IC509(IR)	BLOCK-1271
IC510	BLOCK-1272
IC510(IR)	BLOCK-1272
IC511	BLOCK-1275
IC512	BLOCK-1277
IC512(IR)	BLOCK-1277
IC6	BLOCK-1267
IC7	BLOCK-1212
IC724	BLOCK-2193
IC7420	BLOCK-1365
IC7475	BLOCK-1412
IC8	BLOCK-1421
IC8B	BLOCK-1421
IC903	BLOCK-1256
IC91	BLOCK-1212
ICF-1	BLOCK-1212
ICL7106CPL	BLOCK-966
ICL7107CDL	BLOCK-965
ICL7107CPL	BLOCK-965
ICL7116CPL	BLOCK-976
ICL7126CPL	BLOCK-964
ICL7126RCPL	BLOCK-977
ICL7129CPL	BLOCK-978
ICL8038	BLOCK-1919
ICM7555CBA	BLOCK-2082
ICM7555IPA	BLOCK-2080
ID20-0308	BLOCK-1298
ID2732A-3	BLOCK-1018
IDH0742	BLOCK-1093
IDH0802	BLOCK-1088
IDUPC7912H	BLOCK-2108
IFSI-3122P	BLOCK-907
IGL78M05	BLOCK-2087
IGL78M06	BLOCK-2093
IGLA3150	BLOCK-447
IGLA6324	BLOCK-2171
IGLA6458DF	BLOCK-1801
IGLB1290	BLOCK-944
IGLB1649	BLOCK-755
IGLC4011	BLOCK-1039
IGLC4011B	BLOCK-1039
IGLC4013B	BLOCK-1041
IGLC4069UB	BLOCK-1096
IGLC4081B	BLOCK-525
IGSTK5322	BLOCK-305
IGSTK5431	BLOCK-714
IGSTK5431SL	BLOCK-714
IGSTK5431ST	BLOCK-714
IGSTK6962	BLOCK-729
IGSTK6972	BLOCK-806
IHHA11711	BLOCK-578
IHHA11714	BLOCK-587
IHHA11715	BLOCK-541
IHHA11747ANT	BLOCK-686
IHHA11749	BLOCK-672
IJNJM4558DM	BLOCK-1801
ILA7621	BLOCK-752
ILD1020	BLOCK-12
ILH0111	BLOCK-459
ILL0052	BLOCK-453
ILM0292	BLOCK-903
ILM340T12	BLOCK-2097
ILMC14013B	BLOCK-1041
ILMC14081B	BLOCK-1105
ILT0014	BLOCK-85
IMAN262	BLOCK-250
IMC1358P	BLOCK-1269
IMDN6838	BLOCK-685
IN4066BP	BLOCK-1093

DEVICE TYPE	REPL CODE
INMPC1363C	BLOCK-396
INMPC1363CA	BLOCK-396
INMPC1373H	BLOCK-695
INMPC1373HA	BLOCK-695
INMPC1513HA	BLOCK-768
INMPD14027BC	BLOCK-1066
INMPD4011BC	BLOCK-1039
INMPD4013BC	BLOCK-1041
INMPD4066BC	BLOCK-1093
INMPD4081C	BLOCK-525
INMPD4093BC	BLOCK-1110
INS82LS05N	BLOCK-1591
INUPC1373H	BLOCK-695
INUPD4011BC	BLOCK-1039
INUPD4013BC	BLOCK-1041
INUPD4066BC	BLOCK-1093
INUPD4081C	BLOCK-525
INUPD4093BC	BLOCK-1110
IP20-0161	BLOCK-145
IP20-0174	BLOCK-1212
IP20-0205	BLOCK-1293
IP20-0206	BLOCK-1411
IP20-0207	BLOCK-1426
IP20-0208	BLOCK-1486
IP20-0209	BLOCK-1338
IP20-0210	BLOCK-2140
IP20-0220	BLOCK-2173
IP20-0253	BLOCK-2173
IP20-0281	BLOCK-229
IP20-0308	BLOCK-1298
IP20-0310	BLOCK-1294
IP20-0315	BLOCK-1426
IP20-0316	BLOCK-1411
IP20-0317	BLOCK-248
IP20-0429	BLOCK-180
IP20-0447	BLOCK-973
IP20-2010	BLOCK-2140
IP317	BLOCK-887
IP340-12	BLOCK-890
IP340-5	BLOCK-1022
IP7805	BLOCK-2087
IP7805A	BLOCK-2087
IP7812	BLOCK-2097
IP7815	BLOCK-2119
IP7905	BLOCK-2091
IP7912	BLOCK-2108
IP7915	BLOCK-2128
IP8255A-5	BLOCK-1855
IR3N05	BLOCK-1868
IR3N06	BLOCK-1897
IR9358	BLOCK-1993
IR9393	BLOCK-2063
ISSRM211403	BLOCK-1002
ISSRM2114C3	BLOCK-1002
ITA7176AP	BLOCK-1269
ITA7607AP	BLOCK-532
ITA7608CP5	BLOCK-519
ITA7609P-2	BLOCK-484
ITA7609P2	BLOCK-484
ITT1330	BLOCK-1407
ITT1352	BLOCK-1422
ITT1800-5	BLOCK-2150
ITT1800N	BLOCK-2150
ITT1801N	BLOCK-2151
ITT1806-5	BLOCK-2156
ITT1806N	BLOCK-2156
ITT1807-5	BLOCK-2156
ITT1807N	BLOCK-2157
ITT1808-5	BLOCK-2158
ITT1808N	BLOCK-2158
ITT1809-5	BLOCK-2159
ITT1809N	BLOCK-2159
ITT1810-5	BLOCK-2161
ITT1810N	BLOCK-2161
ITT1811-5	BLOCK-2162
ITT1811N	BLOCK-2162
ITT3064	BLOCK-1808
ITT3065	BLOCK-1269
ITT4116-3	BLOCK-1003
ITT4116-4	BLOCK-1003
ITT7400N	BLOCK-1293
ITT7401N	BLOCK-1294
ITT7402N	BLOCK-1295
ITT7403N	BLOCK-1296
ITT7404N	BLOCK-1297
ITT7405N	BLOCK-1298
ITT7406N	BLOCK-1299
ITT7407N	BLOCK-1300
ITT7408N	BLOCK-1301
ITT7409N	BLOCK-1302
ITT74107N	BLOCK-1305
ITT74109N	BLOCK-1306
ITT7410N	BLOCK-1304
ITT7411N	BLOCK-1307

DEVICE TYPE	REPL CODE
ITT74121N	BLOCK-1311
ITT74122N	BLOCK-1312
ITT74123N	BLOCK-1313
ITT7412N	BLOCK-1310
ITT7413N	BLOCK-1317
ITT74141N	BLOCK-1321
ITT74145N	BLOCK-1325
ITT74150N	BLOCK-1327
ITT74151N	BLOCK-1328
ITT74152N	BLOCK-1329
ITT74153N	BLOCK-1330
ITT74154N	BLOCK-1331
ITT74155N	BLOCK-1332
ITT74156N	BLOCK-1333
ITT74160N	BLOCK-1337
ITT74161N	BLOCK-1338
ITT74162N	BLOCK-1339
ITT74163N	BLOCK-1340
ITT74164N	BLOCK-1341
ITT74165N	BLOCK-1342
ITT7416N	BLOCK-1336
ITT74174N	BLOCK-1346
ITT74175N	BLOCK-1347
ITT7417N	BLOCK-1343
ITT74180N	BLOCK-1352
ITT74181N	BLOCK-1353
ITT74182N	BLOCK-1354
ITT74190N	BLOCK-1355
ITT74192N	BLOCK-1357
ITT74193N	BLOCK-1358
ITT74195N	BLOCK-1359
ITT7420N	BLOCK-1365
ITT7421N	BLOCK-1366
ITT7428N	BLOCK-1373
ITT7430N	BLOCK-1377
ITT7432N	BLOCK-1378
ITT7433N	BLOCK-1379
ITT7437N	BLOCK-1384
ITT7438N	BLOCK-1385
ITT7440N	BLOCK-1390
ITT7442N	BLOCK-1392
ITT7443N	BLOCK-1393
ITT7444N	BLOCK-1394
ITT7445N	BLOCK-1395
ITT7446AN	BLOCK-1396
ITT7447AN	BLOCK-1397
ITT7448N	BLOCK-1398
ITT7450N	BLOCK-1401
ITT7451N	BLOCK-1402
ITT7453N	BLOCK-1403
ITT7454N	BLOCK-1404
ITT7460N	BLOCK-1406
ITT7470N	BLOCK-1408
ITT7472N	BLOCK-1409
ITT7473N	BLOCK-1410
ITT7475N	BLOCK-1412
ITT7476N	BLOCK-1413
ITT7480N	BLOCK-1414
ITT7482N	BLOCK-1416
ITT7483N	BLOCK-1417
ITT7486N	BLOCK-1419
ITT7490N	BLOCK-1423
ITT7491AN	BLOCK-1424
ITT7492N	BLOCK-1425
ITT7493N	BLOCK-1426
ITT7494N	BLOCK-1427
ITT7495AN	BLOCK-1428
ITT7496N	BLOCK-1429
ITT74H00N	BLOCK-1471
ITT74H01N	BLOCK-1472
ITT74H04N	BLOCK-1473
ITT74H05N	BLOCK-1474
ITT74H10N	BLOCK-1476
ITT74H11N	BLOCK-1482
ITT74H20N	BLOCK-1484
ITT74H21N	BLOCK-1485
ITT74H30N	BLOCK-1487
ITT74H40N	BLOCK-1488
ITT74H50N	BLOCK-1489
ITT74H51N	BLOCK-1490
ITT74H53N	BLOCK-1492
ITT74H54N	BLOCK-1493
ITT74H60N	BLOCK-1495
ITT74H72N	BLOCK-1499
ITT74H73N	BLOCK-1500
ITT74H74N	BLOCK-1501
ITT74H76N	BLOCK-1502
ITT9093-5	BLOCK-1950
ITT9093N	BLOCK-1950
ITT9094-5	BLOCK-1951
ITT9094N	BLOCK-1951
ITT9097-5	BLOCK-1952
ITT9097N	BLOCK-1952
ITT9099-5	BLOCK-1953

DEVICE TYPE	REPL CODE
ITT9099N	BLOCK-1953
ITT930-5	BLOCK-2197
ITT930N	BLOCK-2197
ITT932-5	BLOCK-2199
ITT932N	BLOCK-2199
ITT933-5	BLOCK-2200
ITT933N	BLOCK-2200
ITT935-5	BLOCK-2201
ITT935N	BLOCK-2201
ITT936-5	BLOCK-2202
ITT936N	BLOCK-2202
ITT937-5	BLOCK-2203
ITT937N	BLOCK-2203
ITT941-5	BLOCK-2205
ITT941N	BLOCK-2205
ITT944-5	BLOCK-2206
ITT944N	BLOCK-2206
ITT945-5	BLOCK-2207
ITT945N	BLOCK-2207
ITT946-5	BLOCK-2208
ITT946N	BLOCK-2208
ITT948-5	BLOCK-2209
ITT948N	BLOCK-2209
ITT949-5	BLOCK-2210
ITT949N	BLOCK-2210
ITT951-5	BLOCK-2213
ITT951N	BLOCK-2213
ITT961-5	BLOCK-2216
ITT961N	BLOCK-2216
ITT962-5	BLOCK-2217
ITT962N	BLOCK-2217
ITT963-5	BLOCK-2218
ITT963N	BLOCK-2218
ITTA75902P	BLOCK-1265
ITTC4027B	BLOCK-1066
ITTC4030BP	BLOCK-1069
ITTC4049BP	BLOCK-1084
ITTC4052BP	BLOCK-1087
ITTC5012BP	BLOCK-1035
IUPC1363C	BLOCK-396
IUPD1937C	BLOCK-740
IUPD1986C	BLOCK-739
IX0037CE	BLOCK-1157
IX0043CE	BLOCK-1269
IX0054CE	BLOCK-136
IX0069CE	BLOCK-1029
IX0086TA	BLOCK-643
IX0137CE	BLOCK-732
IX0178CE	BLOCK-526
IX0199CE	BLOCK-740
IX0212CE	BLOCK-657
IX0212CEZZ	BLOCK-657
IX0213CE	BLOCK-603
IX0214CE	BLOCK-563
IX0225CE	BLOCK-364
IX0232CE	BLOCK-1020
IX0238CE	BLOCK-662
IX0243CE	BLOCK-1269
IX0252CE	BLOCK-401
IX0252CEZZ	BLOCK-401
IX0260CE	BLOCK-644
IX0275CE	BLOCK-559
IX0325C	BLOCK-851
IX0416CE	BLOCK-773
IX0487CE	BLOCK-775
IX0488CE	BLOCK-205
IX0499CE	BLOCK-776
IX0600CE	BLOCK-771
IX0633CE	BLOCK-1195
IX0638CE	BLOCK-2087
IX0948CE	BLOCK-763
IX1190CE	BLOCK-1191
IYBA5102AALC	BLOCK-545
IYBA6304A	BLOCK-665
J1000-7400	BLOCK-1293
J1000-7402	BLOCK-1295
J1000-7404	BLOCK-1297
J1000-7410	BLOCK-1304
J1000-74121	BLOCK-1311
J1000-7447	BLOCK-1397
J1000-7476	BLOCK-1413
J1000-7490	BLOCK-1423
J1000-7492	BLOCK-1425
J1000-NE555	BLOCK-2079
J241219	BLOCK-1422
J241221	BLOCK-87
J241222	BLOCK-1269
J241270	BLOCK-1421
J4-1000	BLOCK-1293
J4-1002	BLOCK-1295
J4-1004	BLOCK-1297
J4-1010	BLOCK-1304
J4-1047	BLOCK-1397
J4-1075	BLOCK-1412

DEVICE TYPE	REPL CODE
J4-1076	BLOCK-1413
J4-1090	BLOCK-1423
J4-1092	BLOCK-1425
J4-1121	BLOCK-1311
J4-1203	BLOCK-1303
J4-1215	BLOCK-2058
J4-1555	BLOCK-2079
JA-1555	BLOCK-2079
JB8255AP-5	BLOCK-1855
JRC386D	BLOCK-1851
K0201E	BLOCK-2144
K18046	BLOCK-74
K26044	BLOCK-89
K26174	BLOCK-84
K4-590	BLOCK-2058
K4-598	BLOCK-1303
K76026	BLOCK-242
KA2101	BLOCK-1269
KA2131	BLOCK-660
KA2153	BLOCK-534
KA2155	BLOCK-824
KA2181	BLOCK-695
KA2224	BLOCK-621
KA2261	BLOCK-237
KA2911	BLOCK-532
KA2914A	BLOCK-559
KA2919	BLOCK-709
KA33V	BLOCK-1157
KA431CD	BLOCK-2228
KA7812	BLOCK-2097
KB4409	BLOCK-1825
KC383C	BLOCK-246
KC580C1	BLOCK-184
KC581C	BLOCK-247
KC582C	BLOCK-240
KC583C	BLOCK-246
KC850C1	BLOCK-92
KD2114	BLOCK-1944
KD6311	BLOCK-44
KGE41013	BLOCK-91
KGE46441	BLOCK-85
KGE46442	BLOCK-180
KIA6930P	BLOCK-563
KIA7205AP	BLOCK-145
KIA7205P	BLOCK-145
KIA7310P	BLOCK-180
KIA7313AP	BLOCK-453
KIA7373AP	BLOCK-453
KIA7630	BLOCK-563
KIA7630P	BLOCK-563
KIA7640AP	BLOCK-821
KIA78012AP	BLOCK-2097
KIA78012AP(KEC)	BLOCK-2097
KIA7805PI	BLOCK-912
KIA7809PI	BLOCK-918
KIA7812P	BLOCK-2097
KIA78L005AP	BLOCK-2144
KIA78L05AP	BLOCK-2144
KIA78L05BP	BLOCK-2144
KLH5489	BLOCK-1275
KM41256AP-15	BLOCK-1116
KM4164-15	BLOCK-1006
KM5624	BLOCK-242
KM858C	BLOCK-186
KS10969-L5	BLOCK-1336
KS20967-L1	BLOCK-1293
KS20967-L2	BLOCK-1297
KS20967-L3	BLOCK-1419
KS20969-L2	BLOCK-1417
KS20969-L3	BLOCK-1425
KS20969-L4	BLOCK-1428
KS21282-L1	BLOCK-1301
KS21282-L3	BLOCK-1378
KIA7805P	BLOCK-2087
KIA7809PI	BLOCK-918
L-612099	BLOCK-1301
L-612107	BLOCK-1378
L-612150	BLOCK-1330
L-612158	BLOCK-1336
L-612161	BLOCK-1341
L123CB	BLOCK-1984
L2114-550	BLOCK-1002
L272	BLOCK-833
L272B	BLOCK-833
L272M	BLOCK-834
L272MB	BLOCK-834
L274	BLOCK-43
L292	BLOCK-1186
L292V	BLOCK-1186
L293B	BLOCK-730
L295	BLOCK-731
L296	BLOCK-667
L296V	BLOCK-667

DEVICE TYPE	REPL CODE
L5630	BLOCK-1157
L5631	BLOCK-1157
L5631-AA	BLOCK-1157
L7805CV	BLOCK-2087
L7808CY	BLOCK-2095
L7812CV	BLOCK-2097
L7815CV	BLOCK-2119
L7818CT	BLOCK-896
L7824CT	BLOCK-898
L78M05	BLOCK-2087
L78M05-A	BLOCK-2087
L78M05-LU	BLOCK-2087
L78M05-RA	BLOCK-2087
L78M05ABV	BLOCK-2087
L78M05C-V	BLOCK-2087
L78M05CV	BLOCK-2087
L78M05P	BLOCK-2087
L78M06	BLOCK-2093
L78M06SA	BLOCK-2093
L78M09	BLOCK-881
L78M09-RA	BLOCK-881
L78M09-SA	BLOCK-881
L78M09SA	BLOCK-881
L78M12	BLOCK-2097
L78M12-LU	BLOCK-2097
L78M12-RA	BLOCK-2097
L78M12-SA	BLOCK-2097
L78M12CV	BLOCK-2097
L78M12RL	BLOCK-2097
L78N05	BLOCK-2144
L78N06	BLOCK-2093
L78N06-SE	BLOCK-2093
L78N12	BLOCK-2097
L78N12-RA	BLOCK-2097
L7912CV	BLOCK-2108
L7918CT	BLOCK-897
L7924CT	BLOCK-899
LA-1201	BLOCK-4
LA-1201T	BLOCK-4
LA-3300	BLOCK-6
LA-3301	BLOCK-7
LA-7530	BLOCK-807
LA-7601W	BLOCK-659
LA1041	BLOCK-13
LA1110	BLOCK-14
LA1111	BLOCK-3
LA1111P	BLOCK-3
LA1140	BLOCK-591
LA1152N	BLOCK-467
LA1200	BLOCK-4
LA1201	BLOCK-4
LA1201(B)	BLOCK-4
LA1201(C)	BLOCK-4
LA1201B	BLOCK-4
LA1201C	BLOCK-4
LA1201C-W	BLOCK-4
LA1201T	BLOCK-4
LA1201W	BLOCK-4
LA1222	BLOCK-215
LA1230	BLOCK-1813
LA1240	BLOCK-202
LA1245	BLOCK-595
LA1320A	BLOCK-442
LA13301	BLOCK-7
LA1342	BLOCK-1265
LA1352	BLOCK-175
LA1353	BLOCK-78
LA1354	BLOCK-176
LA1357	BLOCK-471
LA1357AB	BLOCK-471
LA1357B	BLOCK-471
LA1357N	BLOCK-471
LA1357NB	BLOCK-471
LA1363	BLOCK-1269
LA1363W	BLOCK-1269
LA1364	BLOCK-5
LA1364N	BLOCK-5
LA1365	BLOCK-1269
LA1365N	BLOCK-1269
LA1366	BLOCK-147
LA1366N	BLOCK-147
LA1367	BLOCK-148
LA1368	BLOCK-1291
LA1368BP	BLOCK-1291
LA1368CW	BLOCK-1291
LA1369	BLOCK-172
LA1373	BLOCK-92
LA1374	BLOCK-119
LA1375	BLOCK-120
LA1376	BLOCK-120
LA1385	BLOCK-234
LA1390	BLOCK-204
LA1390B	BLOCK-204
LA1390C	BLOCK-204

DEVICE TYPE	REPL CODE
LA1460	BLOCK-433
LA1461	BLOCK-433
LA2600S	BLOCK-427
LA3018	BLOCK-1944
LA3086N	BLOCK-1963
LA3115	BLOCK-441
LA3148	BLOCK-1283
LA3150	BLOCK-447
LA3155	BLOCK-203
LA3160	BLOCK-159
LA3161	BLOCK-159
LA3166N	BLOCK-147
LA3190	BLOCK-204
LA3201	BLOCK-168
LA3210	BLOCK-454
LA3210ALC	BLOCK-454
LA3220	BLOCK-621
LA3300	BLOCK-6
LA3301	BLOCK-7
LA3310	BLOCK-218
LA3311	BLOCK-213
LA3350	BLOCK-205
LA3350A	BLOCK-205
LA3361	BLOCK-237
LA3361A	BLOCK-237
LA3370	BLOCK-449
LA350	BLOCK-205
LA4030-7	BLOCK-8
LA4030P	BLOCK-9
LA4031P	BLOCK-10
LA4032P	BLOCK-11
LA4100	BLOCK-169
LA4101	BLOCK-169
LA4102	BLOCK-216
LA4112	BLOCK-451
LA4125	BLOCK-369
LA4125T	BLOCK-369
LA4126	BLOCK-369
LA4126T	BLOCK-369
LA4140	BLOCK-453
LA4170	BLOCK-465
LA4182	BLOCK-653
LA4183	BLOCK-653
LA4190	BLOCK-653
LA4192	BLOCK-653
LA4192S	BLOCK-653
LA4201	BLOCK-458
LA4220	BLOCK-199
LA4230	BLOCK-373
LA4250	BLOCK-373
LA4260	BLOCK-671
LA4261	BLOCK-671
LA4270	BLOCK-779
LA4400	BLOCK-181
LA4400FR	BLOCK-181
LA4400Y	BLOCK-199
LA4420	BLOCK-199
LA4440	BLOCK-364
LA4440-R	BLOCK-364
LA4440R	BLOCK-364
LA4445	BLOCK-687
LA4505	BLOCK-661
LA5112	BLOCK-467
LA5112N	BLOCK-467
LA5112N-G	BLOCK-467
LA5115N	BLOCK-467
LA6324	BLOCK-2171
LA6339	BLOCK-1873
LA6358	BLOCK-1993
LA6358S	BLOCK-1994
LA6393D	BLOCK-2063
LA6458D	BLOCK-1801
LA6458DF	BLOCK-1801
LA6458S	BLOCK-516
LA7016	BLOCK-762
LA7016//-1	BLOCK-762
LA703E	BLOCK-1212
LA7040	BLOCK-545
LA7220	BLOCK-780
LA7222	BLOCK-1252
LA7222-TV	BLOCK-1252
LA7507	BLOCK-529
LA7510	BLOCK-1253
LA7520	BLOCK-709
LA7530	BLOCK-807
LA7530N	BLOCK-807
LA7601	BLOCK-659
LA7601G	BLOCK-659
LA7601W	BLOCK-659
LA7620	BLOCK-824
LA7621	BLOCK-752
LA7625	BLOCK-836
LA7626	BLOCK-817
LA7629	BLOCK-1188

DEVICE TYPE	REPL CODE
LA7650K	BLOCK-843
LA7650KN	BLOCK-843
LA7655	BLOCK-843
LA7655N	BLOCK-843
LA7670	BLOCK-1238
LA7760	BLOCK-781
LA7761	BLOCK-782
LA7800	BLOCK-525
LA7800R	BLOCK-525
LA7802	BLOCK-526
LA7806	BLOCK-527
LA7811	BLOCK-753
LA7823	BLOCK-633
LA7830	BLOCK-754
LA7831	BLOCK-778
LA7835	BLOCK-835
LA7835-TV	BLOCK-835
LA7835K	BLOCK-835
LA7836	BLOCK-1263
LA7836-TV	BLOCK-1263
LA7837	BLOCK-1266
LA7838	BLOCK-1222
LA7910	BLOCK-644
LA7913	BLOCK-816
LA7915	BLOCK-815
LA7916	BLOCK-1184
LA7920	BLOCK-688
LA7940	BLOCK-396
LAB1363	BLOCK-1269
LAD-011	BLOCK-118
LAD010	BLOCK-117
LAD011	BLOCK-118
LAP-011	BLOCK-1
LAP-012	BLOCK-2
LAP011	BLOCK-1
LAS1505	BLOCK-1022
LAS723	BLOCK-1983
LAS723B	BLOCK-1983
LB-2000	BLOCK-2208
LB1274	BLOCK-986
LB1274K	BLOCK-986
LB1275	BLOCK-985
LB1287	BLOCK-993
LB1288	BLOCK-993
LB1290	BLOCK-944
LB1331	BLOCK-543
LB1332	BLOCK-543
LB1403	BLOCK-548
LB1405	BLOCK-503
LB1409	BLOCK-501
LB1416	BLOCK-500
LB1649	BLOCK-755
LB2000	BLOCK-2208
LB2001	BLOCK-2217
LB2002	BLOCK-2197
LB2003	BLOCK-2199
LB2004	BLOCK-2206
LB2005	BLOCK-2200
LB2006	BLOCK-2202
LB2007	BLOCK-2201
LB2030	BLOCK-2207
LB2031	BLOCK-1950
LB2032	BLOCK-1953
LB2100	BLOCK-2210
LB2101	BLOCK-2218
LB2102	BLOCK-2216
LB2106	BLOCK-2203
LB2130	BLOCK-2209
LB2131	BLOCK-1951
LB2132	BLOCK-1952
LB3000	BLOCK-1293
LB3001	BLOCK-1304
LB3001(FANON)	BLOCK-16
LB3002	BLOCK-1365
LB3003	BLOCK-1377
LB3004	BLOCK-1401
LB3005	BLOCK-1406
LB3006	BLOCK-1297
LB3008	BLOCK-1295
LB3009	BLOCK-1390
LB3060	BLOCK-1414
LB3150	BLOCK-1423
LB3175	BLOCK-1424
LC-289(ELCOM)	BLOCK-1822
LC4001B	BLOCK-1029
LC4001BP	BLOCK-1029
LC4011	BLOCK-1039
LC4011B	BLOCK-1039
LC4013B	BLOCK-1041
LC4066B	BLOCK-1093
LC4069UB	BLOCK-1096
LC4071B	BLOCK-1098
LC4081B	BLOCK-1105
LC7120	BLOCK-523

DEVICE TYPE	REPL CODE
LC7131	BLOCK-463
LCS1008P	BLOCK-1269
LD-1110	BLOCK-14
LD-3150	BLOCK-20
LD1020	BLOCK-12
LD1020A	BLOCK-12
LD1041	BLOCK-13
LD1110	BLOCK-14
LD300	BLOCK-16
LD3000	BLOCK-15
LD3001	BLOCK-16
LD3001W	BLOCK-16
LD3001X	BLOCK-16
LD3040	BLOCK-17
LD3061	BLOCK-25
LD3080	BLOCK-18
LD3110	BLOCK-142
LD3110A	BLOCK-142
LD3120	BLOCK-19
LD3150	BLOCK-20
LD3150C	BLOCK-20
LED-15005	BLOCK-2189
LED21085	BLOCK-2185
LED21092	BLOCK-2213
LF13331N	BLOCK-1915
LF347	BLOCK-1912
LF347N	BLOCK-1912
LF351N	BLOCK-1908
LF353	BLOCK-1910
LF353N	BLOCK-1910
LF355	BLOCK-1908
LF355AH	BLOCK-2031
LF355BN	BLOCK-2036
LF355H	BLOCK-2031
LF355J	BLOCK-2036
LF355N	BLOCK-2036
LF356	BLOCK-2036
LF356AH	BLOCK-2031
LF356BN	BLOCK-2036
LF356H	BLOCK-2031
LF356J	BLOCK-2036
LF356N	BLOCK-2036
LF357AH	BLOCK-2031
LF357BN	BLOCK-2036
LF357H	BLOCK-2031
LF357J	BLOCK-2036
LF357N	BLOCK-2036
LH21256-12	BLOCK-1116
LK1352P	BLOCK-1422
LKB52P	BLOCK-1422
LM-1310N	BLOCK-1825
LM-2111	BLOCK-1262
LM-2111N	BLOCK-1262
LM-377N	BLOCK-1828
LM-703LN	BLOCK-1212
LM1011AN	BLOCK-932
LM1011N	BLOCK-932
LM101AH	BLOCK-160
LM101H	BLOCK-160
LM1304	BLOCK-1275
LM1304N	BLOCK-1275
LM1305	BLOCK-1277
LM1305M	BLOCK-1277
LM1305N	BLOCK-1277
LM1305N01	BLOCK-1277
LM1310	BLOCK-1825
LM1310N	BLOCK-1825
LM1310N01	BLOCK-1825
LM131ON	BLOCK-1825
LM1329	BLOCK-1821
LM1351#	BLOCK-1421
LM1351N	BLOCK-1421
LM13600N	BLOCK-1925
LM1391N	BLOCK-1843
LM1408N-6	BLOCK-971
LM1408N-7	BLOCK-971
LM140K-15	BLOCK-894
LM1458	BLOCK-1801
LM1458C	BLOCK-18
LM1458M	BLOCK-1801
LM1458N	BLOCK-1801
LM1496	BLOCK-2138
LM1496H	BLOCK-2138
LM1496J	BLOCK-2139
LM1496N	BLOCK-2139
LM1596H	BLOCK-2138
LM1800	BLOCK-1376
LM1800A	BLOCK-1376
LM1800N	BLOCK-1376
LM1819N	BLOCK-656
LM1820	BLOCK-1389
LM1820A	BLOCK-1389
LM1820N	BLOCK-1389
LM1823	BLOCK-1933

DEVICE TYPE	REPL CODE
LM1823N	BLOCK-1933
LM1829	BLOCK-1821
LM1841	BLOCK-1290
LM1845M	BLOCK-1287
LM1870	BLOCK-844
LM1877	BLOCK-2175
LM1877N	BLOCK-2175
LM1877N-10	BLOCK-2175
LM1877N-3	BLOCK-2175
LM1877N-9	BLOCK-2175
LM1889	BLOCK-1889
LM1889N	BLOCK-1889
LM1946H	BLOCK-2138
LM201AH	BLOCK-160
LM201AN	BLOCK-2141
LM201H	BLOCK-160
LM207N	BLOCK-2058
LM2111	BLOCK-1262
LM2111A	BLOCK-1262
LM2111M	BLOCK-1262
LM2111N	BLOCK-1262
LM2111N01	BLOCK-1262
LM2113N01	BLOCK-1264
LM211H	BLOCK-1980
LM223K	BLOCK-2004
LM224AD	BLOCK-2171
LM224AF	BLOCK-2171
LM224AJ	BLOCK-2171
LM224AN	BLOCK-2171
LM224D	BLOCK-2171
LM224J	BLOCK-2171
LM224N	BLOCK-2171
LM239	BLOCK-1873
LM239A	BLOCK-1873
LM239AJ	BLOCK-1873
LM239AN	BLOCK-1873
LM239J	BLOCK-1873
LM239N	BLOCK-1873
LM258AH	BLOCK-1992
LM258AN	BLOCK-1992
LM258AT	BLOCK-1992
LM258H	BLOCK-1992
LM258JG	BLOCK-1993
LM258N	BLOCK-1993
LM258P	BLOCK-1993
LM258T	BLOCK-1992
LM2896	BLOCK-673
LM2896P1	BLOCK-673
LM2900	BLOCK-2191
LM2900N	BLOCK-2191
LM2901	BLOCK-1873
LM2901D	BLOCK-1874
LM2901J	BLOCK-1873
LM2901N	BLOCK-1873
LM2902	BLOCK-2171
LM2902D	BLOCK-2172
LM2902J	BLOCK-2171
LM2902N	BLOCK-2171
LM2904J	BLOCK-1993
LM2904JG	BLOCK-1993
LM2904L	BLOCK-1992
LM2904M	BLOCK-1995
LM2904N	BLOCK-1993
LM2904P	BLOCK-1993
LM2917	BLOCK-2211
LM2917-8	BLOCK-2211
LM2917J	BLOCK-2211
LM2917N	BLOCK-2211
LM2917N-8	BLOCK-2214
LM2917N8	BLOCK-2214
LM301AD	BLOCK-2142
LM301ADE	BLOCK-2141
LM301AH	BLOCK-160
LM301AN	BLOCK-2141
LM301AP	BLOCK-2141
LM301AT	BLOCK-2141
LM301AV	BLOCK-2141
LM3028AH	BLOCK-1280
LM3028BH	BLOCK-1280
LM302H	BLOCK-1985
LM3045D	BLOCK-1963
LM3045J	BLOCK-1963
LM3045N	BLOCK-1963
LM3046N	BLOCK-1963
LM3053H	BLOCK-1280
LM305AP	BLOCK-903
LM305H	BLOCK-902
LM305T	BLOCK-902
LM3064M	BLOCK-1805
LM3064N	BLOCK-1808
LM3064N-01	BLOCK-1808
LM3065	BLOCK-1269
LM3065N	BLOCK-1269
LM3065N01	BLOCK-1269

DEVICE TYPE	REPL CODE
LM3067N	BLOCK-1285
LM3070A	BLOCK-1271
LM3070N	BLOCK-1271
LM3070N01	BLOCK-1271
LM3071A	BLOCK-1272
LM3071N	BLOCK-1272
LM3071N01	BLOCK-1272
LM3072	BLOCK-1270
LM3072N	BLOCK-1270
LM3075N01	BLOCK-1279
LM3075N01A	BLOCK-1279
LM307DE	BLOCK-2143
LM307F	BLOCK-2143
LM307H	BLOCK-2031
LM307J	BLOCK-2143
LM307N	BLOCK-2143
LM307P	BLOCK-2143
LM3080AN	BLOCK-2215
LM3086N	BLOCK-1963
LM3086W	BLOCK-1963
LM3089N	BLOCK-1813
LM308AH	BLOCK-2037
LM308AJ-8	BLOCK-2037
LM308AN	BLOCK-2042
LM308AT	BLOCK-2037
LM308D	BLOCK-2042
LM308H	BLOCK-2037
LM308J-8	BLOCK-2042
LM308N	BLOCK-2042
LM308T	BLOCK-2037
LM309DA	BLOCK-1022
LM309K	BLOCK-1022
LM3105	BLOCK-1277
LM310H	BLOCK-1985
LM310N	BLOCK-1986
LM311	BLOCK-1980
LM311D	BLOCK-1981
LM311F	BLOCK-1981
LM311H	BLOCK-1980
LM311N	BLOCK-1981
LM311NDS	BLOCK-1981
LM311P	BLOCK-1981
LM311T	BLOCK-1980
LM317	BLOCK-887
LM317K	BLOCK-1983
LM317KC	BLOCK-2083
LM317LZ	BLOCK-879
LM317T	BLOCK-2083
LM318	BLOCK-1974
LM3189N	BLOCK-1197
LM318H	BLOCK-1972
LM318J-8	BLOCK-1973
LM318JG	BLOCK-1973
LM318M	BLOCK-1974
LM318N	BLOCK-1973
LM318P	BLOCK-1973
LM319F	BLOCK-1976
LM319H	BLOCK-1975
LM319J	BLOCK-1976
LM319K	BLOCK-1975
LM319N	BLOCK-1976
LM320K12	BLOCK-891
LM320K15	BLOCK-895
LM320K5.0	BLOCK-889
LM320L-12	BLOCK-882
LM320LZ-12	BLOCK-882
LM320LZ12	BLOCK-882
LM320LZ15	BLOCK-883
LM320MP-12	BLOCK-2108
LM320MP-24	BLOCK-2136
LM320MP-5.2	BLOCK-2091
LM320T-12	BLOCK-2108
LM320T-15	BLOCK-2128
LM320T-24	BLOCK-2136
LM320T-5.0	BLOCK-2091
LM320T-5.2	BLOCK-2091
LM320T-6.0	BLOCK-2094
LM320T12	BLOCK-2108
LM320T15	BLOCK-2128
LM320T24	BLOCK-2136
LM320T5.0	BLOCK-2091
LM320T6.0	BLOCK-2094
LM320T8.0	BLOCK-2095
LM323K	BLOCK-2004
LM324	BLOCK-2171
LM324A	BLOCK-2171
LM324AD	BLOCK-2171
LM324AF	BLOCK-2171
LM324AJ	BLOCK-2171
LM324AN	BLOCK-2171
LM324D	BLOCK-2171
LM324J	BLOCK-2171
LM324N	BLOCK-2171
LM3301N	BLOCK-2191

DEVICE TYPE	REPL CODE
LM3302	BLOCK-1873
LM3302J	BLOCK-1873
LM3302N	BLOCK-1873
LM337K	BLOCK-887
LM337LZ	BLOCK-880
LM337MT	BLOCK-2084
LM337T	BLOCK-2084
LM338K	BLOCK-2024
LM339	BLOCK-1873
LM339A	BLOCK-1873
LM339AJ	BLOCK-1873
LM339AN	BLOCK-1873
LM339D	BLOCK-1874
LM339DP	BLOCK-1873
LM339J	BLOCK-1873
LM339N	BLOCK-1873
LM340	BLOCK-2087
LM340-12	BLOCK-890
LM340-12DA	BLOCK-898
LM340-12KC	BLOCK-2097
LM340-12U	BLOCK-2097
LM340-15DA	BLOCK-892
LM340-15U	BLOCK-2119
LM340-18KC	BLOCK-2085
LM340-18U	BLOCK-2085
LM340-24KC	BLOCK-2137
LM340-24U	BLOCK-2137
LM340-5	BLOCK-1022
LM340-5DA	BLOCK-1022
LM340-5KC	BLOCK-2087
LM340-5U	BLOCK-2087
LM340-6KC	BLOCK-2093
LM340-6U	BLOCK-2093
LM340-8KC	BLOCK-2095
LM340-8U	BLOCK-2095
LM3401N	BLOCK-2191
LM340AK12	BLOCK-890
LM340AK15	BLOCK-892
LM340AT5.0	BLOCK-2087
LM340K-12	BLOCK-890
LM340K-15	BLOCK-892
LM340K-18	BLOCK-896
LM340K-24	BLOCK-898
LM340K12	BLOCK-890
LM340K15	BLOCK-892
LM340K5.0	BLOCK-1022
LM340KC12	BLOCK-890
LM340KC15	BLOCK-892
LM340T	BLOCK-2087
LM340T-12	BLOCK-2097
LM340T-12R	BLOCK-2097
LM340T-15	BLOCK-2119
LM340T-15R	BLOCK-2119
LM340T-18	BLOCK-2085
LM340T-24	BLOCK-2137
LM340T-24R	BLOCK-2137
LM340T-5	BLOCK-2087
LM340T-5.0	BLOCK-2087
LM340T-5.0R	BLOCK-2087
LM340T-6	BLOCK-2093
LM340T-6.0	BLOCK-2093
LM340T-6.0R	BLOCK-2093
LM340T-8.0	BLOCK-2095
LM340T12	BLOCK-2097
LM340T15	BLOCK-2119
LM340T24	BLOCK-2137
LM340T5	BLOCK-2087
LM340T5.0	BLOCK-2087
LM340T6.0	BLOCK-2093
LM340T8.0	BLOCK-2095
LM340U12	BLOCK-2097
LM340U15	BLOCK-2119
LM340U24	BLOCK-2137
LM340U5	BLOCK-2087
LM340U6	BLOCK-2093
LM340U8	BLOCK-2095
LM341-12	BLOCK-2097
LM341-15	BLOCK-2119
LM341-24	BLOCK-2137
LM341-5	BLOCK-2087
LM341-6	BLOCK-2093
LM341P-12	BLOCK-2097
LM341P-15	BLOCK-2119
LM341P-24	BLOCK-2137
LM341P-5.0	BLOCK-2087
LM341P-6.0	BLOCK-2093
LM341P-8.0	BLOCK-2095
LM341P12	BLOCK-2097
LM341P15	BLOCK-2119
LM341P5.0	BLOCK-2087
LM341P8.0	BLOCK-2095
LM342P12	BLOCK-2097
LM342P15	BLOCK-2119
LM342P5.0	BLOCK-2087

ORIGINAL DEVICE TYPES AND REPLACEMENT CODES 37

DEVICE TYPE	REPL CODE
LM348	BLOCK-2071
LM348D	BLOCK-2072
LM348J	BLOCK-2071
LM348N	BLOCK-2071
LM350K	BLOCK-2135
LM3524N	BLOCK-701
LM358	BLOCK-1993
LM358AH	BLOCK-1992
LM358AN	BLOCK-1993
LM358AT	BLOCK-1993
LM358D	BLOCK-1995
LM358H	BLOCK-1992
LM358J	BLOCK-1993
LM358JG	BLOCK-1993
LM358L	BLOCK-1992
LM358M	BLOCK-1995
LM358N	BLOCK-1993
LM358NB	BLOCK-1992
LM358P	BLOCK-1993
LM358S	BLOCK-1994
LM358T	BLOCK-1992
LM359J	BLOCK-1924
LM359N	BLOCK-1924
LM370	BLOCK-1024
LM373N	BLOCK-1824
LM375N	BLOCK-1824
LM376N	BLOCK-903
LM377-N	BLOCK-2175
LM377KC	BLOCK-2084
LM377N	BLOCK-2175
LM377N10	BLOCK-1828
LM378N	BLOCK-1828
LM380	BLOCK-1303
LM380N	BLOCK-1303
LM381A	BLOCK-2061
LM381AN	BLOCK-2061
LM381N	BLOCK-2061
LM3820N	BLOCK-1389
LM383	BLOCK-221
LM383AT	BLOCK-221
LM383T	BLOCK-221
LM384N	BLOCK-1917
LM386	BLOCK-1851
LM3862M	BLOCK-1851
LM386A	BLOCK-1851
LM386LM	BLOCK-1851
LM386N	BLOCK-1851
LM386N-1	BLOCK-1851
LM386N-3	BLOCK-1851
LM387	BLOCK-1852
LM387AN	BLOCK-1852
LM387N	BLOCK-1852
LM388N	BLOCK-1858
LM388N-1	BLOCK-1858
LM3900	BLOCK-2191
LM3900N	BLOCK-2191
LM3909	BLOCK-1931
LM3909N	BLOCK-1931
LM390N	BLOCK-1853
LM3914	BLOCK-496
LM3914N	BLOCK-496
LM3915	BLOCK-497
LM3916	BLOCK-536
LM393JG	BLOCK-2063
LM393M	BLOCK-2064
LM393N	BLOCK-2063
LM393NB	BLOCK-2063
LM393P	BLOCK-2063
LM4250CH	BLOCK-2065
LM4250CN	BLOCK-2066
LM555	BLOCK-2079
LM555C	BLOCK-2079
LM555CJ	BLOCK-2079
LM555CN	BLOCK-2079
LM555CN#1	BLOCK-2079
LM555CN#2	BLOCK-2079
LM556CN	BLOCK-2145
LM556D	BLOCK-2147
LM556M	BLOCK-2147
LM55SCN	BLOCK-2145
LM565CN	BLOCK-2174
LM566CH	BLOCK-2204
LM566H	BLOCK-2204
LM567CM	BLOCK-1871
LM703	BLOCK-1212
LM703E	BLOCK-1212
LM703L	BLOCK-1212
LM703LH	BLOCK-1212
LM703LN	BLOCK-1212
LM709CH	BLOCK-1949
LM709CN	BLOCK-1954
LM710CH	BLOCK-1955
LM710CN	BLOCK-1957
LM711CH	BLOCK-1958
LM711CN	BLOCK-1962
LM723CG	BLOCK-1983
LM723CH	BLOCK-1983
LM723CN	BLOCK-1984
LM723H	BLOCK-1983
LM725CH	BLOCK-1987
LM733CH	BLOCK-1989
LM733CN	BLOCK-1990
LM741	BLOCK-2058
LM741AH	BLOCK-2056
LM741C	BLOCK-2056
LM741CH	BLOCK-2056
LM741CJ	BLOCK-2058
LM741CM	BLOCK-2060
LM741CN	BLOCK-2058
LM741EH	BLOCK-2056
LM741EN	BLOCK-2058
LM741N	BLOCK-2058
LM746	BLOCK-1270
LM746H	BLOCK-1256
LM746N	BLOCK-1815
LM747AH	BLOCK-2069
LM747AJ	BLOCK-2070
LM747CH	BLOCK-2069
LM747CJ	BLOCK-2070
LM747CN	BLOCK-2070
LM747EH	BLOCK-2069
LM747EJ	BLOCK-2070
LM747H	BLOCK-2069
LM747J	BLOCK-2070
LM748CH	BLOCK-160
LM748CN7	BLOCK-2141
LM748H	BLOCK-160
LM748N	BLOCK-2141
LM7805ACZ	BLOCK-2144
LM7805CT	BLOCK-2087
LM7805CV	BLOCK-2087
LM7808A-8	BLOCK-2160
LM7812	BLOCK-2097
LM7812CT	BLOCK-2097
LM78L05A	BLOCK-2144
LM78L05ACH	BLOCK-2144
LM78L05ACI	BLOCK-2144
LM78L05ACZ	BLOCK-2144
LM78L08ACH	BLOCK-2160
LM78L08ACZ	BLOCK-2160
LM78L08CH	BLOCK-2160
LM78L12	BLOCK-2074
LM78L12ACZ	BLOCK-2074
LM7905CT	BLOCK-2091
LM7912CT	BLOCK-2108
LM8360	BLOCK-973
LM8361	BLOCK-973
LM8361D	BLOCK-973
LM8361DH	BLOCK-974
LM8560	BLOCK-975
LMC555CM	BLOCK-2082
LN1304N01	BLOCK-1275
LN1310	BLOCK-1825
LN2064B	BLOCK-994
LR40992	BLOCK-677
LR40993	BLOCK-678
LS-0142	BLOCK-1290
LS00	BLOCK-1568
LS02	BLOCK-1570
LS05	BLOCK-1573
LS125	BLOCK-1585
LS139	BLOCK-1592
LS14	BLOCK-1593
LS15	BLOCK-1597
LS155	BLOCK-1600
LS157N	BLOCK-1602
LS174	BLOCK-1615
LS20	BLOCK-1626
LS240	BLOCK-1630
LS244	BLOCK-1634
LS245	BLOCK-1635
LS301AT	BLOCK-160
LS307B	BLOCK-2143
LS32	BLOCK-1659
LS32N	BLOCK-1659
LS373	BLOCK-1671
LS374	BLOCK-1672
LS40	BLOCK-1683
LS51N	BLOCK-1689
LS703L	BLOCK-1212
LS74	BLOCK-1706
LS86	BLOCK-1713
LSC1008	BLOCK-1269
LSC1008P	BLOCK-1269
M083B1	BLOCK-959
M086B1	BLOCK-960
M1358P	BLOCK-1422
M2002	BLOCK-1280
M2003	BLOCK-1963
M2009	BLOCK-1797
M2016	BLOCK-1963
M2032	BLOCK-1970
M2128-20	BLOCK-1004
M2716	BLOCK-1017
M2716F1	BLOCK-1017
M2716M	BLOCK-1017
M2732AF1	BLOCK-1018
M304/4050	BLOCK-1786
M3764-15RS	BLOCK-1006
M3764-20RS	BLOCK-1006
M380	BLOCK-1303
M4001BP	BLOCK-1029
M4002BP	BLOCK-1030
M4011BP	BLOCK-1039
M4012BP	BLOCK-1040
M4015B	BLOCK-1043
M4016BP	BLOCK-1048
M40175BP	BLOCK-1050
M4020BP	BLOCK-1059
M4023BP	BLOCK-1062
M4024BP	BLOCK-1063
M4025BP	BLOCK-1064
M4040BP	BLOCK-1076
M4042BP	BLOCK-1077
M4043BP	BLOCK-1078
M4049BP	BLOCK-1084
M4050BP	BLOCK-1085
M4051BP	BLOCK-1086
M4052BP	BLOCK-1087
M4053BP	BLOCK-1088
M4066BP	BLOCK-1093
M4069UBP	BLOCK-1096
M4071BP	BLOCK-1098
M4081BP	BLOCK-1105
M4508B	BLOCK-1120
M4510BP	BLOCK-1121
M4512BP	BLOCK-1123
M4514BP	BLOCK-1125
M4515BP	BLOCK-1126
M4516BP	BLOCK-1127
M4518BP	BLOCK-1129
M4520BP	BLOCK-1130
M4528BP	BLOCK-1135
M4539BP	BLOCK-1141
M5101	BLOCK-93
M5101P	BLOCK-93
M5106	BLOCK-95
M5106P	BLOCK-95
M5108	BLOCK-97
M5108-9170	BLOCK-97
M5108P	BLOCK-97
M5109	BLOCK-1971
M5109P	BLOCK-1971
M5112	BLOCK-96
M51124	BLOCK-96
M5112Y	BLOCK-96
M5113	BLOCK-1223
M5113T	BLOCK-1223
M5115	BLOCK-108
M5115P	BLOCK-108
M5115P-9085	BLOCK-108
M5115PA	BLOCK-108
M5115PR	BLOCK-108
M5115PRA	BLOCK-146
M5115RP	BLOCK-146
M51171L	BLOCK-288
M51172P	BLOCK-424
M5117L	BLOCK-288
M51182	BLOCK-455
M51182L	BLOCK-455
M51204	BLOCK-422
M51204L	BLOCK-422
M5124	BLOCK-96
M51247	BLOCK-162
M51247P	BLOCK-162
M51307BSP	BLOCK-825
M5131P	BLOCK-94
M5134	BLOCK-94
M5134-8266	BLOCK-94
M5134P	BLOCK-94
M51354AP	BLOCK-642
M51354APO	BLOCK-642
M51354P	BLOCK-642
M51355P	BLOCK-639
M51356	BLOCK-642
M51356P	BLOCK-642
M51358P	BLOCK-532
M5135P	BLOCK-5
M51365P	BLOCK-1196
M51365SP	BLOCK-1196
M51366P	BLOCK-1828
M51366SP	BLOCK-826
M51375P	BLOCK-406
M51376BSP	BLOCK-1195
M51376SP	BLOCK-1195
M51389P	BLOCK-425
M5143	BLOCK-1269
M5143P	BLOCK-1269
M5144BP	BLOCK-1269
M5144P	BLOCK-1269
M5146P	BLOCK-295
M51513L	BLOCK-245
M51513LB	BLOCK-245
M51515L	BLOCK-278
M51516L	BLOCK-348
M51517L	BLOCK-351
M51518L	BLOCK-349
M51521L	BLOCK-159
M5152L	BLOCK-159
M5153L	BLOCK-1825
M5153P	BLOCK-1825
M51544L	BLOCK-505
M5155P	BLOCK-124
M5169	BLOCK-1407
M5169P	BLOCK-1407
M51709T	BLOCK-1949
M5176P	BLOCK-1407
M5183	BLOCK-1422
M5183P	BLOCK-1422
M51841P	BLOCK-2079
M51848L	BLOCK-2081
M51848P	BLOCK-2079
M5185AP	BLOCK-399
M5185P	BLOCK-399
M5186AP	BLOCK-418
M5186BP	BLOCK-418
M5186P	BLOCK-418
M51903L	BLOCK-498
M5190P	BLOCK-1291
M5191P	BLOCK-48
M5192P	BLOCK-172
M5193P	BLOCK-284
M5195	BLOCK-285
M5195P	BLOCK-285
M519OP	BLOCK-1291
M5218L	BLOCK-1802
M5218P	BLOCK-1801
M5223P	BLOCK-1801
M5224P	BLOCK-2171
M5233P	BLOCK-2063
M5236L	BLOCK-849
M5236ML	BLOCK-850
M5278L-05	BLOCK-2144
M5278L05	BLOCK-2144
M5304	BLOCK-1406
M5304P	BLOCK-1406
M5310	BLOCK-1377
M5310P	BLOCK-1377
M53200	BLOCK-1293
M53200P	BLOCK-1293
M53201	BLOCK-1294
M53201P	BLOCK-1294
M53202	BLOCK-1295
M53202P	BLOCK-1295
M53203	BLOCK-1296
M53203P	BLOCK-1296
M53204	BLOCK-1297
M53204P	BLOCK-1297
M53205	BLOCK-1298
M53205P	BLOCK-1298
M53206P	BLOCK-1299
M53207P	BLOCK-1300
M53208P	BLOCK-1301
M53209P	BLOCK-1302
M53210	BLOCK-1304
M53210P	BLOCK-1304
M53213P	BLOCK-1317
M53214P	BLOCK-1320
M53216P	BLOCK-1336
M53217P	BLOCK-1343
M53220	BLOCK-1365
M53220P	BLOCK-1365
M53227P	BLOCK-1372
M53230	BLOCK-1377
M53230P	BLOCK-1377
M53237P	BLOCK-1384
M53238P	BLOCK-1385
M53238BP	BLOCK-1385
M5323P	BLOCK-1393
M53240	BLOCK-1390
M53240P	BLOCK-1390
M53241	BLOCK-1391
M53241P	BLOCK-1391
M53242	BLOCK-1392
M53242P	BLOCK-1392
M53243	BLOCK-1393
M53243P	BLOCK-1393
M53244	BLOCK-1394
M53244P	BLOCK-1394
M53245P	BLOCK-1395
M53247	BLOCK-1397
M53247P	BLOCK-1397
M53248	BLOCK-1398
M53248A	BLOCK-1398
M53248P	BLOCK-1398
M53250	BLOCK-1401
M53250P	BLOCK-1401
M53253	BLOCK-1403
M53253P	BLOCK-1403
M53260	BLOCK-1406
M53260P	BLOCK-1406
M53270	BLOCK-1408
M53270P	BLOCK-1408
M53272	BLOCK-1409
M53272P	BLOCK-1409
M53273	BLOCK-1410
M53273P	BLOCK-1410
M53274	BLOCK-1411
M53274P	BLOCK-1411
M53275	BLOCK-1412
M53275P	BLOCK-1412
M53276	BLOCK-1413
M53276P	BLOCK-1413
M53280	BLOCK-1414
M53280P	BLOCK-1414
M53283	BLOCK-1417
M53283P	BLOCK-1417
M53285P	BLOCK-1418
M53286	BLOCK-1419
M53286P	BLOCK-1419
M53289P	BLOCK-1420
M53290	BLOCK-1423
M53290P	BLOCK-1423
M53291	BLOCK-1424
M53291P	BLOCK-1424
M53292	BLOCK-1425
M53292P	BLOCK-1425
M53293	BLOCK-1426
M53293P	BLOCK-1426
M53295	BLOCK-1428
M53295P	BLOCK-1428
M53296	BLOCK-1429
M53296P	BLOCK-1429
M53307	BLOCK-1305
M53307P	BLOCK-1305
M53321	BLOCK-1311
M53321P	BLOCK-1311
M53322P	BLOCK-1312
M53323P	BLOCK-1313
M53325P	BLOCK-1314
M53326P	BLOCK-1315
M53332P	BLOCK-1318
M53345P	BLOCK-1325
M53350P	BLOCK-1327
M53351	BLOCK-1328
M53351P	BLOCK-1328
M53353P	BLOCK-1330
M53354P	BLOCK-1331
M53355P	BLOCK-1332
M53356P	BLOCK-1333
M53360P	BLOCK-1337
M53361P	BLOCK-1338
M53362P	BLOCK-1339
M53363P	BLOCK-1340
M53364P	BLOCK-1341
M53365P	BLOCK-1342
M53370P	BLOCK-1344
M53374P	BLOCK-1346
M53375P	BLOCK-1347
M53380	BLOCK-1352
M53380P	BLOCK-1352
M53381P	BLOCK-1353
M53382P	BLOCK-1354
M53390P	BLOCK-1355
M53391P	BLOCK-1356
M53392	BLOCK-1357
M53392P	BLOCK-1357
M53393	BLOCK-1358
M53393P	BLOCK-1358
M53398P	BLOCK-1362
M53399P	BLOCK-1363
M5352	BLOCK-1401
M5352P	BLOCK-1401
M5362	BLOCK-1392
M5362P	BLOCK-1392
M5372	BLOCK-1409
M5372P	BLOCK-1409
M5374	BLOCK-1411
M5374P	BLOCK-1411
M5375	BLOCK-1408
M5375P	BLOCK-1408
M5395	BLOCK-1428
M5395P	BLOCK-1428
M54459L	BLOCK-726
M54516P	BLOCK-993
M54517P	BLOCK-988
M54519P	BLOCK-988
M54521P	BLOCK-987
M54522P	BLOCK-942
M54523P	BLOCK-936
M54524P	BLOCK-934
M54525P	BLOCK-935
M54526P	BLOCK-937
M54527P	BLOCK-986
M54528P	BLOCK-985
M54529P	BLOCK-984
M54530P	BLOCK-983
M54531P	BLOCK-983
M54533P	BLOCK-982
M54534P	BLOCK-981
M54535P	BLOCK-980
M54536P	BLOCK-979
M54543L	BLOCK-615
M54543L-B	BLOCK-615
M54543LB	BLOCK-615
M54549L	BLOCK-871
M58418P	BLOCK-568
M58472P	BLOCK-1410
M58478P	BLOCK-568
M58485P	BLOCK-719
M58725P	BLOCK-1004
M5930	BLOCK-2197
M5930P	BLOCK-2197
M5932	BLOCK-2199
M5932P	BLOCK-2199
M5933	BLOCK-2200
M5933P	BLOCK-2200
M5935	BLOCK-2201
M5935P	BLOCK-2201
M5936	BLOCK-2202
M5936P	BLOCK-2202
M5937	BLOCK-2203
M5937P	BLOCK-2203
M5944	BLOCK-2206
M5944P	BLOCK-2206
M5945	BLOCK-2207
M5945P	BLOCK-2207
M5946	BLOCK-2208
M5946P	BLOCK-2208
M5948	BLOCK-2209
M5948P	BLOCK-2209
M5949	BLOCK-2210
M5949P	BLOCK-2210
M5953	BLOCK-1950
M5953P	BLOCK-1950
M5955P	BLOCK-1952
M5956	BLOCK-1951
M5956P	BLOCK-1951
M5961	BLOCK-2216
M5961P	BLOCK-2216
M5962	BLOCK-2217
M5962P	BLOCK-2217
M5963	BLOCK-2218
M5963P	BLOCK-2218
M5F7805	BLOCK-2087
M5G1400P	BLOCK-1020
M5K4164ANP-12	BLOCK-1006
M5K4164ANP-15	BLOCK-1006
M5K4164S-15	BLOCK-1163
M5K4164S-20	BLOCK-1163
M5L2114LP-3	BLOCK-1002
M5L2716K	BLOCK-1017
M5L2732K	BLOCK-1018
M5L2732K-6	BLOCK-1018
M5L2764K	BLOCK-1019
M5L2764K-2	BLOCK-1019
M5L2764K-FA552	BLOCK-1019
M5L8255AP-5	BLOCK-1855
M5R4558P	BLOCK-1801
M74LS00P	BLOCK-1568
M74LS02P	BLOCK-1570
M74LS03P	BLOCK-1571
M74LS04P	BLOCK-1572
M74LS05P	BLOCK-1573
M74LS08P	BLOCK-1574
M74LS09P	BLOCK-1575
M74LS107AP	BLOCK-1577
M74LS107P	BLOCK-1577
M74LS109AP	BLOCK-1578
M74LS109P	BLOCK-1578
M74LS10P	BLOCK-1576
M74LS112AP	BLOCK-1580
M74LS114AP	BLOCK-1581
M74LS11P	BLOCK-1579

DEVICE TYPE	REPL CODE
M74LS122P	BLOCK-1583
M74LS123P	BLOCK-1584
M74LS125AP	BLOCK-1585
M74LS126AP	BLOCK-1586
M74LS12P	BLOCK-1582
M74LS132P	BLOCK-1588
M74LS133P	BLOCK-1589
M74LS136P	BLOCK-1590
M74LS138BP	BLOCK-1591
M74LS138P	BLOCK-1591
M74LS139P	BLOCK-1592
M74LS13P	BLOCK-1587
M74LS145P	BLOCK-1594
M74LS148P	BLOCK-1596
M74LS14P	BLOCK-1593
M74LS151P	BLOCK-1598
M74LS153P	BLOCK-1599
M74LS155P	BLOCK-1600
M74LS156P	BLOCK-1601
M74LS157P	BLOCK-1602
M74LS158P	BLOCK-1603
M74LS15P	BLOCK-1597
M74LS160AP	BLOCK-1604
M74LS161AP	BLOCK-1605
M74LS161P	BLOCK-1605
M74LS162AP	BLOCK-1606
M74LS163AP	BLOCK-1607
M74LS163P	BLOCK-1607
M74LS164P	BLOCK-1608
M74LS165P	BLOCK-1609
M74LS166AP	BLOCK-1610
M74LS166P	BLOCK-1610
M74LS170P	BLOCK-1613
M74LS173AP	BLOCK-1614
M74LS174P	BLOCK-1615
M74LS175P	BLOCK-1616
M74LS190P	BLOCK-1618
M74LS191P	BLOCK-1619
M74LS192P	BLOCK-1620
M74LS193P	BLOCK-1621
M74LS194AP	BLOCK-1622
M74LS195AP	BLOCK-1623
M74LS196P	BLOCK-1624
M74LS197P	BLOCK-1625
M74LS20P	BLOCK-1626
M74LS21P	BLOCK-1627
M74LS221P	BLOCK-1629
M74LS22P	BLOCK-1628
M74LS240P	BLOCK-1630
M74LS241P	BLOCK-1631
M74LS242P	BLOCK-1632
M74LS243P	BLOCK-1633
M74LS244P	BLOCK-1634
M74LS245P	BLOCK-1635
M74LS247P	BLOCK-1636
M74LS248P	BLOCK-1637
M74LS251P	BLOCK-1639
M74LS253P	BLOCK-1640
M74LS257AP	BLOCK-1641
M74LS258AP	BLOCK-1642
M74LS259P	BLOCK-1643
M74LS266P	BLOCK-1646
M74LS273P	BLOCK-1648
M74LS279P	BLOCK-1649
M74LS27P	BLOCK-1647
M74LS280P	BLOCK-1651
M74LS283P	BLOCK-1652
M74LS290P	BLOCK-1653
M74LS293P	BLOCK-1654
M74LS295AP	BLOCK-1655
M74LS295BP	BLOCK-1655
M74LS298P	BLOCK-1656
M74LS299P	BLOCK-1657
M74LS30P	BLOCK-1658
M74LS32P	BLOCK-1659
M74LS352P	BLOCK-1662
M74LS353P	BLOCK-1663
M74LS365AP	BLOCK-1666
M74LS366AP	BLOCK-1667
M74LS367AP	BLOCK-1668
M74LS368AP	BLOCK-1669
M74LS373P	BLOCK-1671
M74LS374P	BLOCK-1672
M74LS377P	BLOCK-1673
M74LS37P	BLOCK-1670
M74LS386P	BLOCK-1677
M74LS38P	BLOCK-1676
M74LS390P	BLOCK-1678
M74LS393P	BLOCK-1679
M74LS395AP	BLOCK-1680
M74LS395P	BLOCK-1680
M74LS40P	BLOCK-1683
M74LS42P	BLOCK-1684
M74LS47P	BLOCK-1685
M74LS48P	BLOCK-1686
M74LS490P	BLOCK-1688
M74LS51P	BLOCK-1689
M74LS640-1P	BLOCK-1699
M74LS641-1P	BLOCK-1700
M74LS642-1P	BLOCK-1701
M74LS643-1P	BLOCK-1702
M74LS645-1P	BLOCK-1703
M74LS670P	BLOCK-1704
M74LS73AP	BLOCK-1705
M74LS73P	BLOCK-1705
M74LS74AP	BLOCK-1706
M74LS74P	BLOCK-1706
M74LS75P	BLOCK-1707
M74LS76AP	BLOCK-1708
M74LS83AP	BLOCK-1711
M74LS83P	BLOCK-1711
M74LS85P	BLOCK-1712
M74LS86BP	BLOCK-1713
M74LS86P	BLOCK-1713
M74LS90P	BLOCK-1714
M74LS91P	BLOCK-1715
M74LS92P	BLOCK-1716
M74LS93P	BLOCK-1717
M7611AP	BLOCK-401
M7641	BLOCK-1410
M7644BP	BLOCK-534
MA1125	BLOCK-1269
MA1128	BLOCK-1269
MA1306W	BLOCK-28
MA3065	BLOCK-1269
MA356	BLOCK-2036
MA5113	BLOCK-1282
MA5186AP	BLOCK-418
MA6301	BLOCK-118
MA6301T	BLOCK-118
MA7300	BLOCK-931
MA741	BLOCK-2058
MA7805	BLOCK-2087
MB3202	BLOCK-80
MB3710	BLOCK-182
MB3712	BLOCK-411
MB3712HM	BLOCK-411
MB3713	BLOCK-412
MB3713HM	BLOCK-412
MB3756	BLOCK-258
MB3759	BLOCK-710
MB3759P	BLOCK-710
MB400	BLOCK-1293
MB401	BLOCK-1304
MB402	BLOCK-1365
MB403	BLOCK-1377
MB404	BLOCK-1390
MB405	BLOCK-1401
MB406	BLOCK-1406
MB407	BLOCK-1409
MB408	BLOCK-1414
MB410	BLOCK-1305
MB411	BLOCK-1403
MB416	BLOCK-1294
MB417	BLOCK-1295
MB418	BLOCK-1297
MB420	BLOCK-1411
MB4204	BLOCK-1873
MB4204C	BLOCK-1873
MB4204M	BLOCK-1873
MB424	BLOCK-1173
MB433	BLOCK-1385
MB434	BLOCK-1385
MB435	BLOCK-1384
MB440	BLOCK-1313
MB442	BLOCK-1392
MB443	BLOCK-1325
MB445	BLOCK-1328
MB447	BLOCK-1352
MB448	BLOCK-1418
MB449	BLOCK-1419
MB450	BLOCK-1338
MB451	BLOCK-1337
MB452	BLOCK-1429
MB453	BLOCK-1428
MB454	BLOCK-1424
MB455	BLOCK-1362
MB456	BLOCK-1356
MB457	BLOCK-1355
MB458	BLOCK-1353
MB459	BLOCK-1354
MB460	BLOCK-1344
MB601	BLOCK-1293
MB602	BLOCK-1304
MB603	BLOCK-1365
MB604	BLOCK-1377
MB605	BLOCK-1390
MB606	BLOCK-1401
MB607	BLOCK-1406
MB609	BLOCK-1409
MB613	BLOCK-1492
MB614	BLOCK-1485
MB618	BLOCK-1503
MB7051C	BLOCK-1750
MB7052Z	BLOCK-1749
MB7053Z	BLOCK-1763
MB7056C	BLOCK-1743
MB7057Z	BLOCK-1752
MB7058Z	BLOCK-1762
MB7121EC	BLOCK-1764
MB7122EC	BLOCK-1765
MB7123EZ	BLOCK-1756
MB7124EZ	BLOCK-1755
MB7128EZ	BLOCK-1754
MB7128HZ	BLOCK-1754
MB7131EC	BLOCK-1760
MB7131HC	BLOCK-1760
MB7132EC	BLOCK-1759
MB7132HC	BLOCK-1759
MB74LS00	BLOCK-1568
MB74LS00M	BLOCK-1568
MB74LS01	BLOCK-1569
MB74LS01M	BLOCK-1569
MB74LS02	BLOCK-1570
MB74LS02M	BLOCK-1570
MB74LS03	BLOCK-1571
MB74LS03M	BLOCK-1571
MB74LS04	BLOCK-1572
MB74LS04M	BLOCK-1572
MB74LS05	BLOCK-1573
MB74LS05M	BLOCK-1573
MB74LS08	BLOCK-1574
MB74LS08M	BLOCK-1574
MB74LS09	BLOCK-1575
MB74LS09M	BLOCK-1575
MB74LS10	BLOCK-1576
MB74LS107	BLOCK-1577
MB74LS107A	BLOCK-1577
MB74LS107AM	BLOCK-1577
MB74LS107M	BLOCK-1577
MB74LS109A	BLOCK-1578
MB74LS109AM	BLOCK-1578
MB74LS10M	BLOCK-1576
MB74LS11	BLOCK-1579
MB74LS112A	BLOCK-1580
MB74LS112AM	BLOCK-1580
MB74LS114A	BLOCK-1581
MB74LS114AM	BLOCK-1581
MB74LS11M	BLOCK-1579
MB74LS12	BLOCK-1582
MB74LS122	BLOCK-1583
MB74LS122M	BLOCK-1583
MB74LS123	BLOCK-1584
MB74LS123M	BLOCK-1584
MB74LS125A	BLOCK-1585
MB74LS125AM	BLOCK-1585
MB74LS126A	BLOCK-1586
MB74LS126AM	BLOCK-1586
MB74LS12M	BLOCK-1582
MB74LS13	BLOCK-1587
MB74LS132	BLOCK-1588
MB74LS132M	BLOCK-1588
MB74LS136	BLOCK-1590
MB74LS136M	BLOCK-1590
MB74LS138	BLOCK-1591
MB74LS138M	BLOCK-1591
MB74LS139	BLOCK-1592
MB74LS139M	BLOCK-1592
MB74LS13M	BLOCK-1587
MB74LS14	BLOCK-1593
MB74LS145	BLOCK-1594
MB74LS145M	BLOCK-1594
MB74LS14M	BLOCK-1593
MB74LS15	BLOCK-1597
MB74LS151	BLOCK-1598
MB74LS151M	BLOCK-1598
MB74LS153	BLOCK-1599
MB74LS153M	BLOCK-1599
MB74LS155	BLOCK-1600
MB74LS155M	BLOCK-1600
MB74LS156	BLOCK-1601
MB74LS156M	BLOCK-1601
MB74LS157	BLOCK-1602
MB74LS157M	BLOCK-1602
MB74LS158	BLOCK-1603
MB74LS158M	BLOCK-1603
MB74LS15M	BLOCK-1597
MB74LS160A	BLOCK-1604
MB74LS160AM	BLOCK-1604
MB74LS161A	BLOCK-1605
MB74LS161AM	BLOCK-1605
MB74LS162A	BLOCK-1606
MB74LS162AM	BLOCK-1606
MB74LS163A	BLOCK-1607
MB74LS163AM	BLOCK-1607
MB74LS174	BLOCK-1615
MB74LS174M	BLOCK-1615
MB74LS175	BLOCK-1616
MB74LS175M	BLOCK-1616
MB74LS181	BLOCK-1617
MB74LS181M	BLOCK-1617
MB74LS190	BLOCK-1618
MB74LS190M	BLOCK-1618
MB74LS191	BLOCK-1619
MB74LS191M	BLOCK-1619
MB74LS192	BLOCK-1620
MB74LS192M	BLOCK-1620
MB74LS193	BLOCK-1621
MB74LS193M	BLOCK-1621
MB74LS20	BLOCK-1626
MB74LS20M	BLOCK-1626
MB74LS21	BLOCK-1627
MB74LS21M	BLOCK-1627
MB74LS22	BLOCK-1628
MB74LS221	BLOCK-1629
MB74LS221M	BLOCK-1629
MB74LS22M	BLOCK-1628
MB74LS240	BLOCK-1630
MB74LS240M	BLOCK-1630
MB74LS241	BLOCK-1631
MB74LS242	BLOCK-1632
MB74LS242M	BLOCK-1632
MB74LS243	BLOCK-1633
MB74LS243M	BLOCK-1633
MB74LS244	BLOCK-1634
MB74LS244M	BLOCK-1634
MB74LS245	BLOCK-1635
MB74LS245M	BLOCK-1635
MB74LS247	BLOCK-1636
MB74LS247M	BLOCK-1636
MB74LS248	BLOCK-1637
MB74LS248M	BLOCK-1637
MB74LS249	BLOCK-1638
MB74LS249M	BLOCK-1638
MB74LS251	BLOCK-1639
MB74LS251M	BLOCK-1639
MB74LS253	BLOCK-1640
MB74LS253M	BLOCK-1640
MB74LS257	BLOCK-1641
MB74LS257M	BLOCK-1641
MB74LS258	BLOCK-1642
MB74LS258M	BLOCK-1642
MB74LS26	BLOCK-1644
MB74LS266	BLOCK-1646
MB74LS266M	BLOCK-1646
MB74LS26M	BLOCK-1644
MB74LS27	BLOCK-1647
MB74LS273M	BLOCK-1648
MB74LS27M	BLOCK-1647
MB74LS28	BLOCK-1650
MB74LS280	BLOCK-1651
MB74LS280M	BLOCK-1651
MB74LS283	BLOCK-1652
MB74LS283M	BLOCK-1652
MB74LS28M	BLOCK-1650
MB74LS30	BLOCK-1658
MB74LS30M	BLOCK-1658
MB74LS32	BLOCK-1659
MB74LS32M	BLOCK-1659
MB74LS33	BLOCK-1660
MB74LS33M	BLOCK-1660
MB74LS352	BLOCK-1662
MB74LS352M	BLOCK-1662
MB74LS353	BLOCK-1663
MB74LS353M	BLOCK-1663
MB74LS365A	BLOCK-1666
MB74LS365AM	BLOCK-1666
MB74LS366A	BLOCK-1667
MB74LS366AM	BLOCK-1667
MB74LS367A	BLOCK-1668
MB74LS367AM	BLOCK-1668
MB74LS368A	BLOCK-1669
MB74LS368AM	BLOCK-1669
MB74LS37	BLOCK-1670
MB74LS373	BLOCK-1671
MB74LS373P	BLOCK-1671
MB74LS374	BLOCK-1672
MB74LS37M	BLOCK-1670
MB74LS38	BLOCK-1676
MB74LS386	BLOCK-1677
MB74LS386M	BLOCK-1677
MB74LS38M	BLOCK-1676
MB74LS40	BLOCK-1683
MB74LS40M	BLOCK-1683
MB74LS42	BLOCK-1684
MB74LS42M	BLOCK-1684
MB74LS47	BLOCK-1685
MB74LS47M	BLOCK-1685
MB74LS48	BLOCK-1686
MB74LS48M	BLOCK-1686
MB74LS49	BLOCK-1687
MB74LS49M	BLOCK-1687
MB74LS51	BLOCK-1689
MB74LS51M	BLOCK-1689
MB74LS54	BLOCK-1690
MB74LS54M	BLOCK-1690
MB74LS55	BLOCK-1693
MB74LS55M	BLOCK-1693
MB74LS640	BLOCK-1699
MB74LS640M	BLOCK-1699
MB74LS641	BLOCK-1700
MB74LS641M	BLOCK-1700
MB74LS642	BLOCK-1701
MB74LS642M	BLOCK-1701
MB74LS643	BLOCK-1702
MB74LS643M	BLOCK-1702
MB74LS645	BLOCK-1703
MB74LS645M	BLOCK-1703
MB74LS73A	BLOCK-1705
MB74LS73AM	BLOCK-1705
MB74LS74A	BLOCK-1706
MB74LS74AM	BLOCK-1706
MB74LS76A	BLOCK-1708
MB74LS76AM	BLOCK-1708
MB74LS76P	BLOCK-1708
MB74LS78	BLOCK-1710
MB74LS78AM	BLOCK-1710
MB74LS83A	BLOCK-1711
MB74LS83AM	BLOCK-1711
MB74LS85	BLOCK-1712
MB74LS85M	BLOCK-1712
MB74LS86	BLOCK-1713
MB74LS86M	BLOCK-1713
MB74LS86P	BLOCK-1713
MB74S00	BLOCK-1719
MB75LS54	BLOCK-1690
MB75LS54M	BLOCK-1690
MB8116E	BLOCK-1003
MB8116EC	BLOCK-1003
MB8116EP	BLOCK-1003
MB81256-15	BLOCK-1116
MB8128-15	BLOCK-1004
MB8128-15C	BLOCK-1004
MB84001B	BLOCK-1029
MB84001BM	BLOCK-1029
MB84002B	BLOCK-1030
MB84002BM	BLOCK-1030
MB84008B	BLOCK-1034
MB84008BM	BLOCK-1034
MB84011	BLOCK-1039
MB84011-U	BLOCK-1039
MB84011B	BLOCK-1039
MB84011BM	BLOCK-1039
MB84011M	BLOCK-1039
MB84011U	BLOCK-1039
MB84011V	BLOCK-1029
MB84012B	BLOCK-1040
MB84012BM	BLOCK-1040
MB84013B	BLOCK-1041
MB84013BM	BLOCK-1041
MB84016B	BLOCK-1048
MB84016BM	BLOCK-1048
MB84017B	BLOCK-1051
MB84017BM	BLOCK-1051
MB84019B	BLOCK-1058
MB84019BM	BLOCK-1058
MB84020B	BLOCK-1059
MB84020BM	BLOCK-1059
MB84022B	BLOCK-1061
MB84022BM	BLOCK-1061
MB84023B	BLOCK-1062
MB84023BM	BLOCK-1062
MB84025B	BLOCK-1064
MB84025BM	BLOCK-1064
MB84027B	BLOCK-1066
MB84028B	BLOCK-1067
MB84028BM	BLOCK-1067
MB84029B	BLOCK-1068
MB84029BM	BLOCK-1068
MB84040B	BLOCK-1076
MB84040BM	BLOCK-1076
MB84049B	BLOCK-1084
MB84049BM	BLOCK-1084
MB84050B	BLOCK-1085
MB84050BM	BLOCK-1085
MB84051B	BLOCK-1086
MB84051BM	BLOCK-1086
MB84052B	BLOCK-1087
MB84052BM	BLOCK-1087
MB84053B	BLOCK-1088
MB84053BM	BLOCK-1088
MB84060B	BLOCK-1091
MB84060BM	BLOCK-1091
MB84066B	BLOCK-1093
MB84066BM	BLOCK-1093
MB84068B	BLOCK-1093
MB84068BM	BLOCK-1095
MB84069B	BLOCK-1096
MB84069BM	BLOCK-1096
MB84070B	BLOCK-1097
MB84070BM	BLOCK-1097
MB84071B	BLOCK-1098
MB84071BM	BLOCK-1098
MB84072B	BLOCK-1099
MB84072BM	BLOCK-1099
MB84073B	BLOCK-1100
MB84073BM	BLOCK-1100
MB84075B	BLOCK-1101
MB84075BM	BLOCK-1101
MB84077B	BLOCK-1103
MB84077BM	BLOCK-1103
MB84078B	BLOCK-1104
MB84078BM	BLOCK-1104
MB84081B	BLOCK-1105
MB84081BM	BLOCK-1105
MB84082B	BLOCK-1106
MB84082BM	BLOCK-1106
MB84520B	BLOCK-1130
MBL6821N	BLOCK-1169
MBM2147E	BLOCK-1005
MBM2147H55Z	BLOCK-1005
MBM2147H70Z	BLOCK-1005
MBM2732-35Z	BLOCK-1018
MBM2732-45Z	BLOCK-1018
MBM2732A-20-QGA1	BLOCK-1018
MBM2732A20Z	BLOCK-1018
MBM2732A25	BLOCK-1166
MBM2732A25Z	BLOCK-1018
MBM2732A30	BLOCK-1018
MBM2732A30Z	BLOCK-1018
MBM2732A35Z	BLOCK-1018
MBM2764-20	BLOCK-1019
MBM2764-25	BLOCK-1019
MBM2764-30	BLOCK-1019
MC-1310P	BLOCK-1825
MC-1326	BLOCK-1292
MC-1328G	BLOCK-1256
MC-1356P	BLOCK-1262
MC-1364	BLOCK-1808
MC-1375P	BLOCK-1279
MC-14011CP	BLOCK-1039
MC-1550P	BLOCK-1845
MC-3340P	BLOCK-1859
MC-7402	BLOCK-1295
MC-858	BLOCK-1969
MC1044P	BLOCK-2206
MC1303	BLOCK-1281
MC1303L	BLOCK-1281
MC1303P	BLOCK-1281
MC1304	BLOCK-1275
MC1304P	BLOCK-1275
MC1304PQ	BLOCK-1275
MC1305	BLOCK-1277
MC1305P	BLOCK-1277
MC1305P-C	BLOCK-1277
MC1305PC	BLOCK-1277
MC1305PQ	BLOCK-1277
MC1306P	BLOCK-1400
MC1309	BLOCK-214
MC1309P	BLOCK-214
MC1310	BLOCK-1825
MC1310A	BLOCK-1825
MC1310P	BLOCK-1825
MC1311	BLOCK-1212
MC1312P	BLOCK-1823
MC1314G	BLOCK-1223
MC1314P	BLOCK-1826
MC1315P	BLOCK-1827
MC1316P	BLOCK-1842
MC1324	BLOCK-1292
MC1324P	BLOCK-1292
MC1326	BLOCK-1256
MC1326P	BLOCK-1292
MC1326PQ	BLOCK-1292
MC1328	BLOCK-1270
MC1328G	BLOCK-1256
MC1328P	BLOCK-1815
MC1328PQ	BLOCK-1815
MC1329P	BLOCK-1270
MC1330	BLOCK-1277
MC13301P	BLOCK-563
MC1330A	BLOCK-1407
MC1330A1P	BLOCK-1407

ORIGINAL DEVICE TYPES AND REPLACEMENT CODES 39

DEVICE TYPE	REPL CODE
MC1330P	BLOCK-1407
MC133OP	BLOCK-1407
MC1344	BLOCK-1804
MC1344P	BLOCK-1804
MC1345	BLOCK-1804
MC1345P	BLOCK-1804
MC1345PQ	BLOCK-1804
MC1349P	BLOCK-1820
MC1350	BLOCK-1405
MC1350P	BLOCK-1405
MC1351	BLOCK-1421
MC1351N	BLOCK-1421
MC1351P	BLOCK-1421
MC1352	BLOCK-1422
MC1352P	BLOCK-1422
MC1353P	BLOCK-78
MC1355P	BLOCK-1770
MC1356P	BLOCK-1290
MC1357	BLOCK-1262
MC1357A	BLOCK-1262
MC1357P	BLOCK-1262
MC1357PQ	BLOCK-1262
MC1358	BLOCK-1269
MC1358P	BLOCK-1269
MC1358PQ	BLOCK-1269
MC1364	BLOCK-1808
MC1364G	BLOCK-1805
MC1364P	BLOCK-1808
MC1364PQ	BLOCK-1808
MC1370	BLOCK-1271
MC1370P	BLOCK-1271
MC1371	BLOCK-1272
MC1371P	BLOCK-1272
MC1372	BLOCK-1901
MC1372P	BLOCK-1901
MC1373P	BLOCK-1907
MC1375P	BLOCK-1279
MC1375PQ	BLOCK-1279
MC1377P	BLOCK-1934
MC1391P	BLOCK-1843
MC1393AP	BLOCK-1819
MC1393P	BLOCK-1819
MC1394	BLOCK-1843
MC1394P	BLOCK-1876
MC1398	BLOCK-1291
MC1398P	BLOCK-1291
MC1399P	BLOCK-1875
MC14000	BLOCK-1028
MC14000UBCL	BLOCK-1028
MC14000UBCP	BLOCK-1028
MC14001	BLOCK-1029
MC14001B	BLOCK-1029
MC14001BCL	BLOCK-1029
MC14001BCP	BLOCK-1029
MC14001CP	BLOCK-1029
MC14001UBCP	BLOCK-1029
MC14002	BLOCK-1030
MC14002BAL	BLOCK-1030
MC14002BCL	BLOCK-1030
MC14002BCP	BLOCK-1030
MC14002CP	BLOCK-1030
MC14002P	BLOCK-1030
MC14006BAL	BLOCK-1031
MC14006BCL	BLOCK-1031
MC14006BCP	BLOCK-1031
MC14007	BLOCK-1032
MC14007UBAL	BLOCK-1032
MC14007UBCP	BLOCK-1032
MC14008BAL	BLOCK-1034
MC14008BCL	BLOCK-1034
MC14008BCP	BLOCK-1034
MC14011	BLOCK-1039
MC14011B	BLOCK-1039
MC14011BAL	BLOCK-1039
MC14011BCL	BLOCK-1039
MC14011BCP	BLOCK-1039
MC14011CP	BLOCK-1039
MC14012	BLOCK-1040
MC14012BAL	BLOCK-1040
MC14012BCL	BLOCK-1040
MC14012BCP	BLOCK-1040
MC14012CP	BLOCK-1040
MC14013	BLOCK-1041
MC14013A1	BLOCK-1041
MC14013B	BLOCK-1041
MC14013BAL	BLOCK-1041
MC14013BCL	BLOCK-1041
MC14013BCP	BLOCK-1041
MC14013CP	BLOCK-1041
MC14014BAL	BLOCK-1042
MC14014BCL	BLOCK-1042
MC14014BCP	BLOCK-1042
MC14014CP	BLOCK-1042
MC14015	BLOCK-1043
MC14015BAL	BLOCK-1043
MC14015BCL	BLOCK-1043
MC14015BCP	BLOCK-1043
MC14015CP	BLOCK-1043
MC14016	BLOCK-1093
MC14016BAL	BLOCK-1048
MC14016BCL	BLOCK-1048
MC14016BCP	BLOCK-1048
MC14016CP	BLOCK-1048
MC14017	BLOCK-1051
MC14017BAL	BLOCK-1051
MC14017BCL	BLOCK-1051
MC14017BCP	BLOCK-1051
MC14017CP	BLOCK-1051
MC14018BAL	BLOCK-1053
MC14018BCL	BLOCK-1053
MC14018BCP	BLOCK-1053
MC14018BL	BLOCK-1053
MC14020	BLOCK-1059
MC14020BAL	BLOCK-1059
MC14020BCL	BLOCK-1059
MC14020BCP	BLOCK-1059
MC14020CP	BLOCK-1059
MC14021	BLOCK-1060
MC14021BAL	BLOCK-1060
MC14021BCL	BLOCK-1060
MC14021BCP	BLOCK-1060
MC14021CP	BLOCK-1060
MC14022BAL	BLOCK-1061
MC14022BCL	BLOCK-1061
MC14022BCP	BLOCK-1061
MC14023	BLOCK-1062
MC14023B	BLOCK-1062
MC14023BAL	BLOCK-1062
MC14023BCL	BLOCK-1062
MC14023BCP	BLOCK-1062
MC14023CP	BLOCK-1062
MC14024	BLOCK-1063
MC14024B	BLOCK-1063
MC14024BAL	BLOCK-1063
MC14024BCL	BLOCK-1063
MC14024BCP	BLOCK-1063
MC14024CP	BLOCK-1063
MC14025	BLOCK-1064
MC14025AL	BLOCK-1064
MC14025B	BLOCK-1064
MC14025BAL	BLOCK-1064
MC14025BCL	BLOCK-1064
MC14025BCP	BLOCK-1064
MC14025CP	BLOCK-1064
MC14027	BLOCK-1066
MC14027BAL	BLOCK-1066
MC14027BCL	BLOCK-1066
MC14027BCP	BLOCK-1066
MC14027CP	BLOCK-1066
MC14028	BLOCK-1067
MC14028BAL	BLOCK-1067
MC14028BCL	BLOCK-1067
MC14028BCP	BLOCK-1067
MC14028CP	BLOCK-1067
MC14029BAL	BLOCK-1068
MC14029BCL	BLOCK-1068
MC14029BCP	BLOCK-1068
MC14029CP	BLOCK-1068
MC14032BAL	BLOCK-1071
MC14032BCL	BLOCK-1071
MC14032BCP	BLOCK-1071
MC14034BAL	BLOCK-1073
MC14034BCL	BLOCK-1073
MC14034BCP	BLOCK-1073
MC14035BAL	BLOCK-1074
MC14035BCL	BLOCK-1074
MC14035BCP	BLOCK-1074
MC14038BAL	BLOCK-1075
MC14038BCL	BLOCK-1075
MC14038BCP	BLOCK-1075
MC14040	BLOCK-1076
MC14040B	BLOCK-1076
MC14040BAL	BLOCK-1076
MC14040BCL	BLOCK-1076
MC14040BCP	BLOCK-1076
MC14040CP	BLOCK-1076
MC14042	BLOCK-1077
MC14042BAL	BLOCK-1077
MC14042BCL	BLOCK-1077
MC14042BCP	BLOCK-1077
MC14042CP	BLOCK-1077
MC14043BAL	BLOCK-1078
MC14043BCL	BLOCK-1078
MC14043BCP	BLOCK-1078
MC14044BAL	BLOCK-1079
MC14044BCL	BLOCK-1079
MC14044BCP	BLOCK-1079
MC14046BAL	BLOCK-2149
MC14046BCL	BLOCK-1081
MC14046BCP	BLOCK-1081
MC14046CP	BLOCK-1076
MC14049	BLOCK-1084
MC14049B	BLOCK-1084
MC14049CP	BLOCK-1084
MC14049UB	BLOCK-1084
MC14049UBAL	BLOCK-1084
MC14049UBCL	BLOCK-1084
MC14049UBCP	BLOCK-1084
MC14050	BLOCK-1085
MC14050B	BLOCK-1085
MC14050BAL	BLOCK-1085
MC14050BCL	BLOCK-1085
MC14050BCP	BLOCK-1085
MC14050CP	BLOCK-1085
MC14051	BLOCK-1086
MC14051BAL	BLOCK-1086
MC14051BCL	BLOCK-1086
MC14051BCP	BLOCK-1086
MC14052	BLOCK-1087
MC14052B	BLOCK-1087
MC14052BAL	BLOCK-1087
MC14052BCL	BLOCK-1087
MC14052BCP	BLOCK-1087
MC14053	BLOCK-1088
MC14053B	BLOCK-1088
MC14053BAL	BLOCK-1088
MC14053BCL	BLOCK-1088
MC14053BCP	BLOCK-1088
MC14060B	BLOCK-1091
MC14060BAL	BLOCK-1091
MC14060BCP	BLOCK-1091
MC14066	BLOCK-1093
MC14066B	BLOCK-1093
MC14066BAL	BLOCK-1093
MC14066BCL	BLOCK-1093
MC14066BCP	BLOCK-1093
MC14067BAL	BLOCK-1094
MC14068B	BLOCK-1095
MC14068BAL	BLOCK-1095
MC14068BCL	BLOCK-1095
MC14068BCP	BLOCK-1095
MC14069U	BLOCK-1096
MC14069UBAL	BLOCK-1096
MC14069UBCL	BLOCK-1096
MC14069UBCP	BLOCK-1096
MC14070BAL	BLOCK-1097
MC14070BCL	BLOCK-1097
MC14070BCP	BLOCK-1097
MC14070BL	BLOCK-1097
MC14071	BLOCK-1098
MC14071BAL	BLOCK-1098
MC14071BCL	BLOCK-1098
MC14071BCP	BLOCK-1098
MC14071CP	BLOCK-1098
MC14072BAL	BLOCK-1099
MC14072BCL	BLOCK-1099
MC14072BCP	BLOCK-1099
MC14073B	BLOCK-1100
MC14073BAL	BLOCK-1100
MC14073BCL	BLOCK-1100
MC14073BCP	BLOCK-1100
MC14075BAL	BLOCK-1101
MC14075BCL	BLOCK-1101
MC14075BCP	BLOCK-1101
MC14076BAL	BLOCK-1102
MC14076BCL	BLOCK-1102
MC14076BCP	BLOCK-1102
MC14077BAL	BLOCK-1103
MC14077BCL	BLOCK-1103
MC14077BCP	BLOCK-1103
MC14078BAL	BLOCK-1104
MC14078BCL	BLOCK-1104
MC14078BCP	BLOCK-1104
MC1408-7N	BLOCK-971
MC14081	BLOCK-1105
MC14081B	BLOCK-1105
MC14081BAL	BLOCK-1105
MC14081BCL	BLOCK-1105
MC14081BCP	BLOCK-525
MC14081CP	BLOCK-1105
MC14082B	BLOCK-1106
MC14082BAL	BLOCK-1106
MC14082BCL	BLOCK-1106
MC14082BCP	BLOCK-1106
MC14093	BLOCK-1110
MC14093B	BLOCK-1110
MC14093BAL	BLOCK-1110
MC14093BCL	BLOCK-1110
MC14093BCP	BLOCK-1110
MC14097BAL	BLOCK-1113
MC14099BAL	BLOCK-1115
MC14099BCL	BLOCK-1115
MC14099BCP	BLOCK-1115
MC1411P	BLOCK-934
MC1412P	BLOCK-935
MC1413P	BLOCK-936
MC14160BAL	BLOCK-1044
MC14160BCL	BLOCK-1044
MC14160BCP	BLOCK-1044
MC14161BAL	BLOCK-1045
MC14161BCL	BLOCK-1045
MC14161BCP	BLOCK-1045
MC14162BAL	BLOCK-1046
MC14162BCL	BLOCK-1046
MC14162BCP	BLOCK-1046
MC14163BAL	BLOCK-1047
MC14163BCL	BLOCK-1047
MC14163BCP	BLOCK-1047
MC1416P	BLOCK-937
MC14174B	BLOCK-1049
MC14174BAL	BLOCK-1049
MC14174BCL	BLOCK-1049
MC14174BCP	BLOCK-1049
MC14175BCL	BLOCK-1050
MC14175BCP	BLOCK-1050
MC14194BAL	BLOCK-1056
MC14194BCL	BLOCK-1056
MC14194BCP	BLOCK-1056
MC1422P1	BLOCK-1872
MC14412FL	BLOCK-963
MC14412FP	BLOCK-963
MC14412L	BLOCK-963
MC14412P	BLOCK-963
MC14412VL	BLOCK-963
MC14412VP	BLOCK-963
MC14433L	BLOCK-970
MC14433P	BLOCK-970
MC14501UBCL	BLOCK-1117
MC14501UBCP	BLOCK-1117
MC14502BAL	BLOCK-1118
MC14502BCL	BLOCK-1118
MC14502BCP	BLOCK-1118
MC14503BAL	BLOCK-1035
MC14503BCL	BLOCK-1837
MC14503BCP	BLOCK-1035
MC14506BCL	BLOCK-1119
MC14506BCP	BLOCK-1119
MC14506UBCP	BLOCK-1119
MC14508BAL	BLOCK-1120
MC14508BCL	BLOCK-1120
MC14508BCP	BLOCK-1120
MC145104	BLOCK-244
MC145104P	BLOCK-244
MC145109	BLOCK-156
MC145109P	BLOCK-156
MC14510BAL	BLOCK-1121
MC14510BCL	BLOCK-1121
MC14510BCP	BLOCK-1121
MC14511	BLOCK-1122
MC14511B	BLOCK-1122
MC14511B1	BLOCK-1122
MC14511BAL	BLOCK-1122
MC14511BCL	BLOCK-1122
MC14511BCP	BLOCK-1122
MC14511BP	BLOCK-1122
MC14511L	BLOCK-1122
MC14512BAL	BLOCK-1123
MC14512BCL	BLOCK-1123
MC14512BCP	BLOCK-1123
MC14512CP	BLOCK-1123
MC14513BCL	BLOCK-1124
MC14513BCP	BLOCK-1124
MC14514	BLOCK-1125
MC14514B	BLOCK-1125
MC14514BAL	BLOCK-1125
MC14514BCL	BLOCK-1125
MC14514BCP	BLOCK-1125
MC14515	BLOCK-1126
MC14515BAL	BLOCK-1126
MC14515BCL	BLOCK-1126
MC14515BCP	BLOCK-1126
MC14516BAL	BLOCK-1127
MC14516BCL	BLOCK-1127
MC14516BCP	BLOCK-1127
MC14516CP	BLOCK-1127
MC14517BAL	BLOCK-1128
MC14517BCL	BLOCK-1128
MC14517BCP	BLOCK-1128
MC14518	BLOCK-1129
MC14518B	BLOCK-1129
MC14518BAL	BLOCK-1129
MC14518BCL	BLOCK-1129
MC14518BCP	BLOCK-1129
MC14518CP	BLOCK-1129
MC14519BAL	BLOCK-1058
MC14519BCP	BLOCK-1058
MC14520BAL	BLOCK-1130
MC14520BCL	BLOCK-1130
MC14520BCP	BLOCK-1130
MC14521BCL	BLOCK-1131
MC14521BCP	BLOCK-1131
MC14522BCL	BLOCK-1132
MC14522BCP	BLOCK-1132
MC14526	BLOCK-1133
MC14526BCL	BLOCK-1133
MC14526BCP	BLOCK-1133
MC14527BAL	BLOCK-1134
MC14527BCL	BLOCK-1134
MC14527BCP	BLOCK-1134
MC14527CP	BLOCK-1134
MC14528	BLOCK-1114
MC14528BAL	BLOCK-1135
MC14528BCL	BLOCK-1135
MC14528BCP	BLOCK-1135
MC14529B	BLOCK-1136
MC14529BCL	BLOCK-1136
MC14529BCP	BLOCK-1136
MC14531BCL	BLOCK-1137
MC14531BCP	BLOCK-1137
MC14532BAL	BLOCK-1138
MC14532BCL	BLOCK-1138
MC14532BCP	BLOCK-1138
MC14536BAL	BLOCK-1139
MC14536BCP	BLOCK-1139
MC14538	BLOCK-1114
MC14538B	BLOCK-1140
MC14538BAL	BLOCK-1140
MC14538BCP	BLOCK-1140
MC14538BP	BLOCK-1140
MC14539BCL	BLOCK-1141
MC14539BCP	BLOCK-1141
MC14539CP	BLOCK-1141
MC14541BAL	BLOCK-1142
MC14541BCL	BLOCK-1142
MC14541BCP	BLOCK-1142
MC14543BAL	BLOCK-1143
MC14543BCL	BLOCK-1143
MC14543BCP	BLOCK-1143
MC14547BCL	BLOCK-1144
MC14547BCP	BLOCK-1144
MC14553BCL	BLOCK-1145
MC14553BCP	BLOCK-1145
MC14555BAL	BLOCK-1146
MC14555BCL	BLOCK-1146
MC14555BCP	BLOCK-1146
MC14555CP	BLOCK-1146
MC14556	BLOCK-1147
MC14556BAL	BLOCK-1147
MC14556BCL	BLOCK-1147
MC14556BCP	BLOCK-1147
MC14556CP	BLOCK-1147
MC14558BCL	BLOCK-1148
MC14558BCP	BLOCK-1148
MC1455D	BLOCK-2079
MC1455P1	BLOCK-2079
MC14562BCL	BLOCK-1149
MC14562BCP	BLOCK-1149
MC14566BCL	BLOCK-1150
MC14566BCP	BLOCK-1150
MC14568	BLOCK-1151
MC14568B	BLOCK-1151
MC14568BCL	BLOCK-1151
MC14568BCP	BLOCK-1151
MC14568CP	BLOCK-1151
MC14569BCL	BLOCK-1152
MC14569BCP	BLOCK-1152
MC1458	BLOCK-1801
MC14582BAL	BLOCK-1052
MC14582BCL	BLOCK-1052
MC14582BCP	BLOCK-1052
MC14583BCL	BLOCK-1153
MC14583BCP	BLOCK-1153
MC14584	BLOCK-1038
MC14584B	BLOCK-1038
MC14584BAL	BLOCK-1038
MC14584BCL	BLOCK-1038
MC14584BCP	BLOCK-1038
MC14585BAL	BLOCK-1154
MC14585BCL	BLOCK-1154
MC14585BCP	BLOCK-1092
MC1458CG	BLOCK-2073
MC1458CP1	BLOCK-1801
MC1458G	BLOCK-1801
MC1458N	BLOCK-1801
MC1458P	BLOCK-1801
MC1458P1	BLOCK-1801
MC1458V	BLOCK-1801
MC14597BCL	BLOCK-1155
MC14597BCP	BLOCK-1155
MC14598BCL	BLOCK-1156
MC14598BCP	BLOCK-1156
MC1461R	BLOCK-2068
MC1468L	BLOCK-1979
MC1469R	BLOCK-2068
MC1488	BLOCK-1771
MC1488F	BLOCK-1771
MC1488L	BLOCK-1771
MC1488N	BLOCK-1771
MC1488P	BLOCK-1771
MC1488PD	BLOCK-1771
MC1489	BLOCK-1772
MC1489/SN75189N	BLOCK-1772
MC1489A	BLOCK-1772
MC1489AP	BLOCK-1772
MC1489F	BLOCK-1772
MC1489L	BLOCK-1772
MC1489N	BLOCK-1772
MC1489P	BLOCK-1772
MC1496	BLOCK-2138
MC1496A	BLOCK-2139
MC1496G	BLOCK-2138
MC1496K	BLOCK-2138
MC1496L	BLOCK-2139
MC1496N	BLOCK-2139
MC1496P	BLOCK-2139
MC1550	BLOCK-1844
MC1550G	BLOCK-1844
MC1550P	BLOCK-1845
MC1558G	BLOCK-1801
MC1558P	BLOCK-43
MC155DG	BLOCK-1844
MC1709CG	BLOCK-1949
MC1709CL	BLOCK-1954
MC1709CP1	BLOCK-1949
MC1709CP2	BLOCK-1954
MC1709G	BLOCK-1949
MC1710CG	BLOCK-1955
MC1710CL	BLOCK-1957
MC1710CP	BLOCK-1957
MC1710P	BLOCK-1957
MC1711CG	BLOCK-1958
MC1711CL	BLOCK-1962
MC1711CP	BLOCK-1962
MC1711L	BLOCK-1962
MC1723CG	BLOCK-1983
MC1723CL	BLOCK-1984
MC1723CP	BLOCK-1984
MC1723G	BLOCK-1983
MC1731G	BLOCK-2056
MC1733CG	BLOCK-1989
MC1733CL	BLOCK-1990
MC1733CP	BLOCK-1990
MC1733G	BLOCK-1989
MC1741	BLOCK-2056
MC1741-5C	BLOCK-2056
MC1741C2P	BLOCK-2057
MC1741CD	BLOCK-2060
MC1741CG	BLOCK-2056
MC1741CL	BLOCK-2057
MC1741CP	BLOCK-2058
MC1741CP1	BLOCK-2058
MC1741CP2	BLOCK-2057
MC1741G	BLOCK-2056
MC1741NCG	BLOCK-2056
MC1741NCP1	BLOCK-2058
MC1741P1	BLOCK-2058
MC1741SCG	BLOCK-2056
MC1741SCP1	BLOCK-1973
MC1747CG	BLOCK-2069
MC1747CL	BLOCK-2070
MC1747CP2	BLOCK-2070
MC1747G	BLOCK-2069
MC1747L	BLOCK-2070
MC1748G	BLOCK-160
MC1776CP1	BLOCK-1938
MC1800P	BLOCK-2150
MC1801P	BLOCK-2151
MC1802P	BLOCK-2152
MC1803P	BLOCK-2153
MC1804P	BLOCK-2154
MC1805P	BLOCK-2155
MC1806P	BLOCK-2156
MC1807P	BLOCK-2157
MC1808P	BLOCK-2158
MC1809P	BLOCK-2159
MC1810P	BLOCK-2161
MC1811P	BLOCK-2162
MC1812P	BLOCK-2163
MC1813P	BLOCK-2164
MC1814P	BLOCK-2165
MC1816L	BLOCK-1960
MC1816P	BLOCK-1960
MC1820P	BLOCK-1961

DEVICE TYPE	REPL CODE
MC1946A	BLOCK-2138
MC3000L	BLOCK-1471
MC3000P	BLOCK-1471
MC3001	BLOCK-180
MC3001L	BLOCK-1475
MC3001P	BLOCK-1475
MC3004L	BLOCK-1472
MC3004P	BLOCK-1472
MC3005L	BLOCK-1476
MC3005P	BLOCK-1476
MC3006L	BLOCK-1482
MC3006P	BLOCK-1482
MC3008L	BLOCK-1473
MC3008P	BLOCK-1473
MC3009L	BLOCK-1474
MC3009P	BLOCK-1474
MC3010L	BLOCK-1484
MC3010P	BLOCK-1484
MC3011L	BLOCK-1485
MC3011P	BLOCK-1485
MC3012L	BLOCK-1486
MC3012P	BLOCK-1486
MC3016L	BLOCK-1487
MC3016P	BLOCK-1487
MC3018L	BLOCK-1497
MC3018P	BLOCK-1497
MC3019L	BLOCK-1496
MC3019P	BLOCK-1496
MC3020L	BLOCK-1489
MC3020P	BLOCK-1489
MC3021L	BLOCK-1504
MC3021P	BLOCK-1504
MC3023L	BLOCK-1490
MC3023P	BLOCK-1490
MC3024L	BLOCK-1488
MC3024P	BLOCK-1488
MC3030L	BLOCK-1495
MC3030P	BLOCK-1495
MC3031L	BLOCK-1491
MC3031P	BLOCK-1491
MC3032L	BLOCK-1492
MC3032P	BLOCK-1492
MC3033L	BLOCK-1493
MC3033P	BLOCK-1493
MC3034L	BLOCK-1494
MC3034P	BLOCK-1494
MC303P	BLOCK-1500
MC3054L	BLOCK-1498
MC3054P	BLOCK-1498
MC3055L	BLOCK-1499
MC3055P	BLOCK-1499
MC3062P	BLOCK-1729
MC3063L	BLOCK-1500
MC3301P	BLOCK-2191
MC3302	BLOCK-1873
MC3302P	BLOCK-1873
MC3303J	BLOCK-1265
MC3303L	BLOCK-1265
MC3303N	BLOCK-1265
MC3303P	BLOCK-1265
MC3310P	BLOCK-1881
MC3320P	BLOCK-1882
MC3340	BLOCK-1859
MC3340P	BLOCK-1859
MC3340P1	BLOCK-1794
MC3340PA	BLOCK-1859
MC3346P	BLOCK-1963
MC3357P	BLOCK-1897
MC3359P	BLOCK-1914
MC3360P	BLOCK-1773
MC3370P	BLOCK-1800
MC3373	BLOCK-694
MC3386P	BLOCK-1963
MC34001P	BLOCK-2036
MC34002P	BLOCK-1910
MC34004P	BLOCK-1912
MC3401L	BLOCK-2191
MC3401P	BLOCK-2191
MC3403J	BLOCK-1265
MC3403L	BLOCK-1265
MC3403N	BLOCK-2171
MC3403P	BLOCK-2171
MC34060L	BLOCK-734
MC34060P	BLOCK-734
MC34065P	BLOCK-746
MC3456P	BLOCK-2145
MC3458G	BLOCK-1932
MC3476G	BLOCK-2065
MC3476P1	BLOCK-2066
MC3479P	BLOCK-837
MC3490P	BLOCK-952
MC3491P	BLOCK-953
MC3492P	BLOCK-953
MC3494P	BLOCK-954

DEVICE TYPE	REPL CODE
MC4004	BLOCK-1415
MC4012P	BLOCK-1040
MC4016P	BLOCK-1048
MC4018L	BLOCK-1053
MC4018P	BLOCK-1053
MC4040P	BLOCK-2140
MC4042P	BLOCK-1077
MC4044	BLOCK-2140
MC4044CP	BLOCK-2140
MC4044D	BLOCK-2140
MC4044L	BLOCK-2140
MC4044P	BLOCK-2140
MC4047	BLOCK-1082
MC4050	BLOCK-1786
MC4051B	BLOCK-1086
MC4558ACP1	BLOCK-1801
MC4558CD	BLOCK-1803
MC4558CP	BLOCK-1801
MC4558CP1	BLOCK-1801
MC4558CU	BLOCK-1801
MC4741CL	BLOCK-2071
MC4741CP	BLOCK-2071
MC4741L	BLOCK-2071
MC5201	BLOCK-473
MC53200	BLOCK-1293
MC660P	BLOCK-2098
MC661P	BLOCK-2099
MC662P	BLOCK-2100
MC663P	BLOCK-2101
MC664P	BLOCK-2102
MC665P	BLOCK-2103
MC666P	BLOCK-2104
MC667P	BLOCK-2105
MC668P	BLOCK-2106
MC669P	BLOCK-2107
MC670P	BLOCK-2109
MC671P	BLOCK-2110
MC672	BLOCK-2111
MC672P	BLOCK-2111
MC673P	BLOCK-2112
MC674P	BLOCK-2113
MC675P	BLOCK-2114
MC676P	BLOCK-2115
MC677P	BLOCK-2116
MC678P	BLOCK-2117
MC679P	BLOCK-2118
MC6800CL	BLOCK-1164
MC6800CP	BLOCK-1164
MC6800P	BLOCK-1164
MC6802P	BLOCK-1165
MC6809E	BLOCK-1167
MC6809EL	BLOCK-1167
MC6809EP	BLOCK-1167
MC6809ES	BLOCK-1167
MC6809L	BLOCK-1166
MC6809P	BLOCK-1166
MC6809S	BLOCK-1166
MC680P	BLOCK-2120
MC681P	BLOCK-2121
MC6820	BLOCK-1169
MC6821	BLOCK-1169
MC6821CP	BLOCK-1169
MC6821P	BLOCK-1169
MC682P	BLOCK-2122
MC683P	BLOCK-2123
MC684P	BLOCK-2124
MC6850CP	BLOCK-1170
MC6850P	BLOCK-1170
MC685P	BLOCK-2125
MC6860L	BLOCK-1171
MC6860P	BLOCK-1171
MC686P	BLOCK-2126
MC6875	BLOCK-1172
MC6875L	BLOCK-1172
MC6880AL	BLOCK-1173
MC6880AP	BLOCK-1173
MC6885L	BLOCK-1174
MC6885P	BLOCK-1174
MC6886L	BLOCK-1175
MC6886P	BLOCK-1175
MC6887L	BLOCK-1176
MC6887P	BLOCK-1176
MC6888L	BLOCK-1177
MC6888P	BLOCK-1177
MC6889L	BLOCK-1178
MC6889P	BLOCK-1178
MC689P	BLOCK-2127
MC690P	BLOCK-2129
MC691P	BLOCK-2130
MC696P	BLOCK-2131
MC700G	BLOCK-2176
MC703G	BLOCK-2177
MC704G	BLOCK-2178
MC705G	BLOCK-2179

DEVICE TYPE	REPL CODE
MC706G	BLOCK-2180
MC707G	BLOCK-2181
MC708G	BLOCK-2182
MC709G	BLOCK-2183
MC710G	BLOCK-2185
MC711G	BLOCK-2186
MC712G	BLOCK-2187
MC713G	BLOCK-2188
MC714G	BLOCK-2189
MC715G	BLOCK-2190
MC721G	BLOCK-2192
MC724P	BLOCK-2193
MC726G	BLOCK-2194
MC727G	BLOCK-2195
MC7400	BLOCK-1293
MC7400F	BLOCK-1293
MC7400L	BLOCK-1293
MC7400N	BLOCK-1293
MC7400P	BLOCK-1293
MC7401F	BLOCK-1294
MC7401L	BLOCK-1294
MC7401P	BLOCK-1294
MC7402F	BLOCK-1295
MC7402L	BLOCK-1295
MC7402P	BLOCK-1295
MC7403L	BLOCK-1296
MC7403P	BLOCK-1296
MC7404L	BLOCK-1297
MC7404P	BLOCK-1297
MC7405L	BLOCK-1298
MC7405P	BLOCK-1298
MC7406L	BLOCK-1299
MC7406P	BLOCK-1299
MC7407L	BLOCK-1300
MC7407P	BLOCK-1300
MC7408L	BLOCK-1301
MC7408P	BLOCK-1301
MC7409L	BLOCK-1302
MC7409P	BLOCK-1302
MC74107P	BLOCK-1305
MC7410F	BLOCK-1304
MC7410L	BLOCK-1304
MC7410P	BLOCK-1304
MC74121P	BLOCK-1311
MC74145P	BLOCK-1325
MC74150P	BLOCK-1327
MC74152P	BLOCK-1329
MC74153P	BLOCK-1330
MC74155P	BLOCK-1332
MC74156P	BLOCK-1333
MC74164AP	BLOCK-1341
MC74165P	BLOCK-1342
MC7416L	BLOCK-1336
MC7416P	BLOCK-1336
MC74176P	BLOCK-1348
MC74177P	BLOCK-1349
MC7417L	BLOCK-1343
MC7417P	BLOCK-1343
MC74180P	BLOCK-1352
MC74181P	BLOCK-1353
MC74182P	BLOCK-1354
MC74192P	BLOCK-1357
MC74193P	BLOCK-1358
MC74195P	BLOCK-1359
MC7420F	BLOCK-1365
MC7420L	BLOCK-1365
MC7420P	BLOCK-1365
MC7430F	BLOCK-1377
MC7430L	BLOCK-1377
MC7430P	BLOCK-1377
MC7437F	BLOCK-1384
MC7437L	BLOCK-1384
MC7437P	BLOCK-1384
MC7438F	BLOCK-1385
MC7438L	BLOCK-1385
MC7438P	BLOCK-1385
MC7440F	BLOCK-1390
MC7440L	BLOCK-1390
MC7440P	BLOCK-1390
MC74415P	BLOCK-1328
MC7441AL	BLOCK-1391
MC7441AP	BLOCK-1391
MC7442L	BLOCK-1392
MC7442P	BLOCK-1392
MC7443L	BLOCK-1393
MC7443P	BLOCK-1393
MC7444L	BLOCK-1394
MC7444P	BLOCK-1394
MC7445L	BLOCK-1395
MC7445P	BLOCK-1395
MC7446L	BLOCK-1396
MC7446P	BLOCK-1396
MC7447L	BLOCK-1397
MC7447P	BLOCK-1397

DEVICE TYPE	REPL CODE
MC7448L	BLOCK-1398
MC7448P	BLOCK-1398
MC7450F	BLOCK-1401
MC7450L	BLOCK-1401
MC7450P	BLOCK-1401
MC7451F	BLOCK-1402
MC7451L	BLOCK-1402
MC7451P	BLOCK-1402
MC7453F	BLOCK-1403
MC7453L	BLOCK-1403
MC7453P	BLOCK-1403
MC7454F	BLOCK-1404
MC7454L	BLOCK-1404
MC7454P	BLOCK-1404
MC7460F	BLOCK-1406
MC7460L	BLOCK-1406
MC7460P	BLOCK-1406
MC7470L	BLOCK-1408
MC7470P	BLOCK-1408
MC7472F	BLOCK-1409
MC7472L	BLOCK-1409
MC7472P	BLOCK-1409
MC7473F	BLOCK-1410
MC7473L	BLOCK-1410
MC7473P	BLOCK-1410
MC7474P	BLOCK-1411
MC7475L	BLOCK-1412
MC7475P	BLOCK-1412
MC7476L	BLOCK-1413
MC7476P	BLOCK-1413
MC7480L	BLOCK-1414
MC7480P	BLOCK-1414
MC7483L	BLOCK-1417
MC7483P	BLOCK-1417
MC7486F	BLOCK-1419
MC7486L	BLOCK-1419
MC7486P	BLOCK-1419
MC7490AP	BLOCK-1423
MC7490F	BLOCK-1423
MC7490L	BLOCK-1423
MC7490P	BLOCK-1423
MC7491AL	BLOCK-1424
MC7491AP	BLOCK-1424
MC7492F	BLOCK-1425
MC7492L	BLOCK-1425
MC7492P	BLOCK-1425
MC7493F	BLOCK-1426
MC7493L	BLOCK-1426
MC7493P	BLOCK-1426
MC7494L	BLOCK-1427
MC7494P	BLOCK-1427
MC7495L	BLOCK-1428
MC7495P	BLOCK-1428
MC7496L	BLOCK-1429
MC7496P	BLOCK-1429
MC74H74AF	BLOCK-1501
MC74H74AL	BLOCK-1501
MC74H74P	BLOCK-1501
MC74H87F	BLOCK-1505
MC74H87L	BLOCK-1505
MC74H87P	BLOCK-1505
MC74HC00N	BLOCK-1506
MC74HC02N	BLOCK-1507
MC74HC04F	BLOCK-1508
MC74HC08	BLOCK-1509
MC74HC109N	BLOCK-1511
MC74HC10N	BLOCK-1510
MC74HC132N	BLOCK-1516
MC74HC138N	BLOCK-1517
MC74HC139N	BLOCK-1518
MC74HC151N	BLOCK-1520
MC74HC161	BLOCK-1523
MC74HC163N	BLOCK-1524
MC74HC164N	BLOCK-1525
MC74HC165N	BLOCK-1526
MC74HC173N	BLOCK-1527
MC74HC174N	BLOCK-1528
MC74HC175N	BLOCK-1529
MC74HC240N	BLOCK-1530
MC74HC244N	BLOCK-1531
MC74HC257N	BLOCK-1532
MC74HC273N	BLOCK-1534
MC74HC373N	BLOCK-1537
MC74HC374N	BLOCK-1538
MC74HC390N	BLOCK-1540
MC74HC393N	BLOCK-1541
MC74HC4020N	BLOCK-1543
MC74HC4040N	BLOCK-1544
MC74HC4053N	BLOCK-1545
MC74HC4060N	BLOCK-1546
MC74HC86N	BLOCK-1550
MC75452P	BLOCK-1777
MC75452P1	BLOCK-1777
MC75491P	BLOCK-1780

DEVICE TYPE	REPL CODE
MC774G	BLOCK-2220
MC776P	BLOCK-2221
MC7805	BLOCK-2087
MC7805ACK	BLOCK-2004
MC7805ACT	BLOCK-2087
MC7805AK	BLOCK-2004
MC7805BK	BLOCK-2004
MC7805BT	BLOCK-2087
MC7805C	BLOCK-2087
MC7805CK	BLOCK-2004
MC7805CP	BLOCK-2087
MC7805CT	BLOCK-2087
MC7805CTDS	BLOCK-2087
MC7805K	BLOCK-2004
MC7805UC	BLOCK-2087
MC7806ACT	BLOCK-2093
MC7806BT	BLOCK-2093
MC7806CT	BLOCK-2093
MC7808BT	BLOCK-2095
MC7808C	BLOCK-2095
MC7808CP	BLOCK-2095
MC7808CT	BLOCK-2095
MC7812	BLOCK-2097
MC7812AC	BLOCK-2097
MC7812ACK	BLOCK-2014
MC7812ACT	BLOCK-2097
MC7812AK	BLOCK-2014
MC7812BK	BLOCK-2014
MC7812BT	BLOCK-2097
MC7812C	BLOCK-2097
MC7812CK	BLOCK-2014
MC7812CT	BLOCK-2097
MC7812K	BLOCK-2014
MC7815	BLOCK-2119
MC7815ACT	BLOCK-2119
MC7815BT	BLOCK-2119
MC7815CK	BLOCK-892
MC7815CP	BLOCK-2119
MC7815CT	BLOCK-2119
MC7815P	BLOCK-2119
MC7818ACT	BLOCK-2085
MC7818BT	BLOCK-2085
MC7818CK	BLOCK-896
MC7818CT	BLOCK-2085
MC7819CT	BLOCK-2086
MC7824	BLOCK-2137
MC7824ACT	BLOCK-2137
MC7824BT	BLOCK-2137
MC7824CK	BLOCK-898
MC7824CT	BLOCK-2137
MC789	BLOCK-2224
MC789P	BLOCK-2224
MC78L05	BLOCK-2144
MC78L05,CP	BLOCK-2144
MC78L05ACG	BLOCK-2144
MC78L05ACP	BLOCK-2144
MC78L05ACPRP	BLOCK-2144
MC78L05C	BLOCK-2144
MC78L05CG	BLOCK-2144
MC78L05CP	BLOCK-2144
MC78L06AV	BLOCK-2173
MC78L08ACG	BLOCK-2160
MC78L08ACP	BLOCK-2160
MC78L08CG	BLOCK-2160
MC78L08CP	BLOCK-2160
MC78L12ACG	BLOCK-2074
MC78L12ACP	BLOCK-2074
MC78L12CG	BLOCK-2074
MC78L12CP	BLOCK-2074
MC78L15ACG	BLOCK-2075
MC78L15ACP	BLOCK-2075
MC78L15CG	BLOCK-2075
MC78L15CP	BLOCK-2075
MC78L18ACP	BLOCK-884
MC78L18CP	BLOCK-884
MC78L24ACP	BLOCK-886
MC78L24CP	BLOCK-886
MC78M05CT	BLOCK-2087
MC78M06CT	BLOCK-2093
MC78M08CT	BLOCK-2095
MC78M12	BLOCK-2097
MC78M12CT	BLOCK-2097
MC78M15CT	BLOCK-2119
MC78M18CT	BLOCK-2085
MC78M24CP	BLOCK-2137
MC78M24CT	BLOCK-2137
MC78T05CK	BLOCK-2004
MC78T05K	BLOCK-2004
MC78T12CK	BLOCK-888
MC78T12CT	BLOCK-2097
MC78T12K	BLOCK-2014
MC78T15CK	BLOCK-894
MC7905	BLOCK-2091
MC7905.2CT	BLOCK-2091

DEVICE TYPE	REPL CODE
MC7905ACT	BLOCK-2091
MC7905C	BLOCK-2091
MC7905CK	BLOCK-889
MC7905CT	BLOCK-2091
MC7906CT	BLOCK-2094
MC7908CT	BLOCK-2096
MC7912ACT	BLOCK-2108
MC7912CK	BLOCK-891
MC7912CT	BLOCK-2108
MC7915ACT	BLOCK-2128
MC7915CK	BLOCK-895
MC7915CP	BLOCK-2128
MC7915CT	BLOCK-2128
MC7918CK	BLOCK-2128
MC7918CT	BLOCK-2086
MC7924CK	BLOCK-899
MC7924CT	BLOCK-2136
MC79L05AC	BLOCK-893
MC79L05ACG	BLOCK-893
MC79L05ACP	BLOCK-893
MC79L05CG	BLOCK-893
MC79L05CP	BLOCK-2091
MC79L12AC	BLOCK-882
MC79L12ACG	BLOCK-882
MC79L12ACP	BLOCK-882
MC79L12CG	BLOCK-882
MC79L12CP	BLOCK-882
MC79L15ACG	BLOCK-883
MC79L15ACP	BLOCK-883
MC79L15CG	BLOCK-883
MC79L15CP	BLOCK-883
MC79L18ACP	BLOCK-885
MC79L18CP	BLOCK-885
MC800G	BLOCK-2176
MC803G	BLOCK-2177
MC804G	BLOCK-2178
MC805G	BLOCK-2179
MC806G	BLOCK-2180
MC807G	BLOCK-2181
MC808G	BLOCK-2182
MC809G	BLOCK-2183
MC810G	BLOCK-2185
MC811G	BLOCK-2186
MC812G	BLOCK-2187
MC813G	BLOCK-2188
MC814G	BLOCK-2189
MC815G	BLOCK-2190
MC821G	BLOCK-2192
MC824P	BLOCK-2193
MC826G	BLOCK-2194
MC827G	BLOCK-2195
MC830L	BLOCK-2197
MC830P	BLOCK-2197
MC831L	BLOCK-2198
MC831P	BLOCK-2198
MC832L	BLOCK-2199
MC832N	BLOCK-2197
MC832P	BLOCK-2199
MC833L	BLOCK-2200
MC833P	BLOCK-2200
MC835L	BLOCK-2201
MC835P	BLOCK-2201
MC836L	BLOCK-2202
MC836P	BLOCK-2202
MC837L	BLOCK-2203
MC837P	BLOCK-2203
MC840L	BLOCK-2201
MC840P	BLOCK-2201
MC841P	BLOCK-2205
MC844L	BLOCK-2206
MC844P	BLOCK-2206
MC845L	BLOCK-2207
MC845P	BLOCK-2207
MC846L	BLOCK-2208
MC846P	BLOCK-2208
MC848L	BLOCK-2209
MC848P	BLOCK-2209
MC849L	BLOCK-2210
MC849P	BLOCK-2210
MC850L	BLOCK-2212
MC850P	BLOCK-2212
MC851L	BLOCK-2213
MC851P	BLOCK-2213
MC852L	BLOCK-1953
MC852P	BLOCK-1953
MC853L	BLOCK-1950
MC853P	BLOCK-1950
MC855L	BLOCK-1952
MC855P	BLOCK-1952
MC856L	BLOCK-1951
MC856P	BLOCK-1951
MC857L	BLOCK-1968
MC857P	BLOCK-1968
MC858	BLOCK-1969

ORIGINAL DEVICE TYPES AND REPLACEMENT CODES 41

DEVICE TYPE	REPL CODE
MC858L	BLOCK-1969
MC858P	BLOCK-1969
MC8601P	BLOCK-2089
MC861L	BLOCK-2216
MC861P	BLOCK-2216
MC862L	BLOCK-2217
MC862P	BLOCK-2217
MC863L	BLOCK-2218
MC863P	BLOCK-2218
MC874G	BLOCK-2220
MC876P	BLOCK-2221
MC889P	BLOCK-2224
MC8T26AL	BLOCK-1173
MC8T26AP	BLOCK-1173
MC8T28L	BLOCK-1178
MC8T28P	BLOCK-1178
MC8T95L	BLOCK-1174
MC8T95P	BLOCK-1174
MC8T96L	BLOCK-1175
MC8T96P	BLOCK-1175
MC8T97L	BLOCK-1176
MC8T97P	BLOCK-1176
MC8T98L	BLOCK-1177
MC8T98P	BLOCK-1177
MC911G	BLOCK-2186
MC944L	BLOCK-2206
MC9601L	BLOCK-2089
MC9930	BLOCK-2197
MCC555A	BLOCK-2079
MCC555B	BLOCK-2079
MCF6021	BLOCK-1791
MCM2114	BLOCK-1002
MCM2114-30	BLOCK-1002
MCM2114-45	BLOCK-1002
MCM2114P-30	BLOCK-1002
MCM2114P-45	BLOCK-1002
MCM2114P20	BLOCK-1002
MCM2114P45	BLOCK-1002
MCM2147C100	BLOCK-1005
MCM2147C55	BLOCK-1005
MCM2147C70	BLOCK-1005
MCM2147C85	BLOCK-1005
MCM2532C	BLOCK-1015
MCM2708C	BLOCK-1016
MCM2708L	BLOCK-1016
MCM2716C	BLOCK-1017
MCM2716L	BLOCK-1017
MCM4027AC3	BLOCK-1000
MCM4027AC4	BLOCK-1000
MCM4116AC20	BLOCK-1003
MCM4116AC25	BLOCK-1003
MCM4116AC30	BLOCK-1003
MCM4116BC20	BLOCK-1003
MCM4116BC25	BLOCK-1003
MCM4116BC30	BLOCK-1003
MCM4116BP20	BLOCK-1003
MCM4116BP25	BLOCK-1003
MCM4116BP30	BLOCK-1003
MCM4116BP35	BLOCK-1003
MCM65116C	BLOCK-1004
MCM65116P	BLOCK-1004
MCM65116P15	BLOCK-1004
MCM65116P20	BLOCK-1004
MCM6664AL20	BLOCK-1163
MCM6664AP	BLOCK-1163
MCM6664AP15	BLOCK-1163
MCM6664AP20	BLOCK-1163
MCM6665	BLOCK-1006
MCM6665AL15	BLOCK-1006
MCM6665AL20	BLOCK-1006
MCM6665AP	BLOCK-1006
MCM6665AP15	BLOCK-1006
MCM6665AP20	BLOCK-1006
MCM6665L25	BLOCK-1006
MCM6810CL	BLOCK-1168
MCM6810CP	BLOCK-1168
MCM6810P	BLOCK-1168
MCM7641D	BLOCK-1757
MCM7643D	BLOCK-1765
MCM7681	BLOCK-1759
MCM7681D	BLOCK-1759
MCM7681P	BLOCK-1759
MCM7685D	BLOCK-1754
MCS2114-30	BLOCK-1002
MCS2114-35	BLOCK-1002
MCS2114L-45	BLOCK-1002
MCT492P	BLOCK-1425
MCT7808CT	BLOCK-2095
MD74LS37M	BLOCK-1670
ME32865-00001-B	BLOCK-1809
MEM4001	BLOCK-1029
MEM4007	BLOCK-1032
MEM4011	BLOCK-1039
MEM4013	BLOCK-1041
MEM4016	BLOCK-1048
MEM4049	BLOCK-1084
MEM4050	BLOCK-1085
MEM4051	BLOCK-1086
MFC4000	BLOCK-1773
MFC4000A	BLOCK-1773
MFC4000B	BLOCK-1773
MFC4010	BLOCK-1774
MFC4010A	BLOCK-1774
MFC4050	BLOCK-1786
MFC4052	BLOCK-1786
MFC4060	BLOCK-1787
MFC4060A	BLOCK-1787
MFC4062	BLOCK-1787
MFC4062A	BLOCK-1787
MFC4063	BLOCK-1788
MFC4063A	BLOCK-1788
MFC4064	BLOCK-1789
MFC4064A	BLOCK-1789
MFC6010	BLOCK-1790
MFC6020	BLOCK-1791
MFC6032	BLOCK-1792
MFC6032A	BLOCK-1792
MFC6033	BLOCK-1793
MFC6033A	BLOCK-1793
MFC6034	BLOCK-1793
MFC6034A	BLOCK-1793
MFC6040	BLOCK-1859
MFC6050	BLOCK-1794
MFC6060	BLOCK-1795
MFC6080	BLOCK-1796
MFC8020	BLOCK-1797
MFC8021	BLOCK-1797
MFC8021A	BLOCK-1797
MFC8030	BLOCK-1798
MFC8070	BLOCK-1800
MH745	BLOCK-2091
MH746	BLOCK-1638
MIC723-1	BLOCK-1983
MIC7400J	BLOCK-1293
MIC7400N	BLOCK-1293
MIC7401J	BLOCK-1294
MIC7401N	BLOCK-1294
MIC7402J	BLOCK-1295
MIC7402N	BLOCK-1295
MIC7403J	BLOCK-1296
MIC7403N	BLOCK-1296
MIC7404J	BLOCK-1297
MIC7404N	BLOCK-1297
MIC7405J	BLOCK-1298
MIC7405N	BLOCK-1298
MIC741-5C	BLOCK-2056
MIC74107J	BLOCK-1305
MIC74107N	BLOCK-1305
MIC7410J	BLOCK-1304
MIC7410N	BLOCK-1304
MIC74121J	BLOCK-1311
MIC74121N	BLOCK-1311
MIC7413J	BLOCK-1317
MIC7413N	BLOCK-1317
MIC74145J	BLOCK-1325
MIC74145N	BLOCK-1325
MIC74150J	BLOCK-1327
MIC74150N	BLOCK-1327
MIC74151J	BLOCK-1328
MIC74151N	BLOCK-1328
MIC74154J	BLOCK-1331
MIC74154N	BLOCK-1331
MIC74155J	BLOCK-1332
MIC74155N	BLOCK-1332
MIC74156J	BLOCK-1333
MIC74156N	BLOCK-1333
MIC74180J	BLOCK-1352
MIC74180N	BLOCK-1352
MIC7420J	BLOCK-1365
MIC7420N	BLOCK-1365
MIC7428J	BLOCK-1373
MIC7428N	BLOCK-1373
MIC7430J	BLOCK-1377
MIC7430N	BLOCK-1377
MIC7440J	BLOCK-1390
MIC7440N	BLOCK-1390
MIC7441AJ	BLOCK-1391
MIC7441AN	BLOCK-1391
MIC7442J	BLOCK-1392
MIC7442N	BLOCK-1392
MIC7443J	BLOCK-1393
MIC7443N	BLOCK-1393
MIC7444J	BLOCK-1394
MIC7444N	BLOCK-1394
MIC7445J	BLOCK-1395
MIC7445N	BLOCK-1395
MIC7450J	BLOCK-1401
MIC7450N	BLOCK-1401
MIC7451J	BLOCK-1402
MIC7451N	BLOCK-1402
MIC7453J	BLOCK-1403
MIC7453N	BLOCK-1403
MIC7454J	BLOCK-1404
MIC7454N	BLOCK-1404
MIC7460J	BLOCK-1406
MIC7460N	BLOCK-1406
MIC7470J	BLOCK-1408
MIC7470N	BLOCK-1408
MIC7472J	BLOCK-1409
MIC7472N	BLOCK-1409
MIC7473J	BLOCK-1410
MIC7473N	BLOCK-1410
MIC7474J	BLOCK-1411
MIC7474N	BLOCK-1411
MIC7475J	BLOCK-1412
MIC7475N	BLOCK-1412
MIC7476J	BLOCK-1413
MIC7476N	BLOCK-1413
MIC7481J	BLOCK-1415
MIC7481N	BLOCK-1415
MIC7482J	BLOCK-1416
MIC7482N	BLOCK-1416
MIC7483J	BLOCK-1417
MIC7483N	BLOCK-1417
MIC7486J	BLOCK-1419
MIC7486N	BLOCK-1419
MIC7490J	BLOCK-1423
MIC7490N	BLOCK-1423
MIC7491AJ	BLOCK-1424
MIC7491AN	BLOCK-1424
MIC7492J	BLOCK-1425
MIC7492N	BLOCK-1425
MIC7493J	BLOCK-1426
MIC7493N	BLOCK-1426
MIC7494J	BLOCK-1427
MIC7494N	BLOCK-1427
MIC7495J	BLOCK-1428
MIC7495N	BLOCK-1428
MIC7496J	BLOCK-1429
MIC77413J	BLOCK-1317
MIC9093-5D	BLOCK-1950
MIC9093-5P	BLOCK-1950
MIC9094-5D	BLOCK-1951
MIC9094-5P	BLOCK-1951
MIC9097-5D	BLOCK-1952
MIC9097-5P	BLOCK-1952
MIC9099-5D	BLOCK-1953
MIC9099-5P	BLOCK-1953
MIC930-5D	BLOCK-2197
MIC930-5P	BLOCK-2197
MIC931-5D	BLOCK-2198
MIC931-5P	BLOCK-2198
MIC932-5D	BLOCK-2199
MIC932-5P	BLOCK-2199
MIC933-5D	BLOCK-2200
MIC933-5P	BLOCK-2200
MIC935-5D	BLOCK-2201
MIC935-5P	BLOCK-2201
MIC936-5D	BLOCK-2202
MIC936-5P	BLOCK-2202
MIC937-5D	BLOCK-2203
MIC937-5P	BLOCK-2203
MIC941-5D	BLOCK-2205
MIC941-5P	BLOCK-2205
MIC944-5D	BLOCK-2206
MIC944-5P	BLOCK-2206
MIC945-5D	BLOCK-2207
MIC945-5P	BLOCK-2207
MIC946-5D	BLOCK-2208
MIC946-5P	BLOCK-2208
MIC948-5D	BLOCK-2209
MIC948-5P	BLOCK-2209
MIC949-5D	BLOCK-2210
MIC949-5P	BLOCK-2210
MIC950-5D	BLOCK-2212
MIC950-5P	BLOCK-2212
MIC951-5D	BLOCK-2213
MIC951-5P	BLOCK-2213
MIC961-5D	BLOCK-2216
MIC961-5P	BLOCK-2216
MIC962-5D	BLOCK-2217
MIC962-5P	BLOCK-2217
MIC963-5D	BLOCK-2218
MIC963-5P	BLOCK-2218
MIS-18101-3	BLOCK-2197
MIS-18101-5	BLOCK-2207
MIS-181C1-4	BLOCK-2199
MJC574J	BLOCK-1157
MK3881N	BLOCK-1026
MK3881N4	BLOCK-1026
MK3881P	BLOCK-1026
MK3882N	BLOCK-1027
MK3882P	BLOCK-1027
MK4027J-3	BLOCK-1000
MK4027J-4	BLOCK-1000
MK4027N-3	BLOCK-1000
MK4027N-4	BLOCK-1000
MK4027P-3	BLOCK-1000
MK4116	BLOCK-1003
MK4116-4	BLOCK-1003
MK41164-3GP	BLOCK-1003
MK4116E-2	BLOCK-1003
MK4116E-3	BLOCK-1003
MK4116E-4	BLOCK-1003
MK4116J-2	BLOCK-1003
MK4116J-3	BLOCK-1003
MK4116J-4	BLOCK-1003
MK4116J-53GP	BLOCK-1003
MK4116N-2	BLOCK-1003
MK4116N-3	BLOCK-1003
MK4116N-3GP	BLOCK-1003
MK4116N-4	BLOCK-1003
MK4116N-44GP	BLOCK-1003
MK4116P-2	BLOCK-1003
MK4116P-3	BLOCK-1003
MK4116P-4	BLOCK-1003
MK4564N-15	BLOCK-1006
MK4802P-3	BLOCK-1004
MK50240N	BLOCK-959
MK50240P	BLOCK-959
MK50241N	BLOCK-959
MK50241P	BLOCK-959
MK5089J	BLOCK-676
MK5089K	BLOCK-676
MK5089P	BLOCK-676
MK50981N	BLOCK-679
MK50992N	BLOCK-677
ML1458S	BLOCK-1801
ML201AT	BLOCK-160
ML301T	BLOCK-160
ML307S	BLOCK-2143
ML709CT	BLOCK-1949
ML723CM	BLOCK-1984
ML723CP	BLOCK-1984
ML723CT	BLOCK-1983
ML723T	BLOCK-1983
ML741CS	BLOCK-2058
ML741CT	BLOCK-2056
ML747CP	BLOCK-2070
ML747CT	BLOCK-2069
ML748CS	BLOCK-2141
ML7815P	BLOCK-2119
ML78L05A	BLOCK-2144
ML78L18A	BLOCK-884
ML78L24A	BLOCK-886
ML924	BLOCK-738
MLC74HC00A	BLOCK-1506
MLC74HC02A	BLOCK-1507
MLC74HC04A	BLOCK-1508
MLC74HC08A	BLOCK-1509
MLC74HC109	BLOCK-1511
MLC74HC11	BLOCK-1512
MLC74HC14A	BLOCK-1519
MLC74HC32A	BLOCK-1536
MLC74HC86	BLOCK-1550
MLM101AG	BLOCK-2141
MLM111AG	BLOCK-1980
MLM139	BLOCK-1873
MLM139AL	BLOCK-1873
MLM139L	BLOCK-1873
MLM201AG	BLOCK-160
MLM211AG	BLOCK-1980
MLM224L	BLOCK-2171
MLM224P	BLOCK-2171
MLM239AL	BLOCK-1873
MLM239L	BLOCK-1873
MLM2901P	BLOCK-1873
MLM301AG	BLOCK-160
MLM301AU	BLOCK-2141
MLM307P1	BLOCK-2143
MLM307U	BLOCK-2143
MLM309K	BLOCK-1022
MLM311P1	BLOCK-1981
MLM324	BLOCK-2171
MLM324L	BLOCK-2171
MLM324P	BLOCK-2171
MLM324P1	BLOCK-2171
MLM339(P)	BLOCK-1873
MLM339AL	BLOCK-1873
MLM339L	BLOCK-1873
MLM339P	BLOCK-1873
MM14511BCN	BLOCK-1122
MM14511CN	BLOCK-1122
MM2114N-3	BLOCK-1002
MM2708Q	BLOCK-1016
MM2716Q	BLOCK-1017
MM4000(IC)	BLOCK-1028
MM4007	BLOCK-1032
MM4009	BLOCK-1084
MM4011	BLOCK-1039
MM4012	BLOCK-1040
MM4013	BLOCK-1041
MM4015	BLOCK-1043
MM4016A	BLOCK-1048
MM4017	BLOCK-1051
MM4019(IC)	BLOCK-1058
MM4020(IC)	BLOCK-1059
MM4024	BLOCK-1063
MM4025	BLOCK-1064
MM4027	BLOCK-1066
MM4030	BLOCK-1069
MM4040	BLOCK-1076
MM4042	BLOCK-1077
MM4049	BLOCK-1084
MM4050	BLOCK-1085
MM4051	BLOCK-1086
MM4052(IC)	BLOCK-1087
MM4511	BLOCK-1122
MM4518	BLOCK-1129
MM5290N-4	BLOCK-1003
MM53113N	BLOCK-974
MM5316	BLOCK-973
MM5316N	BLOCK-974
MM5369AA/N	BLOCK-962
MM5387	BLOCK-973
MM5387AA	BLOCK-973
MM5387AA-N	BLOCK-973
MM5387AA/N	BLOCK-973
MM5387AB/N	BLOCK-973
MM5387N	BLOCK-973
MM5402	BLOCK-973
MM5402N	BLOCK-973
MM55104	BLOCK-244
MM55104N	BLOCK-244
MM5601AN	BLOCK-1029
MM5602AN	BLOCK-1030
MM5611AN	BLOCK-1039
MM5612AN	BLOCK-1040
MM5617AN	BLOCK-1051
MM5619AN	BLOCK-1058
MM5620AN	BLOCK-1059
MM5622AN	BLOCK-1061
MM5623AN	BLOCK-1062
MM5624AN	BLOCK-1063
MM5625AN	BLOCK-1064
MM5627AN	BLOCK-1066
MM5630AN	BLOCK-1069
MM5660BN	BLOCK-1085
MM741	BLOCK-2058
MM74C00	BLOCK-1431
MM74C00N	BLOCK-1431
MM74C02	BLOCK-1432
MM74C02N	BLOCK-1432
MM74C04	BLOCK-1433
MM74C04J	BLOCK-1433
MM74C04N	BLOCK-1433
MM74C08	BLOCK-1434
MM74C08N	BLOCK-1434
MM74C10	BLOCK-1435
MM74C107N	BLOCK-1436
MM74C10N	BLOCK-1435
MM74C14J	BLOCK-1437
MM74C14N	BLOCK-1437
MM74C151J	BLOCK-1438
MM74C151N	BLOCK-1438
MM74C154J	BLOCK-1439
MM74C154N	BLOCK-1439
MM74C157J	BLOCK-1440
MM74C157N	BLOCK-1440
MM74C161N	BLOCK-1441
MM74C164N	BLOCK-1442
MM74C173N	BLOCK-1443
MM74C174N	BLOCK-1444
MM74C175N	BLOCK-1445
MM74C192N	BLOCK-1446
MM74C193N	BLOCK-1447
MM74C20	BLOCK-1448
MM74C20N	BLOCK-1448
MM74C221	BLOCK-1449
MM74C221N	BLOCK-1449
MM74C240N	BLOCK-1450
MM74C244N	BLOCK-1451
MM74C30N	BLOCK-1452
MM74C32N	BLOCK-1453
MM74C373N	BLOCK-1454
MM74C374N	BLOCK-1455
MM74C48N	BLOCK-1456
MM74C73N	BLOCK-1457
MM74C74	BLOCK-1458
MM74C74N	BLOCK-1458
MM74C76N	BLOCK-1459
MM74C85	BLOCK-1460
MM74C85N	BLOCK-1460
MM74C86	BLOCK-1097
MM74C90	BLOCK-1461
MM74C901N	BLOCK-1462
MM74C902N	BLOCK-1463
MM74C903N	BLOCK-1464
MM74C904N	BLOCK-1465
MM74C90N	BLOCK-1461
MM74C922N	BLOCK-1466
MM74C923N	BLOCK-1467
MM74C925N	BLOCK-1468
MM74C93	BLOCK-1469
MM74C93N	BLOCK-1469
MM74C95N	BLOCK-1470
MM74HC240N	BLOCK-1530
MN1206A	BLOCK-1014
MN1301	BLOCK-626
MN3001	BLOCK-627
MN3007	BLOCK-628
MN3011	BLOCK-627
MN3101	BLOCK-626
MN380	BLOCK-1303
MN380H	BLOCK-1303
MN4001B	BLOCK-1029
MN40098B	BLOCK-1036
MN4011B	BLOCK-1039
MN4013B	BLOCK-1041
MN4027B	BLOCK-1066
MN4052B	BLOCK-1087
MN4053B	BLOCK-1088
MN4066B	BLOCK-1093
MN4066BP	BLOCK-1093
MN4069UB	BLOCK-1096
MN4069UBS	BLOCK-1096
MN4071B	BLOCK-1098
MN4081B	BLOCK-1105
MN4116	BLOCK-1003
MN41256-15	BLOCK-1116
MN4164P-15A	BLOCK-1006
MN4503B	BLOCK-1035
MN4528B	BLOCK-1135
MN5101	BLOCK-1161
MN5613AN	BLOCK-1041
MN6016A	BLOCK-395
MN6040	BLOCK-244
MN6040A	BLOCK-244
MN6049	BLOCK-828
MN6061A	BLOCK-395
MN6076	BLOCK-402
MN6163	BLOCK-795
MN6163A	BLOCK-795
MOS6502	BLOCK-1158
MOS6502A	BLOCK-1158
MOS7712	BLOCK-1574
MOS8712	BLOCK-1574
MOS8713	BLOCK-1572
MP0574J	BLOCK-1157
MP1304P	BLOCK-1275
MP1304PQ	BLOCK-1275
MP5106P	BLOCK-95
MP5190P	BLOCK-1291
MPB121D	BLOCK-2106
MPB123D	BLOCK-2099
MPB124D	BLOCK-2117
MPB125D	BLOCK-2100
MPC-31C	BLOCK-46
MPC-574J	BLOCK-1157
MPC1001H2	BLOCK-125
MPC1352C	BLOCK-404
MPC1355C	BLOCK-1422
MPC1356C	BLOCK-471
MPC1356C2	BLOCK-471
MPC1363	BLOCK-396
MPC1363C	BLOCK-396
MPC1367C	BLOCK-539
MPC1372C	BLOCK-398
MPC1373H	BLOCK-695
MPC1373HA	BLOCK-695
MPC1382C	BLOCK-603
MPC14305	BLOCK-2087
MPC1458C	BLOCK-1801
MPC20C	BLOCK-73
MPC23C	BLOCK-44
MPC29C	BLOCK-87
MPC29C2	BLOCK-87
MPC30C	BLOCK-47
MPC31C	BLOCK-46
MPC46C	BLOCK-75
MPC47C	BLOCK-74
MPC48C	BLOCK-89
MPC558C	BLOCK-49

DEVICE TYPE	REPL CODE
MPC561C	BLOCK-72
MPC562C	BLOCK-48
MPC566HB	BLOCK-101
MPC566HC	BLOCK-101
MPC566HD	BLOCK-101
MPC570C	BLOCK-84
MPC574J	BLOCK-1157
MPC575C	BLOCK-137
MPC575C2	BLOCK-136
MPC577H	BLOCK-80
MPC595C	BLOCK-175
MPC596C	BLOCK-176
MPC596C2	BLOCK-176
MPC596C2B	BLOCK-176
MPC7812H	BLOCK-2097
MPD14027BC	BLOCK-1066
MPD1937C	BLOCK-740
MPD4011BC	BLOCK-1039
MPD4066BC	BLOCK-1093
MPD4093BC	BLOCK-1110
MPD858	BLOCK-186
MPD858C	BLOCK-186
MPD861C	BLOCK-243
MPD861CE	BLOCK-243
MPS2114-30	BLOCK-1002
MPS2114-35	BLOCK-1002
MPS2114-45	BLOCK-1002
MPS2114-50	BLOCK-1002
MPS6502	BLOCK-1158
MPS6502A	BLOCK-1158
MR2525	BLOCK-163
MS115P	BLOCK-108
MSM2114	BLOCK-1002
MSM2114L-3RS	BLOCK-1002
MSM2114L3RS	BLOCK-1002
MSM2764RS	BLOCK-1019
MSM311EL	BLOCK-1981
MSM4001	BLOCK-1029
MSM4002	BLOCK-1030
MSM4008	BLOCK-1034
MSM4011	BLOCK-1039
MSM4011RS	BLOCK-1039
MSM4012	BLOCK-1040
MSM4013	BLOCK-1041
MSM4013RS	BLOCK-1041
MSM4016	BLOCK-1048
MSM4016RS	BLOCK-1048
MSM4017	BLOCK-1051
MSM4017AN	BLOCK-1051
MSM4019	BLOCK-1058
MSM40192	BLOCK-1054
MSM40193	BLOCK-1055
MSM4020	BLOCK-1059
MSM4023	BLOCK-1062
MSM4025	BLOCK-1064
MSM4027	BLOCK-1066
MSM4028	BLOCK-1067
MSM4030	BLOCK-1069
MSM4040	BLOCK-1076
MSM4042	BLOCK-1077
MSM4043	BLOCK-1078
MSM4044	BLOCK-1079
MSM4050	BLOCK-1085
MSM4050RS	BLOCK-1085
MSM4064	BLOCK-1096
MSM4066RS	BLOCK-1093
MSM4068	BLOCK-1095
MSM4069	BLOCK-1096
MSM4069RS	BLOCK-1096
MSM4069UBRU	BLOCK-1096
MSM4071	BLOCK-1098
MSM4071BN	BLOCK-1098
MSM4072	BLOCK-1099
MSM4073	BLOCK-1100
MSM4075	BLOCK-1101
MSM4078	BLOCK-1104
MSM4081	BLOCK-1105
MSM4081RS	BLOCK-1105
MSM4082	BLOCK-1106
MSM4085	BLOCK-1107
MSM4086	BLOCK-1108
MSM4514RS	BLOCK-1125
MSM4518	BLOCK-1129
MSM4520	BLOCK-1130
MT4264-15	BLOCK-1006
MT4264-20	BLOCK-1006
MTC1370PQ	BLOCK-1272
MWA110	BLOCK-749
MWA120	BLOCK-750
MWA220	BLOCK-751
MWS5101D	BLOCK-1161
MWS5101E	BLOCK-1161
MX-2269	BLOCK-644
MX-2273	BLOCK-695

DEVICE TYPE	REPL CODE
MX-3013	BLOCK-4
MX-3198	BLOCK-2144
MX-3235	BLOCK-243
MX-3240	BLOCK-288
MX-3243	BLOCK-1825
MX-3256	BLOCK-182
MX-3260	BLOCK-180
MX-3364	BLOCK-112
MX-3369	BLOCK-2139
MX-3370	BLOCK-183
MX-3372	BLOCK-182
MX-3379	BLOCK-192
MX-3389	BLOCK-80
MX-3393	BLOCK-136
MX-3399	BLOCK-186
MX-3452	BLOCK-2095
MX-3540	BLOCK-242
MX-3545	BLOCK-180
MX-3587	BLOCK-182
MX-3634	BLOCK-1039
MX-3808	BLOCK-1706
MX-3948	BLOCK-159
MX-3976	BLOCK-237
MX-3977	BLOCK-368
MX-3979	BLOCK-552
MX-4295	BLOCK-1648
MX-4313	BLOCK-1336
MX-4314	BLOCK-1592
MX-4339	BLOCK-904
MX-4376	BLOCK-1087
MX-4404	BLOCK-563
MX-4465	BLOCK-500
MX-4476	BLOCK-550
MX-4478	BLOCK-1157
MX-4479	BLOCK-235
MX-4542	BLOCK-544
MX-4712	BLOCK-558
MX-4821	BLOCK-936
MX-4875	BLOCK-504
MX-4969	BLOCK-1200
MX-4974	BLOCK-214
MX-5200	BLOCK-607
MX-5201	BLOCK-473
MX-5202	BLOCK-223
MX-5279	BLOCK-404
MX-5280	BLOCK-532
MX-5281	BLOCK-525
MX-5367	BLOCK-975
MX-5375	BLOCK-398
MX-5389	BLOCK-975
MX-5421	BLOCK-364
MX-5429	BLOCK-544
MX-5430	BLOCK-237
MX-5436	BLOCK-454
MX-5451	BLOCK-593
MX-5475	BLOCK-653
MX-5476	BLOCK-505
MX-5594	BLOCK-237
MX-5595	BLOCK-621
MX-5596	BLOCK-453
MX-5679	BLOCK-1993
MX-6092	BLOCK-732
MX-6093	BLOCK-657
MX-6159	BLOCK-762
MX-6160	BLOCK-662
MX-6387	BLOCK-732
MX-6406	BLOCK-1727
MX-6452	BLOCK-537
MX-6536	BLOCK-725
MX-6829	BLOCK-559
MX3139	BLOCK-1338
MX3256	BLOCK-180
MX3336	BLOCK-1405
MX3587	BLOCK-182
MX3634	BLOCK-1039
MX3979	BLOCK-552
MX4340	BLOCK-771
MX5201	BLOCK-473
MX5202	BLOCK-223
MX5560	BLOCK-1167
MX5741	BLOCK-2079
MXC-1312	BLOCK-1823
MXC1312A	BLOCK-1823
MXF-5375	BLOCK-398
MZT33-01	BLOCK-1157
N-7400A	BLOCK-1293
N-7404A	BLOCK-1297
N-7408A	BLOCK-1301
N-7473A	BLOCK-1410
N4000	BLOCK-1028
N4001	BLOCK-1029
N4002	BLOCK-1030
N4007	BLOCK-1032
N4009	BLOCK-1084

DEVICE TYPE	REPL CODE
N4011	BLOCK-1039
N4012	BLOCK-1040
N4013	BLOCK-1041
N4015	BLOCK-1043
N4016	BLOCK-1048
N4019	BLOCK-1058
N4021	BLOCK-1060
N4023	BLOCK-1062
N4025	BLOCK-1064
N4027	BLOCK-1066
N4030	BLOCK-1069
N4049	BLOCK-1084
N4050	BLOCK-1085
N4081	BLOCK-1105
N5065A	BLOCK-1269
N5070B	BLOCK-1271
N5070N	BLOCK-1271
N5071A	BLOCK-1272
N5072A	BLOCK-1270
N5111	BLOCK-1262
N5111A	BLOCK-1262
N5596K	BLOCK-2138
N5723T	BLOCK-1983
N5741V	BLOCK-2058
N6121860001	BLOCK-1048
N6123270001	BLOCK-553
N6123700001	BLOCK-567
N6124120002	BLOCK-784
N6124440001	BLOCK-735
N6124670001	BLOCK-831
N6124790001	BLOCK-2087
N6124930001	BLOCK-1088
N6125070001	BLOCK-1199
N6125080001	BLOCK-870
N6125450001	BLOCK-767
N6125630001	BLOCK-781
N6125640001	BLOCK-782
N7400A	BLOCK-1293
N7400F	BLOCK-1293
N7400N	BLOCK-1293
N7401	BLOCK-1294
N7401A	BLOCK-1293
N7401F	BLOCK-1293
N7401N	BLOCK-1294
N7402A	BLOCK-1295
N7402F	BLOCK-1295
N7402N	BLOCK-1295
N7403A	BLOCK-1296
N7403F	BLOCK-1296
N7403N	BLOCK-1296
N7404A	BLOCK-1297
N7404F	BLOCK-1297
N7404N	BLOCK-1297
N7405A	BLOCK-1298
N7405F	BLOCK-1298
N7405N	BLOCK-1298
N7406A	BLOCK-1299
N7406F	BLOCK-1299
N7406N	BLOCK-1299
N7407A	BLOCK-1300
N7407F	BLOCK-1300
N7407N	BLOCK-1300
N7408A	BLOCK-1301
N7408F	BLOCK-1301
N7408N	BLOCK-1301
N7409A	BLOCK-1302
N7409F	BLOCK-1302
N7409N	BLOCK-1302
N74107A	BLOCK-1305
N74107F	BLOCK-1305
N74107N	BLOCK-1305
N74109B	BLOCK-1306
N74109F	BLOCK-1306
N74109N	BLOCK-1306
N7410A	BLOCK-1304
N7410F	BLOCK-1304
N7410N	BLOCK-1304
N74116N	BLOCK-1862
N7411A	BLOCK-1307
N7411F	BLOCK-1307
N7411N	BLOCK-1307
N74121A	BLOCK-1311
N74121F	BLOCK-1311
N74121N	BLOCK-1311
N74122A	BLOCK-1312
N74122F	BLOCK-1312
N74122N	BLOCK-1312
N74123A	BLOCK-1313
N74123B	BLOCK-1313
N74123F	BLOCK-1313
N74123N	BLOCK-1313
N74125F	BLOCK-1314
N74125N	BLOCK-1314
N74126F	BLOCK-1315

DEVICE TYPE	REPL CODE
N74126N	BLOCK-1315
N74128F	BLOCK-1316
N7412F	BLOCK-1310
N7412N	BLOCK-1310
N74132B	BLOCK-1318
N74132F	BLOCK-1318
N74132N	BLOCK-1318
N7413A	BLOCK-1317
N7413F	BLOCK-1317
N7413N	BLOCK-1317
N74141B	BLOCK-1321
N74145	BLOCK-1325
N74145A	BLOCK-1325
N74145B	BLOCK-1325
N74145F	BLOCK-1325
N74145N	BLOCK-1325
N74147F	BLOCK-1326
N74147N	BLOCK-1326
N7414B	BLOCK-1320
N7414F	BLOCK-1320
N7414N	BLOCK-1320
N74150A	BLOCK-1327
N74150B	BLOCK-1327
N74150F	BLOCK-1327
N74150N	BLOCK-1327
N74151A	BLOCK-1328
N74151B	BLOCK-1328
N74151F	BLOCK-1328
N74151N	BLOCK-1328
N74152F	BLOCK-1329
N74152N	BLOCK-1329
N74153B	BLOCK-1330
N74153F	BLOCK-1330
N74153N	BLOCK-1330
N74154A	BLOCK-1331
N74154F	BLOCK-1331
N74154N	BLOCK-1331
N74155B	BLOCK-1332
N74155F	BLOCK-1332
N74155N	BLOCK-1332
N74156B	BLOCK-1333
N74156F	BLOCK-1333
N74156N	BLOCK-1333
N74157F	BLOCK-1334
N74157N	BLOCK-1334
N74158F	BLOCK-1335
N74158N	BLOCK-1335
N74160B	BLOCK-1337
N74160F	BLOCK-1337
N74160N	BLOCK-1337
N74161B	BLOCK-1338
N74161F	BLOCK-1338
N74161N	BLOCK-1338
N74162B	BLOCK-1339
N74162F	BLOCK-1339
N74162N	BLOCK-1339
N74163B	BLOCK-1340
N74163F	BLOCK-1340
N74163N	BLOCK-1340
N74164A	BLOCK-1341
N74164F	BLOCK-1341
N74164N	BLOCK-1341
N74165B	BLOCK-1342
N74165F	BLOCK-1342
N74165N	BLOCK-1342
N7416A	BLOCK-1336
N7416F	BLOCK-1336
N7416N	BLOCK-1336
N74170B	BLOCK-1344
N74170F	BLOCK-1344
N74170N	BLOCK-1344
N74173N	BLOCK-1345
N74174	BLOCK-1346
N74174F	BLOCK-1346
N74174N	BLOCK-1346
N74175B	BLOCK-1347
N74175F	BLOCK-1347
N74175N	BLOCK-1347
N74178A	BLOCK-1350
N74179B	BLOCK-1351
N74179F	BLOCK-1351
N7417A	BLOCK-1343
N7417F	BLOCK-1343
N7417N	BLOCK-1343
N74180A	BLOCK-1352
N74180F	BLOCK-1352
N74180N	BLOCK-1352
N74181F	BLOCK-1353
N74181N	BLOCK-1353
N74182B	BLOCK-1354
N74182F	BLOCK-1354
N74182N	BLOCK-1354
N74190B	BLOCK-1355
N74190F	BLOCK-1355

DEVICE TYPE	REPL CODE
N74190N	BLOCK-1355
N74191B	BLOCK-1356
N74191F	BLOCK-1356
N74191N	BLOCK-1356
N74192A	BLOCK-1357
N74192B	BLOCK-1357
N74192F	BLOCK-1357
N74192N	BLOCK-1357
N74192P	BLOCK-1357
N74193A	BLOCK-1358
N74193B	BLOCK-1358
N74193F	BLOCK-1358
N74193N	BLOCK-1358
N74195B	BLOCK-1359
N74195F	BLOCK-1359
N74195N	BLOCK-1359
N74196A	BLOCK-1360
N74198F	BLOCK-1362
N74198N	BLOCK-1362
N74199F	BLOCK-1363
N74199N	BLOCK-1363
N7420A	BLOCK-1365
N7420F	BLOCK-1365
N7420N	BLOCK-1365
N7421A	BLOCK-1366
N7421F	BLOCK-1366
N7421N	BLOCK-1366
N74221F	BLOCK-1368
N74221N	BLOCK-1368
N7423F	BLOCK-1369
N7423N	BLOCK-1369
N7427A	BLOCK-1372
N7427F	BLOCK-1372
N7427N	BLOCK-1372
N7428F	BLOCK-1373
N7428N	BLOCK-1373
N7430A	BLOCK-1377
N7430F	BLOCK-1377
N7430N	BLOCK-1377
N7432A	BLOCK-1378
N7432F	BLOCK-1378
N7432N	BLOCK-1378
N7433F	BLOCK-1379
N7433N	BLOCK-1379
N74365AF	BLOCK-1380
N74365AN	BLOCK-1380
N74366AF	BLOCK-1381
N74366AN	BLOCK-1381
N74367AF	BLOCK-1381
N74367AN	BLOCK-1382
N74368AF	BLOCK-1383
N74368AN	BLOCK-1383
N7437A	BLOCK-1384
N7437F	BLOCK-1384
N7437N	BLOCK-1384
N7438A	BLOCK-1385
N7438F	BLOCK-1385
N7438N	BLOCK-1385
N7439A	BLOCK-1386
N7439F	BLOCK-1386
N7439N	BLOCK-1386
N7440A	BLOCK-1390
N7440F	BLOCK-1390
N7440N	BLOCK-1390
N7441B	BLOCK-1391
N7441F	BLOCK-1391
N7442A	BLOCK-1392
N7442BA	BLOCK-1392
N7442BF	BLOCK-1392
N7442F	BLOCK-1392
N7442N	BLOCK-1392
N7443A	BLOCK-1393
N7443F	BLOCK-1393
N7443N	BLOCK-1393
N7444B	BLOCK-1394
N7444F	BLOCK-1394
N7444N	BLOCK-1394
N7445B	BLOCK-1395
N7445F	BLOCK-1395
N7445N	BLOCK-1395
N7446AF	BLOCK-1396
N7446AN	BLOCK-1396
N7446B	BLOCK-1396
N7447A	BLOCK-1397
N7447AF	BLOCK-1397
N7447B	BLOCK-1397
N7447F	BLOCK-1397
N7448A	BLOCK-1398
N7448B	BLOCK-1398
N7448F	BLOCK-1398
N7448N	BLOCK-1398
N7450A	BLOCK-1401
N7450F	BLOCK-1401
N7450N	BLOCK-1401

DEVICE TYPE	REPL CODE
N7451A	BLOCK-1402
N7451F	BLOCK-1402
N7451N	BLOCK-1402
N7453A	BLOCK-1403
N7453F	BLOCK-1403
N7453N	BLOCK-1403
N7454A	BLOCK-1404
N7454F	BLOCK-1404
N7454N	BLOCK-1404
N7460A	BLOCK-1406
N7460F	BLOCK-1406
N7460N	BLOCK-1406
N7470A	BLOCK-1408
N7470F	BLOCK-1408
N7470N	BLOCK-1408
N7472A	BLOCK-1409
N7472F	BLOCK-1409
N7472N	BLOCK-1409
N7473	BLOCK-1410
N7473A	BLOCK-1410
N7473F	BLOCK-1410
N7473N	BLOCK-1410
N7474A	BLOCK-1411
N7474F	BLOCK-1411
N7474N	BLOCK-1411
N7475A	BLOCK-1412
N7475B	BLOCK-1412
N7475F	BLOCK-1412
N7475N	BLOCK-1412
N7476A	BLOCK-1413
N7476B	BLOCK-1413
N7476F	BLOCK-1413
N7476N	BLOCK-1413
N7480A	BLOCK-1414
N7480F	BLOCK-1414
N7480N	BLOCK-1414
N7483B	BLOCK-1417
N7483F	BLOCK-1417
N7483N	BLOCK-1417
N7485A	BLOCK-1418
N7485B	BLOCK-1418
N7485F	BLOCK-1418
N7485N	BLOCK-1418
N7486A	BLOCK-1419
N7486F	BLOCK-1419
N7486N	BLOCK-1419
N7489B	BLOCK-1420
N7490A	BLOCK-1423
N7490F	BLOCK-1423
N7490N	BLOCK-1423
N7491A	BLOCK-1424
N7491AF	BLOCK-1424
N7491AN	BLOCK-1424
N7491F	BLOCK-1424
N7492A	BLOCK-1425
N7492F	BLOCK-1425
N7492N	BLOCK-1425
N7493A	BLOCK-1426
N7493F	BLOCK-1426
N7493N	BLOCK-1426
N7494B	BLOCK-1427
N7494F	BLOCK-1427
N7494N	BLOCK-1427
N7495A	BLOCK-1428
N7495AF	BLOCK-1428
N7495AN	BLOCK-1428
N7495F	BLOCK-1428
N7495N	BLOCK-1428
N7496B	BLOCK-1429
N7496F	BLOCK-1429
N7496N	BLOCK-1429
N74H00A	BLOCK-1471
N74H00F	BLOCK-1471
N74H00N	BLOCK-1471
N74H01A	BLOCK-1472
N74H01F	BLOCK-1472
N74H01N	BLOCK-1472
N74H04A	BLOCK-1473
N74H04F	BLOCK-1473
N74H04N	BLOCK-1473
N74H05A	BLOCK-1474
N74H05F	BLOCK-1474
N74H05N	BLOCK-1474
N74H08A	BLOCK-1475
N74H08F	BLOCK-1475
N74H08N	BLOCK-1475
N74H101A	BLOCK-1477
N74H101F	BLOCK-1477
N74H101N	BLOCK-1477
N74H102A	BLOCK-1478
N74H102F	BLOCK-1478
N74H102N	BLOCK-1478
N74H103A	BLOCK-1479
N74H103F	BLOCK-1479

ORIGINAL DEVICE TYPES AND REPLACEMENT CODES 43

DEVICE TYPE	REPL CODE
N74H103N	BLOCK-1479
N74H106B	BLOCK-1480
N74H106F	BLOCK-1480
N74H106N	BLOCK-1480
N74H108A	BLOCK-1481
N74H108F	BLOCK-1481
N74H108N	BLOCK-1481
N74H10A	BLOCK-1476
N74H10F	BLOCK-1476
N74H10N	BLOCK-1476
N74H11A	BLOCK-1482
N74H11F	BLOCK-1482
N74H11N	BLOCK-1482
N74H20A	BLOCK-1484
N74H20F	BLOCK-1484
N74H20N	BLOCK-1484
N74H21A	BLOCK-1485
N74H21F	BLOCK-1485
N74H21N	BLOCK-1485
N74H22A	BLOCK-1486
N74H22F	BLOCK-1486
N74H22N	BLOCK-1486
N74H30A	BLOCK-1487
N74H30F	BLOCK-1487
N74H30N	BLOCK-1487
N74H30W	BLOCK-1487
N74H40A	BLOCK-1488
N74H40F	BLOCK-1488
N74H40N	BLOCK-1488
N74H50A	BLOCK-1489
N74H50F	BLOCK-1489
N74H50N	BLOCK-1489
N74H51A	BLOCK-1490
N74H51F	BLOCK-1490
N74H51N	BLOCK-1490
N74H52A	BLOCK-1491
N74H52F	BLOCK-1491
N74H52N	BLOCK-1491
N74H53A	BLOCK-1492
N74H53F	BLOCK-1492
N74H53N	BLOCK-1492
N74H54A	BLOCK-1493
N74H54F	BLOCK-1493
N74H54N	BLOCK-1493
N74H55A	BLOCK-1494
N74H55F	BLOCK-1494
N74H55N	BLOCK-1494
N74H60A	BLOCK-1495
N74H60F	BLOCK-1495
N74H60N	BLOCK-1495
N74H61A	BLOCK-1496
N74H61F	BLOCK-1496
N74H61N	BLOCK-1496
N74H62A	BLOCK-1497
N74H62F	BLOCK-1497
N74H62N	BLOCK-1497
N74H71A	BLOCK-1498
N74H71F	BLOCK-1498
N74H71N	BLOCK-1498
N74H72A	BLOCK-1499
N74H72F	BLOCK-1499
N74H72N	BLOCK-1499
N74H73A	BLOCK-1500
N74H73F	BLOCK-1500
N74H73N	BLOCK-1500
N74H74A	BLOCK-1501
N74H74F	BLOCK-1501
N74H74N	BLOCK-1501
N74H76B	BLOCK-1502
N74H76F	BLOCK-1502
N74H76N	BLOCK-1502
N74LS00	BLOCK-1568
N74LS00F	BLOCK-1568
N74LS00N	BLOCK-1568
N74LS01F	BLOCK-1569
N74LS01N	BLOCK-1569
N74LS02F	BLOCK-1570
N74LS02N	BLOCK-1570
N74LS03F	BLOCK-1571
N74LS03N	BLOCK-1571
N74LS04F	BLOCK-1572
N74LS04N	BLOCK-1572
N74LS05F	BLOCK-1573
N74LS05N	BLOCK-1573
N74LS08F	BLOCK-1574
N74LS08N	BLOCK-1574
N74LS09F	BLOCK-1575
N74LS09N	BLOCK-1575
N74LS107F	BLOCK-1577
N74LS107N	BLOCK-1577
N74LS109AF	BLOCK-1578
N74LS109AN	BLOCK-1578
N74LS109F	BLOCK-1578
N74LS109N	BLOCK-1578
N74LS10F	BLOCK-1576
N74LS10N	BLOCK-1576
N74LS112F	BLOCK-1580
N74LS112N	BLOCK-1580
N74LS114F	BLOCK-1581
N74LS114N	BLOCK-1581
N74LS11F	BLOCK-1579
N74LS11N	BLOCK-1579
N74LS123AF	BLOCK-1584
N74LS123AN	BLOCK-1584
N74LS125F	BLOCK-1585
N74LS125N	BLOCK-1585
N74LS126F	BLOCK-1586
N74LS126N	BLOCK-1586
N74LS12F	BLOCK-1582
N74LS12N	BLOCK-1582
N74LS132F	BLOCK-1588
N74LS132N	BLOCK-1588
N74LS136F	BLOCK-1590
N74LS136N	BLOCK-1590
N74LS138F	BLOCK-1591
N74LS138N	BLOCK-1591
N74LS139F	BLOCK-1592
N74LS139N	BLOCK-1592
N74LS13F	BLOCK-1587
N74LS13N	BLOCK-1587
N74LS145F	BLOCK-1594
N74LS145N	BLOCK-1594
N74LS14F	BLOCK-1593
N74LS14N	BLOCK-1593
N74LS151F	BLOCK-1598
N74LS151N	BLOCK-1598
N74LS153F	BLOCK-1599
N74LS153N	BLOCK-1599
N74LS155F	BLOCK-1600
N74LS155N	BLOCK-1600
N74LS156F	BLOCK-1601
N74LS156N	BLOCK-1601
N74LS157F	BLOCK-1602
N74LS157N	BLOCK-1602
N74LS158F	BLOCK-1603
N74LS158N	BLOCK-1603
N74LS15F	BLOCK-1597
N74LS15N	BLOCK-1597
N74LS160AF	BLOCK-1604
N74LS160AN	BLOCK-1604
N74LS161AF	BLOCK-1605
N74LS161AN	BLOCK-1605
N74LS162AF	BLOCK-1606
N74LS162AN	BLOCK-1606
N74LS163AF	BLOCK-1607
N74LS163AN	BLOCK-1607
N74LS164F	BLOCK-1608
N74LS164N	BLOCK-1608
N74LS168AF	BLOCK-1611
N74LS168AN	BLOCK-1611
N74LS169AF	BLOCK-1612
N74LS169AN	BLOCK-1612
N74LS170B	BLOCK-1613
N74LS170F	BLOCK-1613
N74LS170N	BLOCK-1613
N74LS174B	BLOCK-1615
N74LS174F	BLOCK-1615
N74LS174N	BLOCK-1615
N74LS175B	BLOCK-1616
N74LS175F	BLOCK-1616
N74LS175N	BLOCK-1616
N74LS181F	BLOCK-1617
N74LS181N	BLOCK-1617
N74LS190F	BLOCK-1618
N74LS190N	BLOCK-1618
N74LS191F	BLOCK-1619
N74LS191N	BLOCK-1619
N74LS192F	BLOCK-1620
N74LS192N	BLOCK-1620
N74LS193F	BLOCK-1621
N74LS193N	BLOCK-1621
N74LS194AF	BLOCK-1622
N74LS194AN	BLOCK-1622
N74LS195AF	BLOCK-1623
N74LS195AN	BLOCK-1623
N74LS196F	BLOCK-1624
N74LS196N	BLOCK-1624
N74LS197F	BLOCK-1625
N74LS197N	BLOCK-1625
N74LS20F	BLOCK-1626
N74LS20N	BLOCK-1626
N74LS21F	BLOCK-1627
N74LS21N	BLOCK-1627
N74LS221F	BLOCK-1629
N74LS221N	BLOCK-1629
N74LS22F	BLOCK-1628
N74LS22N	BLOCK-1628
N74LS240F	BLOCK-1630
N74LS240N	BLOCK-1630
N74LS241F	BLOCK-1631
N74LS241N	BLOCK-1631
N74LS242F	BLOCK-1632
N74LS242N	BLOCK-1632
N74LS243F	BLOCK-1633
N74LS243N	BLOCK-1633
N74LS244F	BLOCK-1634
N74LS244N	BLOCK-1634
N74LS245F	BLOCK-1635
N74LS245N	BLOCK-1635
N74LS251AF	BLOCK-1639
N74LS251AN	BLOCK-1639
N74LS253F	BLOCK-1640
N74LS253N	BLOCK-1640
N74LS257AF	BLOCK-1641
N74LS257AN	BLOCK-1641
N74LS258AF	BLOCK-1642
N74LS258AN	BLOCK-1642
N74LS259F	BLOCK-1643
N74LS259N	BLOCK-1643
N74LS260F	BLOCK-1645
N74LS260N	BLOCK-1645
N74LS266F	BLOCK-1646
N74LS266N	BLOCK-1646
N74LS26F	BLOCK-1644
N74LS26N	BLOCK-1644
N74LS273F	BLOCK-1648
N74LS273N	BLOCK-1648
N74LS279F	BLOCK-1649
N74LS279N	BLOCK-1649
N74LS27F	BLOCK-1647
N74LS27N	BLOCK-1647
N74LS283F	BLOCK-1652
N74LS283N	BLOCK-1652
N74LS28F	BLOCK-1650
N74LS28N	BLOCK-1650
N74LS290F	BLOCK-1653
N74LS290N	BLOCK-1653
N74LS293F	BLOCK-1654
N74LS293N	BLOCK-1654
N74LS295BF	BLOCK-1655
N74LS295BN	BLOCK-1655
N74LS298F	BLOCK-1656
N74LS298N	BLOCK-1656
N74LS299F	BLOCK-1657
N74LS299N	BLOCK-1657
N74LS30F	BLOCK-1658
N74LS30N	BLOCK-1658
N74LS32F	BLOCK-1659
N74LS32N	BLOCK-1659
N74LS33F	BLOCK-1660
N74LS33N	BLOCK-1660
N74LS352N	BLOCK-1662
N74LS353N	BLOCK-1663
N74LS363F	BLOCK-1664
N74LS363N	BLOCK-1664
N74LS364F	BLOCK-1665
N74LS364N	BLOCK-1665
N74LS365AF	BLOCK-1666
N74LS365AN	BLOCK-1666
N74LS366AN	BLOCK-1667
N74LS367AF	BLOCK-1668
N74LS367AN	BLOCK-1668
N74LS368AF	BLOCK-1669
N74LS368AN	BLOCK-1669
N74LS373F	BLOCK-1671
N74LS373N	BLOCK-1671
N74LS374F	BLOCK-1672
N74LS374N	BLOCK-1672
N74LS377F	BLOCK-1673
N74LS377N	BLOCK-1673
N74LS378F	BLOCK-1674
N74LS378N	BLOCK-1674
N74LS379F	BLOCK-1675
N74LS379N	BLOCK-1675
N74LS37F	BLOCK-1670
N74LS37N	BLOCK-1670
N74LS386F	BLOCK-1677
N74LS386N	BLOCK-1677
N74LS38F	BLOCK-1676
N74LS38N	BLOCK-1676
N74LS390F	BLOCK-1678
N74LS390N	BLOCK-1678
N74LS393F	BLOCK-1679
N74LS393N	BLOCK-1679
N74LS395AF	BLOCK-1680
N74LS395AN	BLOCK-1680
N74LS398F	BLOCK-1681
N74LS398N	BLOCK-1681
N74LS399F	BLOCK-1682
N74LS399N	BLOCK-1682
N74LS40F	BLOCK-1683
N74LS40N	BLOCK-1683
N74LS42F	BLOCK-1684
N74LS42N	BLOCK-1684
N74LS490F	BLOCK-1688
N74LS490N	BLOCK-1688
N74LS51F	BLOCK-1689
N74LS51N	BLOCK-1689
N74LS54F	BLOCK-1690
N74LS54N	BLOCK-1690
N74LS55F	BLOCK-1693
N74LS55N	BLOCK-1693
N74LS640-1F	BLOCK-1699
N74LS640-1N	BLOCK-1699
N74LS640F	BLOCK-1699
N74LS640N	BLOCK-1699
N74LS641-1F	BLOCK-1700
N74LS641-1N	BLOCK-1700
N74LS641F	BLOCK-1700
N74LS641N	BLOCK-1700
N74LS642-1N	BLOCK-1701
N74LS642F	BLOCK-1701
N74LS642N	BLOCK-1701
N74LS645-1F	BLOCK-1703
N74LS645F	BLOCK-1703
N74LS645N	BLOCK-1703
N74LS670B	BLOCK-1704
N74LS670F	BLOCK-1704
N74LS670N	BLOCK-1704
N74LS73F	BLOCK-1705
N74LS73N	BLOCK-1705
N74LS74AF	BLOCK-1706
N74LS74AN	BLOCK-1706
N74LS74F	BLOCK-1706
N74LS74N	BLOCK-1706
N74LS75F	BLOCK-1707
N74LS75N	BLOCK-1707
N74LS76F	BLOCK-1708
N74LS76N	BLOCK-1708
N74LS78F	BLOCK-1710
N74LS78N	BLOCK-1710
N74LS83AF	BLOCK-1711
N74LS83AN	BLOCK-1711
N74LS83F	BLOCK-1711
N74LS85F	BLOCK-1712
N74LS85N	BLOCK-1712
N74LS86F	BLOCK-1713
N74LS86N	BLOCK-1713
N74LS90F	BLOCK-1714
N74LS90N	BLOCK-1714
N74LS92F	BLOCK-1716
N74LS92N	BLOCK-1716
N74LS93F	BLOCK-1717
N74LS93N	BLOCK-1717
N74S00A	BLOCK-1719
N74S00F	BLOCK-1719
N74S00N	BLOCK-1719
N74S02F	BLOCK-1720
N74S02N	BLOCK-1720
N74S03A	BLOCK-1721
N74S03F	BLOCK-1721
N74S03N	BLOCK-1721
N74S04A	BLOCK-1722
N74S04F	BLOCK-1722
N74S04N	BLOCK-1722
N74S05A	BLOCK-1723
N74S05F	BLOCK-1723
N74S05N	BLOCK-1723
N74S08F	BLOCK-1724
N74S08N	BLOCK-1724
N74S09F	BLOCK-1725
N74S09N	BLOCK-1725
N74S10A	BLOCK-1726
N74S10F	BLOCK-1726
N74S10N	BLOCK-1726
N74S112B	BLOCK-1728
N74S112F	BLOCK-1728
N74S112N	BLOCK-1728
N74S113A	BLOCK-1729
N74S113F	BLOCK-1729
N74S113N	BLOCK-1729
N74S11A	BLOCK-1727
N74S11F	BLOCK-1727
N74S11N	BLOCK-1727
N74S133B	BLOCK-1731
N74S133F	BLOCK-1731
N74S133N	BLOCK-1731
N74S133W	BLOCK-1731
N74S134B	BLOCK-1732
N74S134F	BLOCK-1732
N74S134N	BLOCK-1732
N74S134W	BLOCK-1732
N74S138B	BLOCK-1733
N74S138F	BLOCK-1733
N74S138N	BLOCK-1733
N74S140A	BLOCK-1734
N74S140F	BLOCK-1734
N74S140N	BLOCK-1734
N74S151B	BLOCK-1736
N74S151F	BLOCK-1736
N74S151N	BLOCK-1742
N74S153F	BLOCK-1737
N74S153J	BLOCK-1737
N74S153N	BLOCK-1737
N74S157B	BLOCK-1738
N74S157F	BLOCK-1738
N74S157N	BLOCK-1738
N74S158B	BLOCK-1739
N74S158F	BLOCK-1739
N74S158N	BLOCK-1739
N74S15A	BLOCK-1735
N74S15F	BLOCK-1735
N74S15N	BLOCK-1735
N74S174B	BLOCK-1740
N74S174F	BLOCK-1740
N74S174N	BLOCK-1740
N74S175B	BLOCK-1741
N74S175F	BLOCK-1741
N74S175N	BLOCK-1741
N74S181F	BLOCK-1742
N74S181J	BLOCK-1742
N74S181N	BLOCK-1742
N74S181W	BLOCK-1742
N74S194B	BLOCK-1744
N74S194F	BLOCK-1744
N74S194J	BLOCK-1744
N74S194N	BLOCK-1744
N74S194W	BLOCK-1744
N74S20A	BLOCK-1745
N74S20F	BLOCK-1745
N74S20N	BLOCK-1745
N74S22A	BLOCK-1746
N74S22F	BLOCK-1746
N74S22N	BLOCK-1746
N74S251B	BLOCK-1747
N74S251F	BLOCK-1747
N74S251N	BLOCK-1747
N74S258B	BLOCK-1748
N74S258F	BLOCK-1748
N74S258J	BLOCK-1748
N74S258N	BLOCK-1748
N74S40A	BLOCK-1753
N74S40F	BLOCK-1753
N74S40N	BLOCK-1753
N74S51F	BLOCK-1761
N74S51N	BLOCK-1761
N74S551F	BLOCK-1761
N74S64A	BLOCK-1766
N74S64F	BLOCK-1766
N74S64N	BLOCK-1766
N74S65A	BLOCK-1767
N74S65F	BLOCK-1767
N74S65N	BLOCK-1767
N74S74A	BLOCK-1768
N74S74F	BLOCK-1768
N74S74N	BLOCK-1768
N74S86A	BLOCK-1769
N74S86F	BLOCK-1769
N74S86N	BLOCK-1769
N74S86W	BLOCK-1769
N8250F	BLOCK-1854
N8250N	BLOCK-1854
N8252F	BLOCK-1861
N8252N	BLOCK-1861
N8266F	BLOCK-1857
N8266N	BLOCK-1857
N8268F	BLOCK-1414
N8270F	BLOCK-1350
N8270N	BLOCK-1350
N8271N	BLOCK-1351
N8280F	BLOCK-1348
N8280N	BLOCK-1348
N8281F	BLOCK-1349
N8281N	BLOCK-1349
N8290F	BLOCK-1360
N8290N	BLOCK-1360
N8291F	BLOCK-1361
N8293F	BLOCK-1625
N8293N	BLOCK-1625
N82S123N	BLOCK-1750
N82S129N	BLOCK-1749
N82S137BN	BLOCK-1765
N8455F	BLOCK-1390
N8470F	BLOCK-1304
N8828F	BLOCK-1411
N8840F	BLOCK-1401
N8848F	BLOCK-1501
N8875F	BLOCK-1294
N8881F	BLOCK-1294
N8881N	BLOCK-1294
N8H80A	BLOCK-1471
N8H80J	BLOCK-1471
N8T22A	BLOCK-2089
N8T22N	BLOCK-1312
N8T26	BLOCK-1173
N8T26A	BLOCK-1173
N8T26AF	BLOCK-1173
N8T26AN	BLOCK-1173
N8T28F	BLOCK-1178
N8T28N	BLOCK-1178
N8T95B	BLOCK-1174
N8T95F	BLOCK-1174
N8T95N	BLOCK-1174
N8T96B	BLOCK-1174
N8T96F	BLOCK-1175
N8T96N	BLOCK-1175
N8T97	BLOCK-1176
N8T97B	BLOCK-1176
N8T97F	BLOCK-1176
N8T97N	BLOCK-1176
N8T98F	BLOCK-1177
N8T98N	BLOCK-1177
N9308F	BLOCK-1862
N9308N	BLOCK-1862
N9309F	BLOCK-1863
N9309N	BLOCK-1863
N9314F	BLOCK-1865
N9314N	BLOCK-1865
N9334N	BLOCK-1643
N9602B	BLOCK-2090
N9602N	BLOCK-2090
NA1302	BLOCK-1809
NA2002	BLOCK-221
NAM383	BLOCK-221
NB-07014	BLOCK-1269
ND-07001	BLOCK-1293
ND74LS01F	BLOCK-1569
NE532AN	BLOCK-1993
NE532AT	BLOCK-1992
NE532N	BLOCK-1993
NE532T	BLOCK-1992
NE536T	BLOCK-2051
NE542N	BLOCK-1852
NE545B	BLOCK-930
NE550A	BLOCK-1984
NE550F	BLOCK-1984
NE550N	BLOCK-1984
NE555	BLOCK-2079
NE555JG	BLOCK-2079
NE555N	BLOCK-2079
NE555P	BLOCK-2079
NE555V	BLOCK-2079
NE556	BLOCK-2145
NE5560F	BLOCK-658
NE5560N	BLOCK-658
NE556A	BLOCK-2145
NE556N	BLOCK-2145
NE558	BLOCK-1988
NE558CP	BLOCK-1988
NE558N	BLOCK-1988
NE565A	BLOCK-2174
NE566T	BLOCK-2204
NE567N	BLOCK-1868
NE592	BLOCK-1990
NE592A	BLOCK-1990
NE592D	BLOCK-1990
NE592D14	BLOCK-1991
NE592F	BLOCK-1990
NE592H	BLOCK-1989
NE592K	BLOCK-1989
NE592N	BLOCK-1990
NE592N14	BLOCK-1990
NE645(B)	BLOCK-933
NE645BN	BLOCK-933
NE646N	BLOCK-933
NIM4558D	BLOCK-1801
NJ703N	BLOCK-1212
NJM-4558S	BLOCK-516
NJM-703N	BLOCK-1212
NJM072D	BLOCK-1910
NJM082M	BLOCK-1911
NJM2201	BLOCK-80
NJM2901	BLOCK-2063
NJM2901D	BLOCK-2063
NJM2901M	BLOCK-1874
NJM2901N	BLOCK-1873
NJM2902M	BLOCK-2172
NJM2902N	BLOCK-2171
NJM2903D	BLOCK-2063
NJM2903S	BLOCK-699
NJM2904	BLOCK-1993
NJM2904D	BLOCK-1993
NJM2904M	BLOCK-1995
NJM386D	BLOCK-1851

DEVICE TYPE	REPL CODE
NJM4558C	BLOCK-1801
NJM4558D	BLOCK-1801
NJM4558D-1	BLOCK-1801
NJM4558D-K	BLOCK-1801
NJM4558DD	BLOCK-1801
NJM4558DM	BLOCK-1801
NJM4558DMA	BLOCK-1801
NJM4558DV	BLOCK-1801
NJM4558DX	BLOCK-1801
NJM4558K	BLOCK-1801
NJM4558M	BLOCK-1803
NJM4558M-TE2	BLOCK-1803
NJM4558MDA	BLOCK-1801
NJM4558S	BLOCK-516
NJM4558SD	BLOCK-516
NJM4559D	BLOCK-1801
NJM555D	BLOCK-2079
NJM58L05	BLOCK-2219
NJM703	BLOCK-1212
NJM703N	BLOCK-1212
NJM7805A	BLOCK-2087
NJM7805FA	BLOCK-2087
NJM7812A	BLOCK-2097
NJM7812FA	BLOCK-921
NJM7818FA	BLOCK-2085
NJM78L05	BLOCK-2144
NJM78L05A	BLOCK-2144
NJM78L05A(T3)	BLOCK-2144
NJM78L05AV	BLOCK-2144
NJM78L06A	BLOCK-2173
NJM78L09	BLOCK-881
NJM78L09A	BLOCK-881
NJM78L12A	BLOCK-2074
NJM78M05	BLOCK-2087
NJM78M05A	BLOCK-2087
NJM78M05FA	BLOCK-2087
NJM78M09FA	BLOCK-918
NJM78M12	BLOCK-2097
NJM78M12A	BLOCK-2097
NJM78M18A	BLOCK-2085
NJM7906A	BLOCK-2094
NJM79L05A	BLOCK-893
NJM79L12A	BLOCK-882
NJM79M05A	BLOCK-2091
NJM79M06A	BLOCK-2094
NJM79M12A	BLOCK-2108
NMC2147H-3	BLOCK-1005
NMC6508J-5	BLOCK-1160
NMC6508J-9	BLOCK-1160
NMC6508N-5	BLOCK-1160
NMC6508N-9	BLOCK-1160
NP4510	BLOCK-244
NPC544	BLOCK-115
NPC7629	BLOCK-156
NPC76315	BLOCK-222
NPC76365	BLOCK-244
NS142	BLOCK-1269
NS746	BLOCK-1270
NT826AB	BLOCK-1173
NTC-21	BLOCK-1269
NTE1000	BLOCK-1
NTE1002	BLOCK-3
NTE1003	BLOCK-4
NTE1004	BLOCK-5
NTE1005	BLOCK-6
NTE1006	BLOCK-7
NTE1009	BLOCK-9
NTE1010	BLOCK-10
NTE1011	BLOCK-11
NTE1012	BLOCK-12
NTE1013	BLOCK-13
NTE1014	BLOCK-14
NTE1015	BLOCK-15
NTE1016	BLOCK-16
NTE1017	BLOCK-17
NTE1018	BLOCK-18
NTE1019	BLOCK-19
NTE1022	BLOCK-21
NTE1023	BLOCK-22
NTE1024	BLOCK-23
NTE1025	BLOCK-24
NTE1027	BLOCK-26
NTE1028	BLOCK-27
NTE1029	BLOCK-28
NTE1030	BLOCK-29
NTE1031	BLOCK-30
NTE1032	BLOCK-31
NTE1033	BLOCK-32
NTE1034	BLOCK-33
NTE1035	BLOCK-34
NTE1036	BLOCK-35
NTE1037	BLOCK-36
NTE1038	BLOCK-37
NTE1039	BLOCK-38
NTE1040	BLOCK-39
NTE1041	BLOCK-40
NTE1042	BLOCK-41
NTE1043	BLOCK-42
NTE1045	BLOCK-43
NTE1046	BLOCK-44
NTE1047	BLOCK-45
NTE1048	BLOCK-46
NTE1049	BLOCK-47
NTE1050	BLOCK-48
NTE1051	BLOCK-49
NTE1052	BLOCK-50
NTE1053	BLOCK-51
NTE1054	BLOCK-52
NTE1055	BLOCK-53
NTE1056	BLOCK-54
NTE1057	BLOCK-55
NTE1058	BLOCK-56
NTE1060	BLOCK-58
NTE1061	BLOCK-59
NTE1062	BLOCK-60
NTE1064	BLOCK-62
NTE1065	BLOCK-63
NTE1066	BLOCK-64
NTE1067	BLOCK-65
NTE1068	BLOCK-66
NTE1069	BLOCK-67
NTE1070	BLOCK-68
NTE1071	BLOCK-69
NTE1072	BLOCK-70
NTE1073	BLOCK-71
NTE1074	BLOCK-72
NTE1075A	BLOCK-73
NTE1078	BLOCK-76
NTE1079	BLOCK-77
NTE1080	BLOCK-78
NTE1081A	BLOCK-79
NTE1082	BLOCK-80
NTE1083	BLOCK-81
NTE1084	BLOCK-82
NTE1085	BLOCK-83
NTE1086	BLOCK-84
NTE1087	BLOCK-85
NTE1089	BLOCK-87
NTE1090	BLOCK-88
NTE1091	BLOCK-89
NTE1092	BLOCK-90
NTE1093	BLOCK-91
NTE1094	BLOCK-92
NTE1096	BLOCK-94
NTE1097	BLOCK-95
NTE1098	BLOCK-96
NTE1099	BLOCK-97
NTE1100	BLOCK-98
NTE1101	BLOCK-99
NTE1102	BLOCK-100
NTE1103	BLOCK-101
NTE1104	BLOCK-102
NTE1105	BLOCK-103
NTE1106	BLOCK-104
NTE1107	BLOCK-105
NTE1108	BLOCK-106
NTE1109	BLOCK-107
NTE1110	BLOCK-108
NTE1112	BLOCK-110
NTE1115	BLOCK-112
NTE1115A	BLOCK-113
NTE1116	BLOCK-114
NTE1117	BLOCK-115
NTE1119	BLOCK-117
NTE1120	BLOCK-118
NTE1121	BLOCK-119
NTE1122	BLOCK-120
NTE1123	BLOCK-121
NTE1124	BLOCK-122
NTE1126	BLOCK-124
NTE1127	BLOCK-125
NTE1128	BLOCK-126
NTE1130	BLOCK-127
NTE1131	BLOCK-128
NTE1132	BLOCK-129
NTE1133	BLOCK-130
NTE1134	BLOCK-131
NTE1135	BLOCK-132
NTE1137	BLOCK-134
NTE1139	BLOCK-135
NTE1140	BLOCK-136
NTE1141	BLOCK-137
NTE1142	BLOCK-138
NTE1148	BLOCK-139
NTE1149	BLOCK-140
NTE1150	BLOCK-141
NTE1152	BLOCK-142
NTE1153	BLOCK-143
NTE1154	BLOCK-144
NTE1155	BLOCK-145
NTE1156	BLOCK-146
NTE1158	BLOCK-147
NTE1159	BLOCK-148
NTE1160	BLOCK-149
NTE1161	BLOCK-150
NTE1162	BLOCK-151
NTE1163	BLOCK-152
NTE1164	BLOCK-153
NTE1165	BLOCK-154
NTE1166	BLOCK-155
NTE1167	BLOCK-156
NTE1168	BLOCK-157
NTE1169	BLOCK-158
NTE1170	BLOCK-159
NTE1171	BLOCK-160
NTE1172	BLOCK-161
NTE1173	BLOCK-162
NTE1174	BLOCK-163
NTE1175	BLOCK-164
NTE1176	BLOCK-165
NTE1177	BLOCK-166
NTE1178	BLOCK-167
NTE1179	BLOCK-168
NTE1180	BLOCK-169
NTE1181	BLOCK-170
NTE1183	BLOCK-172
NTE1184	BLOCK-173
NTE1185	BLOCK-174
NTE1186	BLOCK-175
NTE1187	BLOCK-176
NTE1188	BLOCK-177
NTE1189	BLOCK-178
NTE1191	BLOCK-179
NTE1192	BLOCK-180
NTE1193	BLOCK-181
NTE1194	BLOCK-182
NTE1195	BLOCK-183
NTE1196	BLOCK-184
NTE1197	BLOCK-185
NTE1198	BLOCK-186
NTE1199	BLOCK-187
NTE1200	BLOCK-188
NTE1201	BLOCK-189
NTE1202	BLOCK-190
NTE1203	BLOCK-191
NTE1204	BLOCK-192
NTE1205	BLOCK-193
NTE1206	BLOCK-194
NTE1209	BLOCK-197
NTE1210	BLOCK-198
NTE1211	BLOCK-199
NTE1212	BLOCK-200
NTE1213	BLOCK-201
NTE1214	BLOCK-202
NTE1215	BLOCK-203
NTE1216	BLOCK-204
NTE1217	BLOCK-205
NTE1218	BLOCK-206
NTE1219	BLOCK-207
NTE1221	BLOCK-209
NTE1223	BLOCK-211
NTE1224	BLOCK-212
NTE1226	BLOCK-214
NTE1227	BLOCK-215
NTE1228	BLOCK-216
NTE1229	BLOCK-217
NTE1230	BLOCK-218
NTE1231	BLOCK-219
NTE1231A	BLOCK-220
NTE1232	BLOCK-221
NTE1233	BLOCK-222
NTE1234	BLOCK-223
NTE1235	BLOCK-224
NTE1236	BLOCK-225
NTE1237	BLOCK-226
NTE1239	BLOCK-228
NTE1240	BLOCK-229
NTE1241	BLOCK-230
NTE1242	BLOCK-231
NTE1243	BLOCK-232
NTE1245	BLOCK-234
NTE1246	BLOCK-235
NTE1247	BLOCK-236
NTE1248	BLOCK-237
NTE1249	BLOCK-238
NTE1250	BLOCK-239
NTE1251	BLOCK-240
NTE1252	BLOCK-241
NTE1254	BLOCK-243
NTE1255	BLOCK-244
NTE1256	BLOCK-245
NTE1257	BLOCK-246
NTE1258	BLOCK-247
NTE1261	BLOCK-248
NTE1262	BLOCK-249
NTE1263	BLOCK-250
NTE1264	BLOCK-251
NTE1265	BLOCK-252
NTE1266	BLOCK-253
NTE1267	BLOCK-254
NTE1268	BLOCK-255
NTE1269	BLOCK-256
NTE1271	BLOCK-258
NTE1272	BLOCK-259
NTE1273	BLOCK-260
NTE1278	BLOCK-265
NTE1281	BLOCK-268
NTE1282	BLOCK-269
NTE1283	BLOCK-270
NTE1284	BLOCK-271
NTE1285	BLOCK-272
NTE1286	BLOCK-273
NTE1287	BLOCK-274
NTE1288	BLOCK-275
NTE1289	BLOCK-276
NTE1290	BLOCK-277
NTE1291	BLOCK-278
NTE1292	BLOCK-279
NTE1293	BLOCK-280
NTE1294	BLOCK-281
NTE1295	BLOCK-282
NTE1296	BLOCK-283
NTE1297	BLOCK-284
NTE1298	BLOCK-285
NTE1299	BLOCK-286
NTE1300	BLOCK-287
NTE1301	BLOCK-288
NTE1302	BLOCK-289
NTE1303	BLOCK-290
NTE1304	BLOCK-291
NTE1305	BLOCK-292
NTE1306	BLOCK-293
NTE1307	BLOCK-294
NTE1308	BLOCK-295
NTE1309	BLOCK-296
NTE1310	BLOCK-297
NTE1311	BLOCK-298
NTE1312	BLOCK-299
NTE1313	BLOCK-300
NTE1320	BLOCK-307
NTE1321	BLOCK-308
NTE1322	BLOCK-309
NTE1323	BLOCK-310
NTE1324	BLOCK-311
NTE1325	BLOCK-312
NTE1326	BLOCK-313
NTE1327	BLOCK-314
NTE1328	BLOCK-315
NTE1329	BLOCK-316
NTE1330	BLOCK-317
NTE1331	BLOCK-318
NTE1332	BLOCK-319
NTE1333	BLOCK-320
NTE1334	BLOCK-321
NTE1335	BLOCK-322
NTE1336	BLOCK-323
NTE1337	BLOCK-324
NTE1338	BLOCK-325
NTE1339	BLOCK-326
NTE1340	BLOCK-327
NTE1341	BLOCK-328
NTE1342	BLOCK-329
NTE1343	BLOCK-330
NTE1344	BLOCK-331
NTE1345	BLOCK-332
NTE1346	BLOCK-333
NTE1347	BLOCK-334
NTE1348	BLOCK-335
NTE1349	BLOCK-336
NTE1350	BLOCK-337
NTE1351	BLOCK-338
NTE1352	BLOCK-339
NTE1354	BLOCK-341
NTE1355	BLOCK-342
NTE1356	BLOCK-343
NTE1357	BLOCK-344
NTE1358	BLOCK-345
NTE1359	BLOCK-346
NTE1361	BLOCK-348
NTE1362	BLOCK-349
NTE1363	BLOCK-350
NTE1364	BLOCK-351
NTE1365	BLOCK-352
NTE1366	BLOCK-353
NTE1367	BLOCK-354
NTE1368	BLOCK-355
NTE1369	BLOCK-356
NTE1370	BLOCK-357
NTE1371	BLOCK-358
NTE1372	BLOCK-359
NTE1373	BLOCK-360
NTE1374	BLOCK-361
NTE1375	BLOCK-362
NTE1376	BLOCK-363
NTE1377	BLOCK-364
NTE1378	BLOCK-365
NTE1379	BLOCK-366
NTE1380	BLOCK-367
NTE1381	BLOCK-368
NTE1382	BLOCK-369
NTE1383	BLOCK-370
NTE1384	BLOCK-371
NTE1385	BLOCK-372
NTE1386	BLOCK-373
NTE1387	BLOCK-374
NTE1388	BLOCK-375
NTE1389	BLOCK-376
NTE1390	BLOCK-377
NTE1391	BLOCK-378
NTE1392	BLOCK-379
NTE1393	BLOCK-380
NTE1394	BLOCK-381
NTE1395	BLOCK-382
NTE1396	BLOCK-383
NTE1397	BLOCK-384
NTE1398	BLOCK-385
NTE1399	BLOCK-386
NTE1400	BLOCK-387
NTE1401	BLOCK-388
NTE1402	BLOCK-389
NTE1403	BLOCK-390
NTE1404	BLOCK-391
NTE1405	BLOCK-392
NTE1406	BLOCK-393
NTE1407	BLOCK-394
NTE1408	BLOCK-395
NTE1409	BLOCK-396
NTE1409N	BLOCK-396
NTE1410	BLOCK-398
NTE1411	BLOCK-399
NTE1412	BLOCK-400
NTE1413	BLOCK-401
NTE1414	BLOCK-402
NTE1415	BLOCK-403
NTE1416	BLOCK-404
NTE1417	BLOCK-405
NTE1418	BLOCK-406
NTE1419	BLOCK-407
NTE1420	BLOCK-408
NTE1421	BLOCK-409
NTE1422	BLOCK-410
NTE1423	BLOCK-411
NTE1424	BLOCK-412
NTE1425	BLOCK-413
NTE1426	BLOCK-414
NTE1427	BLOCK-415
NTE1428	BLOCK-416
NTE1429	BLOCK-417
NTE1430	BLOCK-418
NTE1431	BLOCK-419
NTE1432	BLOCK-420
NTE1433	BLOCK-421
NTE1434	BLOCK-422
NTE1435	BLOCK-423
NTE1436	BLOCK-424
NTE1437	BLOCK-425
NTE1438	BLOCK-426
NTE1439	BLOCK-427
NTE1440	BLOCK-428
NTE1441	BLOCK-429
NTE1442	BLOCK-430
NTE1443	BLOCK-431
NTE1444	BLOCK-432
NTE1445	BLOCK-433
NTE1446	BLOCK-434
NTE1447	BLOCK-435
NTE1448	BLOCK-436
NTE1449	BLOCK-437
NTE1450	BLOCK-438
NTE1451	BLOCK-439
NTE1452	BLOCK-440
NTE1453	BLOCK-441
NTE1454	BLOCK-442
NTE1455	BLOCK-443
NTE1457	BLOCK-445
NTE1458	BLOCK-446
NTE1459	BLOCK-447
NTE1460	BLOCK-448
NTE1461	BLOCK-449
NTE1462	BLOCK-450
NTE1463	BLOCK-451
NTE1464	BLOCK-452
NTE1465	BLOCK-453
NTE1466	BLOCK-454
NTE1467	BLOCK-455
NTE1468	BLOCK-456
NTE1469	BLOCK-457
NTE1470	BLOCK-458
NTE1471	BLOCK-459
NTE1472	BLOCK-460
NTE1473	BLOCK-461
NTE1474	BLOCK-462
NTE1475	BLOCK-463
NTE1476	BLOCK-464
NTE1477	BLOCK-465
NTE1478	BLOCK-466
NTE1479	BLOCK-467
NTE1480	BLOCK-468
NTE1481	BLOCK-469
NTE1482	BLOCK-470
NTE1483	BLOCK-471
NTE1484	BLOCK-472
NTE1485	BLOCK-473
NTE1486	BLOCK-474
NTE1487	BLOCK-475
NTE1488	BLOCK-476
NTE1489	BLOCK-477
NTE1490	BLOCK-478
NTE1491	BLOCK-479
NTE1492	BLOCK-480
NTE1493	BLOCK-481
NTE1494	BLOCK-482
NTE1495	BLOCK-483
NTE1496	BLOCK-484
NTE1497	BLOCK-485
NTE1498	BLOCK-486
NTE1499	BLOCK-487
NTE1500	BLOCK-488
NTE15005	BLOCK-768
NTE1501	BLOCK-489
NTE1502	BLOCK-490
NTE1503	BLOCK-491
NTE1504	BLOCK-492
NTE15040	BLOCK-818
NTE15041	BLOCK-876
NTE15042	BLOCK-1200
NTE15045	BLOCK-853
NTE15046	BLOCK-841
NTE1505	BLOCK-493
NTE1507	BLOCK-495
NTE1508	BLOCK-496
NTE1509	BLOCK-497
NTE1510	BLOCK-498
NTE1511	BLOCK-499
NTE1512	BLOCK-500
NTE1513	BLOCK-501
NTE1514	BLOCK-502
NTE1515	BLOCK-503
NTE1516	BLOCK-504
NTE1517	BLOCK-505
NTE1518	BLOCK-506
NTE1519	BLOCK-507
NTE1520	BLOCK-508
NTE1521	BLOCK-509
NTE1522	BLOCK-510
NTE1523	BLOCK-511
NTE1525	BLOCK-512
NTE1526	BLOCK-513
NTE1527	BLOCK-514
NTE1528	BLOCK-515
NTE1529	BLOCK-516
NTE1530	BLOCK-517
NTE1531	BLOCK-518
NTE1532	BLOCK-519
NTE1533	BLOCK-520
NTE1534	BLOCK-521
NTE1535	BLOCK-522
NTE1536	BLOCK-523
NTE1537	BLOCK-524
NTE1538	BLOCK-525
NTE1539	BLOCK-526
NTE1540	BLOCK-527
NTE1541	BLOCK-528
NTE1542	BLOCK-529
NTE1543	BLOCK-530
NTE1544	BLOCK-531
NTE1545	BLOCK-532
NTE1546	BLOCK-533
NTE1547	BLOCK-534
NTE1548	BLOCK-535
NTE1549	BLOCK-536
NTE1550	BLOCK-537
NTE1551	BLOCK-538
NTE1552	BLOCK-539

ORIGINAL DEVICE TYPES AND REPLACEMENT CODES 45

DEVICE TYPE	REPL CODE	DEVICE TYPE	REPL CODE	DEVICE TYPE	REPL CODE	DEVICE TYPE	REPL CODE	DEVICE TYPE	REPL CODE	DEVICE TYPE	REPL CODE
NTE1553	BLOCK-540	NTE1679	BLOCK-665	NTE1791	BLOCK-772	NTE2004	BLOCK-933	NTE4017B	BLOCK-1051	NTE4547B	BLOCK-1144
NTE1554	BLOCK-541	NTE1680	BLOCK-666	NTE1792	BLOCK-773	NTE2011	BLOCK-934	NTE40182B	BLOCK-1052	NTE4553B	BLOCK-1145
NTE1556	BLOCK-543	NTE1681	BLOCK-667	NTE1794	BLOCK-775	NTE2012	BLOCK-935	NTE4018B	BLOCK-1053	NTE4555B	BLOCK-1146
NTE1557	BLOCK-544	NTE1682	BLOCK-668	NTE1795	BLOCK-776	NTE2013	BLOCK-936	NTE40192B	BLOCK-1054	NTE4556B	BLOCK-1147
NTE1558	BLOCK-545	NTE1683	BLOCK-669	NTE1796	BLOCK-777	NTE2014	BLOCK-937	NTE40193B	BLOCK-1055	NTE4558B	BLOCK-1148
NTE1559	BLOCK-546	NTE1684	BLOCK-670	NTE1797	BLOCK-778	NTE2015	BLOCK-938	NTE40194B	BLOCK-1056	NTE4566B	BLOCK-1150
NTE1560	BLOCK-547	NTE1685	BLOCK-671	NTE1798	BLOCK-779	NTE2016	BLOCK-939	NTE40195B	BLOCK-1057	NTE4569B	BLOCK-1152
NTE1561	BLOCK-548	NTE1687	BLOCK-673	NTE1799	BLOCK-780	NTE2017	BLOCK-940	NTE4019B	BLOCK-1058	NTE4583B	BLOCK-1153
NTE1562	BLOCK-549	NTE1688	BLOCK-674	NTE1800	BLOCK-781	NTE2018	BLOCK-941	NTE4020B	BLOCK-1059	NTE4585B	BLOCK-1154
NTE1563	BLOCK-550	NTE1689	BLOCK-675	NTE1801	BLOCK-782	NTE2019	BLOCK-942	NTE4021B	BLOCK-1060	NTE4597B	BLOCK-1155
NTE1565	BLOCK-552	NTE1690	BLOCK-676	NTE1803	BLOCK-784	NTE2020	BLOCK-943	NTE4022B	BLOCK-1061	NTE4598B	BLOCK-1156
NTE1566	BLOCK-553	NTE1691	BLOCK-677	NTE1804	BLOCK-785	NTE2021	BLOCK-944	NTE4023B	BLOCK-1062	NTE615	BLOCK-1157
NTE1567	BLOCK-554	NTE1692	BLOCK-678	NTE1805	BLOCK-786	NTE2022	BLOCK-945	NTE4024B	BLOCK-1063	NTE615P	BLOCK-1157
NTE1568	BLOCK-555	NTE1693	BLOCK-679	NTE1808	BLOCK-789	NTE2023	BLOCK-946	NTE4025B	BLOCK-1064	NTE6502	BLOCK-1158
NTE1569	BLOCK-556	NTE1700	BLOCK-680	NTE1809	BLOCK-790	NTE2024	BLOCK-947	NTE4026B	BLOCK-1065	NTE6507	BLOCK-1159
NTE1570	BLOCK-557	NTE1701	BLOCK-681	NTE1810	BLOCK-791	NTE2025	BLOCK-948	NTE4027B	BLOCK-1066	NTE6508	BLOCK-1160
NTE1571	BLOCK-558	NTE1702	BLOCK-682	NTE1811	BLOCK-792	NTE2026	BLOCK-949	NTE4028B	BLOCK-1067	NTE65101	BLOCK-1161
NTE1572	BLOCK-559	NTE1703	BLOCK-683	NTE1812	BLOCK-793	NTE2027	BLOCK-950	NTE4029B	BLOCK-1068	NTE6532	BLOCK-1162
NTE1573	BLOCK-560	NTE1704	BLOCK-684	NTE1813	BLOCK-794	NTE2028	BLOCK-951	NTE4030B	BLOCK-1069	NTE6800	BLOCK-1164
NTE1575	BLOCK-562	NTE1705	BLOCK-685	NTE1814	BLOCK-795	NTE2029	BLOCK-952	NTE4031B	BLOCK-1070	NTE6802	BLOCK-1165
NTE1576	BLOCK-563	NTE1707	BLOCK-687	NTE1815	BLOCK-796	NTE2030	BLOCK-953	NTE4032B	BLOCK-1071	NTE6809	BLOCK-1166
NTE1577	BLOCK-564	NTE1708	BLOCK-688	NTE1816	BLOCK-797	NTE2031	BLOCK-954	NTE4033B	BLOCK-1072	NTE6809E	BLOCK-1167
NTE1578	BLOCK-565	NTE1709	BLOCK-689	NTE1817	BLOCK-798	NTE2032	BLOCK-955	NTE4034B	BLOCK-1073	NTE6810	BLOCK-1168
NTE1579	BLOCK-566	NTE1710	BLOCK-690	NTE1818	BLOCK-799	NTE2047	BLOCK-963	NTE4035B	BLOCK-1074	NTE6821	BLOCK-1169
NTE1580	BLOCK-567	NTE1711	BLOCK-691	NTE1819	BLOCK-800	NTE2050	BLOCK-965	NTE4038B	BLOCK-1075	NTE684	BLOCK-1919
NTE1581	BLOCK-568	NTE1712	BLOCK-692	NTE1820	BLOCK-801	NTE2051	BLOCK-966	NTE4040B	BLOCK-1076	NTE6850	BLOCK-1170
NTE1582	BLOCK-569	NTE1713	BLOCK-693	NTE1822	BLOCK-803	NTE2053	BLOCK-968	NTE4042B	BLOCK-1077	NTE6860	BLOCK-1171
NTE1585	BLOCK-572	NTE1714M	BLOCK-694	NTE1823	BLOCK-804	NTE2054	BLOCK-969	NTE4043B	BLOCK-1078	NTE6875	BLOCK-1172
NTE1589	BLOCK-576	NTE1714S	BLOCK-695	NTE1824	BLOCK-805	NTE2055	BLOCK-970	NTE4044B	BLOCK-1079	NTE6880	BLOCK-1173
NTE1591	BLOCK-578	NTE1715	BLOCK-696	NTE1827	BLOCK-807	NTE2056	BLOCK-971	NTE4045B	BLOCK-1080	NTE6885	BLOCK-1174
NTE1593	BLOCK-580	NTE1716	BLOCK-697	NTE1828	BLOCK-808	NTE2060	BLOCK-973	NTE4046B	BLOCK-1081	NTE6886	BLOCK-1175
NTE1596	BLOCK-583	NTE1717	BLOCK-698	NTE1829	BLOCK-809	NTE2061	BLOCK-974	NTE4047B	BLOCK-1082	NTE6887	BLOCK-1176
NTE1597	BLOCK-584	NTE1718	BLOCK-699	NTE1830	BLOCK-810	NTE2062	BLOCK-975	NTE4048B	BLOCK-1083	NTE6888	BLOCK-1177
NTE1600	BLOCK-587	NTE1719	BLOCK-700	NTE1831	BLOCK-811	NTE2070	BLOCK-979	NTE4049	BLOCK-1084	NTE6889	BLOCK-1178
NTE1602	BLOCK-589	NTE1720	BLOCK-701	NTE1834	BLOCK-813	NTE2071	BLOCK-980	NTE4050B	BLOCK-1085	NTE700	BLOCK-1179
NTE1604	BLOCK-591	NTE1721	BLOCK-702	NTE1835	BLOCK-814	NTE2072	BLOCK-981	NTE4051B	BLOCK-1086	NTE701	BLOCK-1190
NTE1606	BLOCK-593	NTE1722	BLOCK-703	NTE1836	BLOCK-815	NTE2073	BLOCK-982	NTE4052B	BLOCK-1087	NTE7010	BLOCK-1191
NTE1607	BLOCK-594	NTE1723	BLOCK-704	NTE1838	BLOCK-817	NTE2074	BLOCK-983	NTE4053B	BLOCK-1088	NTE702	BLOCK-1201
NTE1608	BLOCK-595	NTE1724	BLOCK-705	NTE1839	BLOCK-818	NTE2075	BLOCK-984	NTE4055B	BLOCK-1089	NTE703	BLOCK-1212
NTE1609	BLOCK-596	NTE1725	BLOCK-706	NTE1840	BLOCK-819	NTE2076	BLOCK-985	NTE4056B	BLOCK-1090	NTE703A	BLOCK-1212
NTE1610	BLOCK-597	NTE1726	BLOCK-707	NTE1842	BLOCK-821	NTE2077	BLOCK-986	NTE4060B	BLOCK-1091	NTE704	BLOCK-1223
NTE1611	BLOCK-598	NTE1727	BLOCK-708	NTE1843	BLOCK-822	NTE2078	BLOCK-987	NTE4063B	BLOCK-1092	NTE705A	BLOCK-1244
NTE1612	BLOCK-599	NTE1728	BLOCK-709	NTE1844	BLOCK-823	NTE2079	BLOCK-988	NTE4066B	BLOCK-1093	NTE706	BLOCK-1245
NTE1613	BLOCK-600	NTE1729	BLOCK-710	NTE1845	BLOCK-824	NTE2080	BLOCK-989	NTE4067B	BLOCK-1094	NTE707	BLOCK-1256
NTE1615	BLOCK-602	NTE1730	BLOCK-711	NTE1847	BLOCK-826	NTE2081	BLOCK-990	NTE4068B	BLOCK-1095	NTE708	BLOCK-1262
NTE1616	BLOCK-603	NTE1731	BLOCK-712	NTE1848	BLOCK-827	NTE2082	BLOCK-991	NTE4069	BLOCK-1096	NTE709	BLOCK-1264
NTE1618	BLOCK-605	NTE1732	BLOCK-713	NTE1849	BLOCK-828	NTE2083	BLOCK-992	NTE4070B	BLOCK-1097	NTE710	BLOCK-1265
NTE1619	BLOCK-606	NTE1733	BLOCK-714	NTE1852	BLOCK-831	NTE2084	BLOCK-993	NTE4071B	BLOCK-1098	NTE711	BLOCK-1267
NTE1620	BLOCK-607	NTE1734	BLOCK-715	NTE1855	BLOCK-835	NTE2085	BLOCK-994	NTE4072B	BLOCK-1099	NTE712	BLOCK-1269
NTE1621	BLOCK-608	NTE1735	BLOCK-716	NTE1856	BLOCK-836	NTE2086	BLOCK-995	NTE4073B	BLOCK-1100	NTE713	BLOCK-1270
NTE1624	BLOCK-611	NTE1736	BLOCK-717	NTE1859	BLOCK-839	NTE2087	BLOCK-996	NTE4075B	BLOCK-1101	NTE714	BLOCK-1271
NTE1625	BLOCK-612	NTE1737	BLOCK-718	NTE1860	BLOCK-840	NTE2088	BLOCK-997	NTE4076B	BLOCK-1102	NTE715	BLOCK-1272
NTE1626	BLOCK-613	NTE1738	BLOCK-719	NTE1861	BLOCK-841	NTE2102	BLOCK-999	NTE4077B	BLOCK-1103	NTE716	BLOCK-1273
NTE1627	BLOCK-614	NTE1739	BLOCK-720	NTE1863	BLOCK-843	NTE2104	BLOCK-1000	NTE4078B	BLOCK-1104	NTE718	BLOCK-1275
NTE1628	BLOCK-615	NTE1740	BLOCK-721	NTE1872	BLOCK-852	NTE2107	BLOCK-1001	NTE4081B	BLOCK-1105	NTE720	BLOCK-1277
NTE1629	BLOCK-616	NTE1741	BLOCK-722	NTE1874	BLOCK-854	NTE2114	BLOCK-1002	NTE4082B	BLOCK-1106	NTE7214	BLOCK-1278
NTE1630	BLOCK-617	NTE1742	BLOCK-723	NTE1875	BLOCK-855	NTE2117	BLOCK-1003	NTE4085B	BLOCK-1107	NTE723	BLOCK-1279
NTE1631	BLOCK-618	NTE1743	BLOCK-724	NTE1876	BLOCK-856	NTE2128	BLOCK-1004	NTE4086B	BLOCK-1108	NTE724	BLOCK-1280
NTE1632	BLOCK-619	NTE1744	BLOCK-725	NTE1877	BLOCK-857	NTE2147	BLOCK-1005	NTE4089B	BLOCK-1109	NTE725	BLOCK-1281
NTE1633	BLOCK-620	NTE1745	BLOCK-726	NTE1879	BLOCK-859	NTE2164	BLOCK-1006	NTE4093B	BLOCK-1110	NTE726	BLOCK-1282
NTE1634	BLOCK-621	NTE1747	BLOCK-728	NTE1900	BLOCK-879	NTE2532	BLOCK-1015	NTE4095B	BLOCK-1111	NTE727	BLOCK-1283
NTE1635	BLOCK-622	NTE1748	BLOCK-729	NTE1901	BLOCK-880	NTE2708	BLOCK-1016	NTE4096B	BLOCK-1112	NTE728	BLOCK-1284
NTE1636	BLOCK-623	NTE1749	BLOCK-730	NTE1902	BLOCK-881	NTE2716	BLOCK-1017	NTE4097B	BLOCK-1113	NTE729	BLOCK-1285
NTE1637	BLOCK-624	NTE1750	BLOCK-731	NTE1903	BLOCK-882	NTE2732	BLOCK-1018	NTE4098B	BLOCK-1114	NTE730	BLOCK-1286
NTE1639	BLOCK-626	NTE1751	BLOCK-732	NTE1905	BLOCK-883	NTE2764	BLOCK-1019	NTE4099B	BLOCK-1115	NTE731	BLOCK-1287
NTE1641	BLOCK-628	NTE1752	BLOCK-733	NTE1906	BLOCK-884	NTE2800	BLOCK-1020	NTE4164	BLOCK-1006	NTE736	BLOCK-1289
NTE1648	BLOCK-634	NTE1753	BLOCK-734	NTE1907	BLOCK-885	NTE3092	BLOCK-1021	NTE4501	BLOCK-1117	NTE737	BLOCK-1290
NTE1649	BLOCK-635	NTE1754	BLOCK-735	NTE1908	BLOCK-886	NTE309K	BLOCK-1022	NTE4502B	BLOCK-1118	NTE738	BLOCK-1291
NTE1650	BLOCK-636	NTE1755	BLOCK-736	NTE1911	BLOCK-887	NTE3880	BLOCK-1025	NTE4503B	BLOCK-1035	NTE739	BLOCK-1292
NTE1651	BLOCK-637	NTE1757	BLOCK-738	NTE1912	BLOCK-888	NTE3881	BLOCK-1026	NTE4506B	BLOCK-1119	NTE7400	BLOCK-1293
NTE1654	BLOCK-640	NTE1758	BLOCK-739	NTE1913	BLOCK-889	NTE3882	BLOCK-1027	NTE4508B	BLOCK-1120	NTE7401	BLOCK-1294
NTE1655	BLOCK-641	NTE1759	BLOCK-740	NTE1914	BLOCK-890	NTE4000	BLOCK-1028	NTE4510B	BLOCK-1121	NTE7402	BLOCK-1295
NTE1656	BLOCK-642	NTE1761	BLOCK-742	NTE1915	BLOCK-891	NTE4001B	BLOCK-1029	NTE4511B	BLOCK-1122	NTE7403	BLOCK-1296
NTE1657	BLOCK-643	NTE1762	BLOCK-743	NTE1916	BLOCK-892	NTE4002B	BLOCK-1030	NTE4512B	BLOCK-1123	NTE7404	BLOCK-1297
NTE1658	BLOCK-644	NTE1764	BLOCK-745	NTE1917	BLOCK-893	NTE4006B	BLOCK-1031	NTE4513B	BLOCK-1124	NTE7405	BLOCK-1298
NTE1659	BLOCK-645	NTE1771	BLOCK-752	NTE1918	BLOCK-894	NTE4007	BLOCK-1032	NTE4514B	BLOCK-1125	NTE7406	BLOCK-1299
NTE1660	BLOCK-646	NTE1772	BLOCK-753	NTE1919	BLOCK-895	NTE40085B	BLOCK-1033	NTE4515B	BLOCK-1126	NTE7407	BLOCK-1300
NTE1661	BLOCK-647	NTE1773	BLOCK-754	NTE1920	BLOCK-896	NTE4008B	BLOCK-1034	NTE4516B	BLOCK-1127	NTE7408	BLOCK-1301
NTE1662	BLOCK-648	NTE1774	BLOCK-755	NTE1923	BLOCK-897	NTE40097B	BLOCK-1035	NTE4517B	BLOCK-1128	NTE7409	BLOCK-1302
NTE1663	BLOCK-649	NTE1775	BLOCK-756	NTE1924	BLOCK-898	NTE40098B	BLOCK-1036	NTE4518B	BLOCK-1129	NTE740A	BLOCK-1303
NTE1664	BLOCK-650	NTE1776	BLOCK-757	NTE1925	BLOCK-899	NTE40100B	BLOCK-1037	NTE4520B	BLOCK-1130	NTE7410	BLOCK-1304
NTE1666	BLOCK-652	NTE1777	BLOCK-758	NTE1926	BLOCK-900	NTE40106B	BLOCK-1038	NTE4521B	BLOCK-1131	NTE74107	BLOCK-1305
NTE1667	BLOCK-653	NTE1778	BLOCK-759	NTE1927	BLOCK-901	NTE4011B	BLOCK-1039	NTE4522B	BLOCK-1132	NTE74109	BLOCK-1306
NTE1668	BLOCK-654	NTE1779	BLOCK-760	NTE1928	BLOCK-902	NTE4012B	BLOCK-1040	NTE4526B	BLOCK-1133	NTE7411	BLOCK-1307
NTE1669	BLOCK-655	NTE1780	BLOCK-761	NTE1930	BLOCK-903	NTE4013B	BLOCK-1041	NTE4527B	BLOCK-1134	NTE74110	BLOCK-1308
NTE1670	BLOCK-656	NTE1781	BLOCK-762	NTE1932	BLOCK-904	NTE4014B	BLOCK-1042	NTE4528B	BLOCK-1135	NTE74111	BLOCK-1309
NTE1671	BLOCK-657	NTE1782	BLOCK-763	NTE1934	BLOCK-905	NTE4015B	BLOCK-1043	NTE4529B	BLOCK-1136	NTE7412	BLOCK-1310
NTE1672	BLOCK-658	NTE1783	BLOCK-764	NTE1934X	BLOCK-906	NTE40160B	BLOCK-1044	NTE4531B	BLOCK-1137	NTE74121	BLOCK-1311
NTE1673	BLOCK-659	NTE1785	BLOCK-766	NTE1936	BLOCK-907	NTE40161B	BLOCK-1045	NTE4532B	BLOCK-1138	NTE74122	BLOCK-1312
NTE1674	BLOCK-660	NTE1786	BLOCK-767	NTE1938	BLOCK-908	NTE40162B	BLOCK-1046	NTE4536B	BLOCK-1139	NTE74123	BLOCK-1313
NTE1675	BLOCK-661	NTE1787	BLOCK-768	NTE1940	BLOCK-909	NTE40163B	BLOCK-1047	NTE4538B	BLOCK-1140	NTE74125	BLOCK-1314
NTE1676	BLOCK-662	NTE1788	BLOCK-769	NTE2000	BLOCK-929	NTE4016B	BLOCK-1048	NTE4539B	BLOCK-1141	NTE74126	BLOCK-1315
NTE1677	BLOCK-663	NTE1789	BLOCK-770	NTE2001	BLOCK-930	NTE40174B	BLOCK-1049	NTE4541B	BLOCK-1142	NTE74128	BLOCK-1316
NTE1678	BLOCK-664	NTE1790	BLOCK-771	NTE2003	BLOCK-932	NTE40175B	BLOCK-1050	NTE4543B	BLOCK-1143	NTE7413	BLOCK-1317

DEVICE TYPE	REPL CODE
NTE74132	BLOCK-1318
NTE74136	BLOCK-1319
NTE7414	BLOCK-1320
NTE74141	BLOCK-1321
NTE74142	BLOCK-1322
NTE74143	BLOCK-1323
NTE74144	BLOCK-1324
NTE74145	BLOCK-1325
NTE74147	BLOCK-1326
NTE74150	BLOCK-1327
NTE74151	BLOCK-1328
NTE74152	BLOCK-1329
NTE74153	BLOCK-1330
NTE74154	BLOCK-1331
NTE74155	BLOCK-1332
NTE74156	BLOCK-1333
NTE74157	BLOCK-1334
NTE74158	BLOCK-1335
NTE7416	BLOCK-1336
NTE74160	BLOCK-1337
NTE74161	BLOCK-1338
NTE74162	BLOCK-1339
NTE74163	BLOCK-1340
NTE74164	BLOCK-1341
NTE74165	BLOCK-1342
NTE7417	BLOCK-1343
NTE74170	BLOCK-1344
NTE74173	BLOCK-1345
NTE74174	BLOCK-1346
NTE74175	BLOCK-1347
NTE74176	BLOCK-1348
NTE74177	BLOCK-1349
NTE74178	BLOCK-1350
NTE74179	BLOCK-1351
NTE74180	BLOCK-1352
NTE74181	BLOCK-1353
NTE74182	BLOCK-1354
NTE74190	BLOCK-1355
NTE74191	BLOCK-1356
NTE74192	BLOCK-1357
NTE74193	BLOCK-1358
NTE74195	BLOCK-1359
NTE74196	BLOCK-1360
NTE74197	BLOCK-1361
NTE74198	BLOCK-1362
NTE74199	BLOCK-1363
NTE742	BLOCK-1364
NTE7420	BLOCK-1365
NTE7421	BLOCK-1366
NTE7422	BLOCK-1367
NTE74221	BLOCK-1368
NTE7423	BLOCK-1369
NTE74249	BLOCK-1370
NTE74251	BLOCK-1371
NTE7427	BLOCK-1372
NTE7428	BLOCK-1373
NTE74290	BLOCK-1374
NTE74293	BLOCK-1375
NTE743	BLOCK-1376
NTE7430	BLOCK-1377
NTE7432	BLOCK-1378
NTE7433	BLOCK-1379
NTE74365	BLOCK-1380
NTE74366	BLOCK-1381
NTE74367	BLOCK-1382
NTE74368	BLOCK-1383
NTE7437	BLOCK-1384
NTE7438	BLOCK-1385
NTE7439	BLOCK-1386
NTE74390	BLOCK-1387
NTE74393	BLOCK-1388
NTE744	BLOCK-1389
NTE7440	BLOCK-1390
NTE7441	BLOCK-1391
NTE7442	BLOCK-1392
NTE7443	BLOCK-1393
NTE7444	BLOCK-1394
NTE7445	BLOCK-1395
NTE7446	BLOCK-1396
NTE7447	BLOCK-1397
NTE7448	BLOCK-1398
NTE74490	BLOCK-1399
NTE745	BLOCK-1400
NTE7450	BLOCK-1401
NTE7451	BLOCK-1402
NTE7453	BLOCK-1403
NTE7454	BLOCK-1404
NTE746	BLOCK-1405
NTE7460	BLOCK-1406
NTE747	BLOCK-1407
NTE7470	BLOCK-1408
NTE7472	BLOCK-1409
NTE7473	BLOCK-1410
NTE7474	BLOCK-1411
NTE7475	BLOCK-1412
NTE7476	BLOCK-1413
NTE748	BLOCK-1421
NTE7480	BLOCK-1414
NTE7481	BLOCK-1415
NTE7482	BLOCK-1416
NTE7483	BLOCK-1417
NTE7485	BLOCK-1418
NTE7486	BLOCK-1419
NTE7489	BLOCK-1420
NTE749	BLOCK-1422
NTE7490	BLOCK-1423
NTE7491	BLOCK-1424
NTE7492	BLOCK-1425
NTE7493A	BLOCK-1426
NTE7494	BLOCK-1427
NTE7495	BLOCK-1428
NTE7496	BLOCK-1429
NTE7497	BLOCK-1430
NTE74C00	BLOCK-1431
NTE74C02	BLOCK-1432
NTE74C04	BLOCK-1433
NTE74C08	BLOCK-1434
NTE74C10	BLOCK-1435
NTE74C107	BLOCK-1436
NTE74C14	BLOCK-1437
NTE74C151	BLOCK-1438
NTE74C154	BLOCK-1439
NTE74C157	BLOCK-1440
NTE74C161	BLOCK-1441
NTE74C164	BLOCK-1442
NTE74C173	BLOCK-1443
NTE74C174	BLOCK-1444
NTE74C175	BLOCK-1445
NTE74C192	BLOCK-1446
NTE74C193	BLOCK-1447
NTE74C20	BLOCK-1448
NTE74C221	BLOCK-1449
NTE74C240	BLOCK-1450
NTE74C244	BLOCK-1451
NTE74C30	BLOCK-1452
NTE74C32	BLOCK-1453
NTE74C373	BLOCK-1454
NTE74C374	BLOCK-1455
NTE74C48	BLOCK-1456
NTE74C73	BLOCK-1457
NTE74C74	BLOCK-1458
NTE74C76	BLOCK-1459
NTE74C85	BLOCK-1460
NTE74C90	BLOCK-1461
NTE74C901	BLOCK-1462
NTE74C902	BLOCK-1463
NTE74C903	BLOCK-1464
NTE74C904	BLOCK-1465
NTE74C922	BLOCK-1466
NTE74C923	BLOCK-1467
NTE74C925	BLOCK-1468
NTE74C93	BLOCK-1469
NTE74C95	BLOCK-1470
NTE74H00	BLOCK-1471
NTE74H01	BLOCK-1472
NTE74H04	BLOCK-1473
NTE74H05	BLOCK-1474
NTE74H08	BLOCK-1475
NTE74H10	BLOCK-1476
NTE74H101	BLOCK-1477
NTE74H102	BLOCK-1478
NTE74H103	BLOCK-1479
NTE74H106	BLOCK-1480
NTE74H108	BLOCK-1481
NTE74H11	BLOCK-1482
NTE74H183	BLOCK-1483
NTE74H20	BLOCK-1484
NTE74H21	BLOCK-1485
NTE74H22	BLOCK-1486
NTE74H30	BLOCK-1487
NTE74H40	BLOCK-1488
NTE74H50	BLOCK-1489
NTE74H51	BLOCK-1490
NTE74H52	BLOCK-1491
NTE74H53	BLOCK-1492
NTE74H54	BLOCK-1493
NTE74H55	BLOCK-1494
NTE74H60	BLOCK-1495
NTE74H61	BLOCK-1496
NTE74H62	BLOCK-1497
NTE74H71	BLOCK-1498
NTE74H72	BLOCK-1499
NTE74H73	BLOCK-1500
NTE74H74	BLOCK-1501
NTE74H76	BLOCK-1502
NTE74H78	BLOCK-1503
NTE74H87	BLOCK-1505
NTE74L93	BLOCK-1567
NTE74LS00	BLOCK-1568
NTE74LS01	BLOCK-1569
NTE74LS02	BLOCK-1570
NTE74LS03	BLOCK-1571
NTE74LS04	BLOCK-1572
NTE74LS05	BLOCK-1573
NTE74LS08	BLOCK-1574
NTE74LS09	BLOCK-1575
NTE74LS10	BLOCK-1576
NTE74LS107	BLOCK-1577
NTE74LS109A	BLOCK-1578
NTE74LS11	BLOCK-1579
NTE74LS112A	BLOCK-1580
NTE74LS114	BLOCK-1581
NTE74LS12	BLOCK-1582
NTE74LS122	BLOCK-1583
NTE74LS123	BLOCK-1584
NTE74LS125A	BLOCK-1585
NTE74LS126A	BLOCK-1586
NTE74LS13	BLOCK-1587
NTE74LS132	BLOCK-1588
NTE74LS133	BLOCK-1589
NTE74LS136	BLOCK-1590
NTE74LS138	BLOCK-1591
NTE74LS139	BLOCK-1592
NTE74LS14	BLOCK-1593
NTE74LS145	BLOCK-1594
NTE74LS147	BLOCK-1595
NTE74LS148	BLOCK-1596
NTE74LS15	BLOCK-1597
NTE74LS151	BLOCK-1598
NTE74LS153	BLOCK-1599
NTE74LS155	BLOCK-1600
NTE74LS156	BLOCK-1601
NTE74LS157	BLOCK-1602
NTE74LS158	BLOCK-1603
NTE74LS160A	BLOCK-1604
NTE74LS161A	BLOCK-1605
NTE74LS162A	BLOCK-1606
NTE74LS163A	BLOCK-1607
NTE74LS164	BLOCK-1608
NTE74LS165	BLOCK-1609
NTE74LS166	BLOCK-1610
NTE74LS168A	BLOCK-1611
NTE74LS169A	BLOCK-1612
NTE74LS170	BLOCK-1613
NTE74LS173A	BLOCK-1614
NTE74LS174	BLOCK-1615
NTE74LS175	BLOCK-1616
NTE74LS181	BLOCK-1617
NTE74LS190	BLOCK-1618
NTE74LS191	BLOCK-1619
NTE74LS192	BLOCK-1620
NTE74LS193	BLOCK-1621
NTE74LS194	BLOCK-1622
NTE74LS194A	BLOCK-1622
NTE74LS195A	BLOCK-1623
NTE74LS196	BLOCK-1624
NTE74LS197	BLOCK-1625
NTE74LS20	BLOCK-1626
NTE74LS21	BLOCK-1627
NTE74LS22	BLOCK-1628
NTE74LS221	BLOCK-1629
NTE74LS240	BLOCK-1630
NTE74LS241	BLOCK-1631
NTE74LS242	BLOCK-1632
NTE74LS243	BLOCK-1633
NTE74LS244	BLOCK-1634
NTE74LS245	BLOCK-1635
NTE74LS247	BLOCK-1636
NTE74LS248	BLOCK-1637
NTE74LS249	BLOCK-1638
NTE74LS251	BLOCK-1639
NTE74LS253	BLOCK-1640
NTE74LS257	BLOCK-1641
NTE74LS258	BLOCK-1642
NTE74LS259	BLOCK-1643
NTE74LS26	BLOCK-1644
NTE74LS260	BLOCK-1645
NTE74LS266	BLOCK-1646
NTE74LS27	BLOCK-1647
NTE74LS273	BLOCK-1648
NTE74LS279	BLOCK-1649
NTE74LS28	BLOCK-1650
NTE74LS280	BLOCK-1651
NTE74LS283	BLOCK-1652
NTE74LS290	BLOCK-1653
NTE74LS293	BLOCK-1654
NTE74LS295	BLOCK-1655
NTE74LS295A	BLOCK-1655
NTE74LS298	BLOCK-1656
NTE74LS299	BLOCK-1657
NTE74LS30	BLOCK-1658
NTE74LS32	BLOCK-1659
NTE74LS33	BLOCK-1660
NTE74LS348	BLOCK-1661
NTE74LS352	BLOCK-1662
NTE74LS353	BLOCK-1663
NTE74LS363	BLOCK-1664
NTE74LS364	BLOCK-1665
NTE74LS365A	BLOCK-1666
NTE74LS366A	BLOCK-1667
NTE74LS367	BLOCK-1668
NTE74LS368	BLOCK-1669
NTE74LS37	BLOCK-1670
NTE74LS373	BLOCK-1671
NTE74LS374	BLOCK-1672
NTE74LS377	BLOCK-1673
NTE74LS378	BLOCK-1674
NTE74LS379	BLOCK-1675
NTE74LS38	BLOCK-1676
NTE74LS386	BLOCK-1677
NTE74LS390	BLOCK-1678
NTE74LS393	BLOCK-1679
NTE74LS395A	BLOCK-1680
NTE74LS398	BLOCK-1681
NTE74LS40	BLOCK-1683
NTE74LS42	BLOCK-1684
NTE74LS47	BLOCK-1685
NTE74LS48	BLOCK-1686
NTE74LS49	BLOCK-1687
NTE74LS490	BLOCK-1688
NTE74LS51	BLOCK-1689
NTE74LS54	BLOCK-1690
NTE74LS540	BLOCK-1691
NTE74LS541	BLOCK-1692
NTE74LS55	BLOCK-1693
NTE74LS624	BLOCK-1694
NTE74LS625	BLOCK-1695
NTE74LS626	BLOCK-1696
NTE74LS627	BLOCK-1697
NTE74LS629	BLOCK-1698
NTE74LS640	BLOCK-1699
NTE74LS641	BLOCK-1700
NTE74LS642	BLOCK-1701
NTE74LS643	BLOCK-1702
NTE74LS645	BLOCK-1703
NTE74LS670	BLOCK-1704
NTE74LS73	BLOCK-1705
NTE74LS74A	BLOCK-1706
NTE74LS75	BLOCK-1707
NTE74LS76A	BLOCK-1708
NTE74LS77	BLOCK-1709
NTE74LS78	BLOCK-1710
NTE74LS83A	BLOCK-1711
NTE74LS85	BLOCK-1712
NTE74LS86	BLOCK-1713
NTE74LS90	BLOCK-1714
NTE74LS91	BLOCK-1715
NTE74LS92	BLOCK-1716
NTE74LS93	BLOCK-1717
NTE74S00	BLOCK-1719
NTE74S02	BLOCK-1720
NTE74S03	BLOCK-1721
NTE74S04	BLOCK-1722
NTE74S05	BLOCK-1723
NTE74S08	BLOCK-1724
NTE74S09	BLOCK-1725
NTE74S10	BLOCK-1726
NTE74S11	BLOCK-1727
NTE74S112	BLOCK-1728
NTE74S113	BLOCK-1729
NTE74S124	BLOCK-1730
NTE74S133	BLOCK-1731
NTE74S134	BLOCK-1732
NTE74S138	BLOCK-1733
NTE74S140	BLOCK-1734
NTE74S15	BLOCK-1735
NTE74S151	BLOCK-1736
NTE74S153	BLOCK-1737
NTE74S157	BLOCK-1738
NTE74S158	BLOCK-1739
NTE74S174	BLOCK-1740
NTE74S175	BLOCK-1741
NTE74S181	BLOCK-1742
NTE74S194	BLOCK-1744
NTE74S20	BLOCK-1745
NTE74S22	BLOCK-1746
NTE74S251	BLOCK-1747
NTE74S258	BLOCK-1748
NTE74S287	BLOCK-1749
NTE74S288	BLOCK-1750
NTE74S30	BLOCK-1751
NTE74S387	BLOCK-1752
NTE74S40	BLOCK-1753
NTE74S472	BLOCK-1755
NTE74S474	BLOCK-1757
NTE74S475	BLOCK-1758
NTE74S51	BLOCK-1761
NTE74S570	BLOCK-1762
NTE74S571	BLOCK-1763
NTE74S572	BLOCK-1764
NTE74S573	BLOCK-1765
NTE74S64	BLOCK-1766
NTE74S65	BLOCK-1767
NTE74S74	BLOCK-1768
NTE74S86	BLOCK-1769
NTE750	BLOCK-1770
NTE75188	BLOCK-1771
NTE75189	BLOCK-1772
NTE75450B	BLOCK-1775
NTE75451B	BLOCK-1776
NTE75452B	BLOCK-1777
NTE75453B	BLOCK-1778
NTE75454B	BLOCK-1779
NTE75491B	BLOCK-1780
NTE75492B	BLOCK-1781
NTE75493	BLOCK-1782
NTE75494	BLOCK-1783
NTE75498	BLOCK-1785
NTE760	BLOCK-1790
NTE778A	BLOCK-1801
NTE778S	BLOCK-1802
NTE778SM	BLOCK-1803
NTE780	BLOCK-1805
NTE781	BLOCK-1806
NTE783	BLOCK-1808
NTE784	BLOCK-1809
NTE786	BLOCK-1811
NTE787	BLOCK-1812
NTE788	BLOCK-1813
NTE789	BLOCK-1814
NTE790	BLOCK-1815
NTE791	BLOCK-1816
NTE793	BLOCK-1818
NTE795	BLOCK-1820
NTE797	BLOCK-1821
NTE798	BLOCK-1822
NTE799	BLOCK-1823
NTE801	BLOCK-1825
NTE802	BLOCK-1826
NTE803	BLOCK-1827
NTE804	BLOCK-1828
NTE806	BLOCK-1829
NTE807	BLOCK-1830
NTE8080A	BLOCK-1832
NTE809	BLOCK-1833
NTE80C95	BLOCK-1835
NTE80C96	BLOCK-1836
NTE80C97	BLOCK-1837
NTE810A	BLOCK-1838
NTE8123	BLOCK-1840
NTE815	BLOCK-1843
NTE818	BLOCK-1846
NTE819	BLOCK-1847
NTE820	BLOCK-1848
NTE821	BLOCK-1849
NTE822	BLOCK-1850
NTE823	BLOCK-1851
NTE824	BLOCK-1852
NTE825	BLOCK-1853
NTE8255	BLOCK-1855
NTE826	BLOCK-1856
NTE8266	BLOCK-1857
NTE828	BLOCK-1858
NTE829	BLOCK-1859
NTE8301	BLOCK-1861
NTE8308	BLOCK-1862
NTE8309	BLOCK-1863
NTE8314	BLOCK-1865
NTE8316	BLOCK-1866
NTE8318	BLOCK-1867
NTE832	BLOCK-1868
NTE8321	BLOCK-1869
NTE8328	BLOCK-1870
NTE832SM	BLOCK-1871
NTE834	BLOCK-1873
NTE834SM	BLOCK-1874
NTE835	BLOCK-1875
NTE836	BLOCK-1876
NTE8368	BLOCK-1877
NTE8370	BLOCK-1879
NTE8374	BLOCK-1880
NTE841	BLOCK-1884
NTE842	BLOCK-1885
NTE843	BLOCK-1886
NTE844	BLOCK-1887
NTE845	BLOCK-1888
NTE846	BLOCK-1889
NTE848	BLOCK-1891
NTE849	BLOCK-1892
NTE850	BLOCK-1893
NTE851	BLOCK-1894
NTE852	BLOCK-1895
NTE8520	BLOCK-1896
NTE853	BLOCK-1897
NTE854	BLOCK-1898
NTE8542	BLOCK-1899
NTE8546	BLOCK-1900
NTE855	BLOCK-1901
NTE8556	BLOCK-1906
NTE856	BLOCK-1907
NTE857M	BLOCK-1908
NTE857SM	BLOCK-1909
NTE858M	BLOCK-1910
NTE858SM	BLOCK-1911
NTE859	BLOCK-1912
NTE859SM	BLOCK-1913
NTE860	BLOCK-1914
NTE861	BLOCK-1915
NTE8613	BLOCK-1916
NTE862	BLOCK-1917
NTE863	BLOCK-1918
NTE864	BLOCK-1919
NTE867	BLOCK-1922
NTE868	BLOCK-1923
NTE869	BLOCK-1924
NTE871	BLOCK-1926
NTE873	BLOCK-1928
NTE874	BLOCK-1929
NTE875	BLOCK-1930
NTE876	BLOCK-1931
NTE877	BLOCK-1932
NTE878	BLOCK-1933
NTE887M	BLOCK-1937
NTE888M	BLOCK-1938
NTE889M	BLOCK-1939
NTE900	BLOCK-1940
NTE901	BLOCK-1941
NTE902	BLOCK-1942
NTE903	BLOCK-1943
NTE904	BLOCK-1944
NTE905	BLOCK-1945
NTE906	BLOCK-1946
NTE907	BLOCK-1947
NTE908	BLOCK-1948
NTE909	BLOCK-1949
NTE9093	BLOCK-1950
NTE9094	BLOCK-1951
NTE9097	BLOCK-1952
NTE9099	BLOCK-1953
NTE909D	BLOCK-1954
NTE910	BLOCK-1955
NTE910D	BLOCK-1957
NTE911	BLOCK-1958
NTE911D	BLOCK-1962
NTE912	BLOCK-1963
NTE913	BLOCK-1964
NTE9135	BLOCK-1965
NTE914	BLOCK-1966
NTE915	BLOCK-1967
NTE9157	BLOCK-1968
NTE9158	BLOCK-1969
NTE916	BLOCK-1970
NTE917	BLOCK-1971
NTE918	BLOCK-1972
NTE918M	BLOCK-1973
NTE918SM	BLOCK-1974
NTE919	BLOCK-1975
NTE919D	BLOCK-1976
NTE9200	BLOCK-1978
NTE921	BLOCK-1979
NTE922	BLOCK-1980
NTE922M	BLOCK-1981
NTE922SM	BLOCK-1982
NTE923	BLOCK-1983
NTE923D	BLOCK-1984
NTE924	BLOCK-1985
NTE924M	BLOCK-1986
NTE925	BLOCK-1987
NTE926	BLOCK-1988
NTE927	BLOCK-1989
NTE927D	BLOCK-1990
NTE927SM	BLOCK-1991
NTE928	BLOCK-1992
NTE928M	BLOCK-1993
NTE928S	BLOCK-1994
NTE928SM	BLOCK-1995
NTE929	BLOCK-1996
NTE930	BLOCK-1997
NTE931	BLOCK-2004
NTE932	BLOCK-2007
NTE933	BLOCK-2014
NTE934	BLOCK-2020
NTE935	BLOCK-2024
NTE936	BLOCK-2025

ORIGINAL DEVICE TYPES AND REPLACEMENT CODES 47

DEVICE TYPE	REPL CODE
NTE937	BLOCK-2031
NTE9370	BLOCK-2032
NTE937M	BLOCK-2036
NTE938	BLOCK-2037
NTE938M	BLOCK-2042
NTE939	BLOCK-2043
NTE93L08	BLOCK-2049
NTE93L16	BLOCK-2050
NTE940	BLOCK-2051
NTE9402	BLOCK-2054
NTE9403	BLOCK-2055
NTE941	BLOCK-2056
NTE941D	BLOCK-2057
NTE941M	BLOCK-2058
NTE941S	BLOCK-2059
NTE941SM	BLOCK-2060
NTE942	BLOCK-2061
NTE943	BLOCK-2062
NTE943M	BLOCK-2063
NTE943SM	BLOCK-2064
NTE944	BLOCK-2065
NTE944M	BLOCK-2066
NTE946	BLOCK-2068
NTE947	BLOCK-2069
NTE947D	BLOCK-2070
NTE948	BLOCK-2071
NTE948SM	BLOCK-2072
NTE949	BLOCK-2073
NTE950	BLOCK-2074
NTE951	BLOCK-2075
NTE952	BLOCK-2076
NTE953	BLOCK-2077
NTE954	BLOCK-2078
NTE955M	BLOCK-2079
NTE955MC	BLOCK-2080
NTE955S	BLOCK-2081
NTE955SM	BLOCK-2082
NTE956	BLOCK-2083
NTE957	BLOCK-2084
NTE958	BLOCK-2085
NTE959	BLOCK-2086
NTE960	BLOCK-2087
NTE9601	BLOCK-2089
NTE9602	BLOCK-2090
NTE961	BLOCK-2091
NTE962	BLOCK-2093
NTE963	BLOCK-2094
NTE964	BLOCK-2095
NTE965	BLOCK-2096
NTE966	BLOCK-2097
NTE9660	BLOCK-2098
NTE9661	BLOCK-2099
NTE9663	BLOCK-2101
NTE9664	BLOCK-2102
NTE9666	BLOCK-2104
NTE9668	BLOCK-2106
NTE9669	BLOCK-2107
NTE967	BLOCK-2108
NTE9670	BLOCK-2109
NTE9671	BLOCK-2110
NTE9672	BLOCK-2111
NTE9675	BLOCK-2114
NTE9676	BLOCK-2115
NTE9678	BLOCK-2117
NTE9679	BLOCK-2118
NTE968	BLOCK-2119
NTE9680	BLOCK-2120
NTE9681	BLOCK-2121
NTE9682	BLOCK-2122
NTE9689	BLOCK-2127
NTE969	BLOCK-2128
NTE9691	BLOCK-2130
NTE96L02	BLOCK-2132
NTE96LS02	BLOCK-2133
NTE96S02	BLOCK-2134
NTE970	BLOCK-2135
NTE971	BLOCK-2136
NTE972	BLOCK-2137
NTE973	BLOCK-2138
NTE973D	BLOCK-2139
NTE974	BLOCK-2140
NTE975	BLOCK-2141
NTE975SM	BLOCK-2142
NTE976	BLOCK-2143
NTE977	BLOCK-2144
NTE978	BLOCK-2145
NTE978C	BLOCK-2146
NTE978SM	BLOCK-2147
NTE980	BLOCK-2149
NTE9800	BLOCK-2150
NTE9801	BLOCK-2151
NTE9802	BLOCK-2152
NTE9803	BLOCK-2153
NTE9804	BLOCK-2154

DEVICE TYPE	REPL CODE
NTE9805	BLOCK-2155
NTE9806	BLOCK-2156
NTE9807	BLOCK-2157
NTE9808	BLOCK-2158
NTE9809	BLOCK-2159
NTE981	BLOCK-2160
NTE9810	BLOCK-2161
NTE9811	BLOCK-2162
NTE9812	BLOCK-2163
NTE9813	BLOCK-2164
NTE9814	BLOCK-2165
NTE982	BLOCK-2166
NTE983	BLOCK-2167
NTE984	BLOCK-2168
NTE985	BLOCK-2169
NTE986	BLOCK-2170
NTE987	BLOCK-2171
NTE987SM	BLOCK-2172
NTE988	BLOCK-2173
NTE989	BLOCK-2174
NTE990	BLOCK-2175
NTE991	BLOCK-2184
NTE9910	BLOCK-2185
NTE992	BLOCK-2191
NTE9924	BLOCK-2193
NTE9926	BLOCK-2194
NTE993	BLOCK-2196
NTE9930	BLOCK-2197
NTE9932	BLOCK-2199
NTE9933	BLOCK-2200
NTE9935	BLOCK-2201
NTE9936	BLOCK-2202
NTE9937	BLOCK-2203
NTE9944	BLOCK-2206
NTE9945	BLOCK-2207
NTE9946	BLOCK-2208
NTE9948	BLOCK-2209
NTE9949	BLOCK-2210
NTE995	BLOCK-2211
NTE9950	BLOCK-2212
NTE9951	BLOCK-2213
NTE995M	BLOCK-2214
NTE996	BLOCK-2215
NTE9961	BLOCK-2216
NTE9962	BLOCK-2217
NTE9963	BLOCK-2218
NTE997	BLOCK-2219
NTE998	BLOCK-2222
NTE9989	BLOCK-2224
OB2764K-2	BLOCK-1019
ON7407N	BLOCK-1300
ON7438N	BLOCK-1385
ON74LS03N	BLOCK-1571
ON74LS08N	BLOCK-1574
ON74LS145N	BLOCK-1594
ON74LS163AN	BLOCK-1607
ON74LS174N	BLOCK-1615
ON74LS175N	BLOCK-1616
ON74LS240N	BLOCK-1630
ON74LS244N	BLOCK-1634
ON74LS245N	BLOCK-1635
ON74LS373N	BLOCK-1671
ON74LS374N	BLOCK-1672
ON74LS670N	BLOCK-1704
ON74S00N	BLOCK-1719
ON74S10N	BLOCK-1726
ON74S138N	BLOCK-1591
ON74S20N	BLOCK-1745
OP-05CDE	BLOCK-2042
OP-05CNB	BLOCK-2042
OP-05CT	BLOCK-2037
OP-05EDE	BLOCK-2042
OP-05ENB	BLOCK-2042
OP-05ET	BLOCK-2037
OP-07CDE	BLOCK-2042
OP-07CNB	BLOCK-2042
OP-07CT	BLOCK-2037
OP-07DDE	BLOCK-2042
OP-07DNB	BLOCK-2042
OP-07DT	BLOCK-2037
OP-07EDE	BLOCK-2042
OP-07ENB	BLOCK-2042
OP-07ET	BLOCK-2037
OP-27EDE	BLOCK-2042
OP-27ENB	BLOCK-2042
OP-27ET	BLOCK-2037
OP-27FDE	BLOCK-2042
OP-27FNB	BLOCK-2042
OP-27FT	BLOCK-2037
OP-27GDE	BLOCK-2042
OP-27GNB	BLOCK-2042
OP-27GT	BLOCK-2037
OP-37EDE	BLOCK-2042
OP-37ENB	BLOCK-2042

DEVICE TYPE	REPL CODE
OP-37ET	BLOCK-2037
OP-37FDE	BLOCK-2042
OP-37FNB	BLOCK-2042
OP-37FT	BLOCK-2037
OP-37GDE	BLOCK-2042
OP-37GNB	BLOCK-2042
OP-37GT	BLOCK-2037
OP-47FT	BLOCK-2037
OP-47GNB	BLOCK-2042
OP-47GT	BLOCK-2037
OV2764KFA552	BLOCK-1019
P0165STD01E	BLOCK-1018
P055I	BLOCK-1019
P056I	BLOCK-1019
P1103	BLOCK-259
P2114	BLOCK-1002
P2114-3	BLOCK-1002
P61AM6116H	BLOCK-1004
P61LS00**H	BLOCK-1568
P61LS139*H	BLOCK-1592
P61LS14**H	BLOCK-1593
P61LS32**H	BLOCK-1659
P61LS373*H	BLOCK-1671
P61LS374*H	BLOCK-1672
P61LS74**H	BLOCK-1706
P61XX0002	BLOCK-2079
P61XX0027	BLOCK-1873
P6502B	BLOCK-1158
P8255A-5	BLOCK-1855
PA-500	BLOCK-88
PA234	BLOCK-1274
PA277	BLOCK-1828
PA306	BLOCK-1828
PA401	BLOCK-88
PA401X	BLOCK-88
PA500	BLOCK-88
PA501	BLOCK-139
PA501X	BLOCK-88
PA53C	BLOCK-993
PA7001/518	BLOCK-1377
PA7001/519	BLOCK-1365
PA7001/520	BLOCK-1304
PA7001/521	BLOCK-1293
PA7001/522	BLOCK-1390
PA7001/523	BLOCK-1402
PA7001/524	BLOCK-1403
PA7001/525	BLOCK-1295
PA7001/526	BLOCK-1294
PA7001/527	BLOCK-1297
PA7001/528	BLOCK-1298
PA7001/529	BLOCK-1411
PA7001/530	BLOCK-1408
PA7001/531	BLOCK-1410
PA7001/532	BLOCK-1409
PA7001/533	BLOCK-1406
PA7001/539	BLOCK-1404
PA7001/593	BLOCK-1325
PA7703	BLOCK-1212
PA7703E	BLOCK-1212
PBM74LS244P	BLOCK-1634
PBM74LS30P	BLOCK-1658
PBM74LS393P	BLOCK-1679
PBM74LS74AP	BLOCK-1706
PBM74LS86P	BLOCK-1713
PC-20004	BLOCK-20
PC-20005	BLOCK-26
PC-20006	BLOCK-23
PC-20007	BLOCK-7
PC-20008	BLOCK-1277
PC-20012	BLOCK-29
PC-20015	BLOCK-1839
PC-20018	BLOCK-1277
PC-20023	BLOCK-10
PC-20045	BLOCK-1825
PC-20046	BLOCK-1852
PC-20051	BLOCK-80
PC-20069	BLOCK-58
PC-20071	BLOCK-132
PC-20082	BLOCK-205
PC-20083	BLOCK-207
PC-20085	BLOCK-223
PC-20087	BLOCK-973
PC-20110	BLOCK-453
PC-20143	BLOCK-973
PC-20151	BLOCK-214
PC1026C	BLOCK-214
PC1167C2	BLOCK-476
PC1363C	BLOCK-396
PC17805H	BLOCK-2087
PC20003	BLOCK-4
PC20004	BLOCK-20
PC20018	BLOCK-7
PC339C	BLOCK-1873
PC393C	BLOCK-2063

DEVICE TYPE	REPL CODE
PC4558C	BLOCK-1801
PC563H2	BLOCK-79
PC574J	BLOCK-1157
PC7327C	BLOCK-2189
PC7805H	BLOCK-2087
PC7805HF	BLOCK-2087
PC7812HF	BLOCK-2097
PC78L05	BLOCK-2144
PCMB74LS00M	BLOCK-1568
PCMB74LS02M	BLOCK-1570
PCMB74LS04M	BLOCK-1572
PCMB74LS05M	BLOCK-1573
PCMB74LS10M	BLOCK-1576
PCMB74LS32M	BLOCK-1659
PD4066BC	BLOCK-1093
PD9093-59	BLOCK-1950
PD9094-59	BLOCK-1951
PD9097-59	BLOCK-1952
PD9099-59	BLOCK-1953
PD9930-59	BLOCK-2197
PD9932-59	BLOCK-2199
PD9933-59	BLOCK-2200
PD9935-59	BLOCK-2201
PD9936-59	BLOCK-2202
PD9937-59	BLOCK-2203
PD9944-59	BLOCK-2206
PD9945-59	BLOCK-2207
PD9946-59	BLOCK-2208
PD9948-59	BLOCK-2209
PD9949-59	BLOCK-2210
PD9950-59	BLOCK-2212
PD9951-59	BLOCK-2213
PD9961-59	BLOCK-2216
PD9962-59	BLOCK-2217
PD9963-59	BLOCK-2218
PE9093-59	BLOCK-1950
PE9094-59	BLOCK-1951
PE9097-59	BLOCK-1952
PE9099-59	BLOCK-1953
PE9930-59	BLOCK-2197
PE9932-59	BLOCK-2199
PE9933-59	BLOCK-2200
PE9935-59	BLOCK-2201
PE9936-59	BLOCK-2202
PE9937-59	BLOCK-2203
PE9944-59	BLOCK-2206
PE9945-59	BLOCK-2207
PE9946-59	BLOCK-2208
PE9948-59	BLOCK-2209
PE9949-59	BLOCK-2210
PE9950-59	BLOCK-2212
PE9951-59	BLOCK-2213
PE9961-59	BLOCK-2216
PE9962-59	BLOCK-2217
PE9963-59	BLOCK-2218
PL-307-047-9-001	BLOCK-2138
PL-307-047-9-002	BLOCK-1280
PL102	BLOCK-156
PL20015	BLOCK-1839
PLL02	BLOCK-156
PLL02A	BLOCK-156
PLL02A-C	BLOCK-156
PLL02A-F	BLOCK-156
PLL02A-G	BLOCK-222
PLL02AG	BLOCK-222
PLL03	BLOCK-259
PLL03A	BLOCK-259
PM308AP	BLOCK-2042
PM355AJ	BLOCK-2031
PM355J	BLOCK-2031
PM356AJ	BLOCK-2031
PM356J	BLOCK-2031
PM357AJ	BLOCK-2031
PM357J	BLOCK-2031
PM725CJ	BLOCK-1987
PM741CJ	BLOCK-2056
PM741CY	BLOCK-2057
PMC1367P	BLOCK-539
PO74LS04	BLOCK-1572
PO74LS368A	BLOCK-1669
POSN7407N	BLOCK-1300
POSN74145N	BLOCK-1325
POSN7416N	BLOCK-1336
POSN7417N	BLOCK-1343
POSN7445N	BLOCK-1395
POSN74LS123N	BLOCK-1584
POSN74LS138N	BLOCK-1591
POSN74LS14N	BLOCK-1593
POSN74LS157N	BLOCK-1602
POSN74LS161AN	BLOCK-1605
POSN74LS166N	BLOCK-1610
POSN74LS221N	BLOCK-1629
POSN74LS245N	BLOCK-1635
POSN74LS368AN	BLOCK-1669

DEVICE TYPE	REPL CODE
POSN74LS373N	BLOCK-1671
POSN74S153N	BLOCK-1737
PS2006B	BLOCK-1021
PTC1101	BLOCK-1405
PTC1623	BLOCK-1405
PTC1956	BLOCK-1806
PTC701	BLOCK-1262
PTC703	BLOCK-1264
PTC705	BLOCK-1270
PTC707	BLOCK-1256
PTC708	BLOCK-1244
PTC709	BLOCK-1275
PTC711	BLOCK-1276
PTC713	BLOCK-1277
PTC715	BLOCK-1271
PTC719	BLOCK-1272
PTC723	BLOCK-1279
PTC726	BLOCK-1269
PTC728	BLOCK-1815
PTC729	BLOCK-1376
PTC730	BLOCK-1808
PTC731	BLOCK-1287
PTC732	BLOCK-1828
PTC733	BLOCK-1816
PTC734	BLOCK-1389
PTC735	BLOCK-1825
PTC736	BLOCK-1801
PTC737	BLOCK-1290
PTC738	BLOCK-1828
PTC739	BLOCK-1292
PTC740	BLOCK-172
PTC741	BLOCK-1291
PTC742	BLOCK-1828
PTC743	BLOCK-1212
PTC744	BLOCK-1280
PTC745	BLOCK-1421
PTC746	BLOCK-1422
PTC754	BLOCK-5
PTC757	BLOCK-56
PTC765	BLOCK-1281
PTC767	BLOCK-1813
PTC769	BLOCK-1821
PTC776	BLOCK-1285
PTC780	BLOCK-145
PTC781	BLOCK-98
PTC787	BLOCK-1813
PTSN7417N	BLOCK-1343
Q-01369	BLOCK-102
Q-013696	BLOCK-102
Q7219	BLOCK-1801
Q7227	BLOCK-525
Q7227R	BLOCK-525
Q7256	BLOCK-644
Q7293	BLOCK-535
Q7296	BLOCK-535
Q7327A	BLOCK-1088
Q7350	BLOCK-1713
Q7368	BLOCK-727
Q7372A	BLOCK-1088
Q7393	BLOCK-1041
Q7402	BLOCK-1713
Q7406	BLOCK-1087
Q7448	BLOCK-1088
Q7478A	BLOCK-1093
Q7652	BLOCK-1093
Q7653A	BLOCK-1093
Q7668	BLOCK-205
Q7682	BLOCK-563
Q7748	BLOCK-215
QA703E	BLOCK-1212
QB400428	BLOCK-2189
QC0011	BLOCK-762
QC0031A	BLOCK-642
QC0032	BLOCK-866
QC0042	BLOCK-2097
QC0065	BLOCK-722
QC0069	BLOCK-1993
QC0073	BLOCK-753
QC0074	BLOCK-763
QC0080	BLOCK-671
QC0081	BLOCK-1087
QC0085	BLOCK-709
QC0093	BLOCK-659
QC0234	BLOCK-1098
QC0235	BLOCK-1098
QC0239	BLOCK-1098
QC0275	BLOCK-824
QC0291	BLOCK-663
QC0300	BLOCK-754
QC0343	BLOCK-1041
QC0352	BLOCK-779
QC0380	BLOCK-725
QC0543	BLOCK-759
QC0554	BLOCK-752

DEVICE TYPE	REPL CODE
QC0573	BLOCK-778
QC0680	BLOCK-839
QC0713	BLOCK-826
QC0846	BLOCK-761
QC0852	BLOCK-817
QQ-0PPL02A0	BLOCK-156
QQ-M07205AT	BLOCK-145
QQ-MAN612AN	BLOCK-238
QQ-MBA511BX	BLOCK-154
QQ-MBA521AX	BLOCK-155
QQ-MC3001AN	BLOCK-180
QQ-MC3001AT	BLOCK-180
QQ-MC3001DT	BLOCK-180
QQ-MO7205AT	BLOCK-145
QQ-OPLL020A	BLOCK-222
QQ-OPLL02A0	BLOCK-156
QQ-OPLL02AN	BLOCK-156
QQ-OPLL02AO	BLOCK-156
QQ-OPLLO2AN	BLOCK-156
QQ-OPPL02A0	BLOCK-156
QQ0PLL02AN	BLOCK-156
QQM07205AT	BLOCK-145
QQMAN612AN	BLOCK-238
QQOPLLO2AN	BLOCK-222
QRT-272	BLOCK-1958
QSBLA0ZSY003	BLOCK-1266
QSBLAOZSQ019	BLOCK-759
QSI-5022	BLOCK-1841
QTVCM-16	BLOCK-1279
QTVCM-44	BLOCK-1212
QTVCM-45	BLOCK-1281
QTVCM-48	BLOCK-1280
QTVCM-5	BLOCK-1264
QTVCM-500	BLOCK-1293
QTVCM-502	BLOCK-1411
QTVCM-505	BLOCK-1423
QTVCM-59	BLOCK-1405
QTVCM-73	BLOCK-1825
QTVCM-75	BLOCK-4
QTVCM-77	BLOCK-102
QTVCM-78	BLOCK-98
QTVCM-79	BLOCK-101
QTVCM-81	BLOCK-145
R10254P936	BLOCK-2202
R10254P945	BLOCK-2207
R10254P945B	BLOCK-2207
R10254P946	BLOCK-2208
R10255P946B	BLOCK-2208
R10256P962	BLOCK-2217
R10256P962B	BLOCK-2217
R2432-1	BLOCK-1265
R2434-1	BLOCK-1245
R2445-1	BLOCK-1267
R24451	BLOCK-1267
R2516	BLOCK-1279
R2516-1	BLOCK-1269
R4295-1	BLOCK-1787
R4295-1(RCA)	BLOCK-1787
R4295-2	BLOCK-1787
R4439-1	BLOCK-1814
R4439-1(RCA)	BLOCK-1814
R4439-2	BLOCK-1814
R4439-2(RCA)	BLOCK-1814
R6502-40	BLOCK-1158
R6502C	BLOCK-1158
R6502P	BLOCK-1158
R6507C	BLOCK-1159
R6507P	BLOCK-1159
R6532P	BLOCK-1162
RC1458DE	BLOCK-1801
RC1458DN	BLOCK-1801
RC1458NB	BLOCK-1801
RC2041M	BLOCK-1801
RC2043NB	BLOCK-1801
RC2403NB	BLOCK-2063
RC3302DB	BLOCK-1873
RC4136	BLOCK-2219
RC4136DB	BLOCK-2219
RC4136DC	BLOCK-2219
RC4156DB	BLOCK-2071
RC4156DC	BLOCK-2071
RC4195NB	BLOCK-910
RC4558	BLOCK-1801
RC4558DE	BLOCK-1801
RC4558DQ	BLOCK-1801
RC4558JG	BLOCK-1801
RC4558NB	BLOCK-1801
RC4558P	BLOCK-1801
RC4559DE	BLOCK-1801
RC4559NB	BLOCK-1801
RC4739DB	BLOCK-1281
RC4739DP	BLOCK-1281
RC555	BLOCK-2079
RC555DE	BLOCK-2079

DEVICE TYPE	REPL CODE
RC555NB	BLOCK-2079
RC555T	BLOCK-2079
RC556DB	BLOCK-2145
RC556DC	BLOCK-2145
RC709D	BLOCK-1954
RC709DC	BLOCK-1954
RC709T	BLOCK-1949
RC710DC	BLOCK-1957
RC710T	BLOCK-1955
RC714CDE	BLOCK-2042
RC714CH	BLOCK-2037
RC714EDE	BLOCK-2042
RC714EH	BLOCK-2037
RC714LDE	BLOCK-2042
RC714LH	BLOCK-2037
RC723D	BLOCK-1984
RC723T	BLOCK-1983
RC741D	BLOCK-2057
RC741DC	BLOCK-2057
RC741DE	BLOCK-2058
RC741DN	BLOCK-2058
RC741H	BLOCK-2056
RC741N	BLOCK-2058
RC741NB	BLOCK-2058
RC741T	BLOCK-2056
RC747DB	BLOCK-2070
RC747DC	BLOCK-2070
RC747T	BLOCK-2069
RC7812FA	BLOCK-2097
RC78L05A	BLOCK-2144
RC78L12A	BLOCK-2074
RC78M05FA	BLOCK-2087
RE-300-IC	BLOCK-1212
RE-303-IC	BLOCK-1808
RE-304-IC	BLOCK-1422
RE-305-IC	BLOCK-1269
RE-306-IC	BLOCK-1270
RE-307-IC	BLOCK-1271
RE-310-IC	BLOCK-1407
RE-311-IC	BLOCK-1262
RE-312-IC	BLOCK-1292
RE-313-IC	BLOCK-1291
RE-314-IC	BLOCK-1364
RE-315-IC	BLOCK-1290
RE-316-IC	BLOCK-1289
RE-317-IC	BLOCK-1376
RE-319-IC	BLOCK-1275
RE-320-IC	BLOCK-1277
RE-321-IC	BLOCK-1303
RE-322-IC	BLOCK-1828
RE-323-IC	BLOCK-1815
RE300-IC	BLOCK-1212
RE301-IC	BLOCK-1264
RE302-IC	BLOCK-1244
RE303-IC	BLOCK-1808
RE304-IC	BLOCK-1422
RE305-IC	BLOCK-1269
RE306-IC	BLOCK-1270
RE307-IC	BLOCK-1271
RE308-IC	BLOCK-1272
RE310-IC	BLOCK-1407
RE311-IC	BLOCK-1262
RE312-IC	BLOCK-1292
RE313-IC	BLOCK-1291
RE314-IC	BLOCK-1364
RE315-IC	BLOCK-1290
RE316-IC	BLOCK-1289
RE317-IC	BLOCK-1376
RE319-IC	BLOCK-1275
RE320-IC	BLOCK-1277
RE321-IC	BLOCK-1303
RE322-IC	BLOCK-1828
RE323-IC	BLOCK-1815
RE324-IC	BLOCK-1816
RE325-IC	BLOCK-5
RE326-IC	BLOCK-7
RE327-IC	BLOCK-50
RE328-IC	BLOCK-78
RE329-IC	BLOCK-101
RE330-IC	BLOCK-138
RE331-IC	BLOCK-1421
RE332-IC	BLOCK-1813
RE333-IC	BLOCK-1821
RE334-IC	BLOCK-1825
RE335-IC	BLOCK-4
RE336-IC	BLOCK-6
RE337-IC	BLOCK-44
RE338-IC	BLOCK-56
RE339-IC	BLOCK-79
RE340-M	BLOCK-85
RE340M	BLOCK-85
RE341-M	BLOCK-80
RE342-M	BLOCK-90
RE343-IC	BLOCK-94

DEVICE TYPE	REPL CODE
RE344-M	BLOCK-98
RE345-IC	BLOCK-1279
RE346-IC	BLOCK-1285
RE347M	BLOCK-23
RE348-IC	BLOCK-24
RE348-M	BLOCK-24
RE348M	BLOCK-24
RE349-IC	BLOCK-26
RE349-M	BLOCK-26
RE351-IC	BLOCK-36
RE352-IC	BLOCK-71
RE353-M	BLOCK-88
RE354-IC	BLOCK-127
RE355-IC	BLOCK-130
RE356-IC	BLOCK-131
RE357-IC	BLOCK-145
RE358-IC	BLOCK-154
RE359-IC	BLOCK-155
RE360-IC	BLOCK-222
RE361-IC	BLOCK-1287
RE362-IC	BLOCK-1389
RE364-IC	BLOCK-1805
RE367-IC	BLOCK-48
RE368-IC	BLOCK-54
RE370-IC	BLOCK-100
RE372-IC	BLOCK-126
RE373-IC	BLOCK-128
RE374-IC	BLOCK-129
RE375-IC	BLOCK-132
RE376-IC	BLOCK-143
RE377-IC	BLOCK-150
RE378-IC	BLOCK-1269
RE380-IC	BLOCK-153
RE382-IC	BLOCK-1293
RE383-IC	BLOCK-1410
RE384-IC	BLOCK-1411
RE385-IC	BLOCK-1410
RE386-IC	BLOCK-1338
RE387-IC	BLOCK-2144
RE388-IC	BLOCK-1313
RE389-IC	BLOCK-163
RE390-IC	BLOCK-180
REN1003	BLOCK-4
REN1004	BLOCK-5
REN1005	BLOCK-6
REN1006	BLOCK-7
REN1024	BLOCK-23
REN1025	BLOCK-24
REN1027	BLOCK-26
REN1028	BLOCK-27
REN1037	BLOCK-36
REN1046	BLOCK-44
REN1050	BLOCK-48
REN1052	BLOCK-50
REN1056	BLOCK-54
REN1058	BLOCK-56
REN1073	BLOCK-71
REN1080	BLOCK-78
REN1081A	BLOCK-79
REN1082	BLOCK-80
REN1087	BLOCK-85
REN1090	BLOCK-88
REN1092	BLOCK-90
REN1096	BLOCK-94
REN1099	BLOCK-97
REN1100	BLOCK-98
REN1102	BLOCK-100
REN1103	BLOCK-101
REN1127	BLOCK-125
REN1128	BLOCK-126
REN1130	BLOCK-127
REN1131	BLOCK-128
REN1132	BLOCK-129
REN1133	BLOCK-130
REN1134	BLOCK-131
REN1135	BLOCK-132
REN1142	BLOCK-138
REN1153	BLOCK-143
REN1155	BLOCK-145
REN1161	BLOCK-150
REN1162	BLOCK-151
REN1163	BLOCK-152
REN1164	BLOCK-153
REN1165	BLOCK-154
REN1166	BLOCK-155
REN1167	BLOCK-156
REN1174	BLOCK-163
REN1192	BLOCK-180
REN703A	BLOCK-1212
REN705A	BLOCK-1244
REN708	BLOCK-1262
REN709	BLOCK-1264
REN712	BLOCK-1269
REN713	BLOCK-1270

DEVICE TYPE	REPL CODE
REN714	BLOCK-1271
REN715	BLOCK-1272
REN718	BLOCK-1275
REN720	BLOCK-1277
REN723	BLOCK-1279
REN729	BLOCK-1285
REN731	BLOCK-1287
REN736	BLOCK-1289
REN737	BLOCK-1290
REN738	BLOCK-1291
REN739	BLOCK-1292
REN740	BLOCK-1303
REN7400	BLOCK-1293
REN74123	BLOCK-1313
REN74161	BLOCK-1338
REN742	BLOCK-1364
REN743	BLOCK-1376
REN744	BLOCK-1389
REN747	BLOCK-1407
REN7473	BLOCK-1410
REN7474	BLOCK-1411
REN748	BLOCK-1421
REN749	BLOCK-1422
REN7493A	BLOCK-1426
REN755	BLOCK-1786
REN780	BLOCK-1805
REN783	BLOCK-1808
REN788	BLOCK-1813
REN790	BLOCK-1815
REN791	BLOCK-1816
REN797	BLOCK-1821
REN801	BLOCK-1825
REN804	BLOCK-1828
REN977	BLOCK-2144
RG602	BLOCK-1830
RG602T	BLOCK-1830
RG604	BLOCK-1833
RG604T	BLOCK-1833
RH-1X0001TAZZ	BLOCK-1421
RH-1X0004CEZZ	BLOCK-1422
RH-1X0005PAZZ	BLOCK-1413
RH-1X0015TAZZ	BLOCK-234
RH-1X0018TAZZ	BLOCK-225
RH-1X0020CEZZ	BLOCK-5
RH-1X0021CEZZ	BLOCK-78
RH-1X0022CEZZ	BLOCK-49
RH-1X0032CEZZ	BLOCK-87
RH-1X0038CEZZ	BLOCK-1269
RH-1X0043CEZZ	BLOCK-1269
RH-1X0047CEZZ	BLOCK-178
RH-1X0092CEZZ	BLOCK-532
RH-1X1020AFZZ	BLOCK-112
RH-1X1039AFZZ	BLOCK-185
RH-FX0008CEZZ	BLOCK-1021
RH-FX0014CEZZ	BLOCK-1021
RH-IX00004CEZZ	BLOCK-1422
RH-IX0001PAZZ	BLOCK-176
RH-IX0001TAZZ	BLOCK-1421
RH-IX0004CE	BLOCK-1422
RH-IX0004CEZZ	BLOCK-1422
RH-IX0005PAZZ	BLOCK-1413
RH-IX0012PAZZ	BLOCK-1297
RH-IX0014PAZZ	BLOCK-2087
RH-IX0015TAZZ	BLOCK-185
RH-IX0017TAZZ	BLOCK-1269
RH-IX0018TAZZ	BLOCK-225
RH-IX0020CE	BLOCK-5
RH-IX0020CEZZ	BLOCK-5
RH-IX0021CEZZ	BLOCK-78
RH-IX0022CEZZ	BLOCK-49
RH-IX0023CEZZ	BLOCK-72
RH-IX0024CEZZ	BLOCK-557
RH-IX0024PAZZ	BLOCK-561
RH-IX0025CEZZ	BLOCK-48
RH-IX0025PAZZ	BLOCK-1883
RH-IX0028PAZZ	BLOCK-480
RH-IX0032CEZZ	BLOCK-87
RH-IX0035GAZZ	BLOCK-234
RH-IX0035TAZZ	BLOCK-234
RH-IX0037CEZZ	BLOCK-1157
RH-IX0038PAZZ	BLOCK-1299
RH-IX0039PAZZ	BLOCK-1305
RH-IX0040PAZZ	BLOCK-1311
RH-IX0041PAZZ	BLOCK-1313
RH-IX0042PAZZ	BLOCK-1318
RH-IX0043CE	BLOCK-1269
RH-IX0043CEZZ	BLOCK-1269
RH-IX0047CE	BLOCK-178
RH-IX0047CEZZ	BLOCK-178
RH-IX0054CEZZ	BLOCK-136
RH-IX0065CEZZ	BLOCK-537
RH-IX0069CEZZ	BLOCK-1029
RH-IX0072CEZZ	BLOCK-1039
RH-IX0072PAZZ	BLOCK-1571

DEVICE TYPE	REPL CODE
RH-IX0092CEZZ	BLOCK-401
RH-IX0093CEZZ	BLOCK-519
RH-IX0094CEZZ	BLOCK-484
RH-IX0111CEZZ	BLOCK-2144
RH-IX0114PAZZ	BLOCK-1125
RH-IX0116CEZZ	BLOCK-1300
RH-IX0137CEZZ	BLOCK-732
RH-IX0199CEZZ	BLOCK-740
RH-IX0203GEZZ	BLOCK-644
RH-IX0212CEZZ	BLOCK-657
RH-IX0213CEZZ	BLOCK-603
RH-IX0214CEZZ	BLOCK-563
RH-IX0223CEZZ	BLOCK-559
RH-IX0224CEZZ	BLOCK-557
RH-IX0225CEZZ	BLOCK-364
RH-IX0238CEZZ	BLOCK-662
RH-IX0252CEZZ	BLOCK-401
RH-IX0260CEZZ	BLOCK-644
RH-IX0275CEZZ	BLOCK-559
RH-IX0325CEZZ	BLOCK-851
RH-IX0416CEZZ	BLOCK-773
RH-IX0450CEZZ	BLOCK-814
RH-IX0487CEZZ	BLOCK-775
RH-IX0488CEZZ	BLOCK-205
RH-IX0499CEZZ	BLOCK-776
RH-IX0600CEZZ	BLOCK-771
RH-IX0633CEZZ	BLOCK-1195
RH-IX0934CEZZ	BLOCK-779
RH-IX0948CEZZ	BLOCK-763
RH-IX1018AFZZ	BLOCK-464
RH-IX1020AFZZ	BLOCK-112
RH-IX1030AFZZ	BLOCK-226
RH-IX1039AFZZ	BLOCK-196
RH-IX1068AFZZ	BLOCK-180
RH-IX1112CEZZ	BLOCK-854
RH-IX1190CEZZ	BLOCK-1191
RH-IX1712CEZZ	BLOCK-1191
RL709T	BLOCK-1949
RLH111	BLOCK-1304
RM4558	BLOCK-1801
RM723T	BLOCK-1983
RM741TE	BLOCK-2056
RS-862S	BLOCK-72
RS1458	BLOCK-1801
RS380	BLOCK-1303
RS4001	BLOCK-1029
RS4011	BLOCK-1039
RS4013	BLOCK-1041
RS4017	BLOCK-1051
RS4020	BLOCK-1059
RS4027	BLOCK-1066
RS4049	BLOCK-1084
RS4050	BLOCK-1085
RS4518	BLOCK-1129
RS555	BLOCK-2079
RS723	BLOCK-1983
RS7400	BLOCK-1293
RS7402	BLOCK-1295
RS7404	BLOCK-1297
RS7406(IC)	BLOCK-1299
RS7408(IC)	BLOCK-1301
RS741	BLOCK-2056
RS7410(IC)	BLOCK-1304
RS74123	BLOCK-1313
RS7413(IC)	BLOCK-1317
RS74145	BLOCK-1325
RS74150	BLOCK-1327
RS74154	BLOCK-1331
RS74192	BLOCK-1357
RS74193	BLOCK-1358
RS74196	BLOCK-1360
RS7420	BLOCK-1365
RS7427	BLOCK-1372
RS7432	BLOCK-1378
RS7441	BLOCK-1391
RS7447	BLOCK-1397
RS7448	BLOCK-1398
RS7451	BLOCK-1402
RS7473	BLOCK-1410
RS7474	BLOCK-1411
RS7475	BLOCK-1412
RS7476	BLOCK-1413
RS7485	BLOCK-1418
RS7486	BLOCK-1419
RS7490	BLOCK-1423
RS7492	BLOCK-1425
RT-271	BLOCK-1958
RT-272	BLOCK-1958
RT-7399	BLOCK-4
RT6790	BLOCK-4
RT7399	BLOCK-4
RV1BA1330	BLOCK-205
RV1LA3301	BLOCK-7
RV1LA4126	BLOCK-369

DEVICE TYPE	REPL CODE
RV1S1998	BLOCK-973
RV1SM5104	BLOCK-244
RV4136DB	BLOCK-2219
RV4136DC	BLOCK-2219
RV4156DC	BLOCK-2071
RV4558DE	BLOCK-1801
RV4558NB	BLOCK-1801
RV4559DE	BLOCK-1801
RV4559NB	BLOCK-1801
RV555NB	BLOCK-2079
RV556DB	BLOCK-2145
RV741NB	BLOCK-2058
RVIBA1330	BLOCK-205
RVIBA328MR	BLOCK-159
RVIBA532S	BLOCK-362
RVIHD38980A	BLOCK-974
RVILM703	BLOCK-1212
RVIM51182	BLOCK-455
RVIMC4080	BLOCK-81
RVIS1998	BLOCK-973
RVITMS1943	BLOCK-973
RVITMS1943N2	BLOCK-973
RVIUPC1018CE	BLOCK-550
RVIUPC20C2	BLOCK-73
RVIUPC22C	BLOCK-76
RVIUPC575	BLOCK-136
RVIUPC78L05	BLOCK-2144
RVIUPD861	BLOCK-243
RX-IX0035GAZZ	BLOCK-234
S-2041-8661	BLOCK-2206
S-2200-1135	BLOCK-1473
S1998	BLOCK-973
S1998A	BLOCK-973
S380	BLOCK-1303
S4LS132N	BLOCK-1588
S4LS221N	BLOCK-1629
S4LS374N	BLOCK-1672
S5723T	BLOCK-1983
S6508	BLOCK-1160
S6508-1	BLOCK-1160
S6508A	BLOCK-1160
S6508E	BLOCK-1160
S6508P	BLOCK-1160
S6800	BLOCK-1164
S6802	BLOCK-1165
S6810P	BLOCK-1168
S6821P	BLOCK-1169
S6850	BLOCK-1170
S6850P	BLOCK-1170
S8H80J	BLOCK-1471
S8T28F	BLOCK-1178
S9308F	BLOCK-1862
S9314F	BLOCK-1865
S9602F	BLOCK-2090
SA723CN	BLOCK-1984
SA7805CDA	BLOCK-2004
SA7805CU	BLOCK-2087
SA7812CDA	BLOCK-2014
SA7812CU	BLOCK-2097
SA78HV05CDA	BLOCK-2004
SA78HV05CU	BLOCK-2087
SA78HV12CDA	BLOCK-2014
SA78HV12CU	BLOCK-2097
SAA1024	BLOCK-561
SAA1025	BLOCK-1883
SAA1124	BLOCK-561
SAA7220	BLOCK-832
SAA7220N	BLOCK-832
SAB3037	BLOCK-767
SAB3037N	BLOCK-767
SAF1039P	BLOCK-741
SAF1039PN	BLOCK-741
SAJ110	BLOCK-957
SAJ2160	BLOCK-84
SAJ72155	BLOCK-74
SAJ72156	BLOCK-89
SAJ72157	BLOCK-44
SAJ72158	BLOCK-45
SAJ72159	BLOCK-75
SAJ72160	BLOCK-84
SAJ72161	BLOCK-87
SAJ72162	BLOCK-46
SC4001BH	BLOCK-1029
SC4001UBC	BLOCK-1029
SC42502P	BLOCK-2184
SC5117	BLOCK-1275
SC5118P	BLOCK-1277
SC5150P	BLOCK-1288
SC5172P	BLOCK-1276
SC5172PC	BLOCK-1276
SC5172PQ	BLOCK-1276
SC51750	BLOCK-2056
SC5175G	BLOCK-2056
SC5177P	BLOCK-1275

DEVICE TYPE	REPL CODE
SC5199P	BLOCK-1275
SC5204P	BLOCK-1822
SC5219P	BLOCK-1272
SC5220P	BLOCK-1269
SC5221P	BLOCK-1786
SC5245P	BLOCK-1421
SC5282P	BLOCK-1845
SC5858	BLOCK-1190
SC5896	BLOCK-1846
SC5898	BLOCK-219
SC8709P	BLOCK-1270
SC8723P	BLOCK-1815
SC9314P	BLOCK-1275
SC9430P	BLOCK-1407
SC9431P	BLOCK-1422
SC9436P	BLOCK-1269
SC9438P	BLOCK-1269
SC9963P	BLOCK-2193
SC9964P	BLOCK-2221
SC9965P	BLOCK-2224
SCD-1062013	BLOCK-1969
SCF40501	BLOCK-1786
SCL4000	BLOCK-1028
SCL4000B	BLOCK-1028
SCL4000BC	BLOCK-1028
SCL4000BD	BLOCK-1028
SCL4000BE	BLOCK-1028
SCL4000BF	BLOCK-1028
SCL4000BH	BLOCK-1028
SCL4001	BLOCK-1029
SCL4001B	BLOCK-1029
SCL4001BC	BLOCK-1029
SCL4001BD	BLOCK-1029
SCL4001BE	BLOCK-1029
SCL4001BF	BLOCK-1029
SCL4001UBD	BLOCK-1029
SCL4001UBE	BLOCK-1029
SCL4001UBF	BLOCK-1029
SCL4001UBH	BLOCK-1029
SCL4002	BLOCK-1030
SCL4002B	BLOCK-1030
SCL4002BC	BLOCK-1030
SCL4002BD	BLOCK-1030
SCL4002BE	BLOCK-1030
SCL4002BF	BLOCK-1030
SCL4002BH	BLOCK-1030
SCL4006BC	BLOCK-1031
SCL4006BD	BLOCK-1031
SCL4006BE	BLOCK-1031
SCL4006BF	BLOCK-1031
SCL4007	BLOCK-1032
SCL4007UBC	BLOCK-1032
SCL4007UBD	BLOCK-1032
SCL4007UBE	BLOCK-1032
SCL4008B	BLOCK-1034
SCL4008BC	BLOCK-1034
SCL4008BD	BLOCK-1034
SCL4008BE	BLOCK-1034
SCL4008BF	BLOCK-1034
SCL4008BH	BLOCK-1034
SCL4008BP	BLOCK-1034
SCL4009	BLOCK-1084
SCL4009UBC	BLOCK-1084
SCL4009UBD	BLOCK-1084
SCL4009UBE	BLOCK-1084
SCL4010BE	BLOCK-1085
SCL4011	BLOCK-1039
SCL4011B	BLOCK-1039
SCL4011BC	BLOCK-1039
SCL4011BD	BLOCK-1039
SCL4011BE	BLOCK-1039
SCL4011BF	BLOCK-1039
SCL4011BH	BLOCK-1039
SCL4011UBC	BLOCK-1039
SCL4011UBD	BLOCK-1039
SCL4011UBE	BLOCK-1039
SCL4011UBF	BLOCK-1039
SCL4011UBH	BLOCK-1039
SCL4012	BLOCK-1040
SCL4012B	BLOCK-1040
SCL4012BC	BLOCK-1040
SCL4012BD	BLOCK-1040
SCL4012BE	BLOCK-1040
SCL4012BF	BLOCK-1040
SCL4012BH	BLOCK-1040
SCL4013	BLOCK-1041
SCL4013AC	BLOCK-1041
SCL4013AD	BLOCK-1041
SCL4013AE	BLOCK-1041
SCL4013AF	BLOCK-1041
SCL4013AH	BLOCK-1041
SCL4013BC	BLOCK-1041
SCL4013BD	BLOCK-1041
SCL4013BE	BLOCK-1041

DEVICE TYPE	REPL CODE
SCL4013BF	BLOCK-1041
SCL4013BH	BLOCK-1041
SCL4013BP	BLOCK-1041
SCL4014BC	BLOCK-1042
SCL4014BD	BLOCK-1042
SCL4014BE	BLOCK-1042
SCL4015	BLOCK-1043
SCL4015BC	BLOCK-1043
SCL4015BD	BLOCK-1043
SCL4015BE	BLOCK-1043
SCL4016	BLOCK-1048
SCL4017	BLOCK-1051
SCL4017ABC	BLOCK-1051
SCL4017ABD	BLOCK-1051
SCL4017ABE	BLOCK-1051
SCL4017ABF	BLOCK-1051
SCL4017ABH	BLOCK-1051
SCL4017AC	BLOCK-1051
SCL4017AD	BLOCK-1051
SCL4017AE	BLOCK-1051
SCL4017AF	BLOCK-1051
SCL4017AH	BLOCK-1051
SCL4017B	BLOCK-1051
SCL4017BC	BLOCK-1051
SCL4017BD	BLOCK-1051
SCL4017BE	BLOCK-1051
SCL4017BF	BLOCK-1051
SCL4017BH	BLOCK-1051
SCL4018AC	BLOCK-1053
SCL4018AD	BLOCK-1053
SCL4018AE	BLOCK-1053
SCL4018AF	BLOCK-1053
SCL4018AH	BLOCK-1053
SCL4018B	BLOCK-1053
SCL4018BD	BLOCK-1053
SCL4018BE	BLOCK-1053
SCL4018BF	BLOCK-1053
SCL4018BH	BLOCK-1053
SCL4019	BLOCK-1058
SCL4020	BLOCK-1059
SCL4020ABC	BLOCK-1059
SCL4020ABD	BLOCK-1059
SCL4020ABE	BLOCK-1059
SCL4020ABF	BLOCK-1059
SCL4020ABH	BLOCK-1059
SCL4020AC	BLOCK-1059
SCL4020AD	BLOCK-1059
SCL4020AE	BLOCK-1059
SCL4020AF	BLOCK-1059
SCL4020AH	BLOCK-1059
SCL4020B	BLOCK-1059
SCL4021	BLOCK-1060
SCL4021BC	BLOCK-1060
SCL4021BD	BLOCK-1060
SCL4021BE	BLOCK-1060
SCL4022ABC	BLOCK-1061
SCL4022ABD	BLOCK-1061
SCL4022ABE	BLOCK-1061
SCL4022ABF	BLOCK-1061
SCL4022ABH	BLOCK-1061
SCL4022AC	BLOCK-1061
SCL4022AD	BLOCK-1061
SCL4022AE	BLOCK-1061
SCL4022AF	BLOCK-1061
SCL4022AH	BLOCK-1061
SCL4022B	BLOCK-1061
SCL4022BC	BLOCK-1061
SCL4022BD	BLOCK-1061
SCL4022BE	BLOCK-1061
SCL4022BF	BLOCK-1061
SCL4022BH	BLOCK-1061
SCL4023	BLOCK-1062
SCL4023AC	BLOCK-1062
SCL4023AD	BLOCK-1062
SCL4023AE	BLOCK-1062
SCL4023AF	BLOCK-1062
SCL4023AH	BLOCK-1062
SCL4023B	BLOCK-1062
SCL4023BC	BLOCK-1062
SCL4023BD	BLOCK-1062
SCL4023BE	BLOCK-1062
SCL4023BF	BLOCK-1062
SCL4023BH	BLOCK-1062
SCL4024	BLOCK-1063
SCL4024AC	BLOCK-1063
SCL4024AD	BLOCK-1063
SCL4024AE	BLOCK-1063
SCL4024AF	BLOCK-1063
SCL4024AH	BLOCK-1063
SCL4024AT	BLOCK-1063
SCL4024B	BLOCK-1063
SCL4024BC	BLOCK-1063
SCL4024BD	BLOCK-1063
SCL4024BE	BLOCK-1063

DEVICE TYPE	REPL CODE
SCL4024BF	BLOCK-1063
SCL4024BH	BLOCK-1063
SCL4025	BLOCK-1064
SCL4025AC	BLOCK-1064
SCL4025AD	BLOCK-1064
SCL4025AE	BLOCK-1064
SCL4025AF	BLOCK-1064
SCL4025AH	BLOCK-1064
SCL4025B	BLOCK-1064
SCL4025BC	BLOCK-1064
SCL4025BD	BLOCK-1064
SCL4025BE	BLOCK-1064
SCL4025BF	BLOCK-1064
SCL4025BH	BLOCK-1064
SCL4026AB	BLOCK-1065
SCL4026ABC	BLOCK-1065
SCL4026ABD	BLOCK-1065
SCL4026ABE	BLOCK-1065
SCL4027	BLOCK-1066
SCL4027B	BLOCK-1066
SCL4027BC	BLOCK-1066
SCL4027BD	BLOCK-1066
SCL4027BE	BLOCK-1066
SCL4027BF	BLOCK-1066
SCL4027BH	BLOCK-1066
SCL4028AC	BLOCK-1067
SCL4028AD	BLOCK-1067
SCL4028AE	BLOCK-1067
SCL4028AF	BLOCK-1067
SCL4028AH	BLOCK-1067
SCL4028B	BLOCK-1067
SCL4028BC	BLOCK-1067
SCL4028BD	BLOCK-1067
SCL4028BE	BLOCK-1067
SCL4028BF	BLOCK-1067
SCL4028BH	BLOCK-1067
SCL4029AC	BLOCK-1068
SCL4029AD	BLOCK-1068
SCL4029AE	BLOCK-1068
SCL4029AF	BLOCK-1068
SCL4029AH	BLOCK-1068
SCL4029B	BLOCK-1068
SCL4029BC	BLOCK-1068
SCL4029BD	BLOCK-1068
SCL4029BE	BLOCK-1068
SCL4029BF	BLOCK-1068
SCL4029BH	BLOCK-1068
SCL4030	BLOCK-1069
SCL4030AC	BLOCK-1069
SCL4030AD	BLOCK-1069
SCL4030AE	BLOCK-1069
SCL4030AF	BLOCK-1069
SCL4030AH	BLOCK-1069
SCL4030B	BLOCK-1069
SCL4030BC	BLOCK-1069
SCL4030BD	BLOCK-1069
SCL4030BE	BLOCK-1069
SCL4030BF	BLOCK-1069
SCL4030BH	BLOCK-1069
SCL4033AB	BLOCK-1072
SCL4033ABC	BLOCK-1072
SCL4033ABD	BLOCK-1072
SCL4033ABE	BLOCK-1072
SCL4034BC	BLOCK-1073
SCL4034BD	BLOCK-1073
SCL4034BE	BLOCK-1073
SCL4035BC	BLOCK-1074
SCL4035BD	BLOCK-1074
SCL4035BE	BLOCK-1074
SCL4040	BLOCK-1076
SCL4040ABC	BLOCK-1076
SCL4040ABD	BLOCK-1076
SCL4040AC	BLOCK-1076
SCL4040AD	BLOCK-1076
SCL4040AE	BLOCK-1076
SCL4040AF	BLOCK-1076
SCL4040AH	BLOCK-1076
SCL4040B	BLOCK-1076
SCL4042	BLOCK-1077
SCL4042B	BLOCK-1077
SCL4042BC	BLOCK-1077
SCL4042BD	BLOCK-1077
SCL4042BE	BLOCK-1077
SCL4042BF	BLOCK-1077
SCL4042BH	BLOCK-1077
SCL4043ABC	BLOCK-1078
SCL4043ABD	BLOCK-1078
SCL4043ABE	BLOCK-1078
SCL4043ABF	BLOCK-1078
SCL4043ABH	BLOCK-1078
SCL4043B	BLOCK-1078
SCL4043BC	BLOCK-1078
SCL4043BD	BLOCK-1078
SCL4043BE	BLOCK-1078

DEVICE TYPE	REPL CODE
SCL4043BF	BLOCK-1078
SCL4043BH	BLOCK-1078
SCL4044ABC	BLOCK-1079
SCL4044ABE	BLOCK-1079
SCL4044ABF	BLOCK-1079
SCL4044ABH	BLOCK-1079
SCL4044B	BLOCK-1079
SCL4044BC	BLOCK-1079
SCL4044BD	BLOCK-1079
SCL4044BE	BLOCK-1079
SCL4044BF	BLOCK-1079
SCL4044BH	BLOCK-1079
SCL4047B	BLOCK-1082
SCL4047BC	BLOCK-1082
SCL4047BD	BLOCK-1082
SCL4047BE	BLOCK-1082
SCL4047BF	BLOCK-1082
SCL4047BH	BLOCK-1082
SCL4049	BLOCK-1084
SCL4050	BLOCK-1085
SCL4051	BLOCK-1086
SCL4052	BLOCK-1087
SCL4060ABC	BLOCK-1091
SCL4060ABD	BLOCK-1091
SCL4060ABE	BLOCK-1091
SCL4060ABF	BLOCK-1091
SCL4060ABH	BLOCK-1091
SCL4060AC	BLOCK-1091
SCL4060AD	BLOCK-1091
SCL4060AE	BLOCK-1091
SCL4060AF	BLOCK-1091
SCL4060AH	BLOCK-1091
SCL4060B	BLOCK-1091
SCL4068B	BLOCK-1095
SCL4068BC	BLOCK-1095
SCL4068BD	BLOCK-1095
SCL4068BE	BLOCK-1095
SCL4068BF	BLOCK-1095
SCL4068BH	BLOCK-1095
SCL4070B	BLOCK-1097
SCL4070BC	BLOCK-1097
SCL4070BD	BLOCK-1097
SCL4070BE	BLOCK-1097
SCL4070BF	BLOCK-1097
SCL4070BH	BLOCK-1097
SCL4071AC	BLOCK-1098
SCL4071AD	BLOCK-1098
SCL4071AE	BLOCK-1098
SCL4071AF	BLOCK-1098
SCL4071AH	BLOCK-1098
SCL4071B	BLOCK-1098
SCL4071BC	BLOCK-1098
SCL4071BD	BLOCK-1098
SCL4071BE	BLOCK-1098
SCL4071BF	BLOCK-1098
SCL4071BH	BLOCK-1098
SCL4072AC	BLOCK-1099
SCL4072AD	BLOCK-1099
SCL4072AE	BLOCK-1099
SCL4072AF	BLOCK-1099
SCL4072AH	BLOCK-1099
SCL4072B	BLOCK-1099
SCL4072BC	BLOCK-1099
SCL4072BD	BLOCK-1099
SCL4072BE	BLOCK-1099
SCL4072BF	BLOCK-1099
SCL4072BH	BLOCK-1099
SCL4073AC	BLOCK-1100
SCL4073AD	BLOCK-1100
SCL4073AE	BLOCK-1100
SCL4073AF	BLOCK-1100
SCL4073AH	BLOCK-1100
SCL4073B	BLOCK-1100
SCL4073BC	BLOCK-1100
SCL4073BD	BLOCK-1100
SCL4073BE	BLOCK-1100
SCL4073BF	BLOCK-1100
SCL4073BH	BLOCK-1100
SCL4075AC	BLOCK-1101
SCL4075AD	BLOCK-1101
SCL4075AE	BLOCK-1101
SCL4075AF	BLOCK-1101
SCL4075AH	BLOCK-1101
SCL4075B	BLOCK-1101
SCL4075BC	BLOCK-1101
SCL4075BD	BLOCK-1101
SCL4075BE	BLOCK-1101
SCL4075BF	BLOCK-1101
SCL4075BH	BLOCK-1101
SCL4076BC	BLOCK-1102
SCL4076BD	BLOCK-1102
SCL4076BE	BLOCK-1102
SCL4077B	BLOCK-1103
SCL4077BC	BLOCK-1103

DEVICE TYPE	REPL CODE
SCL4077BD	BLOCK-1103
SCL4077BE	BLOCK-1103
SCL4077BF	BLOCK-1103
SCL4077BH	BLOCK-1103
SCL4078B	BLOCK-1104
SCL4078BC	BLOCK-1104
SCL4078BD	BLOCK-1104
SCL4078BE	BLOCK-1104
SCL4078BF	BLOCK-1104
SCL4078BH	BLOCK-1104
SCL4081	BLOCK-1105
SCL4081AC	BLOCK-1105
SCL4081AD	BLOCK-1105
SCL4081AE	BLOCK-1105
SCL4081AF	BLOCK-1105
SCL4081AH	BLOCK-1105
SCL4081B	BLOCK-1105
SCL4081BC	BLOCK-1105
SCL4081BD	BLOCK-1105
SCL4081BE	BLOCK-1105
SCL4081BF	BLOCK-1105
SCL4081BH	BLOCK-1105
SCL4082B	BLOCK-1106
SCL4082BC	BLOCK-1106
SCL4082BD	BLOCK-1106
SCL4082BE	BLOCK-1106
SCL4082BF	BLOCK-1106
SCL4082BH	BLOCK-1106
SCL4085B	BLOCK-1107
SCL4085BC	BLOCK-1106
SCL4085BD	BLOCK-1107
SCL4085BE	BLOCK-1107
SCL4085BF	BLOCK-1107
SCL4085BH	BLOCK-1107
SCL4086B	BLOCK-1108
SCL4086BC	BLOCK-1108
SCL4086BD	BLOCK-1108
SCL4086BE	BLOCK-1108
SCL4086BF	BLOCK-1108
SCL4086BH	BLOCK-1108
SCL4160B	BLOCK-1044
SCL4161B	BLOCK-1045
SCL4162B	BLOCK-1046
SCL4163B	BLOCK-1047
SCL4174B	BLOCK-1049
SCL4449UBE	BLOCK-1084
SCL4502BE	BLOCK-1118
SCL4510AC	BLOCK-1121
SCL4510AD	BLOCK-1121
SCL4510AE	BLOCK-1121
SCL4510AF	BLOCK-1121
SCL4510AH	BLOCK-1121
SCL4510B	BLOCK-1121
SCL4510BC	BLOCK-1121
SCL4510BD	BLOCK-1121
SCL4510BE	BLOCK-1121
SCL4510BF	BLOCK-1121
SCL4510BH	BLOCK-1121
SCL4511AC	BLOCK-1122
SCL4511AD	BLOCK-1122
SCL4511AE	BLOCK-1122
SCL4511AF	BLOCK-1122
SCL4511AH	BLOCK-1122
SCL4511BC	BLOCK-1122
SCL4511BD	BLOCK-1122
SCL4511BE	BLOCK-1122
SCL4511BF	BLOCK-1122
SCL4511BH	BLOCK-1122
SCL4512BC	BLOCK-1123
SCL4512BD	BLOCK-1123
SCL4512BE	BLOCK-1123
SCL4514B	BLOCK-1125
SCL4514BD	BLOCK-1125
SCL4514BE	BLOCK-1125
SCL4514BH	BLOCK-1125
SCL4515B	BLOCK-1126
SCL4515BD	BLOCK-1126
SCL4515BE	BLOCK-1126
SCL4515BH	BLOCK-1126
SCL4516AC	BLOCK-1127
SCL4516AD	BLOCK-1127
SCL4516AE	BLOCK-1127
SCL4516AF	BLOCK-1127
SCL4516AH	BLOCK-1127
SCL4516B	BLOCK-1127
SCL4516BC	BLOCK-1127
SCL4516BD	BLOCK-1127
SCL4516BE	BLOCK-1127
SCL4516BF	BLOCK-1127
SCL4516BH	BLOCK-1127
SCL4518	BLOCK-1129
SCL4518AC	BLOCK-1129
SCL4518AD	BLOCK-1129
SCL4518AE	BLOCK-1129

DEVICE TYPE	REPL CODE
SCL4518AF	BLOCK-1129
SCL4518AH	BLOCK-1129
SCL4518B	BLOCK-1129
SCL4518BC	BLOCK-1129
SCL4518BD	BLOCK-1129
SCL4518BE	BLOCK-1129
SCL4518BF	BLOCK-1129
SCL4518BH	BLOCK-1129
SCL4520AC	BLOCK-1130
SCL4520AD	BLOCK-1130
SCL4520AE	BLOCK-1130
SCL4520AF	BLOCK-1130
SCL4520AH	BLOCK-1130
SCL4520B	BLOCK-1130
SCL4520BC	BLOCK-1130
SCL4520BD	BLOCK-1130
SCL4520BE	BLOCK-1130
SCL4520BF	BLOCK-1130
SCL4520BH	BLOCK-1130
SCL4522BC	BLOCK-1132
SCL4522BD	BLOCK-1132
SCL4522BE	BLOCK-1132
SCL4522BF	BLOCK-1132
SCL4522BH	BLOCK-1132
SCL4526B	BLOCK-1133
SCL4527B	BLOCK-1134
SCL4527BC	BLOCK-1134
SCL4527BD	BLOCK-1134
SCL4527BE	BLOCK-1134
SCL4527BF	BLOCK-1134
SCL4527BH	BLOCK-1134
SCL4528B	BLOCK-1114
SCL4543BC	BLOCK-1142
SCL4543BE	BLOCK-1143
SCL4555B	BLOCK-1146
SCL4555BC	BLOCK-1146
SCL4555BD	BLOCK-1146
SCL4555BE	BLOCK-1146
SCL4555BF	BLOCK-1146
SCL4555BH	BLOCK-1146
SCL4556BC	BLOCK-1147
SCL4556BD	BLOCK-1147
SCL4556BE	BLOCK-1147
SCL4556BF	BLOCK-1147
SCL4556BH	BLOCK-1147
SCM90072C	BLOCK-1003
SCM90072P	BLOCK-1003
SE5-0930	BLOCK-2166
SE5-0933	BLOCK-1269
SE555V	BLOCK-2079
SFC2109RM	BLOCK-2004
SFC2209R	BLOCK-2004
SFC2301A	BLOCK-160
SFC2309R	BLOCK-2004
SFC2458DC	BLOCK-1801
SFC2709C	BLOCK-1949
SFC2710C	BLOCK-1955
SFC2711C	BLOCK-1958
SFC2741C	BLOCK-2056
SFC2748C	BLOCK-160
SFC2748M	BLOCK-160
SFC2805EC	BLOCK-2087
SFC2805RC	BLOCK-1022
SFC2806EC	BLOCK-2093
SFC2808EC	BLOCK-2095
SFC2812EC	BLOCK-2097
SFC2812RC	BLOCK-2014
SFC2815EC	BLOCK-2119
SFC2824EC	BLOCK-2137
SFC400E	BLOCK-1293
SFC401E	BLOCK-1294
SFC402E	BLOCK-1295
SFC4050	BLOCK-1786
SFC4052	BLOCK-1786
SFC408E	BLOCK-1301
SFC409E	BLOCK-1302
SFC4107E	BLOCK-1305
SFC4121E	BLOCK-1311
SFC4122E	BLOCK-1312
SFC4123E	BLOCK-1313
SFC4141E	BLOCK-1321
SFC4155E	BLOCK-1332
SFC4156E	BLOCK-1333
SFC4180E	BLOCK-1352
SFC4181E	BLOCK-1353
SFC4182E	BLOCK-1354
SFC4192E	BLOCK-1357
SFC4193E	BLOCK-1358
SFC420E	BLOCK-1365
SFC430E	BLOCK-1377
SFC437E	BLOCK-1384
SFC438E	BLOCK-1385
SFC440E	BLOCK-1390
SFC442E	BLOCK-1392

DEVICE TYPE	REPL CODE
SFC450E	BLOCK-1401
SFC451E	BLOCK-1402
SFC453E	BLOCK-1403
SFC454E	BLOCK-1404
SFC460E	BLOCK-1406
SFC472E	BLOCK-1409
SFC473E	BLOCK-1410
SFC474E	BLOCK-1411
SFC475E	BLOCK-1412
SFC476E	BLOCK-1413
SFC485E	BLOCK-1418
SFC486E	BLOCK-1419
SFC492E	BLOCK-1425
SFC6050	BLOCK-1794
SG1458M	BLOCK-1801
SG1488J	BLOCK-1771
SG1489J	BLOCK-1772
SG1496	BLOCK-2138
SG1496T	BLOCK-2138
SG201AT	BLOCK-160
SG224J	BLOCK-2171
SG224N	BLOCK-2171
SG301AM	BLOCK-2141
SG301AT	BLOCK-160
SG3081N	BLOCK-1970
SG308AM	BLOCK-2042
SG308AT	BLOCK-2037
SG308AY	BLOCK-2042
SG308M	BLOCK-2042
SG308T	BLOCK-2037
SG308Y	BLOCK-2042
SG309K	BLOCK-1022
SG310M	BLOCK-1986
SG310T	BLOCK-1985
SG323K	BLOCK-2004
SG324N	BLOCK-2171
SG340-05K	BLOCK-2004
SG340-12K	BLOCK-2014
SG3524	BLOCK-701
SG3525A	BLOCK-702
SG3526N	BLOCK-703
SG3527A	BLOCK-704
SG555CM	BLOCK-2079
SG555M	BLOCK-2079
SG710CN	BLOCK-1957
SG710CT	BLOCK-1955
SG711CN	BLOCK-1962
SG711CT	BLOCK-1958
SG723CN	BLOCK-1984
SG723CT	BLOCK-1983
SG723T	BLOCK-1983
SG733CJ	BLOCK-1990
SG733CT	BLOCK-1990
SG7400N	BLOCK-1293
SG7402N	BLOCK-1295
SG7410N	BLOCK-1304
SG741CM	BLOCK-2058
SG741CN	BLOCK-2057
SG741CT	BLOCK-2056
SG741SCT	BLOCK-2056
SG7420N	BLOCK-1365
SG7430N	BLOCK-1377
SG7440N	BLOCK-1390
SG7450N	BLOCK-1401
SG7451N	BLOCK-1402
SG7453N	BLOCK-1403
SG7454N	BLOCK-1404
SG7460N	BLOCK-1406
SG747CT	BLOCK-2069
SG7812	BLOCK-2097
SG7812CK	BLOCK-2014
SG7812T	BLOCK-2074
SG7815T	BLOCK-2075
SG7818ACP	BLOCK-2085
SG7818CK	BLOCK-896
SG7818CP	BLOCK-2085
SG7824CK	BLOCK-898
SG7908ACP	BLOCK-2096
SG7908CP	BLOCK-2096
SG7918ACP	BLOCK-2086
SG7918CK	BLOCK-897
SGS17139XRH	BLOCK-111
SH323SC	BLOCK-2004
SI-3052P	BLOCK-905
SI-3052V	BLOCK-906
SI-3122P	BLOCK-907
SI-3122V	BLOCK-907
SI-3152P	BLOCK-908
SI-3242P	BLOCK-909
SI-7115B	BLOCK-827
SI-CA3064E	BLOCK-1808
SI-CA3065E	BLOCK-1269
SI-HA11580	BLOCK-184
SI-LM3065N	BLOCK-1269

DEVICE TYPE	REPL CODE
SI-MC1352P	BLOCK-1422
SI-MC1358P	BLOCK-1269
SI-MC1364P	BLOCK-1808
SI-SN76650N	BLOCK-1422
SI-TA7074P	BLOCK-1422
SI-TA7075P	BLOCK-78
SI-UPC1031H	BLOCK-234
SI-UPC574J	BLOCK-1157
SI-UPC580C	BLOCK-184
SI3052V	BLOCK-906
SI7115B	BLOCK-827
SIIR3N05XXXXX	BLOCK-1868
SILC4001BXXXX	BLOCK-1029
SILC4069UBXXX	BLOCK-1096
SILM8560BXXXX	BLOCK-975
SITA7640APIXX	BLOCK-821
SK10001	BLOCK-730
SK10002	BLOCK-735
SK10003	BLOCK-740
SK10007	BLOCK-309
SK10008	BLOCK-681
SK10010	BLOCK-600
SK10012	BLOCK-697
SK10013	BLOCK-682
SK10016	BLOCK-683
SK10017	BLOCK-688
SK10018	BLOCK-687
SK10019	BLOCK-694
SK10021	BLOCK-696
SK10022	BLOCK-698
SK10023	BLOCK-699
SK10024	BLOCK-702
SK10025	BLOCK-703
SK10026	BLOCK-704
SK10028	BLOCK-707
SK10029	BLOCK-717
SK10030	BLOCK-718
SK10031	BLOCK-720
SK10033	BLOCK-726
SK10035	BLOCK-731
SK10036	BLOCK-733
SK10037	BLOCK-734
SK10038	BLOCK-736
SK10041	BLOCK-739
SK10042	BLOCK-741
SK10044	BLOCK-743
SK10045	BLOCK-744
SK10047	BLOCK-880
SK10048	BLOCK-884
SK10049	BLOCK-885
SK10050	BLOCK-886
SK10052	BLOCK-887
SK10053	BLOCK-896
SK10054	BLOCK-897
SK10055	BLOCK-898
SK10056	BLOCK-899
SK10058	BLOCK-910
SK10061	BLOCK-1803
SK10062	BLOCK-1871
SK10063	BLOCK-1874
SK10064	BLOCK-1911
SK10065	BLOCK-1913
SK10066	BLOCK-1924
SK10067	BLOCK-1931
SK10069	BLOCK-1974
SK10070	BLOCK-1982
SK10071	BLOCK-1991
SK10072	BLOCK-1994
SK10073	BLOCK-1995
SK10074	BLOCK-2060
SK10075	BLOCK-2064
SK10076	BLOCK-2065
SK10077	BLOCK-2066
SK10079	BLOCK-2072
SK10080	BLOCK-689
SK10081	BLOCK-556
SK10082	BLOCK-824
SK10083	BLOCK-825
SK10084	BLOCK-826
SK10085	BLOCK-835
SK10086	BLOCK-866
SK10087	BLOCK-906
SK10139	BLOCK-1802
SK10141	BLOCK-854
SK10142	BLOCK-812
SK10143	BLOCK-818
SK10144	BLOCK-819
SK10145	BLOCK-820
SK10146	BLOCK-822
SK10147	BLOCK-823
SK10148	BLOCK-827
SK10149	BLOCK-828
SK10150	BLOCK-829
SK10151	BLOCK-831

DEVICE TYPE	REPL CODE
SK10152	BLOCK-832
SK10153	BLOCK-833
SK10154	BLOCK-834
SK10155	BLOCK-836
SK10156	BLOCK-837
SK10158	BLOCK-841
SK10159	BLOCK-842
SK10161	BLOCK-851
SK10164	BLOCK-852
SK10166	BLOCK-855
SK10168	BLOCK-857
SK10169	BLOCK-859
SK10170	BLOCK-860
SK10171	BLOCK-861
SK10172	BLOCK-862
SK10173	BLOCK-863
SK10174	BLOCK-864
SK10175	BLOCK-865
SK10190	BLOCK-972
SK10191	BLOCK-1116
SK10221	BLOCK-587
SK10223	BLOCK-589
SK10226	BLOCK-596
SK10230	BLOCK-608
SK10232	BLOCK-612
SK10233	BLOCK-618
SK10234	BLOCK-621
SK10238	BLOCK-627
SK10239	BLOCK-628
SK10241	BLOCK-656
SK10242	BLOCK-658
SK10244	BLOCK-665
SK10246	BLOCK-667
SK10247	BLOCK-672
SK10252	BLOCK-638
SK10274	BLOCK-529
SK10276	BLOCK-949
SK10277	BLOCK-950
SK10278	BLOCK-964
SK10279	BLOCK-965
SK10280	BLOCK-966
SK10281	BLOCK-967
SK10282	BLOCK-970
SK10283	BLOCK-976
SK10284	BLOCK-977
SK10285	BLOCK-978
SK10286	BLOCK-979
SK10287	BLOCK-980
SK10288	BLOCK-981
SK10289	BLOCK-982
SK10290	BLOCK-983
SK10291	BLOCK-984
SK10292	BLOCK-985
SK10293	BLOCK-986
SK10294	BLOCK-987
SK10295	BLOCK-975
SK10296	BLOCK-1005
SK10297	BLOCK-1006
SK10298	BLOCK-1014
SK10447	BLOCK-1978
SK10448	BLOCK-2053
SK10449	BLOCK-2080
SK10450	BLOCK-2082
SK10451	BLOCK-2214
SK10452	BLOCK-2172
SK10453	BLOCK-2147
SK10454	BLOCK-2146
SK10455	BLOCK-2142
SK10456	BLOCK-1167
SK10469	BLOCK-1180
SK10470	BLOCK-1181
SK10471	BLOCK-1182
SK10472	BLOCK-1183
SK10473	BLOCK-1184
SK10474	BLOCK-1185
SK10475	BLOCK-1186
SK10476	BLOCK-1187
SK10477	BLOCK-1188
SK10478	BLOCK-1189
SK10479	BLOCK-1191
SK10480	BLOCK-1192
SK10481	BLOCK-1193
SK10482	BLOCK-1194
SK10483	BLOCK-1195
SK10484	BLOCK-1196
SK10485	BLOCK-1197
SK10486	BLOCK-1198
SK10488	BLOCK-1200
SK10489	BLOCK-1202
SK10493	BLOCK-633
SK10494	BLOCK-747
SK10496	BLOCK-846
SK10499	BLOCK-850
SK10500	BLOCK-849

DEVICE TYPE	REPL CODE
SK10501	BLOCK-869
SK10502	BLOCK-871
SK10505	BLOCK-875
SK10507	BLOCK-877
SK10508	BLOCK-878
SK10509	BLOCK-845
SK10514	BLOCK-1934
SK10515	BLOCK-2227
SK10516	BLOCK-2228
SK10519	BLOCK-1004
SK10520	BLOCK-870
SK1900	BLOCK-1169
SK1901	BLOCK-1164
SK1911	BLOCK-968
SK1912	BLOCK-1018
SK1914	BLOCK-1738
SK1915	BLOCK-1737
SK1917	BLOCK-1720
SK1918	BLOCK-1672
SK1919	BLOCK-1639
SK1920	BLOCK-1627
SK1921	BLOCK-1625
SK1922	BLOCK-1614
SK1923	BLOCK-1603
SK1924	BLOCK-1575
SK2214	BLOCK-1002
SK2716	BLOCK-1017
SK2880	BLOCK-1025
SK2881	BLOCK-1026
SK2882	BLOCK-1027
SK3022	BLOCK-177
SK3023	BLOCK-1223
SK3070	BLOCK-1267
SK3071	BLOCK-1283
SK3072	BLOCK-1269
SK3073	BLOCK-1284
SK3074	BLOCK-1285
SK3075	BLOCK-1271
SK3076	BLOCK-1272
SK3077	BLOCK-1270
SK3078	BLOCK-1814
SK3101	BLOCK-1245
SK3102	BLOCK-1265
SK3129	BLOCK-1282
SK3134	BLOCK-1244
SK3135	BLOCK-1262
SK3140	BLOCK-1811
SK3141	BLOCK-1805
SK3143	BLOCK-1286
SK3144	BLOCK-1279
SK3146	BLOCK-1812
SK3147	BLOCK-1813
SK3149	BLOCK-1816
SK3152	BLOCK-23
SK3153	BLOCK-26
SK3154	BLOCK-310
SK3155	BLOCK-24
SK3157	BLOCK-1212
SK3158	BLOCK-1821
SK3159	BLOCK-1275
SK3160	BLOCK-1825
SK3162	BLOCK-1281
SK3163	BLOCK-1289
SK3164	BLOCK-1983
SK3165	BLOCK-1984
SK3166	BLOCK-2073
SK3167	BLOCK-1291
SK3168	BLOCK-1422
SK3169	BLOCK-1806
SK3170	BLOCK-1287
SK3171	BLOCK-1389
SK3172	BLOCK-1376
SK3184	BLOCK-112
SK3185	BLOCK-2168
SK3186	BLOCK-163
SK3204	BLOCK-1287
SK3205	BLOCK-2166
SK3206	BLOCK-1876
SK3207	BLOCK-1846
SK3208	BLOCK-1886
SK3209	BLOCK-1816
SK3210	BLOCK-165
SK3211	BLOCK-1821
SK3212	BLOCK-164
SK3213	BLOCK-166
SK3214	BLOCK-2169
SK3215	BLOCK-1808
SK3216	BLOCK-1822
SK3223	BLOCK-98
SK3224	BLOCK-100
SK3225	BLOCK-102
SK3226	BLOCK-69
SK3227	BLOCK-67
SK3228	BLOCK-59

DEVICE TYPE	REPL CODE
SK3229	BLOCK-108
SK3230	BLOCK-144
SK3231	BLOCK-145
SK3233	BLOCK-2138
SK3234	BLOCK-1405
SK3235	BLOCK-1292
SK3236	BLOCK-1421
SK3237	BLOCK-1820
SK3238	BLOCK-1823
SK3240	BLOCK-1804
SK3242	BLOCK-1844
SK3243	BLOCK-149
SK3249	BLOCK-50
SK3254	BLOCK-1810
SK3255	BLOCK-1843
SK3276	BLOCK-1400
SK3277	BLOCK-1826
SK3278	BLOCK-1827
SK3279	BLOCK-1407
SK3280	BLOCK-1770
SK3281	BLOCK-101
SK3282	BLOCK-143
SK3283	BLOCK-99
SK3284	BLOCK-78
SK3285	BLOCK-103
SK3286	BLOCK-128
SK3287	BLOCK-129
SK3288	BLOCK-4
SK3289	BLOCK-147
SK3290	BLOCK-148
SK3291	BLOCK-88
SK3292	BLOCK-139
SK3295	BLOCK-70
SK3328	BLOCK-1303
SK3358	BLOCK-7
SK3361	BLOCK-40
SK3362	BLOCK-39
SK3363	BLOCK-42
SK3364	BLOCK-646
SK3365	BLOCK-5
SK3366	BLOCK-38
SK3367	BLOCK-41
SK3368	BLOCK-28
SK3369	BLOCK-34
SK3370	BLOCK-35
SK3371	BLOCK-36
SK3372	BLOCK-29
SK3373	BLOCK-227
SK3375	BLOCK-1290
SK3376	BLOCK-10
SK342	BLOCK-1944
SK3435	BLOCK-27
SK3436	BLOCK-27
SK3445	BLOCK-180
SK3446	BLOCK-95
SK3451	BLOCK-1830
SK3453	BLOCK-1364
SK3454	BLOCK-1815
SK3455	BLOCK-1828
SK3456	BLOCK-1917
SK3457	BLOCK-52
SK3458	BLOCK-54
SK3459	BLOCK-56
SK3460	BLOCK-58
SK3461	BLOCK-80
SK3462	BLOCK-2144
SK3465	BLOCK-1801
SK3468	BLOCK-174
SK3469	BLOCK-183
SK3470	BLOCK-47
SK3471	BLOCK-44
SK3472	BLOCK-90
SK3473	BLOCK-136
SK3474	BLOCK-79
SK3475	BLOCK-48
SK3476	BLOCK-83
SK3477	BLOCK-85
SK3478	BLOCK-127
SK3480	BLOCK-167
SK3482	BLOCK-193
SK3483	BLOCK-231
SK3484	BLOCK-182
SK3485	BLOCK-138
SK3486	BLOCK-173
SK3487	BLOCK-223
SK3488	BLOCK-126
SK3489	BLOCK-131
SK3490	BLOCK-130
SK3491	BLOCK-1264
SK3492	BLOCK-11
SK3493	BLOCK-211
SK3494	BLOCK-53
SK3495	BLOCK-72
SK3496	BLOCK-71

DEVICE TYPE	REPL CODE
SK3497	BLOCK-237
SK3498	BLOCK-76
SK3499	BLOCK-9
SK3514	BLOCK-2056
SK3524	BLOCK-1809
SK3525	BLOCK-1280
SK3526	BLOCK-2069
SK3539	BLOCK-1948
SK3540	BLOCK-1943
SK3541	BLOCK-1966
SK3542	BLOCK-1944
SK3543	BLOCK-1963
SK3544	BLOCK-1971
SK3545	BLOCK-1947
SK3546	BLOCK-1945
SK3547	BLOCK-1940
SK3548	BLOCK-1946
SK3549	BLOCK-1941
SK3550	BLOCK-1970
SK3552	BLOCK-2058
SK3553	BLOCK-2056
SK3556	BLOCK-2070
SK3564	BLOCK-2079
SK3565	BLOCK-160
SK3567	BLOCK-1980
SK3567A	BLOCK-1980
SK3568	BLOCK-1997
SK3569	BLOCK-1873
SK3590	BLOCK-1954
SK3591	BLOCK-2087
SK3592	BLOCK-2097
SK3593	BLOCK-2119
SK3595	BLOCK-2174
SK3596	BLOCK-2143
SK3629	BLOCK-1022
SK3630	BLOCK-2095
SK3637	BLOCK-224
SK3641	BLOCK-2141
SK3643	BLOCK-2171
SK3644	BLOCK-2141
SK3645	BLOCK-160
SK3646	BLOCK-1926
SK3667	BLOCK-947
SK3668	BLOCK-1981
SK3669	BLOCK-2093
SK3670	BLOCK-2137
SK3671	BLOCK-2091
SK3672	BLOCK-2094
SK3673	BLOCK-2108
SK3674	BLOCK-2128
SK3675	BLOCK-2136
SK3688	BLOCK-2191
SK3689	BLOCK-2145
SK3691	BLOCK-1992
SK3692	BLOCK-1993
SK3694	BLOCK-946
SK3695	BLOCK-1996
SK3696	BLOCK-1997
SK3699	BLOCK-2085
SK3700	BLOCK-205
SK3701	BLOCK-181
SK3702	BLOCK-441
SK3703	BLOCK-94
SK3704	BLOCK-201
SK3705	BLOCK-178
SK3706	BLOCK-170
SK3707	BLOCK-226
SK3708	BLOCK-158
SK3711	BLOCK-107
SK3712	BLOCK-196
SK3713	BLOCK-195
SK3714	BLOCK-188
SK3723	BLOCK-6
SK3724	BLOCK-2160
SK3725	BLOCK-184
SK3726	BLOCK-265
SK3727	BLOCK-153
SK3728	BLOCK-157
SK3729	BLOCK-162
SK3730	BLOCK-210
SK3731	BLOCK-232
SK3732	BLOCK-156
SK3733	BLOCK-185
SK3734	BLOCK-1829
SK3735	BLOCK-204
SK3736	BLOCK-202
SK3737	BLOCK-168
SK3738	BLOCK-203
SK3739	BLOCK-199
SK3740	BLOCK-206
SK3741	BLOCK-207
SK3742	BLOCK-176
SK3743	BLOCK-121
SK3762	BLOCK-215

DEVICE TYPE	REPL CODE
SK3763	BLOCK-214
SK3827	BLOCK-155
SK3828	BLOCK-230
SK3829	BLOCK-1813
SK3832	BLOCK-219
SK3833	BLOCK-295
SK3852	BLOCK-221
SK3853	BLOCK-361
SK3872	BLOCK-248
SK3873	BLOCK-245
SK3874	BLOCK-159
SK3875	BLOCK-278
SK3876	BLOCK-132
SK3877	BLOCK-73
SK3878	BLOCK-234
SK3879	BLOCK-235
SK3880	BLOCK-243
SK3881	BLOCK-515
SK3882	BLOCK-1849
SK3884	BLOCK-523
SK3885	BLOCK-425
SK3887	BLOCK-2167
SK3888	BLOCK-169
SK3889	BLOCK-216
SK3890	BLOCK-141
SK3891	BLOCK-1859
SK3892	BLOCK-2139
SK3914	BLOCK-453
SK3916	BLOCK-260
SK3918	BLOCK-2170
SK3919	BLOCK-1850
SK3920	BLOCK-250
SK3921	BLOCK-255
SK3922	BLOCK-272
SK3923	BLOCK-273
SK3924	BLOCK-2061
SK3927	BLOCK-1848
SK3928	BLOCK-1847
SK3962	BLOCK-881
SK3963	BLOCK-238
SK3964	BLOCK-427
SK3965	BLOCK-2140
SK3966	BLOCK-973
SK3967	BLOCK-431
SK3968	BLOCK-150
SK3969	BLOCK-114
SK3973	BLOCK-2173
SK3974	BLOCK-1819
SK3975	BLOCK-934
SK4000	BLOCK-1028
SK4000B	BLOCK-1028
SK4000UB	BLOCK-1028
SK4001	BLOCK-1029
SK4001B	BLOCK-1029
SK4002	BLOCK-1030
SK4002B	BLOCK-1030
SK4006B	BLOCK-1031
SK4007	BLOCK-1032
SK4007UB	BLOCK-1032
SK4008B	BLOCK-1034
SK4009	BLOCK-1084
SK40098B	BLOCK-1036
SK4009UB	BLOCK-1084
SK40100B	BLOCK-1037
SK40106B	BLOCK-1038
SK4011	BLOCK-1039
SK4011B	BLOCK-1039
SK4012	BLOCK-1040
SK4012B	BLOCK-1040
SK4013	BLOCK-1041
SK4013B	BLOCK-1041
SK4014B	BLOCK-1042
SK4015	BLOCK-1043
SK4015B	BLOCK-1043
SK4016	BLOCK-1048
SK40160B	BLOCK-1044
SK40161	BLOCK-1045
SK40161B	BLOCK-1045
SK40162B	BLOCK-1046
SK40163B	BLOCK-1047
SK4016B	BLOCK-1048
SK4017	BLOCK-1051
SK40174B	BLOCK-1049
SK40175B	BLOCK-1050
SK4017B	BLOCK-1051
SK40182B	BLOCK-1052
SK4018B	BLOCK-1053
SK4019	BLOCK-1058
SK40192B	BLOCK-1054
SK40193B	BLOCK-1055
SK40194B	BLOCK-1056
SK40195B	BLOCK-1057
SK4019B	BLOCK-1058
SK4020	BLOCK-1059

DEVICE TYPE	REPL CODE
SK4020B	BLOCK-1059
SK4021	BLOCK-1060
SK4021B	BLOCK-1060
SK4022B	BLOCK-1061
SK4023	BLOCK-1062
SK4023B	BLOCK-1062
SK4024	BLOCK-1063
SK4024B	BLOCK-1063
SK4025	BLOCK-1064
SK4025B	BLOCK-1064
SK4026B	BLOCK-1065
SK4027	BLOCK-1066
SK4027B	BLOCK-1066
SK4028B	BLOCK-1067
SK4029	BLOCK-1068
SK4029B	BLOCK-1068
SK4030	BLOCK-1069
SK4030B	BLOCK-1069
SK4031B	BLOCK-1070
SK4032B	BLOCK-1071
SK4033B	BLOCK-1072
SK4034B	BLOCK-1073
SK4035	BLOCK-1074
SK4035B	BLOCK-1074
SK4038B	BLOCK-1075
SK4040	BLOCK-1076
SK4040B	BLOCK-1076
SK4042	BLOCK-1077
SK4042B	BLOCK-1077
SK4043B	BLOCK-1078
SK4044B	BLOCK-1079
SK4045B	BLOCK-1080
SK4046	BLOCK-2149
SK4046B	BLOCK-2149
SK4047	BLOCK-1082
SK4047B	BLOCK-1082
SK4048B	BLOCK-1083
SK4049	BLOCK-1084
SK4049UB	BLOCK-1084
SK4050	BLOCK-1085
SK4050B	BLOCK-1085
SK4051	BLOCK-1086
SK4051B	BLOCK-1086
SK4052	BLOCK-1087
SK4052B	BLOCK-1087
SK4053B	BLOCK-1088
SK4055B	BLOCK-1089
SK4056	BLOCK-1090
SK4056B	BLOCK-1090
SK4060	BLOCK-1091
SK4060B	BLOCK-1091
SK4063B	BLOCK-1092
SK4066	BLOCK-1093
SK4066B	BLOCK-1093
SK4067B	BLOCK-1094
SK4068	BLOCK-1095
SK4068B	BLOCK-1095
SK4069	BLOCK-1096
SK4069UB	BLOCK-1096
SK4070	BLOCK-1097
SK4070B	BLOCK-1097
SK4071	BLOCK-1098
SK4071B	BLOCK-1098
SK4072	BLOCK-1099
SK4072B	BLOCK-1099
SK4073	BLOCK-1100
SK4073B	BLOCK-1100
SK4075	BLOCK-1101
SK4075B	BLOCK-1101
SK4076	BLOCK-1102
SK4076B	BLOCK-1102
SK4077	BLOCK-1103
SK4077B	BLOCK-1103
SK4078	BLOCK-1104
SK4078B	BLOCK-1104
SK4081	BLOCK-1105
SK4081B	BLOCK-1105
SK4082	BLOCK-1106
SK4082B	BLOCK-1106
SK4085B	BLOCK-1107
SK4086	BLOCK-1108
SK4086B	BLOCK-1108
SK4089B	BLOCK-1109
SK4093	BLOCK-1110
SK4093B	BLOCK-1110
SK4095B	BLOCK-1111
SK4096B	BLOCK-1112
SK4097B	BLOCK-1113
SK4098	BLOCK-1114
SK4098B	BLOCK-1114
SK4099	BLOCK-1115
SK4099B	BLOCK-1115
SK4368	BLOCK-1877
SK4501UB	BLOCK-1117
SK4502B	BLOCK-1118
SK4503	BLOCK-1035
SK4503B	BLOCK-1035
SK4506UB	BLOCK-1119
SK4508B	BLOCK-1120
SK4510	BLOCK-1121
SK4510B	BLOCK-1121
SK4511	BLOCK-1122
SK4511B	BLOCK-1122
SK4512	BLOCK-1123
SK4512B	BLOCK-1123
SK4513B	BLOCK-1124
SK4514	BLOCK-1125
SK4514B	BLOCK-1125
SK4515	BLOCK-1126
SK4515B	BLOCK-1126
SK4516B	BLOCK-1127
SK4517B	BLOCK-1128
SK4518	BLOCK-1129
SK4518B	BLOCK-1129
SK4520	BLOCK-1130
SK4520B	BLOCK-1130
SK4521B	BLOCK-1131
SK4522B	BLOCK-1132
SK4526B	BLOCK-1133
SK4527B	BLOCK-1134
SK4529B	BLOCK-1136
SK4531B	BLOCK-1137
SK4532B	BLOCK-1138
SK4536B	BLOCK-1139
SK4538B	BLOCK-1140
SK4539B	BLOCK-1141
SK4541B	BLOCK-1142
SK4543B	BLOCK-1143
SK4547B	BLOCK-1144
SK4553B	BLOCK-1145
SK4555B	BLOCK-1146
SK4556B	BLOCK-1147
SK4558B	BLOCK-1148
SK4562B	BLOCK-1149
SK4566B	BLOCK-1150
SK4568B	BLOCK-1151
SK4569B	BLOCK-1152
SK4583B	BLOCK-1153
SK4584B	BLOCK-1038
SK4585B	BLOCK-1154
SK4597B	BLOCK-1155
SK4598B	BLOCK-1156
SK4801	BLOCK-205
SK4802	BLOCK-749
SK4803	BLOCK-751
SK4804	BLOCK-750
SK4812	BLOCK-619
SK4814	BLOCK-1933
SK4815	BLOCK-306
SK4816	BLOCK-647
SK4821	BLOCK-357
SK4822	BLOCK-360
SK4825	BLOCK-559
SK4826	BLOCK-1912
SK4828	BLOCK-593
SK4831	BLOCK-653
SK4833	BLOCK-650
SK4834	BLOCK-615
SK4835	BLOCK-591
SK4837	BLOCK-527
SK4838	BLOCK-463
SK4839	BLOCK-455
SK4843	BLOCK-1915
SK4910	BLOCK-993
SK5188	BLOCK-1771
SK5189	BLOCK-1772
SK6810	BLOCK-1168
SK7400	BLOCK-1293
SK7401	BLOCK-1294
SK7402	BLOCK-1295
SK7403	BLOCK-1296
SK7404	BLOCK-1297
SK7405	BLOCK-1298
SK7406	BLOCK-1299
SK7407	BLOCK-1300
SK7408	BLOCK-1301
SK7409	BLOCK-1302
SK7410	BLOCK-1304
SK74107	BLOCK-1305
SK74109	BLOCK-1306
SK7411	BLOCK-1307
SK74110	BLOCK-1308
SK74111	BLOCK-1309
SK7412	BLOCK-1310
SK74121	BLOCK-1311
SK74122	BLOCK-1312
SK74123	BLOCK-1313
SK74125	BLOCK-1314
SK74126	BLOCK-1315
SK74128	BLOCK-1316
SK7413	BLOCK-1317
SK74132	BLOCK-1318
SK7414	BLOCK-1320
SK74141	BLOCK-1321
SK74142	BLOCK-1322
SK74143	BLOCK-1323
SK74144	BLOCK-1324
SK74145	BLOCK-1325
SK74146	BLOCK-1348
SK74147	BLOCK-1326
SK74150	BLOCK-1327
SK74151	BLOCK-1328
SK74153	BLOCK-1330
SK74154	BLOCK-1331
SK74155	BLOCK-1332
SK74156	BLOCK-1333
SK74157	BLOCK-1334
SK74158	BLOCK-1335
SK7416	BLOCK-1336
SK74160	BLOCK-1337
SK74161	BLOCK-1338
SK74162	BLOCK-1339
SK74163	BLOCK-1340
SK74164	BLOCK-1341
SK74165	BLOCK-1342
SK7417	BLOCK-1343
SK74170	BLOCK-1344
SK74173	BLOCK-1345
SK74174	BLOCK-1346
SK74175	BLOCK-1347
SK74176	BLOCK-1348
SK74177	BLOCK-1349
SK74178	BLOCK-1350
SK74179	BLOCK-1351
SK74180	BLOCK-1352
SK74181	BLOCK-1353
SK74182	BLOCK-1354
SK74190	BLOCK-1355
SK74191	BLOCK-1356
SK74192	BLOCK-1357
SK74193	BLOCK-1358
SK74195	BLOCK-1359
SK74196	BLOCK-1360
SK74197	BLOCK-1361
SK74198	BLOCK-1362
SK74199	BLOCK-1363
SK7420	BLOCK-1365
SK7422	BLOCK-1367
SK74221	BLOCK-1368
SK7423	BLOCK-1369
SK74251	BLOCK-1371
SK7427	BLOCK-1372
SK7428	BLOCK-1373
SK74290	BLOCK-1374
SK74293	BLOCK-1375
SK7430	BLOCK-1377
SK7432	BLOCK-1378
SK7433	BLOCK-1379
SK74365	BLOCK-1380
SK74366	BLOCK-1381
SK74367	BLOCK-1382
SK74368	BLOCK-1383
SK7437	BLOCK-1384
SK7438	BLOCK-1385
SK7439	BLOCK-1386
SK74390	BLOCK-1387
SK7440	BLOCK-1390
SK7441	BLOCK-1391
SK7442	BLOCK-1392
SK7443	BLOCK-1393
SK7444	BLOCK-1394
SK7445	BLOCK-1395
SK7446	BLOCK-1396
SK7447	BLOCK-1397
SK7448	BLOCK-1398
SK7450	BLOCK-1401
SK7451	BLOCK-1402
SK7453	BLOCK-1403
SK7454	BLOCK-1404
SK7460	BLOCK-1406
SK7470	BLOCK-1408
SK7472	BLOCK-1409
SK7473	BLOCK-1410
SK7474	BLOCK-1411
SK7475	BLOCK-1412
SK7476	BLOCK-1413
SK7480	BLOCK-1414
SK7481	BLOCK-1415
SK7483	BLOCK-1417
SK7485	BLOCK-1418
SK7486	BLOCK-1419
SK7489	BLOCK-1420
SK7490	BLOCK-1423
SK7492	BLOCK-1425
SK7493	BLOCK-1426
SK7494	BLOCK-1427
SK7495	BLOCK-1428
SK7496	BLOCK-1429
SK7497	BLOCK-1430
SK74C02	BLOCK-1432
SK74C04	BLOCK-1433
SK74C10	BLOCK-1435
SK74C107	BLOCK-1436
SK74C151	BLOCK-1438
SK74C154	BLOCK-1439
SK74C157	BLOCK-1440
SK74C161	BLOCK-1441
SK74C164	BLOCK-1442
SK74C173	BLOCK-1443
SK74C174	BLOCK-1444
SK74C175	BLOCK-1445
SK74C192	BLOCK-1446
SK74C193	BLOCK-1447
SK74C20	BLOCK-1448
SK74C221	BLOCK-1449
SK74C240	BLOCK-1450
SK74C244	BLOCK-1451
SK74C30	BLOCK-1452
SK74C32	BLOCK-1453
SK74C373	BLOCK-1454
SK74C374	BLOCK-1455
SK74C48	BLOCK-1456
SK74C73	BLOCK-1457
SK74C76	BLOCK-1459
SK74C901	BLOCK-1462
SK74C902	BLOCK-1463
SK74C903	BLOCK-1464
SK74C904	BLOCK-1465
SK74C922	BLOCK-1466
SK74C923	BLOCK-1467
SK74C925	BLOCK-1468
SK74C93	BLOCK-1469
SK74C95	BLOCK-1470
SK74H00	BLOCK-1471
SK74H04	BLOCK-1473
SK74LS00	BLOCK-1568
SK74LS01	BLOCK-1569
SK74LS02	BLOCK-1570
SK74LS03	BLOCK-1571
SK74LS04	BLOCK-1572
SK74LS05	BLOCK-1573
SK74LS08	BLOCK-1574
SK74LS10	BLOCK-1576
SK74LS107	BLOCK-1577
SK74LS109	BLOCK-1578
SK74LS11	BLOCK-1579
SK74LS114	BLOCK-1581
SK74LS12	BLOCK-1582
SK74LS122	BLOCK-1583
SK74LS123	BLOCK-1584
SK74LS123P	BLOCK-1584
SK74LS125A	BLOCK-1585
SK74LS13	BLOCK-1587
SK74LS133	BLOCK-1589
SK74LS136	BLOCK-1590
SK74LS138	BLOCK-1591
SK74LS139	BLOCK-1592
SK74LS14	BLOCK-1593
SK74LS145	BLOCK-1594
SK74LS148	BLOCK-1596
SK74LS15	BLOCK-1597
SK74LS151	BLOCK-1598
SK74LS153	BLOCK-1599
SK74LS155	BLOCK-1600
SK74LS156	BLOCK-1601
SK74LS157	BLOCK-1602
SK74LS161	BLOCK-1605
SK74LS163	BLOCK-1607
SK74LS164	BLOCK-1608
SK74LS168A	BLOCK-1611
SK74LS169A	BLOCK-1612
SK74LS174	BLOCK-1615
SK74LS175	BLOCK-1616
SK74LS191	BLOCK-1619
SK74LS193	BLOCK-1621
SK74LS194	BLOCK-1622
SK74LS194A	BLOCK-1622
SK74LS195	BLOCK-1623
SK74LS196	BLOCK-1624
SK74LS20	BLOCK-1626
SK74LS22	BLOCK-1628
SK74LS221	BLOCK-1629
SK74LS240	BLOCK-1630
SK74LS241	BLOCK-1631
SK74LS242	BLOCK-1632
SK74LS243	BLOCK-1633
SK74LS244	BLOCK-1634
SK74LS245	BLOCK-1635
SK74LS247	BLOCK-1636
SK74LS248	BLOCK-1637
SK74LS249	BLOCK-1638
SK74LS253	BLOCK-1640
SK74LS257	BLOCK-1641
SK74LS258	BLOCK-1642
SK74LS259	BLOCK-1643
SK74LS26	BLOCK-1644
SK74LS260	BLOCK-1645
SK74LS266	BLOCK-1646
SK74LS27	BLOCK-1647
SK74LS273	BLOCK-1648
SK74LS279	BLOCK-1649
SK74LS28	BLOCK-1650
SK74LS280	BLOCK-1651
SK74LS290	BLOCK-1653
SK74LS293	BLOCK-1654
SK74LS298	BLOCK-1656
SK74LS30	BLOCK-1658
SK74LS32	BLOCK-1659
SK74LS33	BLOCK-1660
SK74LS348	BLOCK-1661
SK74LS352	BLOCK-1662
SK74LS353	BLOCK-1663
SK74LS364	BLOCK-1665
SK74LS367	BLOCK-1668
SK74LS37	BLOCK-1670
SK74LS373	BLOCK-1671
SK74LS377	BLOCK-1673
SK74LS378	BLOCK-1674
SK74LS379	BLOCK-1675
SK74LS38	BLOCK-1676
SK74LS386	BLOCK-1677
SK74LS393	BLOCK-1679
SK74LS398	BLOCK-1681
SK74LS399	BLOCK-1682
SK74LS40	BLOCK-1683
SK74LS42	BLOCK-1684
SK74LS47	BLOCK-1685
SK74LS48	BLOCK-1686
SK74LS49	BLOCK-1687
SK74LS490	BLOCK-1688
SK74LS51	BLOCK-1689
SK74LS54	BLOCK-1690
SK74LS55	BLOCK-1693
SK74LS624	BLOCK-1694
SK74LS625	BLOCK-1695
SK74LS626	BLOCK-1696
SK74LS627	BLOCK-1697
SK74LS629	BLOCK-1698
SK74LS641	BLOCK-1700
SK74LS642	BLOCK-1701
SK74LS645	BLOCK-1703
SK74LS73	BLOCK-1705
SK74LS73A	BLOCK-1705
SK74LS74	BLOCK-1706
SK74LS74A	BLOCK-1706
SK74LS75	BLOCK-1707
SK74LS76A	BLOCK-1708
SK74LS77	BLOCK-1709
SK74LS78	BLOCK-1710
SK74LS83	BLOCK-1711
SK74LS83A	BLOCK-1711
SK74LS85	BLOCK-1712
SK74LS86	BLOCK-1713
SK74LS90	BLOCK-1714
SK74LS92	BLOCK-1716
SK74LS93	BLOCK-1717
SK74S00	BLOCK-1719
SK74S04	BLOCK-1722
SK74S11	BLOCK-1727
SK74S112	BLOCK-1728
SK74S174	BLOCK-1740
SK74S20	BLOCK-1745
SK74S22	BLOCK-1746
SK74S74	BLOCK-1768
SK74S86	BLOCK-1769
SK75454B	BLOCK-1779
SK75493	BLOCK-1782
SK7600	BLOCK-411
SK7601	BLOCK-412
SK7602	BLOCK-578
SK7603	BLOCK-384
SK7604	BLOCK-512
SK7605	BLOCK-544
SK7606	BLOCK-636
SK7607	BLOCK-474
SK7608	BLOCK-362
SK7610	BLOCK-548
SK7611	BLOCK-611
SK7612	BLOCK-974
SK7614	BLOCK-1882
SK7615	BLOCK-1979
SK7616	BLOCK-1901
SK7617	BLOCK-1990
SK7618	BLOCK-708
SK7619	BLOCK-728
SK7620	BLOCK-547
SK7622	BLOCK-368
SK7626	BLOCK-669
SK7627	BLOCK-387
SK7630	BLOCK-350
SK7632	BLOCK-639
SK7633	BLOCK-642
SK7634	BLOCK-422
SK7635	BLOCK-401
SK7637	BLOCK-497
SK7638	BLOCK-496
SK7641	BLOCK-1910
SK7642	BLOCK-902
SK7643	BLOCK-903
SK7644	BLOCK-879
SK7645	BLOCK-1897
SK7646	BLOCK-535
SK7647	BLOCK-533
SK7648	BLOCK-550
SK7649	BLOCK-280
SK7650	BLOCK-705
SK7652	BLOCK-613
SK7653	BLOCK-662
SK7654	BLOCK-373
SK7656	BLOCK-503
SK7657	BLOCK-543
SK7658	BLOCK-369
SK7659	BLOCK-313
SK7660	BLOCK-314
SK7661	BLOCK-268
SK7662	BLOCK-321
SK7663	BLOCK-366
SK7665	BLOCK-374
SK7667	BLOCK-486
SK7670	BLOCK-640
SK7671	BLOCK-509
SK7672	BLOCK-563
SK7673	BLOCK-643
SK7674	BLOCK-821
SK7676	BLOCK-534
SK7677	BLOCK-558
SK7678	BLOCK-560
SK7679	BLOCK-637
SK7681	BLOCK-557
SK7686	BLOCK-1885
SK7687	BLOCK-641
SK7692	BLOCK-1923
SK7703	BLOCK-713
SK7704	BLOCK-505
SK7705	BLOCK-706
SK7706	BLOCK-365
SK7707	BLOCK-363
SK7708	BLOCK-1907
SK7711	BLOCK-673
SK7713	BLOCK-668
SK7717	BLOCK-1929
SK7718	BLOCK-1928
SK7719	BLOCK-700
SK7720	BLOCK-540
SK7721	BLOCK-1988
SK7722	BLOCK-247
SK7723	BLOCK-240
SK7724	BLOCK-246
SK7725	BLOCK-1930
SK7726	BLOCK-553
SK7727	BLOCK-1833
SK7728	BLOCK-60
SK7729	BLOCK-620
SK7730	BLOCK-1838
SK7731	BLOCK-1914
SK7734	BLOCK-642
SK7735	BLOCK-603
SK7736	BLOCK-644
SK7737	BLOCK-1898
SK7739	BLOCK-905
SK7740	BLOCK-907
SK7741	BLOCK-908
SK7742	BLOCK-909
SK7743	BLOCK-567
SK7751	BLOCK-1872
SK7752	BLOCK-2076
SK7753	BLOCK-1
SK7754	BLOCK-15
SK7755	BLOCK-16
SK7756	BLOCK-19
SK7757	BLOCK-22
SK7758	BLOCK-31
SK7759	BLOCK-32
SK7760	BLOCK-45

52 ORIGINAL DEVICE TYPES AND REPLACEMENT CODES

DEVICE TYPE	REPL CODE
SK7761	BLOCK-51
SK7762	BLOCK-55
SK7763	BLOCK-61
SK7764	BLOCK-64
SK7765	BLOCK-65
SK7766	BLOCK-77
SK7767	BLOCK-87
SK7768	BLOCK-89
SK7769	BLOCK-92
SK7770	BLOCK-96
SK7771	BLOCK-104
SK7772	BLOCK-105
SK7773	BLOCK-119
SK7774	BLOCK-124
SK7775	BLOCK-125
SK7776	BLOCK-134
SK7778	BLOCK-152
SK7779	BLOCK-171
SK7781	BLOCK-209
SK7782	BLOCK-212
SK7783	BLOCK-217
SK7784	BLOCK-225
SK7785	BLOCK-229
SK7786	BLOCK-241
SK7787	BLOCK-252
SK7789	BLOCK-258
SK7790	BLOCK-262
SK7791	BLOCK-292
SK7792	BLOCK-409
SK7793	BLOCK-419
SK7794	BLOCK-423
SK7795	BLOCK-429
SK7796	BLOCK-440
SK7797	BLOCK-447
SK7798	BLOCK-448
SK7799	BLOCK-458
SK7800	BLOCK-460
SK7801	BLOCK-468
SK7802	BLOCK-491
SK7803	BLOCK-507
SK7804	BLOCK-516
SK7805	BLOCK-554
SK7806	BLOCK-2077
SK7807	BLOCK-2078
SK7808	BLOCK-364
SK7809	BLOCK-714
SK7810	BLOCK-715
SK7811	BLOCK-716
SK7813	BLOCK-1856
SK7814	BLOCK-378
SK7815	BLOCK-377
SK7817	BLOCK-721
SK7827	BLOCK-617
SK7841	BLOCK-657
SK7C00	BLOCK-1506
SK7C02	BLOCK-1507
SK7C04	BLOCK-1508
SK7C08	BLOCK-1509
SK7C10	BLOCK-1510
SK7C109	BLOCK-1511
SK7C11	BLOCK-1512
SK7C123	BLOCK-1513
SK7C125	BLOCK-1514
SK7C126	BLOCK-1515
SK7C132	BLOCK-1516
SK7C138	BLOCK-1517
SK7C139	BLOCK-1518
SK7C14	BLOCK-1519
SK7C151	BLOCK-1520
SK7C153	BLOCK-1521
SK7C154	BLOCK-1522
SK7C161	BLOCK-1523
SK7C163	BLOCK-1524
SK7C164	BLOCK-1525
SK7C165	BLOCK-1526
SK7C173	BLOCK-1527
SK7C174	BLOCK-1528
SK7C175	BLOCK-1529
SK7C240	BLOCK-1530
SK7C299	BLOCK-1535
SK7C32	BLOCK-1536
SK7C373	BLOCK-1537
SK7C374	BLOCK-1538
SK7C377	BLOCK-1539
SK7C390	BLOCK-1540
SK7C393	BLOCK-1541
SK7C40105	BLOCK-1542
SK7C4020	BLOCK-1543
SK7C4040	BLOCK-1544
SK7C4053	BLOCK-1545
SK7C4060	BLOCK-1546
SK7C4067	BLOCK-1547
SK7C573	BLOCK-1548
SK7C574	BLOCK-1549
SK7C86	BLOCK-1550
SK7CT00	BLOCK-1551
SK7CT04	BLOCK-1552
SK7CT08	BLOCK-1553
SK7CT138	BLOCK-1554
SK7CT14	BLOCK-1555
SK7CT158	BLOCK-1554
SK7CT161	BLOCK-1556
SK7CT163	BLOCK-1557
SK7CT174	BLOCK-1558
SK7CT240	BLOCK-1559
SK7CT244	BLOCK-1560
SK7CT273	BLOCK-1561
SK7CT32	BLOCK-1562
SK7CT373	BLOCK-1563
SK7CT573	BLOCK-1565
SK7CT574	BLOCK-1566
SK9012	BLOCK-2175
SK9013	BLOCK-1852
SK9014	BLOCK-1277
SK9015	BLOCK-539
SK9016	BLOCK-398
SK9017	BLOCK-1989
SK9018	BLOCK-392
SK9019	BLOCK-945
SK9020	BLOCK-944
SK9025	BLOCK-1201
SK9026	BLOCK-1190
SK9027	BLOCK-1892
SK9028	BLOCK-1179
SK9029	BLOCK-641
SK9030	BLOCK-1860
SK9043	BLOCK-389
SK9044	BLOCK-609
SK9045	BLOCK-400
SK9046	BLOCK-407
SK9048	BLOCK-416
SK9049	BLOCK-413
SK9050	BLOCK-415
SK9051	BLOCK-409
SK9054	BLOCK-256
SK9055	BLOCK-257
SK9056	BLOCK-428
SK9058	BLOCK-403
SK9059	BLOCK-393
SK9061	BLOCK-473
SK9062	BLOCK-2054
SK9063	BLOCK-2055
SK9064	BLOCK-395
SK9067	BLOCK-2004
SK9068	BLOCK-1828
SK9069	BLOCK-844
SK9070	BLOCK-1958
SK9071	BLOCK-1962
SK9074	BLOCK-198
SK9077	BLOCK-939
SK9078	BLOCK-940
SK9079	BLOCK-941
SK9080	BLOCK-942
SK9081	BLOCK-943
SK9089	BLOCK-1868
SK9090	BLOCK-1967
SK9092	BLOCK-935
SK9093	BLOCK-936
SK9094	BLOCK-937
SK9095	BLOCK-938
SK9126	BLOCK-1853
SK9127	BLOCK-1858
SK9128	BLOCK-1985
SK9129	BLOCK-1986
SK9130	BLOCK-565
SK9144	BLOCK-1972
SK9145	BLOCK-1973
SK9146	BLOCK-2031
SK9147	BLOCK-2036
SK9166	BLOCK-2037
SK9167	BLOCK-2042
SK9168	BLOCK-2096
SK9169	BLOCK-2074
SK9170	BLOCK-2075
SK9171	BLOCK-2051
SK9172	BLOCK-2219
SK9173	BLOCK-2071
SK9174	BLOCK-1987
SK9175	BLOCK-1955
SK9176	BLOCK-1957
SK9177	BLOCK-1949
SK9178	BLOCK-1889
SK9179	BLOCK-1157
SK9180	BLOCK-408
SK9181	BLOCK-279
SK9182	BLOCK-276
SK9183	BLOCK-277
SK9184	BLOCK-442
SK9185	BLOCK-234
SK9186	BLOCK-375
SK9187	BLOCK-433
SK9188	BLOCK-467
SK9189	BLOCK-318
SK9190	BLOCK-111
SK9191	BLOCK-385
SK9192	BLOCK-552
SK9193	BLOCK-531
SK9194	BLOCK-459
SK9195	BLOCK-562
SK9196	BLOCK-480
SK9197	BLOCK-510
SK9198	BLOCK-396
SK9199	BLOCK-414
SK9200	BLOCK-1942
SK9201	BLOCK-2215
SK9202	BLOCK-1884
SK9203	BLOCK-955
SK9204	BLOCK-969
SK9206	BLOCK-1894
SK9207	BLOCK-1918
SK9208	BLOCK-451
SK9209	BLOCK-2211
SK9210	BLOCK-1851
SK9211	BLOCK-471
SK9212	BLOCK-454
SK9214	BLOCK-406
SK9215	BLOCK-2083
SK9216	BLOCK-2084
SK9217	BLOCK-1975
SK9218	BLOCK-1976
SK9219	BLOCK-893
SK9221	BLOCK-882
SK9222	BLOCK-883
SK9225	BLOCK-383
SK9226	BLOCK-115
SK9227	BLOCK-765
SK9228	BLOCK-274
SK9244	BLOCK-545
SK9245	BLOCK-623
SK9246	BLOCK-399
SK9247	BLOCK-418
SK9248	BLOCK-1895
SK9249	BLOCK-537
SK9250	BLOCK-501
SK9251	BLOCK-367
SK9252	BLOCK-275
SK9257	BLOCK-1932
SK9265	BLOCK-284
SK9266	BLOCK-151
SK9267	BLOCK-270
SK9268	BLOCK-271
SK9269	BLOCK-386
SK9276	BLOCK-2168
SK9281	BLOCK-1875
SK9282	BLOCK-276
SK9283	BLOCK-2086
SK9298	BLOCK-538
SK9299	BLOCK-282
SK9311	BLOCK-283
SK9312	BLOCK-478
SK9313	BLOCK-382
SK9314	BLOCK-286
SK9315	BLOCK-405
SK9320	BLOCK-519
SK9321	BLOCK-484
SK9322	BLOCK-456
SK9323	BLOCK-2059
SK9324	BLOCK-649
SK9325	BLOCK-660
SK9326	BLOCK-402
SK9331	BLOCK-890
SK9332	BLOCK-892
SK9333	BLOCK-887
SK9334	BLOCK-889
SK9335	BLOCK-891
SK9336	BLOCK-895
SK9337	BLOCK-888
SK9338	BLOCK-894
SK9339	BLOCK-2135
SK9340	BLOCK-2007
SK9341	BLOCK-2014
SK9342	BLOCK-2020
SK9344	BLOCK-2024
SK9353	BLOCK-390
SK9354	BLOCK-388
SK9355	BLOCK-461
SK9356	BLOCK-457
SK9357	BLOCK-988
SK9359	BLOCK-719
SK9360	BLOCK-568
SK9361	BLOCK-348
SK9379	BLOCK-532
SK9380	BLOCK-401
SK9381	BLOCK-1887
SK9382	BLOCK-1888
SK9383	BLOCK-1893
SK9384	BLOCK-220
SK9386	BLOCK-1922
SK9393	BLOCK-459
SK9394	BLOCK-281
SK9395	BLOCK-351
SK9396	BLOCK-349
SK9397	BLOCK-2081
SK9398	BLOCK-285
SK9402	BLOCK-659
SK9403	BLOCK-525
SK9404	BLOCK-526
SK9406	BLOCK-317
SK9700	BLOCK-1176
SK9701	BLOCK-1173
SK9702	BLOCK-1170
SK9703	BLOCK-1759
SK9704	BLOCK-1729
SK9706	BLOCK-622
SK9707	BLOCK-595
SK9715	BLOCK-695
SK9716	BLOCK-774
SK9717	BLOCK-772
SK9721	BLOCK-2063
SK9722	BLOCK-789
SK9724	BLOCK-790
SK9725	BLOCK-674
SK9726	BLOCK-760
SK9727	BLOCK-791
SK9728	BLOCK-814
SK9729	BLOCK-670
SK9730	BLOCK-763
SK9731	BLOCK-761
SK9732	BLOCK-725
SK9733	BLOCK-792
SK9734	BLOCK-793
SK9735	BLOCK-794
SK9736	BLOCK-808
SK9737	BLOCK-809
SK9739	BLOCK-795
SK9740	BLOCK-813
SK9743	BLOCK-757
SK9744	BLOCK-759
SK9745	BLOCK-779
SK9746	BLOCK-762
SK9748	BLOCK-709
SK9749	BLOCK-752
SK9750	BLOCK-817
SK9752	BLOCK-754
SK9753	BLOCK-778
SK9755	BLOCK-816
SK9756	BLOCK-815
SK9757	BLOCK-325
SK9761	BLOCK-713
SK9762	BLOCK-729
SK9764	BLOCK-853
SK9766	BLOCK-768
SK9767	BLOCK-756
SK9769	BLOCK-663
SK9770	BLOCK-1021
SK9781	BLOCK-1461
SK9782	BLOCK-1460
SK9783	BLOCK-1458
SK9785	BLOCK-1434
SK9786	BLOCK-1780
SK9787	BLOCK-1035
SK9788	BLOCK-1952
SK9789	BLOCK-1951
SK9791	BLOCK-1177
SK9792	BLOCK-1175
SK9793	BLOCK-1174
SK9794	BLOCK-1172
SK9795	BLOCK-1171
SK9796	BLOCK-1166
SK9797	BLOCK-1165
SK9799	BLOCK-1909
SK9800	BLOCK-1908
SK9805	BLOCK-1437
SK9806	BLOCK-1431
SK9807	BLOCK-1776
SK9809	BLOCK-1178
SK9813	BLOCK-996
SK9814	BLOCK-951
SK9815	BLOCK-661
SK9816	BLOCK-614
SK9817	BLOCK-599
SK9818	BLOCK-476
SK9822	BLOCK-1917
SK9825	BLOCK-1783
SK9827	BLOCK-1778
SK9828	BLOCK-1775
SK9829	BLOCK-1161
SK9831	BLOCK-1867
SK9832	BLOCK-1003
SK9833	BLOCK-1319
SK9834	BLOCK-1777
SK9835	BLOCK-1388
SK9837	BLOCK-1366
SK9838	BLOCK-786
SK9839	BLOCK-788
SK9850	BLOCK-771
SK9851	BLOCK-1919
SK9862	BLOCK-1937
SK9863	BLOCK-1938
SK9864	BLOCK-1939
SK9866	BLOCK-753
SK9867	BLOCK-755
SK9868	BLOCK-756
SK9870	BLOCK-758
SK9871	BLOCK-759
SK9873	BLOCK-764
SK9874	BLOCK-767
SK9875	BLOCK-769
SK9876	BLOCK-770
SK9877	BLOCK-773
SK9878	BLOCK-775
SK9879	BLOCK-776
SK9880	BLOCK-777
SK9881	BLOCK-780
SK9882	BLOCK-781
SK9883	BLOCK-782
SK9884	BLOCK-784
SK9885	BLOCK-807
SK9886	BLOCK-788
SK9888	BLOCK-787
SK9889	BLOCK-801
SK9890	BLOCK-798
SK9891	BLOCK-796
SK9892	BLOCK-800
SK9893	BLOCK-799
SK9894	BLOCK-785
SK9895	BLOCK-811
SK9896	BLOCK-810
SK9897	BLOCK-797
SK9898	BLOCK-806
SK9899	BLOCK-805
SK9902	BLOCK-803
SK9903	BLOCK-991
SK9906	BLOCK-992
SK9907	BLOCK-995
SK9908	BLOCK-997
SK9911	BLOCK-576
SK9912	BLOCK-710
SK9913	BLOCK-994
SK9945	BLOCK-840
SK9946	BLOCK-839
SK9947	BLOCK-664
SK9948	BLOCK-671
SK9949	BLOCK-675
SK9950	BLOCK-685
SK9951	BLOCK-690
SK9952	BLOCK-691
SK9953	BLOCK-692
SK9954	BLOCK-693
SK9955	BLOCK-711
SK9976	BLOCK-1157
SK9977	BLOCK-602
SK9978	BLOCK-654
SK9980	BLOCK-680
SK9981	BLOCK-684
SK9982	BLOCK-686
SK9983	BLOCK-701
SK9987	BLOCK-345
SK9989	BLOCK-449
SK9990	BLOCK-598
SK9992	BLOCK-1925
SK9993	BLOCK-2063
SK9994	BLOCK-722
SK9995	BLOCK-723
SK9996	BLOCK-724
SK9997	BLOCK-326
SK9998	BLOCK-332
SL01640	BLOCK-2189
SL02518	BLOCK-2176
SL02519	BLOCK-2189
SL02734	BLOCK-2189
SL02779	BLOCK-2189
SL03019	BLOCK-2189
SL03667	BLOCK-2188
SL03911	BLOCK-2210
SL03912	BLOCK-2216
SL03913	BLOCK-2218
SL03914	BLOCK-2199
SL03915	BLOCK-2206
SL03916	BLOCK-2203
SL03917	BLOCK-2213
SL04194	BLOCK-2185
SL04217	BLOCK-2176
SL04218	BLOCK-2189
SL04563	BLOCK-2197
SL04567	BLOCK-2202
SL04568	BLOCK-2199
SL04570	BLOCK-2213
SL04732	BLOCK-2189
SL11877	BLOCK-2197
SL11878	BLOCK-2202
SL11879	BLOCK-2208
SL11880	BLOCK-2209
SL11881	BLOCK-2217
SL14959	BLOCK-1476
SL14960	BLOCK-1482
SL14971	BLOCK-1301
SL14972	BLOCK-1385
SL16122	BLOCK-2185
SL16201	BLOCK-2197
SL16203	BLOCK-2213
SL16204	BLOCK-2208
SL16206	BLOCK-2199
SL16208	BLOCK-2203
SL16209	BLOCK-1953
SL16210	BLOCK-2202
SL16211	BLOCK-2207
SL16212	BLOCK-2217
SL16215	BLOCK-2206
SL16216	BLOCK-2210
SL16218	BLOCK-2218
SL1622C	BLOCK-1338
SL1626C	BLOCK-1338
SL16516	BLOCK-2197
SL16517	BLOCK-2200
SL16518	BLOCK-2202
SL16519	BLOCK-2208
SL16520	BLOCK-2217
SL16521	BLOCK-2207
SL16522	BLOCK-1953
SL16584	BLOCK-2197
SL16585	BLOCK-2198
SL16586	BLOCK-2199
SL16587	BLOCK-2202
SL16588	BLOCK-2207
SL16589	BLOCK-2208
SL16590	BLOCK-2209
SL16591	BLOCK-2217
SL16793	BLOCK-1293
SL16794	BLOCK-1294
SL16795	BLOCK-1295
SL16796	BLOCK-1297
SL16797	BLOCK-1298
SL16798	BLOCK-1301
SL16799	BLOCK-1307
SL16800	BLOCK-1365
SL16801	BLOCK-1304
SL16802	BLOCK-1377
SL16803	BLOCK-1390
SL16804	BLOCK-1401
SL16805	BLOCK-1403
SL16806	BLOCK-1410
SL16807	BLOCK-1411
SL16808	BLOCK-1413
SL16809	BLOCK-1425
SL16810	BLOCK-1429
SL16811	BLOCK-2206
SL17242	BLOCK-1410
SL17284	BLOCK-2199
SL17289	BLOCK-2202
SL17869	BLOCK-1301
SL17887	BLOCK-1307
SL18386	BLOCK-1417
SL18387	BLOCK-1305
SL18699	BLOCK-2186
SL20721	BLOCK-1244
SL20755	BLOCK-1270
SL21017	BLOCK-1272
SL21122	BLOCK-1271
SL21384	BLOCK-1967
SL21441	BLOCK-1256
SL21577	BLOCK-1967
SL21654	BLOCK-1269
SL53424	BLOCK-2197
SL53425	BLOCK-2199
SL53426	BLOCK-2201
SL53427	BLOCK-2202
SL53428	BLOCK-2206
SL53429	BLOCK-2208
SL53431	BLOCK-1294
SL7059	BLOCK-1212
SL7283	BLOCK-1212
SL7308	BLOCK-1212
SL7531	BLOCK-1212

ORIGINAL DEVICE TYPES AND REPLACEMENT CODES 53

DEVICE TYPE	REPL CODE
SL7593	BLOCK-1212
SL8020	BLOCK-1212
SM5104	BLOCK-244
SM5104F	BLOCK-244
SM5104G	BLOCK-244
SM5104P	BLOCK-244
SM5109	BLOCK-156
SM63	BLOCK-1412
SM73(I.C.)	BLOCK-1412
SMC7400N	BLOCK-1293
SMC7402N	BLOCK-1295
SMC7408N	BLOCK-1301
SMC7410N	BLOCK-1304
SMC7420N	BLOCK-1365
SMC7430N	BLOCK-1377
SMC7440N	BLOCK-1390
SMC7451N	BLOCK-1402
SMC7473N	BLOCK-1410
SMC7474N	BLOCK-1411
SMC7475N	BLOCK-1412
SMC7476N	BLOCK-1413
SMC7490N	BLOCK-1423
SMC7493N	BLOCK-1426
SN-7474	BLOCK-1411
SN-76001N	BLOCK-111
SN151800J	BLOCK-2150
SN151800N	BLOCK-2150
SN151801J	BLOCK-2151
SN151801N	BLOCK-2151
SN151802J	BLOCK-2152
SN151802N	BLOCK-2152
SN151803J	BLOCK-2153
SN151803N	BLOCK-2153
SN151804J	BLOCK-2154
SN151804N	BLOCK-2154
SN151805J	BLOCK-2155
SN151805N	BLOCK-2155
SN151806J	BLOCK-2156
SN151806N	BLOCK-2156
SN151807J	BLOCK-2157
SN151807N	BLOCK-2157
SN151808J	BLOCK-2158
SN151808N	BLOCK-2158
SN151809J	BLOCK-2159
SN151809N	BLOCK-2159
SN151810J	BLOCK-2161
SN151810N	BLOCK-2161
SN151811J	BLOCK-2162
SN151811N	BLOCK-2162
SN151812J	BLOCK-2163
SN151812N	BLOCK-2163
SN158093J	BLOCK-1950
SN158093N	BLOCK-1950
SN158094J	BLOCK-1951
SN158094N	BLOCK-1951
SN158097J	BLOCK-1952
SN158097N	BLOCK-1952
SN158099J	BLOCK-1953
SN158099N	BLOCK-1953
SN15830J	BLOCK-2197
SN15830N	BLOCK-2197
SN15831J	BLOCK-2198
SN15831N	BLOCK-2198
SN15832J	BLOCK-2199
SN15832N	BLOCK-2199
SN15833J	BLOCK-2200
SN15833N	BLOCK-2200
SN15835J	BLOCK-2201
SN15835N	BLOCK-2201
SN15836J	BLOCK-2202
SN15836N	BLOCK-2202
SN15837J	BLOCK-2203
SN15837N	BLOCK-2203
SN15838J	BLOCK-1965
SN15838N	BLOCK-1965
SN15844J	BLOCK-2206
SN15844N	BLOCK-2206
SN15845J	BLOCK-2207
SN15845N	BLOCK-2207
SN15846J	BLOCK-2208
SN15846N	BLOCK-2208
SN15848J	BLOCK-2209
SN15848N	BLOCK-2209
SN15849J	BLOCK-2210
SN15849N	BLOCK-2210
SN15850J	BLOCK-2212
SN15850N	BLOCK-2212
SN15851J	BLOCK-2213
SN15851N	BLOCK-2213
SN15857J	BLOCK-1968
SN15857N	BLOCK-1968
SN15858J	BLOCK-1969
SN15858N	BLOCK-1969
SN15861J	BLOCK-2216
SN15861N	BLOCK-2216
SN15862J	BLOCK-2217
SN15862N	BLOCK-2217
SN15863J	BLOCK-2218
SN15863N	BLOCK-2218
SN16704C	BLOCK-1157
SN16706N	BLOCK-1269
SN16726N	BLOCK-1272
SN16727N	BLOCK-1815
SN16840C	BLOCK-1816
SN16859	BLOCK-1816
SN16869A	BLOCK-1816
SN16869A#	BLOCK-1838
SN16890	BLOCK-1794
SN52101AL	BLOCK-160
SN52101L	BLOCK-160
SN5211P	BLOCK-1981
SN52301	BLOCK-2141
SN52723L	BLOCK-1983
SN52723N	BLOCK-1984
SN52741	BLOCK-2056
SN52741L	BLOCK-2056
SN52741P	BLOCK-2056
SN52747L	BLOCK-2069
SN52748J	BLOCK-2141
SN52748L	BLOCK-160
SN52748N	BLOCK-2141
SN52748P	BLOCK-2141
SN72301	BLOCK-2141
SN72301AL	BLOCK-160
SN72301AN	BLOCK-160
SN72301AP	BLOCK-2141
SN72307	BLOCK-2143
SN72307JA	BLOCK-2143
SN72307JP	BLOCK-2143
SN72307N	BLOCK-2143
SN72307P	BLOCK-2143
SN72311P	BLOCK-1981
SN7248L	BLOCK-160
SN72555JP	BLOCK-2079
SN72555P	BLOCK-2079
SN72558L	BLOCK-1958
SN72709	BLOCK-1954
SN72709J	BLOCK-1954
SN72709T	BLOCK-1949
SN72723J	BLOCK-1984
SN72723L	BLOCK-1983
SN72723N	BLOCK-1984
SN72741J	BLOCK-2057
SN72741P	BLOCK-2058
SN72747	BLOCK-2069
SN72747J	BLOCK-2070
SN72747JA	BLOCK-2070
SN72747L	BLOCK-2069
SN72747N	BLOCK-2070
SN72748J	BLOCK-2141
SN72748JG	BLOCK-2141
SN72748L	BLOCK-160
SN72748N	BLOCK-2141
SN72748P	BLOCK-2141
SN72905	BLOCK-2091
SN72906	BLOCK-2094
SN72912	BLOCK-2108
SN72915	BLOCK-2128
SN72924	BLOCK-2137
SN7400	BLOCK-1293
SN7400A	BLOCK-1293
SN7400J	BLOCK-1293
SN7400N	BLOCK-1293
SN7400N-10	BLOCK-1293
SN7401	BLOCK-1294
SN7401J	BLOCK-1294
SN7401N	BLOCK-1294
SN7402	BLOCK-1295
SN7402J	BLOCK-1295
SN7402N	BLOCK-1295
SN7402N-10	BLOCK-1295
SN7403	BLOCK-1296
SN7403J	BLOCK-1296
SN7403N	BLOCK-1296
SN7404	BLOCK-1297
SN7404J	BLOCK-1297
SN7404N	BLOCK-1297
SN7404N-10	BLOCK-1297
SN7405	BLOCK-1298
SN7405A	BLOCK-1298
SN7405J	BLOCK-1298
SN7405N	BLOCK-1298
SN7406	BLOCK-1299
SN7406J	BLOCK-1299
SN7406N	BLOCK-1299
SN7406N-10	BLOCK-1299
SN7407	BLOCK-1300
SN7407J	BLOCK-1300
SN7407N	BLOCK-1300
SN7408	BLOCK-1301
SN7408J	BLOCK-1301
SN7408N	BLOCK-1301
SN7408N-10	BLOCK-1301
SN7409	BLOCK-1302
SN7409J	BLOCK-1302
SN7409N	BLOCK-1302
SN7410	BLOCK-1304
SN74107	BLOCK-1305
SN74107N	BLOCK-1305
SN74109	BLOCK-1306
SN74109J	BLOCK-1306
SN74109N	BLOCK-1306
SN7410J	BLOCK-1304
SN7410N	BLOCK-1304
SN7410N-10	BLOCK-1304
SN74110J	BLOCK-1308
SN74110N	BLOCK-1308
SN74111	BLOCK-1309
SN74111J	BLOCK-1309
SN74111N	BLOCK-1309
SN74116	BLOCK-1862
SN74116N	BLOCK-1862
SN7412	BLOCK-1310
SN74121	BLOCK-1311
SN74121J	BLOCK-1311
SN74121N	BLOCK-1311
SN74122	BLOCK-1312
SN74122J	BLOCK-1312
SN74122N	BLOCK-1312
SN74123	BLOCK-1313
SN74123J	BLOCK-1313
SN74123N	BLOCK-1313
SN74123N-10	BLOCK-1313
SN74125	BLOCK-1314
SN74125J	BLOCK-1314
SN74125N	BLOCK-1314
SN74126	BLOCK-1315
SN74126J	BLOCK-1315
SN74126N	BLOCK-1315
SN74128	BLOCK-1316
SN74128J	BLOCK-1316
SN74128N	BLOCK-1316
SN7412J	BLOCK-1310
SN7412N	BLOCK-1310
SN7412W	BLOCK-1310
SN7413	BLOCK-1317
SN74132	BLOCK-1318
SN74132J	BLOCK-1318
SN74132N	BLOCK-1318
SN74136	BLOCK-1319
SN74136J	BLOCK-1319
SN74136N	BLOCK-1319
SN7413J	BLOCK-1317
SN7413N	BLOCK-1317
SN7413N-10	BLOCK-1317
SN7414	BLOCK-1320
SN74141	BLOCK-1321
SN74141J	BLOCK-1321
SN74141N	BLOCK-1321
SN74142	BLOCK-1322
SN74142J	BLOCK-1322
SN74142N	BLOCK-1322
SN74143	BLOCK-1323
SN74143J	BLOCK-1323
SN74143N	BLOCK-1323
SN74144	BLOCK-1324
SN74144J	BLOCK-1324
SN74144N	BLOCK-1324
SN74145	BLOCK-1325
SN74145J	BLOCK-1325
SN74145N	BLOCK-1325
SN74145N-10	BLOCK-1325
SN74147	BLOCK-1326
SN74147J	BLOCK-1326
SN74147N	BLOCK-1326
SN7414J	BLOCK-1320
SN7414N	BLOCK-1320
SN74150	BLOCK-1327
SN74150J	BLOCK-1327
SN74150N	BLOCK-1327
SN74150N-10	BLOCK-1327
SN74151A	BLOCK-1328
SN74151AJ	BLOCK-1328
SN74151AN	BLOCK-1328
SN74151N	BLOCK-1328
SN74152A	BLOCK-1329
SN74152N	BLOCK-1329
SN74152S	BLOCK-1329
SN74153	BLOCK-1330
SN74153J	BLOCK-1330
SN74153N	BLOCK-1330
SN74154	BLOCK-1331
SN74154J	BLOCK-1331
SN74154N	BLOCK-1331
SN74154N-10	BLOCK-1331
SN74155	BLOCK-1332
SN74155J	BLOCK-1332
SN74155N	BLOCK-1332
SN74156	BLOCK-1333
SN74156J	BLOCK-1333
SN74156N	BLOCK-1333
SN74157	BLOCK-1334
SN74157J	BLOCK-1334
SN74157N	BLOCK-1334
SN7416	BLOCK-1336
SN74160	BLOCK-1337
SN74160J	BLOCK-1337
SN74160N	BLOCK-1337
SN74161	BLOCK-1338
SN74161J	BLOCK-1338
SN74161N	BLOCK-1338
SN74162	BLOCK-1339
SN74162J	BLOCK-1339
SN74162N	BLOCK-1339
SN74163	BLOCK-1340
SN74163J	BLOCK-1340
SN74163N	BLOCK-1340
SN74164	BLOCK-1341
SN74164J	BLOCK-1341
SN74164N	BLOCK-1341
SN74165	BLOCK-1342
SN74165J	BLOCK-1342
SN74165N	BLOCK-1342
SN7416J	BLOCK-1336
SN7416N	BLOCK-1336
SN7417	BLOCK-1343
SN74170	BLOCK-1344
SN74170J	BLOCK-1344
SN74170N	BLOCK-1344
SN74173	BLOCK-1345
SN74173J	BLOCK-1345
SN74173N	BLOCK-1345
SN74174	BLOCK-1346
SN74174J	BLOCK-1346
SN74174N	BLOCK-1346
SN74175	BLOCK-1347
SN74175J	BLOCK-1347
SN74175N	BLOCK-1347
SN74176	BLOCK-1348
SN74176J	BLOCK-1348
SN74176N	BLOCK-1348
SN74177J	BLOCK-1349
SN74177N	BLOCK-1349
SN74178	BLOCK-1350
SN74178J	BLOCK-1350
SN74178N	BLOCK-1350
SN74179	BLOCK-1351
SN74179J	BLOCK-1351
SN74179N	BLOCK-1351
SN7417J	BLOCK-1343
SN7417N	BLOCK-1343
SN74180	BLOCK-1352
SN74180J	BLOCK-1352
SN74180N	BLOCK-1352
SN74181	BLOCK-1353
SN74181J	BLOCK-1353
SN74181N	BLOCK-1353
SN74182	BLOCK-1354
SN74182J	BLOCK-1354
SN74182N	BLOCK-1354
SN74190	BLOCK-1355
SN74190J	BLOCK-1355
SN74190N	BLOCK-1355
SN74191J	BLOCK-1356
SN74191N	BLOCK-1356
SN74192	BLOCK-1357
SN74192J	BLOCK-1357
SN74192N	BLOCK-1357
SN74192N-10	BLOCK-1357
SN74193	BLOCK-1358
SN74193J	BLOCK-1358
SN74193N	BLOCK-1358
SN74193N-10	BLOCK-1358
SN74195	BLOCK-1359
SN74195J	BLOCK-1359
SN74195N	BLOCK-1359
SN74196	BLOCK-1360
SN74196J	BLOCK-1360
SN74196N	BLOCK-1360
SN74196N-10	BLOCK-1360
SN74197	BLOCK-1361
SN74197J	BLOCK-1361
SN74197N	BLOCK-1361
SN74198	BLOCK-1362
SN74198J	BLOCK-1362
SN74198N	BLOCK-1362
SN74199	BLOCK-1363
SN74199J	BLOCK-1363
SN74199N	BLOCK-1363
SN7420	BLOCK-1365
SN7420J	BLOCK-1365
SN7420N	BLOCK-1365
SN7420N-10	BLOCK-1365
SN7422	BLOCK-1367
SN74221	BLOCK-1368
SN74221J	BLOCK-1368
SN74221N	BLOCK-1368
SN7422J	BLOCK-1367
SN7422N	BLOCK-1367
SN7423	BLOCK-1369
SN7423J	BLOCK-1369
SN7423N	BLOCK-1369
SN74249	BLOCK-1370
SN74249J	BLOCK-1370
SN74249N	BLOCK-1370
SN74251	BLOCK-1371
SN74251J	BLOCK-1371
SN74251N	BLOCK-1371
SN7427	BLOCK-1372
SN7427J	BLOCK-1372
SN7427N	BLOCK-1372
SN7427N-10	BLOCK-1372
SN7428	BLOCK-1373
SN7428J	BLOCK-1373
SN7428N	BLOCK-1373
SN74290	BLOCK-1374
SN74290J	BLOCK-1374
SN74290N	BLOCK-1374
SN74293	BLOCK-1375
SN74293J	BLOCK-1375
SN74293N	BLOCK-1375
SN7430	BLOCK-1377
SN7430J	BLOCK-1377
SN7430N	BLOCK-1377
SN7432	BLOCK-1378
SN7432J	BLOCK-1378
SN7432N	BLOCK-1378
SN7432N-10	BLOCK-1378
SN7433	BLOCK-1379
SN7433J	BLOCK-1379
SN7433N	BLOCK-1379
SN74365A	BLOCK-1380
SN74365AJ	BLOCK-1380
SN74365AN	BLOCK-1380
SN74365J	BLOCK-1380
SN74365N	BLOCK-1380
SN74366A	BLOCK-1381
SN74366AJ	BLOCK-1381
SN74366AN	BLOCK-1381
SN74366J	BLOCK-1381
SN74366N	BLOCK-1381
SN74367A	BLOCK-1382
SN74367AJ	BLOCK-1382
SN74367AN	BLOCK-1382
SN74367J	BLOCK-1382
SN74367N	BLOCK-1382
SN74368A	BLOCK-1383
SN74368AJ	BLOCK-1383
SN74368AN	BLOCK-1383
SN74368J	BLOCK-1383
SN74368N	BLOCK-1383
SN7437	BLOCK-1384
SN7437J	BLOCK-1384
SN7437N	BLOCK-1384
SN7438	BLOCK-1385
SN7438J	BLOCK-1385
SN7438N	BLOCK-1385
SN74390	BLOCK-1387
SN74390J	BLOCK-1387
SN74390N	BLOCK-1387
SN74393	BLOCK-1388
SN74393J	BLOCK-1388
SN74393N	BLOCK-1388
SN7439N	BLOCK-1386
SN7440	BLOCK-1390
SN7440J	BLOCK-1390
SN7440N	BLOCK-1390
SN7441N-10	BLOCK-1391
SN7442A	BLOCK-1392
SN7442AJ	BLOCK-1392
SN7442AN	BLOCK-1392
SN7442N	BLOCK-1392
SN7443A	BLOCK-1393
SN7443AJ	BLOCK-1393
SN7443AN	BLOCK-1393
SN7443N	BLOCK-1393
SN7444A	BLOCK-1394
SN7444AJ	BLOCK-1394
SN7444AN	BLOCK-1394
SN7444N	BLOCK-1394
SN7445	BLOCK-1395
SN7445J	BLOCK-1395
SN7445N	BLOCK-1395
SN7446A	BLOCK-1396
SN7446AJ	BLOCK-1396
SN7446AN	BLOCK-1396
SN7447A	BLOCK-1397
SN7447AJ	BLOCK-1397
SN7447AN	BLOCK-1397
SN7447J	BLOCK-1397
SN7447N	BLOCK-1397
SN7447N-10	BLOCK-1397
SN7448	BLOCK-1398
SN7448J	BLOCK-1398
SN7448N	BLOCK-1398
SN7448N-10	BLOCK-1398
SN74490	BLOCK-1399
SN74490J	BLOCK-1399
SN74490N	BLOCK-1399
SN7450	BLOCK-1401
SN74500N	BLOCK-1719
SN745064NE	BLOCK-994
SN745065NE	BLOCK-994
SN745066NE	BLOCK-995
SN745067NE	BLOCK-995
SN745068NE	BLOCK-996
SN745069NE	BLOCK-996
SN7450J	BLOCK-1401
SN7450N	BLOCK-1401
SN7451	BLOCK-1402
SN7451J	BLOCK-1402
SN7451N	BLOCK-1402
SN7451N-10	BLOCK-1402
SN7453	BLOCK-1403
SN7453J	BLOCK-1403
SN7453N	BLOCK-1403
SN7454	BLOCK-1404
SN7454J	BLOCK-1404
SN7454N	BLOCK-1404
SN7460	BLOCK-1406
SN7460J	BLOCK-1406
SN7460N	BLOCK-1406
SN7470	BLOCK-1408
SN7470J	BLOCK-1408
SN7470N	BLOCK-1408
SN7472	BLOCK-1409
SN7472J	BLOCK-1409
SN7472N	BLOCK-1409
SN7473	BLOCK-1410
SN7473J	BLOCK-1410
SN7473N	BLOCK-1410
SN7473N-10	BLOCK-1410
SN7474	BLOCK-1411
SN74741L	BLOCK-2056
SN7474J	BLOCK-1411
SN7474M	BLOCK-1411
SN7474N	BLOCK-1411
SN7474N-10	BLOCK-1411
SN7475	BLOCK-1412
SN7475J	BLOCK-1412
SN7475N	BLOCK-1412
SN7475N-10	BLOCK-1412
SN7476	BLOCK-1413
SN7476B	BLOCK-1413
SN7476J	BLOCK-1413
SN7476N	BLOCK-1413
SN7476N-10	BLOCK-1413
SN7480	BLOCK-1414
SN7480J	BLOCK-1414
SN7480N	BLOCK-1414
SN7481A	BLOCK-1415
SN7481AN	BLOCK-1415
SN7482	BLOCK-1416
SN7482J	BLOCK-1416
SN7482N	BLOCK-1416
SN7483A	BLOCK-1417
SN7483AJ	BLOCK-1415
SN7483AN	BLOCK-1417
SN7485	BLOCK-1418
SN7485J	BLOCK-1418
SN7485N	BLOCK-1418
SN7485N-10	BLOCK-1418
SN7486	BLOCK-1419
SN7486J	BLOCK-1419
SN7486N	BLOCK-1419
SN7486N-10	BLOCK-1419
SN7489	BLOCK-1420
SN7489N	BLOCK-1420
SN7490A	BLOCK-1423
SN7490AJ	BLOCK-1423
SN7490AN	BLOCK-1423
SN7490N	BLOCK-1423
SN7490N-10	BLOCK-1423
SN7491A	BLOCK-1424

54 ORIGINAL DEVICE TYPES AND REPLACEMENT CODES

DEVICE TYPE	REPL CODE
SN7491AJ	BLOCK-1424
SN7491AN	BLOCK-1424
SN74921-10	BLOCK-1425
SN7492A	BLOCK-1425
SN7492AJ	BLOCK-1425
SN7492AN	BLOCK-1425
SN7492N	BLOCK-1425
SN7492N-10	BLOCK-1425
SN7493	BLOCK-1426
SN7493A	BLOCK-1426
SN7493AJ	BLOCK-1426
SN7493AN	BLOCK-1426
SN7493N	BLOCK-1426
SN7494	BLOCK-1427
SN7494J	BLOCK-1427
SN7494N	BLOCK-1427
SN7495AJ	BLOCK-1428
SN7495AN	BLOCK-1428
SN7496J	BLOCK-1429
SN7496N	BLOCK-1429
SN7497	BLOCK-1430
SN7497J	BLOCK-1430
SN7497N	BLOCK-1430
SN74H00	BLOCK-1471
SN74H00J	BLOCK-1471
SN74H00N	BLOCK-1471
SN74H01	BLOCK-1472
SN74H01J	BLOCK-1472
SN74H01N	BLOCK-1472
SN74H04	BLOCK-1473
SN74H04J	BLOCK-1473
SN74H04N	BLOCK-1473
SN74H05	BLOCK-1474
SN74H05J	BLOCK-1474
SN74H05N	BLOCK-1474
SN74H10	BLOCK-1476
SN74H101	BLOCK-1477
SN74H101J	BLOCK-1477
SN74H101N	BLOCK-1477
SN74H102	BLOCK-1478
SN74H102J	BLOCK-1478
SN74H102N	BLOCK-1478
SN74H103	BLOCK-1479
SN74H103J	BLOCK-1479
SN74H103N	BLOCK-1479
SN74H106	BLOCK-1480
SN74H106J	BLOCK-1480
SN74H106N	BLOCK-1480
SN74H108	BLOCK-1481
SN74H108J	BLOCK-1481
SN74H108N	BLOCK-1481
SN74H10J	BLOCK-1476
SN74H10N	BLOCK-1476
SN74H11	BLOCK-1482
SN74H11J	BLOCK-1482
SN74H11N	BLOCK-1482
SN74H183	BLOCK-1483
SN74H183J	BLOCK-1483
SN74H183N	BLOCK-1483
SN74H20	BLOCK-1484
SN74H20J	BLOCK-1484
SN74H20N	BLOCK-1484
SN74H21	BLOCK-1485
SN74H21J	BLOCK-1485
SN74H21N	BLOCK-1485
SN74H22	BLOCK-1486
SN74H22J	BLOCK-1486
SN74H22N	BLOCK-1486
SN74H30	BLOCK-1487
SN74H30J	BLOCK-1487
SN74H30N	BLOCK-1487
SN74H40	BLOCK-1488
SN74H40J	BLOCK-1488
SN74H40N	BLOCK-1488
SN74H50	BLOCK-1489
SN74H50J	BLOCK-1489
SN74H50N	BLOCK-1489
SN74H51	BLOCK-1490
SN74H51J	BLOCK-1490
SN74H51N	BLOCK-1490
SN74H52	BLOCK-1491
SN74H52J	BLOCK-1491
SN74H52N	BLOCK-1491
SN74H53	BLOCK-1492
SN74H53J	BLOCK-1492
SN74H53N	BLOCK-1492
SN74H54	BLOCK-1493
SN74H54DC	BLOCK-1493
SN74H54J	BLOCK-1493
SN74H54N	BLOCK-1493
SN74H55	BLOCK-1494
SN74H55J	BLOCK-1494
SN74H55N	BLOCK-1494
SN74H60	BLOCK-1495

DEVICE TYPE	REPL CODE
SN74H60J	BLOCK-1495
SN74H60N	BLOCK-1495
SN74H61	BLOCK-1496
SN74H61J	BLOCK-1496
SN74H61N	BLOCK-1496
SN74H62	BLOCK-1497
SN74H62J	BLOCK-1497
SN74H62N	BLOCK-1497
SN74H71	BLOCK-1498
SN74H71J	BLOCK-1498
SN74H71N	BLOCK-1498
SN74H72	BLOCK-1499
SN74H72J	BLOCK-1499
SN74H72N	BLOCK-1499
SN74H73	BLOCK-1500
SN74H73J	BLOCK-1500
SN74H73N	BLOCK-1500
SN74H74	BLOCK-1501
SN74H74J	BLOCK-1501
SN74H74N	BLOCK-1501
SN74H76	BLOCK-1502
SN74H76J	BLOCK-1502
SN74H76N	BLOCK-1502
SN74H78	BLOCK-1503
SN74H78J	BLOCK-1503
SN74H78N	BLOCK-1503
SN74H87	BLOCK-1505
SN74H87J	BLOCK-1505
SN74H87N	BLOCK-1505
SN74HC02N	BLOCK-1507
SN74HC04ANS	BLOCK-1508
SN74HC240N	BLOCK-1530
SN74HC32N	BLOCK-1536
SN74HC86N	BLOCK-1550
SN74L03	BLOCK-1571
SN74L93J	BLOCK-1567
SN74L93N	BLOCK-1567
SN74LS00	BLOCK-1568
SN74LS00J	BLOCK-1568
SN74LS00JD	BLOCK-1568
SN74LS00JDS	BLOCK-1568
SN74LS00JS	BLOCK-1568
SN74LS00N	BLOCK-1568
SN74LS00ND	BLOCK-1568
SN74LS00NDS	BLOCK-1568
SN74LS00NS	BLOCK-1568
SN74LS00W	BLOCK-1568
SN74LS01	BLOCK-1569
SN74LS01J	BLOCK-1569
SN74LS01JD	BLOCK-1569
SN74LS01JDS	BLOCK-1569
SN74LS01JS	BLOCK-1569
SN74LS01N	BLOCK-1569
SN74LS01ND	BLOCK-1569
SN74LS01NDS	BLOCK-1569
SN74LS01NS	BLOCK-1569
SN74LS02	BLOCK-1570
SN74LS02J	BLOCK-1570
SN74LS02JD	BLOCK-1570
SN74LS02JDS	BLOCK-1570
SN74LS02JS	BLOCK-1570
SN74LS02N	BLOCK-1570
SN74LS02NDS	BLOCK-1570
SN74LS02NS	BLOCK-1570
SN74LS02W	BLOCK-1570
SN74LS03	BLOCK-1492
SN74LS03J	BLOCK-1571
SN74LS03JD	BLOCK-1571
SN74LS03JDS	BLOCK-1571
SN74LS03JS	BLOCK-1571
SN74LS03N	BLOCK-1571
SN74LS03ND	BLOCK-1571
SN74LS03NDS	BLOCK-1571
SN74LS03NS	BLOCK-1571
SN74LS03W	BLOCK-1571
SN74LS04	BLOCK-1572
SN74LS04J	BLOCK-1572
SN74LS04JD	BLOCK-1572
SN74LS04JDS	BLOCK-1572
SN74LS04JS	BLOCK-1572
SN74LS04N	BLOCK-1572
SN74LS04ND	BLOCK-1572
SN74LS04NDS	BLOCK-1572
SN74LS04NS	BLOCK-1572
SN74LS05	BLOCK-1573
SN74LS05J	BLOCK-1573
SN74LS05JD	BLOCK-1573
SN74LS05JDS	BLOCK-1573
SN74LS05JS	BLOCK-1573
SN74LS05N	BLOCK-1573
SN74LS05ND	BLOCK-1573
SN74LS05NDS	BLOCK-1573
SN74LS05NS	BLOCK-1573
SN74LS08	BLOCK-1574

DEVICE TYPE	REPL CODE
SN74LS08J	BLOCK-1574
SN74LS08JD	BLOCK-1574
SN74LS08JDS	BLOCK-1574
SN74LS08JS	BLOCK-1574
SN74LS08N	BLOCK-1574
SN74LS08ND	BLOCK-1574
SN74LS08NDS	BLOCK-1574
SN74LS08NS	BLOCK-1574
SN74LS08W	BLOCK-1574
SN74LS09	BLOCK-1575
SN74LS09J	BLOCK-1575
SN74LS09JD	BLOCK-1575
SN74LS09JDS	BLOCK-1575
SN74LS09JS	BLOCK-1575
SN74LS09N	BLOCK-1575
SN74LS09ND	BLOCK-1575
SN74LS09NDS	BLOCK-1575
SN74LS09NS	BLOCK-1575
SN74LS10	BLOCK-1576
SN74LS107A	BLOCK-1577
SN74LS107AJ	BLOCK-1577
SN74LS107AJD	BLOCK-1577
SN74LS107AJDS	BLOCK-1577
SN74LS107AJS	BLOCK-1577
SN74LS107AN	BLOCK-1577
SN74LS107AND	BLOCK-1577
SN74LS107ANDS	BLOCK-1577
SN74LS107ANS	BLOCK-1577
SN74LS107AW	BLOCK-1577
SN74LS107N	BLOCK-1577
SN74LS109A	BLOCK-1578
SN74LS109AJ	BLOCK-1578
SN74LS109AJD	BLOCK-1578
SN74LS109AJDS	BLOCK-1578
SN74LS109AJS	BLOCK-1578
SN74LS109AN	BLOCK-1578
SN74LS109AND	BLOCK-1578
SN74LS109ANDS	BLOCK-1578
SN74LS109ANS	BLOCK-1578
SN74LS109AW	BLOCK-1578
SN74LS109N	BLOCK-1578
SN74LS10J	BLOCK-1576
SN74LS10JD	BLOCK-1576
SN74LS10JDS	BLOCK-1576
SN74LS10JS	BLOCK-1576
SN74LS10N	BLOCK-1576
SN74LS10ND	BLOCK-1576
SN74LS10NDS	BLOCK-1576
SN74LS10NS	BLOCK-1576
SN74LS10W	BLOCK-1576
SN74LS11	BLOCK-1579
SN74LS112A	BLOCK-1580
SN74LS112AJ	BLOCK-1580
SN74LS112AJD	BLOCK-1580
SN74LS112AJDS	BLOCK-1580
SN74LS112AJS	BLOCK-1580
SN74LS112AN	BLOCK-1580
SN74LS112AND	BLOCK-1580
SN74LS112ANDS	BLOCK-1580
SN74LS112ANS	BLOCK-1580
SN74LS112N	BLOCK-1728
SN74LS114A	BLOCK-1581
SN74LS114AJ	BLOCK-1581
SN74LS114AJD	BLOCK-1581
SN74LS114AJDS	BLOCK-1581
SN74LS114AJS	BLOCK-1581
SN74LS114AN	BLOCK-1581
SN74LS114AND	BLOCK-1581
SN74LS114ANDS	BLOCK-1581
SN74LS114ANS	BLOCK-1581
SN74LS11J	BLOCK-1579
SN74LS11JD	BLOCK-1579
SN74LS11JDS	BLOCK-1579
SN74LS11JS	BLOCK-1579
SN74LS11N	BLOCK-1579
SN74LS11ND	BLOCK-1579
SN74LS11NDS	BLOCK-1579
SN74LS11NS	BLOCK-1579
SN74LS11W	BLOCK-1579
SN74LS12	BLOCK-1582
SN74LS122	BLOCK-1583
SN74LS122J	BLOCK-1583
SN74LS122JD	BLOCK-1583
SN74LS122JDS	BLOCK-1583
SN74LS122JS	BLOCK-1583
SN74LS122N	BLOCK-1583
SN74LS122ND	BLOCK-1583
SN74LS122NDS	BLOCK-1583
SN74LS122NS	BLOCK-1583
SN74LS123	BLOCK-1584
SN74LS123J	BLOCK-1584
SN74LS123JD	BLOCK-1584
SN74LS123JDS	BLOCK-1584
SN74LS123JS	BLOCK-1584

DEVICE TYPE	REPL CODE
SN74LS123N	BLOCK-1584
SN74LS123ND	BLOCK-1584
SN74LS123NDS	BLOCK-1584
SN74LS123NS	BLOCK-1584
SN74LS123W	BLOCK-1584
SN74LS124J	BLOCK-1698
SN74LS124N	BLOCK-1698
SN74LS125A	BLOCK-1585
SN74LS125AJ	BLOCK-1585
SN74LS125AJD	BLOCK-1585
SN74LS125AJDS	BLOCK-1585
SN74LS125AJS	BLOCK-1585
SN74LS125AN	BLOCK-1585
SN74LS125AND	BLOCK-1585
SN74LS125ANDS	BLOCK-1585
SN74LS125ANS	BLOCK-1585
SN74LS126A	BLOCK-1586
SN74LS126AJ	BLOCK-1586
SN74LS126AJD	BLOCK-1586
SN74LS126AJDS	BLOCK-1586
SN74LS126AJS	BLOCK-1586
SN74LS126AN	BLOCK-1586
SN74LS126AND	BLOCK-1586
SN74LS126ANDS	BLOCK-1586
SN74LS126ANS	BLOCK-1586
SN74LS126J	BLOCK-1586
SN74LS126N	BLOCK-1586
SN74LS12J	BLOCK-1582
SN74LS12JD	BLOCK-1582
SN74LS12JDS	BLOCK-1582
SN74LS12JS	BLOCK-1582
SN74LS12N	BLOCK-1582
SN74LS12ND	BLOCK-1582
SN74LS12NDS	BLOCK-1582
SN74LS12NS	BLOCK-1582
SN74LS13	BLOCK-1587
SN74LS132	BLOCK-1588
SN74LS132J	BLOCK-1588
SN74LS132JD	BLOCK-1588
SN74LS132JDS	BLOCK-1588
SN74LS132JS	BLOCK-1588
SN74LS132N	BLOCK-1588
SN74LS132ND	BLOCK-1588
SN74LS132NDS	BLOCK-1588
SN74LS132NS	BLOCK-1588
SN74LS133J	BLOCK-1589
SN74LS133JD	BLOCK-1589
SN74LS133JDS	BLOCK-1589
SN74LS133JS	BLOCK-1589
SN74LS133N	BLOCK-1589
SN74LS133ND	BLOCK-1589
SN74LS133NDS	BLOCK-1589
SN74LS133NS	BLOCK-1589
SN74LS136	BLOCK-1590
SN74LS136J	BLOCK-1590
SN74LS136JD	BLOCK-1590
SN74LS136JDS	BLOCK-1590
SN74LS136JS	BLOCK-1590
SN74LS136N	BLOCK-1590
SN74LS136ND	BLOCK-1590
SN74LS136NDS	BLOCK-1590
SN74LS136NS	BLOCK-1590
SN74LS136W	BLOCK-1590
SN74LS138	BLOCK-1591
SN74LS138J	BLOCK-1591
SN74LS138JD	BLOCK-1591
SN74LS138JDS	BLOCK-1591
SN74LS138JS	BLOCK-1591
SN74LS138N	BLOCK-1591
SN74LS138ND	BLOCK-1591
SN74LS138NDS	BLOCK-1591
SN74LS138NS	BLOCK-1591
SN74LS138X	BLOCK-1591
SN74LS139	BLOCK-1592
SN74LS139AN	BLOCK-1592
SN74LS139J	BLOCK-1592
SN74LS139JD	BLOCK-1592
SN74LS139JDS	BLOCK-1592
SN74LS139JS	BLOCK-1592
SN74LS139N	BLOCK-1592
SN74LS139ND	BLOCK-1592
SN74LS139NDS	BLOCK-1592
SN74LS139NS	BLOCK-1592
SN74LS139X	BLOCK-1592
SN74LS13J	BLOCK-1587
SN74LS13JD	BLOCK-1587
SN74LS13JDS	BLOCK-1587
SN74LS13JS	BLOCK-1587
SN74LS13N	BLOCK-1587
SN74LS13ND	BLOCK-1587
SN74LS13NDS	BLOCK-1587
SN74LS13NS	BLOCK-1587
SN74LS14	BLOCK-1593
SN74LS145	BLOCK-1594

DEVICE TYPE	REPL CODE
SN74LS145J	BLOCK-1594
SN74LS145JD	BLOCK-1594
SN74LS145JDS	BLOCK-1594
SN74LS145JS	BLOCK-1594
SN74LS145N	BLOCK-1594
SN74LS145ND	BLOCK-1594
SN74LS145NDS	BLOCK-1594
SN74LS145NS	BLOCK-1594
SN74LS147	BLOCK-1595
SN74LS147J	BLOCK-1595
SN74LS147JD	BLOCK-1595
SN74LS147JDS	BLOCK-1595
SN74LS147JS	BLOCK-1595
SN74LS147N	BLOCK-1595
SN74LS147ND	BLOCK-1595
SN74LS147NDS	BLOCK-1595
SN74LS147NS	BLOCK-1595
SN74LS148	BLOCK-1596
SN74LS148AJ	BLOCK-1596
SN74LS148AJD	BLOCK-1596
SN74LS148AJDS	BLOCK-1596
SN74LS148AJS	BLOCK-1596
SN74LS148AN	BLOCK-1596
SN74LS148AND	BLOCK-1596
SN74LS148ANDS	BLOCK-1596
SN74LS148ANS	BLOCK-1596
SN74LS148J	BLOCK-1596
SN74LS148JD	BLOCK-1596
SN74LS148JDS	BLOCK-1596
SN74LS148JS	BLOCK-1596
SN74LS148N	BLOCK-1596
SN74LS148ND	BLOCK-1596
SN74LS148NDS	BLOCK-1596
SN74LS148NS	BLOCK-1596
SN74LS14J	BLOCK-1593
SN74LS14JD	BLOCK-1593
SN74LS14JDS	BLOCK-1593
SN74LS14JS	BLOCK-1593
SN74LS14N	BLOCK-1593
SN74LS14ND	BLOCK-1593
SN74LS14NDS	BLOCK-1593
SN74LS14NS	BLOCK-1593
SN74LS15	BLOCK-1597
SN74LS151	BLOCK-1598
SN74LS151J	BLOCK-1598
SN74LS151JD	BLOCK-1598
SN74LS151JDS	BLOCK-1598
SN74LS151JS	BLOCK-1598
SN74LS151N	BLOCK-1598
SN74LS151ND	BLOCK-1598
SN74LS151NDS	BLOCK-1598
SN74LS151NS	BLOCK-1598
SN74LS151X	BLOCK-1598
SN74LS153	BLOCK-1599
SN74LS153J	BLOCK-1599
SN74LS153JD	BLOCK-1599
SN74LS153JDS	BLOCK-1599
SN74LS153JS	BLOCK-1599
SN74LS153N	BLOCK-1599
SN74LS153ND	BLOCK-1599
SN74LS153NDS	BLOCK-1599
SN74LS153NS	BLOCK-1599
SN74LS155	BLOCK-1600
SN74LS155AN	BLOCK-1600
SN74LS155J	BLOCK-1600
SN74LS155JD	BLOCK-1600
SN74LS155JDS	BLOCK-1600
SN74LS155JS	BLOCK-1600
SN74LS155N	BLOCK-1600
SN74LS155ND	BLOCK-1600
SN74LS155NDS	BLOCK-1600
SN74LS155NS	BLOCK-1600
SN74LS156	BLOCK-1601
SN74LS156J	BLOCK-1601
SN74LS156JD	BLOCK-1601
SN74LS156JDS	BLOCK-1601
SN74LS156JS	BLOCK-1601
SN74LS156N	BLOCK-1601
SN74LS156ND	BLOCK-1601
SN74LS156NDS	BLOCK-1601
SN74LS156NS	BLOCK-1601
SN74LS157	BLOCK-1602
SN74LS157J	BLOCK-1602
SN74LS157JD	BLOCK-1602
SN74LS157JDS	BLOCK-1602
SN74LS157JS	BLOCK-1602
SN74LS157N	BLOCK-1602
SN74LS157ND	BLOCK-1602
SN74LS157NDS	BLOCK-1602
SN74LS157NS	BLOCK-1602
SN74LS158	BLOCK-1603
SN74LS158J	BLOCK-1603
SN74LS158JD	BLOCK-1603
SN74LS158JDS	BLOCK-1603

DEVICE TYPE	REPL CODE
SN74LS158JS	BLOCK-1603
SN74LS158N	BLOCK-1603
SN74LS158ND	BLOCK-1603
SN74LS158NDS	BLOCK-1603
SN74LS158NS	BLOCK-1603
SN74LS15J	BLOCK-1597
SN74LS15JD	BLOCK-1597
SN74LS15JDS	BLOCK-1597
SN74LS15JS	BLOCK-1597
SN74LS15N	BLOCK-1597
SN74LS15ND	BLOCK-1597
SN74LS15NDS	BLOCK-1597
SN74LS15NS	BLOCK-1597
SN74LS160A	BLOCK-1604
SN74LS160AJ	BLOCK-1604
SN74LS160AJD	BLOCK-1604
SN74LS160AJDS	BLOCK-1604
SN74LS160AJS	BLOCK-1604
SN74LS160AN	BLOCK-1604
SN74LS160AND	BLOCK-1604
SN74LS160ANDS	BLOCK-1604
SN74LS160ANS	BLOCK-1604
SN74LS160J	BLOCK-1604
SN74LS160N	BLOCK-1604
SN74LS161A	BLOCK-1605
SN74LS161AJ	BLOCK-1605
SN74LS161AJD	BLOCK-1605
SN74LS161AJDS	BLOCK-1605
SN74LS161AJS	BLOCK-1605
SN74LS161AN	BLOCK-1605
SN74LS161AND	BLOCK-1605
SN74LS161ANDS	BLOCK-1605
SN74LS161ANS	BLOCK-1605
SN74LS161AW	BLOCK-1605
SN74LS161J	BLOCK-1605
SN74LS161N	BLOCK-1605
SN74LS162A	BLOCK-1606
SN74LS162AJ	BLOCK-1606
SN74LS162AJD	BLOCK-1606
SN74LS162AJDS	BLOCK-1606
SN74LS162AJS	BLOCK-1606
SN74LS162AN	BLOCK-1606
SN74LS162AND	BLOCK-1606
SN74LS162ANDS	BLOCK-1606
SN74LS162ANS	BLOCK-1606
SN74LS162J	BLOCK-1606
SN74LS162N	BLOCK-1606
SN74LS163A	BLOCK-1607
SN74LS163AJ	BLOCK-1607
SN74LS163AJD	BLOCK-1607
SN74LS163AJDS	BLOCK-1607
SN74LS163AJS	BLOCK-1607
SN74LS163AN	BLOCK-1607
SN74LS163AND	BLOCK-1607
SN74LS163ANDS	BLOCK-1607
SN74LS163ANS	BLOCK-1607
SN74LS163AW	BLOCK-1607
SN74LS163J	BLOCK-1607
SN74LS163N	BLOCK-1607
SN74LS164	BLOCK-1608
SN74LS164J	BLOCK-1608
SN74LS164JD	BLOCK-1608
SN74LS164JDS	BLOCK-1608
SN74LS164JS	BLOCK-1608
SN74LS164N	BLOCK-1608
SN74LS164ND	BLOCK-1608
SN74LS164NDS	BLOCK-1608
SN74LS164NS	BLOCK-1608
SN74LS165A	BLOCK-1609
SN74LS165J	BLOCK-1609
SN74LS165JD	BLOCK-1609
SN74LS165JDS	BLOCK-1609
SN74LS165JS	BLOCK-1609
SN74LS165N	BLOCK-1609
SN74LS165ND	BLOCK-1609
SN74LS165NDS	BLOCK-1609
SN74LS165NS	BLOCK-1609
SN74LS166	BLOCK-1610
SN74LS166A	BLOCK-1610
SN74LS166AN	BLOCK-1610
SN74LS166J	BLOCK-1610
SN74LS166JD	BLOCK-1610
SN74LS166JDS	BLOCK-1610
SN74LS166JS	BLOCK-1610
SN74LS166N	BLOCK-1610
SN74LS166ND	BLOCK-1610
SN74LS166NDS	BLOCK-1610
SN74LS166NS	BLOCK-1610
SN74LS168N	BLOCK-1611
SN74LS169AJ	BLOCK-1612
SN74LS169AN	BLOCK-1612
SN74LS169B	BLOCK-1612
SN74LS169J	BLOCK-1612
SN74LS169N	BLOCK-1612

ORIGINAL DEVICE TYPES AND REPLACEMENT CODES 55

DEVICE TYPE	REPL CODE
SN74LS170	BLOCK-1613
SN74LS170J	BLOCK-1613
SN74LS170JD	BLOCK-1613
SN74LS170JDS	BLOCK-1613
SN74LS170JS	BLOCK-1613
SN74LS170N	BLOCK-1613
SN74LS170ND	BLOCK-1613
SN74LS170NDS	BLOCK-1613
SN74LS170NS	BLOCK-1613
SN74LS173A	BLOCK-1614
SN74LS173AJ	BLOCK-1614
SN74LS173AJD	BLOCK-1614
SN74LS173AJDS	BLOCK-1614
SN74LS173AJS	BLOCK-1614
SN74LS173AN	BLOCK-1614
SN74LS173AND	BLOCK-1614
SN74LS173ANDS	BLOCK-1614
SN74LS173ANS	BLOCK-1614
SN74LS174	BLOCK-1615
SN74LS174J	BLOCK-1615
SN74LS174JD	BLOCK-1615
SN74LS174JDS	BLOCK-1615
SN74LS174JS	BLOCK-1615
SN74LS174N	BLOCK-1615
SN74LS174ND	BLOCK-1615
SN74LS174NDS	BLOCK-1615
SN74LS174NS	BLOCK-1615
SN74LS175	BLOCK-1616
SN74LS175J	BLOCK-1616
SN74LS175JD	BLOCK-1616
SN74LS175JDS	BLOCK-1616
SN74LS175JS	BLOCK-1616
SN74LS175N	BLOCK-1616
SN74LS175ND	BLOCK-1616
SN74LS175NDS	BLOCK-1616
SN74LS175NS	BLOCK-1616
SN74LS181	BLOCK-1617
SN74LS181J	BLOCK-1617
SN74LS181JD	BLOCK-1617
SN74LS181JDS	BLOCK-1617
SN74LS181JS	BLOCK-1617
SN74LS181N	BLOCK-1617
SN74LS181ND	BLOCK-1617
SN74LS181NDS	BLOCK-1617
SN74LS181NS	BLOCK-1617
SN74LS190	BLOCK-1618
SN74LS190J	BLOCK-1618
SN74LS190JD	BLOCK-1618
SN74LS190JDS	BLOCK-1618
SN74LS190JS	BLOCK-1618
SN74LS190N	BLOCK-1618
SN74LS190ND	BLOCK-1618
SN74LS190NDS	BLOCK-1618
SN74LS190NS	BLOCK-1618
SN74LS191	BLOCK-1619
SN74LS191J	BLOCK-1619
SN74LS191JD	BLOCK-1619
SN74LS191JDS	BLOCK-1619
SN74LS191JS	BLOCK-1619
SN74LS191N	BLOCK-1619
SN74LS191ND	BLOCK-1619
SN74LS191NDS	BLOCK-1619
SN74LS191NS	BLOCK-1619
SN74LS191W	BLOCK-1619
SN74LS192	BLOCK-1620
SN74LS192J	BLOCK-1620
SN74LS192JD	BLOCK-1620
SN74LS192JDS	BLOCK-1620
SN74LS192JS	BLOCK-1620
SN74LS192N	BLOCK-1620
SN74LS192ND	BLOCK-1620
SN74LS192NDS	BLOCK-1620
SN74LS192NS	BLOCK-1620
SN74LS193	BLOCK-1621
SN74LS193J	BLOCK-1621
SN74LS193JD	BLOCK-1621
SN74LS193JDS	BLOCK-1621
SN74LS193JS	BLOCK-1621
SN74LS193N	BLOCK-1621
SN74LS193ND	BLOCK-1621
SN74LS193NDS	BLOCK-1621
SN74LS193NS	BLOCK-1621
SN74LS193W	BLOCK-1621
SN74LS194A	BLOCK-1622
SN74LS194AJ	BLOCK-1622
SN74LS194AJD	BLOCK-1622
SN74LS194AJDS	BLOCK-1622
SN74LS194AJS	BLOCK-1622
SN74LS194AN	BLOCK-1622
SN74LS194AND	BLOCK-1622
SN74LS194ANDS	BLOCK-1622
SN74LS194ANS	BLOCK-1622
SN74LS195	BLOCK-1623
SN74LS195A	BLOCK-1623
SN74LS195AJ	BLOCK-1623
SN74LS195AJD	BLOCK-1623
SN74LS195AJDS	BLOCK-1623
SN74LS195AJS	BLOCK-1623
SN74LS195AN	BLOCK-1623
SN74LS195AND	BLOCK-1623
SN74LS195ANDS	BLOCK-1623
SN74LS195ANS	BLOCK-1623
SN74LS195N	BLOCK-1623
SN74LS196	BLOCK-1624
SN74LS196J	BLOCK-1624
SN74LS196JD	BLOCK-1624
SN74LS196JDS	BLOCK-1624
SN74LS196JS	BLOCK-1624
SN74LS196N	BLOCK-1624
SN74LS196ND	BLOCK-1624
SN74LS196NDS	BLOCK-1624
SN74LS196NS	BLOCK-1624
SN74LS197	BLOCK-1625
SN74LS197J	BLOCK-1625
SN74LS197JD	BLOCK-1625
SN74LS197JDS	BLOCK-1625
SN74LS197JS	BLOCK-1625
SN74LS197N	BLOCK-1625
SN74LS197ND	BLOCK-1625
SN74LS197NDS	BLOCK-1625
SN74LS197NS	BLOCK-1625
SN74LS20	BLOCK-1626
SN74LS20J	BLOCK-1626
SN74LS20JD	BLOCK-1626
SN74LS20JDS	BLOCK-1626
SN74LS20JS	BLOCK-1626
SN74LS20N	BLOCK-1626
SN74LS20ND	BLOCK-1626
SN74LS20NDS	BLOCK-1626
SN74LS20NS	BLOCK-1626
SN74LS20W	BLOCK-1626
SN74LS21	BLOCK-1627
SN74LS21J	BLOCK-1627
SN74LS21JD	BLOCK-1627
SN74LS21JDS	BLOCK-1627
SN74LS21JS	BLOCK-1627
SN74LS21N	BLOCK-1627
SN74LS21ND	BLOCK-1627
SN74LS21NDS	BLOCK-1627
SN74LS21NS	BLOCK-1627
SN74LS22	BLOCK-1628
SN74LS221	BLOCK-1629
SN74LS221J	BLOCK-1629
SN74LS221JD	BLOCK-1629
SN74LS221JDS	BLOCK-1629
SN74LS221JS	BLOCK-1629
SN74LS221N	BLOCK-1629
SN74LS221ND	BLOCK-1629
SN74LS221NDS	BLOCK-1629
SN74LS221NS	BLOCK-1629
SN74LS221W	BLOCK-1629
SN74LS22J	BLOCK-1628
SN74LS22JD	BLOCK-1628
SN74LS22JDS	BLOCK-1628
SN74LS22JS	BLOCK-1628
SN74LS22N	BLOCK-1628
SN74LS22ND	BLOCK-1628
SN74LS22NDS	BLOCK-1628
SN74LS22NS	BLOCK-1628
SN74LS240	BLOCK-1630
SN74LS240J	BLOCK-1630
SN74LS240JD	BLOCK-1630
SN74LS240JDS	BLOCK-1630
SN74LS240JS	BLOCK-1630
SN74LS240N	BLOCK-1630
SN74LS240ND	BLOCK-1630
SN74LS240NDS	BLOCK-1630
SN74LS240NS	BLOCK-1630
SN74LS241	BLOCK-1631
SN74LS241J	BLOCK-1631
SN74LS241JD	BLOCK-1631
SN74LS241JDS	BLOCK-1631
SN74LS241JS	BLOCK-1631
SN74LS241N	BLOCK-1631
SN74LS241ND	BLOCK-1631
SN74LS241NDS	BLOCK-1631
SN74LS241NS	BLOCK-1631
SN74LS242	BLOCK-1632
SN74LS242J	BLOCK-1632
SN74LS242JD	BLOCK-1632
SN74LS242JDS	BLOCK-1632
SN74LS242JS	BLOCK-1632
SN74LS242N	BLOCK-1632
SN74LS242ND	BLOCK-1632
SN74LS242NDS	BLOCK-1632
SN74LS242NS	BLOCK-1632
SN74LS243	BLOCK-1633
SN74LS243J	BLOCK-1633
SN74LS243JD	BLOCK-1633
SN74LS243JDS	BLOCK-1633
SN74LS243JS	BLOCK-1633
SN74LS243N	BLOCK-1633
SN74LS243ND	BLOCK-1633
SN74LS243NDS	BLOCK-1633
SN74LS243NS	BLOCK-1633
SN74LS244	BLOCK-1634
SN74LS244J	BLOCK-1634
SN74LS244JD	BLOCK-1634
SN74LS244JDS	BLOCK-1634
SN74LS244JS	BLOCK-1634
SN74LS244N	BLOCK-1634
SN74LS244ND	BLOCK-1634
SN74LS244NDS	BLOCK-1634
SN74LS244NS	BLOCK-1634
SN74LS245	BLOCK-1635
SN74LS245J	BLOCK-1635
SN74LS245JD	BLOCK-1635
SN74LS245JDS	BLOCK-1635
SN74LS245JS	BLOCK-1635
SN74LS245N	BLOCK-1635
SN74LS245ND	BLOCK-1635
SN74LS245NDS	BLOCK-1635
SN74LS245NS	BLOCK-1635
SN74LS247	BLOCK-1636
SN74LS247J	BLOCK-1636
SN74LS247JD	BLOCK-1636
SN74LS247JDS	BLOCK-1636
SN74LS247JS	BLOCK-1636
SN74LS247N	BLOCK-1636
SN74LS247ND	BLOCK-1636
SN74LS247NDS	BLOCK-1636
SN74LS247NS	BLOCK-1636
SN74LS248	BLOCK-1637
SN74LS248J	BLOCK-1637
SN74LS248JD	BLOCK-1637
SN74LS248JDS	BLOCK-1637
SN74LS248JS	BLOCK-1637
SN74LS248N	BLOCK-1637
SN74LS248ND	BLOCK-1637
SN74LS248NDS	BLOCK-1637
SN74LS248NS	BLOCK-1637
SN74LS249	BLOCK-1638
SN74LS249J	BLOCK-1638
SN74LS249JD	BLOCK-1638
SN74LS249JDS	BLOCK-1638
SN74LS249JS	BLOCK-1638
SN74LS249N	BLOCK-1638
SN74LS249ND	BLOCK-1638
SN74LS249NDS	BLOCK-1638
SN74LS249NS	BLOCK-1638
SN74LS251	BLOCK-1639
SN74LS251J	BLOCK-1639
SN74LS251JD	BLOCK-1639
SN74LS251JDS	BLOCK-1639
SN74LS251JS	BLOCK-1639
SN74LS251N	BLOCK-1639
SN74LS251ND	BLOCK-1639
SN74LS251NDS	BLOCK-1639
SN74LS251NS	BLOCK-1639
SN74LS251X	BLOCK-1639
SN74LS253	BLOCK-1640
SN74LS253J	BLOCK-1640
SN74LS253JD	BLOCK-1640
SN74LS253JDS	BLOCK-1640
SN74LS253JS	BLOCK-1640
SN74LS253N	BLOCK-1640
SN74LS253ND	BLOCK-1640
SN74LS253NDS	BLOCK-1640
SN74LS253NS	BLOCK-1640
SN74LS257	BLOCK-1641
SN74LS257AJ	BLOCK-1641
SN74LS257AJD	BLOCK-1641
SN74LS257AJDS	BLOCK-1641
SN74LS257AJS	BLOCK-1641
SN74LS257AN	BLOCK-1641
SN74LS257AND	BLOCK-1641
SN74LS257ANDS	BLOCK-1641
SN74LS257ANS	BLOCK-1641
SN74LS257BN	BLOCK-1641
SN74LS257J	BLOCK-1641
SN74LS257N	BLOCK-1641
SN74LS257X	BLOCK-1641
SN74LS258	BLOCK-1642
SN74LS258AJ	BLOCK-1642
SN74LS258AJD	BLOCK-1642
SN74LS258AJDS	BLOCK-1642
SN74LS258AJS	BLOCK-1642
SN74LS258AN	BLOCK-1642
SN74LS258AND	BLOCK-1642
SN74LS258ANDS	BLOCK-1642
SN74LS258ANS	BLOCK-1642
SN74LS258J	BLOCK-1642
SN74LS258N	BLOCK-1642
SN74LS258X	BLOCK-1642
SN74LS259	BLOCK-1643
SN74LS259J	BLOCK-1643
SN74LS259JD	BLOCK-1643
SN74LS259JDS	BLOCK-1643
SN74LS259JS	BLOCK-1643
SN74LS259N	BLOCK-1643
SN74LS259ND	BLOCK-1643
SN74LS259NDS	BLOCK-1643
SN74LS259NS	BLOCK-1643
SN74LS259W	BLOCK-1643
SN74LS26	BLOCK-1644
SN74LS260J	BLOCK-1645
SN74LS260JD	BLOCK-1645
SN74LS260JDS	BLOCK-1645
SN74LS260JS	BLOCK-1645
SN74LS260N	BLOCK-1645
SN74LS260ND	BLOCK-1645
SN74LS260NDS	BLOCK-1645
SN74LS260NS	BLOCK-1645
SN74LS266	BLOCK-1646
SN74LS266J	BLOCK-1646
SN74LS266JD	BLOCK-1646
SN74LS266JDS	BLOCK-1646
SN74LS266JS	BLOCK-1646
SN74LS266N	BLOCK-1646
SN74LS266ND	BLOCK-1646
SN74LS266NDS	BLOCK-1646
SN74LS266NS	BLOCK-1646
SN74LS266W	BLOCK-1646
SN74LS26J	BLOCK-1644
SN74LS26JD	BLOCK-1644
SN74LS26JDS	BLOCK-1644
SN74LS26JS	BLOCK-1644
SN74LS26N	BLOCK-1644
SN74LS26ND	BLOCK-1644
SN74LS26NS	BLOCK-1644
SN74LS273	BLOCK-1648
SN74LS273J	BLOCK-1648
SN74LS273JD	BLOCK-1648
SN74LS273JDS	BLOCK-1648
SN74LS273JS	BLOCK-1648
SN74LS273N	BLOCK-1648
SN74LS273ND	BLOCK-1648
SN74LS273NDS	BLOCK-1648
SN74LS273NS	BLOCK-1648
SN74LS279	BLOCK-1649
SN74LS279J	BLOCK-1649
SN74LS279JD	BLOCK-1649
SN74LS279JDS	BLOCK-1649
SN74LS279JS	BLOCK-1649
SN74LS279N	BLOCK-1649
SN74LS279ND	BLOCK-1649
SN74LS279NDS	BLOCK-1649
SN74LS279NS	BLOCK-1649
SN74LS27J	BLOCK-1647
SN74LS27JD	BLOCK-1647
SN74LS27JDS	BLOCK-1647
SN74LS27N	BLOCK-1647
SN74LS27ND	BLOCK-1647
SN74LS27NDS	BLOCK-1647
SN74LS27NS	BLOCK-1647
SN74LS27W	BLOCK-1647
SN74LS28	BLOCK-1650
SN74LS280	BLOCK-1651
SN74LS280J	BLOCK-1651
SN74LS280JD	BLOCK-1651
SN74LS280JDS	BLOCK-1651
SN74LS280JS	BLOCK-1651
SN74LS280N	BLOCK-1651
SN74LS280ND	BLOCK-1651
SN74LS280NDS	BLOCK-1651
SN74LS280NS	BLOCK-1651
SN74LS280W	BLOCK-1651
SN74LS283	BLOCK-1652
SN74LS283J	BLOCK-1652
SN74LS283JD	BLOCK-1652
SN74LS283JDS	BLOCK-1652
SN74LS283JS	BLOCK-1652
SN74LS283N	BLOCK-1652
SN74LS283ND	BLOCK-1652
SN74LS283NDS	BLOCK-1652
SN74LS283NS	BLOCK-1652
SN74LS28J	BLOCK-1650
SN74LS28JD	BLOCK-1650
SN74LS28JDS	BLOCK-1650
SN74LS28JS	BLOCK-1650
SN74LS28N	BLOCK-1650
SN74LS28ND	BLOCK-1650
SN74LS28NDS	BLOCK-1650
SN74LS28NS	BLOCK-1650
SN74LS290	BLOCK-1653
SN74LS290J	BLOCK-1653
SN74LS290JD	BLOCK-1653
SN74LS290JDS	BLOCK-1653
SN74LS290JS	BLOCK-1653
SN74LS290N	BLOCK-1653
SN74LS290ND	BLOCK-1653
SN74LS290NDS	BLOCK-1653
SN74LS290NS	BLOCK-1653
SN74LS293	BLOCK-1654
SN74LS293J	BLOCK-1654
SN74LS293JD	BLOCK-1654
SN74LS293JDS	BLOCK-1654
SN74LS293JS	BLOCK-1654
SN74LS293N	BLOCK-1654
SN74LS293ND	BLOCK-1654
SN74LS293NDS	BLOCK-1654
SN74LS293NS	BLOCK-1654
SN74LS295AJ	BLOCK-1655
SN74LS295AJD	BLOCK-1655
SN74LS295AJDS	BLOCK-1655
SN74LS295AJS	BLOCK-1655
SN74LS295AN	BLOCK-1655
SN74LS295AND	BLOCK-1655
SN74LS295ANDS	BLOCK-1655
SN74LS295ANS	BLOCK-1655
SN74LS295B	BLOCK-1655
SN74LS295BJ	BLOCK-1655
SN74LS295BN	BLOCK-1655
SN74LS298	BLOCK-1656
SN74LS298J	BLOCK-1656
SN74LS298JD	BLOCK-1656
SN74LS298JDS	BLOCK-1656
SN74LS298JS	BLOCK-1656
SN74LS298N	BLOCK-1656
SN74LS298ND	BLOCK-1656
SN74LS298NDS	BLOCK-1656
SN74LS298NS	BLOCK-1656
SN74LS299	BLOCK-1657
SN74LS299J	BLOCK-1657
SN74LS299JD	BLOCK-1657
SN74LS299JDS	BLOCK-1657
SN74LS299JS	BLOCK-1657
SN74LS299N	BLOCK-1657
SN74LS299ND	BLOCK-1657
SN74LS299NDS	BLOCK-1657
SN74LS299NS	BLOCK-1657
SN74LS30	BLOCK-1658
SN74LS30J	BLOCK-1658
SN74LS30JD	BLOCK-1658
SN74LS30JDS	BLOCK-1658
SN74LS30JS	BLOCK-1658
SN74LS30N	BLOCK-1658
SN74LS30ND	BLOCK-1658
SN74LS30NDS	BLOCK-1658
SN74LS30NS	BLOCK-1658
SN74LS30W	BLOCK-1658
SN74LS32	BLOCK-1659
SN74LS32J	BLOCK-1659
SN74LS32JD	BLOCK-1659
SN74LS32JDS	BLOCK-1659
SN74LS32JS	BLOCK-1659
SN74LS32N	BLOCK-1659
SN74LS32ND	BLOCK-1659
SN74LS32NDS	BLOCK-1659
SN74LS32NS	BLOCK-1659
SN74LS32W	BLOCK-1659
SN74LS33	BLOCK-1660
SN74LS33J	BLOCK-1660
SN74LS33N	BLOCK-1660
SN74LS33NS	BLOCK-1660
SN74LS348	BLOCK-1661
SN74LS348AJ	BLOCK-1661
SN74LS348AJD	BLOCK-1661
SN74LS348AJDS	BLOCK-1661
SN74LS348AJS	BLOCK-1661
SN74LS348AN	BLOCK-1661
SN74LS348AND	BLOCK-1661
SN74LS348ANDS	BLOCK-1661
SN74LS348ANS	BLOCK-1661
SN74LS348J	BLOCK-1661
SN74LS348JD	BLOCK-1661
SN74LS348JDS	BLOCK-1661
SN74LS348JS	BLOCK-1661
SN74LS348N	BLOCK-1661
SN74LS348ND	BLOCK-1661
SN74LS348NDS	BLOCK-1661
SN74LS348NS	BLOCK-1661
SN74LS352	BLOCK-1662
SN74LS352J	BLOCK-1662
SN74LS352JD	BLOCK-1662
SN74LS352JDS	BLOCK-1662
SN74LS352JS	BLOCK-1662
SN74LS352N	BLOCK-1662
SN74LS352ND	BLOCK-1662
SN74LS352NDS	BLOCK-1662
SN74LS352NS	BLOCK-1662
SN74LS353	BLOCK-1663
SN74LS353J	BLOCK-1663
SN74LS353JD	BLOCK-1663
SN74LS353JDS	BLOCK-1663
SN74LS353JS	BLOCK-1663
SN74LS353N	BLOCK-1663
SN74LS353ND	BLOCK-1663
SN74LS353NDS	BLOCK-1663
SN74LS353NS	BLOCK-1663
SN74LS365A	BLOCK-1666
SN74LS365AJ	BLOCK-1666
SN74LS365AJD	BLOCK-1666
SN74LS365AJDS	BLOCK-1666
SN74LS365AJS	BLOCK-1666
SN74LS365AN	BLOCK-1666
SN74LS365AND	BLOCK-1666
SN74LS365ANDS	BLOCK-1666
SN74LS365ANS	BLOCK-1666
SN74LS365J	BLOCK-1666
SN74LS365N	BLOCK-1666
SN74LS366A	BLOCK-1667
SN74LS366AJ	BLOCK-1667
SN74LS366AJD	BLOCK-1667
SN74LS366AJDS	BLOCK-1667
SN74LS366AJS	BLOCK-1667
SN74LS366AN	BLOCK-1667
SN74LS366AND	BLOCK-1667
SN74LS366ANDS	BLOCK-1667
SN74LS366ANS	BLOCK-1667
SN74LS366J	BLOCK-1667
SN74LS366N	BLOCK-1667
SN74LS367	BLOCK-1668
SN74LS367A	BLOCK-1668
SN74LS367AJ	BLOCK-1668
SN74LS367AJD	BLOCK-1668
SN74LS367AJDS	BLOCK-1668
SN74LS367AJS	BLOCK-1668
SN74LS367AN	BLOCK-1668
SN74LS367AN-X	BLOCK-1668
SN74LS367AND	BLOCK-1668
SN74LS367ANDS	BLOCK-1668
SN74LS367ANS	BLOCK-1668
SN74LS367J	BLOCK-1668
SN74LS367N	BLOCK-1668
SN74LS368A	BLOCK-1669
SN74LS368AJ	BLOCK-1669
SN74LS368AJD	BLOCK-1669
SN74LS368AJDS	BLOCK-1669
SN74LS368AJS	BLOCK-1669
SN74LS368AN	BLOCK-1669
SN74LS368AND	BLOCK-1669
SN74LS368ANDS	BLOCK-1669
SN74LS368ANS	BLOCK-1669
SN74LS368J	BLOCK-1669
SN74LS368N	BLOCK-1669
SN74LS37	BLOCK-1670
SN74LS373	BLOCK-1671
SN74LS373J	BLOCK-1671
SN74LS373JD	BLOCK-1671
SN74LS373JDS	BLOCK-1671
SN74LS373JS	BLOCK-1671
SN74LS373N	BLOCK-1671
SN74LS373ND	BLOCK-1671
SN74LS373NDS	BLOCK-1671
SN74LS373NS	BLOCK-1671
SN74LS374	BLOCK-1672
SN74LS374J	BLOCK-1672
SN74LS374JD	BLOCK-1672
SN74LS374JDS	BLOCK-1672
SN74LS374JS	BLOCK-1672
SN74LS374N	BLOCK-1672
SN74LS374ND	BLOCK-1672
SN74LS374NDS	BLOCK-1672
SN74LS374NS	BLOCK-1672
SN74LS377	BLOCK-1673
SN74LS377J	BLOCK-1673
SN74LS377JD	BLOCK-1673
SN74LS377JDS	BLOCK-1673
SN74LS377JS	BLOCK-1673
SN74LS377N	BLOCK-1673
SN74LS377ND	BLOCK-1673
SN74LS377NDS	BLOCK-1673
SN74LS377NS	BLOCK-1673
SN74LS378	BLOCK-1674
SN74LS378J	BLOCK-1674
SN74LS378JD	BLOCK-1674
SN74LS378JDS	BLOCK-1674
SN74LS378JS	BLOCK-1674
SN74LS378N	BLOCK-1674
SN74LS378ND	BLOCK-1674
SN74LS378NDS	BLOCK-1674
SN74LS378NS	BLOCK-1674
SN74LS379	BLOCK-1675

56 ORIGINAL DEVICE TYPES AND REPLACEMENT CODES

DEVICE TYPE	REPL CODE
SN74LS379J	BLOCK-1675
SN74LS379JD	BLOCK-1675
SN74LS379JDS	BLOCK-1675
SN74LS379JS	BLOCK-1675
SN74LS379N	BLOCK-1675
SN74LS379ND	BLOCK-1675
SN74LS379NDS	BLOCK-1675
SN74LS379NS	BLOCK-1675
SN74LS37J	BLOCK-1670
SN74LS37JD	BLOCK-1670
SN74LS37JDS	BLOCK-1670
SN74LS37JS	BLOCK-1670
SN74LS37N	BLOCK-1670
SN74LS37ND	BLOCK-1670
SN74LS37NDS	BLOCK-1670
SN74LS37NS	BLOCK-1670
SN74LS38	BLOCK-1676
SN74LS386	BLOCK-1677
SN74LS386J	BLOCK-1677
SN74LS386JD	BLOCK-1677
SN74LS386JDS	BLOCK-1677
SN74LS386JS	BLOCK-1677
SN74LS386N	BLOCK-1677
SN74LS386ND	BLOCK-1677
SN74LS386NDS	BLOCK-1677
SN74LS386NS	BLOCK-1677
SN74LS38J	BLOCK-1676
SN74LS38JD	BLOCK-1676
SN74LS38JDS	BLOCK-1676
SN74LS38JS	BLOCK-1676
SN74LS38N	BLOCK-1676
SN74LS38ND	BLOCK-1676
SN74LS38NDS	BLOCK-1676
SN74LS38NS	BLOCK-1676
SN74LS390	BLOCK-1678
SN74LS390J	BLOCK-1678
SN74LS390JD	BLOCK-1678
SN74LS390JDS	BLOCK-1678
SN74LS390JS	BLOCK-1678
SN74LS390N	BLOCK-1678
SN74LS390ND	BLOCK-1678
SN74LS390NDS	BLOCK-1678
SN74LS390NS	BLOCK-1678
SN74LS393	BLOCK-1679
SN74LS393J	BLOCK-1679
SN74LS393JD	BLOCK-1679
SN74LS393JDS	BLOCK-1679
SN74LS393JS	BLOCK-1679
SN74LS393N	BLOCK-1679
SN74LS393ND	BLOCK-1679
SN74LS393NDS	BLOCK-1679
SN74LS393NS	BLOCK-1679
SN74LS395A	BLOCK-1680
SN74LS395AN	BLOCK-1680
SN74LS395J	BLOCK-1680
SN74LS395JD	BLOCK-1680
SN74LS395JDS	BLOCK-1680
SN74LS395JS	BLOCK-1680
SN74LS395N	BLOCK-1680
SN74LS395ND	BLOCK-1680
SN74LS395NDS	BLOCK-1680
SN74LS395NS	BLOCK-1680
SN74LS398J	BLOCK-1681
SN74LS398JD	BLOCK-1681
SN74LS398JDS	BLOCK-1681
SN74LS398JS	BLOCK-1681
SN74LS398N	BLOCK-1681
SN74LS398ND	BLOCK-1681
SN74LS398NDS	BLOCK-1681
SN74LS398NS	BLOCK-1681
SN74LS399	BLOCK-1682
SN74LS399J	BLOCK-1682
SN74LS399JD	BLOCK-1682
SN74LS399JDS	BLOCK-1682
SN74LS399JS	BLOCK-1682
SN74LS399N	BLOCK-1682
SN74LS399ND	BLOCK-1682
SN74LS399NDS	BLOCK-1682
SN74LS399NS	BLOCK-1682
SN74LS40	BLOCK-1683
SN74LS40J	BLOCK-1683
SN74LS40JD	BLOCK-1683
SN74LS40JDS	BLOCK-1683
SN74LS40JS	BLOCK-1683
SN74LS40N	BLOCK-1683
SN74LS40ND	BLOCK-1683
SN74LS40NDS	BLOCK-1683
SN74LS40NS	BLOCK-1683
SN74LS42	BLOCK-1684
SN74LS42J	BLOCK-1684
SN74LS42JD	BLOCK-1684
SN74LS42JDS	BLOCK-1684
SN74LS42JS	BLOCK-1684
SN74LS42N	BLOCK-1684

DEVICE TYPE	REPL CODE
SN74LS42ND	BLOCK-1684
SN74LS42NDS	BLOCK-1684
SN74LS42NS	BLOCK-1684
SN74LS42W	BLOCK-1684
SN74LS47	BLOCK-1685
SN74LS47J	BLOCK-1685
SN74LS47JD	BLOCK-1685
SN74LS47JDS	BLOCK-1685
SN74LS47JS	BLOCK-1685
SN74LS47N	BLOCK-1685
SN74LS47ND	BLOCK-1685
SN74LS47NDS	BLOCK-1685
SN74LS47NS	BLOCK-1685
SN74LS48J	BLOCK-1686
SN74LS48JD	BLOCK-1686
SN74LS48JDS	BLOCK-1686
SN74LS48JS	BLOCK-1686
SN74LS48N	BLOCK-1686
SN74LS48ND	BLOCK-1686
SN74LS48NDS	BLOCK-1686
SN74LS48NS	BLOCK-1686
SN74LS49	BLOCK-1687
SN74LS490	BLOCK-1688
SN74LS490J	BLOCK-1688
SN74LS490JD	BLOCK-1688
SN74LS490JDS	BLOCK-1688
SN74LS490JS	BLOCK-1688
SN74LS490N	BLOCK-1688
SN74LS490ND	BLOCK-1688
SN74LS490NDS	BLOCK-1688
SN74LS490NS	BLOCK-1688
SN74LS49J	BLOCK-1687
SN74LS49JD	BLOCK-1687
SN74LS49JDS	BLOCK-1687
SN74LS49JS	BLOCK-1687
SN74LS49N	BLOCK-1687
SN74LS49ND	BLOCK-1687
SN74LS49NDS	BLOCK-1687
SN74LS49NS	BLOCK-1687
SN74LS51	BLOCK-1689
SN74LS51J	BLOCK-1689
SN74LS51JD	BLOCK-1689
SN74LS51JDS	BLOCK-1689
SN74LS51JS	BLOCK-1689
SN74LS51N	BLOCK-1689
SN74LS51ND	BLOCK-1689
SN74LS51NDS	BLOCK-1689
SN74LS51NS	BLOCK-1689
SN74LS51W	BLOCK-1689
SN74LS54	BLOCK-1690
SN74LS540	BLOCK-1691
SN74LS540J	BLOCK-1691
SN74LS540JD	BLOCK-1691
SN74LS540JDS	BLOCK-1691
SN74LS540JS	BLOCK-1691
SN74LS540N	BLOCK-1691
SN74LS540ND	BLOCK-1691
SN74LS540NDS	BLOCK-1691
SN74LS540NS	BLOCK-1691
SN74LS541	BLOCK-1692
SN74LS541J	BLOCK-1692
SN74LS541JD	BLOCK-1692
SN74LS541JDS	BLOCK-1692
SN74LS541JS	BLOCK-1692
SN74LS541N	BLOCK-1692
SN74LS541ND	BLOCK-1692
SN74LS541NDS	BLOCK-1692
SN74LS541NS	BLOCK-1692
SN74LS54J	BLOCK-1690
SN74LS54JD	BLOCK-1690
SN74LS54JDS	BLOCK-1690
SN74LS54JS	BLOCK-1690
SN74LS54N	BLOCK-1690
SN74LS54ND	BLOCK-1690
SN74LS54NDS	BLOCK-1690
SN74LS54NS	BLOCK-1690
SN74LS55	BLOCK-1693
SN74LS55J	BLOCK-1693
SN74LS55JD	BLOCK-1693
SN74LS55JDS	BLOCK-1693
SN74LS55JS	BLOCK-1693
SN74LS55N	BLOCK-1693
SN74LS55ND	BLOCK-1693
SN74LS55NDS	BLOCK-1693
SN74LS55NS	BLOCK-1693
SN74LS624	BLOCK-1694
SN74LS624J	BLOCK-1694
SN74LS624N	BLOCK-1694
SN74LS625	BLOCK-1695
SN74LS625J	BLOCK-1695
SN74LS625N	BLOCK-1696
SN74LS626	BLOCK-1696
SN74LS626J	BLOCK-1696
SN74LS626N	BLOCK-1696

DEVICE TYPE	REPL CODE
SN74LS627	BLOCK-1697
SN74LS627J	BLOCK-1697
SN74LS627N	BLOCK-1697
SN74LS629	BLOCK-1698
SN74LS629J	BLOCK-1698
SN74LS629N	BLOCK-1698
SN74LS640	BLOCK-1699
SN74LS640J	BLOCK-1699
SN74LS640JD	BLOCK-1699
SN74LS640JDS	BLOCK-1699
SN74LS640JS	BLOCK-1699
SN74LS640N	BLOCK-1699
SN74LS640ND	BLOCK-1699
SN74LS640NDS	BLOCK-1699
SN74LS640NS	BLOCK-1699
SN74LS641	BLOCK-1700
SN74LS641J	BLOCK-1700
SN74LS641JD	BLOCK-1700
SN74LS641JDS	BLOCK-1700
SN74LS641JS	BLOCK-1700
SN74LS641N	BLOCK-1700
SN74LS641ND	BLOCK-1700
SN74LS641NDS	BLOCK-1700
SN74LS641NS	BLOCK-1700
SN74LS642	BLOCK-1701
SN74LS642J	BLOCK-1701
SN74LS642JD	BLOCK-1701
SN74LS642JDS	BLOCK-1701
SN74LS642JS	BLOCK-1701
SN74LS642N	BLOCK-1701
SN74LS642ND	BLOCK-1701
SN74LS642NDS	BLOCK-1701
SN74LS642NS	BLOCK-1701
SN74LS643	BLOCK-1702
SN74LS643J	BLOCK-1702
SN74LS643JD	BLOCK-1702
SN74LS643JDS	BLOCK-1702
SN74LS643JS	BLOCK-1702
SN74LS643N	BLOCK-1702
SN74LS643ND	BLOCK-1702
SN74LS643NDS	BLOCK-1702
SN74LS643NS	BLOCK-1702
SN74LS645	BLOCK-1703
SN74LS645J	BLOCK-1703
SN74LS645JD	BLOCK-1703
SN74LS645JDS	BLOCK-1703
SN74LS645JS	BLOCK-1703
SN74LS645N	BLOCK-1703
SN74LS645ND	BLOCK-1703
SN74LS645NDS	BLOCK-1703
SN74LS645NS	BLOCK-1703
SN74LS670	BLOCK-1704
SN74LS670J	BLOCK-1704
SN74LS670JD	BLOCK-1704
SN74LS670JDS	BLOCK-1704
SN74LS670JS	BLOCK-1704
SN74LS670N	BLOCK-1704
SN74LS670ND	BLOCK-1704
SN74LS670NDS	BLOCK-1704
SN74LS670NS	BLOCK-1704
SN74LS73A	BLOCK-1705
SN74LS73AJ	BLOCK-1705
SN74LS73AJD	BLOCK-1705
SN74LS73AJDS	BLOCK-1705
SN74LS73AJS	BLOCK-1705
SN74LS73AN	BLOCK-1705
SN74LS73AND	BLOCK-1705
SN74LS73ANDS	BLOCK-1705
SN74LS73ANS	BLOCK-1705
SN74LS73AW	BLOCK-1705
SN74LS73J	BLOCK-1705
SN74LS73N	BLOCK-1705
SN74LS74	BLOCK-1706
SN74LS74A	BLOCK-1706
SN74LS74AJ	BLOCK-1706
SN74LS74AJD	BLOCK-1706
SN74LS74AJDS	BLOCK-1706
SN74LS74AJS	BLOCK-1706
SN74LS74AN	BLOCK-1706
SN74LS74AND	BLOCK-1706
SN74LS74ANDS	BLOCK-1706
SN74LS74ANS	BLOCK-1706
SN74LS74AW	BLOCK-1706
SN74LS74N	BLOCK-1706
SN74LS75	BLOCK-1707
SN74LS75J	BLOCK-1707
SN74LS75JD	BLOCK-1707
SN74LS75JDS	BLOCK-1707
SN74LS75JS	BLOCK-1707
SN74LS75N	BLOCK-1707
SN74LS75ND	BLOCK-1707
SN74LS75NDS	BLOCK-1707
SN74LS75NS	BLOCK-1707
SN74LS75W	BLOCK-1707

DEVICE TYPE	REPL CODE
SN74LS76A	BLOCK-1708
SN74LS76AJ	BLOCK-1708
SN74LS76AJD	BLOCK-1708
SN74LS76AJDS	BLOCK-1708
SN74LS76AJS	BLOCK-1708
SN74LS76AN	BLOCK-1708
SN74LS76AND	BLOCK-1708
SN74LS76ANDS	BLOCK-1708
SN74LS76ANS	BLOCK-1708
SN74LS77J	BLOCK-1709
SN74LS77JD	BLOCK-1709
SN74LS77JDS	BLOCK-1709
SN74LS77JS	BLOCK-1709
SN74LS77N	BLOCK-1709
SN74LS77ND	BLOCK-1709
SN74LS77NDS	BLOCK-1709
SN74LS77NS	BLOCK-1709
SN74LS78A	BLOCK-1710
SN74LS78AJ	BLOCK-1710
SN74LS78AJD	BLOCK-1710
SN74LS78AJDS	BLOCK-1710
SN74LS78AJS	BLOCK-1710
SN74LS78AN	BLOCK-1710
SN74LS78AND	BLOCK-1710
SN74LS78ANDS	BLOCK-1710
SN74LS78ANS	BLOCK-1710
SN74LS83A	BLOCK-1711
SN74LS83AJ	BLOCK-1711
SN74LS83AJD	BLOCK-1711
SN74LS83AJDS	BLOCK-1711
SN74LS83AJS	BLOCK-1711
SN74LS83AN	BLOCK-1711
SN74LS83AND	BLOCK-1711
SN74LS83ANDS	BLOCK-1711
SN74LS83ANS	BLOCK-1711
SN74LS83AW	BLOCK-1711
SN74LS83N	BLOCK-1711
SN74LS85J	BLOCK-1712
SN74LS85JD	BLOCK-1712
SN74LS85JDS	BLOCK-1712
SN74LS85JS	BLOCK-1712
SN74LS85N	BLOCK-1712
SN74LS85ND	BLOCK-1712
SN74LS85NDS	BLOCK-1712
SN74LS85NS	BLOCK-1712
SN74LS85W	BLOCK-1712
SN74LS86	BLOCK-1713
SN74LS86AN	BLOCK-1713
SN74LS86J	BLOCK-1713
SN74LS86JD	BLOCK-1713
SN74LS86JDS	BLOCK-1713
SN74LS86JS	BLOCK-1713
SN74LS86N	BLOCK-1713
SN74LS86ND	BLOCK-1713
SN74LS86NDS	BLOCK-1713
SN74LS86NS	BLOCK-1713
SN74LS86W	BLOCK-1713
SN74LS90	BLOCK-1714
SN74LS90J	BLOCK-1714
SN74LS90JD	BLOCK-1714
SN74LS90JDS	BLOCK-1714
SN74LS90JS	BLOCK-1714
SN74LS90N	BLOCK-1714
SN74LS90ND	BLOCK-1714
SN74LS90NDS	BLOCK-1714
SN74LS90NS	BLOCK-1714
SN74LS91	BLOCK-1715
SN74LS91J	BLOCK-1715
SN74LS91JD	BLOCK-1715
SN74LS91JDS	BLOCK-1715
SN74LS91JS	BLOCK-1715
SN74LS91N	BLOCK-1715
SN74LS91ND	BLOCK-1715
SN74LS91NDS	BLOCK-1715
SN74LS91NS	BLOCK-1715
SN74LS92	BLOCK-1716
SN74LS92J	BLOCK-1716
SN74LS92JD	BLOCK-1716
SN74LS92JDS	BLOCK-1716
SN74LS92JS	BLOCK-1716
SN74LS92N	BLOCK-1716
SN74LS92ND	BLOCK-1716
SN74LS92NDS	BLOCK-1716
SN74LS92NS	BLOCK-1716
SN74LS93	BLOCK-1717
SN74LS93J	BLOCK-1717
SN74LS93JD	BLOCK-1717
SN74LS93JDS	BLOCK-1717
SN74LS93N	BLOCK-1717
SN74LS93ND	BLOCK-1717
SN74LS93NDS	BLOCK-1717
SN74LS93NS	BLOCK-1717
SN74LS93W	BLOCK-1717
SN74S00	BLOCK-1719

DEVICE TYPE	REPL CODE
SN74S00J	BLOCK-1719
SN74S00N	BLOCK-1719
SN74S02	BLOCK-1720
SN74S02J	BLOCK-1720
SN74S02N	BLOCK-1720
SN74S03	BLOCK-1721
SN74S03J	BLOCK-1721
SN74S03N	BLOCK-1721
SN74S04	BLOCK-1722
SN74S04J	BLOCK-1722
SN74S04N	BLOCK-1722
SN74S05	BLOCK-1723
SN74S05J	BLOCK-1723
SN74S05N	BLOCK-1723
SN74S08	BLOCK-1724
SN74S08J	BLOCK-1724
SN74S08N	BLOCK-1724
SN74S09	BLOCK-1725
SN74S09J	BLOCK-1725
SN74S09N	BLOCK-1725
SN74S10	BLOCK-1726
SN74S10J	BLOCK-1726
SN74S10N	BLOCK-1726
SN74S11	BLOCK-1727
SN74S112	BLOCK-1728
SN74S112J	BLOCK-1728
SN74S112N	BLOCK-1728
SN74S113	BLOCK-1729
SN74S113J	BLOCK-1729
SN74S113N	BLOCK-1729
SN74S11J	BLOCK-1727
SN74S11N	BLOCK-1727
SN74S124	BLOCK-1730
SN74S124J	BLOCK-1730
SN74S124N	BLOCK-1730
SN74S133	BLOCK-1731
SN74S133J	BLOCK-1731
SN74S133N	BLOCK-1731
SN74S134	BLOCK-1732
SN74S134J	BLOCK-1732
SN74S134N	BLOCK-1732
SN74S138	BLOCK-1733
SN74S138J	BLOCK-1733
SN74S138N	BLOCK-1733
SN74S138X	BLOCK-1733
SN74S140	BLOCK-1734
SN74S140J	BLOCK-1734
SN74S140N	BLOCK-1734
SN74S15	BLOCK-1735
SN74S151	BLOCK-1736
SN74S151J	BLOCK-1736
SN74S151N	BLOCK-1736
SN74S151X	BLOCK-1736
SN74S153	BLOCK-1737
SN74S153J	BLOCK-1737
SN74S153N	BLOCK-1737
SN74S153X	BLOCK-1737
SN74S157	BLOCK-1738
SN74S157J	BLOCK-1738
SN74S157N	BLOCK-1738
SN74S158	BLOCK-1739
SN74S158J	BLOCK-1739
SN74S158N	BLOCK-1739
SN74S15J	BLOCK-1735
SN74S15N	BLOCK-1735
SN74S163	BLOCK-1607
SN74S163J	BLOCK-1607
SN74S174	BLOCK-1740
SN74S174J	BLOCK-1740
SN74S174N	BLOCK-1740
SN74S175	BLOCK-1741
SN74S175J	BLOCK-1741
SN74S175N	BLOCK-1741
SN74S181	BLOCK-1742
SN74S181J	BLOCK-1742
SN74S181N	BLOCK-1742
SN74S188	BLOCK-1743
SN74S194	BLOCK-1744
SN74S194J	BLOCK-1744
SN74S194N	BLOCK-1744
SN74S20	BLOCK-1745
SN74S20J	BLOCK-1745
SN74S20N	BLOCK-1745
SN74S22	BLOCK-1746
SN74S22J	BLOCK-1746
SN74S22N	BLOCK-1746
SN74S251	BLOCK-1747
SN74S251J	BLOCK-1747
SN74S251N	BLOCK-1747
SN74S258J	BLOCK-1748
SN74S258N	BLOCK-1748
SN74S287	BLOCK-1749
SN74S288	BLOCK-1750
SN74S30	BLOCK-1751

DEVICE TYPE	REPL CODE
SN74S30J	BLOCK-1751
SN74S30N	BLOCK-1751
SN74S387	BLOCK-1752
SN74S40	BLOCK-1753
SN74S40J	BLOCK-1753
SN74S40N	BLOCK-1753
SN74S454	BLOCK-1754
SN74S472	BLOCK-1755
SN74S473	BLOCK-1756
SN74S474	BLOCK-1757
SN74S475	BLOCK-1758
SN74S476	BLOCK-1765
SN74S477	BLOCK-1764
SN74S478	BLOCK-1759
SN74S479	BLOCK-1760
SN74S51	BLOCK-1761
SN74S51J	BLOCK-1761
SN74S51N	BLOCK-1761
SN74S64	BLOCK-1766
SN74S64J	BLOCK-1766
SN74S64N	BLOCK-1766
SN74S65	BLOCK-1767
SN74S65J	BLOCK-1767
SN74S65N	BLOCK-1767
SN74S74	BLOCK-1768
SN74S74J	BLOCK-1768
SN74S74N	BLOCK-1768
SN74S86	BLOCK-1769
SN74S86J	BLOCK-1769
SN74S86N	BLOCK-1769
SN75067NE	BLOCK-995
SN75068NE	BLOCK-996
SN75069NE	BLOCK-996
SN7506NE	BLOCK-994
SN75115J	BLOCK-2092
SN75115N	BLOCK-2092
SN75188	BLOCK-1771
SN75188J	BLOCK-1771
SN75188N	BLOCK-1771
SN75189	BLOCK-1772
SN75189A	BLOCK-1772
SN75189AN	BLOCK-1772
SN75189N	BLOCK-1772
SN75450B	BLOCK-1775
SN75450BN	BLOCK-1775
SN75451B	BLOCK-1776
SN75451BP	BLOCK-1776
SN75452B	BLOCK-1777
SN75452BP	BLOCK-1777
SN75452P	BLOCK-1777
SN75453B	BLOCK-1778
SN75453BP	BLOCK-1778
SN75454BN	BLOCK-1779
SN75480N	BLOCK-951
SN75491AN	BLOCK-1780
SN75491N	BLOCK-1780
SN75492N	BLOCK-1781
SN75493N	BLOCK-1782
SN75494N	BLOCK-1783
SN75497N	BLOCK-1784
SN75498N	BLOCK-1785
SN76001	BLOCK-111
SN76001A	BLOCK-111
SN76001AN	BLOCK-111
SN76001N	BLOCK-111
SN76007	BLOCK-1839
SN76007N	BLOCK-1839
SN76008	BLOCK-1405
SN76104N	BLOCK-1275
SN76105N	BLOCK-1277
SN76111	BLOCK-1276
SN76111N	BLOCK-1276
SN76115	BLOCK-1825
SN76115N	BLOCK-1825
SN76116N	BLOCK-1376
SN76131M	BLOCK-1281
SN76131N	BLOCK-1281
SN76177ND	BLOCK-1828
SN76242N	BLOCK-1271
SN76243AN	BLOCK-1272
SN76243N	BLOCK-1272
SN76246A	BLOCK-1790
SN76246N	BLOCK-1270
SN76298	BLOCK-1291
SN76298N	BLOCK-1291
SN764246N	BLOCK-1815
SN76514	BLOCK-1927
SN76564	BLOCK-1808
SN76565N	BLOCK-1808
SN76591P	BLOCK-1843
SN76594	BLOCK-1876
SN76600	BLOCK-1405
SN766008	BLOCK-1405
SN76600P	BLOCK-1405

DEVICE TYPE	REPL CODE
SN76635N	BLOCK-1389
SN76642	BLOCK-1264
SN76642N	BLOCK-1264
SN76642P	BLOCK-1264
SN76643	BLOCK-1262
SN76643A	BLOCK-1262
SN76643N	BLOCK-1262
SN76650	BLOCK-1422
SN76650N	BLOCK-1422
SN76651N	BLOCK-1421
SN76653N	BLOCK-1262
SN7666	BLOCK-1269
SN76664N	BLOCK-1269
SN76665	BLOCK-1269
SN76665N	BLOCK-1269
SN76666	BLOCK-1269
SN76666N	BLOCK-1269
SN76666N-S	BLOCK-1269
SN76666NS	BLOCK-1269
SN76669N	BLOCK-1290
SN7666N	BLOCK-1269
SN76675	BLOCK-1422
SN76676HC	BLOCK-1806
SN76678	BLOCK-1289
SN76678P	BLOCK-1289
SN76689	BLOCK-1813
SN76689N	BLOCK-1813
SN76873N	BLOCK-542
SN94017P	BLOCK-1289
SN94018N	BLOCK-1279
SN94174	BLOCK-1816
SN962-2M	BLOCK-2217
SNM74S251J	BLOCK-1747
SP180CI	BLOCK-1944
SRM2114C3	BLOCK-1002
SS1001	BLOCK-324
SS4001AE	BLOCK-1029
SS4002AE	BLOCK-1030
SS4011AE	BLOCK-1039
SS4012AE	BLOCK-1040
SS4013AE	BLOCK-1041
SS4023AE	BLOCK-1062
SS4025AE	BLOCK-1064
SS4027AE	BLOCK-1066
SS4028AE	BLOCK-1067
SS4030AE	BLOCK-1069
SSN74LS243	BLOCK-1633
ST984539-006	BLOCK-2209
STK-011	BLOCK-23
STK-015	BLOCK-26
STK-018	BLOCK-135
STK-036	BLOCK-139
STK-075	BLOCK-139
STK-082	BLOCK-314
STK-084	BLOCK-315
STK-154	BLOCK-27
STK-433	BLOCK-206
STK-437	BLOCK-207
STK-563F	BLOCK-713
STK0050	BLOCK-268
STK0080	BLOCK-309
STK011	BLOCK-23
STK013	BLOCK-200
STK015	BLOCK-26
STK016	BLOCK-310
STK020	BLOCK-24
STK020F	BLOCK-24
STK031	BLOCK-319
STK040	BLOCK-311
STK041	BLOCK-312
STK043	BLOCK-312
STK050	BLOCK-323
STK058	BLOCK-327
STK070	BLOCK-323
STK075	BLOCK-328
STK077	BLOCK-313
STK078	BLOCK-313
STK080	BLOCK-314
STK082	BLOCK-314
STK083	BLOCK-315
STK084	BLOCK-315
STK086	BLOCK-324
STK1050	BLOCK-345
STK1050A	BLOCK-345
STK2230	BLOCK-855
STK3042	BLOCK-325
STK3042III	BLOCK-325
STK3082	BLOCK-326
STK3082II	BLOCK-326
STK4021	BLOCK-306
STK4021M	BLOCK-306
STK4024II	BLOCK-796
STK4044II	BLOCK-862
STK4044V	BLOCK-862
STK4121II	BLOCK-854
STK413	BLOCK-206
STK415	BLOCK-206
STK4152II	BLOCK-800
STK4171II	BLOCK-303
STK4273	BLOCK-859
STK430	BLOCK-330
STK4311	BLOCK-857
STK433	BLOCK-206
STK435	BLOCK-206
STK4362	BLOCK-206
STK437	BLOCK-207
STK439	BLOCK-207
STK4392	BLOCK-207
STK441	BLOCK-316
STK443	BLOCK-326
STK457	BLOCK-317
STK459	BLOCK-317
STK460	BLOCK-334
STK461	BLOCK-318
STK463	BLOCK-318
STK465	BLOCK-332
STK465A	BLOCK-332
STK5322	BLOCK-305
STK5332	BLOCK-860
STK5372	BLOCK-863
STK5431	BLOCK-714
STK5431SL	BLOCK-714
STK5431ST	BLOCK-714
STK5451	BLOCK-716
STK5466ST	BLOCK-802
STK5471	BLOCK-803
STK5476	BLOCK-861
STK5479	BLOCK-864
STK5481	BLOCK-804
STK5482	BLOCK-852
STK5490	BLOCK-804
STK561F	BLOCK-713
STK563A	BLOCK-713
STK563F	BLOCK-713
STK583F	BLOCK-713
STK583FST	BLOCK-713
STK6722H	BLOCK-865
STK6940	BLOCK-805
STK6941	BLOCK-805
STK6942	BLOCK-805
STK6960	BLOCK-729
STK6961	BLOCK-729
STK6962	BLOCK-729
STK6962H	BLOCK-729
STK6970	BLOCK-806
STK6971	BLOCK-806
STK6972	BLOCK-806
STK6981B	BLOCK-717
STK6982	BLOCK-717
STK6982B	BLOCK-717
STK7308	BLOCK-851
STK7309	BLOCK-851
STK7563F	BLOCK-718
STK7563FE	BLOCK-718
STL6961	BLOCK-729
STR-30120	BLOCK-700
STR-30123	BLOCK-757
STR-3125	BLOCK-722
STR-380	BLOCK-535
STR-381	BLOCK-533
STR-53041	BLOCK-819
STR-S6301	BLOCK-1260
STR2012	BLOCK-847
STR2012A	BLOCK-847
STR2013	BLOCK-848
STR30115	BLOCK-875
STR30120	BLOCK-818
STR30123	BLOCK-757
STR30125	BLOCK-876
STR30130	BLOCK-758
STR30130-A	BLOCK-758
STR30130A	BLOCK-758
STR30134	BLOCK-759
STR30135	BLOCK-759
STR30135F	BLOCK-759
STR3035	BLOCK-724
STR3115	BLOCK-721
STR3123	BLOCK-722
STR3125	BLOCK-722
STR3130	BLOCK-723
STR3135	BLOCK-724
STR3225	BLOCK-722
STR3230	BLOCK-723
STR380	BLOCK-535
STR381	BLOCK-533
STR381A	BLOCK-533
STR382	BLOCK-540
STR383	BLOCK-700
STR385	BLOCK-540
STR50041	BLOCK-873
STR50041A	BLOCK-873
STR5100	BLOCK-866
STR51041	BLOCK-874
STR53041	BLOCK-819
STR53043	BLOCK-820
STR53043A	BLOCK-820
STR54041	BLOCK-777
STR729	BLOCK-632
STRD3030	BLOCK-758
STRS6301	BLOCK-1260
STRS6301-LF953	BLOCK-1260
SVI-LD3120	BLOCK-19
SVI-STK018	BLOCK-135
SVISTK020F	BLOCK-24
SW-10548	BLOCK-2203
SW-10549	BLOCK-1969
SW4001	BLOCK-1029
SW4002	BLOCK-1030
SW4011	BLOCK-1039
SW4012	BLOCK-1040
SW4013	BLOCK-1041
SW4015	BLOCK-1043
SW4016	BLOCK-1048
SW4017	BLOCK-1051
SW4019	BLOCK-1058
SW4020	BLOCK-1059
SW4021	BLOCK-1060
SW4023	BLOCK-1062
SW4024	BLOCK-1063
SW4025	BLOCK-1064
SW4027	BLOCK-1066
SW4030	BLOCK-1069
SW4049	BLOCK-1084
SW4050	BLOCK-1085
SW705-2M	BLOCK-1950
SW705-2P	BLOCK-1950
SW706-2M	BLOCK-1953
SW706-2P	BLOCK-1953
SW708-2M	BLOCK-1951
SW708-2P	BLOCK-1951
SW709-2M	BLOCK-1952
SW709-2P	BLOCK-1952
SW74LS247	BLOCK-1636
SW930-2M	BLOCK-2197
SW930-2P	BLOCK-2197
SW932-2M	BLOCK-2199
SW932-2P	BLOCK-2199
SW933-2M	BLOCK-2200
SW933-2P	BLOCK-2200
SW935-2M	BLOCK-2201
SW935-2P	BLOCK-2201
SW936-2M	BLOCK-2202
SW936-2P	BLOCK-2202
SW937-2M	BLOCK-2203
SW937-2P	BLOCK-2203
SW941-2M	BLOCK-2205
SW941-2P	BLOCK-2205
SW944-2M	BLOCK-2206
SW944-2P	BLOCK-2206
SW945-2M	BLOCK-2207
SW945-2P	BLOCK-2207
SW946-2M	BLOCK-2208
SW946-2P	BLOCK-2208
SW948-2M	BLOCK-2209
SW948-2P	BLOCK-2209
SW949-2M	BLOCK-2210
SW949-2P	BLOCK-2210
SW950-2M	BLOCK-2212
SW950-2P	BLOCK-2212
SW951-2M	BLOCK-2213
SW951-2P	BLOCK-2213
SW957-2M	BLOCK-1968
SW957-2P	BLOCK-1968
SW958-2M	BLOCK-1969
SW958-2P	BLOCK-1969
SW961-2M	BLOCK-2216
SW961-2P	BLOCK-2216
SW962-2P	BLOCK-2217
SW963-2M	BLOCK-2218
SW963-2P	BLOCK-2218
SY2128	BLOCK-1004
SY6502	BLOCK-1158
SYC2114	BLOCK-1002
SYC2114-3	BLOCK-1002
SYC2147	BLOCK-1005
SYC2147-3	BLOCK-1005
SYC2147-6	BLOCK-1005
SYC2147L	BLOCK-1005
SYC6502	BLOCK-1158
SYC6507	BLOCK-1159
SYC6532	BLOCK-1162
SYD2114	BLOCK-1002
SYD2114-2	BLOCK-1002
SYD2114-3	BLOCK-1002
SYD2114L	BLOCK-1002
SYD2114L-2	BLOCK-1002
SYD2114L-3	BLOCK-1002
SYD2147	BLOCK-1005
SYD2147-3	BLOCK-1005
SYD2147-6	BLOCK-1005
SYD2147L	BLOCK-1005
SYD6502	BLOCK-1158
SYD6507	BLOCK-1159
SYD6532	BLOCK-1162
SYK4332	BLOCK-206
SYP2114	BLOCK-1002
SYP2114-2	BLOCK-1002
SYP2114-3	BLOCK-1002
SYP2114L	BLOCK-1002
SYP2114L-2	BLOCK-1002
SYP2114L-3	BLOCK-1002
SYP6502	BLOCK-1158
SYP6507	BLOCK-1159
SYP6532	BLOCK-1162
SZ-LA4102	BLOCK-216
T-2508	BLOCK-732
T-Q7037	BLOCK-5
T-Q7038	BLOCK-5
T0B	BLOCK-1212
T1A	BLOCK-1223
T1B	BLOCK-1212
T1F	BLOCK-1421
T1H	BLOCK-1265
T1J	BLOCK-1275
T1M	BLOCK-1274
T1N	BLOCK-1275
T1T	BLOCK-1786
T2062	BLOCK-1808
T2370	BLOCK-1846
T2508	BLOCK-732
T2D	BLOCK-1279
T2E	BLOCK-1774
T2G	BLOCK-1279
T2J	BLOCK-1262
T2L	BLOCK-1797
T2M	BLOCK-1806
T2N	BLOCK-1814
T2N-1	BLOCK-1814
T2N1	BLOCK-1814
T2R	BLOCK-1801
T2T	BLOCK-1813
T2T-1	BLOCK-1813
T2T-2	BLOCK-1813
T2V	BLOCK-1289
T3A	BLOCK-1376
T3A-1	BLOCK-1376
T3A-2	BLOCK-1376
T3B	BLOCK-2058
T7400B1	BLOCK-1293
T7400D1	BLOCK-1293
T7400D2	BLOCK-1293
T7401B1	BLOCK-1294
T7401D1	BLOCK-1294
T7401D2	BLOCK-1294
T7402B1	BLOCK-1295
T7402D1	BLOCK-1295
T7402D2	BLOCK-1295
T7403B1	BLOCK-1296
T7403D1	BLOCK-1296
T7403D2	BLOCK-1296
T7404B1	BLOCK-1297
T7405B1	BLOCK-1298
T7406B1	BLOCK-1299
T7407B1	BLOCK-1300
T7408B1	BLOCK-1301
T7408D1	BLOCK-1301
T7408D2	BLOCK-1301
T7409B1	BLOCK-1302
T7409D1	BLOCK-1302
T7409D2	BLOCK-1302
T74107B1	BLOCK-1305
T7410B1	BLOCK-1304
T7410D1	BLOCK-1304
T7410D2	BLOCK-1304
T7416B1	BLOCK-1336
T74180B1	BLOCK-1352
T74193B1	BLOCK-1358
T7420B1	BLOCK-1365
T7420D1	BLOCK-1365
T7420D2	BLOCK-1365
T7428B1	BLOCK-1373
T7428D1	BLOCK-1373
T7428D2	BLOCK-1373
T7430B1	BLOCK-1377
T7430D1	BLOCK-1377
T7430D2	BLOCK-1377
T7433B1	BLOCK-1379
T7433D1	BLOCK-1379
T7433D2	BLOCK-1379
T7440B1	BLOCK-1390
T7440D1	BLOCK-1390
T7440D2	BLOCK-1390
T7441AB1	BLOCK-1391
T7441AD1	BLOCK-1391
T7441AD2	BLOCK-1391
T7442B1	BLOCK-1392
T7442D1	BLOCK-1392
T7442D2	BLOCK-1392
T7443B1	BLOCK-1393
T7443D1	BLOCK-1393
T7443D2	BLOCK-1393
T7444B1	BLOCK-1394
T7444D1	BLOCK-1394
T7444D2	BLOCK-1394
T7450B1	BLOCK-1401
T7450D1	BLOCK-1401
T7450D2	BLOCK-1401
T7451B1	BLOCK-1402
T7451D1	BLOCK-1402
T7451D2	BLOCK-1402
T7453B1	BLOCK-1403
T7453D1	BLOCK-1403
T7453D2	BLOCK-1403
T7454B1	BLOCK-1404
T7454D1	BLOCK-1404
T7454D2	BLOCK-1404
T7460B1	BLOCK-1406
T7460D1	BLOCK-1406
T7460D2	BLOCK-1406
T7472B1	BLOCK-1409
T7472D1	BLOCK-1409
T7472D2	BLOCK-1409
T7473B1	BLOCK-1410
T7473D1	BLOCK-1410
T7473D2	BLOCK-1410
T7474B1	BLOCK-1411
T7474D1	BLOCK-1411
T7474D2	BLOCK-1411
T7475B1	BLOCK-1412
T7476B1	BLOCK-1413
T7476D1	BLOCK-1413
T7476D2	BLOCK-1413
T7481B1	BLOCK-1415
T7483B1	BLOCK-1417
T7486B1	BLOCK-1419
T7486D1	BLOCK-1419
T7486D2	BLOCK-1419
T7490B1	BLOCK-1423
T7490D1	BLOCK-1423
T7490D2	BLOCK-1423
T7492B1	BLOCK-1425
T7492D1	BLOCK-1425
T7492D2	BLOCK-1425
T7493B1	BLOCK-1426
T7493D1	BLOCK-1426
T7493D2	BLOCK-1426
T74H00B1	BLOCK-1471
T74H00D1	BLOCK-1471
T74H00D2	BLOCK-1471
T74H10B1	BLOCK-1476
T74H10D1	BLOCK-1476
T74H10D2	BLOCK-1476
T74H11B1	BLOCK-1482
T74H11D1	BLOCK-1482
T74H11D2	BLOCK-1482
T74H20B1	BLOCK-1484
T74H20D1	BLOCK-1484
T74H20D2	BLOCK-1484
T74H21B1	BLOCK-1485
T74H21D1	BLOCK-1485
T74H21D2	BLOCK-1485
T74H40B1	BLOCK-1488
T74H40D1	BLOCK-1488
T74H40D2	BLOCK-1488
T74H50B1	BLOCK-1489
T74H50D1	BLOCK-1489
T74H50D2	BLOCK-1489
T74H51B1	BLOCK-1490
T74H51D1	BLOCK-1490
T74H51D2	BLOCK-1490
T74H52B1	BLOCK-1491
T74H52D1	BLOCK-1491
T74H52D2	BLOCK-1491
T74H53B1	BLOCK-1492
T74H53D1	BLOCK-1492
T74H53D2	BLOCK-1492
T74H54B1	BLOCK-1493
T74H54D1	BLOCK-1493
T74H54D2	BLOCK-1493
T74H61B1	BLOCK-1496
T74H61D1	BLOCK-1496
T74H61D2	BLOCK-1496
T74H62B1	BLOCK-1497
T74H62D1	BLOCK-1497
T74H62D2	BLOCK-1497
T74H71B1	BLOCK-1498
T74H71D2	BLOCK-1498
T74H72B1	BLOCK-1499
T74H72D1	BLOCK-1499
T74H72D2	BLOCK-1499
T74LS00B1	BLOCK-1568
T74LS00D1	BLOCK-1568
T74LS01B1N	BLOCK-1569
T74LS02B1	BLOCK-1570
T74LS02D1	BLOCK-1570
T74LS03B1	BLOCK-1571
T74LS03D1	BLOCK-1571
T74LS04B1	BLOCK-1572
T74LS08B1	BLOCK-1574
T74LS08D1	BLOCK-1574
T74LS09B1	BLOCK-1575
T74LS09D1	BLOCK-1575
T74LS109B1	BLOCK-1578
T74LS109D1	BLOCK-1578
T74LS10B1	BLOCK-1576
T74LS10D1	BLOCK-1576
T74LS112B1	BLOCK-1580
T74LS112D1	BLOCK-1580
T74LS114B1	BLOCK-1581
T74LS114D1	BLOCK-1581
T74LS11B1	BLOCK-1579
T74LS11D1	BLOCK-1579
T74LS136B1	BLOCK-1590
T74LS136D1	BLOCK-1590
T74LS138B1	BLOCK-1591
T74LS13B1	BLOCK-1587
T74LS14B1	BLOCK-1593
T74LS14BI	BLOCK-1593
T74LS155B1	BLOCK-1600
T74LS157B1	BLOCK-1602
T74LS15B1	BLOCK-1597
T74LS15D1	BLOCK-1597
T74LS161B1	BLOCK-1605
T74LS174B1	BLOCK-1615
T74LS175B1	BLOCK-1616
T74LS20B1	BLOCK-1626
T74LS20D1	BLOCK-1626
T74LS21B1	BLOCK-1627
T74LS21D1	BLOCK-1627
T74LS22B1	BLOCK-1628
T74LS22D1	BLOCK-1628
T74LS244B1	BLOCK-1634
T74LS257B1	BLOCK-1641
T74LS26B1	BLOCK-1644
T74LS26D1	BLOCK-1644
T74LS279B1	BLOCK-1649
T74LS27B1	BLOCK-1647
T74LS27BI	BLOCK-1647
T74LS27D1	BLOCK-1647
T74LS28B1	BLOCK-1650
T74LS30B1	BLOCK-1658
T74LS30D1	BLOCK-1658
T74LS32B1	BLOCK-1659
T74LS32BI	BLOCK-1659
T74LS32D1	BLOCK-1659
T74LS352B1	BLOCK-1662
T74LS353B1	BLOCK-1663
T74LS373B1	BLOCK-1671
T74LS374B1	BLOCK-1672
T74LS393B1	BLOCK-1679
T74LS398B1	BLOCK-1681
T74LS399B1	BLOCK-1682
T74LS40B1	BLOCK-1683
T74LS40D1	BLOCK-1683
T74LS51B1	BLOCK-1689
T74LS51D1	BLOCK-1689
T74LS54B1	BLOCK-1690
T74LS54D1	BLOCK-1690
T74LS55B1	BLOCK-1693
T74LS55B1N	BLOCK-1693
T74LS55D1	BLOCK-1693
T74LS670B1	BLOCK-1704
T74LS74B1	BLOCK-1706
T74LS74D1	BLOCK-1706
T74LS83B1	BLOCK-1711
T74LS83D1	BLOCK-1711
T74LS86B1	BLOCK-1713
T74LS86D1	BLOCK-1713
T78010AP	BLOCK-904
T900B1	BLOCK-611
T911B1-T	BLOCK-611
T911BIT	BLOCK-611
TA-5702B	BLOCK-1272
TA-7027M	BLOCK-1282

DEVICE TYPE	REPL CODE
TA-7045	BLOCK-1280
TA-7045M	BLOCK-1280
TA-7046	BLOCK-106
TA-7050M	BLOCK-1267
TA-7053M	BLOCK-1256
TA-7055P	BLOCK-193
TA-7055P-E	BLOCK-193
TA-705M	BLOCK-1267
TA-7060	BLOCK-102
TA-7060P	BLOCK-102
TA-7061AP	BLOCK-98
TA-7061B	BLOCK-98
TA-7061P	BLOCK-98
TA-7063	BLOCK-101
TA-7069P	BLOCK-99
TA-7076P	BLOCK-107
TA-7092-P	BLOCK-105
TA-7102P	BLOCK-103
TA-7120P	BLOCK-85
TA-7205P	BLOCK-145
TA-7358P	BLOCK-822
TA-7640AP	BLOCK-821
TA2124P	BLOCK-126
TA310P	BLOCK-180
TA5628	BLOCK-1813
TA5649	BLOCK-1271
TA5649A	BLOCK-1271
TA5702	BLOCK-1272
TA5702B	BLOCK-1272
TA5814	BLOCK-1269
TA5912	BLOCK-1270
TA6472	BLOCK-2168
TA6480	BLOCK-164
TA6699B	BLOCK-1886
TA6896	BLOCK-1846
TA7027M	BLOCK-1282
TA7028M	BLOCK-177
TA7038M#	BLOCK-1223
TA7045	BLOCK-1280
TA7045M	BLOCK-1280
TA7046P	BLOCK-106
TA7047	BLOCK-1944
TA7047M	BLOCK-1944
TA7050M	BLOCK-1267
TA7051P	BLOCK-1265
TA7053M	BLOCK-1256
TA7054	BLOCK-104
TA7054P	BLOCK-104
TA7055P	BLOCK-193
TA7055P-D	BLOCK-193
TA7055P-E	BLOCK-193
TA7055P-F	BLOCK-193
TA7057M	BLOCK-1945
TA7060	BLOCK-102
TA7060P	BLOCK-102
TA7061AP	BLOCK-98
TA7062P	BLOCK-100
TA7063	BLOCK-101
TA70639	BLOCK-50
TA7063P	BLOCK-101
TA7063P-C	BLOCK-101
TA7064P	BLOCK-224
TA7064P-JA	BLOCK-224
TA7066P	BLOCK-456
TA7070A	BLOCK-5
TA7070B	BLOCK-5
TA7070FA-1	BLOCK-5
TA7070P	BLOCK-5
TA7070PFA-1	BLOCK-5
TA7070PFA1	BLOCK-5
TA7070PGL	BLOCK-5
TA7071GL	BLOCK-1269
TA7071P	BLOCK-1269
TA7072P	BLOCK-1421
TA7073AP	BLOCK-1421
TA7073P	BLOCK-1421
TA7074GL	BLOCK-1422
TA7074P	BLOCK-1422
TA7074PGL	BLOCK-1422
TA7075B	BLOCK-78
TA7075P	BLOCK-78
TA7075PL	BLOCK-78
TA7076	BLOCK-107
TA7076AP	BLOCK-107
TA7076P	BLOCK-107
TA7092AP-B	BLOCK-22
TA7092C	BLOCK-105
TA7092P	BLOCK-105
TA7092P-C	BLOCK-105
TA70L9&	BLOCK-99
TA7102P	BLOCK-103
TA7103	BLOCK-167
TA7103P	BLOCK-60
TA7108P	BLOCK-444
TA7115P	BLOCK-233
TA7117P	BLOCK-21
TA7120P	BLOCK-85
TA7120P-B	BLOCK-85
TA7120P-C	BLOCK-85
TA7120P-D	BLOCK-85
TA7120P-E	BLOCK-85
TA7120PB	BLOCK-85
TA7120PC	BLOCK-85
TA7122	BLOCK-83
TA7122AP	BLOCK-83
TA7122AR	BLOCK-83
TA7124P	BLOCK-126
TA7124PFA-3	BLOCK-126
TA7124PFA-4	BLOCK-126
TA7129P	BLOCK-450
TA7130P	BLOCK-223
TA7130P-B	BLOCK-223
TA7130P-C	BLOCK-223
TA7130PB	BLOCK-223
TA7130PC	BLOCK-223
TA7137(IC)	BLOCK-198
TA7137P	BLOCK-454
TA7137P-ST	BLOCK-454
TA7140P	BLOCK-456
TA7140PC	BLOCK-456
TA7145P	BLOCK-131
TA7146P	BLOCK-130
TA7148P	BLOCK-128
TA7149P	BLOCK-129
TA7150P	BLOCK-127
TA7155P	BLOCK-205
TA7157P	BLOCK-194
TA7159P	BLOCK-232
TA7167	BLOCK-188
TA7167P	BLOCK-188
TA7176AP	BLOCK-1269
TA7176AP(FA-1)	BLOCK-225
TA7176APFA	BLOCK-1269
TA7176APFA-1	BLOCK-225
TA7176P	BLOCK-225
TA7176PFA-1	BLOCK-1269
TA7192P	BLOCK-184
TA7200P	BLOCK-143
TA7203P	BLOCK-144
TA7204P	BLOCK-143
TA7205	BLOCK-145
TA7205A	BLOCK-145
TA7205AP	BLOCK-145
TA7205AT	BLOCK-145
TA7205P	BLOCK-145
TA7207P	BLOCK-464
TA7209P	BLOCK-210
TA7215P	BLOCK-366
TA7220P	BLOCK-85
TA7222	BLOCK-265
TA7222AP	BLOCK-265
TA7222P	BLOCK-265
TA7230P	BLOCK-374
TA7270P	BLOCK-810
TA7274P	BLOCK-812
TA7280P	BLOCK-877
TA7281P	BLOCK-878
TA7302P	BLOCK-462
TA7303P	BLOCK-512
TA7310	BLOCK-180
TA7310P	BLOCK-180
TA7310P-U	BLOCK-180
TA7310P-Y	BLOCK-180
TA7313(IC)	BLOCK-453
TA7313AP	BLOCK-453
TA7313F	BLOCK-453
TA7313P	BLOCK-453
TA7313P(IC)	BLOCK-453
TA7314P	BLOCK-509
TA7318P	BLOCK-486
TA7319P	BLOCK-453
TA7325P	BLOCK-711
TA7343AP	BLOCK-643
TA7343P	BLOCK-643
TA7343PS	BLOCK-643
TA7347	BLOCK-853
TA7347P	BLOCK-853
TA7358AP	BLOCK-822
TA7358P	BLOCK-822
TA7361	BLOCK-768
TA7361AP	BLOCK-768
TA7361P	BLOCK-768
TA7504	BLOCK-2058
TA7504M	BLOCK-2056
TA7504P	BLOCK-2058
TA7504S	BLOCK-2058
TA75339	BLOCK-1873
TA75339P	BLOCK-1873
TA75358P	BLOCK-1993
TA75393	BLOCK-2063
TA75393P	BLOCK-2063
TA75458P	BLOCK-1801
TA75557P	BLOCK-1801
TA75557S	BLOCK-516
TA75558P	BLOCK-1801
TA75558S	BLOCK-1993
TA75558S-1	BLOCK-1993
TA75559P	BLOCK-1801
TA75559P(FA-1)	BLOCK-1801
TA75559PFA-1	BLOCK-1801
TA75902P	BLOCK-1265
TA7607	BLOCK-532
TA7607A	BLOCK-532
TA7607AP	BLOCK-532
TA7607BP	BLOCK-532
TA7607P	BLOCK-532
TA7608AP	BLOCK-519
TA7608BP	BLOCK-519
TA7608BR	BLOCK-519
TA7608CP	BLOCK-519
TA7608CP(FA-5)	BLOCK-519
TA7608CP-GS-4	BLOCK-519
TA7608CP-GS-5	BLOCK-519
TA7608CPFA-5	BLOCK-519
TA7608CPFA-6	BLOCK-519
TA7608P	BLOCK-519
TA7608PFA-5	BLOCK-519
TA7608PFA-6	BLOCK-519
TA7608PGS-4	BLOCK-519
TA7608PGS-5	BLOCK-519
TA7609	BLOCK-484
TA7609AP	BLOCK-484
TA7609GS-2	BLOCK-484
TA7609P	BLOCK-484
TA7609P(FA-2)	BLOCK-484
TA7609P-GS-2	BLOCK-484
TA7609PFA-2	BLOCK-484
TA7609PFA2	BLOCK-484
TA7609PGS-2	BLOCK-484
TA7611AP	BLOCK-401
TA7611AP-S	BLOCK-401
TA7614AP	BLOCK-640
TA7614AP-Y	BLOCK-640
TA7614P	BLOCK-640
TA7630	BLOCK-563
TA7630P	BLOCK-563
TA7640AP	BLOCK-821
TA7640P	BLOCK-821
TA7643P	BLOCK-604
TA7644AP	BLOCK-534
TA7644BP	BLOCK-534
TA7644BPFA-1	BLOCK-534
TA7658P	BLOCK-558
TA7660	BLOCK-756
TA7660P	BLOCK-756
TA7660P(FA-1)	BLOCK-756
TA7660PFA	BLOCK-756
TA7660PFA-1	BLOCK-756
TA7662P	BLOCK-184
TA7664P	BLOCK-560
TA7664P(FA-1)	BLOCK-560
TA7664PFA-1	BLOCK-560
TA7664PFA1	BLOCK-560
TA7668AP	BLOCK-637
TA766P	BLOCK-756
TA7670P	BLOCK-657
TA7671P	BLOCK-225
TA7671P(FA-1)	BLOCK-225
TA7671PFA-1	BLOCK-225
TA7673	BLOCK-598
TA7673P	BLOCK-598
TA7680AP	BLOCK-559
TA7680P	BLOCK-559
TA7681AP	BLOCK-557
TA7681P	BLOCK-557
TA7750P	BLOCK-1193
TA7774P	BLOCK-1185
TA7777N	BLOCK-771
TA7777P	BLOCK-771
TA7777P(FA-1)	BLOCK-771
TA7777P(FA-3)	BLOCK-771
TA7777PFA-1	BLOCK-771
TA7777PFA-3	BLOCK-771
TA78005AP	BLOCK-2087
TA78005P	BLOCK-2087
TA78010AP	BLOCK-904
TA78012A	BLOCK-2097
TA78012AP	BLOCK-2097
TA78012M	BLOCK-890
TA7805S	BLOCK-912
TA7809S	BLOCK-918
TA7812S	BLOCK-921
TA78L	BLOCK-2144
TA78L005	BLOCK-2144
TA78L005AP	BLOCK-2144
TA78L005AP-Y	BLOCK-2144
TA78L005P	BLOCK-2144
TA78L008P	BLOCK-2160
TA78L009AP	BLOCK-881
TA78L009P	BLOCK-881
TA78L012AP	BLOCK-2074
TA78L018AP	BLOCK-884
TA78L024P	BLOCK-886
TA78L05S	BLOCK-2144
TA78L09S	BLOCK-881
TA8200AH	BLOCK-1254
TA8677N	BLOCK-771
TA8680AN	BLOCK-1191
TA8680BN	BLOCK-1191
TA8680BNFA-1	BLOCK-1191
TA8680N	BLOCK-1191
TA8703S	BLOCK-1241
TAA521	BLOCK-1949
TAA521A	BLOCK-1954
TAA522	BLOCK-1949
TAA550-B	BLOCK-1157
TAA611A12	BLOCK-111
TAA611B	BLOCK-111
TAA611B12	BLOCK-111
TAA611C11	BLOCK-116
TAA621	BLOCK-109
TAA621A11	BLOCK-109
TAA621A12	BLOCK-110
TASCA3137	BLOCK-165
TBA-800	BLOCK-114
TBA-810S	BLOCK-112
TBA-820	BLOCK-115
TBA10S-H	BLOCK-112
TBA120AS	BLOCK-279
TBA120S	BLOCK-279
TBA120U	BLOCK-567
TBA221	BLOCK-2056
TBA2214	BLOCK-2057
TBA222	BLOCK-2056
TBA281	BLOCK-1983
TBA800	BLOCK-114
TBA810	BLOCK-112
TBA810AS	BLOCK-112
TBA810DS	BLOCK-112
TBA810P	BLOCK-113
TBA810S	BLOCK-112
TBA810S-H	BLOCK-112
TBA820	BLOCK-115
TBA820L	BLOCK-115
TBA820M	BLOCK-281
TBA820MT	BLOCK-398
TBB0747	BLOCK-2069
TBB0747A	BLOCK-2070
TBB0748	BLOCK-160
TBB0748B	BLOCK-2141
TBB1458B	BLOCK-1801
TBP14S10	BLOCK-1749
TBP14S10J	BLOCK-1749
TBP14S10N	BLOCK-1749
TBP14SA10	BLOCK-1752
TBP14SA10J	BLOCK-1752
TBP14SA10N	BLOCK-1752
TBP18S030	BLOCK-1750
TBP18S030J	BLOCK-1750
TBP18S030N	BLOCK-1750
TBP18S42	BLOCK-1755
TBP18S42J	BLOCK-1755
TBP18S42N	BLOCK-1755
TBP18S46	BLOCK-1757
TBP18S46J	BLOCK-1757
TBP18S46N	BLOCK-1757
TBP18SA030	BLOCK-1743
TBP18SA030J	BLOCK-1743
TBP18SA030N	BLOCK-1743
TBP18SA42	BLOCK-1756
TBP18SA42J	BLOCK-1756
TBP18SA42N	BLOCK-1756
TBP18SA46	BLOCK-1758
TBP18SA46J	BLOCK-1758
TBP18SA46N	BLOCK-1758
TBP24S10N	BLOCK-1749
TBP24S41	BLOCK-1765
TBP24S41J	BLOCK-1765
TBP24S41N	BLOCK-1765
TBP24S81	BLOCK-1754
TBP24S81J	BLOCK-1754
TBP24S81N	BLOCK-1754
TBP24SA41	BLOCK-1764
TBP24SA41J	BLOCK-1764
TBP24SA41N	BLOCK-1764
TBP28S86J	BLOCK-1759
TBP28S86N	BLOCK-1759
TBP28SA86J	BLOCK-1760
TBP28SA86N	BLOCK-1760
TC3001	BLOCK-180
TC4000BP	BLOCK-1028
TC4001BP	BLOCK-1029
TC4001P	BLOCK-1029
TC4001UBP	BLOCK-1029
TC4002BP	BLOCK-1030
TC4002P	BLOCK-1030
TC4007BP	BLOCK-1275
TC4008BP	BLOCK-1034
TC4009UBP	BLOCK-1084
TC4010BP	BLOCK-1085
TC4011	BLOCK-1039
TC4011B	BLOCK-1039
TC4011BP	BLOCK-1039
TC4011P	BLOCK-1039
TC4011UBP	BLOCK-1039
TC4012BP	BLOCK-1040
TC4013B	BLOCK-1041
TC4013BCP	BLOCK-1041
TC4013BP	BLOCK-1041
TC4013P	BLOCK-1041
TC4015BP	BLOCK-1043
TC4015P	BLOCK-1043
TC40160BP	BLOCK-1044
TC40161BP	BLOCK-1045
TC40162BP	BLOCK-1046
TC40163BP	BLOCK-1047
TC4016BB	BLOCK-1048
TC4016BP	BLOCK-1093
TC40174BP	BLOCK-1049
TC40175BP	BLOCK-1050
TC4017BP	BLOCK-1051
TC4017P	BLOCK-1051
TC4018BP	BLOCK-1053
TC40193BP	BLOCK-1055
TC4019BP	BLOCK-1058
TC4019P	BLOCK-1058
TC4020BP	BLOCK-1059
TC4021P	BLOCK-1060
TC4022BP	BLOCK-1061
TC4023BP	BLOCK-1062
TC4023P	BLOCK-1062
TC4024BP	BLOCK-1063
TC4025BP	BLOCK-1064
TC4025P	BLOCK-1064
TC4027	BLOCK-1066
TC4027B	BLOCK-1066
TC4027BP	BLOCK-1066
TC4027P	BLOCK-1066
TC4028B	BLOCK-1067
TC4028BP	BLOCK-1067
TC4028P	BLOCK-1067
TC4029BP	BLOCK-1068
TC4030BP	BLOCK-1069
TC4030P	BLOCK-1069
TC4032BP	BLOCK-1071
TC4038BP	BLOCK-1075
TC4040B	BLOCK-1076
TC4040BP	BLOCK-1076
TC4040P	BLOCK-1076
TC4042BP	BLOCK-1077
TC4042P	BLOCK-1077
TC4043BP	BLOCK-1078
TC4044BP	BLOCK-1079
TC4047BP	BLOCK-1082
TC4049	BLOCK-1084
TC4049BP	BLOCK-1084
TC4049P	BLOCK-1084
TC4050BP	BLOCK-1085
TC4050CP	BLOCK-1085
TC4050P	BLOCK-1085
TC4051BP	BLOCK-1086
TC4052	BLOCK-1087
TC4052BP	BLOCK-1087
TC4052BP-1	BLOCK-1087
TC4053BP	BLOCK-1088
TC4055BP	BLOCK-1089
TC4056BP	BLOCK-1090
TC4063BP	BLOCK-1092
TC4066	BLOCK-1093
TC4066B	BLOCK-1093
TC4066BC	BLOCK-1093
TC4066BP	BLOCK-1093
TC4066P	BLOCK-1093
TC4068BP	BLOCK-1095
TC4069BP	BLOCK-1096
TC4069P	BLOCK-1096
TC4069UBP	BLOCK-1096
TC4069UPB	BLOCK-1096
TC4069UPN	BLOCK-1096
TC4071BP	BLOCK-1098
TC4071P.JA	BLOCK-1098
TC4072BP	BLOCK-1099
TC4073BP	BLOCK-1100
TC4075BP	BLOCK-1101
TC4078BP	BLOCK-1104
TC4081	BLOCK-1105
TC4081BP	BLOCK-525
TC4081P	BLOCK-1105
TC4082BP	BLOCK-1106
TC4082P	BLOCK-1106
TC4085BP	BLOCK-1107
TC4086BP	BLOCK-1108
TC4093BP	BLOCK-1110
TC4099BP	BLOCK-1115
TC4508BP	BLOCK-1120
TC4510BP	BLOCK-1121
TC4511BP	BLOCK-1122
TC4512BP	BLOCK-1123
TC4512P	BLOCK-1123
TC4514BP	BLOCK-1125
TC4514P	BLOCK-1125
TC4516BP	BLOCK-1127
TC4518BP	BLOCK-1129
TC4520BP	BLOCK-1130
TC4522BP	BLOCK-1132
TC4527BP	BLOCK-1134
TC4528BP	BLOCK-1135
TC4531BP	BLOCK-1137
TC4538BP	BLOCK-1140
TC4539BP	BLOCK-1141
TC4539P	BLOCK-1141
TC4555BP	BLOCK-1146
TC4556BP	BLOCK-1147
TC4584	BLOCK-1038
TC4584BP	BLOCK-1038
TC5012BP	BLOCK-1035
TC5080	BLOCK-185
TC5080P	BLOCK-195
TC5081	BLOCK-196
TC5081P	BLOCK-196
TC5082	BLOCK-185
TC5082P	BLOCK-185
TC5082P-L	BLOCK-185
TC5508P	BLOCK-1160
TC5508P-1	BLOCK-1160
TC5565	BLOCK-1004
TC7400	BLOCK-1293
TC7400BP	BLOCK-1431
TC7476BP	BLOCK-1459
TC74HC02	BLOCK-1507
TC74HC02P	BLOCK-1507
TC74HC04	BLOCK-1508
TC74HC04P	BLOCK-1508
TC74HC123A	BLOCK-1513
TC74HC123F	BLOCK-1513
TC74HC123F-TP1	BLOCK-1513
TC9100C	BLOCK-156
TC9100P	BLOCK-156
TCA380	BLOCK-1303
TCA830	BLOCK-113
TCA830A	BLOCK-112
TCA830S	BLOCK-113
TCG1000	BLOCK-1
TCG1002	BLOCK-3
TCG1003	BLOCK-4
TCG1004	BLOCK-5
TCG1005	BLOCK-6
TCG1006	BLOCK-7
TCG1009	BLOCK-9
TCG1010	BLOCK-10
TCG1011	BLOCK-11
TCG1012	BLOCK-12
TCG1013	BLOCK-13
TCG1014	BLOCK-14
TCG1015	BLOCK-15
TCG1016	BLOCK-16
TCG1017	BLOCK-17
TCG1018	BLOCK-18
TCG1019	BLOCK-19
TCG1021	BLOCK-20
TCG1022	BLOCK-21
TCG1023	BLOCK-22
TCG1024	BLOCK-23
TCG1025	BLOCK-24
TCG1027	BLOCK-26
TCG1028	BLOCK-27
TCG1029	BLOCK-28
TCG1030	BLOCK-29
TCG1031	BLOCK-30
TCG1032	BLOCK-31
TCG1033	BLOCK-32
TCG1034	BLOCK-33
TCG1035	BLOCK-34
TCG1036	BLOCK-35

ORIGINAL DEVICE TYPES AND REPLACEMENT CODES 59

DEVICE TYPE	REPL CODE
TCG1037	BLOCK-36
TCG1038	BLOCK-37
TCG1039	BLOCK-38
TCG1040	BLOCK-39
TCG1041	BLOCK-40
TCG1042	BLOCK-41
TCG1043	BLOCK-42
TCG1045	BLOCK-43
TCG1046	BLOCK-44
TCG1047	BLOCK-45
TCG1048	BLOCK-46
TCG1049	BLOCK-47
TCG1050	BLOCK-48
TCG1051	BLOCK-49
TCG1052	BLOCK-50
TCG1053	BLOCK-51
TCG1054	BLOCK-52
TCG1055	BLOCK-53
TCG1056	BLOCK-54
TCG1057	BLOCK-55
TCG1058	BLOCK-56
TCG1060	BLOCK-58
TCG1061	BLOCK-59
TCG1062	BLOCK-60
TCG1064	BLOCK-62
TCG1065	BLOCK-63
TCG1066	BLOCK-64
TCG1067	BLOCK-65
TCG1068	BLOCK-66
TCG1069	BLOCK-67
TCG1070	BLOCK-68
TCG1071	BLOCK-69
TCG1072	BLOCK-70
TCG1073	BLOCK-71
TCG1074	BLOCK-72
TCG1075A	BLOCK-73
TCG1077	BLOCK-75
TCG1078	BLOCK-76
TCG1079	BLOCK-77
TCG1080	BLOCK-78
TCG1081A	BLOCK-79
TCG1082	BLOCK-80
TCG1083	BLOCK-81
TCG1084	BLOCK-82
TCG1085	BLOCK-83
TCG1086	BLOCK-84
TCG1087	BLOCK-85
TCG1089	BLOCK-87
TCG1090	BLOCK-88
TCG1091	BLOCK-89
TCG1092	BLOCK-90
TCG1093	BLOCK-91
TCG1094	BLOCK-92
TCG1096	BLOCK-94
TCG1097	BLOCK-95
TCG1098	BLOCK-96
TCG1099	BLOCK-97
TCG1100	BLOCK-98
TCG1101	BLOCK-99
TCG1102	BLOCK-100
TCG1103	BLOCK-101
TCG1104	BLOCK-102
TCG1105	BLOCK-103
TCG1106	BLOCK-104
TCG1107	BLOCK-105
TCG1108	BLOCK-106
TCG1109	BLOCK-107
TCG1110	BLOCK-108
TCG1115	BLOCK-112
TCG1115A	BLOCK-113
TCG1116	BLOCK-114
TCG1117	BLOCK-115
TCG1119	BLOCK-117
TCG1120	BLOCK-118
TCG1121	BLOCK-119
TCG1122	BLOCK-120
TCG1123	BLOCK-121
TCG1124	BLOCK-122
TCG1126	BLOCK-124
TCG1127	BLOCK-125
TCG1128	BLOCK-126
TCG1130	BLOCK-127
TCG1131	BLOCK-128
TCG1132	BLOCK-129
TCG1133	BLOCK-130
TCG1134	BLOCK-131
TCG1135	BLOCK-132
TCG1137	BLOCK-134
TCG1139	BLOCK-135
TCG1140	BLOCK-136
TCG1141	BLOCK-137
TCG1142	BLOCK-138
TCG1148	BLOCK-139
TCG1149	BLOCK-140

DEVICE TYPE	REPL CODE
TCG1150	BLOCK-141
TCG1152	BLOCK-142
TCG1153	BLOCK-143
TCG1154	BLOCK-144
TCG1155	BLOCK-145
TCG1156	BLOCK-146
TCG1158	BLOCK-147
TCG1159	BLOCK-148
TCG1160	BLOCK-149
TCG1161	BLOCK-150
TCG1162	BLOCK-151
TCG1163	BLOCK-152
TCG1164	BLOCK-153
TCG1165	BLOCK-154
TCG1166	BLOCK-155
TCG1167	BLOCK-156
TCG1168	BLOCK-157
TCG1169	BLOCK-158
TCG1170	BLOCK-159
TCG1171	BLOCK-160
TCG1172	BLOCK-161
TCG1173	BLOCK-162
TCG1174	BLOCK-163
TCG1175	BLOCK-164
TCG1176	BLOCK-165
TCG1178	BLOCK-167
TCG1179	BLOCK-168
TCG1180	BLOCK-169
TCG1181	BLOCK-170
TCG1183	BLOCK-172
TCG1184	BLOCK-173
TCG1185	BLOCK-174
TCG1186	BLOCK-175
TCG1187	BLOCK-176
TCG1188	BLOCK-177
TCG1189	BLOCK-178
TCG1191	BLOCK-179
TCG1192	BLOCK-180
TCG1193	BLOCK-181
TCG1194	BLOCK-182
TCG1195	BLOCK-183
TCG1196	BLOCK-184
TCG1197	BLOCK-185
TCG1198	BLOCK-186
TCG1199	BLOCK-187
TCG1200	BLOCK-188
TCG1201	BLOCK-189
TCG1202	BLOCK-190
TCG1203	BLOCK-191
TCG1204	BLOCK-192
TCG1205	BLOCK-193
TCG1206	BLOCK-194
TCG1208	BLOCK-196
TCG1209	BLOCK-197
TCG1210	BLOCK-198
TCG1211	BLOCK-199
TCG1212	BLOCK-200
TCG1213	BLOCK-201
TCG1214	BLOCK-202
TCG1215	BLOCK-203
TCG1216	BLOCK-204
TCG1217	BLOCK-205
TCG1218	BLOCK-206
TCG1219	BLOCK-207
TCG1221	BLOCK-209
TCG1223	BLOCK-211
TCG1224	BLOCK-212
TCG1226	BLOCK-214
TCG1227	BLOCK-215
TCG1228	BLOCK-216
TCG1229	BLOCK-217
TCG1230	BLOCK-218
TCG1231	BLOCK-219
TCG1231A	BLOCK-220
TCG1232	BLOCK-221
TCG1233	BLOCK-222
TCG1234	BLOCK-223
TCG1235	BLOCK-224
TCG1236	BLOCK-225
TCG1237	BLOCK-226
TCG1238	BLOCK-227
TCG1239	BLOCK-228
TCG1240	BLOCK-229
TCG1241	BLOCK-230
TCG1242	BLOCK-231
TCG1243	BLOCK-232
TCG1245	BLOCK-234
TCG1246	BLOCK-235
TCG1247	BLOCK-236
TCG1248	BLOCK-237
TCG1249	BLOCK-238
TCG1250	BLOCK-239
TCG1251	BLOCK-240
TCG1252	BLOCK-241

DEVICE TYPE	REPL CODE
TCG1253	BLOCK-242
TCG1254	BLOCK-243
TCG1256	BLOCK-245
TCG1257	BLOCK-246
TCG1258	BLOCK-247
TCG1261	BLOCK-248
TCG1262	BLOCK-249
TCG1263	BLOCK-250
TCG1264	BLOCK-251
TCG1265	BLOCK-252
TCG1266	BLOCK-253
TCG1267	BLOCK-254
TCG1268	BLOCK-255
TCG1269	BLOCK-256
TCG1271	BLOCK-258
TCG1272	BLOCK-259
TCG1273	BLOCK-260
TCG1278	BLOCK-265
TCG1281	BLOCK-268
TCG1282	BLOCK-269
TCG1283	BLOCK-270
TCG1284	BLOCK-271
TCG1285	BLOCK-272
TCG1286	BLOCK-273
TCG1287	BLOCK-274
TCG1288	BLOCK-275
TCG1289	BLOCK-276
TCG1290	BLOCK-277
TCG1291	BLOCK-278
TCG1292	BLOCK-279
TCG1293	BLOCK-280
TCG1300	BLOCK-287
TCG1301	BLOCK-288
TCG1302	BLOCK-289
TCG1303	BLOCK-290
TCG1304	BLOCK-291
TCG1305	BLOCK-292
TCG1306	BLOCK-293
TCG1307	BLOCK-294
TCG1308	BLOCK-295
TCG1309	BLOCK-296
TCG1310	BLOCK-297
TCG1311	BLOCK-298
TCG1312	BLOCK-299
TCG1313	BLOCK-300
TCG1320	BLOCK-307
TCG1321	BLOCK-308
TCG1322	BLOCK-309
TCG1323	BLOCK-310
TCG1324	BLOCK-311
TCG1325	BLOCK-312
TCG1326	BLOCK-313
TCG1327	BLOCK-314
TCG1328	BLOCK-315
TCG1329	BLOCK-316
TCG1330	BLOCK-317
TCG1331	BLOCK-318
TCG1332	BLOCK-319
TCG1333	BLOCK-320
TCG1334	BLOCK-321
TCG1335	BLOCK-322
TCG1336	BLOCK-323
TCG1337	BLOCK-324
TCG1338	BLOCK-325
TCG1339	BLOCK-326
TCG1340	BLOCK-327
TCG1341	BLOCK-328
TCG1342	BLOCK-329
TCG1343	BLOCK-330
TCG1344	BLOCK-331
TCG1345	BLOCK-332
TCG1346	BLOCK-333
TCG1347	BLOCK-334
TCG1348	BLOCK-335
TCG1349	BLOCK-336
TCG1350	BLOCK-337
TCG1351	BLOCK-338
TCG1352	BLOCK-339
TCG1354	BLOCK-341
TCG1355	BLOCK-342
TCG1361	BLOCK-348
TCG1363	BLOCK-350
TCG1364	BLOCK-351
TCG1365	BLOCK-352
TCG1366	BLOCK-353
TCG1367	BLOCK-354
TCG1368	BLOCK-355
TCG1369	BLOCK-356
TCG1370	BLOCK-357
TCG1371	BLOCK-358
TCG1372	BLOCK-359
TCG1373	BLOCK-360
TCG1374	BLOCK-361
TCG1375	BLOCK-362

DEVICE TYPE	REPL CODE
TCG1376	BLOCK-363
TCG1377	BLOCK-364
TCG1378	BLOCK-365
TCG1379	BLOCK-366
TCG1380	BLOCK-367
TCG1381	BLOCK-368
TCG1382	BLOCK-369
TCG1383	BLOCK-370
TCG1384	BLOCK-371
TCG1385	BLOCK-372
TCG1386	BLOCK-373
TCG1387	BLOCK-374
TCG1388	BLOCK-375
TCG1389	BLOCK-376
TCG1391	BLOCK-378
TCG1392	BLOCK-379
TCG1393	BLOCK-380
TCG1394	BLOCK-381
TCG1395	BLOCK-382
TCG1396	BLOCK-383
TCG1400	BLOCK-387
TCG1401	BLOCK-388
TCG1403	BLOCK-390
TCG1404	BLOCK-391
TCG1405	BLOCK-392
TCG1406	BLOCK-393
TCG1407	BLOCK-394
TCG1408	BLOCK-395
TCG1409	BLOCK-396
TCG1410	BLOCK-398
TCG1411	BLOCK-399
TCG1412	BLOCK-400
TCG1413	BLOCK-401
TCG1414	BLOCK-402
TCG1415	BLOCK-403
TCG1416	BLOCK-404
TCG1417	BLOCK-405
TCG1419	BLOCK-407
TCG1420	BLOCK-408
TCG1421	BLOCK-409
TCG1422	BLOCK-410
TCG1423	BLOCK-411
TCG1424	BLOCK-412
TCG1425	BLOCK-413
TCG1426	BLOCK-414
TCG1427	BLOCK-415
TCG1428	BLOCK-416
TCG1429	BLOCK-417
TCG1430	BLOCK-418
TCG1431	BLOCK-419
TCG1433	BLOCK-421
TCG1434	BLOCK-422
TCG1435	BLOCK-423
TCG1436	BLOCK-424
TCG1437	BLOCK-425
TCG1438	BLOCK-426
TCG1439	BLOCK-427
TCG1440	BLOCK-428
TCG1441	BLOCK-429
TCG1442	BLOCK-430
TCG1445	BLOCK-433
TCG1446	BLOCK-434
TCG1447	BLOCK-435
TCG1448	BLOCK-436
TCG1449	BLOCK-437
TCG1450	BLOCK-438
TCG1451	BLOCK-439
TCG1452	BLOCK-440
TCG1453	BLOCK-441
TCG1454	BLOCK-442
TCG1455	BLOCK-443
TCG1457	BLOCK-445
TCG1458	BLOCK-446
TCG1459	BLOCK-447
TCG1460	BLOCK-448
TCG1461	BLOCK-449
TCG1462	BLOCK-450
TCG1463	BLOCK-451
TCG1464	BLOCK-452
TCG1465	BLOCK-453
TCG1466	BLOCK-454
TCG1467	BLOCK-455
TCG1468	BLOCK-456
TCG1469	BLOCK-457
TCG1470	BLOCK-458
TCG1471	BLOCK-459
TCG1472	BLOCK-460
TCG1473	BLOCK-461
TCG1474	BLOCK-462
TCG1475	BLOCK-463
TCG1476	BLOCK-464
TCG1477	BLOCK-465
TCG1478	BLOCK-466
TCG1479	BLOCK-467

DEVICE TYPE	REPL CODE
TCG1480	BLOCK-468
TCG1481	BLOCK-469
TCG1482	BLOCK-470
TCG1483	BLOCK-471
TCG1484	BLOCK-472
TCG1486	BLOCK-474
TCG1487	BLOCK-475
TCG1488	BLOCK-476
TCG1489	BLOCK-477
TCG1490	BLOCK-478
TCG1491	BLOCK-479
TCG1492	BLOCK-480
TCG1493	BLOCK-481
TCG1494	BLOCK-482
TCG1495	BLOCK-483
TCG1496	BLOCK-484
TCG1497	BLOCK-485
TCG1498	BLOCK-486
TCG1499	BLOCK-487
TCG1500	BLOCK-488
TCG1501	BLOCK-489
TCG1502	BLOCK-490
TCG1503	BLOCK-491
TCG1504	BLOCK-492
TCG1505	BLOCK-493
TCG1507	BLOCK-495
TCG1508	BLOCK-496
TCG1509	BLOCK-497
TCG1510	BLOCK-498
TCG1511	BLOCK-499
TCG1512	BLOCK-500
TCG1513	BLOCK-501
TCG1514	BLOCK-502
TCG1515	BLOCK-503
TCG1516	BLOCK-504
TCG1517	BLOCK-505
TCG1518	BLOCK-506
TCG1519	BLOCK-507
TCG1520	BLOCK-508
TCG1521	BLOCK-509
TCG1522	BLOCK-510
TCG1523	BLOCK-511
TCG1525	BLOCK-512
TCG1526	BLOCK-513
TCG1527	BLOCK-514
TCG1528	BLOCK-515
TCG1529	BLOCK-516
TCG1530	BLOCK-517
TCG1531	BLOCK-518
TCG1532	BLOCK-519
TCG1533	BLOCK-520
TCG1534	BLOCK-521
TCG1536	BLOCK-523
TCG1537	BLOCK-524
TCG1538	BLOCK-525
TCG1539	BLOCK-526
TCG1540	BLOCK-527
TCG1541	BLOCK-528
TCG1543	BLOCK-530
TCG1544	BLOCK-531
TCG1545	BLOCK-532
TCG2003	BLOCK-932
TCG2004	BLOCK-933
TCG2011	BLOCK-934
TCG2012	BLOCK-935
TCG2013	BLOCK-936
TCG2014	BLOCK-937
TCG2015	BLOCK-938
TCG2016	BLOCK-939
TCG2017	BLOCK-940
TCG2018	BLOCK-941
TCG2019	BLOCK-942
TCG2020	BLOCK-943
TCG2021	BLOCK-944
TCG2022	BLOCK-945
TCG2023	BLOCK-946
TCG2025	BLOCK-948
TCG2026	BLOCK-949
TCG2027	BLOCK-950
TCG2028	BLOCK-951
TCG2029	BLOCK-952
TCG2030	BLOCK-953
TCG2031	BLOCK-954
TCG2032	BLOCK-955
TCG2050	BLOCK-965
TCG2051	BLOCK-966
TCG2053	BLOCK-968
TCG2054	BLOCK-969
TCG2060	BLOCK-973
TCG2061	BLOCK-974
TCG2070	BLOCK-979
TCG2071	BLOCK-980
TCG2072	BLOCK-981
TCG2073	BLOCK-982

DEVICE TYPE	REPL CODE
TCG2074	BLOCK-983
TCG2075	BLOCK-984
TCG2076	BLOCK-985
TCG2077	BLOCK-986
TCG2078	BLOCK-987
TCG2079	BLOCK-988
TCG2080	BLOCK-989
TCG2081	BLOCK-990
TCG2082	BLOCK-991
TCG2083	BLOCK-992
TCG2084	BLOCK-993
TCG2085	BLOCK-994
TCG2087	BLOCK-996
TCG2102	BLOCK-999
TCG2104	BLOCK-1000
TCG2114	BLOCK-1002
TCG2117	BLOCK-1003
TCG2147	BLOCK-1005
TCG2708	BLOCK-1016
TCG2716	BLOCK-1017
TCG309K	BLOCK-1022
TCG3880	BLOCK-1025
TCG3881	BLOCK-1026
TCG3882	BLOCK-1027
TCG40-45B	BLOCK-1080
TCG4000	BLOCK-1028
TCG4001B	BLOCK-1029
TCG4002B	BLOCK-1030
TCG4006B	BLOCK-1031
TCG4007	BLOCK-1032
TCG4007B	BLOCK-1032
TCG40085B	BLOCK-1033
TCG4008B	BLOCK-1034
TCG40097B	BLOCK-1035
TCG40098B	BLOCK-1036
TCG40100B	BLOCK-1037
TCG40106B	BLOCK-1038
TCG4011B	BLOCK-1039
TCG4012B	BLOCK-1040
TCG4013B	BLOCK-1041
TCG4014B	BLOCK-1042
TCG4015B	BLOCK-1043
TCG40160B	BLOCK-1044
TCG40161B	BLOCK-1045
TCG40162B	BLOCK-1046
TCG40163B	BLOCK-1047
TCG4016B	BLOCK-1048
TCG40174B	BLOCK-1049
TCG40175B	BLOCK-1050
TCG4017B	BLOCK-1051
TCG40182B	BLOCK-1052
TCG4018B	BLOCK-1053
TCG40192B	BLOCK-1054
TCG40193B	BLOCK-1055
TCG40194B	BLOCK-1056
TCG40195B	BLOCK-1057
TCG4019B	BLOCK-1058
TCG4020B	BLOCK-1059
TCG4021B	BLOCK-1060
TCG4022B	BLOCK-1061
TCG4023B	BLOCK-1062
TCG4024B	BLOCK-1063
TCG4025B	BLOCK-1064
TCG4026B	BLOCK-1065
TCG4027B	BLOCK-1066
TCG4028B	BLOCK-1067
TCG4029B	BLOCK-1068
TCG4030B	BLOCK-1069
TCG4031B	BLOCK-1070
TCG4032B	BLOCK-1071
TCG4033B	BLOCK-1072
TCG4034B	BLOCK-1073
TCG4035B	BLOCK-1074
TCG4038B	BLOCK-1075
TCG4040B	BLOCK-1076
TCG4042B	BLOCK-1077
TCG4043B	BLOCK-1078
TCG4044B	BLOCK-1079
TCG4047B	BLOCK-1082
TCG4048B	BLOCK-1083
TCG4049	BLOCK-1084
TCG4049B	BLOCK-1084
TCG4050B	BLOCK-1085
TCG4051B	BLOCK-1086
TCG4052B	BLOCK-1087
TCG4053B	BLOCK-1088
TCG4055B	BLOCK-1089
TCG4056B	BLOCK-1090
TCG4060B	BLOCK-1091
TCG4063B	BLOCK-1092
TCG4066B	BLOCK-1093
TCG4067B	BLOCK-1094
TCG4068B	BLOCK-1095
TCG4069	BLOCK-1096

DEVICE TYPE	REPL CODE
TCG4069B	BLOCK-1096
TCG4070B	BLOCK-1097
TCG4071B	BLOCK-1098
TCG4072B	BLOCK-1099
TCG4073B	BLOCK-1100
TCG4075B	BLOCK-1101
TCG4076B	BLOCK-1102
TCG4077B	BLOCK-1103
TCG4078B	BLOCK-1104
TCG4081	BLOCK-1105
TCG4081B	BLOCK-1105
TCG4082B	BLOCK-1106
TCG4085B	BLOCK-1107
TCG4086B	BLOCK-1108
TCG4089B	BLOCK-1109
TCG4093B	BLOCK-1110
TCG4095B	BLOCK-1111
TCG4096B	BLOCK-1112
TCG4097B	BLOCK-1113
TCG4098B	BLOCK-1114
TCG4099B	BLOCK-1115
TCG4501	BLOCK-1117
TCG4502B	BLOCK-1118
TCG4503B	BLOCK-1035
TCG4506B	BLOCK-1119
TCG4508B	BLOCK-1120
TCG4510B	BLOCK-1121
TCG4511B	BLOCK-1122
TCG4512B	BLOCK-1123
TCG4514B	BLOCK-1125
TCG4515B	BLOCK-1126
TCG4516B	BLOCK-1127
TCG4517B	BLOCK-1128
TCG4518B	BLOCK-1129
TCG4520B	BLOCK-1130
TCG4521B	BLOCK-1131
TCG4522B	BLOCK-1132
TCG4526B	BLOCK-1133
TCG4527B	BLOCK-1134
TCG4528B	BLOCK-1135
TCG4529B	BLOCK-1136
TCG4531B	BLOCK-1137
TCG4532B	BLOCK-1138
TCG4536B	BLOCK-463
TCG4539B	BLOCK-1141
TCG4541B	BLOCK-1142
TCG4543B	BLOCK-1143
TCG4547B	BLOCK-1144
TCG4555B	BLOCK-1146
TCG4556B	BLOCK-1147
TCG4558B	BLOCK-1148
TCG4566B	BLOCK-1150
TCG4583B	BLOCK-1153
TCG4585B	BLOCK-1154
TCG6502	BLOCK-1158
TCG6507	BLOCK-1159
TCG6508	BLOCK-1160
TCG65101	BLOCK-1161
TCG6532	BLOCK-1162
TCG6800	BLOCK-1164
TCG6802	BLOCK-1165
TCG6810	BLOCK-1168
TCG6821	BLOCK-1169
TCG6850	BLOCK-1170
TCG6875	BLOCK-1172
TCG6880	BLOCK-1173
TCG6885	BLOCK-1174
TCG6886	BLOCK-1175
TCG6887	BLOCK-1176
TCG6888	BLOCK-1177
TCG6889	BLOCK-1178
TCG700	BLOCK-1179
TCG701	BLOCK-1190
TCG702	BLOCK-1201
TCG703A	BLOCK-1212
TCG704	BLOCK-1223
TCG705A	BLOCK-1244
TCG706	BLOCK-1245
TCG707	BLOCK-1256
TCG708	BLOCK-1262
TCG709	BLOCK-1264
TCG710	BLOCK-1265
TCG711	BLOCK-1267
TCG712	BLOCK-1269
TCG714	BLOCK-1271
TCG715	BLOCK-1272
TCG716	BLOCK-1273
TCG717	BLOCK-1274
TCG718	BLOCK-1275
TCG720	BLOCK-1277
TCG723	BLOCK-1279
TCG724	BLOCK-1280
TCG725	BLOCK-1281
TCG726	BLOCK-1282
TCG727	BLOCK-1283
TCG728	BLOCK-1284
TCG729	BLOCK-1285
TCG730	BLOCK-1286
TCG731	BLOCK-1287
TCG736	BLOCK-1289
TCG737	BLOCK-1290
TCG738	BLOCK-1291
TCG7400	BLOCK-1293
TCG7401	BLOCK-1294
TCG7402	BLOCK-1295
TCG7403	BLOCK-1296
TCG7404	BLOCK-1297
TCG7405	BLOCK-1298
TCG7406	BLOCK-1299
TCG7407	BLOCK-1300
TCG7408	BLOCK-1301
TCG7409	BLOCK-1302
TCG740A	BLOCK-1303
TCG741	BLOCK-1917
TCG7410	BLOCK-1304
TCG74107	BLOCK-1305
TCG74109	BLOCK-1306
TCG7411	BLOCK-1307
TCG74110	BLOCK-1308
TCG74111	BLOCK-1309
TCG7412	BLOCK-1310
TCG74121	BLOCK-1311
TCG74122	BLOCK-1312
TCG74123	BLOCK-1313
TCG74125	BLOCK-1314
TCG74126	BLOCK-1315
TCG74128	BLOCK-1316
TCG7413	BLOCK-1317
TCG74132	BLOCK-1318
TCG74136	BLOCK-1319
TCG7414	BLOCK-1320
TCG74141	BLOCK-1321
TCG74144	BLOCK-1324
TCG74145	BLOCK-1325
TCG74147	BLOCK-1326
TCG74150	BLOCK-1327
TCG74151	BLOCK-1328
TCG74152	BLOCK-1329
TCG74153	BLOCK-1330
TCG74154	BLOCK-1331
TCG74155	BLOCK-1332
TCG74156	BLOCK-1333
TCG74157	BLOCK-1334
TCG74158	BLOCK-1335
TCG7416	BLOCK-1336
TCG74160	BLOCK-1337
TCG74161	BLOCK-1338
TCG74162	BLOCK-1339
TCG74163	BLOCK-1340
TCG74164	BLOCK-1341
TCG74165	BLOCK-1342
TCG7417	BLOCK-1343
TCG74170	BLOCK-1346
TCG74173	BLOCK-1345
TCG74174	BLOCK-1346
TCG74175	BLOCK-1347
TCG74176	BLOCK-1348
TCG74177	BLOCK-1349
TCG74178	BLOCK-1350
TCG74179	BLOCK-1351
TCG74180	BLOCK-1352
TCG74181	BLOCK-1353
TCG74182	BLOCK-1354
TCG74190	BLOCK-1355
TCG74191	BLOCK-1356
TCG74192	BLOCK-1357
TCG74193	BLOCK-1358
TCG74195	BLOCK-1359
TCG74196	BLOCK-1360
TCG74197	BLOCK-1361
TCG74198	BLOCK-1362
TCG74199	BLOCK-1363
TCG742	BLOCK-1364
TCG7420	BLOCK-1365
TCG7421	BLOCK-1366
TCG7422	BLOCK-1367
TCG74221	BLOCK-1368
TCG7423	BLOCK-1369
TCG74249	BLOCK-1370
TCG74251	BLOCK-1371
TCG7427	BLOCK-1372
TCG7428	BLOCK-1373
TCG74290	BLOCK-1374
TCG74293	BLOCK-1375
TCG743	BLOCK-1376
TCG7430	BLOCK-1377
TCG7432	BLOCK-1378
TCG7433	BLOCK-1379
TCG74365	BLOCK-1380
TCG74366	BLOCK-1381
TCG74367	BLOCK-1382
TCG74368	BLOCK-1383
TCG7437	BLOCK-1384
TCG7438	BLOCK-1385
TCG7439	BLOCK-1386
TCG74390	BLOCK-1387
TCG74393	BLOCK-1388
TCG744	BLOCK-1389
TCG7440	BLOCK-1390
TCG7441	BLOCK-1391
TCG7442	BLOCK-1392
TCG7443	BLOCK-1393
TCG7444	BLOCK-1394
TCG7445	BLOCK-1395
TCG7446	BLOCK-1396
TCG7447	BLOCK-1397
TCG7448	BLOCK-1398
TCG74490	BLOCK-1399
TCG745	BLOCK-1400
TCG7450	BLOCK-1401
TCG7451	BLOCK-1402
TCG7453	BLOCK-1403
TCG7454	BLOCK-1404
TCG746	BLOCK-1405
TCG7460	BLOCK-1406
TCG747	BLOCK-1407
TCG7470	BLOCK-1408
TCG7472	BLOCK-1409
TCG7473	BLOCK-1410
TCG7474	BLOCK-1411
TCG7475	BLOCK-1412
TCG7476	BLOCK-1413
TCG748	BLOCK-1421
TCG7480	BLOCK-1414
TCG7481	BLOCK-1415
TCG7482	BLOCK-1416
TCG7483	BLOCK-1417
TCG7485	BLOCK-1418
TCG7486	BLOCK-1419
TCG7486A	BLOCK-1419
TCG7489	BLOCK-1420
TCG749	BLOCK-1422
TCG7490	BLOCK-1423
TCG7491	BLOCK-1424
TCG7492	BLOCK-1425
TCG7493A	BLOCK-1426
TCG7494	BLOCK-1427
TCG7495	BLOCK-1428
TCG7496	BLOCK-1429
TCG7497	BLOCK-1430
TCG74C00	BLOCK-1431
TCG74C02	BLOCK-1432
TCG74C04	BLOCK-1433
TCG74C08	BLOCK-1434
TCG74C10	BLOCK-1435
TCG74C107	BLOCK-1436
TCG74C14	BLOCK-1437
TCG74C154	BLOCK-1439
TCG74C157	BLOCK-1440
TCG74C161	BLOCK-1441
TCG74C164	BLOCK-1442
TCG74C173	BLOCK-1443
TCG74C174	BLOCK-1444
TCG74C175	BLOCK-1445
TCG74C192	BLOCK-1446
TCG74C193	BLOCK-1447
TCG74C20	BLOCK-1448
TCG74C221	BLOCK-1449
TCG74C240	BLOCK-1450
TCG74C244	BLOCK-1451
TCG74C30	BLOCK-1452
TCG74C32	BLOCK-1453
TCG74C48	BLOCK-1456
TCG74C74	BLOCK-1458
TCG74C76	BLOCK-1459
TCG74C85	BLOCK-1459
TCG74C90	BLOCK-1461
TCG74C901	BLOCK-1462
TCG74C902	BLOCK-1463
TCG74C903	BLOCK-1464
TCG74C904	BLOCK-1465
TCG74C925	BLOCK-1468
TCG74C95	BLOCK-1470
TCG74H00	BLOCK-1471
TCG74H01	BLOCK-1472
TCG74H04	BLOCK-1473
TCG74H05	BLOCK-1474
TCG74H08	BLOCK-1475
TCG74H10	BLOCK-1476
TCG74H101	BLOCK-1477
TCG74H102	BLOCK-1478
TCG74H103	BLOCK-1479
TCG74H106	BLOCK-1480
TCG74H108	BLOCK-1481
TCG74H11	BLOCK-1482
TCG74H183	BLOCK-1483
TCG74H20	BLOCK-1484
TCG74H21	BLOCK-1485
TCG74H22	BLOCK-1486
TCG74H30	BLOCK-1487
TCG74H40	BLOCK-1488
TCG74H50	BLOCK-1489
TCG74H51	BLOCK-1490
TCG74H52	BLOCK-1491
TCG74H53	BLOCK-1492
TCG74H54	BLOCK-1493
TCG74H55	BLOCK-1494
TCG74H60	BLOCK-1495
TCG74H61	BLOCK-1496
TCG74H62	BLOCK-1497
TCG74H71	BLOCK-1498
TCG74H72	BLOCK-1499
TCG74H73	BLOCK-1500
TCG74H74	BLOCK-1501
TCG74H76	BLOCK-1502
TCG74H78	BLOCK-1503
TCG74H87	BLOCK-1505
TCG74L93	BLOCK-1567
TCG74LS00	BLOCK-1568
TCG74LS01	BLOCK-1569
TCG74LS02	BLOCK-1570
TCG74LS03	BLOCK-1571
TCG74LS04	BLOCK-1572
TCG74LS05	BLOCK-1573
TCG74LS08	BLOCK-1574
TCG74LS09	BLOCK-1575
TCG74LS10	BLOCK-1576
TCG74LS107	BLOCK-1577
TCG74LS109A	BLOCK-1578
TCG74LS11	BLOCK-1579
TCG74LS112A	BLOCK-1580
TCG74LS114	BLOCK-1581
TCG74LS12	BLOCK-1582
TCG74LS123	BLOCK-1584
TCG74LS125A	BLOCK-1585
TCG74LS126A	BLOCK-1586
TCG74LS13	BLOCK-1587
TCG74LS132	BLOCK-1588
TCG74LS136	BLOCK-1590
TCG74LS138	BLOCK-1591
TCG74LS139	BLOCK-1592
TCG74LS14	BLOCK-1593
TCG74LS145	BLOCK-1594
TCG74LS15	BLOCK-1597
TCG74LS151	BLOCK-1598
TCG74LS153	BLOCK-1599
TCG74LS155	BLOCK-1600
TCG74LS156	BLOCK-1601
TCG74LS157	BLOCK-1602
TCG74LS158	BLOCK-1603
TCG74LS160A	BLOCK-1604
TCG74LS161A	BLOCK-1605
TCG74LS162A	BLOCK-1606
TCG74LS163A	BLOCK-1607
TCG74LS164	BLOCK-1608
TCG74LS170	BLOCK-1613
TCG74LS173A	BLOCK-1614
TCG74LS174	BLOCK-1615
TCG74LS175	BLOCK-1616
TCG74LS190	BLOCK-1618
TCG74LS191	BLOCK-1619
TCG74LS192	BLOCK-1620
TCG74LS193	BLOCK-1621
TCG74LS194	BLOCK-1622
TCG74LS195A	BLOCK-1623
TCG74LS196	BLOCK-1624
TCG74LS197	BLOCK-1625
TCG74LS20	BLOCK-1626
TCG74LS21	BLOCK-1627
TCG74LS22	BLOCK-1628
TCG74LS221	BLOCK-1629
TCG74LS240	BLOCK-1630
TCG74LS241	BLOCK-1631
TCG74LS242	BLOCK-1632
TCG74LS243	BLOCK-1633
TCG74LS244	BLOCK-1634
TCG74LS245	BLOCK-1635
TCG74LS247	BLOCK-1636
TCG74LS248	BLOCK-1637
TCG74LS249	BLOCK-1638
TCG74LS251	BLOCK-1639
TCG74LS253	BLOCK-1640
TCG74LS257	BLOCK-1641
TCG74LS258	BLOCK-1642
TCG74LS259	BLOCK-1643
TCG74LS26	BLOCK-1644
TCG74LS266	BLOCK-1646
TCG74LS27	BLOCK-1647
TCG74LS273	BLOCK-1648
TCG74LS279	BLOCK-1649
TCG74LS280	BLOCK-1651
TCG74LS283	BLOCK-1652
TCG74LS290	BLOCK-1653
TCG74LS293	BLOCK-1654
TCG74LS298	BLOCK-1656
TCG74LS30	BLOCK-1658
TCG74LS32	BLOCK-1659
TCG74LS348	BLOCK-1661
TCG74LS352	BLOCK-1662
TCG74LS353	BLOCK-1663
TCG74LS363	BLOCK-1664
TCG74LS364	BLOCK-1665
TCG74LS365A	BLOCK-1666
TCG74LS366A	BLOCK-1667
TCG74LS367	BLOCK-1668
TCG74LS368	BLOCK-1669
TCG74LS37	BLOCK-1670
TCG74LS373	BLOCK-1671
TCG74LS374	BLOCK-1672
TCG74LS377	BLOCK-1673
TCG74LS378	BLOCK-1674
TCG74LS379	BLOCK-1675
TCG74LS38	BLOCK-1676
TCG74LS393	BLOCK-1679
TCG74LS395A	BLOCK-1680
TCG74LS40	BLOCK-1683
TCG74LS42	BLOCK-1684
TCG74LS47	BLOCK-1685
TCG74LS48	BLOCK-1686
TCG74LS49	BLOCK-1687
TCG74LS490	BLOCK-1688
TCG74LS51	BLOCK-1689
TCG74LS54	BLOCK-1690
TCG74LS540	BLOCK-1691
TCG74LS541	BLOCK-1692
TCG74LS55	BLOCK-1693
TCG74LS624	BLOCK-1694
TCG74LS625	BLOCK-1695
TCG74LS626	BLOCK-1696
TCG74LS627	BLOCK-1697
TCG74LS629	BLOCK-1698
TCG74LS640	BLOCK-1699
TCG74LS641	BLOCK-1700
TCG74LS642	BLOCK-1701
TCG74LS643	BLOCK-1702
TCG74LS645	BLOCK-1703
TCG74LS670	BLOCK-1704
TCG74LS73	BLOCK-1705
TCG74LS74A	BLOCK-1706
TCG74LS75	BLOCK-1707
TCG74LS76A	BLOCK-1708
TCG74LS78	BLOCK-1710
TCG74LS83A	BLOCK-1711
TCG74LS85	BLOCK-1712
TCG74LS86	BLOCK-1713
TCG74LS90	BLOCK-1714
TCG74LS92	BLOCK-1716
TCG74LS93	BLOCK-1717
TCG74S00	BLOCK-1719
TCG74S02	BLOCK-1720
TCG74S03	BLOCK-1721
TCG74S04	BLOCK-1722
TCG74S05	BLOCK-1723
TCG74S08	BLOCK-1724
TCG74S09	BLOCK-1725
TCG74S10	BLOCK-1726
TCG74S11	BLOCK-1727
TCG74S112	BLOCK-1728
TCG74S113	BLOCK-1729
TCG74S124	BLOCK-1730
TCG74S133	BLOCK-1731
TCG74S134	BLOCK-1732
TCG74S138	BLOCK-1733
TCG74S140	BLOCK-1734
TCG74S15	BLOCK-1735
TCG74S151	BLOCK-1736
TCG74S153	BLOCK-1737
TCG74S157	BLOCK-1738
TCG74S158	BLOCK-1739
TCG74S163	BLOCK-1607
TCG74S174	BLOCK-1740
TCG74S175	BLOCK-1741
TCG74S181	BLOCK-1742
TCG74S194	BLOCK-1744
TCG74S20	BLOCK-1745
TCG74S22	BLOCK-1746
TCG74S251	BLOCK-1747
TCG74S258	BLOCK-1748
TCG74S280	BLOCK-1651
TCG74S30	BLOCK-1751
TCG74S40	BLOCK-1753
TCG74S51	BLOCK-1761
TCG74S64	BLOCK-1766
TCG74S65	BLOCK-1767
TCG74S74	BLOCK-1768
TCG74S86	BLOCK-1769
TCG750	BLOCK-1770
TCG75188	BLOCK-1771
TCG75189	BLOCK-1772
TCG75450B	BLOCK-1775
TCG75451B	BLOCK-1776
TCG75452B	BLOCK-1777
TCG75453B	BLOCK-1778
TCG75454B	BLOCK-1779
TCG75491B	BLOCK-1780
TCG75492B	BLOCK-1781
TCG75493	BLOCK-1782
TCG75494	BLOCK-1783
TCG768	BLOCK-1822
TCG778A	BLOCK-1801
TCG780	BLOCK-1805
TCG781	BLOCK-1806
TCG783	BLOCK-1808
TCG784	BLOCK-1809
TCG785	BLOCK-1810
TCG786	BLOCK-1811
TCG787	BLOCK-1812
TCG788	BLOCK-1813
TCG789	BLOCK-1814
TCG790	BLOCK-1815
TCG791	BLOCK-1816
TCG797	BLOCK-1821
TCG798	BLOCK-1822
TCG799	BLOCK-1823
TCG801	BLOCK-1825
TCG802	BLOCK-1826
TCG803	BLOCK-1827
TCG804	BLOCK-1828
TCG807	BLOCK-1830
TCG8080A	BLOCK-1832
TCG80C95	BLOCK-1835
TCG80C96	BLOCK-1836
TCG80C97	BLOCK-1837
TCG812	BLOCK-1839
TCG8123	BLOCK-1840
TCG815	BLOCK-1843
TCG818	BLOCK-1846
TCG819	BLOCK-1847
TCG820	BLOCK-1848
TCG821	BLOCK-1849
TCG822	BLOCK-1850
TCG823	BLOCK-1851
TCG824	BLOCK-1852
TCG825	BLOCK-1853
TCG8255	BLOCK-1855
TCG828	BLOCK-1858
TCG829	BLOCK-1859
TCG8301	BLOCK-1861
TCG8308	BLOCK-1862
TCG8309	BLOCK-1863
TCG8314	BLOCK-1865
TCG8316	BLOCK-1866
TCG8318	BLOCK-1867
TCG832	BLOCK-1868
TCG8321	BLOCK-1869
TCG8328	BLOCK-1870
TCG834	BLOCK-1873
TCG836	BLOCK-1876
TCG8368	BLOCK-1877
TCG8370	BLOCK-1879
TCG8374	BLOCK-1880
TCG841	BLOCK-1884
TCG843	BLOCK-1886
TCG846	BLOCK-1889
TCG849	BLOCK-1892
TCG851	BLOCK-1894
TCG852	BLOCK-1895
TCG8520	BLOCK-1896
TCG853	BLOCK-1897
TCG8542	BLOCK-1899
TCG8546	BLOCK-1900
TCG855	BLOCK-1901
TCG8556	BLOCK-1906
TCG856	BLOCK-1907
TCG857M	BLOCK-1908
TCG859	BLOCK-1912
TCG860	BLOCK-1914
TCG900	BLOCK-1940
TCG901	BLOCK-1941
TCG902	BLOCK-1942
TCG903	BLOCK-1943
TCG904	BLOCK-1944
TCG905	BLOCK-1945
TCG906	BLOCK-1946

ORIGINAL DEVICE TYPES AND REPLACEMENT CODES 61

DEVICE TYPE	REPL CODE
TCG907	BLOCK-1947
TCG908	BLOCK-1948
TCG909	BLOCK-1949
TCG9093	BLOCK-1950
TCG9094	BLOCK-1951
TCG9097	BLOCK-1952
TCG9099	BLOCK-1953
TCG909D	BLOCK-1954
TCG910	BLOCK-1955
TCG910D	BLOCK-1955
TCG911	BLOCK-1958
TCG911D	BLOCK-1962
TCG912	BLOCK-1963
TCG913	BLOCK-1964
TCG9135	BLOCK-1965
TCG914	BLOCK-1966
TCG915	BLOCK-1967
TCG9157	BLOCK-1968
TCG9158	BLOCK-1969
TCG916	BLOCK-1970
TCG917	BLOCK-1971
TCG918	BLOCK-1972
TCG918M	BLOCK-1973
TCG919	BLOCK-1975
TCG919D	BLOCK-1976
TCG9200	BLOCK-1978
TCG922	BLOCK-1980
TCG922M	BLOCK-1981
TCG923	BLOCK-1983
TCG923D	BLOCK-1984
TCG924	BLOCK-1985
TCG924M	BLOCK-1986
TCG925	BLOCK-1987
TCG926	BLOCK-1988
TCG927	BLOCK-1989
TCG927D	BLOCK-1990
TCG928	BLOCK-1992
TCG928M	BLOCK-1993
TCG930	BLOCK-1997
TCG931	BLOCK-2004
TCG932	BLOCK-2007
TCG933	BLOCK-2014
TCG934	BLOCK-2020
TCG935	BLOCK-2024
TCG936	BLOCK-2025
TCG937	BLOCK-2031
TCG9370	BLOCK-2032
TCG937M	BLOCK-2036
TCG938	BLOCK-2037
TCG938M	BLOCK-2037
TCG939	BLOCK-2043
TCG93L08	BLOCK-2049
TCG93L16	BLOCK-2050
TCG940	BLOCK-2051
TCG9402	BLOCK-2054
TCG9403	BLOCK-2055
TCG941	BLOCK-2056
TCG941D	BLOCK-2057
TCG941M	BLOCK-2058
TCG942	BLOCK-2061
TCG943	BLOCK-2062
TCG943M	BLOCK-2063
TCG944	BLOCK-2065
TCG944M	BLOCK-2066
TCG946	BLOCK-2068
TCG947	BLOCK-2069
TCG947D	BLOCK-2070
TCG948	BLOCK-2071
TCG949	BLOCK-2073
TCG950	BLOCK-2074
TCG951	BLOCK-2075
TCG952	BLOCK-2076
TCG953	BLOCK-2077
TCG954	BLOCK-2078
TCG955M	BLOCK-2079
TCG956	BLOCK-2083
TCG957	BLOCK-2084
TCG958	BLOCK-2085
TCG959	BLOCK-2086
TCG960	BLOCK-2087
TCG9601	BLOCK-2089
TCG9602	BLOCK-2090
TCG961	BLOCK-2091
TCG962	BLOCK-2093
TCG963	BLOCK-2094
TCG964	BLOCK-2095
TCG965	BLOCK-2096
TCG966	BLOCK-2097
TCG9660	BLOCK-2098
TCG9661	BLOCK-2099
TCG9663	BLOCK-2101
TCG9664	BLOCK-2102
TCG9666	BLOCK-2104
TCG9668	BLOCK-2106

DEVICE TYPE	REPL CODE
TCG9669	BLOCK-2107
TCG967	BLOCK-2108
TCG9670	BLOCK-2109
TCG9671	BLOCK-2110
TCG9672	BLOCK-2111
TCG9673	BLOCK-2112
TCG9675	BLOCK-2114
TCG9676	BLOCK-2115
TCG9678	BLOCK-2117
TCG9679	BLOCK-2118
TCG968	BLOCK-2119
TCG9680	BLOCK-2120
TCG9681	BLOCK-2121
TCG9682	BLOCK-2122
TCG9689	BLOCK-2127
TCG969	BLOCK-2128
TCG9691	BLOCK-2130
TCG96L02	BLOCK-2132
TCG96LS02	BLOCK-2133
TCG96S02	BLOCK-2134
TCG970	BLOCK-2135
TCG971	BLOCK-2136
TCG972	BLOCK-2137
TCG973	BLOCK-2138
TCG973D	BLOCK-2139
TCG974	BLOCK-2140
TCG975	BLOCK-2141
TCG976	BLOCK-2143
TCG977	BLOCK-2144
TCG978	BLOCK-2145
TCG980	BLOCK-2149
TCG9800	BLOCK-2150
TCG9801	BLOCK-2151
TCG9802	BLOCK-2152
TCG9803	BLOCK-2153
TCG9804	BLOCK-2154
TCG9805	BLOCK-2155
TCG9806	BLOCK-2156
TCG9807	BLOCK-2157
TCG9808	BLOCK-2158
TCG9809	BLOCK-2159
TCG981	BLOCK-2160
TCG9810	BLOCK-2161
TCG9811	BLOCK-2162
TCG9812	BLOCK-2163
TCG9813	BLOCK-2164
TCG9814	BLOCK-2165
TCG984	BLOCK-2168
TCG985	BLOCK-2169
TCG986	BLOCK-2170
TCG987	BLOCK-2171
TCG988	BLOCK-2173
TCG989	BLOCK-2174
TCG990	BLOCK-2175
TCG9904	BLOCK-1951
TCG991	BLOCK-2184
TCG9910	BLOCK-2185
TCG992	BLOCK-2191
TCG9924	BLOCK-2193
TCG9926	BLOCK-2191
TCG993	BLOCK-2196
TCG9930	BLOCK-2197
TCG9932	BLOCK-2199
TCG9933	BLOCK-2200
TCG9935	BLOCK-2201
TCG9936	BLOCK-2202
TCG9937	BLOCK-2203
TCG9944	BLOCK-2206
TCG9945	BLOCK-2207
TCG9946	BLOCK-2208
TCG9948	BLOCK-2209
TCG9949	BLOCK-2210
TCG995	BLOCK-2211
TCG9950	BLOCK-2212
TCG9951	BLOCK-2213
TCG996	BLOCK-2215
TCG9961	BLOCK-2216
TCG9962	BLOCK-2217
TCG9963	BLOCK-2218
TCG997	BLOCK-2219
TCG9989	BLOCK-2224
TCM5089	BLOCK-676
TD1060	BLOCK-2197
TD1060P	BLOCK-2197
TD1061	BLOCK-2201
TD1061P	BLOCK-2201
TD1062	BLOCK-2199
TD1062P	BLOCK-2199
TD1063	BLOCK-2200
TD1063P	BLOCK-2200
TD1064	BLOCK-2206
TD1064P	BLOCK-2206
TD1065	BLOCK-2208
TD1065P	BLOCK-2208

DEVICE TYPE	REPL CODE
TD1066	BLOCK-2217
TD1066P	BLOCK-2217
TD1067	BLOCK-2209
TD1067P	BLOCK-2209
TD1070	BLOCK-2207
TD1070P	BLOCK-2207
TD1072	BLOCK-2202
TD1072P	BLOCK-2202
TD1073	BLOCK-1950
TD1073P	BLOCK-1950
TD1074	BLOCK-1953
TD1074P	BLOCK-1953
TD1080	BLOCK-2216
TD1080P	BLOCK-2216
TD1085	BLOCK-2210
TD1085P	BLOCK-2210
TD1086	BLOCK-2218
TD1086P	BLOCK-2218
TD1401	BLOCK-1293
TD1401P	BLOCK-1293
TD1402	BLOCK-1304
TD1402P	BLOCK-1304
TD1403	BLOCK-1365
TD1403P	BLOCK-1365
TD1404	BLOCK-1377
TD1404P	BLOCK-1377
TD1405	BLOCK-1390
TD1405P	BLOCK-1390
TD1406	BLOCK-1401
TD1406P	BLOCK-1401
TD1407	BLOCK-1406
TD1407P	BLOCK-1406
TD1408	BLOCK-1409
TD1408P	BLOCK-1409
TD1409	BLOCK-1410
TD1409P	BLOCK-1410
TD1419	BLOCK-1402
TD1419P	BLOCK-1402
TD2001P	BLOCK-2099
TD2002P	BLOCK-2109
TD2003P	BLOCK-2106
TD2004P	BLOCK-2102
TD2005P	BLOCK-2101
TD2006P	BLOCK-2107
TD2008P	BLOCK-2098
TD2009P	BLOCK-2110
TD2010P	BLOCK-2111
TD2011P	BLOCK-2100
TD2012P	BLOCK-2120
TD2013P	BLOCK-2116
TD2015P	BLOCK-2105
TD3400A	BLOCK-1293
TD3400AP	BLOCK-1293
TD3400P	BLOCK-1293
TD3401A	BLOCK-1294
TD3401AP	BLOCK-1294
TD3402A	BLOCK-1295
TD3402AP	BLOCK-1295
TD3404A	BLOCK-1297
TD3404AP	BLOCK-1297
TD3405A	BLOCK-1298
TD3405AP	BLOCK-1298
TD3409A	BLOCK-1302
TD3409AP	BLOCK-1302
TD3410A	BLOCK-1304
TD3410AP	BLOCK-1304
TD3410P	BLOCK-1304
TD34121A	BLOCK-1311
TD34121AP	BLOCK-1311
TD34192A	BLOCK-1357
TD34192AP	BLOCK-1357
TD3420A	BLOCK-1365
TD3420AP	BLOCK-1365
TD3420P	BLOCK-1365
TD3430A	BLOCK-1377
TD3430AP	BLOCK-1377
TD3430P	BLOCK-1377
TD3440A	BLOCK-1390
TD3440AP	BLOCK-1390
TD3440P	BLOCK-1390
TD3441A	BLOCK-1391
TD3441AP	BLOCK-1391
TD3442A	BLOCK-1392
TD3442AP	BLOCK-1392
TD3447A	BLOCK-1397
TD3447AP	BLOCK-1397
TD3450A	BLOCK-1401
TD3450AP	BLOCK-1401
TD3450P	BLOCK-1401
TD3451A	BLOCK-1402
TD3451AP	BLOCK-1402
TD3451P	BLOCK-1402
TD3460A	BLOCK-1406
TD3460AP	BLOCK-1406

DEVICE TYPE	REPL CODE
TD3460P	BLOCK-1406
TD3472A	BLOCK-1409
TD3472AP	BLOCK-1409
TD3473A	BLOCK-1410
TD3473AP	BLOCK-1410
TD3474A	BLOCK-1411
TD3474AP	BLOCK-1411
TD3474P	BLOCK-1411
TD3475A	BLOCK-1412
TD3475AP	BLOCK-1412
TD3480A	BLOCK-1414
TD3480AP	BLOCK-1414
TD3482P	BLOCK-1416
TD3483P	BLOCK-1417
TD3490A	BLOCK-1423
TD3490AP	BLOCK-1423
TD3490P	BLOCK-1423
TD3491A	BLOCK-1424
TD3491AP	BLOCK-1424
TD3492A	BLOCK-1425
TD3492AP	BLOCK-1425
TD3492P	BLOCK-1425
TD3493A	BLOCK-1426
TD3493AP	BLOCK-1426
TD3493BP	BLOCK-1426
TD3493P	BLOCK-1426
TD3495A	BLOCK-1428
TD3495AP	BLOCK-1428
TD3503A	BLOCK-1341
TD3503AP	BLOCK-1341
TD62003P	BLOCK-936
TD62006P	BLOCK-986
TD62007P	BLOCK-985
TD62104P	BLOCK-988
TD6306P	BLOCK-727
TD6306PFA-1	BLOCK-727
TD6312P	BLOCK-663
TD6350P	BLOCK-663
TDA-1170	BLOCK-276
TDA-2541	BLOCK-428
TDA1013A	BLOCK-831
TDA1013AU	BLOCK-831
TDA1013B	BLOCK-831
TDA1060	BLOCK-658
TDA1083	BLOCK-611
TDA1083A	BLOCK-611
TDA1170	BLOCK-276
TDA1170S	BLOCK-276
TDA11907	BLOCK-219
TDA1190P	BLOCK-220
TDA1190Z	BLOCK-219
TDA1200	BLOCK-1813
TDA1330P	BLOCK-1407
TDA1412	BLOCK-2097
TDA1415	BLOCK-2119
TDA1458D	BLOCK-1801
TDA1521	BLOCK-829
TDA1521U	BLOCK-829
TDA1524	BLOCK-784
TDA1524A	BLOCK-784
TDA1524AN	BLOCK-784
TDA1524PN	BLOCK-784
TDA152AN	BLOCK-784
TDA1541	BLOCK-972
TDA1541AN	BLOCK-972
TDA1541N	BLOCK-972
TDA1670A	BLOCK-842
TDA1675	BLOCK-842
TDA1905	BLOCK-1180
TDA2002	BLOCK-221
TDA2003	BLOCK-275
TDA2003V	BLOCK-275
TDA2004	BLOCK-383
TDA2005M	BLOCK-383
TDA2006	BLOCK-365
TDA2030	BLOCK-367
TDA2540	BLOCK-401
TDA2541	BLOCK-401
TDA2541Q	BLOCK-401
TDA2544	BLOCK-532
TDA2546A	BLOCK-733
TDA2548	BLOCK-532
TDA2577	BLOCK-619
TDA2577A	BLOCK-619
TDA2577AS2	BLOCK-619
TDA2593	BLOCK-1181
TDA2593N	BLOCK-1181
TDA2611	BLOCK-553
TDA2611A	BLOCK-553
TDA2611AQ	BLOCK-553
TDA2653A	BLOCK-785
TDA2653AN	BLOCK-785
TDA2653AU	BLOCK-785
TDA2791	BLOCK-638

DEVICE TYPE	REPL CODE
TDA3047	BLOCK-742
TDA3048	BLOCK-743
TDA3048N	BLOCK-743
TDA3190	BLOCK-220
TDA3564	BLOCK-870
TDA3570	BLOCK-404
TDA3651A	BLOCK-554
TDA3651AQ	BLOCK-554
TDA3652	BLOCK-735
TDA3653B	BLOCK-554
TDA3654	BLOCK-735
TDA3654AQ	BLOCK-735
TDA3654Q	BLOCK-735
TDA3654U	BLOCK-735
TDA4505	BLOCK-1199
TDA4505A	BLOCK-1199
TDA4505B	BLOCK-1199
TDA4600-2	BLOCK-1182
TDA4601	BLOCK-1182
TDA7052	BLOCK-1235
TDA7052/N1	BLOCK-1235
TDA7052N1	BLOCK-1235
TDA7056	BLOCK-1236
TDA8172	BLOCK-769
TDA8174	BLOCK-838
TDA8175	BLOCK-1198
TDAV190Z	BLOCK-219
TDB0117	BLOCK-187
TDB0124DP	BLOCK-2171
TDB0193DP	BLOCK-2063
TDB0555B	BLOCK-2079
TDB0555DP	BLOCK-2079
TDB0556A	BLOCK-2145
TDB0723	BLOCK-1983
TDB0723A	BLOCK-1984
TDB2905SP	BLOCK-2091
TDB2912SP	BLOCK-2108
TDB2915SP	BLOCK-2128
TDB7805	BLOCK-1022
TDB7805T	BLOCK-2087
TDB7806T	BLOCK-2093
TDB7808T	BLOCK-2095
TDB7812	BLOCK-2014
TDB7812T	BLOCK-2097
TDB7815T	BLOCK-2119
TDB7824T	BLOCK-2137
TDD1605S	BLOCK-2087
TDD1606S	BLOCK-2093
TDD1608S	BLOCK-2095
TDD1610S	BLOCK-904
TDD1612S	BLOCK-2097
TDD1615S	BLOCK-2119
TDD1618S	BLOCK-2085
TDD1624S	BLOCK-2137
TEA1009	BLOCK-744
TIA7630	BLOCK-563
TIE	BLOCK-1275
TIJ	BLOCK-1275
TIN	BLOCK-1275
TISM74LS00N	BLOCK-1568
TL061ACP	BLOCK-1937
TL061CP	BLOCK-1937
TL062AC	BLOCK-1939
TL062ACP	BLOCK-1939
TL062C	BLOCK-1939
TL071ACP	BLOCK-1908
TL071CP	BLOCK-1908
TL072CP	BLOCK-1910
TL074CN	BLOCK-1912
TL081	BLOCK-1908
TL081ACP	BLOCK-1908
TL081CFP	BLOCK-1909
TL081CP	BLOCK-1908
TL082CD	BLOCK-1911
TL082CP	BLOCK-1910
TL084	BLOCK-1912
TL084CD	BLOCK-1913
TL084CDP	BLOCK-1912
TL084CN	BLOCK-1912
TL1555P	BLOCK-2079
TL317	BLOCK-879
TL317LP	BLOCK-879
TL431ACD	BLOCK-2228
TL431ACP	BLOCK-2227
TL431CD	BLOCK-2228
TL431CP	BLOCK-2227
TL4558P	BLOCK-1801
TL487C	BLOCK-492
TL487CP	BLOCK-492
TL489C	BLOCK-493
TL489CP	BLOCK-493
TL490C	BLOCK-494
TL490CN	BLOCK-494
TL491C	BLOCK-495

DEVICE TYPE	REPL CODE
TL491CN	BLOCK-495
TL494	BLOCK-710
TL494C	BLOCK-770
TL494CJ	BLOCK-710
TL494CN	BLOCK-770
TL7400N	BLOCK-1293
TL7401N	BLOCK-1294
TL7402N	BLOCK-1295
TL7403N	BLOCK-1296
TL7404N	BLOCK-1297
TL7405N	BLOCK-1298
TL7406N	BLOCK-1299
TL7407N	BLOCK-1300
TL7409N	BLOCK-1302
TL74107N	BLOCK-1305
TL7410N	BLOCK-1304
TL74121N	BLOCK-1311
TL74122N	BLOCK-1312
TL74123N	BLOCK-1313
TL7412N	BLOCK-1310
TL7413N	BLOCK-1317
TL74141N	BLOCK-1321
TL74145N	BLOCK-1325
TL74150N	BLOCK-1327
TL74151N	BLOCK-1328
TL74153N	BLOCK-1330
TL74154N	BLOCK-1331
TL74155N	BLOCK-1332
TL74156N	BLOCK-1333
TL74162N	BLOCK-1339
TL74163N	BLOCK-1340
TL74164N	BLOCK-1341
TL74165N	BLOCK-1342
TL7416N	BLOCK-1336
TL74180N	BLOCK-1352
TL74181N	BLOCK-1353
TL74182N	BLOCK-1354
TL74190N	BLOCK-1355
TL74191N	BLOCK-1356
TL74192N	BLOCK-1357
TL74193N	BLOCK-1358
TL74196N	BLOCK-1360
TL74197N	BLOCK-1361
TL74198N	BLOCK-1362
TL74199N	BLOCK-1363
TL7420N	BLOCK-1365
TL7423N	BLOCK-1369
TL7430N	BLOCK-1377
TL7437N	BLOCK-1384
TL7438N	BLOCK-1385
TL7440N	BLOCK-1390
TL7442N	BLOCK-1392
TL7443N	BLOCK-1393
TL7444N	BLOCK-1394
TL7445N	BLOCK-1395
TL7446AN	BLOCK-1396
TL7447AN	BLOCK-1397
TL7448N	BLOCK-1398
TL7450N	BLOCK-1401
TL7451N	BLOCK-1402
TL7453N	BLOCK-1403
TL7454N	BLOCK-1404
TL7460N	BLOCK-1406
TL7470N	BLOCK-1408
TL7472N	BLOCK-1409
TL7473N	BLOCK-1410
TL7474N	BLOCK-1411
TL7475N	BLOCK-1412
TL7476N	BLOCK-1413
TL7480N	BLOCK-1414
TL7481N	BLOCK-1415
TL7482N	BLOCK-1416
TL7483N	BLOCK-1417
TL7485N	BLOCK-1418
TL7486N	BLOCK-1419
TL7489N	BLOCK-1420
TL7490N	BLOCK-1423
TL7491N	BLOCK-1424
TL7492N	BLOCK-1425
TL7493N	BLOCK-1426
TL7494N	BLOCK-1427
TL7495AN	BLOCK-1428
TL7496N	BLOCK-1429
TL7497N	BLOCK-1430
TL74H183N	BLOCK-1483
TL74H87N	BLOCK-1505
TLA7601G	BLOCK-659
TLA78601G	BLOCK-659
TLC555	BLOCK-2080
TLC555CP	BLOCK-2080
TLC556	BLOCK-2146
TLC556CP	BLOCK-2146
TLP550	BLOCK-1021
TLP551	BLOCK-1021

62 ORIGINAL DEVICE TYPES AND REPLACEMENT CODES

DEVICE TYPE	REPL CODE
TLP551FA-1	BLOCK-1021
TLP551FA-2	BLOCK-1021
TLP651	BLOCK-1021
TM1000	BLOCK-1
TM1001	BLOCK-2
TM1002	BLOCK-3
TM1003	BLOCK-4
TM1004	BLOCK-5
TM1005	BLOCK-6
TM1006	BLOCK-7
TM1008	BLOCK-8
TM1009	BLOCK-9
TM1010	BLOCK-10
TM1011	BLOCK-11
TM1012	BLOCK-12
TM1013	BLOCK-13
TM1014	BLOCK-14
TM1015	BLOCK-15
TM1016	BLOCK-16
TM1017	BLOCK-17
TM1018	BLOCK-18
TM1019	BLOCK-19
TM1021	BLOCK-20
TM1023	BLOCK-22
TM1024	BLOCK-23
TM1025	BLOCK-24
TM1026	BLOCK-25
TM1027	BLOCK-26
TM1028	BLOCK-27
TM1029	BLOCK-28
TM1030	BLOCK-29
TM1031	BLOCK-30
TM1032	BLOCK-31
TM1033	BLOCK-32
TM1034	BLOCK-33
TM1035	BLOCK-34
TM1036	BLOCK-35
TM1037	BLOCK-36
TM1039	BLOCK-38
TM1040	BLOCK-39
TM1041	BLOCK-40
TM1042	BLOCK-41
TM1043	BLOCK-42
TM1045	BLOCK-43
TM1046	BLOCK-44
TM1047	BLOCK-45
TM1048	BLOCK-46
TM1049	BLOCK-47
TM1050	BLOCK-48
TM1051	BLOCK-49
TM1052	BLOCK-50
TM1053	BLOCK-51
TM1054	BLOCK-52
TM1055	BLOCK-53
TM1056	BLOCK-54
TM1057	BLOCK-55
TM1058	BLOCK-56
TM1059	BLOCK-57
TM1060	BLOCK-58
TM1061	BLOCK-59
TM1062	BLOCK-60
TM1063	BLOCK-61
TM1064	BLOCK-62
TM1065	BLOCK-63
TM1066	BLOCK-64
TM1067	BLOCK-65
TM1068	BLOCK-66
TM1069	BLOCK-67
TM1070	BLOCK-68
TM1071	BLOCK-69
TM1072	BLOCK-70
TM1073	BLOCK-71
TM1074	BLOCK-72
TM1075A	BLOCK-73
TM1076	BLOCK-74
TM1078	BLOCK-76
TM1079	BLOCK-77
TM1080	BLOCK-78
TM1081A	BLOCK-79
TM1082	BLOCK-80
TM1083	BLOCK-81
TM1084	BLOCK-82
TM1085	BLOCK-83
TM1086	BLOCK-84
TM1087	BLOCK-85
TM1088	BLOCK-86
TM1089	BLOCK-87
TM1090	BLOCK-88
TM1091	BLOCK-89
TM1092	BLOCK-90
TM1093	BLOCK-91
TM1094	BLOCK-92
TM1095	BLOCK-93
TM1096	BLOCK-94

DEVICE TYPE	REPL CODE
TM1097	BLOCK-95
TM1098	BLOCK-96
TM1099	BLOCK-97
TM1100	BLOCK-98
TM1101	BLOCK-99
TM1102	BLOCK-100
TM1103	BLOCK-101
TM1104	BLOCK-102
TM1105	BLOCK-103
TM1106	BLOCK-104
TM1107	BLOCK-105
TM1108	BLOCK-106
TM1109	BLOCK-107
TM1110	BLOCK-108
TM1111	BLOCK-109
TM1112	BLOCK-110
TM1113	BLOCK-111
TM1115	BLOCK-112
TM1115A	BLOCK-113
TM1116	BLOCK-114
TM1117	BLOCK-115
TM1118	BLOCK-116
TM1119	BLOCK-117
TM1120	BLOCK-118
TM1121	BLOCK-119
TM1122	BLOCK-120
TM1123	BLOCK-121
TM1124	BLOCK-122
TM1125	BLOCK-123
TM1126	BLOCK-124
TM1127	BLOCK-125
TM1128	BLOCK-126
TM1130	BLOCK-127
TM1131	BLOCK-128
TM1132	BLOCK-129
TM1133	BLOCK-130
TM1134	BLOCK-131
TM1135	BLOCK-132
TM1136	BLOCK-133
TM1137	BLOCK-134
TM1139	BLOCK-135
TM1140	BLOCK-136
TM1141	BLOCK-137
TM1142	BLOCK-138
TM1149	BLOCK-140
TM1150	BLOCK-141
TM1152	BLOCK-142
TM1153	BLOCK-143
TM1154	BLOCK-144
TM1155	BLOCK-145
TM1156	BLOCK-146
TM1158	BLOCK-147
TM1159	BLOCK-148
TM1160	BLOCK-149
TM1161	BLOCK-150
TM1162	BLOCK-151
TM1163	BLOCK-152
TM1164	BLOCK-153
TM1165	BLOCK-154
TM1166	BLOCK-155
TM1167	BLOCK-156
TM1168	BLOCK-157
TM1169	BLOCK-158
TM1170	BLOCK-159
TM1171	BLOCK-160
TM1172	BLOCK-161
TM1173	BLOCK-162
TM1174	BLOCK-163
TM1175	BLOCK-164
TM1176	BLOCK-165
TM1177	BLOCK-166
TM1178	BLOCK-167
TM1179	BLOCK-168
TM1180	BLOCK-169
TM1181	BLOCK-170
TM1182	BLOCK-171
TM1183	BLOCK-172
TM1184	BLOCK-173
TM1185	BLOCK-174
TM1186	BLOCK-175
TM1187	BLOCK-176
TM1188	BLOCK-177
TM1189	BLOCK-178
TM1191	BLOCK-179
TM1192	BLOCK-180
TM1193	BLOCK-181
TM1194	BLOCK-182
TM1195	BLOCK-183
TM1196	BLOCK-184
TM1197	BLOCK-185
TM1200	BLOCK-188
TM1201	BLOCK-189
TM1202	BLOCK-190
TM1203	BLOCK-191

DEVICE TYPE	REPL CODE
TM1204	BLOCK-192
TM1205	BLOCK-193
TM1206	BLOCK-194
TM1207	BLOCK-195
TM1208	BLOCK-196
TM1209	BLOCK-197
TM1210	BLOCK-198
TM1211	BLOCK-199
TM1212	BLOCK-200
TM1213	BLOCK-201
TM1214	BLOCK-202
TM1215	BLOCK-203
TM1216	BLOCK-204
TM1217	BLOCK-205
TM1218	BLOCK-206
TM1219	BLOCK-207
TM1221	BLOCK-209
TM1222	BLOCK-210
TM1223	BLOCK-211
TM1224	BLOCK-212
TM1225	BLOCK-213
TM1226	BLOCK-214
TM1227	BLOCK-215
TM1228	BLOCK-216
TM1229	BLOCK-217
TM1230	BLOCK-218
TM1231	BLOCK-219
TM1232	BLOCK-221
TM1233	BLOCK-222
TM1234	BLOCK-223
TM1235	BLOCK-224
TM1236	BLOCK-225
TM1237	BLOCK-226
TM1238	BLOCK-227
TM1239	BLOCK-228
TM1240	BLOCK-229
TM1241	BLOCK-230
TM1242	BLOCK-231
TM1243	BLOCK-232
TM1244	BLOCK-233
TM1245	BLOCK-234
TM1246	BLOCK-235
TM1247	BLOCK-236
TM1248	BLOCK-237
TM1249	BLOCK-238
TM1250	BLOCK-239
TM1251	BLOCK-240
TM1252	BLOCK-241
TM1253	BLOCK-242
TM1254	BLOCK-243
TM1255	BLOCK-244
TM1256	BLOCK-245
TM1257	BLOCK-246
TM1258	BLOCK-247
TM1261	BLOCK-248
TM1262	BLOCK-249
TM1263	BLOCK-250
TM1264	BLOCK-251
TM1265	BLOCK-252
TM1266	BLOCK-253
TM1267	BLOCK-254
TM1268	BLOCK-255
TM1269	BLOCK-256
TM1270	BLOCK-257
TM1271	BLOCK-258
TM1272	BLOCK-259
TM1273	BLOCK-260
TM1274	BLOCK-261
TM1300	BLOCK-287
TM1301	BLOCK-288
TM1302	BLOCK-289
TM1303	BLOCK-290
TM1304	BLOCK-291
TM1305	BLOCK-292
TM1306	BLOCK-293
TM1307	BLOCK-294
TM1308	BLOCK-295
TM1309	BLOCK-296
TM1310	BLOCK-297
TM1311	BLOCK-298
TM1312	BLOCK-299
TM1313	BLOCK-300
TM2004	BLOCK-933
TM2011	BLOCK-934
TM2012	BLOCK-935
TM2013	BLOCK-936
TM2014	BLOCK-937
TM2015	BLOCK-938
TM2016	BLOCK-939
TM2017	BLOCK-940
TM2018	BLOCK-941
TM2019	BLOCK-942
TM2020	BLOCK-943
TM2021	BLOCK-944

DEVICE TYPE	REPL CODE
TM2022	BLOCK-945
TM2023	BLOCK-946
TM2024	BLOCK-947
TM2040	BLOCK-957
TM309K	BLOCK-1022
TM370	BLOCK-1024
TM4000	BLOCK-1028
TM4001B	BLOCK-1029
TM4002B	BLOCK-1030
TM4006B	BLOCK-1031
TM4007B	BLOCK-1032
TM4008B	BLOCK-1034
TM4011B	BLOCK-1039
TM4012B	BLOCK-1040
TM4013B	BLOCK-1041
TM4014B	BLOCK-1042
TM4015B	BLOCK-1043
TM4016B	BLOCK-1048
TM4017B	BLOCK-1051
TM4018B	BLOCK-1053
TM4019B	BLOCK-1058
TM4020B	BLOCK-1059
TM4021B	BLOCK-1060
TM4022B	BLOCK-1061
TM4023B	BLOCK-1062
TM4024B	BLOCK-1063
TM4025B	BLOCK-1064
TM4027B	BLOCK-1066
TM4028B	BLOCK-1067
TM4029B	BLOCK-1068
TM4030B	BLOCK-1069
TM4034B	BLOCK-1073
TM4035B	BLOCK-1074
TM4040B	BLOCK-1076
TM4042B	BLOCK-1077
TM4043B	BLOCK-1078
TM4044B	BLOCK-1079
TM4049B	BLOCK-1084
TM4050B	BLOCK-1085
TM4051B	BLOCK-1086
TM4052B	BLOCK-1087
TM4053B	BLOCK-1088
TM4055B	BLOCK-1089
TM4060B	BLOCK-1091
TM4063B	BLOCK-1092
TM4066B	BLOCK-1093
TM4068B	BLOCK-1095
TM4069	BLOCK-1096
TM4071B	BLOCK-1098
TM4072B	BLOCK-1099
TM4073B	BLOCK-1100
TM4075B	BLOCK-1101
TM4076B	BLOCK-1102
TM4077B	BLOCK-1103
TM4078B	BLOCK-1104
TM4081B	BLOCK-1105
TM4082B	BLOCK-1106
TM4085B	BLOCK-1107
TM4086B	BLOCK-1108
TM4093B	BLOCK-1110
TM4098B	BLOCK-1114
TM4099B	BLOCK-1115
TM4510B	BLOCK-1121
TM4511B	BLOCK-1122
TM4512B	BLOCK-1123
TM4514B	BLOCK-1125
TM4515B	BLOCK-1126
TM4516B	BLOCK-1127
TM4518B	BLOCK-1129
TM4520B	BLOCK-1130
TM4522B	BLOCK-1132
TM4527B	BLOCK-1134
TM4539B	BLOCK-1141
TM4555B	BLOCK-1146
TM4556B	BLOCK-1147
TM703A	BLOCK-1212
TM704	BLOCK-1223
TM705A	BLOCK-1244
TM706	BLOCK-1245
TM707	BLOCK-1256
TM708	BLOCK-1262
TM709	BLOCK-1264
TM710	BLOCK-1265
TM711	BLOCK-1267
TM712	BLOCK-1269
TM713	BLOCK-1270
TM714	BLOCK-1271
TM716	BLOCK-1273
TM717	BLOCK-1274
TM718	BLOCK-1275
TM719	BLOCK-1276
TM720	BLOCK-1277
TM723	BLOCK-1279
TM724	BLOCK-1280

DEVICE TYPE	REPL CODE
TM725	BLOCK-1281
TM726	BLOCK-1282
TM727	BLOCK-1283
TM729	BLOCK-1285
TM730	BLOCK-1286
TM731	BLOCK-1287
TM735	BLOCK-1288
TM736	BLOCK-1289
TM737	BLOCK-1290
TM738	BLOCK-1291
TM739	BLOCK-1292
TM740A	BLOCK-1303
TM741	BLOCK-1917
TM742	BLOCK-1364
TM743	BLOCK-1376
TM744	BLOCK-1389
TM745	BLOCK-1400
TM747	BLOCK-1407
TM748	BLOCK-1421
TM749	BLOCK-1422
TM7492	BLOCK-1425
TM7493A	BLOCK-1426
TM750	BLOCK-1770
TM752	BLOCK-1773
TM753	BLOCK-1774
TM756	BLOCK-1787
TM757	BLOCK-1787
TM758	BLOCK-1788
TM759	BLOCK-1789
TM760	BLOCK-1790
TM761	BLOCK-1791
TM763	BLOCK-1792
TM764	BLOCK-1793
TM765	BLOCK-1793
TM767	BLOCK-1794
TM768	BLOCK-1795
TM770	BLOCK-1796
TM772A	BLOCK-1797
TM773	BLOCK-1798
TM776	BLOCK-1800
TM778	BLOCK-1801
TM778A	BLOCK-1801
TM779-1	BLOCK-1804
TM780	BLOCK-1805
TM781	BLOCK-1806
TM782	BLOCK-1807
TM783	BLOCK-1808
TM784	BLOCK-1809
TM785	BLOCK-1810
TM786	BLOCK-1811
TM787	BLOCK-1812
TM788	BLOCK-1813
TM789	BLOCK-1814
TM790	BLOCK-1815
TM791	BLOCK-1816
TM793	BLOCK-1818
TM795	BLOCK-1820
TM796	BLOCK-1797
TM797	BLOCK-1821
TM799	BLOCK-1823
TM801	BLOCK-1825
TM802	BLOCK-1826
TM803	BLOCK-1827
TM804	BLOCK-1828
TM806	BLOCK-1829
TM807	BLOCK-1830
TM808	BLOCK-1831
TM809	BLOCK-1833
TM810	BLOCK-1838
TM812	BLOCK-1839
TM813	BLOCK-1841
TM814	BLOCK-1842
TM815	BLOCK-1843
TM816	BLOCK-1844
TM817	BLOCK-1845
TM818	BLOCK-1846
TM819	BLOCK-1847
TM820	BLOCK-1848
TM821	BLOCK-1849
TM822	BLOCK-1850
TM823	BLOCK-1851
TM824	BLOCK-1852
TM825	BLOCK-1853
TM826	BLOCK-1856
TM830	BLOCK-1860
TM832	BLOCK-1868
TM833	BLOCK-1872
TM834	BLOCK-1873
TM900	BLOCK-1940
TM901	BLOCK-1941
TM903	BLOCK-1943
TM904	BLOCK-1944
TM905	BLOCK-1945
TM906	BLOCK-1946

DEVICE TYPE	REPL CODE
TM907	BLOCK-1947
TM908	BLOCK-1948
TM909	BLOCK-1949
TM909D	BLOCK-1954
TM910	BLOCK-1955
TM910D	BLOCK-1957
TM911	BLOCK-1958
TM912	BLOCK-1963
TM913	BLOCK-1964
TM914	BLOCK-1966
TM915	BLOCK-1967
TM916	BLOCK-1970
TM917	BLOCK-1971
TM923	BLOCK-1983
TM923D	BLOCK-1984
TM925	BLOCK-1987
TM940	BLOCK-2051
TM941	BLOCK-2056
TM941D	BLOCK-2057
TM941M	BLOCK-2058
TM946	BLOCK-2068
TM947	BLOCK-2069
TM947D	BLOCK-2070
TM949	BLOCK-2073
TM955M	BLOCK-2079
TM960	BLOCK-2087
TM961	BLOCK-2091
TM962	BLOCK-2093
TM963	BLOCK-2094
TM964	BLOCK-2095
TM966	BLOCK-2097
TM967	BLOCK-2108
TM968	BLOCK-2119
TM969	BLOCK-2128
TM971	BLOCK-2136
TM972	BLOCK-2137
TM973	BLOCK-2138
TM973D	BLOCK-2139
TM974	BLOCK-2140
TM975	BLOCK-2141
TM976	BLOCK-2143
TM977	BLOCK-2144
TM978	BLOCK-2145
TM980	BLOCK-2149
TM981	BLOCK-2160
TM982	BLOCK-2166
TM983	BLOCK-2167
TM984	BLOCK-2168
TM985	BLOCK-2169
TM986	BLOCK-2170
TM987	BLOCK-2171
TM988	BLOCK-2173
TM989	BLOCK-2174
TM990	BLOCK-2175
TM991	BLOCK-2184
TM992	BLOCK-2191
TMM2016	BLOCK-1004
TMM2016AP	BLOCK-1004
TMM2016AP-12	BLOCK-1004
TMM2016BP-15	BLOCK-1004
TMM2016P	BLOCK-1004
TMM2016P-1	BLOCK-1004
TMM2016P-2	BLOCK-1004
TMM2016P-7	BLOCK-1004
TMM2764D-2	BLOCK-1019
TMM314AP	BLOCK-1002
TMM314APL-3	BLOCK-1002
TMM315D	BLOCK-1005
TMM315D-1	BLOCK-1005
TMM41256CP-12	BLOCK-1116
TMM4164P-3	BLOCK-1006
TMM416D-3	BLOCK-1003
TMM416D-4	BLOCK-1003
TMM416P-3	BLOCK-1003
TMS-1952	BLOCK-974
TMS1943	BLOCK-973
TMS1943A2	BLOCK-973
TMS1943N2	BLOCK-973
TMS1943N21	BLOCK-973
TMS1943N2L	BLOCK-973
TMS1943NL	BLOCK-973
TMS1952NL	BLOCK-974
TMS2114-20	BLOCK-1002
TMS2114-25	BLOCK-1002
TMS2114-45	BLOCK-1002
TMS2114L-20	BLOCK-1002
TMS2114L-25	BLOCK-1002
TMS2114L-45	BLOCK-1002
TMS2516JL-45	BLOCK-1019
TMS2532-30	BLOCK-1015
TMS2532-35	BLOCK-1015
TMS2532-45	BLOCK-1015
TMS2532JL-45	BLOCK-1015
TMS2564-35	BLOCK-1019

ORIGINAL DEVICE TYPES AND REPLACEMENT CODES 63

DEVICE TYPE	REPL CODE
TMS2564-45	BLOCK-1019
TMS2732AJL-30	BLOCK-1018
TMS2732AJL-35	BLOCK-1018
TMS2764JL-25	BLOCK-1019
TMS3450NL	BLOCK-975
TMS4016-15	BLOCK-1004
TMS4016-20	BLOCK-1004
TMS4016-25	BLOCK-1004
TMS4116-20	BLOCK-1003
TMS4116-25	BLOCK-1003
TMS4164-15	BLOCK-1006
TMS4164-15NLJ	BLOCK-1006
TMS4164-20	BLOCK-1006
TMS4164-20NL	BLOCK-1006
TMS4164-25	BLOCK-1006
TMS4164NLJ	BLOCK-1006
TMS4256-15NL	BLOCK-1116
TMS8080AN	BLOCK-1832
TP4000	BLOCK-1028
TP4001	BLOCK-1029
TP4001AN	BLOCK-1029
TP4001BN	BLOCK-1029
TP4001BP	BLOCK-1029
TP4002	BLOCK-1030
TP4002AN	BLOCK-1030
TP4007	BLOCK-1032
TP4007UBN	BLOCK-1032
TP4008AN	BLOCK-1034
TP4008BN	BLOCK-1034
TP4009	BLOCK-1084
TP4009UBN	BLOCK-1084
TP4010BN	BLOCK-1085
TP4011	BLOCK-1039
TP4011AN	BLOCK-1039
TP4011B	BLOCK-1039
TP4011BN	BLOCK-1039
TP4011BP	BLOCK-1039
TP4012	BLOCK-1040
TP4012AN	BLOCK-1040
TP4013	BLOCK-1041
TP4013AN	BLOCK-1041
TP4013BE	BLOCK-1041
TP4013BN	BLOCK-1041
TP4014AN	BLOCK-1042
TP4015	BLOCK-1043
TP4015AN	BLOCK-1043
TP4016A	BLOCK-1093
TP4016AN	BLOCK-1048
TP4016BN	BLOCK-1048
TP4016UBN	BLOCK-1048
TP4017	BLOCK-1051
TP4017AN	BLOCK-1051
TP4018AN	BLOCK-1053
TP4018BN	BLOCK-1053
TP4019	BLOCK-1058
TP4019AN	BLOCK-1058
TP4019BN	BLOCK-1058
TP4020	BLOCK-1059
TP4020AN	BLOCK-1059
TP4020BN	BLOCK-1059
TP4021	BLOCK-1060
TP4021AN	BLOCK-1060
TP4022AN	BLOCK-1061
TP4023	BLOCK-1062
TP4023AN	BLOCK-1062
TP4024	BLOCK-1063
TP4024AN	BLOCK-1063
TP4024BN	BLOCK-1063
TP4025	BLOCK-1064
TP4025AN	BLOCK-1064
TP4027	BLOCK-1066
TP4027AN	BLOCK-1066
TP4027BN	BLOCK-1066
TP4028AN	BLOCK-1067
TP4028BN	BLOCK-1067
TP4029AN	BLOCK-1068
TP4030AN	BLOCK-1069
TP4040	BLOCK-1076
TP4040AN	BLOCK-1076
TP4040BN	BLOCK-1076
TP4042	BLOCK-1077
TP4042AN	BLOCK-1077
TP4042BN	BLOCK-1077
TP4043AN	BLOCK-1078
TP4043BN	BLOCK-1078
TP4044AN	BLOCK-1079
TP4044BN	BLOCK-1079
TP4049	BLOCK-1084
TP4049AN	BLOCK-1084
TP4049UBN	BLOCK-1084
TP4050	BLOCK-1085
TP4050AN	BLOCK-1085
TP4051	BLOCK-1086
TP4051BN	BLOCK-1086
TP4052	BLOCK-1087
TP4052BN	BLOCK-1087
TP4053BN	BLOCK-1088
TP4068BN	BLOCK-1095
TP4069UBN	BLOCK-1096
TP4069UPN	BLOCK-1096
TP4071BN	BLOCK-1098
TP4072BN	BLOCK-1099
TP4073BN	BLOCK-1100
TP4075BN	BLOCK-1101
TP4078BN	BLOCK-1104
TP4081	BLOCK-1105
TP4081BN	BLOCK-1105
TP4082BN	BLOCK-1106
TP4512	BLOCK-1123
TP4512AN	BLOCK-1123
TP4518	BLOCK-1129
TP4518AN	BLOCK-1129
TP4520AN	BLOCK-1130
TP4522AN	BLOCK-1132
TQ7037	BLOCK-5
TQ7038	BLOCK-5
TR0-9022	BLOCK-88
TR09004	BLOCK-1275
TR09005	BLOCK-1212
TR09006	BLOCK-1212
TR09007	BLOCK-1770
TR09008	BLOCK-1790
TR09009	BLOCK-1790
TR09010	BLOCK-1405
TR09011	BLOCK-1277
TR09018	BLOCK-1813
TR09019	BLOCK-1825
TR09022	BLOCK-88
TR09023	BLOCK-88
TR09024	BLOCK-88
TR0911	BLOCK-1277
TR22873700104	BLOCK-1277
TR228779905696	BLOCK-4
TR2327312	BLOCK-38
TR2327411	BLOCK-40
TR2327422	BLOCK-1277
TR2367151	BLOCK-468
TR2367161	BLOCK-122
TR2367171	BLOCK-1825
TR4083-23273	BLOCK-38
TR4083-2327311	BLOCK-38
TR4083-2327312	BLOCK-38
TRHA1151	BLOCK-226
TRLA1230	BLOCK-1813
TRLA3350	BLOCK-205
TRO-9004	BLOCK-1275
TRO-9005	BLOCK-1212
TRO-9006	BLOCK-1212
TRO-9011	BLOCK-1277
TRO9004	BLOCK-1275
TRO9005	BLOCK-1212
TRO9006	BLOCK-1212
TRO9024	BLOCK-88
TS4078BP	BLOCK-1104
TS555CD	BLOCK-2082
TS555CN	BLOCK-2080
TS556CN	BLOCK-2146
TSC7106CPL	BLOCK-966
TSC7107CPL	BLOCK-965
TSC7116CPL	BLOCK-976
TSC7126CPL	BLOCK-964
TV241072	BLOCK-5
TVC-MK3800	BLOCK-133
TVCM-1	BLOCK-1244
TVCM-11	BLOCK-1269
TVCM-12	BLOCK-1276
TVCM-14	BLOCK-1289
TVCM-15	BLOCK-1287
TVCM-16	BLOCK-1279
TVCM-17	BLOCK-1288
TVCM-18	BLOCK-1290
TVCM-19	BLOCK-1389
TVCM-2	BLOCK-1270
TVCM-20	BLOCK-1289
TVCM-21	BLOCK-1292
TVCM-22	BLOCK-1815
TVCM-23	BLOCK-1828
TVCM-25	BLOCK-1828
TVCM-26	BLOCK-1828
TVCM-27	BLOCK-1291
TVCM-28	BLOCK-1364
TVCM-29	BLOCK-1376
TVCM-3	BLOCK-1256
TVCM-30	BLOCK-1808
TVCM-31	BLOCK-1805
TVCM-32	BLOCK-1284
TVCM-33	BLOCK-1285
TVCM-34	BLOCK-1816
TVCM-35	BLOCK-1303
TVCM-39	BLOCK-1813
TVCM-4	BLOCK-1262
TVCM-41	BLOCK-1421
TVCM-42	BLOCK-1422
TVCM-43	BLOCK-1281
TVCM-44	BLOCK-1212
TVCM-45	BLOCK-1282
TVCM-46	BLOCK-1223
TVCM-47	BLOCK-1809
TVCM-48	BLOCK-1280
TVCM-49	BLOCK-1810
TVCM-5	BLOCK-1264
TVCM-50	BLOCK-1245
TVCM-500	BLOCK-1293
TVCM-501	BLOCK-1486
TVCM-502	BLOCK-1411
TVCM-503	BLOCK-1397
TVCM-504	BLOCK-1426
TVCM-505	BLOCK-1423
TVCM-506	BLOCK-1338
TVCM-51	BLOCK-1265
TVCM-52	BLOCK-1811
TVCM-53	BLOCK-1267
TVCM-54	BLOCK-1283
TVCM-55	BLOCK-1286
TVCM-56	BLOCK-1806
TVCM-57	BLOCK-1812
TVCM-58	BLOCK-1814
TVCM-59	BLOCK-1405
TVCM-6	BLOCK-1275
TVCM-60	BLOCK-138
TVCM-61	BLOCK-1801
TVCM-62	BLOCK-1804
TVCM-65	BLOCK-1828
TVCM-66	BLOCK-1963
TVCM-67	BLOCK-1971
TVCM-7	BLOCK-1277
TVCM-73	BLOCK-1825
TVCM-74	BLOCK-56
TVCM-75	BLOCK-4
TVCM-77	BLOCK-102
TVCM-78	BLOCK-98
TVCM-79	BLOCK-101
TVCM-8	BLOCK-1271
TVCM-81	BLOCK-145
TVCM-82	BLOCK-50
TVCM-9	BLOCK-1272
TVCM21	BLOCK-60
TVCM60	BLOCK-1407
TVCM69	BLOCK-2138
TVCMK3800	BLOCK-133
TVS-MPC23C	BLOCK-44
TVS-UPC23C	BLOCK-44
TVS3153GM1	BLOCK-2168
TVSAN225	BLOCK-60
TVSAN227	BLOCK-61
TVSAN230	BLOCK-63
TVSAN241	BLOCK-151
TVSBA526	BLOCK-599
TVSBN5111	BLOCK-428
TVSBN5416	BLOCK-669
TVSBN5416A	BLOCK-669
TVSCA3126	BLOCK-1821
TVSCA3137	BLOCK-165
TVSCA3139GM1	BLOCK-163
TVSCA3139GMI	BLOCK-163
TVSCA3153GM1	BLOCK-2168
TVSCA3153GMI	BLOCK-2168
TVSCN5310	BLOCK-398
TVSCN5310L	BLOCK-398
TVSCN5310M	BLOCK-398
TVSCN5311	BLOCK-398
TVSCN5311CL	BLOCK-398
TVSCN5311KL	BLOCK-398
TVSCN5411	BLOCK-286
TVSEN11235	BLOCK-537
TVSEN11238	BLOCK-655
TVSEN11301	BLOCK-219
TVSEN11401	BLOCK-388
TVSEN11414	BLOCK-390
TVSEN11436	BLOCK-636
TVSEN11438	BLOCK-2170
TVSEN1364	BLOCK-219
TVSG4011BCM	BLOCK-1039
TVSHA1151	BLOCK-226
TVSHZT33	BLOCK-1157
TVSHZT33S1	BLOCK-1157
TVSJN76664V	BLOCK-1269
TVSL78M12RL	BLOCK-2097
TVSL78N05	BLOCK-2144
TVSLA7530N	BLOCK-807
TVSM51247	BLOCK-162
TVSM51247P	BLOCK-162
TVSM51376SP	BLOCK-1195
TVSM5218L	BLOCK-1802
TVSM53216P	BLOCK-1336
TVSM58725P	BLOCK-1004
TVSMP0574J	BLOCK-1157
TVSMPC1355C	BLOCK-1422
TVSMPC574J	BLOCK-1157
TVSMPC595C	BLOCK-175
TVSMPC596C	BLOCK-176
TVSMPC596C2	BLOCK-176
TVSMPC596C2B	BLOCK-176
TVSMPC596CE	BLOCK-176
TVSMPC596CZ	BLOCK-176
TVSMPS596C2	BLOCK-176
TVSN76665	BLOCK-1269
TVSS4LS221N	BLOCK-1629
TVSS4LS374N	BLOCK-1672
TVSSN7403N	BLOCK-1296
TVSSN7409N	BLOCK-1302
TVSSN74LS09N	BLOCK-1575
TVSSN76642N	BLOCK-1264
TVSSN76664N	BLOCK-1269
TVSSN76665	BLOCK-1269
TVSSN7666N	BLOCK-1269
TVSSTR30130	BLOCK-758
TVSSTR3130	BLOCK-723
TVSSTR3230	BLOCK-723
TVSSTR380	BLOCK-535
TVSSTR381	BLOCK-533
TVSSTR381A	BLOCK-533
TVSTC4011BP	BLOCK-1039
TVSTC4052BP	BLOCK-1087
TVSTC4053BP	BLOCK-1088
TVSTC4066BP	BLOCK-1093
TVSTC4069UBP	BLOCK-1096
TVSTDA1190Z	BLOCK-219
TVSTMS1943A2	BLOCK-973
TVSTMS1943N2	BLOCK-973
TVSUC78M12H2	BLOCK-2097
TVSUP574J	BLOCK-1157
TVSUPC1363CA	BLOCK-396
TVSUPC1380C	BLOCK-184
TVSUPC23	BLOCK-44
TVSUPC23C	BLOCK-44
TVSUPC339C	BLOCK-1873
TVSUPC574J	BLOCK-1157
TVSUPC595C	BLOCK-175
TVSUPC596C2	BLOCK-176
TVSUPC78M-12H2	BLOCK-2097
TVSUPC78M12H2	BLOCK-2097
TVSUPD4011BC	BLOCK-1039
TVSUPD4013BC	BLOCK-1041
TVSUPD4017BC	BLOCK-1051
TVSUPD4027BC	BLOCK-1066
TVSUPD4052BC	BLOCK-1087
TVSUPD4053BC	BLOCK-1088
TVSUPD4066BC	BLOCK-1093
U417B	BLOCK-611
U574	BLOCK-1157
U574(IC)	BLOCK-1157
U574K	BLOCK-1157
U5D7703312	BLOCK-1212
U5D770331X	BLOCK-1212
U5D7703393	BLOCK-1212
U5D7703394	BLOCK-1212
U5D770339X	BLOCK-1212
U5E7746394	BLOCK-1256
U6A7065354	BLOCK-1269
U6A7729394	BLOCK-1277
U6A7732394	BLOCK-1275
U6A7746394	BLOCK-1270
U6A7781394	BLOCK-1272
U6B7780394	BLOCK-1271
U6E7729394	BLOCK-1277
U7F7065354	BLOCK-1269
U7F7729394	BLOCK-1277
U7F7732394	BLOCK-1275
U7F7746394	BLOCK-1270
U7F7780394	BLOCK-1271
U7F7781394	BLOCK-1272
U8B770339	BLOCK-1212
U8B770339-825	BLOCK-1212
U8B7703394	BLOCK-1212
U8B770339X	BLOCK-1212
U8F7737394	BLOCK-1244
U8F7746394	BLOCK-1244
U9A7746394	BLOCK-1815
U9A7781394	BLOCK-1272
U9B7780394	BLOCK-1271
U9C7065354	BLOCK-1269
UA-703E	BLOCK-1212
UA0802BPC	BLOCK-971
UA0802CPC	BLOCK-971
UA1312PC	BLOCK-1823
UA1314PC	BLOCK-1826
UA1315PC	BLOCK-1827
UA1391TC	BLOCK-1843
UA1458CP	BLOCK-1801
UA1458CTC	BLOCK-1801
UA1458P	BLOCK-1801
UA1458TC	BLOCK-1801
UA1488	BLOCK-1771
UA1488D	BLOCK-1771
UA1488DC	BLOCK-1771
UA1488P	BLOCK-1771
UA1488PC	BLOCK-1771
UA1489	BLOCK-1772
UA1489APC	BLOCK-1772
UA1489D	BLOCK-1772
UA1489DC	BLOCK-1772
UA1489P	BLOCK-1772
UA1489PC	BLOCK-1772
UA1489PCQR	BLOCK-1772
UA1748G	BLOCK-160
UA2002	BLOCK-221
UA201AH	BLOCK-160
UA201H	BLOCK-160
UA201HC	BLOCK-160
UA209KM	BLOCK-1022
UA2136P	BLOCK-1290
UA2136PC	BLOCK-1290
UA224DM	BLOCK-2171
UA2901PC	BLOCK-1873
UA2902	BLOCK-2171
UA3018AHM	BLOCK-1944
UA3018H	BLOCK-1944
UA3019HC	BLOCK-1945
UA301AH	BLOCK-160
UA301AHC	BLOCK-160
UA301AT	BLOCK-2141
UA301ATC	BLOCK-2141
UA3026HC	BLOCK-1946
UA3039HC	BLOCK-1947
UA3045DC	BLOCK-1963
UA3045PC	BLOCK-1963
UA3046DC	BLOCK-1963
UA3046PC	BLOCK-1963
UA3054DC	BLOCK-1971
UA3064HC	BLOCK-1805
UA3064PC	BLOCK-1808
UA3064TC	BLOCK-1805
UA3065	BLOCK-1269
UA3065P	BLOCK-1269
UA3065PC	BLOCK-1269
UA3066DC	BLOCK-1284
UA3066PC	BLOCK-1284
UA3067DC	BLOCK-1285
UA3075PC	BLOCK-1279
UA307T	BLOCK-2143
UA307TC	BLOCK-2143
UA3086DC	BLOCK-1963
UA3086DM	BLOCK-1963
UA3086PC	BLOCK-1963
UA3089E	BLOCK-1813
UA308AHC	BLOCK-2037
UA308HC	BLOCK-2037
UA308TC	BLOCK-2042
UA309K	BLOCK-1022
UA309KC	BLOCK-1022
UA311HC	BLOCK-1980
UA311TC	BLOCK-1981
UA317UC	BLOCK-2083
UA324DC	BLOCK-2171
UA324PC	BLOCK-2171
UA3301P	BLOCK-2191
UA3303PC	BLOCK-2171
UA339ADC	BLOCK-1873
UA3401PC	BLOCK-2191
UA3403DC	BLOCK-2171
UA3403PC	BLOCK-2171
UA348DC	BLOCK-2071
UA348PC	BLOCK-2071
UA4136	BLOCK-2219
UA4136PC	BLOCK-2219
UA4188DC	BLOCK-1771
UA4558T	BLOCK-1801
UA555	BLOCK-2079
UA555IC	BLOCK-2079
UA555TC	BLOCK-2079
UA556DC	BLOCK-2145
UA556DM	BLOCK-2145
UA556PC	BLOCK-2145
UA703	BLOCK-1212
UA703A	BLOCK-1212
UA703C	BLOCK-1212
UA703CT	BLOCK-1212
UA703E	BLOCK-1212
UA703HC	BLOCK-1212
UA703L	BLOCK-1212
UA705	BLOCK-1828
UA7094HC	BLOCK-1949
UA709A	BLOCK-1954
UA709C	BLOCK-1954
UA709CJ	BLOCK-1954
UA709CL	BLOCK-1949
UA709CN	BLOCK-1954
UA709CT	BLOCK-1949
UA709DC	BLOCK-1954
UA709HC	BLOCK-1949
UA709HM	BLOCK-1949
UA709N	BLOCK-1954
UA709PC	BLOCK-1954
UA709T	BLOCK-1949
UA710	BLOCK-1955
UA710A	BLOCK-1957
UA710CA	BLOCK-1957
UA710CJ	BLOCK-1957
UA710CL	BLOCK-1955
UA710CN	BLOCK-1957
UA710CT	BLOCK-1955
UA710DC	BLOCK-1957
UA710DM	BLOCK-1957
UA710HC	BLOCK-1955
UA710PC	BLOCK-1957
UA710T	BLOCK-1955
UA711A	BLOCK-1958
UA711CA	BLOCK-1962
UA711CJ	BLOCK-1962
UA711CK	BLOCK-1958
UA711CL	BLOCK-1958
UA711CN	BLOCK-1962
UA711DC	BLOCK-1962
UA711DM	BLOCK-1962
UA711HC	BLOCK-1958
UA711HM	BLOCK-1958
UA711K	BLOCK-1958
UA711PC	BLOCK-1962
UA715(METAL-CAN)	BLOCK-1967
UA720DC	BLOCK-1389
UA723	BLOCK-1983
UA723CA	BLOCK-1984
UA723CJ	BLOCK-1984
UA723CL	BLOCK-1983
UA723CN	BLOCK-1984
UA723CT	BLOCK-1983
UA723DC	BLOCK-1984
UA723DM	BLOCK-1984
UA723HC	BLOCK-1983
UA723HM	BLOCK-1984
UA723L	BLOCK-1983
UA723ML	BLOCK-1984
UA723PC	BLOCK-1984
UA725HC	BLOCK-1987
UA729	BLOCK-1277
UA729CA	BLOCK-1277
UA7300	BLOCK-931
UA732	BLOCK-1275
UA732PC	BLOCK-1275
UA733C	BLOCK-1989
UA733CF	BLOCK-1990
UA733CH	BLOCK-1989
UA733CJ	BLOCK-1990
UA733CN	BLOCK-1990
UA733DC	BLOCK-1990
UA733HC	BLOCK-1989
UA733PC	BLOCK-1990
UA737	BLOCK-1256
UA737E	BLOCK-1244
UA739DC	BLOCK-1281
UA739PC	BLOCK-1281
UA740CT	BLOCK-2051
UA740EHC	BLOCK-2051
UA741	BLOCK-2056
UA741A	BLOCK-2058
UA741C	BLOCK-2057
UA741CA	BLOCK-2057
UA741CD	BLOCK-2060
UA741CJ	BLOCK-2057
UA741CJG	BLOCK-2058
UA741CL	BLOCK-2056
UA741CN	BLOCK-2057
UA741CP	BLOCK-2058
UA741CT	BLOCK-2056
UA741CV	BLOCK-2058
UA741DC	BLOCK-2058
UA741HC	BLOCK-2056
UA741HM	BLOCK-2058
UA741ML	BLOCK-2056
UA741PC	BLOCK-2057

DEVICE TYPE	REPL CODE
UA741TC	BLOCK-2058
UA746	BLOCK-1256
UA746(DIP)	BLOCK-1270
UA746(METAL-CAN)	BLOCK-1256
UA746C	BLOCK-1270
UA746DC	BLOCK-1270
UA746E	BLOCK-1244
UA746HC	BLOCK-1256
UA746PC	BLOCK-1270
UA747C	BLOCK-2069
UA747CA	BLOCK-2070
UA747CF	BLOCK-2070
UA747CJ	BLOCK-2070
UA747CK	BLOCK-2069
UA747CL	BLOCK-2069
UA747CN	BLOCK-2070
UA747DC	BLOCK-2070
UA747HC	BLOCK-2069
UA747HM	BLOCK-2069
UA747K	BLOCK-2069
UA747ML	BLOCK-2069
UA747N	BLOCK-2070
UA747PC	BLOCK-2070
UA748CD	BLOCK-2142
UA748CJ	BLOCK-2141
UA748CJG	BLOCK-2141
UA748CL	BLOCK-160
UA748CN	BLOCK-2141
UA748CP	BLOCK-2141
UA748CT	BLOCK-160
UA748DC	BLOCK-2141
UA748DM	BLOCK-2141
UA748HC	BLOCK-160
UA748HM	BLOCK-160
UA748MJ	BLOCK-2141
UA748MJG	BLOCK-2141
UA748ML	BLOCK-160
UA748N	BLOCK-2141
UA748T	BLOCK-160
UA748TC	BLOCK-2141
UA748V	BLOCK-2141
UA749	BLOCK-2073
UA749D	BLOCK-2073
UA749DHC	BLOCK-2067
UA7508A	BLOCK-2087
UA753TC	BLOCK-1289
UA758	BLOCK-1376
UA758B	BLOCK-1376
UA758BA	BLOCK-1376
UA758N	BLOCK-1376
UA758PC	BLOCK-1376
UA771ARC	BLOCK-1908
UA771ATC	BLOCK-1908
UA771BRC	BLOCK-1908
UA771BTC	BLOCK-1908
UA771LRC	BLOCK-1908
UA771LTC	BLOCK-1908
UA771RC	BLOCK-1908
UA771TC	BLOCK-1908
UA772ARC	BLOCK-1910
UA772ATC	BLOCK-1910
UA772BRC	BLOCK-1910
UA772BTC	BLOCK-1910
UA772RC	BLOCK-1910
UA772TC	BLOCK-1910
UA774LDC	BLOCK-1912
UA774LPC	BLOCK-1912
UA776TC	BLOCK-1938
UA780	BLOCK-1271
UA7805	BLOCK-2087
UA7805A	BLOCK-2087
UA7805C	BLOCK-2087
UA7805CDA	BLOCK-1022
UA7805CKA	BLOCK-1022
UA7805CKC	BLOCK-2087
UA7805CU	BLOCK-2087
UA7805DA	BLOCK-1022
UA7805KC	BLOCK-1022
UA7805KM	BLOCK-1022
UA7805MKA	BLOCK-1022
UA7805UC	BLOCK-2087
UA7805UV	BLOCK-2087
UA7806C	BLOCK-2093
UA7806CKC	BLOCK-2093
UA7806CU	BLOCK-2093
UA7806KC	BLOCK-2093
UA7806UC	BLOCK-2093
UA7806UV	BLOCK-2093
UA7808	BLOCK-2095
UA7808C	BLOCK-2095
UA7808CKC	BLOCK-2095
UA7808CU	BLOCK-2095
UA7808KC	BLOCK-2095

DEVICE TYPE	REPL CODE
UA7808UC	BLOCK-2095
UA7808UV	BLOCK-2095
UA780C	BLOCK-1271
UA780DC	BLOCK-1271
UA780PC	BLOCK-1271
UA781	BLOCK-1272
UA7812C	BLOCK-2097
UA7812CKA	BLOCK-890
UA7812CKC	BLOCK-2097
UA7812CU	BLOCK-2097
UA7812DA	BLOCK-890
UA7812KC	BLOCK-890
UA7812KM	BLOCK-890
UA7812MKA	BLOCK-890
UA7812UC	BLOCK-2097
UA7812UV	BLOCK-2097
UA7815	BLOCK-2119
UA7815C	BLOCK-2119
UA7815CDA	BLOCK-892
UA7815CKA	BLOCK-892
UA7815CKC	BLOCK-2119
UA7815CU	BLOCK-2119
UA7815DA	BLOCK-892
UA7815KC	BLOCK-892
UA7815KM	BLOCK-892
UA7815MKA	BLOCK-892
UA7815UC	BLOCK-2119
UA7818C	BLOCK-2085
UA7818CKA	BLOCK-896
UA7818CKC	BLOCK-2085
UA7818CKG	BLOCK-2085
UA7818CU	BLOCK-2085
UA7818KC	BLOCK-896
UA7818UC	BLOCK-2085
UA7818UV	BLOCK-2085
UA781C	BLOCK-1272
UA781DC	BLOCK-1272
UA781PC	BLOCK-1272
UA7824C	BLOCK-2137
UA7824CKA	BLOCK-898
UA7824CKC	BLOCK-2137
UA7824CU	BLOCK-2137
UA7824KC	BLOCK-898
UA7824UC	BLOCK-2137
UA7824UV	BLOCK-2137
UA787	BLOCK-1821
UA787PC	BLOCK-1821
UA78GKC	BLOCK-900
UA78GU1C	BLOCK-2077
UA78H05KC	BLOCK-2007
UA78H12KC	BLOCK-2014
UA78HGKC	BLOCK-2024
UA78L	BLOCK-2160
UA78L-8.2AWC	BLOCK-2160
UA78L05	BLOCK-2144
UA78L05AC	BLOCK-2144
UA78L05ACLP	BLOCK-2144
UA78L05AWC	BLOCK-2144
UA78L05C	BLOCK-2144
UA78L05CLP	BLOCK-2144
UA78L05S	BLOCK-2144
UA78L05WC	BLOCK-2144
UA78L062WV	BLOCK-2173
UA78L06ACLP	BLOCK-2173
UA78L06AS	BLOCK-2173
UA78L06CLP	BLOCK-2173
UA78L06S	BLOCK-2173
UA78L08	BLOCK-2160
UA78L08AC	BLOCK-2160
UA78L08ACLP	BLOCK-2160
UA78L08AWC	BLOCK-2160
UA78L08C	BLOCK-2160
UA78L08CLP	BLOCK-2160
UA78L08S	BLOCK-2160
UA78L08WC	BLOCK-2160
UA78L09ACLP	BLOCK-881
UA78L09AWC	BLOCK-881
UA78L12AC	BLOCK-2074
UA78L12ACLP	BLOCK-2074
UA78L12AWC	BLOCK-2074
UA78L12C	BLOCK-2074
UA78L12CLP	BLOCK-2074
UA78L15AC	BLOCK-2075
UA78L15ACLP	BLOCK-2075
UA78L15C	BLOCK-2075
UA78L15CLP	BLOCK-2075
UA78L6.2AHC	BLOCK-2173
UA78L62AHC	BLOCK-2173
UA78L62AWC	BLOCK-2173
UA78L62AWV	BLOCK-2173
UA78L82	BLOCK-2160
UA78L82AWC	BLOCK-2160
UA78L82AWV	BLOCK-2160
UA78L82W	BLOCK-2160

DEVICE TYPE	REPL CODE
UA78M05C	BLOCK-2087
UA78M05CKC	BLOCK-2087
UA78M05UC	BLOCK-2087
UA78M06C	BLOCK-2093
UA78M06CKC	BLOCK-2093
UA78M06UC	BLOCK-2093
UA78M08C	BLOCK-2095
UA78M08CKC	BLOCK-2095
UA78M08CKD	BLOCK-2095
UA78M08UC	BLOCK-2095
UA78M12C	BLOCK-2097
UA78M12CKC	BLOCK-2097
UA78M12CKD	BLOCK-2097
UA78M12UC	BLOCK-2097
UA78M15C	BLOCK-2119
UA78M15CKC	BLOCK-2119
UA78M15CKD	BLOCK-2119
UA78M15UC	BLOCK-2119
UA78M24C	BLOCK-2137
UA78M24CKC	BLOCK-2137
UA78M24CKD	BLOCK-2137
UA78M24UC	BLOCK-2137
UA78MGU1C	BLOCK-2077
UA7905	BLOCK-2091
UA7905C	BLOCK-2091
UA7905CKA	BLOCK-889
UA7905CKC	BLOCK-2091
UA7905KC	BLOCK-2091
UA7905KM	BLOCK-889
UA7905MKA	BLOCK-889
UA7905UC	BLOCK-2091
UA7906C	BLOCK-2094
UA7906CKC	BLOCK-2094
UA7906U	BLOCK-2094
UA7906UC	BLOCK-2094
UA7908C	BLOCK-2096
UA7908CKC	BLOCK-2096
UA7908KC	BLOCK-2096
UA7908UC	BLOCK-2096
UA7912	BLOCK-2108
UA7912C	BLOCK-2108
UA7912CKA	BLOCK-891
UA7912CKC	BLOCK-2108
UA7912KC	BLOCK-891
UA7912U	BLOCK-2108
UA7912UC	BLOCK-2108
UA7915C	BLOCK-2128
UA7915CKA	BLOCK-895
UA7915CKC	BLOCK-2128
UA7915KC	BLOCK-895
UA7915KM	BLOCK-895
UA7915U	BLOCK-2128
UA7915UC	BLOCK-2128
UA7918C	BLOCK-2086
UA7918CKA	BLOCK-897
UA7918CKC	BLOCK-2086
UA7918CKG	BLOCK-2086
UA7918UC	BLOCK-2086
UA7924C	BLOCK-2136
UA7924CKA	BLOCK-899
UA7924CKC	BLOCK-2136
UA7924U	BLOCK-2136
UA7924UC	BLOCK-2136
UA796	BLOCK-2138
UA796DC	BLOCK-2139
UA796HC	BLOCK-2138
UA796PC	BLOCK-2139
UA798HM	BLOCK-1992
UA79M05AUC	BLOCK-2091
UA79M05C	BLOCK-2091
UA79M05CKC	BLOCK-2091
UA79M06AUC	BLOCK-2094
UA79M06C	BLOCK-2094
UA79M06CKC	BLOCK-2094
UA79M08AUC	BLOCK-2096
UA79M08C	BLOCK-2096
UA79M12AUC	BLOCK-2108
UA79M12C	BLOCK-2108
UA79M12CKC	BLOCK-2108
UA79M12CKD	BLOCK-2108
UA79M15AUC	BLOCK-2128
UA79M15C	BLOCK-2128
UA79M15CKC	BLOCK-2128
UA79M15CKD	BLOCK-2128
UA79M24AUC	BLOCK-2136
UA79M24C	BLOCK-2136
UA79M24CKC	BLOCK-2136
UA79M24CKD	BLOCK-2136
UA79MGU1C	BLOCK-2078
UA8T26	BLOCK-1173
UA8T26D	BLOCK-1173
UA8T26DC	BLOCK-1173
UA8T26P	BLOCK-1173
UA8T26PC	BLOCK-1173

DEVICE TYPE	REPL CODE
UA8T28	BLOCK-1178
UA8T28D	BLOCK-1178
UA8T28DC	BLOCK-1178
UA8T28DM	BLOCK-1178
UA8T28P	BLOCK-1178
UA8T28PC	BLOCK-1178
UAA4000	BLOCK-736
UBB770339X	BLOCK-1212
UC1406HA	BLOCK-773
UC337K	BLOCK-887
UC3525AN	BLOCK-702
UC4741	BLOCK-2056
UC4741C	BLOCK-2056
UC7805UC	BLOCK-2087
UC78M12H2	BLOCK-2097
UC7933	BLOCK-2097
UDN-3611M	BLOCK-1776
UDN-3612M	BLOCK-1777
UDN-3613M	BLOCK-1778
UDN-3614M	BLOCK-1779
UDN-6118A	BLOCK-944
UDN-6128A	BLOCK-945
UDN-6144A	BLOCK-948
UDN-6164A	BLOCK-949
UDN-6184A	BLOCK-944
UDN-7183A	BLOCK-944
UDN-7186A	BLOCK-945
UDN6116A-1	BLOCK-949
UDN6118A	BLOCK-944
UDN6128A	BLOCK-945
UDN6164A	BLOCK-949
UGH7812	BLOCK-2097
UGJ7109393	BLOCK-1022
UHIC-001	BLOCK-189
UHIC-003	BLOCK-190
UHIC-004	BLOCK-191
UHIC-004E	BLOCK-191
UHIC-005	BLOCK-192
UHIC-007	BLOCK-249
UHIC001	BLOCK-189
UHIC003	BLOCK-190
UHIC004	BLOCK-191
UHIC005	BLOCK-192
UHIC006	BLOCK-236
UHIC007	BLOCK-249
UL-914	BLOCK-2189
ULC3037	BLOCK-1262
ULM2111A	BLOCK-1264
ULM2114A	BLOCK-1815
ULM2274P	BLOCK-1828
ULN-2001A	BLOCK-934
ULN-2002A	BLOCK-935
ULN-2003A	BLOCK-936
ULN-2004A	BLOCK-937
ULN-2046A	BLOCK-1963
ULN-2064A	BLOCK-994
ULN-2081A	BLOCK-1970
ULN-2082A	BLOCK-946
ULN-2086A	BLOCK-1963
ULN-2111A	BLOCK-1262
ULN-2111N	BLOCK-1262
ULN-2113A	BLOCK-1264
ULN-2113N	BLOCK-1264
ULN-2114A	BLOCK-1815
ULN-2114K	BLOCK-1256
ULN-2114N	BLOCK-1815
ULN-2114W	BLOCK-1244
ULN-2120A	BLOCK-1275
ULN-2120N	BLOCK-1275
ULN-2121A	BLOCK-1276
ULN-2121N	BLOCK-1276
ULN-2122A	BLOCK-1277
ULN-2122N	BLOCK-1277
ULN-2124A	BLOCK-1271
ULN-2125A	BLOCK-1287
ULN-2127A	BLOCK-1272
ULN-2127N	BLOCK-1272
ULN-2129A	BLOCK-1279
ULN-2129N	BLOCK-1279
ULN-2136A	BLOCK-1262
ULN-2137A	BLOCK-1389
ULN-2151D	BLOCK-2056
ULN-2151M	BLOCK-2058
ULN-2156D	BLOCK-2056
ULN-2156M	BLOCK-2058
ULN-2157A	BLOCK-2070
ULN-2157H	BLOCK-2070
ULN-2157K	BLOCK-2069
ULN-2158D	BLOCK-1972
ULN-2158M	BLOCK-1973
ULN-2159D	BLOCK-2056
ULN-2159H	BLOCK-2057
ULN-2159M	BLOCK-2058
ULN-2165A	BLOCK-1269

DEVICE TYPE	REPL CODE
ULN-2165N	BLOCK-1269
ULN-2171D	BLOCK-2056
ULN-2171M	BLOCK-2058
ULN-2172D	BLOCK-1972
ULN-2172M	BLOCK-1973
ULN-2173D	BLOCK-2056
ULN-2173M	BLOCK-2058
ULN-2174D	BLOCK-1972
ULN-2174M	BLOCK-1973
ULN-2176D	BLOCK-1973
ULN-2176M	BLOCK-1973
ULN-2177D	BLOCK-2056
ULN-2177H	BLOCK-2057
ULN-2177M	BLOCK-2058
ULN-2178D	BLOCK-1972
ULN-2178M	BLOCK-1973
ULN-2209M	BLOCK-1289
ULN-2210A	BLOCK-1825
ULN-2211P	BLOCK-1364
ULN-2212A	BLOCK-1830
ULN-2212P	BLOCK-1830
ULN-2224A	BLOCK-1292
ULN-2228A	BLOCK-1815
ULN-2244A	BLOCK-1376
ULN-2264A	BLOCK-1808
ULN-2267N	BLOCK-1285
ULN-2269A	BLOCK-1816
ULN-2274P	BLOCK-1828
ULN-2275A	BLOCK-1828
ULN-2275P	BLOCK-1828
ULN-2277P	BLOCK-1828
ULN-2278P	BLOCK-1828
ULN-2280A	BLOCK-1303
ULN-2280B	BLOCK-1303
ULN-2280P	BLOCK-1303
ULN-2281B	BLOCK-1917
ULN-2298A	BLOCK-1818
ULN-2709CM	BLOCK-2058
ULN-2709M	BLOCK-2058
ULN-2723A	BLOCK-1984
ULN-2723K	BLOCK-1983
ULN-2741D	BLOCK-2056
ULN-2747A	BLOCK-2070
ULN2001A	BLOCK-934
ULN2001AJ	BLOCK-934
ULN2001AN	BLOCK-934
ULN2002A	BLOCK-935
ULN2002AJ	BLOCK-935
ULN2002AN	BLOCK-935
ULN2003	BLOCK-936
ULN2003A	BLOCK-936
ULN2003AJ	BLOCK-936
ULN2003AN	BLOCK-936
ULN2004A	BLOCK-937
ULN2004AJ	BLOCK-937
ULN2004AN	BLOCK-937
ULN2005A	BLOCK-938
ULN2011A	BLOCK-934
ULN2012A	BLOCK-935
ULN2013A	BLOCK-936
ULN2014A	BLOCK-937
ULN2015A	BLOCK-938
ULN2046A	BLOCK-1963
ULN2064B	BLOCK-994
ULN2064NE	BLOCK-994
ULN2065B	BLOCK-994
ULN2065NE	BLOCK-994
ULN2066B	BLOCK-995
ULN2066NE	BLOCK-995
ULN2067B	BLOCK-995
ULN2067NE	BLOCK-995
ULN2068B	BLOCK-996
ULN2068NE	BLOCK-996
ULN2069B	BLOCK-996
ULN2071B	BLOCK-997
ULN2081A	BLOCK-1970
ULN2082A	BLOCK-946
ULN2083A	BLOCK-1996
ULN2086A	BLOCK-1963
ULN2110(A)	BLOCK-1825
ULN2111	BLOCK-1262
ULN2111A	BLOCK-1262
ULN2111N	BLOCK-1262
ULN2113	BLOCK-1264
ULN2113A	BLOCK-1264
ULN2113N	BLOCK-1264
ULN2114	BLOCK-1815
ULN2114A	BLOCK-1270
ULN2114K	BLOCK-1256
ULN2114N	BLOCK-1815
ULN2114W	BLOCK-1256
ULN2120A	BLOCK-1275
ULN2121A	BLOCK-1276
ULN2121N	BLOCK-1276

DEVICE TYPE	REPL CODE
ULN2122A	BLOCK-1277
ULN2122N	BLOCK-1277
ULN2124	BLOCK-1271
ULN2124A	BLOCK-1271
ULN2124N	BLOCK-1271
ULN2125A	BLOCK-1287
ULN2127A	BLOCK-1272
ULN2127N	BLOCK-1272
ULN2129A	BLOCK-1279
ULN2136A	BLOCK-1290
ULN2137A	BLOCK-1389
ULN2165A	BLOCK-1269
ULN2165N	BLOCK-1269
ULN2204A	BLOCK-611
ULN2209M	BLOCK-1289
ULN2210	BLOCK-1825
ULN2211	BLOCK-1364
ULN2211A	BLOCK-1364
ULN2211B	BLOCK-1364
ULN2212B	BLOCK-1830
ULN2224A	BLOCK-167
ULN2228A	BLOCK-1815
ULN2231A	BLOCK-2167
ULN2242N	BLOCK-60
ULN2243A	BLOCK-565
ULN2244	BLOCK-1376
ULN2244A	BLOCK-1376
ULN2244N	BLOCK-1376
ULN2244R	BLOCK-1376
ULN2249A	BLOCK-277
ULN2261	BLOCK-1846
ULN2261A	BLOCK-1846
ULN2262	BLOCK-1821
ULN2264	BLOCK-1805
ULN2264A	BLOCK-1808
ULN2264AN	BLOCK-1815
ULN2264K	BLOCK-1805
ULN2266A	BLOCK-1284
ULN2266N	BLOCK-1284
ULN2267	BLOCK-1285
ULN2267A	BLOCK-1285
ULN2267N	BLOCK-1285
ULN2268A	BLOCK-2166
ULN2269A	BLOCK-1816
ULN2274B	BLOCK-1828
ULN2274P	BLOCK-1828
ULN2275	BLOCK-1828
ULN2275B	BLOCK-1828
ULN2275P	BLOCK-1828
ULN2276B	BLOCK-1828
ULN2276P	BLOCK-1828
ULN2276Q	BLOCK-1828
ULN2277	BLOCK-1828
ULN2277B	BLOCK-1828
ULN2277P	BLOCK-1828
ULN2278	BLOCK-1828
ULN2278B	BLOCK-1828
ULN2278P	BLOCK-1828
ULN2278Q	BLOCK-1828
ULN2280A	BLOCK-1303
ULN2280B	BLOCK-1303
ULN2280N	BLOCK-1303
ULN2280P	BLOCK-1303
ULN2280Q	BLOCK-1303
ULN2281B	BLOCK-1917
ULN2289	BLOCK-1813
ULN2289(A)	BLOCK-1813
ULN2289A	BLOCK-1813
ULN2290B	BLOCK-220
ULN2290Q	BLOCK-219
ULN2291M	BLOCK-1843
ULN2294M	BLOCK-1876
ULN2296	BLOCK-1892
ULN2297A	BLOCK-2168
ULN2298	BLOCK-1291
ULN2298A	BLOCK-1291
ULN2298P	BLOCK-1291
ULN2741D	BLOCK-2056
ULN2801A	BLOCK-939
ULN2802A	BLOCK-940
ULN2803A	BLOCK-941
ULN2804A	BLOCK-942
ULN2805A	BLOCK-943
ULN2811A	BLOCK-939
ULN2812A	BLOCK-940
ULN2813A	BLOCK-941
ULN2814A	BLOCK-942
ULN2815A	BLOCK-943
ULN3262A	BLOCK-1821
ULN3701Z	BLOCK-221
ULN3859A	BLOCK-1914
ULN8126A	BLOCK-703
ULN8160A	BLOCK-658
ULN8160R	BLOCK-658

ORIGINAL DEVICE TYPES AND REPLACEMENT CODES 65

DEVICE TYPE	REPL CODE
ULS-2045H	BLOCK-1963
ULS-2151D	BLOCK-2056
ULS-2151M	BLOCK-2058
ULS-2156D	BLOCK-2056
ULS-2156M	BLOCK-2058
ULS-2157A	BLOCK-2070
ULS-2157H	BLOCK-2070
ULS-2157K	BLOCK-2069
ULS-2158D	BLOCK-1972
ULS-2158M	BLOCK-1973
ULS-2159D	BLOCK-2056
ULS-2159H	BLOCK-2057
ULS-2159M	BLOCK-2058
ULS-2171D	BLOCK-2056
ULS-2171M	BLOCK-2058
ULS-2172D	BLOCK-1972
ULS-2172M	BLOCK-1973
ULS-2173D	BLOCK-2056
ULS-2173M	BLOCK-2058
ULS-2174D	BLOCK-1972
ULS-2174M	BLOCK-1973
ULS-2176D	BLOCK-1972
ULS-2176M	BLOCK-1973
ULS-2177D	BLOCK-2056
ULS-2177H	BLOCK-2057
ULS-2177M	BLOCK-2058
ULS-2178D	BLOCK-1973
ULS-2178M	BLOCK-1973
ULS-2723A	BLOCK-1984
ULS-2723K	BLOCK-1983
ULS-2741D	BLOCK-2056
ULS2045H	BLOCK-1963
ULS2741D	BLOCK-2056
ULT2260	BLOCK-1190
ULX-2231A	BLOCK-2167
ULX-2267A	BLOCK-1285
ULX-2289A	BLOCK-1813
ULX2210	BLOCK-1825
ULX2210D	BLOCK-1825
ULX2244	BLOCK-1376
ULX2298	BLOCK-1291
ULX3701Z	BLOCK-221
UN4066B	BLOCK-1093
UP32C	BLOCK-48
UPA2003	BLOCK-936
UPA2003C	BLOCK-936
UPA2004C	BLOCK-937
UPA53C	BLOCK-993
UPA64H	BLOCK-666
UPA64HA	BLOCK-666
UPA79C	BLOCK-998
UPA81C	BLOCK-988
UPB201C	BLOCK-1293
UPB201D	BLOCK-1293
UPB202C	BLOCK-1304
UPB202D	BLOCK-1304
UPB203C	BLOCK-1365
UPB203D	BLOCK-1365
UPB2047D	BLOCK-1397
UPB204C	BLOCK-1377
UPB204D	BLOCK-1377
UPB205C	BLOCK-1390
UPB205D	BLOCK-1390
UPB206C	BLOCK-1401
UPB206D	BLOCK-1401
UPB207C	BLOCK-1402
UPB207D	BLOCK-1402
UPB2080D	BLOCK-1414
UPB2085D	BLOCK-1418
UPB2086D	BLOCK-1419
UPB208D	BLOCK-1403
UPB2091D	BLOCK-1424
UPB209D	BLOCK-1404
UPB210C	BLOCK-1406
UPB210D	BLOCK-1406
UPB211C	BLOCK-1409
UPB211D	BLOCK-1408
UPB212C	BLOCK-1410
UPB212D	BLOCK-1409
UPB213C	BLOCK-1411
UPB214C	BLOCK-1411
UPB214D	BLOCK-1411
UPB2150D	BLOCK-1327
UPB2151D	BLOCK-1328
UPB2154D	BLOCK-1331
UPB215C	BLOCK-1294
UPB215D	BLOCK-1294
UPB2161D	BLOCK-1338
UPB216C	BLOCK-1482
UPB216D	BLOCK-1482
UPB2170D	BLOCK-1344
UPB217C	BLOCK-1412
UPB217D	BLOCK-1412
UPB2180D	BLOCK-1352
UPB2181P	BLOCK-1353
UPB2182D	BLOCK-1354
UPB218C	BLOCK-1391
UPB218D	BLOCK-1391
UPB2192D	BLOCK-1357
UPB2193D	BLOCK-1358
UPB2195D	BLOCK-1359
UPB2198D	BLOCK-1362
UPB219C	BLOCK-1423
UPB219D	BLOCK-1423
UPB222C	BLOCK-1425
UPB222D	BLOCK-1425
UPB223C	BLOCK-1426
UPB223D	BLOCK-1426
UPB224C	BLOCK-1413
UPB224D	BLOCK-1413
UPB225C	BLOCK-1410
UPB225D	BLOCK-1410
UPB226C	BLOCK-1428
UPB226D	BLOCK-1428
UPB227D	BLOCK-1392
UPB230D	BLOCK-1417
UPB232C	BLOCK-1295
UPB232D	BLOCK-1295
UPB233C	BLOCK-1485
UPB233D	BLOCK-1485
UPB234C	BLOCK-1301
UPB234D	BLOCK-1301
UPB235D	BLOCK-1297
UPB236D	BLOCK-1298
UPB237D	BLOCK-1384
UPB238D	BLOCK-1385
UPB7400C	BLOCK-1293
UPB7402C	BLOCK-1295
UPB7404C	BLOCK-1297
UPB7405C	BLOCK-1298
UPB74107C	BLOCK-1305
UPB7410C	BLOCK-1304
UPB74123C	BLOCK-1313
UPB7413C	BLOCK-1317
UPB74141D	BLOCK-1321
UPB74150C	BLOCK-1327
UPB74151C	BLOCK-1328
UPB74153C	BLOCK-1330
UPB74154C	BLOCK-1331
UPB74155C	BLOCK-1332
UPB74156	BLOCK-1333
UPB74156C	BLOCK-1333
UPB74157C	BLOCK-1334
UPB74161C	BLOCK-1338
UPB74164C	BLOCK-1341
UPB74170D	BLOCK-1344
UPB74175C	BLOCK-1347
UPB74180C	BLOCK-1352
UPB74181D	BLOCK-1353
UPB74182C	BLOCK-1354
UPB74192C	BLOCK-1357
UPB74193C	BLOCK-1358
UPB74195C	BLOCK-1359
UPB74198D	BLOCK-1362
UPB7420C	BLOCK-1365
UPB7430C	BLOCK-1377
UPB7437C	BLOCK-1384
UPB7438C	BLOCK-1385
UPB7440C	BLOCK-1390
UPB7442C	BLOCK-1392
UPB7445C	BLOCK-1395
UPB7447C	BLOCK-1397
UPB7450C	BLOCK-1401
UPB7451C	BLOCK-1402
UPB7453C	BLOCK-1403
UPB7454C	BLOCK-1404
UPB7460C	BLOCK-1406
UPB7473C	BLOCK-1410
UPB7474C	BLOCK-1411
UPB7476C	BLOCK-1413
UPB7480C	BLOCK-1414
UPB7485C	BLOCK-1418
UPB7486	BLOCK-1419
UPB7486C	BLOCK-1419
UPB7491C	BLOCK-1424
UPB74H04C	BLOCK-1473
UPB74H40C	BLOCK-1488
UPC-1367C	BLOCK-539
UPC-1379C	BLOCK-647
UPC-16	BLOCK-43
UPC-16A	BLOCK-43
UPC-16C	BLOCK-43
UPC-20C	BLOCK-73
UPC-50C	BLOCK-82
UPC-554C	BLOCK-138
UPC-555A	BLOCK-1212
UPC-571C	BLOCK-76
UPC-573C	BLOCK-173
UPC-574	BLOCK-1157
UPC-574J	BLOCK-1157
UPC-576H	BLOCK-174
UPC-596C	BLOCK-176
UPC-781-05AN	BLOCK-2144
UPC-78L-05AN	BLOCK-2144
UPC1001	BLOCK-114
UPC1001-H2	BLOCK-125
UPC1001H	BLOCK-125
UPC1001H2	BLOCK-125
UPC1008C	BLOCK-161
UPC1018	BLOCK-550
UPC1018C	BLOCK-550
UPC1018C-F	BLOCK-550
UPC1018CE	BLOCK-550
UPC1020	BLOCK-149
UPC1020H	BLOCK-149
UPC1023H	BLOCK-83
UPC1025	BLOCK-79
UPC1025H	BLOCK-149
UPC1026	BLOCK-214
UPC1026C	BLOCK-214
UPC1028H	BLOCK-223
UPC1031H	BLOCK-234
UPC1031H2	BLOCK-234
UPC1032H	BLOCK-159
UPC1057C	BLOCK-598
UPC1154H	BLOCK-182
UPC1155H	BLOCK-182
UPC1156	BLOCK-182
UPC1156H	BLOCK-182
UPC1156H2	BLOCK-182
UPC1158H2	BLOCK-485
UPC1161(C)	BLOCK-477
UPC1161C	BLOCK-477
UPC1171C	BLOCK-479
UPC1173C	BLOCK-448
UPC1181	BLOCK-272
UPC1181H	BLOCK-272
UPC1182	BLOCK-273
UPC1182A	BLOCK-273
UPC1182H	BLOCK-273
UPC1185H	BLOCK-280
UPC1185H2	BLOCK-280
UPC1186H	BLOCK-446
UPC1187V	BLOCK-481
UPC1188H	BLOCK-705
UPC1191V	BLOCK-483
UPC1197C	BLOCK-237
UPC1212C	BLOCK-504
UPC1213C	BLOCK-504
UPC1213C(MS)	BLOCK-504
UPC1216V	BLOCK-506
UPC1230H	BLOCK-376
UPC1230H2	BLOCK-376
UPC1238V	BLOCK-221
UPC1252H2	BLOCK-775
UPC1252HA	BLOCK-775
UPC1253H2	BLOCK-776
UPC1253HA2	BLOCK-776
UPC1277H	BLOCK-613
UPC1350C	BLOCK-517
UPC1352	BLOCK-404
UPC1352C	BLOCK-404
UPC1352S	BLOCK-398
UPC1353	BLOCK-398
UPC1353C	BLOCK-235
UPC1355C	BLOCK-1422
UPC1356C	BLOCK-471
UPC1356C2	BLOCK-471
UPC1358A	BLOCK-408
UPC1358C	BLOCK-408
UPC1358H	BLOCK-408
UPC1358H2	BLOCK-408
UPC1358HZ	BLOCK-408
UPC1360	BLOCK-508
UPC1360C	BLOCK-508
UPC1363C	BLOCK-396
UPC1363CA	BLOCK-396
UPC1366C	BLOCK-510
UPC1366C2	BLOCK-510
UPC1367C	BLOCK-405
UPC1368H	BLOCK-414
UPC1368H2	BLOCK-414
UPC1368HR	BLOCK-414
UPC1368HZ	BLOCK-414
UPC1371	BLOCK-398
UPC1371C	BLOCK-398
UPC1372C	BLOCK-398
UPC1372E	BLOCK-404
UPC1373	BLOCK-695
UPC1373-H	BLOCK-695
UPC1373H	BLOCK-695
UPC1373HA	BLOCK-695
UPC1377C	BLOCK-650
UPC1378	BLOCK-662
UPC1378-H-L	BLOCK-662
UPC1378H	BLOCK-662
UPC1378H-L	BLOCK-662
UPC1378H-P	BLOCK-662
UPC1379C	BLOCK-647
UPC1380	BLOCK-184
UPC1380C	BLOCK-179
UPC1382C	BLOCK-603
UPC1391H	BLOCK-654
UPC1391H-1	BLOCK-654
UPC1391HA	BLOCK-654
UPC1394C	BLOCK-664
UPC1401CA	BLOCK-774
UPC1402CA	BLOCK-774
UPC1406HA	BLOCK-773
UPC143-05	BLOCK-2087
UPC14305	BLOCK-2144
UPC14305H	BLOCK-2087
UPC14308	BLOCK-2095
UPC14308H	BLOCK-2095
UPC14312	BLOCK-2097
UPC14312H	BLOCK-2097
UPC14315D	BLOCK-2119
UPC14315H	BLOCK-2119
UPC143C08	BLOCK-2095
UPC1458	BLOCK-1801
UPC1458C	BLOCK-1801
UPC1470H	BLOCK-823
UPC1474HA	BLOCK-772
UPC1480CA	BLOCK-781
UPC1481CA	BLOCK-782
UPC1486	BLOCK-1200
UPC1486C	BLOCK-1200
UPC1490HA	BLOCK-1194
UPC1498H	BLOCK-748
UPC1513HA	BLOCK-768
UPC151A	BLOCK-2056
UPC151C	BLOCK-2058
UPC1520CA	BLOCK-1202
UPC154A	BLOCK-1987
UPC157A	BLOCK-160
UPC157C	BLOCK-2141
UPC15MA	BLOCK-160
UPC16	BLOCK-43
UPC16A	BLOCK-43
UPC16C	BLOCK-43
UPC177C	BLOCK-1873
UPC17C	BLOCK-45
UPC1880C	BLOCK-184
UPC2002	BLOCK-221
UPC2002V	BLOCK-221
UPC206	BLOCK-73
UPC20C	BLOCK-73
UPC20C1	BLOCK-73
UPC20C2	BLOCK-73
UPC213C	BLOCK-1411
UPC215C	BLOCK-1294
UPC22C	BLOCK-76
UPC23C	BLOCK-44
UPC251A	BLOCK-2069
UPC251C	BLOCK-1801
UPC262C	BLOCK-48
UPC271C	BLOCK-1981
UPC276	BLOCK-122
UPC27C	BLOCK-122
UPC27CI	BLOCK-122
UPC2816C	BLOCK-515
UPC29C	BLOCK-87
UPC29C2	BLOCK-87
UPC29C3	BLOCK-87
UPC301AC	BLOCK-2141
UPC301AN	BLOCK-2141
UPC305(C)	BLOCK-903
UPC305C	BLOCK-903
UPC30C	BLOCK-47
UPC311C	BLOCK-1981
UPC311G	BLOCK-1982
UPC311G2	BLOCK-1982
UPC317H	BLOCK-2083
UPC31C	BLOCK-46
UPC324C	BLOCK-2171
UPC324G2	BLOCK-2172
UPC32C	BLOCK-48
UPC339C	BLOCK-1873
UPC339G2	BLOCK-1874
UPC358	BLOCK-1993
UPC358C	BLOCK-1993
UPC358P	BLOCK-1993
UPC373C	BLOCK-2063
UPC393	BLOCK-2063
UPC393C	BLOCK-2063
UPC4001C	BLOCK-1029
UPC4062	BLOCK-1939
UPC4066	BLOCK-1093
UPC41C	BLOCK-91
UPC451C	BLOCK-2171
UPC4558	BLOCK-1801
UPC4558C	BLOCK-1801
UPC4558D	BLOCK-1801
UPC4570HA	BLOCK-823
UPC46C	BLOCK-75
UPC47C	BLOCK-74
UPC48C1	BLOCK-89
UPC494C	BLOCK-710
UPC49C	BLOCK-84
UPC544C	BLOCK-7
UPC554	BLOCK-138
UPC554C	BLOCK-138
UPC554H	BLOCK-138
UPC555A	BLOCK-1212
UPC555H	,BLOCK-90
UPC558	BLOCK-49
UPC558C	BLOCK-49
UPC561	BLOCK-95
UPC561C	BLOCK-72
UPC562C	BLOCK-48
UPC563H	BLOCK-79
UPC563H2	BLOCK-79
UPC566	BLOCK-50
UPC566H	BLOCK-50
UPC566H-B	BLOCK-50
UPC566H-L	BLOCK-50
UPC566H-M	BLOCK-50
UPC566H-N	BLOCK-50
UPC566H2	BLOCK-50
UPC566H3	BLOCK-50
UPC569C2	BLOCK-178
UPC570C	BLOCK-84
UPC570G	BLOCK-84
UPC571	BLOCK-76
UPC571C	BLOCK-76
UPC573C	BLOCK-173
UPC574	BLOCK-1157
UPC574-J	BLOCK-1157
UPC5742	BLOCK-1157
UPC574I	BLOCK-1157
UPC574J	BLOCK-1157
UPC574J(L)	BLOCK-1157
UPC574J(M)	BLOCK-1157
UPC574J(V)	BLOCK-1157
UPC574J-G	BLOCK-1157
UPC574J-KL	BLOCK-1157
UPC574J-T	BLOCK-1157
UPC574JA	BLOCK-1157
UPC574JAG	BLOCK-1157
UPC574JC	BLOCK-1157
UPC574JK	BLOCK-1157
UPC574JL	BLOCK-1157
UPC574JM	BLOCK-1157
UPC574JT	BLOCK-1157
UPC574K	BLOCK-1157
UPC574K02L	BLOCK-1157
UPC574V	BLOCK-1157
UPC575	BLOCK-136
UPC575C	BLOCK-136
UPC575C2	BLOCK-136
UPC576H	BLOCK-174
UPC577	BLOCK-80
UPC577A	BLOCK-80
UPC577H	BLOCK-80
UPC577N	BLOCK-80
UPC580C	BLOCK-184
UPC587C	BLOCK-214
UPC587C2	BLOCK-214
UPC592A2	BLOCK-183
UPC592H	BLOCK-183
UPC592H2	BLOCK-183
UPC592HZ	BLOCK-183
UPC592N	BLOCK-183
UPC595C	BLOCK-175
UPC595C2	BLOCK-176
UPC596	BLOCK-176
UPC596C	BLOCK-176
UPC596C2	BLOCK-176
UPC71A	BLOCK-1955
UPC741C	BLOCK-2058
UPC7805	BLOCK-2087
UPC7805H	BLOCK-2087
UPC7812	BLOCK-2097
UPC7812H	BLOCK-2097
UPC7812HA	BLOCK-2097
UPC7812HF	BLOCK-2097
UPC7815H	BLOCK-2128
UPC7818HF	BLOCK-2085
UPC78L05	BLOCK-2144
UPC78L05A	BLOCK-2144
UPC78L05J	BLOCK-2144
UPC78L05J-T	BLOCK-2144
UPC78L05J-TP	BLOCK-2144
UPC78L05T	BLOCK-2144
UPC78L08	BLOCK-2160
UPC78L12	BLOCK-2074
UPC78L15	BLOCK-2075
UPC78L24J	BLOCK-886
UPC78M05	BLOCK-2087
UPC78M05AHF	BLOCK-2087
UPC78M05H	BLOCK-2087
UPC78M08H	BLOCK-2095
UPC78M10H	BLOCK-904
UPC78M12	BLOCK-2097
UPC78M12H	BLOCK-2097
UPC78M12H2	BLOCK-2097
UPC78M15H	BLOCK-2119
UPC78M18	BLOCK-2085
UPC78M18H	BLOCK-2085
UPC7912H	BLOCK-2108
UPC81	BLOCK-73
UPC81C	BLOCK-73
UPC858	BLOCK-186
UPC858C	BLOCK-186
UPC861C	BLOCK-243
UPC861CE	BLOCK-243
UPD-858C	BLOCK-186
UPD1937C	BLOCK-740
UPD1986C	BLOCK-739
UPD1987C	BLOCK-740
UPD2102AL-4	BLOCK-999
UPD2102ALC-4	BLOCK-999
UPD2114LC	BLOCK-1002
UPD2114LC-1	BLOCK-1002
UPD2114LC-5	BLOCK-1002
UPD2147A-45	BLOCK-1005
UPD23128EC-088	BLOCK-1570
UPD2732	BLOCK-1018
UPD2732A	BLOCK-1018
UPD2732D	BLOCK-1018
UPD2732D-4	BLOCK-1018
UPD2764D-FC4-A2	BLOCK-1019
UPD2764D-FC5-A3	BLOCK-1019
UPD2814C	BLOCK-242
UPD2816C	BLOCK-515
UPD4001BC	BLOCK-1029
UPD4001C	BLOCK-1029
UPD4002BC	BLOCK-1030
UPD4002C	BLOCK-1030
UPD4011	BLOCK-1039
UPD4011B	BLOCK-1039
UPD4011BC	BLOCK-1039
UPD4011BCM	BLOCK-1039
UPD4011C	BLOCK-1039
UPD4012BC	BLOCK-1040
UPD4012C	BLOCK-1040
UPD4013BC	BLOCK-1041
UPD4013BG	BLOCK-1041
UPD4013C	BLOCK-1041
UPD4015C	BLOCK-1043
UPD4016	BLOCK-1004
UPD4016C	BLOCK-1004
UPD4016C-2	BLOCK-1004
UPD4016CX	BLOCK-1004
UPD4016D-1	BLOCK-1004
UPD4017BC	BLOCK-1051
UPD4017C	BLOCK-1051
UPD4020C	BLOCK-1059
UPD4021C	BLOCK-1060
UPD4023BC	BLOCK-1062
UPD4023C	BLOCK-1062
UPD4025C	BLOCK-1064
UPD4027	BLOCK-1066
UPD4027BC	BLOCK-1066
UPD4027C	BLOCK-1066
UPD4029	BLOCK-1068
UPD4029C	BLOCK-1068
UPD4030BC	BLOCK-1069
UPD4030C	BLOCK-1069
UPD4035C	BLOCK-1074
UPD4040BC	BLOCK-1076
UPD4040C	BLOCK-1076
UPD4042C	BLOCK-1077
UPD4049	BLOCK-1084
UPD4049C	BLOCK-1084
UPD4049UBC	BLOCK-1084
UPD4050C	BLOCK-1085
UPD4051BC	BLOCK-1086
UPD4052	BLOCK-1087
UPD4052BC	BLOCK-1087
UPD4053	BLOCK-1088
UPD4053B	BLOCK-1088

DEVICE TYPE	REPL CODE
UPD4053BC	BLOCK-1088
UPD4053BG	BLOCK-1088
UPD4053BP	BLOCK-1088
UPD4066	BLOCK-1093
UPD4066B	BLOCK-1093
UPD4066B-1	BLOCK-1093
UPD4066BC	BLOCK-1093
UPD4066BP	BLOCK-1093
UPD4066C	BLOCK-1093
UPD4066G	BLOCK-1093
UPD4069	BLOCK-1096
UPD4069C	BLOCK-1096
UPD4069UBC	BLOCK-1096
UPD4071	BLOCK-1098
UPD4071BC	BLOCK-1098
UPD4078C	BLOCK-1104
UPD4081BC	BLOCK-1105
UPD4081C	BLOCK-525
UPD4093BC	BLOCK-1110
UPD416	BLOCK-1003
UPD416-3	BLOCK-1003
UPD4160BC	BLOCK-1044
UPD4164-3	BLOCK-1006
UPD41647H-45	BLOCK-1006
UPD416C	BLOCK-1003
UPD416C-1	BLOCK-1003
UPD416C-2	BLOCK-1003
UPD416D	BLOCK-1003
UPD416D-1	BLOCK-1003
UPD416D-2	BLOCK-1003
UPD4503BC	BLOCK-1035
UPD4503C	BLOCK-1035
UPD4518	BLOCK-1129
UPD4520C	BLOCK-1130
UPD4528B	BLOCK-1135
UPD4528C	BLOCK-1135
UPD4538	BLOCK-1140
UPD4538BC	BLOCK-1140
UPD4539C	BLOCK-1141
UPD4555C	BLOCK-1146
UPD4556C	BLOCK-1147
UPD4584BC	BLOCK-1038
UPD471C	BLOCK-1098
UPD7805	BLOCK-2087
UPD780C	BLOCK-1025
UPD8080AFC	BLOCK-1832
UPD8255AC-5	BLOCK-1855
UPD857C	BLOCK-187
UPD858	BLOCK-186
UPD8585C	BLOCK-186
UPD858C	BLOCK-186
UPD861	BLOCK-243
UPD861C	BLOCK-243
UPD861CE	BLOCK-243
UPD861E	BLOCK-243
US-0908D	BLOCK-2182
US-0909D	BLOCK-2176
US-0910D	BLOCK-2185
US-0911D	BLOCK-2186
US-0912D	BLOCK-2187
US-0913D	BLOCK-2188
US-0921D	BLOCK-2192
US-0960D	BLOCK-2189
US7400A	BLOCK-1293
US7400J	BLOCK-1293
US7401A	BLOCK-1294
US7401J	BLOCK-1294
US7402A	BLOCK-1295
US7402J	BLOCK-1295
US7403A	BLOCK-1296
US7403J	BLOCK-1296
US7404A	BLOCK-1297
US7404J	BLOCK-1297
US7405A	BLOCK-1298
US7405J	BLOCK-1298
US7408A	BLOCK-1301
US7408J	BLOCK-1301
US7409A	BLOCK-1302
US7409J	BLOCK-1302
US74107A	BLOCK-1305
US74107J	BLOCK-1305
US7410A	BLOCK-1304
US7410J	BLOCK-1304
US7411A	BLOCK-1307
US7411J	BLOCK-1307
US74121A	BLOCK-1311
US74121J	BLOCK-1311
US74145A	BLOCK-1325
US74153A	BLOCK-1330
US74154A	BLOCK-1331
US74154AN	BLOCK-1331
US74180A	BLOCK-1352
US74180J	BLOCK-1352
US7420A	BLOCK-1365

DEVICE TYPE	REPL CODE
US7420J	BLOCK-1365
US7427A	BLOCK-1372
US7430A	BLOCK-1377
US7430J	BLOCK-1377
US7432A	BLOCK-1378
US7432J	BLOCK-1378
US7438A	BLOCK-1385
US7438J	BLOCK-1385
US7440A	BLOCK-1390
US7440J	BLOCK-1390
US7441A	BLOCK-1391
US7442A	BLOCK-1392
US7443A	BLOCK-1393
US7444A	BLOCK-1394
US7445A	BLOCK-1395
US7446A	BLOCK-1396
US7447A	BLOCK-1397
US7448A	BLOCK-1398
US7450A	BLOCK-1401
US7450J	BLOCK-1401
US7451A	BLOCK-1402
US7451J	BLOCK-1402
US7453A	BLOCK-1403
US7453J	BLOCK-1403
US7454A	BLOCK-1404
US7454J	BLOCK-1404
US7460A	BLOCK-1406
US7460J	BLOCK-1406
US7470A	BLOCK-1408
US7470J	BLOCK-1408
US7472A	BLOCK-1409
US7472J	BLOCK-1409
US7473A	BLOCK-1410
US7473J	BLOCK-1410
US7474A	BLOCK-1411
US7474J	BLOCK-1411
US7475A	BLOCK-1412
US7475J	BLOCK-1412
US7476A	BLOCK-1413
US7476J	BLOCK-1413
US7480A	BLOCK-1414
US7482A	BLOCK-1416
US7483A	BLOCK-1417
US7486A	BLOCK-1419
US7486J	BLOCK-1419
US7490A	BLOCK-1423
US7490J	BLOCK-1423
US7491A	BLOCK-1424
US7492A	BLOCK-1425
US7492J	BLOCK-1425
US7493A	BLOCK-1426
US7493J	BLOCK-1426
US7494A	BLOCK-1427
US7495A	BLOCK-1428
US7496A	BLOCK-1429
US74H00A	BLOCK-1471
US74H00J	BLOCK-1471
US74H01A	BLOCK-1472
US74H01J	BLOCK-1472
US74H04A	BLOCK-1473
US74H04J	BLOCK-1473
US74H05A	BLOCK-1474
US74H05J	BLOCK-1474
US74H08A	BLOCK-1475
US74H08J	BLOCK-1475
US74H10A	BLOCK-1476
US74H10J	BLOCK-1476
US74H11A	BLOCK-1482
US74H11J	BLOCK-1482
US74H20A	BLOCK-1484
US74H20J	BLOCK-1484
US74H21A	BLOCK-1485
US74H21J	BLOCK-1485
US74H22A	BLOCK-1486
US74H22J	BLOCK-1486
US74H30A	BLOCK-1487
US74H30J	BLOCK-1487
US74H40A	BLOCK-1488
US74H40J	BLOCK-1488
US74H50A	BLOCK-1489
US74H50J	BLOCK-1489
US74H51A	BLOCK-1490
US74H51J	BLOCK-1490
US74H52A	BLOCK-1491
US74H52J	BLOCK-1491
US74H53A	BLOCK-1492
US74H53J	BLOCK-1492
US74H54A	BLOCK-1493
US74H54J	BLOCK-1493
US74H55A	BLOCK-1494
US74H55J	BLOCK-1494
US74H60A	BLOCK-1495
US74H60J	BLOCK-1495
US74H61A	BLOCK-1496

DEVICE TYPE	REPL CODE
US74H61J	BLOCK-1496
US74H62A	BLOCK-1497
US74H62J	BLOCK-1497
US74H71A	BLOCK-1498
US74H71J	BLOCK-1498
US74H72A	BLOCK-1499
US74H72J	BLOCK-1499
US74H73A	BLOCK-1500
US74H73J	BLOCK-1500
US74H74A	BLOCK-1501
US74H74J	BLOCK-1501
US74H76A	BLOCK-1502
US74H76J	BLOCK-1429
US74H78A	BLOCK-1503
US74H78J	BLOCK-1503
USD7703394	BLOCK-1212
USN-7441B	BLOCK-1391
USN-7480A	BLOCK-1414
USN-7482A	BLOCK-1416
USN-7483B	BLOCK-1417
USN-7490A	BLOCK-1423
USN-7491A	BLOCK-1424
USN-7492A	BLOCK-1425
USN-7493A	BLOCK-1426
UZ1-24	BLOCK-1212
V151(BENCO)	BLOCK-2189
V181(BENCO)	BLOCK-2224
VCRS0011	BLOCK-691
VE79092	BLOCK-2210
VE79093	BLOCK-2206
VE79141	BLOCK-2203
VHIAN3211K/-1	BLOCK-786
VHIAN5836//-1	BLOCK-761
VHIAN6291//-1	BLOCK-725
VHIBA6209//1E	BLOCK-697
VHIBA6238AU1E	BLOCK-813
VHIBU4011B/-1	BLOCK-1039
VHICXA1011P-1	BLOCK-839
VHID74LS136P	BLOCK-1590
VHID14042P-1	BLOCK-1077
VHIHD7406//-1	BLOCK-1299
VHIHD7417//-1	BLOCK-1343
VHIHD74LS00P/	BLOCK-1568
VHIHD74LS136P	BLOCK-1590
VHIHD74LS14-1	BLOCK-1593
VHIHD74LS221/	BLOCK-1629
VHIHD74S04/-1	BLOCK-1722
VHIIR9358//-1	BLOCK-1993
VHIIR94558/-1	BLOCK-1801
VHILA7016//-1	BLOCK-762
VHIM5224P//-1	BLOCK-2171
VHIMN3007//-1	BLOCK-628
VHIMN3101//-1	BLOCK-626
VHINJM4558D-1	BLOCK-1801
VHINS74LS32-1	BLOCK-1659
VHISN74LS08-1	BLOCK-1574
VHISN74LS32-1	BLOCK-1659
VHISN74S04/-1	BLOCK-1722
VHISN74S11/-1	BLOCK-1727
VHISTR30130-1	BLOCK-758
VHITA7130P2/F	BLOCK-223
VHITA7343P/-1	BLOCK-643
VHITA7347/-1	BLOCK-853
VHITA7347P/-1	BLOCK-853
VHITA75558S-1	BLOCK-1993
VHITA78010AP1	BLOCK-904
VHITA7809S/-1	BLOCK-918
VHITA8703S/-1	BLOCK-1241
VHITC4001BP-1	BLOCK-1029
VHITC4013BP-1	BLOCK-1041
VHITC4028BP-1	BLOCK-1067
VHITC4030BP-1	BLOCK-1069
VHITC4052BP-1	BLOCK-1087
VHITC4066BP-1	BLOCK-1093
VHITC4081BP-1	BLOCK-525
VHITDA7052/-1	BLOCK-1235
VHITDA7056/-1	BLOCK-1236
VHIUPC1373H-1	BLOCK-695
VHIUPC1373HA/	BLOCK-695
VHIUPC1373HA1	BLOCK-695
VHIUPC1391H-1	BLOCK-654
VHIUPC1486C	BLOCK-1200
VHIUPC1486C-1	BLOCK-1200
VHIUPC1513H-1	BLOCK-768
VHIUPC574JT-1	BLOCK-1157
VHIUPD4066B-1	BLOCK-1093
VHIKA7805PI-1	BLOCK-2087
VHIKA7809PI-1	BLOCK-918
VHISTR30130E	BLOCK-758
VHITA7347P/-1	BLOCK-853
VHITA7809S/-1	BLOCK-918
VHIUPC1406HA1	BLOCK-773
WE4110	BLOCK-1773
WEP1231/1231	BLOCK-219

DEVICE TYPE	REPL CODE
WEP1246/1246	BLOCK-235
WEP1410	BLOCK-398
WEP1410/1410	BLOCK-398
WEP1416L/1416	BLOCK-401
WEP1417L/1417	BLOCK-405
WEP1440L/1440	BLOCK-428
WEP2000	BLOCK-1245
WEP2000/706	BLOCK-1245
WEP2001	BLOCK-154
WEP2001/1165	BLOCK-154
WEP2002	BLOCK-1273
WEP2002/716	BLOCK-1273
WEP2003	BLOCK-155
WEP2003/1166	BLOCK-155
WEP2003/717	BLOCK-1274
WEP2004	BLOCK-1276
WEP2004/719	BLOCK-1276
WEP2006	BLOCK-1279
WEP2006/723	BLOCK-1279
WEP2007	BLOCK-222
WEP2007/1233	BLOCK-156
WEP2008	BLOCK-1284
WEP2008/728	BLOCK-1284
WEP2009	BLOCK-1285
WEP2009/729	BLOCK-1285
WEP2010	BLOCK-233
WEP2010/1244	BLOCK-233
WEP2011	BLOCK-1287
WEP2011/731	BLOCK-1287
WEP2014	BLOCK-1289
WEP2014/736	BLOCK-1289
WEP2015	BLOCK-1290
WEP2015/737	BLOCK-1290
WEP2017	BLOCK-1292
WEP2017/739	BLOCK-1292
WEP2018	BLOCK-1303
WEP2018/740A	BLOCK-1303
WEP2020	BLOCK-1364
WEP2020/742	BLOCK-1364
WEP2021	BLOCK-1376
WEP2021/743	BLOCK-1376
WEP2022	BLOCK-1389
WEP2022/744	BLOCK-1389
WEP2024	BLOCK-1405
WEP2024/746	BLOCK-1405
WEP2026	BLOCK-1421
WEP2026/748	BLOCK-1421
WEP2027	BLOCK-1422
WEP2027/749	BLOCK-1422
WEP2028/370	BLOCK-1024
WEP2030	BLOCK-1267
WEP2030/711	BLOCK-1267
WEP2032	BLOCK-1283
WEP2032/727	BLOCK-1283
WEP2033	BLOCK-1286
WEP2033/730	BLOCK-1286
WEP2035	BLOCK-1291
WEP2035/738	BLOCK-1291
WEP2036	BLOCK-1400
WEP2036/745	BLOCK-1400
WEP2037	BLOCK-1407
WEP2037/747	BLOCK-1407
WEP2039	BLOCK-1810
WEP2039/785	BLOCK-1810
WEP2040	BLOCK-1813
WEP2040/788	BLOCK-1813
WEP2041	BLOCK-1820
WEP2041/795	BLOCK-1820
WEP2043	BLOCK-1821
WEP2043/797	BLOCK-1821
WEP2044	BLOCK-1822
WEP2044/798	BLOCK-1822
WEP2046	BLOCK-1822
WEP2046/804	BLOCK-1828
WEP2053	BLOCK-1801
WEP2053/778	BLOCK-1801
WEP2054	BLOCK-1949
WEP2054/909	BLOCK-1949
WEP2055	BLOCK-1805
WEP2055/780	BLOCK-1805
WEP2056	BLOCK-1954
WEP2056/909D	BLOCK-1954
WEP2057/782	BLOCK-1807
WEP2058	BLOCK-1808
WEP2058/783	BLOCK-1808
WEP2059	BLOCK-1809
WEP2059/784	BLOCK-1809
WEP2060	BLOCK-1955
WEP2060/910	BLOCK-1955
WEP2061	BLOCK-1811
WEP2061/786	BLOCK-1811
WEP2063	BLOCK-2057
WEP2063/941D	BLOCK-2057
WEP2064	BLOCK-1814

DEVICE TYPE	REPL CODE
WEP2064/789	BLOCK-1814
WEP2065	BLOCK-1815
WEP2065/790	BLOCK-1815
WEP2066	BLOCK-1816
WEP2066/791	BLOCK-1816
WEP2067	BLOCK-2138
WEP2067/973	BLOCK-2138
WEP2072	BLOCK-23
WEP2072/1024	BLOCK-23
WEP2073	BLOCK-26
WEP2073/1027	BLOCK-26
WEP2074	BLOCK-27
WEP2074/1028	BLOCK-27
WEP2075	BLOCK-1825
WEP2075/801	BLOCK-1825
WEP2076	BLOCK-28
WEP2076/1029	BLOCK-28
WEP2077	BLOCK-29
WEP2077/1030	BLOCK-29
WEP2079	BLOCK-36
WEP2079/1037	BLOCK-36
WEP2080	BLOCK-43
WEP2080/1045	BLOCK-43
WEP2081	BLOCK-48
WEP2081/1050	BLOCK-48
WEP2082	BLOCK-50
WEP2082/1052	BLOCK-50
WEP2086	BLOCK-60
WEP2086/1062	BLOCK-60
WEP2087	BLOCK-76
WEP2087/1078	BLOCK-76
WEP2089	BLOCK-78
WEP2089/1080	BLOCK-78
WEP2090	BLOCK-79
WEP2090/1081A	BLOCK-79
WEP2093	BLOCK-95
WEP2093/1097	BLOCK-95
WEP2094	BLOCK-107
WEP2094/1109	BLOCK-107
WEP2095	BLOCK-108
WEP2095/1110	BLOCK-108
WEP2098	BLOCK-112
WEP2098/1115	BLOCK-112
WEP2099	BLOCK-113
WEP2099/1115A	BLOCK-113
WEP2100	BLOCK-125
WEP2100/1127	BLOCK-125
WEP2102	BLOCK-130
WEP2102/1133	BLOCK-130
WEP2103	BLOCK-133
WEP2103/1136	BLOCK-133
WEP2104	BLOCK-1963
WEP2104/912	BLOCK-1963
WEP2105	BLOCK-136
WEP2105/1140	BLOCK-136
WEP2106	BLOCK-141
WEP2106/1150	BLOCK-141
WEP2107	BLOCK-144
WEP2107/1154	BLOCK-144
WEP2108	BLOCK-146
WEP2108/1156	BLOCK-146
WEP2109	BLOCK-147
WEP2109/1158	BLOCK-147
WEP2110	BLOCK-1983
WEP2110/923	BLOCK-1983
WEP2111	BLOCK-150
WEP2111/1161	BLOCK-150
WEP2112	BLOCK-153
WEP2112/1164	BLOCK-153
WEP2113	BLOCK-161
WEP2113/1172	BLOCK-161
WEP2114	BLOCK-2056
WEP2114/941	BLOCK-2056
WEP2115	BLOCK-162
WEP2115/1173	BLOCK-162
WEP2117	BLOCK-2069
WEP2117/947	BLOCK-2069
WEP2118	BLOCK-2073
WEP2118/949	BLOCK-2073
WEP2119	BLOCK-2079
WEP2119/955M	BLOCK-2079
WEP2120	BLOCK-180
WEP2120/1192	BLOCK-180
WEP2121	BLOCK-181
WEP2121/1193	BLOCK-181
WEP2122	BLOCK-4
WEP2122/1003	BLOCK-4
WEP2123	BLOCK-5
WEP2123/1004	BLOCK-5
WEP2124	BLOCK-6
WEP2124/1005	BLOCK-6
WEP2125	BLOCK-7
WEP2125/1006	BLOCK-7
WEP2126	BLOCK-182

DEVICE TYPE	REPL CODE
WEP2126/1194	BLOCK-182
WEP2127	BLOCK-186
WEP2127/1198	BLOCK-186
WEP2129	BLOCK-10
WEP2129/1010	BLOCK-10
WEP2130	BLOCK-11
WEP2130/1011	BLOCK-11
WEP2131	BLOCK-189
WEP2131/1201	BLOCK-189
WEP2132	BLOCK-190
WEP2132/1202	BLOCK-190
WEP2133	BLOCK-15
WEP2133/1015	BLOCK-15
WEP2134	BLOCK-17
WEP2134/1017	BLOCK-17
WEP2135	BLOCK-191
WEP2135/1203	BLOCK-191
WEP2136	BLOCK-19
WEP2136/1019	BLOCK-19
WEP2138	BLOCK-20
WEP2138/1021	BLOCK-20
WEP2139	BLOCK-192
WEP2139/1204	BLOCK-192
WEP2141	BLOCK-199
WEP2141/1211	BLOCK-199
WEP2142	BLOCK-200
WEP2142/1212	BLOCK-200
WEP2143	BLOCK-205
WEP2143/1217	BLOCK-205
WEP2144	BLOCK-206
WEP2144/1218	BLOCK-206
WEP2145	BLOCK-207
WEP2145/1219	BLOCK-207
WEP2148	BLOCK-216
WEP2148/1228	BLOCK-216
WEP2150	BLOCK-221
WEP2150/1232	BLOCK-221
WEP2156	BLOCK-38
WEP2156/1039	BLOCK-38
WEP2160	BLOCK-42
WEP2160/1043	BLOCK-42
WEP2175	BLOCK-164
WEP2175/1175	BLOCK-164
WEP2176	BLOCK-59
WEP2176/1061	BLOCK-59
WEP2184	BLOCK-67
WEP2184/1069	BLOCK-67
WEP2185	BLOCK-68
WEP2185/1070	BLOCK-68
WEP2186	BLOCK-69
WEP2186/1071	BLOCK-69
WEP2207	BLOCK-92
WEP2207/1094	BLOCK-92
WEP2214	BLOCK-103
WEP2214/1105	BLOCK-103
WEP2215	BLOCK-104
WEP2215/1106	BLOCK-104
WEP2223	BLOCK-121
WEP2223/1123	BLOCK-121
WEP2228	BLOCK-119
WEP2228/1121	BLOCK-119
WEP2229	BLOCK-120
WEP2229/1122	BLOCK-120
WEP2242	BLOCK-132
WEP2242/1135	BLOCK-132
WEP2245	BLOCK-234
WEP2245/1245	BLOCK-234
WEP2248	BLOCK-138
WEP2248/1142	BLOCK-138
WEP2258	BLOCK-149
WEP2258/1160	BLOCK-149
WEP2259	BLOCK-176
WEP2259/1187	BLOCK-176
WEP2260	BLOCK-58
WEP2260/1060	BLOCK-58
WEP2261	BLOCK-184
WEP2261/1196	BLOCK-184
WEP2262	BLOCK-229
WEP2262/1240	BLOCK-229
WEP2265	BLOCK-248
WEP2266	BLOCK-250
WEP2266/1263	BLOCK-250
WEP2268	BLOCK-256
WEP2268/1269	BLOCK-256
WEP2272	BLOCK-1029
WEP2272/4001B	BLOCK-1029
WEP2275	BLOCK-1048
WEP2275/4016	BLOCK-1048
WEP2279	BLOCK-1426
WEP2279/7493A	BLOCK-1426
WEP2281	BLOCK-1818
WEP2281/793	BLOCK-1818
WEP2285	BLOCK-1846
WEP2285/818	BLOCK-1846

ORIGINAL DEVICE TYPES AND REPLACEMENT CODES 67

DEVICE TYPE	REPL CODE	DEVICE TYPE	REPL CODE	DEVICE TYPE	REPL CODE	DEVICE TYPE	REPL CODE	DEVICE TYPE	REPL CODE	DEVICE TYPE	REPL CODE
WEP2286	BLOCK-175	WEP7490	BLOCK-1423	X0232CE	BLOCK-1020	X440029020	BLOCK-2171	Y130201002	BLOCK-1003	ZN74163E	BLOCK-1340
WEP2286/1186	BLOCK-175	WEP7490/7490	BLOCK-1423	X0238CE	BLOCK-662	X440029040	BLOCK-1993	Y130213002	BLOCK-234	ZN74164E	BLOCK-1341
WEP2289	BLOCK-2095	WEP821/821	BLOCK-1849	X0243CE	BLOCK-1269	X440032060	BLOCK-1299	Y130800701	BLOCK-1018	ZN74165E	BLOCK-1342
WEP2289/964	BLOCK-2095	WEP822/822	BLOCK-1850	X0249CE	BLOCK-1157	X440041040	BLOCK-988	Y130801001	BLOCK-1017	ZN74174E	BLOCK-1346
WEP2293	BLOCK-85	WEP9158	BLOCK-1969	X0252CE	BLOCK-401	X440055560	BLOCK-2145	Y130801401	BLOCK-1017	ZN74191E	BLOCK-1356
WEP2293/1087	BLOCK-85	WEP9158/9158	BLOCK-1969	X0260C	BLOCK-644	X440057330	BLOCK-1990	Y15310157	BLOCK-1269	ZN74192E	BLOCK-1357
WEP2294	BLOCK-151	WEP926	BLOCK-1280	X0275CE	BLOCK-559	X440058120	BLOCK-2097	Y15310158	BLOCK-1157	ZN74193E	BLOCK-1358
WEP2294/1162	BLOCK-151	WEP926/724	BLOCK-1280	X0600CE	BLOCK-771	X440063050	BLOCK-2087	Y15310159	BLOCK-112	ZN7420E	BLOCK-1365
WEP2296	BLOCK-167	WEP927	BLOCK-1282	X1190CE	BLOCK-1191	X440064940	BLOCK-710	Y15377001	BLOCK-1817	ZN7420F	BLOCK-1365
WEP2296/1178	BLOCK-167	WEP927/726	BLOCK-1282	X37C	BLOCK-1157	X440069050	BLOCK-2091	Y15390751	BLOCK-1291	ZN7427E	BLOCK-1372
WEP2297	BLOCK-172	WEP928	BLOCK-1773	X400007801	BLOCK-1025	X440070490	BLOCK-1084	Y15415451	BLOCK-1819	ZN7427J	BLOCK-1372
WEP2297/1183	BLOCK-172	WEP930	BLOCK-1791	X400040161	BLOCK-1004	X440072920	BLOCK-2063	Y15416272	BLOCK-1849	ZN7428E	BLOCK-1373
WEP2308	BLOCK-295	WEP931	BLOCK-1844	X400082550	BLOCK-1855	X440073860	BLOCK-1851	Y15418561	BLOCK-1820	ZN7428J	BLOCK-1373
WEP2308/1308	BLOCK-295	WEP933	BLOCK-2058	X400104162	BLOCK-1003	X440078051	BLOCK-2144	Y15432511	BLOCK-1407	ZN7430E	BLOCK-1377
WEP2319/309K	BLOCK-1022	WEP933/941M	BLOCK-2058	X400141640	BLOCK-1006	X440078054	BLOCK-2144	Y15432512	BLOCK-1407	ZN7430F	BLOCK-1377
WEP2331	BLOCK-1984	WEP934	BLOCK-1	X420100030	BLOCK-1299	X440079120	BLOCK-2108	Y15437031	BLOCK-1291	ZN7432E	BLOCK-1378
WEP2331/923D	BLOCK-1984	WEP934/1000	BLOCK-1	X420100060	BLOCK-1299	X440140050	BLOCK-2087	Y15453001	BLOCK-1269	ZN7432J	BLOCK-1378
WEP4011B	BLOCK-1039	WEP935	BLOCK-12	X420100070	BLOCK-1300	X440150640	BLOCK-666	Y440800101	BLOCK-1750	ZN7437E	BLOCK-1384
WEP4011B/4011B	BLOCK-1039	WEP935/1012	BLOCK-12	X420100160	BLOCK-1336	X440150790	BLOCK-998	Y440800301	BLOCK-1017	ZN7437J	BLOCK-1384
WEP4013B	BLOCK-1041	WEP936	BLOCK-16	X420100170	BLOCK-1343	X440160660	BLOCK-1093	Y440800601	BLOCK-1019	ZN7438E	BLOCK-1385
WEP4013B/4013B	BLOCK-1041	WEP937	BLOCK-56	X420100380	BLOCK-1385	X440193050	BLOCK-903	Y440801101	BLOCK-1019	ZN7438J	BLOCK-1385
WEP4066B/4066B	BLOCK-1093	WEP937/1058	BLOCK-56	X420101230	BLOCK-1313	X440200650	BLOCK-287	Y440801501	BLOCK-1017	ZN7440E	BLOCK-1390
WEP4069	BLOCK-1096	WEP938	BLOCK-73	X420200020	BLOCK-1720	X440223110	BLOCK-1981	Y441800101	BLOCK-1019	ZN7440F	BLOCK-1390
WEP4069/4069	BLOCK-1096	WEP938/1075A	BLOCK-73	X420200040	BLOCK-1722	X440672200	BLOCK-865	Y441800102	BLOCK-1019	ZN7441AE	BLOCK-1391
WEP500	BLOCK-1212	WEP939	BLOCK-80	X420201610	BLOCK-1605	X440672210	BLOCK-865	YEADUPC577H	BLOCK-80	ZN7442E	BLOCK-1392
WEP500/703A	BLOCK-1212	WEP939/1082	BLOCK-80	X420300000	BLOCK-1568	X440751880	BLOCK-1771	YEAM004	BLOCK-96	ZN7442J	BLOCK-1392
WEP501	BLOCK-1223	WEP941	BLOCK-98	X420300020	BLOCK-1570	X440751890	BLOCK-1772	YEAM53274P	BLOCK-1411	ZN7450E	BLOCK-1401
WEP501/704	BLOCK-1223	WEP941/1100	BLOCK-98	X420300040	BLOCK-1572	X440755400	BLOCK-718	YEAMJM78L05	BLOCK-2144	ZN7450F	BLOCK-1401
WEP502	BLOCK-1244	WEP942	BLOCK-100	X420300050	BLOCK-1573	X440756330	BLOCK-718	YEAMLM2900N	BLOCK-2191	ZN7451E	BLOCK-1402
WEP502/705A	BLOCK-1244	WEP942/1102	BLOCK-100	X420300080	BLOCK-1574	X440759820	BLOCK-717	YEAMM5152L	BLOCK-159	ZN7451F	BLOCK-1402
WEP503	BLOCK-1256	WEP943	BLOCK-101	X420300100	BLOCK-1576	X440762120	BLOCK-907	YEAMNJ58L05	BLOCK-2144	ZN7453E	BLOCK-1403
WEP503/707	BLOCK-1256	WEP943/1103	BLOCK-101	X420300140	BLOCK-1593	X460405000	BLOCK-1085	YEAMNJM58L05	BLOCK-2219	ZN7453F	BLOCK-1403
WEP504	BLOCK-1262	WEP944	BLOCK-102	X420300200	BLOCK-1626	X460406901	BLOCK-1096	YEAMNJM78L05	BLOCK-2144	ZN7454E	BLOCK-1404
WEP504/708	BLOCK-1262	WEP944/1104	BLOCK-102	X420300270	BLOCK-1647	X460409300	BLOCK-1110	YEAMUPC577H	BLOCK-80	ZN7454F	BLOCK-1404
WEP505	BLOCK-1264	WEP945	BLOCK-111	X420300300	BLOCK-1658	X460458400	BLOCK-1038	Z80	BLOCK-1025	ZN7460E	BLOCK-1406
WEP505/709	BLOCK-1264	WEP945/1113	BLOCK-111	X420300320	BLOCK-1659	XAA104	BLOCK-1293	Z80A	BLOCK-1025	ZN7460F	BLOCK-1406
WEP506	BLOCK-1265	WEP946	BLOCK-116	X420300510	BLOCK-1689	XAA105	BLOCK-1377	Z8400APS	BLOCK-1025	ZN7470E	BLOCK-1408
WEP506/710	BLOCK-1265	WEP946/1118	BLOCK-116	X420300540	BLOCK-1690	XAA106	BLOCK-1390	ZEN601	BLOCK-1212	ZN7470F	BLOCK-1408
WEP507	BLOCK-1269	WEP948	BLOCK-143	X420300730	BLOCK-1705	XAA107	BLOCK-1410	ZEN603	BLOCK-1270	ZN7472E	BLOCK-1409
WEP507/712	BLOCK-1269	WEP948/1153	BLOCK-143	X420300740	BLOCK-1706	XAA108	BLOCK-1413	ZEN604	BLOCK-1262	ZN7472J	BLOCK-1409
WEP508	BLOCK-1270	WEP949	BLOCK-145	X420300750	BLOCK-1707	XAA109	BLOCK-1423	ZEN605	BLOCK-43	ZN7473E	BLOCK-1410
WEP508/713	BLOCK-1270	WEP949/1155	BLOCK-145	X420300860	BLOCK-1713	XC-1312A	BLOCK-1823	ZEN606	BLOCK-1256	ZN7473F	BLOCK-1410
WEP509	BLOCK-1271	WEP950	BLOCK-1051	X420300930	BLOCK-1717	XC1312	BLOCK-1823	ZEN607	BLOCK-1271	ZN7474E	BLOCK-1411
WEP509/714	BLOCK-1271	WEP950/4017	BLOCK-1051	X420301230	BLOCK-1584	XR-8038CN	BLOCK-1919	ZEN608	BLOCK-1272	ZN7474F	BLOCK-1411
WEP510	BLOCK-1272	WEP966L/966	BLOCK-2097	X420301390	BLOCK-1592	XR-8038CP	BLOCK-1919	ZEN609	BLOCK-1822	ZN7475E	BLOCK-1412
WEP510/715	BLOCK-1272	WEP989	BLOCK-2174	X420301450	BLOCK-1594	XR-8038N	BLOCK-1919	ZEN610	BLOCK-1804	ZN7476E	BLOCK-1413
WEP511	BLOCK-1275	WEP989/989	BLOCK-2174	X420301560	BLOCK-1601	XR-8038P	BLOCK-1919	ZN7400E	BLOCK-1293	ZN7486E	BLOCK-1419
WEP511/718	BLOCK-1275	WEP9914	BLOCK-2189	X420301610	BLOCK-1605	XR082CP	BLOCK-1910	ZN7400F	BLOCK-1293	ZN7486J	BLOCK-1419
WEP512	BLOCK-1277	WEP9924	BLOCK-2193	X420301630	BLOCK-1607	XR084CP	BLOCK-1912	ZN7401E	BLOCK-1294	ZN7490E	BLOCK-1423
WEP512/720	BLOCK-1277	WEP9924/9924	BLOCK-2193	X420301640	BLOCK-1608	XR1458CP	BLOCK-1801	ZN7401F	BLOCK-1294	ZN7490F	BLOCK-1423
WEP514	BLOCK-1281	WEP9946	BLOCK-2208	X420301750	BLOCK-1616	XR1488N	BLOCK-1771	ZN7402E	BLOCK-1295	ZN7492AE	BLOCK-1425
WEP515/725	BLOCK-1281	WEP9946/9946	BLOCK-2208	X420301950	BLOCK-1623	XR1488P	BLOCK-1771	ZN7402F	BLOCK-1295	ZN7492AJ	BLOCK-1425
WEP615/615	BLOCK-1157	WPC23C	BLOCK-44	X420302210	BLOCK-1629	XR1489AN	BLOCK-1772	ZN7403E	BLOCK-1296	ZN7492E	BLOCK-1425
WEP7400	BLOCK-1293	WPD858	BLOCK-186	X420302440	BLOCK-1634	XR2201CP	BLOCK-934	ZN7403J	BLOCK-1296	ZN7492F	BLOCK-1425
WEP7400/7400	BLOCK-1293	WS6945-2	BLOCK-2207	X420302450	BLOCK-1635	XR2202CP	BLOCK-935	ZN7404E	BLOCK-1297	ZN7493AE	BLOCK-1426
WEP7401	BLOCK-1294	X-2408473	BLOCK-2189	X420302730	BLOCK-1648	XR2203CP	BLOCK-936	ZN7405E	BLOCK-1298	ZN7493AJ	BLOCK-1426
WEP7401/7401	BLOCK-1294	X-2408784	BLOCK-2213	X420303730	BLOCK-1671	XR2204CP	BLOCK-937	ZN7408E	BLOCK-1301	ZN7493E	BLOCK-1426
WEP7402	BLOCK-1295	X0043CE	BLOCK-1269	X420303740	BLOCK-1672	XR3403CN	BLOCK-2171	ZN7408J	BLOCK-1301	ZN7493F	BLOCK-1426
WEP7402/7402	BLOCK-1295	X0054CE	BLOCK-136	X420303930	BLOCK-1679	XR3403CP	BLOCK-2171	ZN7409E	BLOCK-1302	ZN7494E	BLOCK-1427
WEP7408	BLOCK-1301	X0065CE	BLOCK-537	X420305410	BLOCK-1692	XR3524	BLOCK-701	ZN7409J	BLOCK-1302	ZN7495AE	BLOCK-1428
WEP7408/7408	BLOCK-1301	X0069CE	BLOCK-1029	X420500000	BLOCK-1568	XR3524CP	BLOCK-701	ZN74107E	BLOCK-1305	ZN7496E	BLOCK-1429
WEP74145	BLOCK-1325	X0087TA	BLOCK-503	X420500020	BLOCK-1570	XR4136CP	BLOCK-2219	ZN7410E	BLOCK-1304	ZNM442E	BLOCK-1392
WEP74145/74145	BLOCK-1325	X0092CE	BLOCK-401	X420500040	BLOCK-1572	XR4195CP	BLOCK-910	ZN7410F	BLOCK-1304	ZPFAK-50701	BLOCK-1018
WEP74162	BLOCK-1339	X0093CE	BLOCK-519	X420500080	BLOCK-1574	XR4558CP	BLOCK-1801	ZN74121E	BLOCK-1311	ZPFAK-50801	BLOCK-1018
WEP74162/74162	BLOCK-1339	X0094CE	BLOCK-519	X420500100	BLOCK-1576	XR8038CN	BLOCK-1919	ZN74122E	BLOCK-1312	ZPFAK-50901	BLOCK-1018
WEP7473	BLOCK-1410	X0137CE	BLOCK-732	X420500110	BLOCK-1579	XR8038CP	BLOCK-1919	ZN74123E	BLOCK-1313	ZPFAK-51001	BLOCK-1018
WEP7473/7473	BLOCK-1410	X0178CE	BLOCK-526	X420500320	BLOCK-1659	XR8038P	BLOCK-1919	ZN7412E	BLOCK-1310	ZTK-33	BLOCK-1157
WEP7474	BLOCK-1411	X0212CE	BLOCK-657	X420500370	BLOCK-1670	XRA6418N	BLOCK-1226	ZN7412J	BLOCK-1310	ZTK33	BLOCK-1157
WEP7474/7474	BLOCK-1411	X0213CE	BLOCK-603	X420500400	BLOCK-1683	XRL555CN	BLOCK-2080	ZN74154E	BLOCK-1736	ZTK33A	BLOCK-1157
WEP7476	BLOCK-1413	X0225C	BLOCK-364	X420500740	BLOCK-1706	XRL556CN	BLOCK-2146	ZN74155E	BLOCK-1332		
WEP7476/7476	BLOCK-1413			X420501120	BLOCK-1580	XRL556CP	BLOCK-2146	ZN74161E	BLOCK-1338		

SECTION 2

REPLACEMENTS

BLOCK-1

52-045-001-0	55810-169	8000-00006-011	87-0022
EA6801	ECG1000	LAP-011	LAP011
NTE1000	SK7753	TCG1000	TM1000
WEP934	WEP934/1000		

BLOCK-2

4001(MODULE)	ECG1001	LAP-012	TM1001

BLOCK-3

09-308050	1111P	166907	ECG1002
ESI-LA1111P	ETI-21	GEIC-69	LA1111
LA1111P	NTE1002	TCG1002	TM1002

BLOCK-4

00001201	000LA1201B	003501	02-121201
0201201	051-0068-02	09-308008	09-308062
09-308063	1002-0006	1077-2382	14LQ007
167900306081	171179-036	172252	19-076001
1C-24	1C-507	221-Z9021	24MW1028
307-005-9-001	41C-407	42-27377	46-1343-3
4IC-407	52-040-009-0	61B042-9	740-8160-190
749-8160-190	8-759-812-01	917-1201-0	917-12010
A-1201B	A1201	A1201B	CXL706
EA16X35	EA3282	EA33X8367	ECG1003
G09-015-B	G09015	LA-1201	LA-1201T
LA1200	LA1201	LA1201(B)	LA1201(C)
LA1201B	LA1201C	LA1201C-W	LA1201T
LA1201W	MX-3013	NTE1003	PC20003
QTVCM-75	RE335-IC	REN1003	RT-7399
RT6790	RT7399	SK3288	TCG1003
TM1003	TR228779905696	TVCM-75	WEP2122
WEP2122/1003			

BLOCK-5

09-308089	111-4-2060-02300	111-4-2060-02400	2002100022
2002120022	200X2100-022	2119-101-1204	221-Z9022
23119012	2360141	2360171	2360361
2360401	2360511	4-2060-02300	4-2060-02400
4206002400	4206004400	4206042400	46-1357-3
46-1369-3	5351361	6644004400	7-759-651-35
	8-759-651-35	8-795-270-70	905-105
905-191	A1364N	AN222	B0306000
B0306004	ECG1004	EP84X74	GEIC-149
GEIC-36	GEIC-96	HA1126	HA1126AW
HA1126D	HA1126DW	HA1146	IC146
LA1364	LA1364N	M5135P	NTE1004
PTC754	RE325-IC	REN1004	RH-
1X0020CEZZ	RH-IX0020CE	RH-IX0020CEZZ	SK3365
T-Q7037	T-Q7038	TA7070A	TA7070B
TA7070FA-1	TA7070P	TA7070PFA-1	TA7070PFA1
TA7070PGL	TCG1004	TM1004	TQ7037
TQ7038	TV241072	WEP2123	WEP2123/1004

BLOCK-6

051-0038-00	2091-50	740-8120-160	A3300
EA33X8356	ECG1005	GE-1005	GEIC-42
HC1000403	LA-3300	LA3300	NTE1005
RE336-IC	REN1005	SK3723	TCG1005
TM1005	WEP2124	WEP2124/1005	

BLOCK-7

000074010	0000LA3301	09-308034	09-308064
1077-2390	16233010	171179-051	172272
2091-49	21M532	221-Z9026	51-41850J01
52-040-010-0	740-5903-301	740-9003-301	741098
8-759-833-01	87-0004	90200090	916072
A3301	ECG1006	G09-009-A	G09009
GEIC-38	HEPC6060P	LA-3301	LA13301
LA3301	NTE1006	PC-20007	PC20018
RE326-IC	REN1006	RV1LA3301	SK3358
TCG1006	TM1006	UPC544C	WEP2125
WEP2125/1006			

BLOCK-8

156018	A4030	ECG1008	ES84X1
GEIC-271	LA4030-7	TM1008	

BLOCK-9

4206105670	46-13131-3	905-152	ECG1009
GEIC-77	LA4030P	NTE1009	SK3499
TCG1009	TM1009		

BLOCK-10

221-Z9046	46-1392-3	916081	ECG1010
GEIC-37	LA4031P	NTE1010	PC-20023
SK3376	TCG1010	TM1010	WEP2129
WEP2129/1010			

BLOCK-11

61K001-12	61K001-9	ECG1011	GEIC-79
LA4032P	NTE1011	SK3492	TCG1011
TM1011	WEP2130	WEP2130/1011	

BLOCK-12

000LD1020A	09-308077	1002-07	ECG1012
IC-549(ELCOM)	ILD1020	LD1020	LD1020A
NTE1012	TCG1012	TM1012	WEP935
WEP935/1012			

BLOCK-13

1041-68	ECG1013	IC-547(ELCOM)	LA1041
LD1041	NTE1013	TCG1013	TM1013

BLOCK-14

740-2001-110	ECG1014	LA1110	LD-1110
LD1110	NTE1014	TCG1014	TM1014

BLOCK-15

0000LD3000	ECG1015	IC-548(ELCOM)	LD3000
NTE1015	SK7754	TCG1015	TM1015
WEP2133	WEP2133/1015		

BLOCK-16

09-308065	1041-69	ECG1016	
LB3001(FANON)	LD300	LD3001	LD3001W
LD3001X	NTE1016	SK7755	TCG1016
TM1016	WEP936		

BLOCK-17

ECG1017	LD3040	NTE1017	TCG1017
TM1017	WEP2134	WEP2134/1017	

BLOCK-18

221-Z9134	46-1383-3	ECG1018	ES84X2
IC-551(ELCOM)	LD3080	LM1458C	NTE1018
TCG1018	TM1018		

BLOCK-19

09-308049	EA33X8352	ECG1019	HC1000503
HC1000603	HC100603	LD3120	NTE1019
SK7756	SVI-LD3120	TCG1019	TM1019
WEP2136	WEP2136/1019		

BLOCK-20

09-308024	1019-25	24MW996	740-1003-150
740-2003-150	8000-00004-308	ECG1021	GE-1021
LD-3150	LD3150	LD3150C	PC-20004
PC20004	TCG1021	TM1021	WEP2138
WEP2138/1021			

BLOCK-21

15-31015-1	ECG1022	GEIC-102	NTE1022
TA7117P	TCG1022		

BLOCK-22

051-0010-00	740-9017-092	ECG1023	GEIC-272
NTE1023	SK7757	TA7092AP-B	TCG1023
TM1023			

BLOCK-23

16740135809	24MW997	905-5	ECG1024
GE-720	NTE1024	PC-20006	RE347M
REN1024	SK3152	STK-011	STK011
TCG1024	TM1024	WEP2072	WEP2072/1024

BLOCK-24

CXL1025	ECG1025	GE-722	NTE1025
RE348-IC	RE348-M	RE348M	REN1025
SK3155	STK020	STK020F	SVISTK020F
TCG1025	TM1025		

BLOCK-25

ECG1026	LD3061	TM1026

BLOCK-26

415200050	905-9-B	CXL1027	EA33X8351
ECG1027	GE-721	HC1000703	NTE1027
PC-20005	RE349-IC	RE349-M	REN1027
SK3153	STK-015	STK015	TCG1027
TM1027	WEP2073	WEP2073/1027	

BLOCK-27

ECG1028	GE-724	NTE1028	REN1028
SK3435	SK3436	STK-154	TCG1028
TM1028	WEP2074	WEP2074/1028	

BLOCK-28

000HA1306U	000LA1306U	00HA1306PU	221-Z9099
266P30601	4-3036	5350121	740-2001-306
916061	ECG1029	G09-006-A	G09-006-B
G09-006-C	G09-010-A	GEIC-162	HA-1306P
HA1306	HA1306P	HA1306PU	HA1306W
HA1306WU	MA1306W	NTE1029	SK3368
TCG1029	TM1029	WEP2076	WEP2076/1029

BLOCK-29

170964	221-Z9100	5350132	5350136
57A32-17	ECG1030	GEIC-163	HA1308
HA1338	NTE1030	PC-20012	SK3372
TCG1030	TM1030	WEP2077	WEP2077/1030

BLOCK-30

5350141	ECG1031	GEIC-273	HA1310
NTE1031	TCG1031	TM1031	

BLOCK-31

24MW1110	5350151	5350152	ECG1032
GEIC-164	HA1311	HA1311W	NTE1032
SK7758	TCG1032	TM1032	

BLOCK-32

5350161	ECG1033	GEIC-165	HA1312
NTE1033	SK7759	TCG1033	TM1033

BLOCK-33

21A101-013	ECG1034	GE1C-166	GEIC-166
HA1313	NTE1034	TCG1034	TM1034

BLOCK-34

221-Z9101	ECG1035	GEIC-167	NTE1035
SK3369	TCG1035	TM1035	

BLOCK-35

221-Z9102	EA33X8368	ECG1036	GEIC-168
HA1316	NTE1036	SK3370	TCG1036
TM1036			

BLOCK-36

051-0036-00	051-0036-01	051-0036-0102	051-0036-02
051-0036-03	221-Z9103	740-9001-322	740781
ECG1037	GEIC-170	HA1322	HA1322C
NTE1037	RE351-IC	REN1037	SK3371
TCG1037	TM1037	WEP2079	WEP2079/1037

BLOCK-37

ECG1038	G09-011-A	G09-012-A	G09-012-B
G09-012-C	GE-1038	GEIC-274	HA1318PU
HA1318PU-1	HA1318PU-2	HA1318PU-3	NTE1038
TCG1038			

BLOCK-38

221-Z9104	2327311	2327312	5351021
57A32-3	ECG1039	GEIC-159	HA1201
IC-512(ELCOM)	NTE1039	SK3366	TCG1039
TM1039	TR2327312	TR4083-23273	
TR4083-2327311	TR4083-2327312	WEP2156	WEP2156/1039

BLOCK-39

ECG1040	GEIC-161	HA1211	NTE1040
SK3362	TCG1040	TM1040	

BLOCK-40

221-Z9105	2327411	5351051	57A32-6
ECG1041	FA-6008T	GEIC-160	HA-1202
HA1202	NTE1041	SK3361	TCG1041
TM1041	TR2327411		

BLOCK-41

170954	221-Z9106	5351031	ECG1042
HA1203	IC-517(ELCOM)	NTE1042	SK3367
TCG1042	TM1042		

BLOCK-42

5350211	905-123	ECG1043	EP84X72
GEIC-169	HA1319	HC1001001	NTE1043
SK3363	TCG1043	TM1043	WEP2160
WEP2160/1043			

BLOCK-43

000074020	09-308033	09-308054	1003-02
1129	1192(IC)	171179-027	1C507
51D70177A02(C)	55810-168	61A023-2	7130A
8-759-101-60	8000-00004-306	8010-171	88X0053-001
916064	A1201-H45	ECG1045	FB274
HEPC6063P	HEPC6083P	IC26	L274
MC1558P	NTE1045	TCG1045	TM1045
UPC-16	UPC-16A	UPC-16C	UPC16
UPC16A	UPC16C	WEP2080	WEP2080/1045
ZEN605			

BLOCK-44

02-300023	02-310023	06300005	09-308009
1038-1804	21A101-005	221-Z9047	489751-175
ECG1046	GEIC-118	KD6311	MPC23C
NTE1046	RE337-IC	REN1046	SAJ72157
SK3471	TCG1046	TM1046	TVS-MPC23C
TVS-UPC23C	TVSUPC23	TVSUPC23C	UPC23C
WPC23C			

BLOCK-45

09-308027	09-308069	21A101-004	37001003
E17336	ECG1047	GEIC-116	NTE1047
SAJ72158	SK7760	TCG1047	TM1047
UPC17C			

BLOCK-46

09-308031	21A101-007	221-Z9156	C31C
ECG1048	GEIC-122	MPC-31C	MPC31C
NTE1048	SAJ72162	TCG1048	TM1048
UPC31C			

BLOCK-47

09-308083	09-308102	ECG1049	ESI-UPC30C
GEIC-121	MPC30C	NTE1049	SK3470
TCG1049	TM1049	UPC30C	

BLOCK-48

21A101-018	ECG1050	GEIC-123	GEIC-133
IC142	M5191P	MPC562C	NTE1050
RE367-IC	REN1050	RH-IX0025CEZZ	SK3475
TCG1050	TM1050	UP32C	UPC262C
UPC32C	UPC562C	WEP2081	WEP2081/1050

BLOCK-49

ECG1051	GEIC-131	IC136	MPC558C
NTE1051	RH-1X0022CEZZ	RH-IX0022CEZZ	TCG1051
TM1051	UPC558	UPC558C	

BLOCK-50

003522	09-308076	1047-25	1077-2408
1089-7619	159947	161193	26810-158
28810-177	37510-165	37510-166	61A0001-12
61A001-12	740-9000-566	916084	BA308
EA33X8585	ECG1052	G09-029-B	GEIC-135
NTE1052	RE327-IC	REN1052	SK3249
TA70639	TCG1052	TM1052	TVCM-82
UPC566	UPC566H	UPC566H-B	UPC566H-L
UPC566H-M	UPC566H-N	UPC566H2	UPC566H3
WEP2082	WEP2082/1052		

BLOCK-51

AN136	ECG1053	GEIC-44	NTE1053
SK7761	TCG1053	TM1053	

BLOCK-52

09-308011	4150002031	905-13	905-13B
AN-203	AN-210	AN203	AN203(C)
AN203AA	AN203BA	AN203BB	ECG1054
GEIC-45	NTE1054	SK3457	TCG1054
TM1054			

BLOCK-53

AN210	AN210B	AN210C	ECG1055
GEIC-47	NTE1055	SK3494	TCG1055
TM1055			

BLOCK-54

1018-25	741853	AN-211	AN211
AN211A	AN211AB	AN211B	ECG1056
GEIC-48	NTE1056	RE368-IC	REN1056
SK3458	TCG1056	TM1056	

BLOCK-55

AN-206	AN206	AN2065	AN206AB
AN206B	AN206S	ECG1057	GEIC-46
NTE1057	SK7762	TCG1057	TM1057

BLOCK-56

18600-156	221-Z9107	5350231	741854
8000-00032-030	88-18920	AN-214	AN-214-G
AN-214-P	AN-214P	AN-214Q	AN214
AN214-QR	AN214D	AN214P	AN214PQR
AN214Q	AN214QR	AN214R	ECG1058
G09-013-A	G09-013-A(1)	GEIC-49	NTE1058
PTC757	RE338-IC	REN1058	SK3459
TCG1058	TM1058	TVCM-74	WEP937
WEP937/1058			

BLOCK-57

AN213	AN213-AB	AN213AB	ECG1059
TM1059			

BLOCK-58

1114-4458	14LN033	221-Z9151	51-90305A60
5632-AN217(BB)	5652-AN217	5652-AN217(BB)	5652-AN217BB
AN217	AN217(BB)	AN217BB	AN217CB
AN217P	AN217PBB	ECG1060	GEIC-50
NTE1060	PC-20069	SK3460	TCG1060
TM1060	WEP2260	WEP2260/1060	

BLOCK-59

221-Z9108	2360021	AN220	ECG1061
GEIC-143	GEIC-51	GEIC-88	HA1108
IC-4	NTE1061	SK3228	TCG1061
TM1061	WEP2176	WEP2176/1061	

BLOCK-60

23119017	46-1364-3	46-5002-31	AN225
CXL1062	ECG1062	GEIC-184	GEIC-52
NTE1062	SK7728	TA7103P	TCG1062
TM1062	TVCM21	TVSAN225	ULN2242N
WEP2086	WEP2086/1062		

BLOCK-61

AN227	CXL1063	ECG1063	GEIC-53
SK7763	TM1063	TVSAN227	

BLOCK-62

AN229	ECG1064	GEIC-54	NTE1064
TCG1064	TM1064		

BLOCK-63

AN228V	AN228W	AN230	ECG1065
GEIC-55	NTE1065	TCG1065	TM1065
TVSAN230			

BLOCK-64

AN231	ECG1066	GEIC-56	NTE1066
SK7764	TCG1066	TM1066	

BLOCK-65

AN234	ECG1067	GEIC-57	NTE1067
SK7765	TCG1067	TM1067	

BLOCK-66

AN242	ECG1068	GEIC-58	NTE1068
TCG1068	TM1068		

BLOCK-67

221-Z9109	AN288	ECG1069	GEIC-64
NTE1069	SK3227	TCG1069	TM1069
WEP2184	WEP2184/1069		

BLOCK-68

AN328	ECG1070	GEIC-65	NTE1070
TCG1070	TM1070	WEP2185	WEP2185/1070

BLOCK-69

221-Z9110	AN342	AN343	ECG1071
GEIC-67	NTE1071	SK3226	TCG1071
TM1071	WEP2186	WEP2186/1071	

BLOCK-70

171982	221-Z9111	5351061	5351062
AN-253BB	AN252	AN253	AN253AB
AN253BB	AN253P	ECG1072	GEIC-59
NTE1072	SK3295	TCG1072	TM1072

BLOCK-71

56-4833	AN277	AN277AB	AN277B
AN277BA	ECG1073	GEIC-63	NTE1073
RE352-IC	REN1073	SK3496	TCG1073
TM1073			

BLOCK-72

AN260	ECG1074	GEIC-60	IC132
MPC561C	NTE1074	RH-IX0023CEZZ	RS-862S
SK3495	TCG1074	TM1074	UPC561C

BLOCK-73

09-308053	1020-25	21A040-068	221-Z9112
266P30201	5350182	57A32-11	8000-00004-307
905-39B	ECG1075	ECG1075A	GE-117
GEIC-117	IC-508(ELCOM)	MPC20C	NTE1075A
RVIUPC20C2	SK3877	TCG1075A	TM1075A
UPC-20C	UPC206	UPC20C	UPC20C1
UPC20C2	UPC81	UPC81C	WEP938
WEP938/1075A			

BLOCK-74

09-308025	21A101-009	ECG1076	GEIC-125
K18046	MPC47C	SAJ72155	TM1076
UPC47C			

BLOCK-75

09-308028	21A101-008	E23036	E25056
ECG1077	GEIC-124	MPC46C	SAJ72159
TCG1077	UPC46C		

BLOCK-76

221-Z9149	ECG1078	GEIC-136	NTE1078
RVIUPC22C	SK3498	TCG1078	TM1078
UPC-571C	UPC22C	UPC571	UPC571C
WEP2087	WEP2087/1078		

BLOCK-77

ECG1079	GEIC-132	NTE1079	SK7766
TCG1079	TM1079		

BLOCK-78

111-4-2060-03900	1C-112	2002120033	200X2120-033
4-2060-02600	4-2060-02900	4-2060-03900	4206002600
4206003900	46-13101-3	54SI-TA7075P	73SI-MC1353P
ECG1080	GEIC-41	GEIC-98	HA1152
HA1353	IC-112	IC112	LA1353
MC1353P	NTE1080	RE328-IC	REN1080
RH-1X0021CEZZ	RH-IX0021CEZZ	SI-TA7075P	SK3284
TA7075B	TA7075P	TA7075PL	TCG1080
TM1080	WEP2089	WEP2089/1080	

BLOCK-79

25810-107	25810-167	25840-167	27840-164
28810-178	37510-167	916083	916098
D369	DDEY027001	ECG1081	ECG1081A
GEIC-197	NTE1081A	PC563H2	RE339-IC
REN1081A	SK3474	TCG1081A	TM1081A
UPC1025	UPC563H	UPC563H2	WEP2090
WEP2090/1081A			

BLOCK-80

02-300577	09-308052	2000-007	2056-04
28810-175	307-112-9-007	307-115-9-001	61A0001-10
61A001-10	741518	741852	C577H
DDEY064001	ECG1082	GEIC-140	MB3202
MPC577H	MX-3389	NJM2201	NTE1082
PC-20051	RE341-M	REN1082	SK3461
TCG1082	TM1082	UPC577	UPC577A
UPC577H	UPC577N	WEP939	WEP939/1082
YEADUPC577H	YEAMUPC577H		

BLOCK-81

223-Z9000	61A030-9	ECG1083	GEIC-275
NTE1083	RVIMC4080	TCG1083	TM1083

BLOCK-82

155892	21A101-011	ECG1084	GEIC-127
NTE1084	TCG1084	TM1084	UPC-50C

BLOCK-83

BA311	ECG1085	GEIC-104	HC1000505
NTE1085	SK3476	TA7122	TA7122AP
TA7122AR	TCG1085	TM1085	UPC1023H

BLOCK-84

09-308029	203743	21A101-012	ECG1086
GEIC-142	K26174	MPC570C	NTE1086
SAJ2160	SAJ72160	TCG1086	TM1086
UPC49C	UPC570C	UPC570G	

BLOCK-85 — BLOCK-111

BLOCK-85

001-0020-02	0207120	051-0020-00	051-0020-02
051-0035-01	051-0035-02	051-0035-03	051-0035-04
09-308095	171179-045	38510-171	45810-169
46-13173-3	58840-203	740-2007-120	740-9037-120
740-937-120	740782	742363	8-759-271-20
905-177	905-201	B0311000	B0311006
B0311007	B0311008	ECG1087	G09-008-B
G09-008-C	G09-008-D	G09-008-E	GEIC-103
ILT0014	KGE46441	NTE1087	RE340-M
RE340M	REN1087	SK3477	TA-7120P
TA7120P	TA7120P-B	TA7120P-C	TA7120P-D
TA7120P-E	TA7120PB	TA7120PC	TA7220P
TCG1087	TM1087	WEP2293	WEP2293/1087

BLOCK-86

ECG1088	TM1088

BLOCK-87

09-308030	21A101-006	221-Z9157	37003001(IC)
ECG1089	GEIC-120	J241221	MPC29C
MPC29C2	NTE1089	RH-1X0032CEZZ	RH-
IX0032CEZZ	SAJ72161	SK7767	TCG1089
TM1089	UPC29C	UPC29C2	UPC29C3

BLOCK-88

223-Z9003	CXL1090	ECG1090	GE-723
NTE1090	PA-500	PA401	PA401X
PA500	PA501X	RE353-M	REN1090
SK3291	TCG1090	TM1090	TR0-9022
TR09022	TR09023	TR09024	TRO9024

BLOCK-89

09-308026	21A101-010	ECG1091	GEIC-126
K26044	MPC48C	NTE1091	SAJ72156
SK7768	TCG1091	TM1091	UPC48C1

BLOCK-90

09-308096	916063	916092	ECG1092
GEIC-130	NTE1092	RE342-M	REN1092
SK3472	TCG1092	TM1092	UPC555H

BLOCK-91

ECG1093	GEIC-201	KGE41013	NTE1093
TCG1093	TM1093	UPC41C	

BLOCK-92

2360221	4-2060-04200	4-2060-07200	46-13103-3
ECG1094	GEIC-157	GEIC-73	HA1118
HA1158	HA1158-O	KC850C1	LA1373
NTE1094	SK7769	TCG1094	TM1094
WEP2207	WEP2207/1094		

BLOCK-93

41C-402	4613443	4IC-402	70270740
ECG1095	M5101	M5101P	TM1095

BLOCK-94

1-805-105-11	203751	266P00301	266P00801
7625	8-750-105-11	8-759-651-34	8-759-651-42
CX089	CX089D	ECG1096	GEIC-312
M5131P	M5134	M5134-8266	M5134P
NTE1096	RE343-IC	REN1096	SK3703
TCG1096	TM1096		

BLOCK-95

09-308061	2093/1097	41C-406	81-46128001-8
ECG1097	M5106	M5106P	MP5106P
NTE1097	SK3446	TCG1097	TM1097
UPC561	WEP2093	WEP2093/1097	

BLOCK-96

004	004(TOYOTA)	ECG1098	GEIC-311
IC-101-004	M5112	M51124	M5112Y
M5124	NTE1098	SK7770	TCG1098
TM1098	YEAM004		

BLOCK-97

266P00602	ECG1099	M5108	M5108-9170
M5108P	NTE1099	REN1099	TCG1099
TM1099			

BLOCK-98

09-308041	09-308059	1021-25	221-Z9023
25810-164	260-10-036	280-0001	307-007-9-001
307-029-1-001	37510-164	58810-170	58840-201
8000-00016-127	8000-0016-127	8000-0016-129	80910-147
88060-147	88510-177	8910-147	90-36
90-73	BA402	DDEY004001	DEEY004001

BLOCK-98 (CONT.)

EA33X8500	ECG1100	GEIC-92	HC10002050
NTE1100	PTC781	QTVCM-78	RE344-M
REN1100	SK3223	TA-7061AP	TA-7061B
TA-7061P	TA7061AP	TCG1100	TM1100
TVCM-78	WEP941	WEP941/1100	

BLOCK-99

223-Z9004	23119011	23119022	46-1362-3
ECG1101	GEIC-95	NTE1101	SK3283
TA-7069P	TA70L9&	TCG1101	TM1101

BLOCK-100

223-Z9005	307-007-9-002	90-35	90-72
ECG1102	GEIC-93	NTE1102	RE370-IC
REN1102	SK3224	TA7062P	TCG1102
TM1102	WEP942	WEP942/1102	

BLOCK-101

051-0011-00	051-0011-00-04	09-308043	223-Z9006
266P30706	57A32-19	740538	740583
916067	ECG1103	EICM-14	G09-007-A
G09-007-B	G09-029-C	G09-029-D	G09-029B
G09007	GEIC-94	MPC566HB	MPC566HC
MPC566HD	NTE1103	QTVCM-79	RE329-IC
REN1103	SK3281	TA-7063	TA7063
TA7063P	TA7063P-C	TCG1103	TM1103
TVCM-79	WEP943	WEP943/1103	

BLOCK-102

223-Z9007	44T-300-99	90-37	90-74
916105	BA401	ECG1104	GEIC-91
NTE1104	Q-01369	Q-013696	QTVCM-77
SK3225	TA-7060	TA-7060P	TA7060
TA7060P	TCG1104	TM1104	TVCM-77
WEP944	WEP944/1104		

BLOCK-103

1C-126	221-Z9113	23119016	46-1363-3
ECG1105	GEIC-101	IC126	NTE1105
SK3285	TA-7102P	TA7102P	TCG1105
TM1105	WEP2214	WEP2214/1105	

BLOCK-104

09-308066	ECG1106	GEIC-90	NTE1106
SK7771	TA7054	TA7054P	TCG1106
TM1106	WEP2215	WEP2215/1106	

BLOCK-105

051-0011-00-05	09-308067	740502	740543
ECG1107	GEIC-276	NTE1107	SK7772
TA-7092-P	TA7092C	TA7092P	TA7092P-C
TCG1107	TM1107		

BLOCK-106

0207046	09-308070	740-9007-046	ECG1108
GEIC-87	NTE1108	TA-7046	TA7046P
TCG1108	TM1108		

BLOCK-107

221-Z9048	23119013	23119023	23119032
23119033	4-2061-05370	40306604	4206105370
46-1370-3	ECG1109	GEIC-99	NTE1109
SK3711	TA-7076P	TA7076	TA7076AP
TA7076P	TCG1109	TM1109	WEP2094
WEP2094/1109			

BLOCK-108

106719	221-Z9114	41C-409	4IC-409
81-46128002-6	ECG1110	GE-1110	GEIC-319
M5115	M5115P	M5115P-9085	M5115PA
M5115PR	MS115P	NTE1110	SK3229
TCG1110	TM1110	WEP2095	WEP2095/1110

BLOCK-109

1200	2042-2	221-64	3531-033-000
42-2	612042-2	87322-04	BB-42-1B
ECG1111	TAA621	TAA621A11	TM1111

BLOCK-110

2042-1	221-64(IC)	221-80	42-1
612042-1	ECG1112	NTE1112	TAA621A12
TM1112			

BLOCK-111

09-308051	53-1487	ECG1113	SGS17139XRH
SK9190	SN-76001N	SN76001	SN76001A
SN76001AN	SN76001N	TAA611A12	TAA611B
TAA611B12	TM1113	WEP945	WEP945/1113

REPLACEMENTS

BLOCK-112

147-7040-01	156025	221-Z9115	3130-3248-801
3597049	3597280	3598173	544-2006-001
6001	905-244	905-59	ECG1115
GEIC-278	MX-3364	NTE1115	RH-
1X1020AFZZ	RH-IX1020AFZZ	SK3184	TBA-810S
TBA10S-H	TBA810	TBA810AS	TBA810DS
TBA810S	TBA810S-H	TCA830A	TCG1115
TM1115	WEP2098	WEP2098/1115	Y15310159

BLOCK-113

221-116	CA810Q	ECG1115A	NTE1115A
TBA810P	TCA830	TCA830S	TCG1115A
TM1115A	WEP2099	WEP2099/1115A	

BLOCK-114

02-540800	221-Z9123	221-Z9146	56A42-1
93T4150133000	ECG1116	GEIC-279	NTE1116
SK3969	TBA-800	TBA800	TCG1116
TM1116	UPC1001		

BLOCK-115

221-Z9049	576-0004-035(IC)	65-20820-00	760522-0012
905-226	905-298	ECG1117	GEIC-280
NPC544	NTE1117	SK9226	TBA-820
TBA820	TBA820L	TCG1117	TM1117

BLOCK-116

ECG1118	TAA611C11	TM1118	WEP946
WEP946/1118			

BLOCK-117

ECG1119	LAD010	NTE1119	TCG1119
TM1119			

BLOCK-118

09-308044	171179-028	55810-170	8010-172
ECG1120	LAD-011	LAD011	MA6301
MA6301T	NTE1120	TCG1120	TM1120

BLOCK-119

2360231	4-2060-04300	4-2060-07300	46-13104-3
ECG1121	GEIC-158	GEIC-39	HA1119
HA1159	LA1374	NTE1121	SK7773
TCG1121	TM1121	WEP2228	WEP2228/1121

BLOCK-120

4-2060-04800	4-2060-04900	46-13105-3	ECG1122
GEIC-156	GEIC-40	HA1117	HA1157
LA1375	LA1376	NTE1122	TCG1122
TM1122	WEP2229	WEP2229/1122	

BLOCK-121

07-28777-40	144011	221-Z9158	51-90305A20
61C001-2	AN236	ECG1123	GEIC-281
NTE1123	SK3743	TCG1123	TM1123
WEP2223	WEP2223/1123		

BLOCK-122

2057A32-25	21M582	221-Z9138	57A32-25
57A32-27	EA33X8371	ECG1124	GEIC-119
NTE1124	TCG1124	TM1124	TR2367161
UPC276	UPC27C	UPC27CI	

BLOCK-123

AN216	ECG1125	GEIC-282	TM1125

BLOCK-124

ECG1126	GEIC-313	M5155P	NTE1126
SK7774	TCG1126	TM1126	

BLOCK-125

09-308084	221-Z9116	916101	ECG1127
GE-1127	MPC1001H2	NTE1127	REN1127
SK7775	TCG1127	TM1127	UPC1001-H2
UPC1001H	UPC1001H2	WEP2100	WEP2100/1127

BLOCK-126

1061-9153	203780	2119-101-1301	221-Z9137
23119981	23119990	2319990	46-1396-3
6644001100	A0311400	B0311400	B0311402
B0311405	ECG1128	GEIC-105	NTE1128
RE372-IC	REN1128	SK3488	TA2124P
TA7124P	TA7124PFA-3	TA7124PFA-4	TCG1128
TM1128			

BLOCK-127

1061-9666	1066-9666	203781	23119993
46-1393-3	6644001900	B0313800	ECG1130
GEIC-111	NTE1130	RE354-IC	REN1130
SK3478	TA7150P	TCG1130	TM1130

BLOCK-128

221-Z9117	23119995	46-1394-3	6644001700
B0313600	ECG1131	GEIC-109	NTE1131
RE373-IC	REN1131	SK3286	TA7148P
TCG1131	TM1131		

BLOCK-129

221-Z9118	23119994	46-1397-3	6644001800
B0313700	ECG1132	GEIC-110	NTE1132
RE374-IC	REN1132	SK3287	TA7149P
TCG1132	TM1132		

BLOCK-130

1061-9856	21A120-071	221-Z9177	23119988
46-1398-3	6644001500	B0313400	ECG1133
GEIC-107	NTE1133	RE355-IC	REN1133
SK3490	TA7146P	TCG1133	TM1133
WEP2102	WEP2102/1133		

BLOCK-131

1061-9161	2119-101-1408	221-Z9050	23119989
46-1395-3	6644001400	B0313300	B0313500
ECG1134	GEIC-106	NTE1134	RE356-IC
REN1134	SK3489	TA7145P	TCG1134
TM1134			

BLOCK-132

171179-070	23119961	BA301	BA301B
ECG1135	EW84X487	GEIC-314	GEIC-318
NTE1135	PC-20071	RE375-IC	REN1135
SK3876	TCG1135	TM1135	WEP2242
WEP2242/1135			

BLOCK-133

AN208	ECG1136	GEIC-284	TM1136
TVC-MK3800	TVCMK3800	WEP2103	WEP2103/1136

BLOCK-134

AN274	ECG1137	GEIC-62	NTE1137
SK7776	TCG1137	TM1137	

BLOCK-135

CXL1139	ECG1139	NTE1139	STK-018
SVI-STK018	TCG1139	TM1139	

BLOCK-136

001-0091	1032-25	1193-0054	1259-6516
178690	2119-202-0803	221-Z9135	36-0083
37001040	4151005753	8-759-157-52	90200100
C575C2	ECG1140	GEIC-138	IX0054CE
MPC575C2	MX-3393	NTE1140	
RH-IX0054CEZZ	RVIUPC575	SK3473	TCG1140
TM1140	UPC575	UPC575C	UPC575C2
WEP2105	WEP2105/1140	X0054CE	

BLOCK-137

ECG1141	MPC575C	NTE1141	TCG1141
TM1141			

BLOCK-138

000074030	051-0039-00	09-308071	18600-155
2057A32-26	21M485	221-Z9024	25810-165
25840-165	28810-176	36-0041	518022S
57A32-2	57A32-21	57A32-28	57A32-29
588-40-202	58810-171	61A0001-11	61A001-11
740-9000-554	740-9000-754	87-0216	8840-171
88510-178	8910-184	916070	C554C
EA33X8372	ECG1142	GEIC-128	HEPC6079P
NTE1142	RE330-IC	REN1142	SK3485
TCG1142	TM1142	TVCM-60	UPC-554C
UPC554	UPC554C	UPC554H	WEP2248
WEP2248/1142			

BLOCK-139

223-Z9008	ECG1148	NTE1148	PA501
SK3292	STK-036	STK-075	TCG1148

BLOCK-140

ECG1149	GEIC-285	NTE1149	TCG1149
TM1149			

BLOCK-141

ECG1150	GEIC-286	NTE1150	SK3890
TCG1150	TM1150	WEP2106	WEP2106/1150

BLOCK-142

ECG1152	IC-550(ELCOM)	LD3110	LD3110A
NTE1152	TCG1152	TM1152	

BLOCK-143

09-308086	221-Z9119	740-9037-204	916109
ECG1153	GEIC-182	NTE1153	RE376-IC
REN1153	SK3282	TA7200P	TA7204P
TCG1153	TM1153	WEP948	WEP948/1153

BLOCK-144

221-Z9120	ECG1154	GEIC-287	NTE1154
SK3230	TA7203P	TCG1154	TM1154
WEP2107	WEP2107/1154		

BLOCK-145

02-124422	02-257205	02-437205	0207205
051-0055-02	051-0055-0203	051-0055-03	09-308080
221-Z9051	26810-159	276-705	307-107-9-003
44T-100-120	44T-300-102	45810-170	5002-031
740-9007-205	740-9607-205	741687	742364
916110	B0319200	BO319200	E1CM-0060
EA33X8389	EA33X8396	ECG1155	EICM-0060
ETI-23	GEIC-179	IP20-0161	KIA7205AP
KIA7205P	NTE1155	PTC780	QQ-M07205AT
QQ-MO7205AT	QQM07205AT	QTVCM-81	RE357-IC
REN1155	SK3231	TA-7205P	TA7205
TA7205A	TA7205AP	TA7205AT	TA7205P
TCG1155	TM1155	TVCM-81	WEP949
WEP949/1155			

BLOCK-146

ECG1156	M5115PRA	M5115RP	NTE1156
TCG1156	TM1156	WEP2108	WEP2108/1156

BLOCK-147

111-4-2060-05100	221-Z9121	4-2060-07500	4206007500
ECG1158	GEIC-71	LA1366	LA1366N
LA3166N	NTE1158	SK3289	TCG1158
TM1158	WEP2109	WEP2109/1158	

BLOCK-148

111-4-2060-05200	221-Z9122	4-2060-05200	4206005200
46-13119-3	ECG1159	GEIC-72	LA1367
NTE1159	SK3290	TCG1159	TM1159

BLOCK-149

2056-06	2091-02	741673	ECG1160
ESI-UPC1020H	GEIC-196	GEIC197	NTE1160
SK3243	TCG1160	TM1160	UPC1020
UPC1020H	UPC1025H	WEP2258	WEP2258/1160

BLOCK-150

1052-6390	221-Z9052	51M33801A06	51S13753A03
AN239	AN239Q	AN239QA	AN239QB
ECG1161	GEIC-300	NTE1161	RE377-IC
REN1161	SK3968	TCG1161	TM1161
WEP2111	WEP2111/1161		

BLOCK-151

02-010241	51-13753A04	51-90433A60	51M33801A01
8-759-424	8-759-425	AN241	AN241A
AN241B	AN241C	AN241D	AN241P
AN241PD	ECG1162	GE-1162	GEIC-320
NTE1162	REN1162	SK9266	TCG1162
TM1162	TVSAN241	WEP2294	WEP2294/1162

BLOCK-152

ECG1163	GEIC-68	NTE1163	REN1163
SK7778	TCG1163	TM1163	

BLOCK-153

02-010245	06300003	1052-6408	221-Z9053
51-13753A10	51M33801A07	51S13753A02	51S13753A10
AN-246	AN245	AN246	ECG1164
GE-1164	GEIC-301	GEIC-321	NTE1164
RE380-IC	REN1164	SK3727	TCG1164
TM1164	WEP2112	WEP2112/1164	

BLOCK-154

AZQQ001GEA	BA511	BA511A	ECG1165
GEIC-315	NTE1165	QQ-MBA511BX	RE358-IC
REN1165	TCG1165	TM1165	WEP2001
WEP2001/1165			

BLOCK-155

106625	1223910	BA521	BA521A
BA521AX	ECG1166	GEIC-316	NTE1166
QQ-MBA521AX	RE359-IC	REN1166	SK3827
TCG1166	TM1166	WEP2003	WEP2003/1166

BLOCK-156

02-360002	051-0100-00	106525	1223909
5002-030	741686	EA33X8388	ECG1167
GEIC-317	MC145109	MC145109P	NPC7629
NTE1167	PL102	PLL02	PLL02A
PLL02A-C	PLL02A-F	QQ-0PPL02A0	QQ-OPLL02A0
QQ-OPLL02AN	QQ-OPLLO2AO	QQ-OPLLO2AN	QQ-OPPL02A0
QQ0PLL02AN	REN1167	SK3732	SM5109
TC9100C	TC9100P	TCG1167	TM1167
WEP2007/1233			

BLOCK-157

02-010331	221-Z9124	221-Z9159	51S13753A08
AN331	ECG1168	GEIC-66	NTE1168
SK3728	TCG1168	TM1168	

BLOCK-158

022-2844-001	022-2844-703	09A10	221-Z9125
5350321	5350491	ECG1169	GE-1169
GEIC-322	HA-1339	HA1139A	HA1339
HA1339A	NTE1169	SK3708	TCG1169
TM1169			

BLOCK-159

003526	02-161521	51-42211P01	741473
905-173	916106	BA328MR	DDEY123001
ECG1170	HC10001200	LA3160	LA3161
M51521L	M5152L	MX-3948	NTE1170
RVIBA328MR	SK3874	TCG1170	TM1170
UPC1032H	YEAMM5152L		

BLOCK-160

1CL-101-TY	1CL-101A-TY	1CL-201-TY	1CL-201A-TY
1CL-301A-PA	1CL-301A-TY	1CL-748-TY	AD101AH
AD201AH	AD201H	AD301AH	AM748HC
AM748HM	AMLM101AH	AMLM101H	AMLM201
AMLM201A	AMLM201AH	AMLM201H	AMLM301
AMLM301A	AMLM301AH	CA101AT	CA101T
CA201T	CA301AS	CA301AT	CA301T
CA3748CT	CA748CT	CA748T	ECG1171
GEIC-173	LM101AH	LM101H	LM201AH
LM201H	LM301AH	LM748CH	LM748H
LS301AT	MC1748G	ML201AT	ML301T
MLM201AG	MLM301AG	NTE1171	SFC2301A
SFC2748C	SFC2748M	SG301AT	SG301AT
SK3565	SK3645	SN52101AL	SN52101L
SN52748L	SN72301AL	SN72301AN	SN7248L
SN72748L	TBB0748	TCG1171	TM1171
UA1748G	UA201AH	UA201H	UA201HC
UA301AH	UA301AHC	UA748CL	UA748CT
UA748HC	UA748HM	UA748ML	UA748T
UPC157A	UPC15MA		

BLOCK-161

159425	C3806P	DDEY017001	DDEY032001
ECG1172	GEIC-199	HEP3806P	HEPC3806P
NTE1172	TCG1172	TM1172	UPC1008C
WEP2113	WEP2113/1172		

BLOCK-162

06300001	1052-6416	1065-2055	221-Z9054
51-13753A18	51M33801A05	51S13753A06	AN247
AN247P	ECG1173	GEIC-302	M51247
M51247P	NTE1173	SK3729	TCG1173
TM1173	TVSM51247	TVSM51247P	WEP2115
WEP2115/1173			

BLOCK-163

1044-7043	1044-7049	141135	143821
1465345-1	221-Z9055	51-13753A46	51S13749A46
51S13753A46	5320500120	612272-1	CA3139
CA3139E	CA3139G	CA3139GM1	CA3139GMI
CA3139GQ	CA3139Q	ECG1174	MR2525
NTE1174	RE389-IC	REN1174	SK3186
TCG1174	TM1174	TVSCA3139GM1	TVSCA3139GMI

BLOCK-164

1063-4939	1081-3541	1183-1633	141270
142718	143033	146052	1465617-2
153684	221-Z9056	86X0086-001	86X86-1
CA3134	CA3134EM	CA3134GQM	CA3134Q
CA3134QM	ECG1175	GE-1175	NTE1175
SK3212	TA6480	TCG1175	TM1175
WEP2175	WEP2175/1175		

BLOCK-165

141280	1465615	1465615-1	1465615-2
221-Z9057	51-13753A20	CA3137	CA3137E
CA3137EM	CA3137EM1	CA3137EMI	ECG1176
NTE1176	SK3210	TASCA3137	TCG1176
TM1176	TVSCA3137		

BLOCK-166

141290	1466701-1	221-Z9126	CA3143
CA3143E	CA3143Q	CA314E	ECG1177
NTE1177	SK3213	TM1177	

BLOCK-167

4-2060-09100	A1369	ECG1178	GE-1178
GEIC-323	NTE1178	SK3480	TA7103
TCG1178	TM1178	ULN2224A	WEP2296
WEP2296/1178			

BLOCK-168

104825	206-5-0723-20110	4152032010	EA33X8374
ECG1179	GE-1179	GEIC-324	GEIC-75
LA3201	NTE1179	SK3737	TCG1179
TM1179			

BLOCK-169

104925	206-5-0824-10110	4152041000	905-214
EA33X8363	ECG1180	GEIC-80	LA4100
LA4101	NTE1180	SK3888	TCG1180
TM1180			

BLOCK-170

14LN034	AN271	AN271B	AN271FB
ECG1181	GEIC-61	NTE1181	SK3706
TCG1181	TM1181		

BLOCK-171

AN289	ECG1182	GEIC-303	SK7779
TM1182			

BLOCK-172

221-Z9058	221-Z9178	266P60304	4206009100
4206009800	46-13125-3	46-1325-3	ECG1183
GE-1183	GEIC-325	LA1369	M5192P
NTE1183	PTC740	TCG1183	TM1183
WEP2297	WEP2297/1183		

BLOCK-173

ECG1184	GEIC-137	NTE1184	SK3486
TCG1184	TM1184	UPC-573C	UPC573C

BLOCK-174

1-800-662-11	221-Z9160	61K001-13	8-759-157-60
87-0217	916002	916102	ECG1185
GEIC-139	NTE1185	SK3468	TCG1185
TM1185	UPC-576H	UPC576H	

BLOCK-175

1001-7150	1001-9677	1140-2302	221-Z9059
40-13184-3	46-13184-3	51-90305A04	ECG1186
EP84X1	EP84X10	GEIC-175	LA1352
MPC595C	NTE1186	TCG1186	TM1186
TVSMPC595C	TVSUPC595C	UPC595C	WEP2286
WEP2286/1186			

BLOCK-176

1001-7168	1001-9685	1116-6691	221-Z9060
46-13185-3	51-90305A05	51-90433A66	56A19-1
C596(IC)	C596C(IC)	ECG1187	GEIC-176
LA1354	MPC596C	MPC596C2	MPC596C2B
NTE1187	RH-IX0001PAZZ	SK3742	TCG1187
TM1187	TVSMPC596C	TVSMPC596C2	TVSMPS596C2
TVSMPC596C2B	TVSMPC596CE	TVSMPC596CZ	
TVSUPC596C2	UPC-596C	UPC595C2	UPC596
UPC596C	UPC596C2	WEP2259	WEP2259/1187

BLOCK-177

ECG1188	EICM-19	GEIC-82	NTE1188
SK3022	TA7028M	TCG1188	TM1188

BLOCK-178

221-Z9127	ECG1189	HA11112	HA1112
NTE1189	RH-1X0047CEZZ	RH-IX0047CE	
RH-IX0047CEZZ	SK3705	TCG1189	TM1189
UPC569C2			

BLOCK-179

221-Z9129	AN380	ECG1191	NTE1191
TCG1191	TM1191	UPC1380C	

BLOCK-180

02-010103	02-257310	02-263001	02-373001
02-437310	10-004#	10-004(IC)	10-004IC
107225	11-113	1224076	2000-017
221-Z9061	266P32501	307-133-9-004	3071339004
39-061-0	640000002	742510	8000-00058-009
AN103	AN103-0	AN103G	C3001
C3001-0	C3001-O	C3001A	C3001M
C3001T	DDEY109001	EA33X8392	EA33X8394
EA33X8509	ECG1192	EX33X8392	GE-1192
GEIC-326	IP20-0429	KGE46442	KIA7310P
MC3001	MX-3260	MX-3545	MX3256
NTE1192	QQ-MC3001AN	QQ-MC3001AT	QQ-MC3001DT
RE390-IC	REN1192	RH-IX1068AFZZ	SK3445
TA310P	TA7310	TA7310P	TA7310P-U
TA7310P-Y	TC3001	TCG1192	TM1192
WEP2120	WEP2120/1192		

BLOCK-181

02-124400	221-Z9062	EA33X8384	ECG1193
GE-1193	LA4400	LA4400FR	NTE1193
SK3701	TCG1193	TM1193	WEP2121
WEP2121/1193			

BLOCK-182

003536	02-173710	1156(TC)	2000-005
221-Z9063	307-112-9-001	916125	C1156H
DDEY061001	DDEY091001	DDF4091001	DDFY091001
ECG1194	GE-1194	GEIC-327	MB3710
MX-3256	MX-3372	MX-3587	MX3587
NTE1194	SK3484	TCG1194	TM1194
UPC1154H	UPC1155H	UPC1156	UPC1156H
UPC1156H2	WEP2126	WEP2126/1194	

BLOCK-183

2000-006	307-112-9-006	307-113-9-004	905-137
905-170	DDEY097001	ECG1195	MX-3370
NTE1195	SK3469	TCG1195	TM1195
UPC592A2	UPC592H	UPC592H2	UPC592HZ
UPC592N			

BLOCK-184

02-301380	02-311380	06300004	1003-0567
1103-7751	2002501708	200X2501-708	2119-101-1602
21A120-044	221-Z9064	23119966	2360431
2360441	35SI-UPC580C	46-13143-3	51-13753A01
51M33801A04	51S13753A01	65SI-HA11580	6644004301
B0318920	ECG1196	EP84X35	GE-1196
GEIC-328	HA11580	HA1177	KC580C1
NTE1196	SI-HA11580	SI-UPC580C	SK3725
TA7192P	TA7662P	TCG1196	TM1196
TVSUPC1380C	UPC1380	UPC1880C	UPC580C
WEP2261	WEP2261/1196		

BLOCK-185

307-107-9-005	ECG1197	NTE1197	
RH-1X1039AFZZ	RH-IX0015TAZZ	SK3733	TC5080
TC5082	TC5082P	TC5082P-L	TCG1197
TM1197			

BLOCK-186

2000-001	2056-105	2065-53	307-095-9-004
307-112-9-009	D858(IC)	DDEY055001	DDFY055001
ECG1198	GE-1198	KM858C	MPD858
MPD858C	MX-3399	NTE1198	TCG1198
UPC858	UPC858C	UPD-858C	UPD858
UPD8585C	UPD858C	WEP2127	WEP2127/1198
WPD858			

BLOCK-187

DDEY037001	ECG1199	NTE1199	TCG1199
TDB0117	UPD857C		

BLOCK-188

1061-9526	23119971	B0315500	ECG1200
NTE1200	SK3714	TA7167	TA7167P
TCG1200	TM1200		

BLOCK-189

2065-50	ECG1201	NTE1201	TCG1201
TM1201	UHIC-001	UHIC001	WEP2131
WEP2131/1201			

BLOCK-190

307-108-9-001	ECG1202	NTE1202	TCG1202
TM1202	UHIC-003	UHIC003	WEP2132
WEP2132/1202			

BLOCK-191

2000-003	307-113-9-002	307-131-9-001	307-131-9001
DDEY087001	DE-087	ECG1203	NTE1203
TCG1203	TM1203	UHIC-004	UHIC-004E
UHIC004	WEP2135	WEP2135/1203	

BLOCK-192

2000-010	2000-011	DDEY082001	ECG1204
MX-3379	NTE1204	TCG1204	TM1204
UHIC-005	UHIC005	WEP2139	WEP2139/1204

BLOCK-193

ECG1205	GEIC-178	NTE1205	SK3482
TA-7055P	TA-7055P-E	TA7055P	TA7055P-D
TA7055P-E	TA7055P-F	TCG1205	TM1205

BLOCK-194

ECG1206	NTE1206	TA7157P	TCG1206
TM1206			

BLOCK-195

307-107-9-006	ECG1207	SK3713	TC5080P
TM1207			

BLOCK-196

307-107-9-004	ECG1208	RH-IX1039AFZZ	SK3712
TC5081	TC5081P	TCG1208	TM1208

BLOCK-197

ECG1209	NTE1209	TCG1209	TM1209

BLOCK-198

0071-0030	BA313	ECG1210	NTE1210
SK9074	TA7137(IC)	TCG1210	TM1210

BLOCK-199

4400Y	ECG1211	GE-1211	GEIC-329
LA4220	LA4400Y	LA4420	NTE1211
SK3739	TCG1211	TM1211	WEP2141
WEP2141/1211			

BLOCK-200

AGH-001	ECG1212	GE-725	NTE1212
STK013	TCG1212	TM1212	WEP2142
WEP2142/1212			

BLOCK-201

2360151	2360631	61L001-3	ECG1213
GEIC-150	HA11228	HA1144	NTE1213
SK3704	TCG1213	TM1213	

BLOCK-202

30900460	39-059-0	668-614A	905-155
905-178	ECG1214	HA1197	HA1197W
LA1240	NTE1214	SK3736	TCG1214
TM1214			

BLOCK-203

ECG1215	GEIC-74	LA3155	NTE1215
SK3738	TCG1215	TM1215	

BLOCK-204

221-Z9166	4206012500	46-13180-3	ECG1216
LA1390	LA1390B	LA1390C	LA3190
NTE1216	SK3735	TCG1216	TM1216

BLOCK-205

003516	051-0088-00	051-0088-00101	266P35001
2911501	46-131238-3	6190000210	740-9003-350
905-121	A3350	AN363N	BA1330
ECG1217	EW84X793	GEIC-76	IX0488CE
LA3350	LA3350A	LA350	NTE1217
PC-20082	Q7668	RH-IX0488CEZZ	RV1BA1330
RVIBA1330	SK3700	SK4801	TA7155P
TCG1217	TM1217	TRLA3350	WEP2143
WEP2143/1217			

BLOCK-206

1077-3844	266P36402	46-131357-3	668-028B
905-122	905-138	ECG1218	GE-1218
GEIC-330	NTE1218	SK3740	STK-433
STK413	STK415	STK433	STK435
STK4362	SYK4332	TCG1218	TM1218
WEP2144	WEP2144/1218		

BLOCK-207

21192050103	266P35801	4835-209-87082	6125040001
668-616A	905-139	905-144	905-171
ECG1219	GE-1219	GEIC-331	NTE1219
PC-20083	SK3741	STK-437	STK437
STK439	STK4392	TCG1219	TM1219
WEP2145	WEP2145/1219		

BLOCK-208

ECG1220			

BLOCK-209

ECG1221	GEIC-202	NTE1221	SK7781
TCG1221	TM1221		

BLOCK-210

ECG1222	SK3730	TA7209P	TM1222

BLOCK-211

07-28816-73	221-Z9161	61C001-12	87-0246
870246	AN360	ECG1223	GEIC-295
NTE1223	SK3493	TCG1223	TM1223

BLOCK-212

221-Z9130	AN374	ECG1224	GEIC-296
NTE1224	SK7782	TCG1224	TM1224

BLOCK-213

ECG1225	GEIC-204	LA3311	TM1225

BLOCK-214

051-0086-00	1099-2949	1155-5570	1181-3201
307-131-9-006	742362	905-325	916155
AN115	BA1310	BA1310F	DDEY179001
EA33X8535	ECG1226	HA12003	MC1309
MC1309P	MX-4974	NTE1226	PC-20151
PC1026C	SK3763	TCG1226	TM1226
UPC1026	UPC1026C	UPC587C	UPC587C2

BLOCK-215

206-5-0131-22210	46-131230-3	ECG1227	GEIC-70
LA1222	NTE1227	Q7748	SK3762
TCG1227	TM1227		

BLOCK-216

1A4102	206-5-0834-10214	905-254	905-291
ECG1228	GE-1228	GEIC-332	LA4102
NTE1228	SK3889	SZ-LA4102	TCG1228
TM1228	WEP2148	WEP2148/1228	

BLOCK-217

ECG1229	GEIC-153	HA1148	NTE1229
SK7783	TCG1229	TM1229	

BLOCK-218

1230	46-1391-3	ECG1230	GEIC-203
LA3310	NTE1230	TCG1230	TM1230

BLOCK-219

1141-3119	15-43636-1	221-98	221-98-05
221-98A05	51-13753A40	51S1373A40	51S13753A40
51X90480A94	56A65-1	56D65-1	AN1364
CA1190G	CA1190GM	CA1190GQ	CA1190GQ/M
ECG1231	EN11301	EN1364	EN1364/1231
EP84X81	GE-1231	GEIC-333	HA1364
NTE1231	SC5898	SK3832	TCG1231
TDA11907	TDA1190Z	TDAV190Z	TM1231
TVSEN11301	TVSEN1364	TVSTDA1190Z	ULN2290Q
WEP1231/1231			

BLOCK-220

15-3017045-1	221-158	221-158-01	221-158-02
221-158-02001	221-158-03	4H20981194	56A122-1
612305	612305-1	612305-2	6123620001
6123620002	A-9982	A-9982-07	A-9982A
ECG1231A	NTE1231A	SK9384	TCG1231A
TDA1190P	TDA3190	ULN2290B	

BLOCK-221

00S08	147-7043-01	15-301513-1	1M383T
2366591	276-703	383(IC)	3L4-9020-1
442-713	51-40000S08	544-2006-011	640000003

BLOCK-221 (CONT.)

8-759-112-38	8520	AE920	CA2002
CA2002M	CA2002V	CA2004	ECG1232
GE-1232	GEIC-334	LM383	LM383AT
LM383T	NA2002	NAM383	NTE1232
SK3852	TCG1232	TDA2002	TM1232
UA2002	ULN3701Z	ULX3701Z	UPC1238V
UPC2002	UPC2002V	WEP2150	WEP2150/1232

BLOCK-222

221-Z9065	ECG1233	NPC76315	NTE1233
PLL02A-G	PLL02AG	QQ-OPLL020A	QQOPLLO2AN
RE360-IC	TCG1233	TM1233	WEP2007

BLOCK-223

003519	0079-0170	02-257130	107255
107625	11119027	1148-9556	1172-5157
154754	221-Z9066	26810-156	51-44789J01
51-44789J02	51-90433A54	51544789J01	51544789J02
5190507A20	51S44789J01	51S44789J02	51X90433A54
51X90458A73	AN5732	916121	AN5215
AN5730	AN5732	ECG1234	EW84X217
GEIC-181	HC10022010	HC10034050	MX-5202
MX5202	NTE1234	PC-20085	SK3487
TA7130P	TA7130P-B	TA7130P-C	TA7130PB
TA7130PC	TCG1234	TM1234	UPC1028H
VHITA7130P2/F			

BLOCK-224

307-107-9-001	B0305401	ECG1235	NTE1235
SK3637	TA7064P	TA7064P-JA	TCG1235
TM1235			

BLOCK-225

1176-7274	2119-101-1505	23119978	46-13121-3
46-13561-3	B0316403	B0356711	ECG1236
GEIC-114	NTE1236	RH-1X0018TAZZ	RH-IX0018TAZZ
SK7784	TA7176AP(FA-1)	TA7176APFA-1	TA7176P
TA7671P	TA7671P(FA-1)	TA7671PFA-1	TCG1236
TM1236			

BLOCK-226

155911	4154011510	ECG1237	GEIC-154
HA1151	HA11516C4	HA11516E1	HA11516J
NTE1237	RH-IX1030AFZZ	SK3707	TCG1237
TM1237	TRHA1151	TVSHA1151	

BLOCK-227

221-Z9131	5652-HA1325	5662-HA1325	ECG1238
HA1325	SK3373	TCG1238	TM1238

BLOCK-228

09A08	51-44837J04	ECG1239	GE-1239
GEIC-335	HA1342	HA1342A	HC10023010
NTE1239	TCG1239	TM1239	

BLOCK-229

221-Z9067	307-120-9-001	AN-315	AN315
ECG1240	IP20-0281	NTE1240	SK7785
TCG1240	TM1240	WEP2262	WEP2262/1240

BLOCK-230

BA318	ECG1241	HC10019210	NTE1241
SK3828	TCG1241	TM1241	

BLOCK-231

87-0233	870233	AN366	ECG1242
GEIC-291	NTE1242	SK3483	TCG1242
TM1242			

BLOCK-232

09-308094	6207159	ECG1243	GEIC-113
NTE1243	SK3731	TA7159P	TCG1243
TM1243			

BLOCK-233

BA501	BA501A	ECG1244	TA7115P
TM1244	WEP2010	WEP2010/1244	

BLOCK-234

2002300012	46-13223-3	56A121-1	8-759-10-31
8-759-110-31	905-165	ECG1245	I02S010310
LA1385	NTE1245	RH-1X0015TAZZ	
RH-IX0035GAZZ	RH-IX0035TAZZ	RX-IX0035GAZZ	SI-UPC1031H
SK3878	SK9185	TCG1245	TM1245
UPC1031H	UPC1031H2	WEP2245	WEP2245/1245
Y130213002			

BLOCK-235

0079-0060	21A120-208	221-Z9179	8-759-113-53
ECG1246	MX-4479	NTE1246	SK3879
TCG1246	TM1246	UPC1353C	WEP1246/1246

BLOCK-236

ECG1247	NTE1247	TCG1247	TM1247
UHIC006			

BLOCK-237

1000-6948	1004-1994	1006-0887	196121
221-Z9132	300012026007	4H-209-80821	51-90305A78
65-33361-00	742723	8-759-833-61	905-253
AN352	AN362	AN362L	BA1320
EA33X8607	EA33X8705	ECG1248	GEIC-292
HA12026	KA2261	LA3361	LA3361A
MX-3976	MX-5430	MX-5594	NTE1248
SK3497	TCG1248	TM1248	UPC1197C

BLOCK-238

02-010612	10-005	10-005(IC)	2000-031
221-Z9068	307-143-9-002	AN612	DDEY130001
EA33X8395	EA33X8508	ECG1249	NTE1249
QQ-MAN612AN	QQMAN612AN	SK3963	TCG1249
TM1249			

BLOCK-239

221-Z9069	AN326	ECG1250	NTE1250
TCG1250	TM1250		

BLOCK-240

2360281	61L001-8	ECG1251	KC582C
NTE1251	SK7723	TCG1251	TM1251

BLOCK-241

ECG1252	NTE1252	SK7786	TCG1252
TM1252			

BLOCK-242

02-435624	2000-052	ECG1253	HD42853
K76026	KM5624	MX-3540	TCG1253
TM1253	UPD2814C		

BLOCK-243

051-0087-00	10-001	1147-161	221-Z9070
5359261	5359262	8000-00049-025	ECG1254
HC10008060	HD42851	HD42851A2	MPD861C
MPD861CE	MX-3235	NTE1254	RVIUPD861
SK3880	TCG1254	TM1254	UPC861C
UPC861CE	UPD861	UPD861C	UPD861CE
UPD861E			

BLOCK-244

02-235104	2906-004	5359251	ECG1255
MC145104	MC145104P	MM55104	MM55104N
MN6040	MN6040A	NP4510	NPC76365
NTE1255	RV1SM5104	SM5104	SM5104F
SM5104G	SM5104P	TM1255	

BLOCK-245

11-102	5652-M51513L	ECG1256	M51513L
M51513LB	NTE1256	SK3873	TCG1256
TM1256			

BLOCK-246

21A120-045	2360291	61L001-5	ECG1257
GE-1257	KC383C	KC583C	NTE1257
SK7724	TCG1257	TM1257	

BLOCK-247

21A120-009	2360271	61L001-6	ECG1258
GE-1258	KC581C	NTE1258	SK7722
TCG1258	TM1258		

BLOCK-248

02-091366	221-Z9071	266P32402	307-131-9-002
39-060-1	ECG1261	HA1366R	HA1366WR
IP20-0317	NTE1261	SK3872	TCG1261
TM1261	WEP2265		

BLOCK-249

02-280007	2000-029	307-143-9-001	DDEY133001
ECG1262	NTE1262	TCG1262	TM1262
UHIC-007	UHIC007		

BLOCK-250

07-28777-42	144012	221-Z9162	46-13172-3
51-90305A21	51X90305A21	61C001-4	70119023
AN262	ECG1263	GEIC-293	IMAN262
NTE1263	SK3920	TCG1263	TM1263
WEP2266	WEP2266/1263		

BLOCK-251

AN302	ECG1264	NTE1264	TCG1264
TM1264			

BLOCK-252

AN303	ECG1265	NTE1265	SK7787
TCG1265	TM1265		

BLOCK-253

07-28777-45	144015	221-Z9180	51-90305A24
61C001-7	AN305	ECG1266	NTE1266
TCG1266	TM1266		

BLOCK-254

AN316	ECG1267	NTE1267	TCG1267
TM1267			

BLOCK-255

70119041	AN318	ECG1268	NTE1268
SK3921	TCG1268	TM1268	

BLOCK-256

07-28777-49	144018	221-Z9181	51-90305A27
61C001-11	AN337	ECG1269	NTE1269
SK9054	TCG1269	TM1269	WEP2268
WEP2268/1269			

BLOCK-257

148715	221-Z9182	51X90458A78	AN640G
ECG1270	SK9055	TM1270	

BLOCK-258

02-173756	2000-033	307-133-9-005	307-133-9001
DDEY131001	ECG1271	MB3756	NTE1271
SK7789	TCG1271	TM1271	

BLOCK-259

02-360003	1224275	EA33X8501	ECG1272
NTE1272	P1103	PLL03	PLL03A
TCG1272	TM1272		

BLOCK-260

ECG1273	NTE1273	SK3916	TCG1273
TM1273			

BLOCK-261

ECG1274	TM1274

BLOCK-262

ECG1275	SK7790

BLOCK-263

544-2006-11	ECG1276

BLOCK-264

DM-59	DM-84	DM-98	ECG1277

BLOCK-265

0070-0120	02-252222	02-257222	2000-032
307-143-9-003	DDEY146001	ECG1278	I05SP72220
NTE1278	SK3726	TA7222	TA7222AP
TA7222P	TCG1278		

BLOCK-266

ECG1279

BLOCK-267

ECG1280

BLOCK-268

ECG1281	NTE1281	SK7661	STK0050
TCG1281			

BLOCK-269

ECG1282	NTE1282	TCG1282

BLOCK-270

ECG1283	NTE1283	SK9267	TCG1283

BLOCK-271

ECG1284	NTE1284	SK9268	TCG1284

BLOCK-272

02-301181	ECG1285	NTE1285	SK3922
TCG1285	UPC1181	UPC1181H	

BLOCK-273

02-301182	DDEY149001	EA33X8545	ECG1286
NTE1286	SK3923	TCG1286	UPC1182
UPC1182A	UPC1182H		

BLOCK-274

ECG1287	NTE1287	SK9228	TCG1287

BLOCK-275

4H-209-81473	ECG1288	NTE1288	SK9252
TCG1288	TDA2003	TDA2003V	

BLOCK-276

15-43637-1	ECG1289	NTE1289	SK9182
SK9282	TCG1289	TDA-1170	TDA1170
TDA1170S			

BLOCK-277

051-0066-00	ECG1290	HA1199	HA1199P
NTE1290	SK9183	TCG1290	ULN2249A

BLOCK-278

ECG1291	M51515L	NTE1291	SK3875
TCG1291			

BLOCK-279

4H2098-0357	56A101-1	ECG1292	NTE1292
SK9181	TBA120AS	TBA120S	TCG1292

BLOCK-280

8-759-101-80	8-759-111-85	ECG1293	NTE1293
SK7649	TCG1293	UPC1185H	UPC1185H2

BLOCK-281

221-202	ECG1294	NTE1294	SK9394
TBA820M			

BLOCK-282

51X90458A52	AN5435	ECG1295	EW84X424
NTE1295	SK9299		

BLOCK-283

2002502202	ECG1296	HA11247	NTE1296
SK9311			

BLOCK-284

266P61401	ECG1297	M5193P	NTE1297
SK9265			

BLOCK-285

266P20402	ECG1298	M5195	M5195P
NTE1298	SK9398		

BLOCK-286

1177-1169	2109-102-5401	2119-102-5401	46-13552-3
51-13753A24	51X13753A24	AN5411	CN5411
ECG1299	NTE1299	SK9314	TVSCN5411

BLOCK-287

CX065	CX065B	ECG1300	NTE1300
TCG1300	TM1300	X440200650	

BLOCK-288

2047-52	CX-075	CX-075B	CX075B
EA33X8333	EA33X8383	ECG1301	M51171L
M5117L	MX-3240	NTE1301	TCG1301
TM1301			

BLOCK-289

8-751-700-10	CX170	ECG1302	NTE1302
TCG1302	TM1302		

BLOCK-290

ECG1303	NTE1303	TCG1303	TM1303

BLOCK-291

CX172	ECG1304	NTE1304	TCG1304
TM1304			

BLOCK-292

CX173	ECG1305	NTE1305	SK7791
TCG1305	TM1305		

BLOCK-293

8-751-771-00	8-751-771-10	905-189	CX-177B
CX177	CX177B	ECG1306	NTE1306
TCG1306	TM1306		

BLOCK-294

CX181	ECG1307	NTE1307	TCG1307
TM1307			

BLOCK-295

095C	8-759-600-95	8-759-601-95	8-759-610-95
8-795-600-95	905-106	905-223	905-234
CX-095C	CX-095D	CX-095E	CX095
CX095A	CX095C	CX095D	CX095E
CX842B	ECG1308	M5146P	NTE1308
SK3833	TCG1308	TM1308	WEP2308
WEP2308/1308			

BLOCK-296

ECG1309	NTE1309	TCG1309	TM1309

BLOCK-297

CX162	ECG1310	NTE1310	TCG1310
TM1310			

BLOCK-298

CX-064-2	CX064	ECG1311	NTE1311
TCG1311	TM1311		

BLOCK-299

CX157	ECG1312	NTE1312	TCG1312
TM1312			

BLOCK-300

CX158	ECG1313	NTE1313	TCG1313
TM1313			

BLOCK-301

148727	51X90458A05	AN6362	ECG1314

BLOCK-302

ECG1315

BLOCK-303

ECG1316	STK4171II

BLOCK-304

ECG1317

BLOCK-305

ECG1318	IGSTK5322	STK5322

BLOCK-306

46-722066-3	6124140001	ECG1319	SK4815
STK4021	STK4021M		

BLOCK-307

ECG1320	NTE1320	TCG1320

BLOCK-308

ECG1321	NTE1321	TCG1321

BLOCK-309

ECG1322	NTE1322	SK10007	STK0080
TCG1322			

BLOCK-310

ECG1323	NTE1323	SK3154	STK016
TCG1323			

BLOCK-311

ECG1324	NTE1324	STK040	TCG1324

BLOCK-312

ECG1325	NTE1325	STK041	STK043
TCG1325			

BLOCK-313

ECG1326	NTE1326	SK7659	STK077
STK078	TCG1326		

BLOCK-314

ECG1327	NTE1327	SK7660	STK-082
STK080	STK082	TCG1327	

BLOCK-315

ECG1328	NTE1328	STK-084	STK083
STK084	TCG1328		

BLOCK-316

ECG1329	NTE1329	STK441	TCG1329

BLOCK-317

ECG1330	NTE1330	SK9406	STK457
STK459	TCG1330		

BLOCK-318

ECG1331	NTE1331	SK9189	STK461
STK463	TCG1331		

BLOCK-319

ECG1332	NTE1332	STK031	TCG1332

BLOCK-320

ECG1333	NTE1333	TCG1333

BLOCK-321

ECG1334	NTE1334	SK7662	TCG1334

BLOCK-322

ECG1335	NTE1335	TCG1335

BLOCK-323

ECG1336	NTE1336	STK050	STK070
TCG1336			

BLOCK-324

ECG1337	NTE1337	SS1001	STK086
TCG1337			

BLOCK-325

ECG1338	NTE1338	SK9757	STK3042
STK3042III	TCG1338		

BLOCK-326

ECG1339	NTE1339	SK9997	STK3082
STK3082II	STK443	TCG1339	

BLOCK-327

ECG1340	NTE1340	STK058	TCG1340

BLOCK-328

ECG1341	NTE1341	STK075	TCG1341

BLOCK-329

ECG1342	NTE1342	TCG1342

BLOCK-330

ECG1343	NTE1343	STK430	TCG1343

BLOCK-331

ECG1344	NTE1344	TCG1344

BLOCK-332

ECG1345	NTE1345	SK9998	STK465
STK465A	TCG1345		

BLOCK-333

ECG1346	NTE1346	TCG1346

BLOCK-334

ECG1347	NTE1347	STK460	TCG1347

BLOCK-335

ECG1348	NTE1348	TCG1348

BLOCK-336

ECG1349	NTE1349	TCG1349

BLOCK-337

ECG1350	NTE1350	TCG1350

BLOCK-338

ECG1351	NTE1351	TCG1351

BLOCK-339

ECG1352	NTE1352	TCG1352

BLOCK-340

ECG1353

BLOCK-341

ECG1354	NTE1354	TCG1354

BLOCK-342

ECG1355	NTE1355	TCG1355

BLOCK-343

ECG1356	NTE1356		

BLOCK-344

ECG1357	NTE1357		

BLOCK-345

ECG1358	NTE1358	SK9987	STK1050
STK1050A			

BLOCK-346

ECG1359	NTE1359		

BLOCK-347

ECG1360			

BLOCK-348

ECG1361	M51516L	NTE1361	SK9361
TCG1361			

BLOCK-349

ECG1362	M51518L	NTE1362	SK9396

BLOCK-350

AN7130	ECG1363	NTE1363	SK7630
TCG1363			

BLOCK-351

ECG1364	M51517L	NTE1364	SK9395
TCG1364			

BLOCK-352

AN7131	AN7140	ECG1365	NTE1365
TCG1365			

BLOCK-353

ECG1366	NTE1366	TCG1366	

BLOCK-354

AN7146H	ECG1367	NTE1367	TCG1367

BLOCK-355

ECG1368	NTE1368	TCG1368	

BLOCK-356

AN7154	ECG1369	NTE1369	TCG1369

BLOCK-357

ECG1370	NTE1370	SK4821	TCG1370

BLOCK-358

AN7156N	ECG1371	NTE1371	TCG1371

BLOCK-359

ECG1372	NTE1372	TCG1372	

BLOCK-360

11119113	1200-3430	1200-3530	1232-7300
1330-9281	2119-102-7401	23119763	37001047
46-13922-3	5113753A67	51S13753A67	AN7158
AN7158N	AN7166	ECG1373	EW84X978
NTE1373	SK4822	TCG1373	

BLOCK-361

15-3015131-2	221-422-01	612332-1	6123320001
6123320003	ECG1374	NTE1374	SK3853
TCG1374			

BLOCK-362

BA532	BA532S	ECG1375	NTE1375
RVIBA532S	SK7608	TCG1375	

BLOCK-363

1006-7445	15-3015131-1	15-3015131-3	221-177
612332-6	6123320006	ECG1376	EP84X109
NTE1376	SK7707	TCG1376	

BLOCK-364

8-759-800-15	ECG1377	IX0225CE	LA4440
LA4440-R	LA4440R	MX-5421	NTE1377
RH-IX0225CEZZ	SK7808	TCG1377	X0225C

BLOCK-365

198609	ECG1378	NTE1378	SK7706
TCG1378	TDA2006		

BLOCK-366

EA33X8707	ECG1379	NTE1379	SK7663
TA7215P	TCG1379		

BLOCK-367

221-316	442-677	ECG1380	HE-442-677
NTE1380	SK9251	TCG1380	TDA2030

BLOCK-368

AN4102	AN7115	AN7115F	ECG1381
MX-3977	NTE1381	SK7622	TCG1381

BLOCK-369

8-759-801-25	8-759-841-25	ECG1382	HC10044030
LA4125	LA4125T	LA4126	LA4126T
NTE1382	RV1LA4126	SK7658	TCG1382

BLOCK-370

51-90305A77	AN7145	AN7145H	AN7145L
AN7145M	AN7146M	ECG1383	NTE1383
TCG1383			

BLOCK-371

ECG1384	NTE1384	TCG1384	

BLOCK-372

ECG1385	NTE1385	TCG1385	

BLOCK-373

ECG1386	LA4230	LA4250	NTE1386
SK7654	TCG1386		

BLOCK-374

1000-6971	2002610109	5652-TA7230P	ECG1387
NTE1387	SK7665	TA7230P	TCG1387

BLOCK-375

ECG1388	NTE1388	SK9186	TCG1388

BLOCK-376

ECG1389	NTE1389	TCG1389	UPC1230H
UPC1230H2			

BLOCK-377

ECG1390	NTE1390	SK7815	

BLOCK-378

ECG1391	NTE1391	SK7814	TCG1391

BLOCK-379

ECG1392	NTE1392	TCG1392	

BLOCK-380

ECG1393	HA1389R	NTE1393	TCG1393

BLOCK-381

ECG1394	NTE1394	TCG1394	

BLOCK-382

1270-1751	37001064	AN7168	AN7168N
ECG1395	NTE1395	SK9313	TCG1395

BLOCK-383

4H-209-80751	ECG1396	GL1010	NTE1396
SK9225	TCG1396	TDA2004	TDA2005M

BLOCK-384

ECG1397	HA1374	HA1374A	NTE1397
SK7603			

BLOCK-385

2004341	ECG1398	HA1377	HA1377A
NTE1398	SK9191		

BLOCK-386

ECG1399	HA1388	NTE1399	SK9269

BLOCK-387

07-28816-71	146894	51-90433A33	51X90433A33
AN304	ECG1400	NTE1400	SK7627
TCG1400			

BLOCK-388

51-13753A48	51S13753A48	51S13753A60	51S137A48
51S3753A48	AN5340	ECG1401	EN11401
HA11401	NTE1401	SK9354	TCG1401
TVSEN11401			

BLOCK-389
07-28815-100	144920	51-90433A07	67-51703-02
AN5010	ECG1402	NTE1402	SK9043

BLOCK-390
266P0501	266P50101	51-13753A49	51S13753A49
AN5440	ECG1403	EN11414	HA11414
NTE1403	SK9353	TCG1403	TVSEN11414

BLOCK-391
8-759-400-01	AN5250	ECG1404	NTE1404
TCG1404			

BLOCK-392
07-28778-70	144848	51-90305A39	61C001-30
ECG1405	HA11220	NTE1405	SK9018
TCG1405			

BLOCK-393
21A120-050	51-13753A29	51-90305A92	AN295
AN295N	AN5510	CN5510	ECG1406
NTE1406	SK9059	TCG1406	

BLOCK-394
ECG1407	NTE1407	TCG1407

BLOCK-395
07-28777-56	144024	51-90305A62	61C001-38
ECG1408	MN6016A	MN6061A	NTE1408
SK9064	TCG1408		

BLOCK-396
06300071	06300508	103D079400	11119032
1191-3340	1231-8341	1266-2425	1311-8641
148741	151333	159970	1UPC1363C
46-131188-3	46-131905-3	51X90480A20	8-719-113-63
8-759-113-63	905-217	C1363CA	ECG1409
EW84X416	I03D079400	INMPC1363C	INMPC1363CA
IUPC1363C	LA7940	MPC1363	MPC1363C
NTE1409	NTE1409N	PC1363C	SK9198
TCG1409	TVSUPC1363CA	UPC1363C	UPC1363CA

BLOCK-397
ECG1409C

BLOCK-398
0019-1313	0079-1313	1177-1151	1181-7137
2109-101-5304	2109-102-5304	21A120-185	37003020
46-13551-3	51-13753A22	51-13753A28	51-13753A54
51-13763A27	51S13753A28	51S13753A56	51X90480A57
5310	56B75-1	56D75-1	8-759-105-56
8-759-115-56	AN5310	AN5310K	AN5310KL
AN5310L	AN5310M	AN5310N	AN5310U
AN5311	AN5311KL	CN5310	CN5310(K)
CN5310(L)	CN5310(M)	CN5310K	CN5310L
CN5310M	CN5311	CN5311CL	CN5311K
CN5311KL	CN5311LM	CX-556	CX556
CX556A	ECG1410	I02DC13710	MPC1372C
MX-5375	MXF-5375	NTE1410	SK9016
TBA820MT	TCG1410	TVSCN5310	TVSCN5310L
TVSCN5310M	TVSCN5311	TVSCN5311CL	TVSCN5311KL
UPC1352S	UPC1353	UPC1371	UPC1371C
UPC1372C	WEP1410	WEP1410/1410	

BLOCK-399
21A120-178	8-759-600-05	8-759-605-55	CX-555A
CX555	CX555A	ECG1411	M5185AP
M5185P	NTE1411	SK9246	TCG1411

BLOCK-400
07-28778-74	144852	51-90305A43	61C001-35
AN6300	ECG1412	NTE1412	SK9045
TCG1412			

BLOCK-401
0073-0180	1138-0979	126-1	15-3015734-1
2002120067	266P11601	394-1	394-2
52TA7611AP-S	56A107-1	6121260001	6123300027
6123390001	6123940001	6123940002	ECG1413
EP84X34	IX0252CE	IX0252CEZZ	M7611AP
NTE1413	RH-IX0092CEZZ	RH-IX0252CEZZ	SK7635
SK9380	TA7611AP	TA7611AP-S	TCG1413
TDA2540	TDA2541	TDA2541Q	
WEP1416L/1416	X0092CE	X0252CE	

BLOCK-402
146595	46-13312-3	51X90433A28	ECG1414
MN6076	NTE1414	SK9326	TCG1414

BLOCK-403
196187	51-13753A23	51S13753A23	AN5320
ECG1415	EW84X394	NTE1415	SK9058
TCG1415			

BLOCK-404
0079-0780	06300034	1184-8108	1195-6570
1352	2119-101-5204	21A120-049	2361281
303-1(IC)	37003012	56A75-1	56A75-1A
56D75-1A	612303-1	6123030001	8-759-125-56
CX556B	ECG1416	EW84X270	MPC1352C
MX-5279	NTE1416	TCG1416	TDA3570
UPC1352	UPC1352C	UPC1372E	

BLOCK-405
21A120-052	8-759-105-57	AN5410	CX-557A
CX-557S	CX557	CX557A	ECG1417
NTE1417	SK9315	TCG1417	UPC1367C
WEP1417L/1417			

BLOCK-406
8-759-600-43	8-759-608-43	CX-843	CX-843A
CX843	CX843A	ECG1418	M51375P
NTE1418	SK9214		

BLOCK-407
07-28778-75	144853	146514	51-90305A44
51-90433A35	51X90433A35	61C001-36	AN6320
AN6320N	ECG1419	NTE1419	SK9046
TCG1419			

BLOCK-408
1231-0256	8-759-103-00	8-759-113-58	ECG1420
NTE1420	SK9180	TCG1420	UPC1358A
UPC1358C	UPC1358H	UPC1358H2	UPC1358HZ

BLOCK-409
07-28778-73	144851	51-90305A73	61C001-34
AN6811	DN6811	ECG1421	NTE1421
SK7792	SK9051	TCG1421	

BLOCK-410
07-28816-72	146895	148720	51-90433A34
51X90458A01	AN6331	AN6332	ECG1422
NTE1422	TCG1422		

BLOCK-411
11-115	2000-037	3071519001	DDEY157001
DDEY171001	ECG1423	MB3712	MB3712HM
NTE1423	SK7600	TCG1423	

BLOCK-412
DDEY172001	ECG1424	MB3713	MB3713HM
NTE1424	SK7601	TCG1424	

BLOCK-413
07-28778-72	144850	51-90305A41	51X90305A41
61C001-33	AN6341	AN6341N	ECG1425
NTE1425	SK9049	TCG1425	

BLOCK-414
8-759-103-68	8-759-113-68	ECG1426	NTE1426
SK9199	TCG1426	UPC1368H	UPC1368H2
UPC1368HR	UPC1368HZ		

BLOCK-415
07-28778-77	144612	150342	51-90305A37
51X90458A56	61C002-2	AN6342	AN6342N
ECG1427	NTE1427	SK9050	TCG1427

BLOCK-416
07-28816-70	144849	146893	51-90305A40
51-90433A32	61C001-32	AN6340	AN6344
ECG1428	NTE1428	SK9048	TCG1428

BLOCK-417
ECG1429	NTE1429	TCG1429

BLOCK-418
226P10601	266P10601	46-13208-3	46-13267-3
ECG1430	EP84X80	M5186AP	M5186BP
M5186P	MA5186AP	NTE1430	SK9247
TCG1430			

BLOCK-419
ECG1431	NTE1431	SK7793	TCG1431

BLOCK-420
ECG1432	NTE1432

BLOCK-421

ECG1433	NTE1433	TCG1433

BLOCK-422

02-151204	02-161204	08200040	ECG1434
M51204	M51204L	NTE1434	SK7634
TCG1434			

BLOCK-423

AN7311	ECG1435	NTE1435	SK7794
TCG1435			

BLOCK-424

916157	ECG1436	M51172P	NTE1436
TCG1436			

BLOCK-425

1140-4118	21A120-074	51-13753A27	AN5330
ECG1437	EP84X60	M51389P	NTE1437
SK3885	TCG1437		

BLOCK-426

ECG1438	NTE1438	TCG1438

BLOCK-427

8-759-826-01	AN829	AN829S	EA33X8522
ECG1439	LA2600S	NTE1439	SK3964
TCG1439			

BLOCK-428

1177-1144	1181-7129	148744	2119-101-5107
2119-102-5100	46-13550-3	51-13753A26	51-13753A35
51-13753A39	51S13753A39	51X90458A30	51X90458A50
56A74-1	56B74-1	AN5111	BN5111
BN5111A	BN5111B	ECG1440	NTE1440
SK9056	TCG1440	TDA-2541	TVSBN5111
WEP1440L/1440			

BLOCK-429

ECG1441	NTE1441	SK7795	TCG1441

BLOCK-430

ECG1442	NTE1442	TCG1442

BLOCK-431

ECG1443	NTE1443	SK3967

BLOCK-432

AN313	ECG1444	GEIC-298	NTE1444

BLOCK-433

2304-1	46-13327-3	612304-1	6123040001
ECG1445	LA1460	LA1461	NTE1445
SK9187	TCG1445		

BLOCK-434

AN301	ECG1446	NTE1446	TCG1446

BLOCK-435

ECG1447	NTE1447	TCG1447

BLOCK-436

AN278	ECG1448	NTE1448	TCG1448

BLOCK-437

ECG1449	NTE1449	TCG1449

BLOCK-438

AN272	ECG1450	GEIC-299	NTE1450
TCG1450			

BLOCK-439

ECG1451	NTE1451	TCG1451

BLOCK-440

AN374P	ECG1452	NTE1452	SK7796
TCG1452			

BLOCK-441

ECG1453	LA3115	NTE1453	SK3702
TCG1453			

BLOCK-442

46-13176-3	46-13341-3	AN340	AN340P
ECG1454	LA1320A	NTE1454	SK9184
TCG1454			

BLOCK-443

ECG1455	NTE1455	TCG1455

BLOCK-444

09-308103	ECG1456	ETI-13	TA7108P

BLOCK-445

ECG1457	NTE1457	TCG1457

BLOCK-446

ECG1458	NTE1458	TCG1458	UPC1186H

BLOCK-447

ECG1459	IGLA3150	LA3150	NTE1459
SK7797	TCG1459		

BLOCK-448

ECG1460	NTE1460	SK7798	TCG1460
UPC1173C			

BLOCK-449

ECG1461	LA3370	NTE1461	SK9989
TCG1461			

BLOCK-450

ECG1462	NTE1462	TA7129P	TCG1462

BLOCK-451

5350702	AN7120	ECG1463	LA4112
NTE1463	SK9208	TCG1463	

BLOCK-452

ECG1464	NTE1464	TCG1464

BLOCK-453

1002-0345	1003-3116	1100-9750	11119030
1122-8863	1172-9878	15-5016T/IC7313AP	16173130
300007313167	415307313P	46-5290-0042	8-759-841-40
905-284	EA33X8559	ECG1465	HA12013
I03SP41400	ILL0052	KIA7313AP	KIA7373AP
LA4140	MX-5596	NTE1465	PC-20110
SK3914	TA7313(IC)	TA7313AP	TA7313F
TA7313P	TA7313P(IC)	TA7319P	TCG1465

BLOCK-454

1006-0903	206-5-0783-21011	46-13322-3	46-5290-0178
8-759-802-10	8-759-832-10	905-235	905-289
93T4150182072	BA333	EA33C8390	ECG1466
LA3210	LA3210ALC	MX-5436	NTE1466
SK9212	TA7137P	TA7137P-ST	TCG1466

BLOCK-455

ECG1467	M51182	M51182L	NTE1467
RVIM51182	SK4839	TCG1467	

BLOCK-456

46-13301-3	ECG1468	NTE1468	SK9322
TA7066P	TA7140P	TA7140PC	TCG1468

BLOCK-457

147437	2360782	ECG1469	HA11215A
HA11251A	NTE1469	SK9356	TCG1469

BLOCK-458

46-13152-3	ECG1470	LA4201	NTE1470
SK7799	TCG1470		

BLOCK-459

06300080	06300090	06300546	1172-0034
1199-7855	1327-1309	2002300023	2002300033
200X2300-033	2364181	46-13538-3	57SI-HA11423
905-368	ECG1471	EP84X215	EW84X310
HA11423	HA114323	ILH0111	NTE1471
SK9194	SK9393	TCG1471	

BLOCK-460

ECG1472	NTE1472	SK7800	TCG1472

BLOCK-461

2361571	2361572	2363261	ECG1473
HA11412	HA11412A	HA11434	NTE1473
SK9355	TCG1473		

BLOCK-462

ECG1474	NTE1474	TA7302P	TCG1474

BLOCK-463

ECG1475	LC7131	NTE1475	SK4838
TCG1475	TCG4536B		

BLOCK-464

ECG1476	NTE1476	RH-IX1018AFZZ	TA7207P
TCG1476			

BLOCK-465

ECG1477	LA4170	NTE1477	TCG1477

BLOCK-466

ECG1478	NTE1478	TCG1478

BLOCK-467

46-13237-3	46-13411	46-13411-3	46-13428-3
ECG1479	LA1152N	LA5112	LA5112N
LA5112N-G	LA5115N	NTE1479	SK9188
TCG1479			

BLOCK-468

ECG1480	HA1452	HA1452W	NTE1480
SK7801	TCG1480	TR2367151	

BLOCK-469

ECG1481	NTE1481	TCG1481

BLOCK-470

442-673	ECG1482	HA11223W	HE-442-673
NTE1482	TCG1482		

BLOCK-471

1172-0026	1186-5888	37007014	46-13302-3
46-13323-3	46-13379-3	A1357N	C1356C2
ECG1483	EP84X154	EP84X213	LA1357
LA1357AB	LA1357B	LA1357N	LA1357NB
MPC1356C	MPC1356C2	NTE1483	SK9211
TCG1483	UPC1356C	UPC1356C2	

BLOCK-472

1KK-HA1196	442-636	ECG1484	HA1196
HE-442-636	NTE1484	TCG1484	

BLOCK-473

0079-0160	11119026	51-13753A25	51-90433A53
51X90433A53	AN5520	AN5720	CN5520
ECG1485	MC5201	MX-5201	MX5201
NTE1485	SK9061		

BLOCK-474

22114626	ECG1486	HA11211	NTE1486
SK7607	TCG1486		

BLOCK-475

AN259	ECG1487	NTE1487	TCG1487

BLOCK-476

1KK-HA11225	30900720	30901120	ECG1488
EI703843	HA11225	NTE1488	PC1167C2
SK9818	TCG1488		

BLOCK-477

30900540	ECG1489	EI307200	NTE1489
TCG1489	UPC1161(C)	UPC1161C	

BLOCK-478

5351411	ECG1490	HA11251	NTE1490
SK9312	TCG1490		

BLOCK-479

ECG1491	NTE1491	TCG1491	UPC1171C

BLOCK-480

8-759-312-21	ECG1492	HA11221	NTE1492
RH-IX0028PAZZ	SK9196	TCG1492	

BLOCK-481

ECG1493	NTE1493	TCG1493	UPC1187V

BLOCK-482

ECG1494	HA1138	NTE1494	TCG1494

BLOCK-483

ECG1495	NTE1495	TCG1495	UPC1191V

BLOCK-484

06300052	1TA7609P-2	46-13190-3	5320200170
56A102-1	6644012700	B0354901	ECG1496
ITA7609P-2	ITA7609P2	NTE1496	
RH-IX0094CEZZ	SK9321	TA7609	TA7609AP
TA7609GS-2	TA7609P	TA7609P(FA-2)	TA7609P-GS-2
TA7609PFA-2	TA7609PFA2	TA7609PGS-2	TCG1496

BLOCK-485

ECG1497	NTE1497	TCG1497	UPC1158H2

BLOCK-486

ECG1498	NTE1498	SK7667	TA7318P
TCG1498			

BLOCK-487

AN6821	ECG1499	NTE1499	TCG1499

BLOCK-488

ECG1500	NTE1500	TCG1500

BLOCK-489

ECG1501	NTE1501	TCG1501

BLOCK-490

ECG1502	NTE1502	TCG1502

BLOCK-491

AN6875	ECG1503	NTE1503	SK7802
TCG1503			

BLOCK-492

ECG1504	NTE1504	TCG1504	TL487C
TL487CP			

BLOCK-493

ECG1505	NTE1505	TCG1505	TL489C
TL489CP			

BLOCK-494

ECG1506	TL490C	TL490CN

BLOCK-495

ECG1507	NTE1507	TCG1507	TL491C
TL491CN			

BLOCK-496

276-1707	ECG1508	LM3914	LM3914N
NTE1508	SK7638	TCG1508	

BLOCK-497

276-1708	ECG1509	LM3915	NTE1509
SK7637	TCG1509		

BLOCK-498

22114634	8-759-619-03	ECG1510	M51903L
NTE1510	TCG1510		

BLOCK-499

ECG1511	NTE1511	TCG1511

BLOCK-500

206-5-2341-41610	905-465	ECG1512	EI703907
LB1416	MX-4465	NTE1512	TCG1512

BLOCK-501

2362991	ECG1513	LB1409	NTE1513
SK9250	TCG1513		

BLOCK-502

ECG1514	NTE1514	TCG1514

BLOCK-503

30900530	44T-300-359	8-759-814-05	905-329
EA33X8521	ECG1515	EW84X18	LB1405
NTE1515	SK7656	TCG1515	X0087TA

BLOCK-504

905-326	ECG1516	I0XDP12130	MX-4875
NTE1516	TCG1516	UPC1212C	UPC1213C
UPC1213C(MS)			

BLOCK-505

4-2069-71710	ECG1517	M51544L	MX-5476
NTE1517	SK7704	TCG1517	

BLOCK-506

ECG1518	NTE1518	TCG1518	UPC1216V

BLOCK-507

ECG1519	NTE1519	SK7803	TCG1519

BLOCK-508

31SI-UPC1360C	ECG1520	EP84X78	NTE1520
TCG1520	UPC1360	UPC1360C	

BLOCK-509

46-13153-3	ECG1521	NTE1521	SK7671
TA7314P	TCG1521		

BLOCK-510

06300066	1001-5956	1192-7852	2002120046
2364172	56A120-1	C1366C2	ECG1522
EW84X115	NTE1522	SK9197	TCG1522
UPC1366C	UPC1366C2		

BLOCK-511

ECG1523	NTE1523	TCG1523	

BLOCK-512

BA404	ECG1525	NTE1525	SK7604
TA7303P	TCG1525		

BLOCK-513

ECG1526	NTE1526	TCG1526	

BLOCK-514

ECG1527	NTE1527	TCG1527	

BLOCK-515

02-392816	2000-036	307-145-9-001	3071459001
DDEY147001	ECG1528	NTE1528	SK3881
TCG1528	UPC2816C	UPD2816C	

BLOCK-516

1229-4625	1231-0165	156545	46-131060-3
5350604	6192100150	A6458S	AN6551
BA715	BA718	ECG1529	EW84X484
EW84X499	I03S064580	LA6458S	NJM-4558S
NJM4558S	NJM4558SD	NTE1529	SK7804
TA75557S	TCG1529		

BLOCK-517

ECG1530	NTE1530	TCG1530	UPC1350C

BLOCK-518

ECG1531	NTE1531	TCG1531	

BLOCK-519

06300050	06300051	1TA7608CP5	46-13216-3
46-13238-3	5320200160	6644012602	A0354805
B0354804	B0354805	B0354806	B0354830
B0354832	B0354833	BO354830	BO356620
ECG1532	ITA7608CP5	NTE1532	
RH-IX0093CEZZ	SK9320	TA7608AP	TA7608BP
TA7608BR	TA7608CP	TA7608CP(FA-5)	
TA7608CP-GS-4	TA7608CP-GS-5	TA7608CPFA-5	TA7608CPFA-6
TA7608P	TA7608PFA-5	TA7608PFA-6	TA7608PGS-4
TA7608PGS-5	TCG1532	X0093CE	X0094CE

BLOCK-520

ECG1533	NTE1533	TCG1533	

BLOCK-521

ECG1534	NTE1534	TCG1534	

BLOCK-522

5190528A53	ECG1535	NTE1535	

BLOCK-523

02-507120	ECG1536	LC7120	NTE1536
SK3884	TCG1536		

BLOCK-524

ECG1537	NTE1537	TCG1537	

BLOCK-525

0079-0510	162922	409-019-4700	46-13257-3
46-13527-3	46-13704-3	46-14367-3	8-759-240-81
905-220	ECG1538	I03DD78000	IGLC4081B
INMPD4081C	INUPD4081C	LA7800	LA7800R
MC14081BCP	MX-5281	NTE1538	Q7227
Q7227R	SK9403	TC4081BP	TCG1538
UPD4081C	VHITC4081BP-1		

BLOCK-526

8-759-878-02	8-759-878-03	8-759-879-03	ECG1539
EW84X269	IX0178CE	LA7802	NTE1539
SK9404	TCG1539	X0178CE	

BLOCK-527

ECG1540	LA7806	NTE1540	SK4837
TCG1540			

BLOCK-528

BA403	ECG1541	NTE1541	TCG1541

BLOCK-529

46-13839-3	ECG1542	LA7507	NTE1542
SK10274			

BLOCK-530

ECG1543	HA11219	NTE1543	TCG1543

BLOCK-531

8-759-312-44	905-287	ECG1544	HA11244
NTE1544	SK9193	TCG1544	

BLOCK-532

0079-0500	06300049	0630049	1124-0239
1126-0239	1145-2786	1178-7454	1205-7832
1214-4945	146857	150325	1756-1
1TA7607AP	2002120051	20SI-TA7607AP	2109-103-2200
2119-101-2607	2119-101-9408	2119-102-2600	2199-101-9408
21A120-051	21A120-099	23119901	266P11301
32119-101-940	32119-102-260	46-13188-3	4H20981787
5320200150	612126	612126-1	6644012500
8-759-276-07	80SI-M51358B	80SI-M51358P	B0345710
B03544710	B0354700	B0354710	B0364710
BO354710	DSD84	ECG1545	EW84X899
I05DA7607A	ITA7607AP	KA2911	M51358P
MX-5280	NTE1545	RH-1X0092CEZZ	SK9379
TA7607	TA7607A	TA7607AP	TA7607BP
TA7607P	TCG1545	TDA2544	TDA2548

BLOCK-533

2002600183	200X2600-183	2364131	267P90205
51X90458A55	51X90480A95	51X90495A14	ECG1546
EW84X427	NTE1546	SK7647	STR-381
STR381	STR381A	TVSSTR381	TVSSTR381A

BLOCK-534

1176-7266	1178-7462	1181-3763	1228-4410
2109-103-3001	2119-101-9602	2119-101-9709	23119796
266P63001	46-13356-3	56A137-1	56SI-TA7644BP
67-32009-01	B0356446	B0356448	BO356446
ECG1547	EW84X900	I05DE76440	KA2153
M7644BP	NTE1547	SK7676	TA7644AP
TA7644BP	TA7644BPFA-1		

BLOCK-535

06300058	223-18	267P90202	46-13588-3
46-13588-9	51-13753A52	51S13753A52	A-13284
ECG1548	EXT95423C	NTE1548	Q7293
Q7296	SK7646	STR-380	STR380
TVSSTR380			

BLOCK-536

276-1709	ECG1549	LM3916	NTE1549

BLOCK-537

06300031	1001-9727	1184-8728	1195-6562
157653	2002300029	21A120-204	2360691
37009010	51-13753A55	51-90305A89	51S13753A55
ECG1550	EN11235	H-IX0065CE	HA11235
MX-6452	NTE1550	RH-IX0065CEZZ	SK9249
TVSEN11235	X0065CE		

BLOCK-538

51X90480A64	AN5132	ECG1551	EW84X422
NTE1551	SK9298		

BLOCK-539

8-729-100-09	8-759-100-07	8-759-100-09	CX-557
CX557B	CX557S	ECG1552	MPC1367C
NTE1552	PMC1367P	SK9015	UPC-1367C

BLOCK-540

1200-2440	1205-7907	2119-601-2709	2119-601-5209
2149-203-0607	46-13546-3	ECG1553	EXT95247C
EXT954-25C	EXT95427C	NTE1553	SK7720
STR382	STR385		

REPLACEMENTS

BLOCK-541

147340	ECG1554	HA11715	IHHA11715
NTE1554			

BLOCK-542

ECG1555	SN76873N		

BLOCK-543

46-13225-3	46-13339-3	ECG1556	LB1331
LB1332	NTE1556	SK7657	

BLOCK-544

1000-7227	1037-7174	300012413001	BA4220
E-HA12413	EA33X8717	ECG1557	HA11413
HA12413	HA12413-03	HC10036010	MX-4542
MX-5429	NTE1557	SK7605	

BLOCK-545

150340	150430	156895	160175
46-13484-3	5352631	8-759-951-02	905-347
BA5102	BA5102A	ECG1558	IYBA5102AALC
LA7040	NTE1558	SK9244	

BLOCK-546

ECG1559	HA12412	NTE1559	

BLOCK-547

1004-2042	1172-9886	5652-AN7410	
5652-AN7410NS	AN7410	AN7410NS	ECG1560
NTE1560	SK7620		

BLOCK-548

1181-3565	5652-BA6124	AN6884	BA3704
BA6124	ECG1561	LB1403	NTE1561
SK7610			

BLOCK-549

ECG1562	NTE1562		

BLOCK-550

0073-0030	051-0011-04	1155-9259	4-2069-70232
46-13392-3	4H-209-80967	5652-AN7218	
5652-UPC1018C	8-759-110-17	AN7218	AN7218H
EA33X8583	ECG1563	MX-4476	NTE1563
RVIUPC1018CE	SK7648	UPC1018	UPC1018C
UPC1018C-F	UPC1018CE		

BLOCK-551

ECG1564			

BLOCK-552

ECG1565	HA11227	MX-3979	MX3979
NTE1565	SK9192		

BLOCK-553

0061-232-70001	1166-5205	150369	327-1
6123270001	ECG1566	N6123270001	NTE1566
SK7726	TDA2611	TDA2611A	TDA2611AQ

BLOCK-554

329-1	4822 209 60955	6123290001	ECG1567
NTE1567	SK7805	TDA3651A	TDA3651AQ
TDA3653B			

BLOCK-555

148722	51X90458A82	AN6345	ECG1568
NTE1568			

BLOCK-556

1259-6532	2002900395	37011063	46-131014-3
AN287	AN287NT	AN6571	ECG1569
NTE1569	SK10081		

BLOCK-557

ECG1570	I05DE76810	NTE1570	RH-
IX0024CEZZ	RH-IX0224CEZZ	SK7681	TA7681AP
TA7681P			

BLOCK-558

1000-6963	4H-209-81252	5652-TA7658P	8-759-276-58
ECG1571	MX-4712	NTE1571	SK7677
TA7658P			

BLOCK-559

2002400752	56A138-1	6123300101	67-32109-02
87SI-TA7680AP	ECG1572	IX0275CE	KA2914A
MX-6829	NTE1572	RH-IX0223CEZZ	
RH-IX0275CEZZ	SK4825	TA7680AP	TA7680P
X0275CE			

BLOCK-560

1180-2055	46-13830-3	B0356640	B0356641
ECG1573	NTE1573	SK7678	TA7664P
TA7664P(FA-1)	TA7664PFA-1	TA7664PFA1	

BLOCK-561

56A34-1	ECG1574	EP84X25	RH-IX0024PAZZ
SAA1024	SAA1124		

BLOCK-562

1157-8572	2360611	ECG1575	HA11229
NTE1575	SK9195		

BLOCK-563

06300160	1043-7275	1180-2089	1259-4917
1287-7940	1316-0951	1393-9285	176226
2002400638	221-230	37001039	46-13590-3
8-759-276-30	B0356190	ECG1576	EP84X229
I05D076300	IX0214CE	KIA6930P	KIA7630
KIA7630P	MC13301P	MX-4404	NTE1576
Q7682	RH-IX0214CEZZ	SK7672	TA7630
TA7630P	TIA7630		

BLOCK-564

ECG1577	NTE1577		

BLOCK-565

DM-37	ECG1578	NTE1578	SK9130
ULN2243A			

BLOCK-566

ECG1579	HA12411	NTE1579	

BLOCK-567

370-2	4835-209-87065	6123700001	6123700002
8-759-001-20	ECG1580	N6123700001	NTE1580
SK7743	TBA120U		

BLOCK-568

8-759-684-78	905-198	ECG1581	M58418P
M58478P	NTE1581	SK9360	

BLOCK-569

46-13299-3	8-751-430-00	905-101	CX-143
CX-143A	CX143	CX143A	ECG1582
NTE1582			

BLOCK-570

ECG1583	HA11702		

BLOCK-571

46-13169-3	8-751-450-00	905-112	CX-145
CX145	ECG1584		

BLOCK-572

5351501	ECG1585	EP84X62	HA11701
NTE1585			

BLOCK-573

46-13174-3	70119049	8-751-410-00	905-103
CX-141	CX141	CX141A	ECG1586

BLOCK-574

ECG1587			

BLOCK-575

46-13164-3	8-751-390-00	905-116	CX-139A
CX139A	ECG1588		

BLOCK-576

1207-6873	2002502317	2369151	23691515
37003023	ECG1589	HA11431	HA11431NT
NTE1589	SK9911		

BLOCK-577

46-13233-3	8-751-380-00	905-115	CX-138
CX-138A	CX138	ECG1590	

BLOCK-578

5359841	ECG1591	EP84X100	HA11711
IHHA11711	NTE1591	SK7602	

BLOCK-579

46-13168-3	8-751-370-00	905-111	CX-137A
CX137A	ECG1592		

BLOCK-580

2362541	ECG1593	HA11409	NTE1593

BLOCK-581

46-13167-3	70119037	8-751-360-00	905-110
CX-136A	CX136	CX136A	ECG1594

BLOCK-582

5351521	ECG1595	EP84X63	HA11703

BLOCK-583

46-13171-3	70119036	8-751-350-00	905-109
CX-135	CX135	ECG1596	NTE1596

BLOCK-584

5351551	ECG1597	EP84X67	HA11706
NTE1597			

BLOCK-585

46-13162-3	70119048	8-751-340-00	8-851-340-00
905-114	CK-134A	CX-134A	CX134A
ECG1598			

BLOCK-586

46-13166-3	8-751-3300-00	905-108	CX-133A
CX133A	ECG1599		

BLOCK-587

147339	ECG1600	HA11417	HA11714
IHHA11714	NTE1600	SK10221	

BLOCK-588

46-13165-3	70119034	8-751-310-00	905-107
CX-131A	CX131A	ECG1601	

BLOCK-589

ECG1602	HA1393	NTE1602	SK10223

BLOCK-590

8-751-300-00	CX-130	CX130	ECG1603

BLOCK-591

BA4110	ECG1604	LA1140	NTE1604
SK4835			

BLOCK-592

46-13175-3	8-751-001-00	905-104	CX-1000
CX-100D	CX100B	CX100D	ECG1605

BLOCK-593

180215	ECG1606	HA1392	MX-5451
NTE1606	SK4828		

BLOCK-594

ECG1607	NTE1607		

BLOCK-595

ECG1608	LA1245	NTE1608	SK9707

BLOCK-596

BA222	ECG1609	EW84X486	NTE1609
SK10226			

BLOCK-597

1148-9523	51X90458A72	AN5722	ECG1610
NTE1610			

BLOCK-598

51X90518A97	ECG1611	NTE1611	SK9990
TA7673	TA7673P	UPC1057C	

BLOCK-599

AN5740	BA526	ECG1612	NTE1612
SK9817	TVSBA526		

BLOCK-600

148729	51X90458A06	AN6677	ECG1613
NTE1613	SK10010		

BLOCK-601

ECG1614			

BLOCK-602

11119024	1128-4239	46-13528-3	51-90305A88
51X90305A88	AN5700	ECG1615	NTE1615
SK9977			

BLOCK-603

0070-0210	1170-0036	1172-0036	1186-5672
2002110345	37001034	6123300024	ECG1616
EP84X216	I02DB13820	IX0213CE	MPC1382C
NTE1616	RH-IX0213CEZZ	SK7735	UPC1382C
X0213CE			

BLOCK-604

46-13338-3	ECG1617	TA7643P	

BLOCK-605

1148-9499	51X90458A71	AN5712	ECG1618
NTE1618			

BLOCK-606

5652-BA5402	BA536	BA5402A	BA5406
ECG1619	NTE1619		

BLOCK-607

0079-0150	11119025	51-90433A52	51X90433A52
AN5710	ECG1620	MX-5200	NTE1620

BLOCK-608

148719	51X90433A99	AN6310	ECG1621
NTE1621	SK10230		

BLOCK-609

07-28778-76	144854	51-90305A45	51X90305A45
AN6321	ECG1622	SK9044	

BLOCK-610

AN6390	ECG1623		

BLOCK-611

1147-8815	EA33X8532	EA33X8537	EA33X8564
ECG1624	NTE1624	SK7611	T900B1
T911B1-T	T911BIT	TDA1083	TDA1083A
U417B	ULN2204A		

BLOCK-612

148717	305-1	51X90458A80	6123050001
AN5703	ECG1625	NTE1625	SK10232

BLOCK-613

8-759-112-77	ECG1626	I02SP12770	NTE1626
SK7652	UPC1277H		

BLOCK-614

BA546	ECG1627	NTE1627	SK9816

BLOCK-615

1033-1874	154652	156725	5190507A87
51X90507A87	ECG1628	EW84X296	M54543L
M54543L-B	M54543LB	NTE1628	SK4834

BLOCK-616

1176-5005	51X90458A75	51X90480A93	AN5752
AN5753	ECG1629	I01SD57530	NTE1629

BLOCK-617

1175-7903	5190507A60	AN6387	ECG1630
EW84X252	NTE1630	SK7827	

BLOCK-618

BA1350	ECG1631	NTE1631	SK10233

BLOCK-619

6123280001	6124450001	ECG1632	NTE1632
SK4812	TDA2577	TDA2577A	TDA2577AS2

BLOCK-620

221-178	A-9982-02A	A9982-02	ECG1633
NTE1633	SK7729		

BLOCK-621

300003220005	EA33X8706	ECG1634	KA2224
LA3220	MX-5595	NTE1634	SK10234

BLOCK-622

ECG1635	HA12002	HA12002-W	NTE1635
SK9706			

BLOCK-623

150401	BA847	ECG1636	NTE1636
SK9245			

REPLACEMENTS

BLOCK-624
ECG1637	NTE1637		

BLOCK-625
ECG1638

BLOCK-626
ECG1639	MN1301	MN3101	NTE1639
VHIMN3101//-1			

BLOCK-627
ECG1640	MN3001	MN3011	SK10238

BLOCK-628
181315	ECG1641	MN3007	NTE1641
SK10239	VHIMN3007//-1		

BLOCK-629
ECG1642

BLOCK-630
ECG1644

BLOCK-631
ECG1645

BLOCK-632
ECG1646	STR729

BLOCK-633
ECG1647	LA7823	SK10493

BLOCK-634
ECG1648	NTE1648

BLOCK-635
ECG1649	NTE1649

BLOCK-636
06300115	06300365	1180-5991	1200-3448
1239-6545	1266-2466	1327-1291	2360501
2365061	2365062	37003027	37003029
46-132638-3	50SI-HA11446	51S13753A65	ECG1650
EN11436	EP84X214	EP84X221	EW84X782
GL3301	GL3301B	HA11436	HA11436A
HA11446	HA11480	HA11480A	NTE1650
SK7606	TVSEN11436		

BLOCK-637
1197-6040	B0356690	ECG1651	NTE1651
SK7679	TA7668AP		

BLOCK-638
1166-5197	1502MH833	150368	649-3
ECG1652	SK10252	TDA2791	

BLOCK-639
2366151	ECG1653	M51355P	SK7632

BLOCK-640
8-759-276-14	B0355431	EA33X8569	ECG1654
NTE1654	SK7670	TA7614AP	TA7614AP-Y
TA7614P			

BLOCK-641
612496-1	6124960001	CA3195	CA3195E
ECG1655	NTE1655	SK7687	SK9029

BLOCK-642
2368282	46-13547-3	46-13912-3	ECG1656
M51354AP	M51354APO	M51354P	M51356
M51356P	NTE1656	QC0031A	SK7633
SK7734			

BLOCK-643
1173-4639	4-2069-71660	46-13645-3	B0325350
B0325355	E-TA7343P	E-TA7343PS	ECG1657
EW84X15	IX0086TA	NTE1657	SK7673
TA7343AP	TA7343P	TA7343PS	VHITA7343P/-1

BLOCK-644
06300121	1200-3463	1346-8905	156730
178695	2366621	266P19201	37011039
409-019-6209	4152079100	46-13581-3	5355611
56A355-1	8-759-800-65	ECG1658	EW84X888
I03D079100	I03S079100	IX0260CE	LA7910
MX-2269	NTE1658	Q7256	
RH-IX0203GEZZ	RH-IX0260CEZZ	SK7736	X0260C

BLOCK-645
160576	160741	8-759-932-80	BA328
BA328LN	ECG1659	NTE1659	

BLOCK-646
147232	148132	5350251	5350961
BA340	ECG1660	EP84X71	G09-017-A
G09-017-B	G09-017-C	G09-017-D	GEIC-171
HA1406	HA1406-2	HA1406-3	HA1406-4
NTE1660	SK3364		

BLOCK-647
1191-4991	185897	415101379C	6123300140
ECG1661	EW84X897	NTE1661	SK4816
UPC-1379C	UPC1379C		

BLOCK-648
ECG1662	NTE1662

BLOCK-649
5113753A68	51S13753A68	AN5255	AN5256
ECG1663	EW84X423	EW84X913	NTE1663
SK9324			

BLOCK-650
1186-5912	37009016	8-759-100-60	ECG1664
NTE1664	SK4833	UPC1377C	

BLOCK-651
8-758-480-00	848	CX-848	CX848
ECG1665			

BLOCK-652
ECG1666	NTE1666

BLOCK-653
E-LA4192	ECG1667	LA4182	LA4183
LA4190	LA4192	LA4192S	MX-5475
NTE1667	SK4831		

BLOCK-654
2002110332	2365021	46-13448-3	ECG1668
NTE1668	SK9978	UPC1391H	UPC1391H-1
UPC1391HA	VHIUPC1391H-1		

BLOCK-655
06300030	1195-6547	51-13753A53	51S13753A53
ECG1669	EN11238	HA11238	HA11238N
NTE1669	TVSEN11238		

BLOCK-656
ECG1670	LM1819N	NTE1670	SK10241

BLOCK-657
6123300023	ECG1671	IX0212CE	IX0212CEZZ
MX-6093	NTE1671	RH-IX0212CEZZ	SK7841
TA7670P	X0212CE		

BLOCK-658
ECG1672	NE5560F	NE5560N	NTE1672
SK10242	TDA1060	ULN8160A	ULN8160R

BLOCK-659
1316-7895	185896	415207601W	46-13548-3
46-13604	46-13604-3	46-13938-3	46-13963-3
6123300334	ECG1673	EW84X910	LA-7601W
LA7601	LA7601G	LA7601W	NTE1673
QC0093	SK9402	TLA7601G	TLA78601G

BLOCK-660
0061-920-60090	1200-2358	1227-8610	1298-0785
2119-102-3003	2119-102-5702	32119-102-300	32119-102-570
4835-209-87612	5190480A96	51X904080A96	51X90480A96
5652-AN5512	6192060090	8-759-994-51	AN5512
ECG1674	EW84X426	KA2131	NTE1674
SK9325			

BLOCK-661
ECG1675	LA4505	NTE1675	SK9815

BLOCK-662
102SD13780	178692	2002300046	2368501
23685015	6123300025	6192080250	8-759-105-82
8-759-113-78	875910582	ECG1676	I02SD13780
IX0238CE	MX-6160	NTE1676	
RH-IX0238CEZZ	SK7653	UPC1378	UPC1378-H-L
UPC1378H	UPC1378H-L	UPC1378H-P	X0238CE

BLOCK-663

1234-1674	1234-5948	2002700175	409-053-7309
46-131203-3	46-13924-3	67-32808-02	B0272120
B0272490	ECG1677	NTE1677	QC0291
SK9769	TD6312P	TD6350P	

BLOCK-664

1232-7573	2119-102-4309	2366721	6192140210
8-759-100-75	ECG1678	NTE1678	SK9947
UPC1394C			

BLOCK-665

156571	156915	BA6304	BA6304A
BA6304AL	ECG1679	IYBA6304A	NTE1679
SK10244			

BLOCK-666

DAP601	ECG1680	NTE1680	UPA64H
UPA64HA	X440150640		

BLOCK-667

ECG1681	L296	L296V	NTE1681
SK10246			

BLOCK-668

221-187	221-187A	2366631	5113753A37
51S13753A37	AN5020	ECG1682	NTE1682
SK7713			

BLOCK-669

5190495A90	5190528A28	5190555A53	AN5416
BN-5416	BN5416	ECG1683	EW84X914
NTE1683	SK7626	TVSBN5416	TVSBN5416A

BLOCK-670

23119548	272P15901	33A17057	46-131245-3
AN5515	AN5515X	ECG1684	NTE1684
SK9729			

BLOCK-671

06300766	0ISA426100A	1368-1846	221-598
409 017 7208	409 017 7505	409-017-7208	46-13969-3
ECG1685	LA4260	LA4261	NTE1685
QC0080	SK9948		

BLOCK-672

156912	ECG1686	HA11749	IHHA11749
SK10247			

BLOCK-673

EA33X8572	ECG1687	LM2896	LM2896P1
NTE1687	SK7711		

BLOCK-674

1027-4264	AN5070	ECG1688	EW84X315
NTE1688	SK9725		

BLOCK-675

1175-7879	154576	5190507A46	51X90507A46
AN6326	AN6326N	ECG1689	EW84X239
EW84X481	NTE1689	SK9949	

BLOCK-676

ECG1690	MK5089J	MK5089K	MK5089P
NTE1690	TCM5089		

BLOCK-677

ECG1691	LR40992	MK50992N	NTE1691

BLOCK-678

ECG1692	LR40993	NTE1692	

BLOCK-679

ECG1693	MK50981N	NTE1693	

BLOCK-680

148721	51X90458A81	AN6343	ECG1700
NTE1700	SK9980		

BLOCK-681

148723	51X90458A02	AN6350	ECG1701
NTE1701	SK10008		

BLOCK-682

148725	51X90458A03	AN6360	ECG1702
NTE1702	SK10013		

BLOCK-683

148726	149585	51X90480A41	AN6361
AN6361N	ECG1703	NTE1703	SK10016

BLOCK-684

0079-0220	185901	51X90518A09	AN7110
AN7110E	E-AN7110E	ECG1704	EW84X738
NTE1704	SK9981		

BLOCK-685

1033-1338	154753	155556	156960
30901350	5190480A92	51X90480A92	5391032
6838	DN6338-A	DN6838	DN6838A
ECG1705	IMDN6838	NTE1705	SK9950

BLOCK-686

156580	156918	160763	ECG1706
HA11747-A	HA11747A	HA11747ANT	HA11747BNT
IHHA11747ANT	SK9982		

BLOCK-687

ECG1707	I03SP44450	LA4445	NTE1707
SK10018			

BLOCK-688

8-759-800-12	ECG1708	LA7920	NTE1708
SK10017			

BLOCK-689

148718	AN6041	ECG1709	NTE1709
SK10080			

BLOCK-690

1231-0116	154577	51X90507A77	6192180061
AN6209	AN6209K	ECG1710	EW84X479
NTE1710	SK9951		

BLOCK-691

1185-0161	155508	155509	5190518A19
51X90518A19	AN6307	ECG1711	EW84X314
NTE1711	SK9952	VCRS0011	

BLOCK-692

1033-1247	154572	5190507A78	51X90507A78
AN6306	ECG1712	EW84X312	NTE1712
SK9953			

BLOCK-693

1033-1262	154574	51X90507A80	AN6328
ECG1713	EW84X375	NTE1713	SK9954

BLOCK-694

ECG1714M	MC3373	NTE1714M	SK10019

BLOCK-695

06300095	1065-2212	11119034	1180-6445
1237-7912	1259-4958	153712	154822
160869	2002700158	2109-301-6008	
34SI-UPC1373HA	37011020	38SI-UPC1373H	46-13815-3
51X90507A29	5355262	5652-UPC1373H	8-759-113-73
C1373HA	ECG1714S	EP84X227	EW84X60
INMPC1373H	INMPC1373HA	INUPC1373H	KA2181
MPC1373H	MPC1373HA	MX-2273	NTE1714S
SK9715	UPC1373	UPC1373-H	UPC1373H
UPC1373HA	VHIUPC1373H-1	VHIUPC1373HA/	
VHIUPC1373HA1			

BLOCK-696

154730	162674	51X90507A28	51X90507A83
AN6873	AN6873N	ECG1715	EW84X385
NTE1715	SK10021		

BLOCK-697

1231-0157	1231-8325	193333	46-13689-3
BA6209	BA6209U	BA6209U1	BA6209UI
ECG1716	EW84X490	NTE1716	SK10012
VHIBA6209//1E			

BLOCK-698

BA6122A	ECG1717	NTE1717	SK10022

BLOCK-699

149586	157971	51X90480A32	AN6913
ECG1718	NJM2903S	NTE1718	SK10023

BLOCK-700

06300111	06300116	1266-2441	ECG1719
EXT954-20C	NTE1719	SK7719	STR-30120
STR383			

BLOCK-701
CA3524E	ECG1720	HA17524P	LM3524N
NTE1720	SG3524	SK9983	XR3524
XR3524CP			

BLOCK-702
ECG1721	NTE1721	SG3525A	SK10024
UC3525AN			

BLOCK-703
ECG1722	NTE1722	SG3526N	SK10025
ULN8126A			

BLOCK-704
ECG1723	NTE1723	SG3527A	SK10026

BLOCK-705
8-759-111-88	ECG1724	NTE1724	SK7650
UPC1188H			

BLOCK-706
06300113	1266-2458	46-132637-3	ECG1725
EW84X784	GL3101A	GL3101B	HA11440
HA11440A	NTE1725	SK7705	

BLOCK-707
5652-AN5316	AN5315	AN5316	ECG1726
EW84X0425	EW84X425	NTE1726	SK10028

BLOCK-708
51X90480A65	5652-AN5436	AN5436	AN5436N
BN5436N	ECG1727	EW84X1009	NTE1727
SK7618			

BLOCK-709
06300357	06300574	103DE75200	1259-6524
1296-1892	1298-0777	1310-8394	2119-102-0503
2119-102-0907	266P12101	32119-102-050	32119-102-090
37007021	409-019-3109	46-131206-3	5652-LA7520
6192020400	A7520(IC)	CA7520	ECG1728
EW84X906	GL3120	I03DE75200	KA2919
LA7520	NTE1728	QC0085	SK9748

BLOCK-710
2368211	4835-209-87803	8-759-937-59	ECG1729
EW84X370	MB3759	MB3759P	NTE1729
SK9912	TL494	TL494CJ	UPC494C
X440064940			

BLOCK-711
ECG1730	EW84X377	NTE1730	SK9955
TA7325P			

BLOCK-712
ECG1731	NTE1731

BLOCK-713
8-749-956-30	8-749-958-30	ECG1732	NTE1732
SK7703	SK9761	STK-563F	STK561F
STK563A	STK563F	STK583F	STK583FST

BLOCK-714
ECG1733	IGSTK5431	IGSTK5431SL	IGSTK5431ST
NTE1733	SK7809	STK5431	STK5431SL
STK5431ST			

BLOCK-715
ECG1734	NTE1734	SK7810

BLOCK-716
163823	5353591	ECG1735	NTE1735
SK7811	STK5451		

BLOCK-717
ECG1736	NTE1736	SK10029	STK6981B
STK6982	STK6982B	X440759820	

BLOCK-718
EAT00-09100	EC-A063	ECG1737	NTE1737
SK10030	STK7563F	STK7563FE	X440755400
X440756330			

BLOCK-719
266P57001	67-01603-01	ECG1738	M58485P
NTE1738	SK9359		

BLOCK-720
1234-1095	159433	781-1	CA3218E
ECG1739	NTE1739	SK10031	

BLOCK-721
06300154	1270-5299	ECG1740	EW84X783
NTE1740	SK7817	STR3115	

BLOCK-722
1217-2284	2119-601-5005	46-13911-3	6125960001
ECG1741	EW15X570	NTE1741	QC0065
SK9994	STR-3125	STR3123	STR3125
STR3225			

BLOCK-723
06300688	1330-9232	1380-6070	2002600213
2370873	409-047-9104	46-131585-3	5190528A41
6192140170	ECG1742	NTE1742	SK9995
STR3130	STR3230	TVSSTR3130	TVSSTR3230

BLOCK-724
2366201	409-047-9203	46-131351-3	56A141-1
8-749-930-35	ECG1743	NTE1743	SK9996
STR3035	STR3135		

BLOCK-725
221-303	23119489	266P28101	2911481
46-131237-3	6124890001	8-759-400-88	AN6291
DBXAN6291	ECG1744	EW84X480	MX-6536
NTE1744	QC0380	SK9732	VHIAN6291//-1

BLOCK-726
ECG1745	M54459L	NTE1745	SK10033

BLOCK-727
1180-2071	2002700101	46-13649-3	B0272050
B0272054	ECG1746	Q7368	TD6306P
TD6306PFA-1			

BLOCK-728
188244	5190495A88	5190528A27	AN5318
AN5318A	AN5318N	ECG1747	EW84X452
NTE1747	SK7619		

BLOCK-729
ECG1748	IGSTK6962	NTE1748	SK9762
STK6960	STK6961	STK6962	STK6962H
STL6961			

BLOCK-730
ECG1749	L293B	NTE1749	SK10001

BLOCK-731
ECG1750	L295	NTE1750	SK10035

BLOCK-732
0070-0500	105S925080	178693	46-133275-3
ECG1751	I05S925080	IX0137CE	MX-6092
MX-6387	NTE1751	RH-IX0137CEZZ	T-2508
T2508	X0137CE		

BLOCK-733
449-1	6124490001	ECG1752	NTE1752
SK10036	TDA2546A		

BLOCK-734
ECG1753	MC34060L	MC34060P	NTE1753
SK10037			

BLOCK-735
1254-2106	1348-1593	32119-103-420	32119-110-016
35683-112-220	3H83-00050-000	444-1(IC)	6124440001
ECG1754	N6124440001	NTE1754	SK10002
TDA3652	TDA3654	TDA3654AQ	TDA3654Q
TDA3654U			

BLOCK-736
ECG1755	NTE1755	SK10038	UAA4000

BLOCK-737
ECG1756

BLOCK-738
ECG1757	ML924	NTE1757

BLOCK-739
119035	1UPD1986C	2109-301-6707	ECG1758
IUPD1986C	NTE1758	SK10041	UPD1986C

BLOCK-740

06300056	06300072	11119033	1200-2200
1266-2433	1UPD1937C	2109-301-6600	2361301
46-132286-3	ECG1759	IUPD1937C	IX0199CE
MPD1937C	NTE1759	RH-IX0199CEZZ	SK10003
UPD1937C	UPD1987C		

BLOCK-741

46SI-SAF1039P	612247-2	6122470002	ECG1760
SAF1039P	SAF1039PN	SK10042	

BLOCK-742

6124500002	ECG1761	NTE1761	TDA3047

BLOCK-743

6124500001	7044260991	ECG1762	NTE1762
SK10044	TDA3048	TDA3048N	

BLOCK-744

ECG1763	SK10045	TEA1009	

BLOCK-745

ECG1764	NTE1764	

BLOCK-746

4835-209-87071	4835-209-87834	6126640001	ECG1765
MC34065P			

BLOCK-747

4835-209-87268	BA6219	BA6219B	ECG1766
SK10494			

BLOCK-748

2916681	2917361	ECG1767	I02SD14980
UPC1498H			

BLOCK-749

ECG1768	MWA110	SK4802

BLOCK-750

ECG1769	MWA120	SK4804

BLOCK-751

ECG1770	MWA220	SK4803

BLOCK-752

2910012	46-131315-3	46-131507-3	ECG1771
ILA7621	LA7621	NTE1771	QC0554
SK9749			

BLOCK-753

46-131099-3	ECG1772	LA7811	NTE1772
QC0073	SK9866		

BLOCK-754

0061-262-50001	06300360	0ISA783000A	11119112
1310-8410	188086	221-796	2910021
409-019-5608	4152078300	46-131208-3	4835-209-87069
6126250001	6192060140	8-759-801-98	ECG1773
GL1130	LA7830	NTE1773	QC0300
SK9752			

BLOCK-755

ECG1774	IGLB1649	LB1649	NTE1774
SK9867			

BLOCK-756

11119118	1234-5922	46-13829-3	B0356602
ECG1775	EW84X267	NTE1775	SK9767
SK9868	TA7660	TA7660P	TA7660P(FA-1)
TA7660PFA	TA7660PFA-1	TA766P	

BLOCK-757

0061-258-60003	1346-8947	223-18-01	267P91003
33A17062	4159301230	6125860001	ECG1776
EW84X909	NTE1776	SK9743	STR-30123
STR30123			

BLOCK-758

0061-258-60004	06300629	0IGL301300A	1313-5546
1348-5024	1374-6847	223-40	223-40-01
267P91002	267P91006	2912177	32119-801-130
32119-901-130	409 243 0806	409-243-0806	4835-209-47056
5653-STR30130	6125860002	905-10085	ECG1777
EW84X1010	NTE1777	SK9870	STR30130
STR30130-A	STR30130A	STRD3030	TVSSTR30130
VHISTR30130-1	VHISTR30130E		

BLOCK-759

23114420	2912175	2912176	2912178
409-047-8602	46-131433-3	46-131438-9	8-749-901-35
ECG1778	NTE1778	QC0543	
QSBLAOZSQ019	SK9744	SK9871	STR30134
STR30135	STR30135F		

BLOCK-760

1027-4389	5190266A68	5190538A20	AN5125
AN5125M	ECG1779	EW84X327	EW84X451
NTE1779	SK9726		

BLOCK-761

11119114	1232-7292	1239-6537	1330-9257
2119-102-7304	23119710	2366392	266P38802
32119-102-730	37001059	409-001-0604	46-13921-3
5190555A50	67-53315-01	AN5836	ECG1780
EW84X976	I01S058360	NTE1780	QC0846
SK9731	VHIAN5836//-1		

BLOCK-762

06300767	0ISA701600A	1368-1853	157677
178476	178746	189152	266P016010
2917611	46-13686-3	4835-209-87074	5369431
6126810001	8-759-800-81	9-901-505-01	A7016
ECG1781	EW84X905	I03S070160	LA7016
LA7016//-1	MX-6159	NTE1781	QC0011
SK9746	VHILA7016//-1		

BLOCK-763

1330-9216	1393-9327	188246	221-468
23119142	2913981	46-131100-3	8-759-402-35
AN5521	ECG1782	IX0948CE	NTE1782
QC0074	RH-IX0948CEZZ	SK9730	

BLOCK-764

06300336	46-131265-3	5190528A43	56A60-1
AN5262	ECG1783	NTE1783	SK9873

BLOCK-765

ECG1784	SK9227

BLOCK-766

ECG1785	NTE1785

BLOCK-767

6124240001	6125450001	ECG1786	N6125450001
NTE1786	SAB3037	SAB3037N	SK9874

BLOCK-768

1027-6244	1217-9065	1255-1198	161715
46-131274-3	5190528A64	5364201	6123300215
6192000040	6192180080	C1513HA	ECG1787
EW84X323	EW84X352	EW84X624	INMPC1513HA
NTE15005	NTE1787	SK9766	TA7361
TA7361AP	TA7361P	UPC1513HA	
VHIUPC1513H-1			

BLOCK-769

215531	221-347	221-347-01	8-759-980-58
ECG1788	NTE1788	SK9875	TDA8172

BLOCK-770

0061-920-01330	1325-1186	188243	33A17056
46-132886-3	46-13493-3	4835-209-87406	5190538A98
5653-AN5265	6192001330	8-759-904-94	AN5265
ECG1789	EW84X1008	NTE1789	SK9876
TL494C	TL494CN		

BLOCK-771

176854	178694	193082	221-467
23119355	46-131432-3	46-131787-3	5653-TA7777N
A23-1101-01A	B0358268	B0358272	B0379400
ECG1790	I05DE7777P	I05DE86770	IX0600CE
MX4340	NTE1790	RH-IX0600CEZZ	SK9850
TA7777N	TA7777P	TA7777P(FA-1)	TA7777P(FA-3)
TA7777PFA-1	TA7777PFA-3	TA8677N	X0600CE

BLOCK-772

1234-5963	23119566	46-131266-3	C1474HA
ECG1791	NTE1791	SK9717	UPC1474HA

BLOCK-773

06300334	1204-000395	14LV233	196122
221-383	23119228	46-132305-3	6126060001
6192000990	8-759-145-27	C1406HA	ECG1792
IX0416CE	NTE1792	RH-IX0416CEZZ	SK9877
UC1406HA	UPC1406HA	VHiUPC1406HA1	

BLOCK-774

102DE14020	2002502338	2002502341	2002502385
6123300102	ECG1793	I02DE14020	SK9716
UPC1401CA	UPC1402CA		

BLOCK-775

161326	2002610039	C1252H2	ECG1794
I02SF12520	IX0487CE	NTE1794	
RH-IX0487CEZZ	SK9878	UPC1252H2	UPC1252HA

BLOCK-776

161325	2002610042	C1253H2	C1353H2
ECG1795	I02SF12530	IX0499CE	NTE1795
RH-IX0499CEZZ	SK9879	UPC1253H2	UPC1253HA2

BLOCK-777

ECG1796	NTE1796	SK9880	STR54041

BLOCK-778

0061-257-40001	06300491	0ISA783100A	1316-1013
176853	192742	409-019-5707	46-131341-3
4835-209-87073	6125740001	ECG1797	LA7831
NTE1797	QC0573	SK9753	

BLOCK-779

272P14001	272P140010	2912561	409-017-7505
46-131343-3	4835 209 87085	4835-209-87085	6126280001
8-759-803-29	ECG1798	LA4270	NTE1798
QC0352	RH-IX0934CEZZ	SK9745	

BLOCK-780

2917601	46-131342-3	ECG1799	I03D072200
LA7220	NTE1799	SK9881	

BLOCK-781

1270-1769	1287-7916	176223	2002610086
37001065	6125630001	67-93401-01	C1480CA
ECG1800	I02DF1480C	LA7760	N6125630001
NTE1800	SK9882	UPC1480CA	

BLOCK-782

1270-1777	1287-7924	176224	2002610091
272P01401	37001066	6125640001	67-93501-01
C1481CA	ECG1801	I02DF1481C	I02DF148C
LA7761	N6125640001	NTE1801	SK9883
UPC1481CA			

BLOCK-783

ECG1802			

BLOCK-784

412-2	4835-209-17148	6124120002	ECG1803
N6124120002	NTE1803	SK9884	TDA1524
TDA1524A	TDA1524AN	TDA1524PN	TDA152AN

BLOCK-785

ECG1804	NTE1804	SK9894	TDA2653A
TDA2653AN	TDA2653AU		

BLOCK-786

1217-8935	1255-1063	6192080080	AN3210K
AN3211K	AN3211NK	ECG1805	EW84X440
EW84X761	EW84X774	NTE1805	SK9838
VHIAN3211K/-1			

BLOCK-787

1229-0862	AN3310K	ECG1806	EW84X589
SK9888			

BLOCK-788

1217-8943	1255-1081	6192180010	AN3312
AN3313	ECG1807	EW84X441	EW84X744
SK9839	SK9886		

BLOCK-789

1217-8950	AN3320K	ECG1808	EW84X442
EW84X536	NTE1808	SK9722	

BLOCK-790

1217-8968	1217-8976	AN3821K	AN3822K
ECG1809	EW84X381	EW84X443	NTE1809
SK9724			

BLOCK-791

1217-8992	1255-1123	612160030	6192160030
6192180160	AN5135K	AN5135NK	AN5153NK
ECG1810	EW84X444	EW84X767	NTE1810
SK9727			

BLOCK-792

1027-4462	5190528A55	AN6356N	ECG1811
EW84X320	NTE1811	SK9733	

BLOCK-793

1027-4850	1217-9008	5190528A56	AN6359
AN6359N	ECG1812	EW84X317	NTE1812
SK9734			

BLOCK-794

1027-4959	1217-9016	5190528A57	6192040031
AN6366	AN6366K	AN6366NK	ECG1813
EW84X445	NTE1813	SK9735	

BLOCK-795

1027-5535	1217-9040	5190528A61	6192040040
ECG1814	EW84X446	EW84Z316	MN6163
MN6163A	NTE1814	SK9739	

BLOCK-796

ECG1815	NTE1815	SK9891	STK4024II

BLOCK-797

ECG1816	NTE1816	SK9897	

BLOCK-798

ECG1817	NTE1817	SK9890	

BLOCK-799

ECG1818	NTE1818	SK9893	

BLOCK-800

ECG1819	NTE1819	SK9892	STK4152II

BLOCK-801

ECG1820	NTE1820	SK9889

BLOCK-802

409-047-5601	ECG1821	STK5466ST

BLOCK-803

ECG1822	NTE1822	SK9902	STK5471

BLOCK-804

905-1224	ECG1823	NTE1823	STK5481
STK5490			

BLOCK-805

ECG1824	NTE1824	SK9899	STK6940
STK6941	STK6942		

BLOCK-806

ECG1825	IGSTK6972	SK9898	STK6970
STK6971	STK6972		

BLOCK-807

1316-7887	185895	193309	415207530
4152075300	415207530N	6123300418	A7530N
ECG1827	EW84X911	I03DA7530N	LA-7530
LA7530	LA7530N	NTE1827	SK9885
TVSLA7530N			

BLOCK-808

1027-5006	5190528A58	6193220010	AN90C21
ECG1828	EW84X319	NTE1828	SK9736

BLOCK-809

1027-5113	5190528A59	AN90C22	ECG1829
EW84X322	NTE1829	SK9737	

BLOCK-810

1298-3920	32119-103-110	ECG1830	NTE1830
SK9896	TA7270P		

BLOCK-811

ECG1831	NTE1831	SK9895

BLOCK-812

ECG1832	SK10142	TA7274P

BLOCK-813

1255-1131	46-131969-3	6192180110	BA6238A
BA6238AU	BA6238AU4	ECG1834	EW84X763
NTE1834	SK9740	VHIBA6238AU1E	

BLOCK-814

1215-8663	1380-4687	178742	23119723
23318409	32109-301-610	46-131131-3	AN5352
AN5352N	ECG1835	NTE1835	
RH-IX0450CEZZ	SK9728		

BLOCK-815

409-019-6605	46-131401-3	ECG1836	LA7915
NTE1836	SK9756	-	

BLOCK-816

6192120110	ECG1837	EW84X810	I03D079130
LA7913	SK9755		

BLOCK-817

06300519	1296-1900	1313-5553	1316-3258
32119-102-010	409-019-4205	46-13156-3	46-131656-3
9-901-504-01	ECG1838	LA7626	NTE1838
QC0852	SK9750		

BLOCK-818

06300403	1271-3111	46-132287-3	56A141-4
ECG1839	NTE15040	NTE1839	SK10143
STR30120			

BLOCK-819

1-424-115-11	223-28	ECG1840	NTE1840
SK10144	STR-53041	STR53041	

BLOCK-820

23314068	46-861739-3	ECG1841	SK10145
STR53043	STR53043A		

BLOCK-821

11119031	1197-6032	185900	4-2069-71730
6192020190	B0356385	ECG1842	EW84X894
KIA7640AP	NTE1842	SITA7640APIXX	SK7674
TA-7640AP	TA7640AP	TA7640P	

BLOCK-822

185899	223883	6192020160	ECG1843
EW84X895	NTE1843	SK10146	TA-7358P
TA7358AP	TA7358P		

BLOCK-823

177612	8-759-106-61	ECG1844	NTE1844
SK10147	UPC1470H	UPC4570HA	

BLOCK-824

06300359	06300616	1298-0793	1310-8402
2119-102-0600	2119-102-1003	2910011	32119-102-060
32119-102-100	4152076200	46-131207-3	4835-209-87094
6192040260	ECG1845	GL3320	KA2155
LA7620	NTE1845	QC0275	SK10082

BLOCK-825

6125750001	ECG1846	M51307BSP	SK10083

BLOCK-826

23318086	272P026010	409-036-0105	46-131629-3
ECG1847	M51366SP	NTE1847	QC0713
SK10084			

BLOCK-827

08200070	08200077	ECG1848	NTE1848
SI-7115B	SI7115B	SK10148	

BLOCK-828

ECG1849	MN6049	NTE1849	SK10149

BLOCK-829

179726	ECG1850	SK10150	TDA1521
TDA1521U			

BLOCK-830

ECG1851

BLOCK-831

4835-209-87066	4835-209-87835	6124670001	ECG1852
N6124670001	NTE1852	SK10151	TDA1013A
TDA1013AU	TDA1013B		

BLOCK-832

ECG1853	SAA7220	SAA7220N	SK10152

BLOCK-833

ECG1854D	L272	L272B	SK10153

BLOCK-834

ECG1854M	L272M	L272MB	SK10154

BLOCK-835

103SD78370	11119226	1346-8939	1374-6821
272P239010	4152078350	46-132648-3	56A360-1
8-759-820-92	8-759-822-02	9-901-507-01	ECG1855
I03SD78350	LA7835	LA7835-TV	LA7835K
NTE1855	SK10085		

BLOCK-836

1330-5313	32119-102-020	ECG1856	LA7625
NTE1856	SK10155		

BLOCK-837

6126880001	ECG1857	MC3479P	SK10156

BLOCK-838

221-992	4835-209-87091	ECG1858	F-51056
TDA8174			

BLOCK-839

06300456	11119140	1316-0973	1330-9273
221-475	221-4757D04	221-475C	2914941
409-004-5705	46-131515-3	70119527	8-752-030-26
A1011(IC)	CXA1011P	ECG1859	I00DF10110
NTE1859	QC0680	SK9946	VHICXA1011P-1

BLOCK-840

06300457	11119139	1316-0981	1330-9265
20112	221-476	221-4767D02	221-476D
2914931	409-004-8409	46-131514-3	46-132398-3
70119526	8-752-011-20	B0358265	CX20112
ECG1860	I00DF01120	NTE1860	SK9945

BLOCK-841

1330-9190	188242	AN5136K	AN5136K-R
AN5136KR	ECG1861	NTE15046	NTE1861
SK10158			

BLOCK-842

ECG1862	SK10159	TDA1670A	TDA1675

BLOCK-843

11119225	1346-8921	221-561	272P238010
272P238020	409-146-7001	409-156-6605	4152076550
46-132647-3	56A359-1	8-759-820-93	ECG1863
FH2000571-4	LA7650K	LA7650KN	LA7655
LA7655N	NTE1863		

BLOCK-844

ECG1864	LM1870	SK9069	

BLOCK-845

AN7062	ECG1865	SK10509	

BLOCK-846

BA6144	ECG1866	SK10496	

BLOCK-847

159249	ECG1867	STR2012	STR2012A

BLOCK-848

ECG1868	STR2013		

BLOCK-849

ECG1869	M5236L	SK10500	

BLOCK-850

ECG1869SM	M5236ML	SK10499	

BLOCK-851

6192140400	ECG1870	I03S073080	IX0325C
RH-IX0325CEZZ	SK10161	STK7308	STK7309

BLOCK-852

ECG1872	NTE1872	SK10164	STK5482

BLOCK-853

1231-0207	1259-4966	192276	196188
37011041	ECG1873	EP84X250	EW84X255
EW84X595	NTE15045	SK9764	TA7347
TA7347P	VHITA7347/-1	VHITA7347P/-1	VHiTA7347P/-1

BLOCK-854

ECG1874	NTE1874	RH-IX1112CEZZ	SK10141
STK4121II			

BLOCK-855

ECG1875	NTE1875	SK10166	STK2230

BLOCK-856

ECG1876	NTE1876

BLOCK-857

ECG1877	NTE1877	SK10168	STK4311

BLOCK-858

ECG1878

BLOCK-859

ECG1879	NTE1879	SK10169	STK4273

BLOCK-860

ECG1880	I23S953320	SK10170	STK5332

BLOCK-861

177619	46-131683-3	ECG1881	SK10171
STK5476			

BLOCK-862

ECG1882	SK10172	STK4044II	STK4044V

BLOCK-863

46-132070-3	ECG1883	SK10173	STK5372

BLOCK-864

ECG1884	SK10174	STK5479

BLOCK-865

ECG1885	SK10175	STK6722H	X440672200
X440672210			

BLOCK-866

46-131098-3	ECG1886	QC0032	SK10086
STR5100			

BLOCK-867

ECG1887

BLOCK-868

ECG1888

BLOCK-869

AN7273	ECG1889	SK10501

BLOCK-870

6105080001	6125080001	6126150001	6126150K01
ECG1890	N6125080001	SK10520	TDA3564

BLOCK-871

161573	173116	5364091	ECG1892
M54549L	SK10502		

BLOCK-872

156546	46-131030-3	ECG1893	HA13403

BLOCK-873

0ISK500410A	1377-2744	23314158	ECG1894
I2B4900410	STR50041	STR50041A	

BLOCK-874

1254-3583	32199-901-200	3283-00160-000	35682-104-620
3H83-00160-000	ECG1895	I2B4910410	STR51041

BLOCK-875

ECG1896	SK10505	STR30115

BLOCK-876

1296-1926	2119-901-0201	32119-901-020	ECG1897
NTE15041	STR30125		

BLOCK-877

ECG1898	SK10507	TA7280P

BLOCK-878

ECG1899	SK10508	TA7281P

BLOCK-879

276-1762	ECG1900	LM317LZ	NTE1900
SK7644	TL317	TL317LP	

BLOCK-880

ECG1901	LM337LZ	NTE1901	SK10047

BLOCK-881

103B98M090	11-117	1380-4729	266P923020
266P931010	39-054-1	46-131516-3	78L09
78L09A	78L09CLP	78M09	78M09A
AN78L09	AN78L09-Y	AN78M09-(LC)	AN78M09F
ECG1902	I01B98M090	I03A98M090	I03B98M090
L78M09	L78M09-RA	L78M09-SA	L78M09SA
NJM78L09	NJM78L09A	NTE1902	SK3962
TA78L009AP	TA78L009P	TA78L09S	UA78L09ACLP
UA78L09AWC			

BLOCK-882

08200050	08200064	79L12	ECG1903
GEVR-112	HEP-C6137P	HEPC6137P	LM320L-12
LM320LZ-12	LM320LZ12	MC79L12AC	MC79L12ACG
MC79L12ACP	MC79L12CG	MC79L12CP	NJM79L12A
NTE1903	SK9221		

BLOCK-883

AN79L15	ECG1905	HEPC6138P	LM320LZ15
MC79L15ACG	MC79L15ACP	MC79L15CG	MC79L15CP
NTE1905	SK9222		

BLOCK-884

4835-209-87552	AN78L18	ECG1906	MC78L18ACP
MC78L18CP	ML78L18A	NTE1906	SK10048
TA78L018AP			

BLOCK-885

AN79L18	ECG1907	MC79L18ACP	MC79L18CP
NTE1907	SK10049		

BLOCK-886

AN78L24	ECG1908	HEPC6135P	MC78L24ACP
MC78L24CP	ML78L24A	NTE1908	SK10050
TA78L024P	UPC78L24J		

BLOCK-887

ECG1911	IP317	LM317	LM337K
NTE1911	SK10052	SK9333	UC337K

BLOCK-888

ECG1912	MC78T12CK	NTE1912	SK9337

BLOCK-889

7905CDA	7905DA	ECG1913	LM320K5.0
MC7905CK	NTE1913	SK9334	UA7905CKA
UA7905KM	UA7905MKA		

BLOCK-890

901528-04	ECG1914	IP340-12	LM340-12
LM340AK12	LM340K-12	LM340K12	LM340KC12
NTE1914	SK9331	TA78012M	UA7812CKA
UA7812DA	UA7812KC	UA7812KM	UA7812MKA

BLOCK-891

ECG1915	LM320K12	MC7912CK	NTE1915
SK9335	UA7912CKA	UA7912KC	

BLOCK-892

7815CDA	ECG1916	LM340-15DA	LM340AK15
LM340K-15	LM340K15	LM340KC15	MC7815CK
NTE1916	SK9332	UA7815CDA	UA7815CKA
UA7815DA	UA7815KC	UA7815KM	UA7815MKA

BLOCK-893

442-665	79L05	79L05AC	AN79L05
ECG1917	HE-442-665	MC79L05AC	MC79L05ACG
MC79L05ACP	MC79L05CG	NJM79L05A	NTE1917
SK9219			

BLOCK-894

ECG1918	LM140K-15	MC78T15CK	NTE1918
SK9338			

BLOCK-895

ECG1919	LM320K15	MC7915CK	NTE1919
SK9336	UA7915CKA	UA7915KC	UA7915KM

BLOCK-896

ECG1920	L7818CT	LM340K-18	MC7818CK
NTE1920	SG7818CK	SK10053	UA7818CKA
UA7818KC			

BLOCK-897

ECG1923	L7918CT	NTE1923	SG7918CK
SK10054	UA7918CKA		

BLOCK-898

ECG1924	L7824CT	LM340-12DA	LM340K-24
MC7824CK	NTE1924	SG7824CK	SK10055
UA7824CKA	UA7824KC		

BLOCK-899

ECG1925	L7924CT	MC7924CK	NTE1925
SK10056	UA7924CKA		

BLOCK-900

ECG1926	NTE1926	UA78GKC

BLOCK-901

ECG1927	NTE1927

BLOCK-902

ECG1928	LM305H	LM305T	NTE1928
SK7642			

BLOCK-903

442-24	C305C(IC)	ECG1930	HE-442-24
ILM0292	LM305AP	LM376N	NTE1930
SK7643	UPC305(C)	UPC305C	X440193050

BLOCK-904

153612	78010AP	AN78M10	ECG1932
MX-4339	NTE1932	T78010AP	TA78010AP
TDD1610S	UPC78M10H	VHITA78010AP1	

BLOCK-905

3052P	ECG1934	NTE1934	SI-3052P
SK7739			

BLOCK-906

08200093	160844	3052V	ECG1934X
NTE1934X	SI-3052V	SI3052V	SK10087

BLOCK-907

1393-9376	3122P	ECG1936	IFSI-3122P
NTE1936	SI-3122P	SI-3122V	SK7740
X440762120			

BLOCK-908

ECG1938	NTE1938	SI-3152P	SK7741

BLOCK-909

ECG1940	NTE1940	SI-3242P	SK7742

BLOCK-910

ECG1941	RC4195NB	SK10058	XR4195CP

BLOCK-911

ECG1942

BLOCK-912

8-759-231-53	ECG1960	KIA7805PI	TA7805S

BLOCK-913

ECG1961

BLOCK-914

ECG1962

BLOCK-915

ECG1963

BLOCK-916

ECG1964

BLOCK-917

ECG1965

BLOCK-918

ECG1966	I0QK98M090	I1KA978090	KIA7809PI
KiA7809Pi	NJM78M09FA	TA7809S	VHITA7809S/-1
VHiKA7809Pi-1	VHiTA7809S/-1		

BLOCK-919

ECG1967

BLOCK-920

ECG1968

BLOCK-921

5653-TA7812S	8-759-701-79	ECG1970	NJM7812FA
TA7812S			

BLOCK-922

ECG1971

BLOCK-923

ECG1972

BLOCK-924

ECG1973

BLOCK-925

ECG1974

BLOCK-926

ECG1975

BLOCK-927

ECG1976

BLOCK-928

ECG1977

BLOCK-929

30901150	ECG2000	HA11226	NTE2000

BLOCK-930

14L0035	4206970020	442-604	722114299
ECG2001	HE-442-604	NE545B	NTE2001

BLOCK-931

905-172	ECG2002	MA7300	UA7300

BLOCK-932

51T40113T01	905-229	ECG2003	LM1011AN
LM1011N	NTE2003	TCG2003	

BLOCK-933

442-672	905-131	ECG2004	HE-442-672
NE645(B)	NE645BN	NE646N	NTE2004
TCG2004	TM2004		

BLOCK-934

552	9665PC	DM-65	ECG2011
M54524P	MC1411P	NTE2011	SK3975
TCG2011	TM2011	ULN-2001A	ULN2001A
ULN2001AJ	ULN2001AN	ULN2011A	XR2201CP

BLOCK-935

9666DC	9666PC	ECG2012	M54525P
MC1412P	NTE2012	SK9092	TCG2012
TM2012	ULN-2002A	ULN2002A	ULN2002AJ
ULN2002AN	ULN2012A	XR2202CP	

BLOCK-936

610010-413	6124920001	75468	76-500-001
9667DC	9667PC	BA12003	ECG2013
M54523P	MC1413P	MX-4821	NTE2013
SK9093	TCG2013	TD62003P	TM2013
ULN-2003A	ULN2003	ULN2003A	ULN2003AJ
ULN2003AN	ULN2013A	UPA2003	UPA2003C
XR2203CP			

BLOCK-937

9668DC	9668PC	ECG2014	M54526P
MC1416P	NTE2014	SK9094	TCG2014
TM2014	ULN-2004A	ULN2004A	ULN2004AJ
ULN2004AN	ULN2014A	UPA2004C	XR2204CP

BLOCK-938

ECG2015	NTE2015	SK9095	TCG2015
TM2015	ULN2005A	ULN2015A	

BLOCK-939

ECG2016	NTE2016	SK9077	TCG2016
TM2016	ULN2801A	ULN2811A	

BLOCK-940

ECG2017	NTE2017	SK9078	TCG2017
TM2017	ULN2802A	ULN2812A	

BLOCK-941

ECG2018	NTE2018	SK9079	TCG2018
TM2018	ULN2803A	ULN2813A	

BLOCK-942

ECG2019	M54522P	NTE2019	SK9080
TCG2019	TM2019	ULN2804A	ULN2814A

BLOCK-943

ECG2020	NTE2020	SK9081	TCG2020
TM2020	ULN2805A	ULN2815A	

BLOCK-944

442-682	ECG2021	HE-442-682	IGLB1290
LB1290	NTE2021	SK9020	TCG2021
TM2021	UDN-6118A	UDN-6184A	UDN-7183A
UDN6118A			

BLOCK-945

442-659	ECG2022	HE-442-659	NTE2022
SK9019	TCG2022	TM2022	UDN-6128A
UDN-7186A	UDN6128A		

BLOCK-946

CA3082	ECG2023	NTE2023	SK3694
TCG2023	TM2023	ULN-2082A	ULN2082A

BLOCK-947

1099-6544	143766	CA3168	CA3168E
ECG2024	NTE2024	SK3667	TM2024

BLOCK-948

ECG2025	NTE2025	TCG2025	UDN-6144A

BLOCK-949

ECG2026	NTE2026	SK10276	TCG2026
UDN-6164A	UDN6116A-1	UDN6164A	

BLOCK-950

DS8870	DS8870J	DS8870N	ECG2027
NTE2027	SK10277	TCG2027	

BLOCK-951

56A16-1	DS8880F	DS8880J	DS8880N
ECG2028	NTE2028	SK9814	SN75480N
TCG2028			

BLOCK-952

ECG2029	MC3490P	NTE2029	TCG2029

BLOCK-953

DS8889J	DS8889N	ECG2030	MC3491P
MC3492P	NTE2030	TCG2030	

BLOCK-954

ECG2031	MC3494P	NTE2031	TCG2031

BLOCK-955

CA3161E	ECG2032	NTE2032	SK9203
TCG2032			

BLOCK-956

ECG2033

BLOCK-957

ECG2040	SAJ110	TM2040

BLOCK-958

ECG2041

BLOCK-959

ECG2042	M083B1	MK50240N	MK50240P
MK50241N	MK50241P		

BLOCK-960

AY1-0212	ECG2043	M086B1

BLOCK-961

ECG2045

BLOCK-962

ECG2046	MM5369AA/N

BLOCK-963

ECG2047	MC14412FL	MC14412FP	MC14412L
MC14412P	MC14412VL	MC14412VP	NTE2047

BLOCK-964

ECG2049	ICL7126CPL	SK10278	TSC7126CPL

BLOCK-965

ECG2050	ICL7107CDL	ICL7107CPL	NTE2050
SK10279	TCG2050	TSC7107CPL	

BLOCK-966

ECG2051	ICL7106CPL	NTE2051	SK10280
TCG2051	TSC7106CPL		

BLOCK-967

ADD3501	ECG2052	SK10281

BLOCK-968

276-1792	ADC0801	ADC0802	ADC0802LCD
ADC0802LCN	ADC0803LCD	ADC0803LCN	ADC0804LCD
ADC0804LCN	ECG2053	NTE2053	SK1911
TCG2053			

BLOCK-969

ECG2054	NTE2054	SK9204	TCG2054

BLOCK-970

C4059P	ECG2055	HEPC4059P	MC14433L
MC14433P	NTE2055	SK10282	

BLOCK-971

AD1408-7D	DAC0806LCN	DAC0807LCN	ECG2056
HA17408P	LM1408N-6	LM1408N-7	MC1408-7N
NTE2056	UA0802BPC	UA0802CPC	

BLOCK-972

ECG2057	SK10190	TDA1541	TDA1541AN
TDA1541N			

BLOCK-973

1-TR-016	1086-9931	1155	1155-6578
1157-4738	16483610	16483611	46-13177-3
46-5291-0164	51-90305A72	51-90433A63	5359701
65-45402-00	8-759-953-87	905-00066	905-00161
905-118	905-118-01	905-210	905-66
905-69	EA-7316B	EA33X8450	EA7316A-1
EA7316B	EA7317B	ECG2060	HD38991
IP20-0447	LM8360	LM8361	LM8361D
MM5316	MM5387	MM5387AA	MM5387AA-N
MM5387AA/N	MM5387AB/N	MM5387N	MM5402
MM5402N	NTE2060	PC-20087	PC-20143
RV1S1998	RVIS1998	RVITMS1943	RVITMS1943N2
S1998	S1998A	SK3966	TCG2060
TMS1943	TMS1943A2	TMS1943N2	TMS1943N21
TMS1943N2L	TMS1943NL	TVSTMS1943A2	
TVSTMS1943N2			

BLOCK-974

02-103898	1172-9894	16453160	5654-HD38991A
905-204	905-301	EA33X8465	EA33X8584
EA33X8600	ECG2061	HD38980A	HD38980C
HD38991A	LM8361DH	MM53113N	MM5316N
NTE2061	RVIHD38980A	SK7612	TCG2061
TMS-1952	TMS1952NL		

BLOCK-975

1007-6537	301003450375	4152085600	ECG2062
LM8560	MX-5367	MX-5389	NTE2062
SILM8560BXXXX	SK10295	TMS3450NL	

BLOCK-976

ECG2063	ICL7116CPL	SK10283	TSC7116CPL

BLOCK-977

ECG2064	ICL7126RCPL	SK10284

BLOCK-978

ECG2065	ICL7129CPL	SK10285

BLOCK-979

ECG2070	M54536P	NTE2070	SK10286
TCG2070			

BLOCK-980

5359522	ECG2071	M54535P	NTE2071
SK10287	TCG2071		

BLOCK-981

13428129	ECG2072	M54534P	NTE2072
SK10288	TCG2072		

BLOCK-982

ECG2073	M54533P	NTE2073	SK10289
TCG2073			

BLOCK-983

2002900042	ECG2074	M54530P	M54531P
NTE2074	SK10290	TCG2074	

BLOCK-984

ECG2075	M54529P	NTE2075	SK10291
TCG2075			

BLOCK-985

46-13502-3	ECG2076	LB1275	M54528P
NTE2076	SK10292	TCG2076	TD62007P

BLOCK-986

46-13307-3	ECG2077	LB1274	LB1274K
M54527P	NTE2077	SK10293	TCG2077
TD62006P			

BLOCK-987

ECG2078	M54521P	NTE2078	SK10294
TCG2078			

BLOCK-988

155711	2002900408	51X90518A71	8-759-645-19
905-221	ECG2079	M54517P	M54519P
NTE2079	SK9357	TCG2079	TD62104P
UPA81C	X440041040		

BLOCK-989

ECG2080	NTE2080	TCG2080

BLOCK-990

ECG2081	NTE2081	TCG2081

BLOCK-991

ECG2082	NTE2082	SK9903	TCG2082

BLOCK-992

ECG2083	NTE2083	SK9906	TCG2083

BLOCK-993

06300107	157972	46-13309-3	59SI-UPA53C
ECG2084	LB1287	LB1288	M54516P
NTE2084	PA53C	SK4910	TCG2084
UPA53C			

BLOCK-994

ECG2085	LN2064B	NTE2085	SK9913
SN745064NE	SN745065NE	SN7506NE	TCG2085
ULN-2064A	ULN2064B	ULN2064NE	ULN2065B
ULN2065NE			

BLOCK-995

ECG2086	NTE2086	SK9907	SN745066NE
SN745067NE	SN75067NE	ULN2066B	ULN2066NE
ULN2067B	ULN2067NE		

BLOCK-996

ECG2087	NTE2087	SK9813	SN745068NE
SN745069NE	SN75068NE	SN75069NE	TCG2087
ULN2068B	ULN2068NE	ULN2069B	

BLOCK-997

ECG2088	NTE2088	SK9908	ULN2071B

BLOCK-998

ECG2090	UPA79C	X440150790

BLOCK-999

2102A	2102AN-4L	2102FDC	2102FPC
21L02B	3108002	4102	AM9102ADC
AM9102APC	AM9102BDC	AM9102BPC	AM9102CDC
AM9102CPC	AM9102DC	AM9102PC	C2102A
ECG2102	NTE2102	TCG2102	UPD2102AL-4
UPD2102ALC-4			

BLOCK-1000

2104	6604	AXX3044	ECG2104
MCM4027AC3	MCM4027AC4	MK4027J-3	MK4027J-4
MK4027N-3	MK4027N-4	MK4027P-3	NTE2104
TCG2104			

BLOCK-1001

2107	6605	AM9060CDC	AM9060CPC
AM9060DDC	AM9060DPC	AM9060EDC	AM9060EPC
ECG2107	NTE2107		

BLOCK-1002

0098-0710	2114	2114-30L	2114L
2114L-3	2114L-30	443-764	8114
901453-01	AC4045NL	AM9114BDC	AM9114BPC
AM9114CDC	AM9114CPC	AXX3038	D2114
D2114-2	D2114A4	D2114A5	D2114AL3
D2114AL4	D2114L	D2114L2	D2114L3
ECG2114	HE-443-764	HM472114-3	HM472114-4
HM472114P-3	HM472114P-4	IC-2114-L3	IC-L2114-550
ISSRM211403	ISSRM2114C3	L2114-550	M5L2114LP-3
MCM2114	MCM2114-30	MCM2114-45	MCM2114P-30
MCM2114P-45	MCM2114P20	MCM2114P45	MCS2114-30
MCS2114-35	MCS2114L-45	MM2114N-3	MPS2114-30
MPS2114-35	MPS2114-45	MPS2114-50	MSM2114
MSM2114L-3RS	MSM2114L3RS	NTE2114	P2114
P2114-3	SK2214	SRM2114C3	SYC2114
SYC2114-3	SYD2114	SYD2114-2	SYD2114-3
SYD2114L	SYD2114L-2	SYD2114L-3	SYP2114
SYP2114-2	SYP2114-3	SYP2114L	SYP2114L-2
SYP2114L-3	TCG2114	TMM314AP	TMM314APL-3
TMS2114-20	TMS2114-25	TMS2114-45	TMS2114L-20
TMS2114L-25	TMS2114L-45	UPD2114LC	UPD2114LC-1
UPD2114LC-5			

BLOCK-1003

2117	221-364	276-2503	276-2505
3108003	3108009	4116	416
443-904	8041016	8041016A	8116
AM9016CDC	AM9016CPC	AM9016DDC	AM9016DPC
AM9016EDC	AM9016EPC	AXX3021	AXX3055
C014331-09	CM4116AD	CM4116AE	CO1433-09
D416	ECG2117	F16K3DC	F16K4DC
F16K5DC	HE-443-904	HYB4116-P2	HYB4116-P3
ITT4116-3	ITT4116-4	MB8116E	MB8116EC
MB8116EP	MCM4116AC20	MCM4116AC25	MCM4116AC30
MCM4116BC20	MCM4116BC25	MCM4116BC30	MCM4116BP20
MCM4116BP25	MCM4116BP30	MCM4116BP35	MK4116
MK4116-4	MK41164-3GP	MK4116E-2	MK4116E-3
MK4116-E4	MK4116J-2	MK4116J-3	MK4116J-4
MK4116J-53GP	MK4116N-2	MK4116N-3	MK4116N-3gp
MK4116N-4	MK4116N-44GP	MK4116P-2	MK4116P-3
MK4116P-4	MM5290N-4	MN4116	NTE2117
SCM90072C	SCM90072P	SK9832	TCG2117
TMM416D-3	TMM416D-4	TMM416P-3	TMS4116-20
TMS4116-25	UPD416	UPD416-3	UPD416C
UPD416C-1	UPD416C-2	UPD416D	UPD416D-1
UPD416D-2	X400104162	Y130201002	

BLOCK-1004

08220027	08220031	2016	2016P
2420TMM16P2	325502-01	325502-03	4016(RAM)
4016CX	8128	D4016CX-15	EA001-02700
EA001-02703	EAO01-00600	ECG2128	HM6116
HM6116-3	HM6116-4	HM6116FP3	HM6116FP4
HM6116LP-4	HM6116P-3	HM6116P-4	HM6116P4
M2128-20	M58725P	MB8128-15	MB8128-15C
MCM65116C	MCM65116P	MCM65116P15	MCM65116P20
MK4802P-3	NTE2128	P61AM6116H	SK10519
SY2128	TC5565	TMM2016	TMM2016AP
TMM2016AP-12	TMM2016BP-15	TMM2016P	TMM2016P-1
TMM2016P-2	TMM2016P-7	TMS4016-15	TMS4016-20
TMS4016-25	TVSM58725P	UPD4016	UPD4016C
UPD4016C-2	UPD4016CX	UPD4016D-1	X400040161

BLOCK-1005

ECG2147	MBM2147E	MBM2147H55Z	MBM2147H70Z
MCM2147C100	MCM2147C55	MCM2147C70	MCM2147C85
NMC2147H-3	NTE2147	SK10296	SYC2147
SYC2147-3	SYC2147-6	SYC2147L	SYD2147
SYD2147-3	SYD2147-6	SYD2147L	TCG2147
TMM315D	TMM315D-1	UPD2147A-45	

BLOCK-1006

4164	4164-2	4164C	5K4164ANP-15
901505-01	AB4164ANP-12	AB4164ANP-15	C060612
D4164C-15	D4164C-2	D4164C-3	ECG2164
HE-443-970	HM4864-2	HM4864-3	HM4864P-2
HYB4164-P2	KM4164-15	M3764-15RS	M3764-20RS
M5K4164ANP-12	M5K4164ANP-15	MCM6665	MCM6665AL15
MCM6665AL20	MCM6665AP	MCM6665AP15	MCM6665AP20
MCM6665L25	MK4564N-15	MN4164P-15A	MT4264-15
MT4264-20	NTE2164	NTE4164	SK10297
TMM4164P-3	TMS4164-15	TMS4164-15NLJ	TMS4164-20
TMS4164-20NL	TMS4164-25	TMS4164NLJ	UPD4164-3
UPD41647H-45	X400141640		

BLOCK-1007

6123420001	ECG2200

BLOCK-1008

ECG2201

BLOCK-1009
6124160003	ECG2202

BLOCK-1010
6123180001	ECG2203

BLOCK-1011
ECG2204

BLOCK-1012
ECG2205

BLOCK-1013
ECG2206

BLOCK-1014
148954	ECG2207	MN1206A	SK10298

BLOCK-1015
1331409	ECG2532	HN462532	HN462532G
HN462532G2	MCM2532C	NTE2532	TMS2532-30
TMS2532-35	TMS2532-45	TMS2532JL-45	

BLOCK-1016
AM2708DC	C2708	ECG2708	MCM2708C
MCM2708L	MM2708Q	NTE2708	TCG2708

BLOCK-1017
2716	2716-SA2-A0	2716-SC2-A1	AM2716DC
D2716D	D2716D-Q1P-4	D2716D-Q1X-0	ECG2716
HN462716G	M2716	M2716F1	M2716M
M5L2716K	MCM2716C	MCM2716L	MM2716Q
NTE2716	SK2716	TCG2716	Y130801001
Y130801401	Y440800301	Y440801501	

BLOCK-1018
08220014	104648D	2732	2732/2332
2732A	2732A-2	2732A-30	2732D
88922010	88940010	88941010	88942010
CLEAK-05710	CLEAK-05903	CLEAK-06002	D2732A
D2732D	ECG2732	HN462732	HN462732G
ID2732A-3	M2732AF1	M5L2732K	M5L2732K-6
MBM2732-35Z	MBM2732-45Z	MBM2732A-20-QGA1	MBM2732A20Z
MBM2732A25Z	MBM2732A30	MBM2732A30Z	MBM2732A35Z
NTE2732	P0165STD01E	SK1912	
TMS2732AJL-30	TMS2732AJL-35	UPD2732	UPD2732A
UPD2732D	UPD2732D-4	Y130800701	ZPFAK-50701
ZPFAK-50801	ZPFAK-50901	ZPFAK-51001	

BLOCK-1019
08220030	08220033	08220090	2516JL-45
276-1251	2764	2764-2	2764-25
2764-250	2764-30	2764-450	2764-FA4-A2
2764-FA5-A3	2764-FC5-A3	88901010	88902010
88921010	88962010	AM2764ADC	CLEAK-06102
CLEAK-06201	CLEAK-12004	CQEAK-01103	D2764
D2764A-2	D2764D	ECG2764	HN482764G
HN482764G-3	I2764	I2764-2	M5L2764K
M5L2764K-2	M5L2764K-FA552	MBM2764-20	MBM2764-25
MBM2764-30	MSM2764RS	NTE2764	OB2764K-2
OV2764KFA552	P055I	P056I	TMM2764-D-2
TMS2516JL-45	TMS2564-35	TMS2564-45	TMS2764JL-25
UPD2764D-FC4-A2	UPD2764D-FC5-A3	Y440800601	Y440801101
Y441800101	Y441800102		

BLOCK-1020
612184-1	ECG2800	ER1400	EW84X272
IX0232CE	M5G1400P	NTE2800	X0232CE

BLOCK-1021
1296-2718	162-37	32309-024-010	408 000 0301
408-000-0301	46-131396-3	46-133389-3	46-133692-3
46-133831-3	4835-130-47042	4835-130-97006	4835-130-97042
5082-4350	5303340001	6N136	6N136-020
6N136-HP	8-719-800-43	A8641301	A8641302
A8641303	ECG3092	FX0014CE	GN13G
HP4510	NTE3092	PS2006B	
RH-FX0008CEZZ	RH-FX0014CEZZ	SK9770	TLP550
TLP551	TLP551FA-1	TLP551FA-2	TLP651

BLOCK-1022
15105300	156-0176-00	157575	1820-0430
2709759	3008321-00	442-30	442-30-2897
7805CDA	901528-03	CXL309K	ECG309K
GE-309K	HE-442-30	IP340-5	LAS1505
LM309DA	LM309K	LM340-5	LM340-5DA
LM340K5.0	MLM309K	NTE309K	SFC2805RC
SG309K	SK3629	TCG309K	TDB7805
TM309K	UA209KM	UA309K	UA309KC
UA7805CDA	UA7805CKA	UA7805DA	UA7805KC
UA7805KM	UA7805MKA	UGJ7109393	WEP2319/309K

BLOCK-1023
35392DC	3539DC	ECG3539

BLOCK-1024
CXL370	ECG370	ECG370A	IC-526(ELCOM)
LM370	TM370	WEP2028/370	

BLOCK-1025
3110001	443-881	AMX3586	D780C-1
ECG3880	HE-443-881	NTE3880	SK2880
TCG3880	UPD780C	X400007801	Z80
Z80A	Z8400APS		

BLOCK-1026
ECG3881	MK3881N	MK3881N4	MK3881P
NTE3881	SK2881	TCG3881	

BLOCK-1027
ECG3882	MK3882N	MK3882P	NTE3882
SK2882	TCG3882		

BLOCK-1028
4000	4000(IC)	67-32720-01	CD4000AD
CD4000AE	CD4000AF	CD4000AZ	CD4000BD
CD4000BE	CD4000BF	CD4000CJ	CD4000CN
CD4000MJ	CD4000MW	CD4000UBD	CD4000UBE
CD4000UBF	CM4000	CM4000AD	CM4000AE
ECG4000	GE-4000	HBC4000AD	HBC4000AF
HBC4000AK	HBF4000AE	HBF4000AF	HCC4000BD
HCC4000BF	HCC4000BK	HCF4000BE	HCF4000BF
HD14000UB	HD4000	HEF4000B	HEF4000BD
HEF4000BP	HEF4000P	HEPC4000P	MC14000
MC14000UBCL	MC14000UBCP	MM4000(IC)	N4000
NTE4000	SCL4000	SCL4000B	SCL4000BC
SCL4000BD	SCL4000BE	SCL4000BF	SCL4000BH
SK4000	SK4000B	SK4000UB	TC4000BP
TCG4000	TM4000	TP4000	

BLOCK-1029
0N188320-2	1181-4910	15-3015569-1	154222
154774	276-2401	34001PC	4001(IC)
4001BDC	4001BDM	4001BFC	4001BFM
4001BPC	443-703	46-13272-3	46-13317-3
51-10655A17	51S10655A17	51X90507A84	544-3001-103
569-0690-500	8-759-140-01	8-759-240-01	905-1106
905-125	991-026025	BO470016	CD4001
CD4001AD	CD4001AE	CD4001AF	CD4001BCJ
CD4001BCN	CD4001BD	CD4001BE	CD4001BF
CD4001BMJ	CD4001BMW	CD4001CJ	CD4001CN
CD4001MJ	CD4001MW	CD4001UBD	CD4001UBE
CD4001UBF	CM4001	CM4001AD	CM4001AE
ECG4001	ECG4001B	F4001	GE-4001
HBC4001AD	HBC4001AF	HBC4001AK	HBF4001
HBF4001A	HBF4001AE	HBF4001AF	HCC4001BD
HCC4001BF	HCC4001BK	HCF4001BE	HCF4001BF
HD14001B	HD4001	HE-443-695	HE-443-703
HEF4001BD	HEF4001BP	HEF4001P	HEPC4001P
I53D040010	I55D040010	IC4001BP	IX0069CE
LC4001B	LC4001BP	M4001BP	MB84001B
MB84001BM	MB84011V	MC14001	MC14001B
MC14001BCL	MC14001BCP	MC14001CP	MC14001UBCP
MEM4001	MM5601AN	MN4001B	MSM4001
N4001	NTE4001B	RH-IX0069CEZZ	RS4001
SC4001BH	SC4001UBC	SCL4001	SCL4001B
SCL4001BC	SCL4001BD	SCL4001BE	SCL4001BF
SCL4001UBD	SCL4001UBE	SCL4001UBF	SCL4001UBH
SILC4001BXXXX	SK4001	SK4001B	SS4001AE
SW4001	TC4001BP	TC4001P	TC4001UBP
TCG4001B	TM4001B	TP4001	TP4001AN
TP4001BN	TP4001BP	UPC4001C	UPD4001BC
UPD4001C	VHITC4001BP-1	WEP2272	
WEP2272/4001B	X0069CE		

BLOCK-1030
154767	34002PC	4002(IC)	4002A
4002BDC	4002BDM	4002BFC	4002BFM
4002BPC	443-704	51X90507A97	6125660001
CD4002	CD4002AD	CD4002AE	CD4002CJ
CD4002BD	CD4002BE	CD4002BF	CD4002CJ
CD4002CN	CD4002MJ	CD4002MW	CD4002UBD
CD4002UBE	CD4002UBF	CM4002	CM4002AD
CM4002AE	D4002BC	ECG4002	ECG4002B
F4002	GE-4002	HBC4002AD	HBC4002AF
HBC4002AK	HBF4002	HBF4002A	HBF4002AE
HBF4002AF	HCC4002BD	HCC4002BF	HCC4002BK
HCF4002BE	HCF4002BF	HD14002B	HD4002
HE-443-704	HEF4002BD	HEF4002BP	HEF4002P
HEPC4002P	M4002BP	MB84002B	MB84002BM
MC14002	MC14002BAL	MC14002BCL	MC14002BCP
MC14002CP	MC14002P	MM5602AN	MSM4002
N4002	NTE4002B	SCL4002	SCL4002B
SCL4002BC	SCL4002BD	SCL4002BE	SCL4002BF

BLOCK-1030 (CONT.)

SCL4002BH	SK4002	SK4002B	SS4002AE
SW4002	TC4002BP	TC4002P	TCG4002B
TM4002B	TP4002	TP4002AN	UPD4002BC
UPD4002C			

BLOCK-1031

4006BDC	4006BPC	CD4006	CD4006AE
CD4006BE	CM4006AD	CM4006AE	ECG4006B
HBF4006AE	HBF4006AF	HCC4006BF	HCF4006BE
HCF4006BF	HD14006B	HEF4006B	HEF4006BD
HEF4006BP	HEF4006P	MC14006BAL	MC14006BCL
MC14006BCP	NTE4006B	SCL4006BC	SCL4006BD
SCL4006BE	SCL4006BF	SK4006B	TCG4006B
TM4006B			

BLOCK-1032

4007(IC)	4007UBDC	4007UBPC	443-604
CD4007AE	CD4007AF	CD4007CJ	CD4007CN
CD4007UBE	CM4007	CM4007AD	CM4007AE
ECG4007	ECG4007B	F4007	GE-4007
HBF4007	HD14007UB	HD4007	HE-443-604
HEPC4005P	HEPC4007P	MC14007	MC14007UBAL
MC14007UBCP	MEM4007	MM4007	N4007
NTE4007	SCL4007	SCL4007UBC	SCL4007UBD
SCL4007UBE	SK4007	SK4007UB	TCG4007
TCG4007B	TM4007B	TP4007	TP4007UBN

BLOCK-1033

340085PC	40085BDC	40085BDM	40085BFC
40085BFM	40085BPC	ECG40085	ECG40085B
NTE40085B	TCG40085B		

BLOCK-1034

4008BDC	4008BDM	4008BFC	4008BFM
4008BPC	443-928	CD4008AD	CD4008AE
CD4008AF	CD4008B	CD4008BCJ	CD4008BCN
CD4008BD	CD4008BE	CD4008BF	CD4008BJ
CD4008BMJ	CD4008BMW	CM4008AD	CM4008AE
ECG4008B	HBC4008AD	HBC4008AF	HBC4008AK
HBF4008AE	HBF4008AF	HCC4008BD	HCC4008BF
HCC4008BK	HCF4008BE	HCF4008BF	HD14008B
HE-443-928	HEF4008B	HEF4008BD	HEF4008BP
HEF4008P	MB84008B	MB84008BM	MC14008BAL
MC14008BCL	MC14008BCP	MSM4008	NTE4008B
SCL4008B	SCL4008BC	SCL4008BD	SCL4008BE
SCL4008BF	SCL4008BH	SCL4008BP	SK4008B
TC4008BP	TCG4008B	TM4008B	TP4008AN
TP4008BN			

BLOCK-1035

148736	154648	155632	340097PC
40097BDC	40097BPC	51X90458A15	51X90480A18
51X90518A34	CD4503B	CD4503BD	CD4503BE
CD4503BF	ECG40097	ECG40097B	EW84X172
HD14503B	HD14503P	HEF40097BD	HEF40097BP
ITTC5012BP	MC14503BAL	MC14503BCP	MN4503B
NTE40097B	NTE4503B	SK4503	SK4503B
SK9787	TC5012BP	TCG40097B	TCG4503B
UPD4503BC	UPD4503C		

BLOCK-1036

340098PC	40098BDC	40098BPC	ECG40098
ECG40098B	HEF40098BD	HEF40098BN	HEF40098BP
HEF40098P	MN40098B	NTE40098B	SK40098B
TCG40098B			

BLOCK-1037

CD40100BD	CD40100BE	CD40100BF	ECG40100B
HCC40100BD	HCC40100BF	HCF40100BE	HCF40100BF
NTE40100B	SK40100B	TCG40100B	

BLOCK-1038

1240-3432	CD40106BCJ	CD40106BCN	CD40106BD
CD40106BE	CD40106BMJ	CD40106BMW	ECG40106B
EP84X240	HCC40106BF	HCF40106BE	HCF40106BF
HD14584BP	MC14584	MC14584B	MC14584BAL
MC14584BCL	MC14584BCP	NTE40106B	SK40106B
SK4584B	TC4584	TC4584BP	TCG40106B
UPD4584BC	X460458400		

BLOCK-1039

1147-09	1157-8564	144968	145818
147234	148730	15-43705-1	15-45141-1
152923	153937	164984	2000-002
2004-70	2032-38	2065-51	221-Z9133
2360851	276-2411	307-095-9-002	307-113-9-001
307-152-9-012	34011PC	37051310	40-035-0
4011(IC)	4011-PC	4011A	4011BDC
4011BDM	4011BFC	4011BFM	4011BPC
4011PC	443-603	46-13258-3	46-133293-3

BLOCK-1039 (CONT.)

51S10655A18	51X90518A36	5362741	
55SI-MC14011BCP	569-0690-501	6123500001	70119062
733W00211	76SI-HD14011BP	8-759-140-11	8-759-240-11
84011	84011U	905-126	991-025515
991-025515-001	991-025515-002	B0470116	BO470116
BU4011B	CD4011	CD4011AD	CD4011AE
CD4011AF	CD4011BCJ	CD4011BCN	CD4011BD
CD4011BE	CD4011BF	CD4011BMJ	CD4011BMW
CD4011CJ	CD4011CN	CD4011MJ	CD4011MW
CD4011UBD	CD4011UBE	CD4011UBF	CM4011
CM4011AD	CM4011AE	D4011BC	DDEY058001
DDEY084001	DDEY089001	EA000-13000	ECG4011
ECG4011B	F4011	F4011PC	GE-4011
HBC4011AD	HBC4011AF	HBC4011AK	HBF4011
HBF4011A	HBF4011AE	HBF4011AF	HCC4011BD
HCC4011BF	HCC4011BK	HCF4011BE	HCF4011BF
HD14011B	HD14011BP	HD4011	HE-443-603
HEF4011BD	HEF4011BP	HEF4011P	IGLC4011
IGLC4011B	INMPD4011BC	INUPD4011BC	LC4011
LC4011B	M4011BP	MB84011	MB84011-U
MB84011B	MB84011M	MB84011M	MB84011U
MC-14011CP	MC14011	MC14011B	MC14011BAL
MC14011BCL	MC14011BCP	MC14011CP	MEM4011
MM4011	MM5611AN	MN4011B	MPD4011BC
MSM4011	MSM4011RS	MX-3634	MX3634
N4011	NTE4011B	RH-IX0072CEZZ	RS4011
SCL4011	SCL4011BC	SCL4011BC	SCL4011BD
SCL4011BE	SCL4011BF	SCL4011BH	SCL4011UBC
SCL4011UBD	SCL4011UBE	SCL4011UBF	SCL4011UBH
SK4011	SK4011B	SS4011AE	SW4011
TC4011	TC4011B	TC4011BP	TC4011P
TC4011UBP	TCG4011B	TM4011B	TP4011
TP4011AN	TP4011B	TP4011BN	TP4011BP
TVSG4011BCM	TVSTC4011BP	TVSUPD4011BC	UPD4011
UPD4011B	UPD4011BC	UPD4011BCM	UPD4011C
VHIBU4011B/-1	WEP4011B	WEP4011B/4011B	

BLOCK-1040

148731	154768	158061	158063
34012PC	4012(IC)	4012A	4012BDC
4012BDM	4012BFC	4012BFM	4012BPC
443-60	443-886	51X90507A98	8-759-240-12
CD4012	CD4012AD	CD4012AE	CD4012AF
CD4012BD	CD4012BE	CD4012CJ	CD4012CJ
CD4012CN	CD4012MJ	CD4012MW	CD4012UBD
CD4012UBE	CD4012UBF	CM4012	CM4012AD
CM4012AE	D4012BC	ECG4012	ECG4012B
F4012	GE-4012	HBC4012AD	HBC4012AE
HBC4012AK	HBF4012	HBF4012A	HBF4012AE
HBF4012AF	HCC4012BD	HCC4012BF	HCC4012BK
HCF4012BE	HCF4012BF	HD14012B	HD4012
HE-443-886	HEF4012	HEF4012BD	HEF4012BP
HEF4012P	M4012BP	MB84012B	MB84012BM
MC14012	MC14012BAL	MC14012BCL	MC14012BCP
MC14012CP	MC4012P	MM4012	MM5612AN
MSM4012	N4012	NTE4012B	SCL4012
SCL4012B	SCL4012BC	SCL4012BD	SCL4012BE
SCL4012BF	SCL4012BH	SK4012	SK4012B
SS4012AE	SW4012	TC4012BP	TCG4012B
TM4012B	TP4012	TP4012AN	UPD4012BC
UPD4012C			

BLOCK-1041

06300316	1191-3357	145586	147887
157654	276-2413	34013PC	4013(IC)
4013B	4013BDC	4013BDM	4013BFC
4013BFM	4013BPC	443-607	46-13313-3
46-13315-3	51-10655A19	51S10655A19	569-0690-502
612074-1	8-759-240-13	905-1239	905-1240
905-1373	905-186	905-662	905-717
991-027358	991-027358-001	B0470130	BU4013B
CD4013	CD4013AD	CD4013AE	CD4013AF
CD4013B	CD4013BCJ	CD4013BCN	CD4013BD
CD4013BE	CD4013BF	CD4013BMJ	CD4013BMW
CD4013BN	CM4013	CM4013AD	CM4013AE
D4013BC	ECG4013	ECG4013B	EW84X863
F4013	GE-4013	HBC4013AD	HBC4013AF
HBC4013AK	HBF4013	HBF4013A	HBF4013AE
HBF4013AF	HCC4013BD	HCC4013BF	HCC4013BK
HCF4013BE	HCF4013BF	HD14013B	HD4013
HE-443-607	HEF4013BD	HEF4013BP	HEF4013P
IGLC4013B	ILMC14013B	INMPD4013BC	INUPD4013BC
LC4013B	MB84013B	MB84013BM	MC14013
MC14013A1	MC14013B	MC14013BAL	MC14013BCL
MC14013BCP	MC14013CP	MEM4013	MM4013
MN4013B	MN5613AN	MSM4013	MSM4013RS
N4013	NTE4013B	Q7393	QC0343
RS4013	SCL4013	SCL4013AC	SCL4013AD
SCL4013AE	SCL4013AF	SCL4013AH	SCL4013BC
SCL4013BD	SCL4013BE	SCL4013BH	SCL4013BH
SCL4013BP	SK4013	SK4013B	SS4013AE
SW4013	TC4013B	TC4013BCP	TC4013BP
TC4013P	TCG4013B	TM4013B	TP4013

BLOCK-1041 (CONT.)

TP4013AN	TP4013BE	TP4013BN	
TVSUPD4013BC	UPD4013BC	UPD4013BG	UPD4013C
VHITC4013BP-1	WEP4013B	WEP4013B/4013B	

BLOCK-1042

34014DC	34014PC	4014BDC	4014BPC
CD4014AD	CD4014AE	CD4014AF	CD4014BD
CD4014BE	CD4014BF	CM4014AD	CM4014AE
ECG4014	ECG4014B	HBF4014A	HBF4014AE
HCF4014BE	HCF4014BF	HD14014B	HEF4014B
HEF4014BD	HEF4014BP	HEF4014P	MC14014BAL
MC14014BCL	MC14014BCP	MC14014CP	NTE4014B
SCL4014BC	SCL4014BD	SCL4014BE	SK4014B
TCG4014B	TM4014B	TP4014AN	

BLOCK-1043

06300315	154771	34015DC	34015DM
34015PC	4015	4015(IC)	4015BDC
4015BPC	51X90507A92	905-147	991-026609
CD4015	CD4015AD	CD4015AE	CD4015AF
CD4015BD	CD4015BE	CD4015BF	CM4015
CM4015AD	CM4015AE	ECG4015	ECG4015B
F4015	GE-4015	H31100127	HBF4015A
HCF4015BE	HCF4015BF	HD14015B	HD4015
HEF4015B	HEF4015P	M4015B	MC14015
MC14015BAL	MC14015BCL	MC14015BCP	MC14015CP
MM4015	N4015	NTE4015B	SCL4015
SCL4015BC	SCL4015BD	SCL4015BE	SK4015
SK4015B	SW4015	TC4015BP	TC4015P
TCG4015B	TM4015B	TP4015	TP4015AN
UPD4015C			

BLOCK-1044

1200-3505	340160PC	37053003	40160BDC
40160BDM	40160BFC	40160BFM	40160BPC
CD40160BCJ	CD40160BCN	CD40160BD	CD40160BE
CD40160BF	CD40160BMJ	CD40160BMW	D4160BC
ECG40160	ECG40160B	HCC40160BD	HCC40160BF
HCC40160BK	HCF40160BE	HCF40160BF	HEF40160B
HEF40160BD	HEF40160BP	MC14160BAL	MC14160BCL
MC14160BCP	NTE40160B	SCL4160B	SK40160B
TC40160BP	TCG40160B	UPD4160BC	

BLOCK-1045

340161PC	40161BDC	40161BDM	40161BFC
40161BFM	40161BPC	CD40161BCJ	CD40161BCN
CD40161BD	CD40161BE	CD40161BF	CD40161BMJ
CD40161BMW	ECG40161	ECG40161B	HCC40161BD
HCC40161BF	HCC40161BK	HCF40161BE	HCF40161BF
HEF40161B	HEF40161BD	HEF40161BP	MC14161BAL
MC14161BCL	MC14161BCP	NTE40161B	SCL4161B
SK40161	SK40161B	TC40161BP	TCG40161B

BLOCK-1046

340162PC	40162BDC	40162BDM	40162BFC
40162BFM	40162BPC	CD40162BCJ	CD40162BCN
CD40162BD	CD40162BE	CD40162BF	CD40162BMJ
CD40162BMW	ECG40162	ECG40162B	HCC40162BF
HCC40162BK	HCF40162BE	HCF40162BF	HEF40162B
HEF40162BD	HEF40162BP	MC14162BAL	MC14162BCL
MC14162BCP	NTE40162B	SCL4162B	SK40162B
TC40162BP	TCG40162B		

BLOCK-1047

340163PC	40163BDC	40163BDM	40163BFC
40163BFM	40163BPC	CD40163BCJ	CD40163BCN
CD40163BD	CD40163BE	CD40163BF	CD40163BMJ
CD40163BMW	ECG40163	ECG40163B	HCC40163BD
HCC40163BF	HCC40163BK	HCF40163BE	HCF40163BF
HEF40163B	HEF40163BD	HEF40163BP	MC14163BAL
MC14163BCL	MC14163BCP	NTE40163B	SCL4163B
SK40163B	TC40163BP	TCG40163B	

BLOCK-1048

058-001138	07-28816-84	0N198382-2	1181-4928
144969	15-45185-1	154027	1710-1
186-1	34016PC	4016	4016(IC)
4016BDC	4016BPC	442-99	443-61
51-10655A21	51-903	51-90433A13	612186-1
6121860001	991-028679-001	AD7516JN	CD4016
CD4016AD	CD4016AE	CD4016AF	CD4016B
CD4016BD	CD4016BE	CD4016BF	CD4016CJ
CD4016CN	CD4016MJ	CM4016	CM4016AD
CM4016AE	DM-86	ECG4016	ECG4016B
EP84X30	F4016	GE-4016	HBF4016
HBF4016A	HCC4016BF	HCF4016BE	HCF4016BF
HD14016B	HD4016	HE-442-99	HEF4016BD
HEF4016BP	HEPC3804P	M4016BP	MB84016B
MB84016BM	MC14016BAL	MC14016BCL	MC14016BCP
MC14016CP	MC4016P	MEM4016	MM4016A
MSM4016	MSM4016RS	N4016	N6121860001

BLOCK-1048 (CONT.)

NTE4016B	SCL4016	SK4016	SK4016B
SW4016	TC4016BB	TCG4016B	TM4016B
TP4016AN	TP4016BN	TP4016UBN	WEP2275
WEP2275/4016			

BLOCK-1049

06300317	1234-5955	1309-1004	263P17402
340174DC	340174DM	340174PC	443-836
B0474370	CD40174BC	CD40174BD	CD40174BE
CD40174BF	ECG40174	ECG40174B	HCC40174BD
HCC40174BF	HCC40174BK	HCF40174BE	HCF40174BF
HD14174B	HE-443-836	HEF40174B	HEF40174BD
HEF40174BP	HEF40174P	MC14174B	MC14174BAL
MC14174BCL	MC14174BCP	NTE40174B	SCL4174B
SK40174B	TC40174BP	TCG40174B	

BLOCK-1050

340175DC	340175DM	340175PC	CD40175BC
ECG40175B	HD14175B	HEF40175B	HEF40175BD
HEF40175BP	M40175BP	MC14175BCL	MC14175BCP
NTE40175B	SK40175B	TC40175BP	TCG40175B

BLOCK-1051

1240-3457	276-2417	34017PC	4017
4017(IC)	4017BDC	4017BDM	4017BFC
4017BFM	4017BPC	443-929	544-3001-117
991-026615	CD4017	CD4017AD	CD4017AE
CD4017AF	CD4017BCJ	CD4017BCN	CD4017BD
CD4017BE	CD4017BF	CD4017BMJ	CD4017BMW
CM4017	CM4017AD	CM4017AE	D4017BC
ECG4017	ECG4017B	EP84X242	EW84X30
F4017	GE-4017	HBC4017AD	HBC4017AF
HBC4017AK	HBF4017	HBF4017A	HBF4017AE
HBF4017AF	HCC4017BD	HCC4017BF	HCC4017BK
HCF4017BE	HCF4017BF	HD14017B	HD4017
HE-443-929	HEF4017BD	HEF4017BP	HEF4017P
MB84017B	MB84017BM	MC14017	MC14017BAL
MC14017BCL	MC14017BCP	MC14017CP	MM4017
MM5617AN	MSM4017	MSM4017AN	NTE4017B
RS4017	SCL4017	SCL4017ABC	SCL4017ABD
SCL4017ABE	SCL4017ABF	SCL4017ABH	SCL4017AC
SCL4017AD	SCL4017AE	SCL4017AF	SCL4017AH
SCL4017B	SCL4017BC	SCL4017BD	SCL4017BE
SCL4017BF	SCL4017BH	SK4017	SK4017B
SW4017	TC4017BP	TC4017P	TCG4017B
TM4017B	TP4017	TP4017AN	
TVSUPD4017BC	UPD4017BC	UPD4017C	WEP950
WEP950/4017			

BLOCK-1052

4582BDC	4582BDM	4582BPC	CD40182BD
CD40182BE	CD40182BF	ECG40182B	ECG4582B
HCC40182BF	HCF40182BE	HCF40182BF	MC14582BAL
MC14582BCL	MC14582BCP	NTE40182B	SK40182B
TCG40182B			

BLOCK-1053

4018BDC	4018BDM	4018BFC	4018BFM
4018BPC	CD4018	CD4018AD	CD4018AE
CD4018AF	CD4018BCJ	CD4018BCN	CD4018BD
CD4018BE	CD4018BF	CD4018BMJ	CD4018BMW
CM4018AD	CM4018AE	ECG4018B	HBC4018AD
HBC4018AF	HBC4018AK	HBF4018AE	HBF4018AF
HCC4018BD	HCC4018BF	HCC4018BK	HCF4018BE
HCF4018BF	HD14018B	HEF4018BD	HEF4018BP
HEF4018P	MC14018BAL	MC14018BCL	MC14018BCP
MC14018BL	MC4018L	MC4018P	NTE4018B
SCL4018AC	SCL4018AD	SCL4018AE	SCL4018AF
SCL4018AH	SCL4018B	SCL4018BD	SCL4018BE
SCL4018BF	SCL4018BH	SK4018B	TC4018BP
TCG4018B	TM4018B	TP4018AN	TP4018BN

BLOCK-1054

340192PC	40192BDC	40192BDM	40192BFC
40192BFM	40192BPC	CD40192BCJ	CD40192BCN
CD40192BD	CD40192BE	CD40192BF	CD40192BMJ
CD40192BMW	CD40192CJ	CD40192CN	ECG40192
ECG40192B	HEF40192BD	HEF40192BP	HEF40192P
MSM40192	NTE40192B	SK40192B	TCG40192B

BLOCK-1055

340193PC	40193BDC	40193BDM	40193BFC
40193BFM	40193BPC	CD40193BCJ	CD40193BCN
CD40193BD	CD40193BE	CD40193BF	CD40193BMJ
CD40193BMW	ECG40193	ECG40193B	HEF40193BD
HEF40193BP	HEF40193P	MSM40193	NTE40193B
SK40193B	TC40193BP	TCG40193B	

BLOCK-1056

340194DC	340194DM	340194PC	40194BDC
40194BPC	CD40194BD	CD40194BE	CD40194BF
ECG40194	ECG40194B	HD14194B	HEF40194B
HEF40194BD	HEF40194BP	HEF40194P	MC14194BAL
MC14194BCL	MC14194BCP	NTE40194B	SK40194B
TCG40194B			

BLOCK-1057

340195DC	340195DM	340195PC	40195BDC
40195BPC	ECG40195	ECG40195B	HEF40195B
HEF40195BD	HEF40195BP	NTE40195B	SK40195B
TCG40195B			

BLOCK-1058

34019PC	4019	4019(IC)	4019BDC
4019BPC	991-028680-001	CD4019	CD4019AD
CD4019AE	CD4019AF	CD4019BC	CD4019BCJ
CD4019BCN	CD4019BD	CD4019BE	CD4019BF
CM4019	CM4019AD	CM4019AE	ECG4019
ECG4019B	F4019	GE-4019	HBF4019A
HCC4019BF	HCF4019BE	HCF4019BF	HD4019
HEF4019BD	HEF4019BP	MB84019B	MB84019BM
MC14519BAL	MC14519BCP	MM4019(IC)	MM5619AN
MSM4019	N4019	NTE4019B	SCL4019
SK4019	SK4019B	SW4019	TC4019BP
TC4019P	TCG4019B	TM4019B	TP4019
TP4019AN	TP4019BN		

BLOCK-1059

09A05	149938	34020PC	4020(CMOS)
4020(IC)	4020BDC	4020BDM	4020BFC
4020BFM	4020BPC	CD4020	CD4020AD
CD4020AE	CD4020AF	CD4020BCJ	CD4020BCN
CD4020BD	CD4020BE	CD4020BF	CD4020BMJ
CD4020BMW	CM4020	CM4020AD	CM4020AE
ECG4020	ECG4020B	F4020	GE-4020
HBC4020AD	HBC4020AF	HBC4020AK	HBF4020
HBF4020A	HBF4020AE	HBF4020AF	HCC4020BD
HCC4020BF	HCC4020BK	HCF4020BE	HCF4020BF
HD14020B	HD4020	HEF4020BD	HEF4020BP
HEF4020P	HEPC4020P	M4020BP	MB84020B
MB84020BM	MC14020	MC14020BAL	MC14020BCL
MC14020BCP	MC14020CP	MM4020(IC)	MM5620AN
MSM4020	NTE4020B	RS4020	SCL4020
SCL4020ABC	SCL4020ABD	SCL4020ABE	SCL4020ABF
SCL4020ABH	SCL4020AC	SCL4020AD	SCL4020AE
SCL4020AF	SCL4020AH	SCL4020B	SK4020
SK4020B	SW4020	TC4020BP	TCG4020B
TM4020B	TP4020	TP4020AN	TP4020BN
UPD4020C			

BLOCK-1060

34021DC	34021PC	4021(IC)	4021BDC
4021BPC	991-029866	CD4021	CD4021AD
CD4021AE	CD4021AF	CD4021BD	CD4021BE
CD4021BF	CM4021	CM4021AD	CM4021AE
ECG4021	ECG4021B	F4021	GE-4021
HBF4021	HBF4021A	HCC4021BF	HCF4021BE
HCF4021BF	HD14021B	HD4021	HEF4021B
HEF4021BD	HEF4021BP	HEF4021P	HEPC4021P
MC14021	MC14021BAL	MC14021BCL	MC14021BCP
MC14021CP	N4021	NTE4021B	SCL4021
SCL4021BC	SCL4021BD	SCL4021BE	SK4021
SK4021B	SW4021	TC4021P	TCG4021B
TM4021B	TP4021	TP4021AN	UPD4021C

BLOCK-1061

4022BDC	4022BDM	4022BFC	4022BFM
4022BPC	443-783	CD4022	CD4022AD
CD4022AE	CD4022AF	CD4022BCJ	CD4022BCN
CD4022BD	CD4022BE	CD4022BF	CD4022BMJ
CD4022BMW	CM4022AD	CM4022AE	ECG4022B
HBC4022AD	HBC4022AE	HBC4022AF	HBC4022AK
HBF4022AE	HBF4022AF	HCC4022BD	HCC4022BE
HCC4022BF	HCC4022BK	HCF4022BE	HCF4022BF
HD14022B	HE-443-783	HEF4022BD	HEF4022BP
HEF4022P	MB84022B	MB84022BM	MC14022BAL
MC14022BCL	MC14022BCP	MM5622AN	NTE4022B
SCL4022ABC	SCL4022ABD	SCL4022ABE	SCL4022ABF
SCL4022ABH	SCL4022AC	SCL4022AD	SCL4022AE
SCL4022AF	SCL4022AH	SCL4022B	SCL4022BC
SCL4022BD	SCL4022BE	SCL4022BF	SCL4022BH
SK4022B	TC4022BP	TCG4022B	TM4022B
TP4022AN			

BLOCK-1062

276-2423	34023PC	4023	4023(IC)
4023A	4023BDC	4023BDM	4023BFC
4023BFM	4023BPC	443-887	991-026605
CD4023	CD4023AD	CD4023AE	CD4023AF
CD4023BCJ	CD4023BCN	CD4023BD	CD4023BE

BLOCK-1062 (CONT.)

CD4023BF	CD4023BMJ	CD4023BMW	CD4023C
CD4023CJ	CD4023CN	CD4023MJ	CD4023MW
CD4023UBD	CD4023UBE	CD4023UBF	CM4023
CM4023AD	CM4023AE	ECG4023	ECG4023B
F4023	GE-4023	HBC4023AD	HBC4023AF
HBC4023AK	HBF4023	HBF4023A	HBF4023AE
HBF4023AF	HCC4023BD	HCC4023BF	HCC4023BK
HCF4023BE	HCF4023BF	HD14023B	HD4023
HE-443-887	HEF4023BD	HEF4023BP	HEF4023P
M4023BP	MB84023B	MB84023BM	MC14023
MC14023B	MC14023BAL	MC14023BCL	MC14023BCP
MC14023CP	MM5623AN	MSM4023	N4023
NTE4023B	SCL4023	SCL4023AC	SCL4023AD
SCL4023AE	SCL4023AF	SCL4023AH	SCL4023B
SCL4023BC	SCL4023BD	SCL4023BE	SS4023AE
SCL4023BH	SK4023	SK4023B	SW4023
SW4023	TC4023BP	TC4023P	TCG4023B
TM4023B	TP4023	TP4023AN	UPD4023BC
UPD4023C			

BLOCK-1063

15-45184-1	154772	221-182-01	4024
4024(IC)	4024BDC	4024BDM	4024BFC
4024BFM	4024BPC	51X90507A93	991-028681-001
CD4024	CD4024AD	CD4024AE	CD4024AF
CD4024BCJ	CD4024BCN	CD4024BD	CD4024BE
CD4024BF	CD4024BMJ	CD4024BMW	CM4024
CM4024AD	CM4024AE	ECG4024	ECG4024B
F4024	F4024B	GE-4024	HBC4024AD
HBC4024AF	HBC4024AK	HBF4024	HBF4024A
HBF4024AE	HBF4024AF	HCC4024BD	HCC4024BF
HCC4024BK	HCF4024BE	HCF4024BF	HD14024B
HD4024	HEF4024BD	HEF4024BP	HEF4024P
M4024BP	MC14024	MC14024B	MC14024BAL
MC14024BCL	MC14024BCP	MC14024CP	MM4024
MM5624AN	NTE4024B	SCL4024	SCL4024AC
SCL4024AD	SCL4024AE	SCL4024AF	SCL4024AH
SCL4024AT	SCL4024B	SCL4024BC	SCL4024BD
SCL4024BE	SCL4024BF	SCL4024BH	SK4024
SK4024B	SW4024	TC4024BP	TCG4024B
TM4024B	TP4024	TP4024AN	TP4024BN

BLOCK-1064

148732	221-Z9072	34025PC	4025
4025(IC)	4025A	4025BDC	4025BDM
4025BFC	4025BFM	4025BPC	443-712
991-028682-001	CD4025	CD4025AD	CD4025AE
CD4025AF	CD4025BCJ	CD4025BCN	CD4025BD
CD4025BE	CD4025BF	CD4025BMJ	CD4025BMW
CD4025CJ	CD4025CN	CD4025MJ	CD4025MW
CD4025UBD	CD4025UBE	CD4025UBF	CM4025
CM4025AD	CM4025AE	ECG4025	ECG4025B
F4025	F4025B	GE-4025	HBC4025AD
HBC4025AF	HBC4025AK	HBF4025	HBF4025A
HBF4025AE	HBF4025AF	HCC4025BD	HCC4025BF
HCC4025BK	HCF4025BE	HCF4025BF	HD14025B
HD4025	HE-443-712	HEF4025BD	HEF4025BP
HEF4025P	HEPC4003P	M4025BP	MB84025B
MB84025BM	MC14025	MC14025AL	MC14025B
MC14025BAL	MC14025BCL	MC14025BCP	MC14025CP
MM4025	MM5625AN	MSM4025	N4025
NTE4025B	SCL4025	SCL4025AC	SCL4025AD
SCL4025AE	SCL4025AF	SCL4025AH	SCL4025B
SCL4025BC	SCL4025BD	SCL4025BE	SCL4025BF
SCL4025BH	SK4025	SK4025B	SS4025AE
SW4025	TC4025BP	TC4025P	TCG4025B
TM4025B	TP4025	TP4025AN	UPD4025C

BLOCK-1065

CD4026AD	CD4026AE	CD4026AF	CD4026BD
CD4026BE	CD4026BF	CM4026AD	CM4026AE
ECG4026B	F4026B	HCC4026BF	HCF4026BE
HCF4026BF	NTE4026B	SCL4026AB	SCL4026ABC
SCL4026ABD	SCL4026ABE	SK4026B	TCG4026B

BLOCK-1066

1143-0584	1207-4707	1207-6907	34027PC
37051246	4027	4027(IC)	4027B
4027BDC	4027BDM	4027BFC	4027BFM
4027BPC	443-606	5190538A18	67-32720-02
8-759-240-27	905-200	991-026022	991-026022-00
CD4027	CD4027AD	CD4027AE	CD4027AF
CD4027BCJ	CD4027BCN	CD4027BD	CD4027BE
CD4027BF	CD4027BMJ	CD4027BMW	CM4027
CM4027AD	CM4027AE	D4027BC	ECG4027
ECG4027B	EW84X464	F4027	GE-4027
HBC4027AD	HBC4027AF	HBC4027AK	HBF4027
HBF4027A	HBF4027AE	HBF4027AF	HCC4027BD
HCC4027BF	HCC4027BK	HCF4027BE	HCF4027BF
HD14027B	HD4027	HE-443-606	HEF4027BD
HEF4027BP	HEF4027P	INMPD14027BC	ITTC4027B
MB84027B	MC14027	MC14027BAL	MC14027BCL

BLOCK-1066 (CONT.)

MC14027BCP	MC14027CP	MM4027	MM5627AN
MN4027B	MPD14027BC	MSM4027	N4027
NTE4027B	RS4027	SCL4027	SCL4027B
SCL4027BC	SCL4027BD	SCL4027BE	SCL4027BF
SCL4027BH	SK4027	SK4027B	SS4027AE
SW4027	TC4027	TC4027B	TC4027BP
TC4027P	TCG4027B	TM4027B	TP4027
TP4027AN	TP4027BN	TVSUPD4027BC	UPD4027
UPD4027BC	UPD4027C		

BLOCK-1067

34028PC	4028BDC	4028BDM	4028BFC
4028BFM	4028BPC	443-713	46-13487-3
70119063	CD4028AD	CD4028AE	CD4028AF
CD4028BCJ	CD4028BCN	CD4028BD	CD4028BE
CD4028BF	CD4028BMJ	CD4028BMW	CD4028F
CM4028AD	CM4028AE	ECG4028	ECG4028B
HBC4028AD	HBC4028AF	HBC4028AK	HBF4028A
HBF4028AE	HBF4028AF	HCC4028BD	HCC4028BF
HCC4028BK	HCF4028BE	HCF4028BF	HD14028B
HE-443-713	HEF4028B	HEF4028BD	HEF4028BP
HEF4028P	MB84028B	MB84028BM	MC14028
MC14028BAL	MC14028BCL	MC14028BCP	MC14028CP
MSM4028	NTE4028B	SCL4028AC	SCL4028AD
SCL4028AE	SCL4028AF	SCL4028AH	SCL4028B
SCL4028BC	SCL4028BD	SCL4028BE	SCL4028BF
SCL4028BH	SK4028B	SS4028AE	TC4028B
TC4028BP	TC4028P	TCG4028B	TM4028B
TP4028AN	TP4028BN	VHITC4028BP-1	

BLOCK-1068

1147-11	15-43704-1	221-Z9073	34029PC
4029	4029BDC	4029BDM	4029BFC
4029BFM	4029BPC	8-759-140-29	8-759-240-29
905-218	CD4029AD	CD4029AE	CD4029AF
CD4029BCJ	CD4029BCN	CD4029BD	CD4029BE
CD4029BF	CD4029BMJ	CD4029BMW	CM4029AD
CM4029AE	ECG4029	ECG4029B	F4029
F4029PC	HBC4029AD	HBC4029AF	HBC4029AK
HBF4029A	HBF4029AE	HBF4029AF	HCC4029BD
HCC4029BF	HCC4029BK	HCF4029BE	HCF4029BF
HEF4029B	HEF4029BD	HEF4029BP	HEF4029P
MB84029B	MB84029BM	MC14029BAL	MC14029BCL
MC14029BCP	MC14029CP	NTE4029B	SCL4029AC
SCL4029AD	SCL4029AE	SCL4029AH	SCL4029AH
SCL4029B	SCL4029BC	SCL4029BD	SCL4029BE
SCL4029BF	SCL4029BH	SK4029	SK4029B
TC4029BP	TCG4029B	TM4029B	TP4029AN
UPD4029	UPD4029C		

BLOCK-1069

1259-5005	191862	34030PC	37051100
4030	4030(IC)	4030BDC	4030BDM
4030BFC	4030BFM	4030BPC	443-784
8-759-240-30	991-026005	CD4030	CD4030AD
CD4030AE	CD4030AF	CD4030BD	CD4030BE
CD4030BF	CD4030BMW	CD4030CJ	CD4030CN
CD4030MJ	CD4030MW	CD4070	CM4030
CM4030AD	CM4030AE	D4030BC	ECG4030
ECG4030B	F4030	GE-4030	HBC4030AD
HBC4030AF	HBC4030AK	HBF4030A	HBF4030AE
HBF4030AF	HCC4030BF	HCF4030BE	HCF4030BF
HD4030	HE-443-784	HEF4030BD	HEF4030BP
HEF4030P	HEPC4030P	ITTC4030BP	MM4030
MM5630AN	MSM4030	N4030	NTE4030B
SCL4030	SCL4030AC	SCL4030AD	SCL4030AE
SCL4030AF	SCL4030AH	SCL4030B	SCL4030BC
SCL4030BD	SCL4030BE	SCL4030BF	SCL4030BH
SK4030	SK4030B	SS4030AE	SW4030
TC4030BP	TC4030P	TCG4030B	TM4030B
TP4030AN	UPD4030BC	UPD4030C	VHITC4030BP-1

BLOCK-1070

4031BDC	4031BPC	CD4031AD	CD4031AE
CD4031AF	CD4031BD	CD4031BE	CD4031BF
ECG4031B	HCC4031BF	HCF4031BE	HCF4031BF
HEF4031B	HEF4031BD	HEF4031BP	HEF4031P
NTE4031B	SK4031B	TCG4031B	

BLOCK-1071

CD4032AD	CD4032AE	CD4032AF	CD4032BD
CD4032BE	CD4032BF	CM4032AD	CM4032AE
ECG4032B	F4032B	HBF4032AE	HBF4032AF
HCC4032BF	HCF4032BE	HCF4032BF	HD14032B
MC14032BAL	MC14032BCL	MC14032BCP	NTE4032B
SK4032B	TC4032BP	TCG4032B	

BLOCK-1072

CD4033	CD4033AD	CD4033AE	CD4033AF
CD4033BD	CD4033BE	CD4033BF	CM4033AD
CM4033AE	ECG4033B	HBF4033AE	HBF4033AF
HCC4033BF	HCF4033BE	HCF4033BF	NTE4033B
SCL4033AB	SCL4033ABC	SCL4033ABD	SCL4033ABE
SK4033B	TCG4033B		

BLOCK-1073

4034BDC	4034BPC	CD4034AD	CD4034AE
CD4034AF	CD4034AY	CD4034BD	CD4034BE
CD4034BF	ECG4034B	HBF4034AE	HCF4034BE
HD14034B	MC14034BAL	MC14034BCL	MC14034BCP
NTE4034B	SCL4034BC	SCL4034BD	SCL4034BE
SK4034B	TCG4034B	TM4034B	

BLOCK-1074

4035BDC	4035BPC	CD4035AD	CD4035AE
CD4035AF	CD4035AY	CD4035BD	CD4035BE
CD4035BF	CM4035AE	ECG4035B	HD14035B
HCC4035BF	HCF4035BE	HCF4035BF	HD14035B
HEF4035B	HEF4035BD	HEF4035BP	HEF4035P
MC14035BAL	MC14035BCL	MC14035BCP	NTE4035B
SCL4035BC	SCL4035BD	SCL4035BE	SK4035
SK4035B	TCG4035B	TM4035B	UPD4035C

BLOCK-1075

CD4038AD	CD4038AE	CD4038AF	CD4038BD
CD4038BE	CD4038BF	CM4038AD	CM4038AE
ECG4038B	HBF4038AE	HBF4038AF	HCC4038BF
HCF4038BF	HD14038B	MC14038BAL	MC14038BCL
MC14038BCP	NTE4038B	SK4038B	TC4038BP
TCG4038B			

BLOCK-1076

154773	34040PC	4040	4040(IC)
4040BDC	4040BDM	4040BFC	4040BFM
4040BPC	443-760	46-13314-3	51X90518A08
AMX4666	CD4040	CD4040AD	CD4040AE
CD4040AF	CD4040BCJ	CD4040BCN	CD4040BD
CD4040BE	CD4040BF	CD4040BMJ	CD4040BMW
CM4040	CM4040AD	CM4040AE	D4040BC
ECG4040	ECG4040B	F4040	GE-4040
HBC4040AD	HBC4040AF	HBC4040AK	HBF4040A
HBF4040AE	HBF4040AF	HCC4040BD	HCC4040BF
HCC4040BK	HCF4040BE	HCF4040BF	HD14040B
HD4040	HEF4040BD	HEF4040BP	HEF4040P
HEPC4040P	M4040BP	MB84040B	MB84040BM
MC14040	MC14040B	MC14040BAL	MC14040BCL
MC14040BCP	MC14040CP	MC14046CP	MM4040
MSM4040	NTE4040B	SCL4040	SCL4040ABC
SCL4040ABD	SCL4040AC	SCL4040AD	SCL4040AE
SCL4040AF	SCL4040AH	SCL4040B	SK4040
SK4040B	TC4040B	TC4040BP	TC4040P
TCG4040B	TM4040B	TP4040	TP4040AN
TP4040BN	UPD4040BC	UPD4040C	

BLOCK-1077

1191-3365	158238	34042PC	4042
4042(IC)	4042BDC	4042BDM	4042BFC
4042BFM	4042BPC	CD4042	CD4042AD
CD4042AE	CD4042AF	CD4042BCJ	CD4042BCN
CD4042BD	CD4042BE	CD4042BF	CD4042BMJ
CD4042BMW	CM4042	CM4042AD	CM4042AE
ECG4042	ECG4042B	F4042	GE-4042
HBC4042AD	HBC4042AF	HBC4042AK	HBF4042A
HBF4042AE	HBF4042AF	HCC4042BD	HCC4042BF
HCC4042BK	HCF4042BE	HCF4042BF	HD14042B
HD14042BP	HD14042P	HEF4042BD	HEF4042BP
HEF4042P	HEPC4042P	M4042BP	MC14042
MC14042BAL	MC14042BCL	MC14042BCP	MC14042CP
MC4042P	MM4042	MSM4042	NTE4042B
SCL4042	SCL4042B	SCL4042BC	SCL4042BD
SCL4042BE	SCL4042BF	SCL4042BH	SK4042
SK4042B	TC4042BP	TC4042P	TCG4042B
TM4042B	TP4042	TP4042AN	TP4042BN
UPD4042C	VHIHD14042P-1		

BLOCK-1078

158240	4043BDC	4043BDM	4043BFC
4043BFM	4043BPC	CD4043AD	CD4043AE
CD4043AF	CD4043BD	CD4043BE	CD4043BF
CD4043CJ	CD4043CN	CD4043MJ	CD4043MW
CM4043AD	CM4043AE	ECG4043B	HBC4043AD
HBC4043AF	HBC4043AK	HBF4043AE	HBF4043AF
HCC4043BD	HCC4043BF	HCC4043BK	HCF4043BE
HCF4043BF	HD14043B	HEF4043BD	HEF4043BP
HEF4043P	M4043BP	MC14043BAL	MC14043BCL
MC14043BCP	MSM4043	NTE4043B	SCL4043ABC
SCL4043ABD	SCL4043ABE	SCL4043ABF	SCL4043ABH
SCL4043B	SCL4043BC	SCL4043BD	SCL4043BE
SCL4043BF	SCL4043BH	SK4043B	TC4043BP
TCG4043B	TM4043B	TP4043AN	TP4043BN

BLOCK-1079

157806	4044BDC	4044BDM	4044BFC
4044BFM	4044BPC	CD4044AD	CD4044AE
CD4044AF	CD4044BD	CD4044BE	CD4044BF
CD4044CJ	CD4044CN	CD4044MJ	CD4044MW
CM4044AD	CM4044AE	ECG4044B	HBC4044AD
HBC4044AF	HBC4044AK	HBF4044AE	HBF4044AF
HCC4044BD	HCC4044BF	HCC4044BK	HCF4044BE
HCF4044BF	HD14044B	HD14044BP	HEF4044B
HEF4044BD	HEF4044BP	HEF4044P	HEPC4008P
MC14044BAL	MC14044BCL	MC14044BCP	MSM4044
NTE4044B	SCL4044ABC	SCL4044ABE	SCL4044ABF
SCL4044ABH	SCL4044B	SCL4044BC	SCL4044BD
SCL4044BE	SCL4044BF	SCL4044BH	SK4044B
TC4044BP	TCG4044B	TM4044B	TP4044AN
TP4044BN			

BLOCK-1080

CD4045AE	CD4045AF	CD4045BD	CD4045BE
CD4045BF	ECG4045B	HCC4045BF	HCF4045BE
HCF4045BF	NTE4045B	SK4045B	TCG40-45B

BLOCK-1081

442-647	8-759-040-46	CD4046A	CD4046AD
CD4046AF	CD4046BD	CD4046BE	CD4046BF
ECG4046B	HE-442-647	HEPC4053P	MC14046BCL
MC14046BCP	NTE4046B		

BLOCK-1082

4047BDC	4047BDM	4047BFC	4047BFM
4047BPC	443-930	CD4047AD	CD4047AE
CD4047AF	CD4047BCJ	CD4047BCN	CD4047BD
CD4047BE	CD4047BF	CD4047BMJ	CD4047BMW
CM4047AD	CM4047AE	CM4047AF	ECG4047B
HBC4047AD	HBC4047AK	HBF4047AE	HCC4047BF
HCF4047BE	HCF4047BF	HE-443-930	HEF4047B
HEF4047BD	HEF4047BP	MC4047	NTE4047B
SCL4047B	SCL4047BC	SCL4047BD	SCL4047BE
SCL4047BF	SCL4047BH	SK4047	SK4047B
TC4047BP	TCG4047B		

BLOCK-1083

CD4048AD	CD4048AE	CD4048AF	CD4048BCJ
CD4048BCN	CD4048BD	CD4048BE	CD4048BF
CD4048BMJ	CM4048AD	CM4048AE	ECG4048B
HBF4048AE	HBF4048AF	HCC4048BF	HCF4048BE
HCF4048BF	NTE4048B	SK4048B	TCG4048B

BLOCK-1084

1287-7932	148236	148733	176225
221-Z9074	276-2449	34049PC	4009(IC)
4049	4049(IC)	4049B	4049BDC
4049BDM	4049BPC	544-3001-140	8-759-240-49
B0470494	CD4009AD	CD4009AE	CD4009AF
CD4009CJ	CD4009CN	CD4009UBD	CD4009UBE
CD4009UBF	CD4049AD	CD4049AE	CD4049AF
CD4049CJ	CD4049CN	CD4049MJ	CD4049UBD
CD4049UBE	CD4049UBF	CM4009	CM4009AD
CM4009AE	CM4049	CM4049AD	CM4049AE
D4049UBC	ECG4049	ECG4049B	F4049
F4049PC	GE-4009	GE-4049	H31100129
HBF4009	HBF4009A	HBF4049A	HCF4049BE
HCF4049BF	HD14049UB	HD4009	HD4049
HEF4049BD	HEF4049BP	HEPC4009P	ITTC4049BP
M4049BP	MB84049B	MB84049BM	MC14049
MC14049B	MC14049CP	MC14049UB	MC14049UBAL
MC14049UBCL	MC14049UBCP	MEM4049	MM4009
MM4049	N4009	N4049	NTE4049
RS4049	SCL4009	SCL4009UBC	SCL4009UBD
SCL4009UBE	SCL4049	SCL4449UBE	SK4009
SK4009UB	SK4049	SK4049UB	SW4049
TC4009UBP	TC4049	TC4049BP	TC4049P
TCG4049	TCG4049B	TM4049B	TP4009
TP4009UBN	TP4049	TP4049AN	TP4049UBN
UPD4049	UPD4049C	UPD4049UBC	X440070490

BLOCK-1085

07-28816-85	144970	221-Z9163	34050PC
4050	4050(IC)	4050BDC	4050BDM
4050BPC	51-90433A11	569-0690-550	70119064
991-028650-001	AMX4584	CO-10816	C010816
C010816-01	CD4010AE	CD4010BE	CD4010CJ
CD4010CN	CD4050	CD4050AD	CD4050AE
CD4050AF	CD4050B	CD4050BCJ	CD4050BCN
CD4050BD	CD4050BE	CD4050BF	CM4010AE
CM4050	CM4050AD	CM4050AE	CO-10816
CO10816-XX	ECG4050	ECG4050B	F4050
GE-4050	HBF4050A	HCF4050BE	HCF4050BF
HD14050B	HD14050BP	HD4050	HE-443-991
HEF4050BD	HEF4050BP	HEPC4050P	M4050BP
MB84050B	MB84050BM	MC14050	MC14050B
MC14050BAL	MC14050BCL	MC14050BCP	MC14050CP
MEM4050	MM4050	MM5660BN	MSM4050

BLOCK-1085 (CONT.)

MSM4050RS	N4050	NTE4050B	RS4050
SCL4010BE	SCL4050	SK4050	SK4050B
SW4050	TC4010BP	TC4050BP	TC4050CP
TC4050P	TCG4050B	TM4050B	TP4010BN
TP4050	TP4050AN	UPD4050C	X460405000

BLOCK-1086

34051PC	4051	4051(IC)	4051BDC
4051BDM	4051BPC	8-759-240-51	905-380
C014336	CD4051	CD4051AE	CD4051B
CD4051BCJ	CD4051BCN	CD4051BD	CD4051BE
CD4051BF	CD4051BMJ	CD4052AE	CM4051
CO14336	ECG4051	ECG4051B	F4051
GE-4051	HBF4051A	HCC4051BF	HCF4051BE
HCF4051BF	HD14051BP	HD14051BP	HEF4051BD
HEF4051BP	HEPC4051P	M4051BP	MB84051B
MB84051BM	MC14051	MC14051BAL	MC14051BCL
MC14051BCP	MC4051B	MEM4051	MM4051
NTE4051B	SCL4051	SK4051	SK4051B
TC4051BP	TCG4051B	TM4051B	TP4051
TP4051BN	UPD4051BC		

BLOCK-1087

06300554	1200-3497	1250-5210	161079
1826-1	263P05209	34052PC	37053002
4052	4052(IC)	4052BDC	4052BDM
4052BPC	409-051-2801	46-131172-3	46-13730-3
4835-209-87277	6124740001	8-759-240-52	99195-1
99195-3	B0470522	BU4052B	CD4052
CD4052BCJ	CD4052BCN	CD4052BD	CD4052BE
CD4052BF	CD4052BMJ	CM4052	D4052BC
ECG4052	ECG4052B	EW84X470	F4052
GD4052B	GE-4052	HBF4052A	HCC4052BF
HCF4052BE	HCF4052BF	HD14052B	HD14052BP
HEF4052BD	HEF4052BP	HEPC4052P	I55D04052B
ITTC4052BP	M4052BP	MB84052B	MB84052BM
MC14052	MC14052B	MC14052BAL	MC14052BCL
MC14052BCP	MM4052(IC)	MN4052B	MX-4376
NTE4052B	Q7406	QC0081	SCL4052
SK4052	SK4052B	TC4052	TC4052BP
TC4052BP-1	TCG4052B	TM4052B	TP4052
TP4052BN	TVSTC4052BP	TVSUPD4052BC	UPD4052
UPD4052BC	VHITC4052BP-1		

BLOCK-1088

06300191	06300192	0801-000961	1207-6931
1313-5793	1316-3217	1365-5113	195791
2002020066	32109-301-890	32119-201-310	37051378
4053	4053BDC	4053BDM	4053BPC
409 051 3006	409-009-2709	409-051-3006	409-132-7909
46-131678-3	46-132422-3	46-13662-3	4835 209 17033
4835-209-17033	483520947083	5190555A60	612493-1
6124930001	8-759-040-53	8-759-140-53	8-759-240-53
905-354	B0470532	BU4053B	CD4053BCJ
CD4053BCN	CD4053BD	CD4053BE	CD4053BF
CD4053BP	D4053BC	ECG4053B	HCC4053BF
HCF4053BE	HCF4053BF	HD14053B	HD14053BP
HEF4053BD	HEF4053BP	I52D040530	I5PD14053B
IDH0802	M4053BP	MB84053B	MB84053BM
MC14053	MC14053B	MC14053BAL	MC14053BCL
MC14053BCP	MN4053B	N6124930001	NTE4053B
Q7327A	Q7372A	Q7448	SK4053B
TC4053BP	TCG4053B	TM4053B	TP4053BN
TVSTC4053BP	TVSUPD4053BC	UPD4053	UPD4053B
UPD4053BC	UPD4053BG	UPD4053BP	

BLOCK-1089

34055PC	4055	4055(IC)	CD4055
CD4055AE	CD4055BD	CD4055BE	CD4055BF
CD4055BY	ECG4055	ECG4055B	GE-4055
HBC4055AD	HBC4055AF	HBC4055AK	HBF4055AE
HBF4055AF	HCC4055BF	HCF4055BE	HCF4055BF
HEPC4055P	NTE4055B	SK4055B	TC4055BP
TCG4055B	TM4055B		

BLOCK-1090

CD4056BD	CD4056BE	CD4056BY	ECG4056B
HCC4056BD	HCC4056BF	HCC4056BK	HCF4056BE
HCF4056BF	NTE4056B	SK4056	SK4056B
TC4056BP	TCG4056B		

BLOCK-1091

443-958	CD4060	CD4060AD	CD4060AE
CD4060AF	CD4060BCJ	CD4060BCN	CD4060BD
CD4060BE	CD4060BF	CD4060BMJ	CD4060BMW
CD4063BE	ECG4060B	HBC4060AD	
HBC4060AF	HBC4060AK	HBF4060AE	HBF4060AF
HCC4060BD	HCC4060BF	HCC4060BK	HCF4060BE
HCF4060BF	HE-443-958	MB84060B	MB84060BM
MC14060B	MC14060BAL	MC14060BCP	NTE4060B
SCL4060ABC	SCL4060ABD	SCL4060ABE	SCL4060ABF

BLOCK-1091 (CONT.)

SCL4060ABH	SCL4060AC	SCL4060AD	SCL4060AE
SCL4060AF	SCL4060AH	SCL4060B	SK4060
SK4060B	TCG4060B	TM4060B	

BLOCK-1092

158249	CD4063BD	CD4063BF	ECG4063B
HCC4063BD	HCC4063BF	HCC4063BK	HCF4063BE
HCF4063BF	MC14585BCP	NTE4063B	SK4063B
TC4063BP	TCG4063B	TM4063B	

BLOCK-1093

0061-238-00001	06300738	0IGS406600A	1033-4340
1225-4975	1239-7766	1296-1918	147478
148734	154649	163755	181860
185233	2002020077	221-173	221-Z9164
2363191	263P06602	276-2466	32109-410-270
32109-410-280	34066PC	37051036	4066(IC)
4066BDC	4066BDM	4066BPC	455947
46-132394-3	46-13296-3	4835-209-87105	51X90458A88
51X90480A16	56A265-1	6123800001	8-759-000-49
8-759-140-66	8-759-240-16	8-759-240-66	8-759-932-33
901502-01	905-1126	905-369	B0470662
BU4066B	BU4066BL	BU4066BP	CD-4066B
CD4066	CD4066AD	CD4066AE	CD4066AF
CD4066B	CD4066BCJ	CD4066BCN	CD4066BD
CD4066BE	CD4066BF	CD4066BMJ	CM4066AD
CM4066AE	D4066BC	ECG4066B	EW84X496
GD4066	GD4066B	GD4066BD	GD4066BP
HCC4066BF	HCF4066BE	HCF4066BF	HD14066B
HD14066BCP	HD14066BP	HEF4066BD	HEF4066BP
I03D040660	I55D04066B	IDH0742	IN4066BP
INMPD4066BC	INUPD4066BC	LC4066B	M4066BP
MB84066B	MB84066BM	MB84068B	MC14016
MC14066	MC14066B	MC14066BAL	MC14066BCL
MC14066BCP	MN4066B	MN4066BP	MPD4066BC
MSM4066RS	NTE4066B	PD4066BC	Q7478A
Q7652	Q7653A	SK4066	SK4066B
TC4016BP	TC4066	TC4066B	TC4066BC
TC4066BP	TC4066P	TCG4066B	TM4066B
TP4016A	TVSTC4066BP	TVSUPD4066BC	UN4066B
UPC4066	UPD4066	UPD4066B	UPD4066B-1
UPD4066BC	UPD4066BP	UPD4066C	UPD4066G
VHITC4066BP-1	VHIUPD4066B-1	WEP4066B/4066B	X440160660

BLOCK-1094

4067BDC	4067BDM	4067BPC	CD4067BD
CD4067BE	CD4067BF	ECG4067B	MC14067BAL
NTE4067B	SK4067B	TCG4067B	

BLOCK-1095

34068PC	4068BDC	4068BDM	4068BFC
4068BFM	4068BPC	CD4068BE	CD4068BE
CD4068BF	ECG4068	ECG4068B	HCC4068BD
HCC4068BF	HCC4068BK	HCF4068BE	HCF4068BF
HD14068B	HEF4068BD	HEF4068BP	HEF4068P
MB84068BM	MC14068B	MC14068BAL	MC14068BCL
MC14068BCP	MSM4068	NTE4068B	SCL4068B
SCL4068BC	SCL4068BD	SCL4068BE	SCL4068BF
SCL4068BH	SK4068	SK4068B	TC4068BP
TCG4068B	TM4068B	TP4068BN	

BLOCK-1096

07-28816-86	144971	152935	207890
34069PC	4069UBDC	4069UBDM	4069UBPC
46-13319-3	483520917279	51-90433A10	51X90507A99
6122500001	8-759-240-69	8-759-904-69	905-260
CD4069BE	CD4069C	CD4069CJ	CD4069CN
CD4069M	CD4069MJ	CD4069N	CD4069UBD
CD4069UBE	CD4069UBF	CM4069B	D4069C
D4069UBC	DM-77	ECG4069	EW84X756
HD14069UB	HEF4069P	HEF4069UBD	HEF4069UBP
IGLC4069UB	LC4069UB	M4069UBP	MB84069B
MB84069BM	MC14069U	MC14069UBAL	MC14069UBCL
MC14069UBCP	MN4069UB	MN4069UBS	MSM4064
MSM4069	MSM4069RS	MSM4069UBRU	NTE4069
SILC4069UBXXX	SK4069	SK4069UB	TC4069BP
TC4069P	TC4069UBP	TC4069UPB	TC4069UPN
TCG4069	TCG4069B	TM4069	TP4069UBN
TP4069UPN	TVSTC4069UBP	UPD4069	UPD4069C
UPD4069UBC	WEP4069	WEP4069/4069	X460406901

BLOCK-1097

34070PC	4070BDC	4070BDM	4070BFC
4070BFM	4070BPC	CD4070BCJ	CD4070BCN
CD4070BD	CD4070BE	CD4070BF	CD4070BMJ
CD4070BMW	ECG4070B	HCC4070BD	HCC4070BF
HCC4070BK	HCF4070BE	HCF4070BF	HD14070B
HEF4070BD	HEF4070BP	HEF4070P	HEPC4004P
MB84070B	MB84070BM	MC14070BAL	MC14070BCL
MC14070BCP	MC14070BL	MM74C86	NTE4070B
SCL4070B	SCL4070BC	SCL4070BD	SCL4070BE
SCL4070BF	SCL4070BH	SK4070	SK4070B
TCG4070B			

BLOCK-1098

1229-4633	1229-4658	154770	161732
34071PC	4071BDC	4071BDM	4071BFC
4071BFM	4071BPC	443-706	46-13816-3
51X90507A85	8-759-240-71	CD4071BCJ	CD4071BCN
CD4071BD	CD4071BE	CD4071BF	CD4071BMJ
CD4071BMW	ECG4071	ECG4071B	HCC4071BD
HCC4071BF	HCC4071BK	HCF4071BE	HCF4071BF
HD14071B	HE-443-706	HEF4071BD	HEF4071BP
HEF4071P	LC4071B	M4071BP	MB84071B
MB84071BM	MC14071	MC14071BAL	MC14071BCL
MC14071BCP	MC14071CP	MN4071B	MSM4071
MSM4071BN	NTE4071B	QC0234	QC0235
QC0239	SCL4071AC	SCL4071AD	SCL4071AE
SCL4071AF	SCL4071AH	SCL4071B	SCL4071BC
SCL4071BD	SCL4071BE	SCL4071BF	SCL4071BH
SK4071	SK4071B	TC4071BP	TC4071P.JA
TCG4071B	TM4071B	TP4071BN	UPD4071
UPD4071BC	UPD471C		

BLOCK-1099

4072BDC	4072BDM	4072BFC	4072BFM
4072BPC	CD4072BD	CD4072BE	CD4072BF
ECG4072B	HCC4072BD	HCC4072BF	HCC4072BK
HCF4072BE	HCF4072BF	HD14072B	HEF4072BD
HEF4072BP	HEF4072P	MB84072B	MB84072BM
MC14072BAL	MC14072BCL	MC14072BCP	MSM4072
NTE4072B	SCL4072AC	SCL4072AD	SCL4072AE
SCL4072AF	SCL4072AH	SCL4072B	SCL4072BC
SCL4072BD	SCL4072BE	SCL4072BF	SCL4072BH
SK4072	SK4072B	TC4072BP	TCG4072B
TM4072B	TP4072BN		

BLOCK-1100

4073BDC	4073BDM	4073BFC	4073BFM
4073BPC	803-0073	CD4073BCJ	CD4073BCN
CD4073BD	CD4073BE	CD4073BF	CD4073BMJ
CD4073BMW	ECG4073B	HCC4073BD	HCC4073BF
HCC4073BK	HCF4073BE	HCF4073BF	HD14073B
HEF4073B	HEF4073BD	HEF4073BP	HEF4073P
MB84073B	MB84073BM	MC14073B	MC14073BAL
MC14073BCL	MC14073BCP	MSM4073	NTE4073B
SCL4073AC	SCL4073AD	SCL4073AE	SCL4073AF
SCL4073AH	SCL4073B	SCL4073BC	SCL4073BD
SCL4073BE	SCL4073BF	SCL4073BH	SK4073
SK4073B	TC4073BP	TCG4073B	TM4073B
TP4073BN			

BLOCK-1101

4075BDC	4075BDM	4075BFC	4075BFM
4075BPC	CD4075BCJ	CD4075BCN	CD4075BD
CD4075BE	CD4075BF	CD4075BMJ	CD4075BMN
CD4075BMW	CD4075BT	ECG4075B	F4075
HCC4075BD	HCC4075BF	HCC4075BK	HCF4075BE
HCF4075BF	HD14075B	HEF4075B	HEF4075BD
HEF4075BP	MB84075B	MB84075BM	MC14075BAL
MC14075BCL	MC14075BCP	MSM4075	NTE4075B
SCL4075AC	SCL4075AD	SCL4075AE	SCL4075AF
SCL4075AH	SCL4075B	SCL4075BC	SCL4075BD
SCL4075BE	SCL4075BF	SCL4075BH	SK4075
SK4075B	TC4075BP	TCG4075B	TM4075B
TP4075BN			

BLOCK-1102

4076BDC	4076BDM	4076BPC	CD4076BC
CD4076BD	CD4076BE	CD4076BF	CD4076BM
ECG4076B	F4076	HCC4076BF	HCF4076BE
HCF4076BF	HD14076B	HEF4076B	HEF4076BD
HEF4076BP	HEF4076P	MC14076BAL	MC14076BCL
MC14076BCP	NTE4076B	SCL4076BC	SCL4076BD
SCL4076BE	SK4076	SK4076B	TCG4076B
TM4076B			

BLOCK-1103

34077PC	4077BDC	4077BDM	4077BFC
4077BFM	4077BPC	CD4077BD	CD4077BE
CD4077BF	ECG4077	ECG4077B	F4077BPC
HCC4077BD	HCC4077BF	HCC4077BK	HCF4077BE
HCF4077BF	HD14077B	HEF4077B	HEF4077BD
HEF4077BP	HEF4077P	MB84077B	MB84077BM
MC14077BAL	MC14077BCL	MC14077BCP	NTE4077B
SCL4077B	SCL4077BC	SCL4077BD	SCL4077BE
SCL4077BF	SCL4077BH	SK4077	SK4077B
TCG4077B	TM4077B		

BLOCK-1104

148735	34078PC	4078BDC	4078BDM
4078BFC	4078BFM	4078BPC	51X90458A89
51X90480A17	CD4078BD	CD4078BE	CD4078BF
ECG4078	ECG4078B	HCC4078BD	HCC4078BF
HCC4078BK	HCF4078BE	HCF4078BF	HD14078B
HEF4078BD	HEF4078BP	HEF4078P	MB84078B
MB84078BM	MC14078BAL	MC14078BCL	MC14078BCP

BLOCK-1104 (CONT.)

MSM4078	NTE4078B	SCL4078B	SCL4078BC
SCL4078BD	SCL4078BE	SCL4078BF	SCL4078BH
SK4078	SK4078B	TC4078BP	TCG4078B
TM4078B	TP4078BN	TS4078BP	UPD4078C

BLOCK-1105

11-103	159435	191985	191986
191989	276-2481	307-152-9-004	34081PC
4081	4081(IC)	4081BDC	4081BDM
4081BFC	4081BFM	4081BPC	443-751
905-268	991-025517	CD4081	CD4081BCJ
CD4081BCN	CD4081BD	CD4081BE	CD4081BF
CD4081BMJ	CD4081BMW	CM4081	D4081
D4081BC	DM-103	ECG4081	ECG4081B
F4081	GE-4081	H31100128	HCC4081BD
HCC4081BF	HCC4081BK	HCF4081BE	HCF4081BF
HD14081B	HD14081BP	HE-443-751	HEF4081BD
HEF4081BP	HEF4081P	ILMC14081B	LC4081B
M4081BP	MB84081B	MB84081BM	MC14081
MC14081B	MC14081BAL	MC14081BCL	MC14081CP
MN4081B	MSM4081	MSM4081RS	N4081
NTE4081B	SCL4081	SCL4081AC	SCL4081AD
SCL4081AE	SCL4081AF	SCL4081AH	SCL4081B
SCL4081BC	SCL4081BD	SCL4081BE	SCL4081BF
SCL4081BH	SK4081	SK4081B	TC4081
TC4081P	TCG4081	TCG4081B	TM4081B
TP4081	TP4081BN	UPD4081BC	

BLOCK-1106

4082BDC	4082BDM	4082BFC	4082BFM
4082BPC	46-13316-3	CD4082	CD4082B
CD4082BD	CD4082BE	CD4082BF	ECG4082B
HCC4082BD	HCC4082BF	HCC4082BK	HCF4082BE
HCF4082BF	HD14082B	HEF4082B	HEF4082BP
HEF4082P	MB84082B	MB84082BM	MC14082B
MC14082BAL	MC14082BCL	MC14082BCP	MSM4082
NTE4082B	SCL4082B	SCL4082BC	SCL4082BD
SCL4082BE	SCL4082BF	SCL4082BH	SCL4085BC
SK4082	SK4082B	TC4082BP	TC4082P
TCG4082B	TM4082B	TP4082BN	

BLOCK-1107

34085PC	4085BDC	4085BDM	4085BFC
4085BFM	4085BPC	CD4085BD	CD4085BE
CD4085BF	ECG4085	ECG4085B	HCC4085BD
HCC4085BF	HCC4085BK	HCF4085BE	HCF4085BF
HEF4085BD	HEF4085BP	HEF4085P	MSM4085
NTE4085B	SCL4085B	SCL4085BD	SCL4085BE
SCL4085BF	SCL4085BH	SK4085B	TC4085BP
TCG4085B	TM4085B		

BLOCK-1108

34086PC	4086BDC	4086BDM	4086BFC
4086BFM	4086BPC	CD4086BD	CD4086BE
CD4086BF	ECG4086	ECG4086BD	HCC4086BD
HCC4086BF	HCC4086BK	HCF4086BE	HCF4086BF
HCF4086BK	HEF4086B	HEF4086BD	HEF4086BP
HEF4086P	MSM4086	NTE4086B	SCL4086B
SCL4086BC	SCL4086BD	SCL4086BE	SCL4086BF
SCL4086BH	SK4086	SK4086B	TC4086BP
TCG4086B	TM4086B		

BLOCK-1109

CD4089BCJ	CD4089BCN	CD4089BD	CD4089BE
CD4089BF	CD4089BMJ	ECG4089B	HCC4089BF
HCF4089BE	HCF4089BF	NTE4089B	SK4089B
TCG4089B			

BLOCK-1110

15-43706-1	4093BDC	4093BDM	443-778
B0470932	CD4093BCJ	CD4093BCN	CD4093BD
CD4093BE	CD4093BMJ	CD4093BMJ	ECG4093B
HD14093B	HE-443-778	HEF4093BD	HEF4093BP
INMPD4093BC	INUPD4093BC	MC14093	MC14093B
MC14093BAL	MC14093BCL	MC14093BCP	MPD4093BC
NTE4093B	SK4093	SK4093B	TC4093BP
TCG4093B	TM4093B	UPD4093BC	X460409300

BLOCK-1111

CD4095BD	CD4095BE	CD4095BF	ECG4095B
HCC4095BF	HCF4095BE	HCF4095BF	NTE4095B
SK4095B	TCG4095B		

BLOCK-1112

CD4096BD	CD4096BE	CD4096BF	ECG4096B
HCC4096BF	HCF4096BE	HCF4096BF	NTE4096B
SK4096B	TCG4096B		

BLOCK-1113

CD4097BD	CD4097BE	CD4097BF	ECG4097B
MC14097BAL	NTE4097B	SK4097B	TCG4097B

BLOCK-1114

15-3015570-1	443-777	443-916	CD4098BD
CD4098BE	CD4098BF	ECG4098B	HCC4098BD
HCC4098BF	HCC4098BK	HCF4098BE	HCF4098BF
HE-443-777	HE-443-916	MC14528	MC14538
NTE4098B	SCL4528B	SK4098	SK4098B
TCG4098B	TM4098B		

BLOCK-1115

34099PC	CD4099	CD4099BCJ	CD4099BCN
CD4099BD	CD4099BE	CD4099BF	CD4099BMJ
CD4099BMW	ECG4099	ECG4099B	HCC4099BD
HCC4099BF	HCC4099BK	HCF4099BE	HCF4099BF
MC14099BAL	MC14099BCL	MC14099BCP	NTE4099B
SK4099	SK4099B	TC4099BP	TCG4099B
TM4099B			

BLOCK-1116

ECG4256	HM50256-15	HM50256-20	HM50256P-12
HM50256P-15	HM50256P-20	I41D41256C	KM41256AP-15
LH21256-12	MB81256-15	MN41256-15	SK10191
TMM41256CP-12	TMS4256-15NL		

BLOCK-1117

ECG4501	HD14501UB	MC14501UBCL	MC14501UBCP
NTE4501	SK4501UB	TCG4501	

BLOCK-1118

CD4502BD	CD4502BE	CD4502BF	ECG4502
ECG4502B	HCF4502BE	HD14502B	HEF4502BD
HEF4502BP	MC14502BAL	MC14502BCL	MC14502BCP
NTE4502B	SCL4502BE	SK4502B	TCG4502B

BLOCK-1119

443-888	4506	ECG4506B	HD14506B
HE-443-888	MC14506BCL	MC14506BCP	MC14506UBCP
NTE4506B	SK4506UB	TCG4506B	

BLOCK-1120

CD4508BE	CD4508BF	ECG4508B	HD14508B
HEF4508BD	HEF4508BP	M4508B	MC14508BAL
MC14508BCL	MC14508BCP	NTE4508B	SK4508B
TC4508BP	TCG4508B		

BLOCK-1121

4510BDC	4510BDM	4510BFC	4510BFM
4510BPC	CD4510BCJ	CD4510BCN	CD4510BD
CD4510BE	CD4510BF	CD4510BMJ	CD4510BMW
ECG4510B	HCC4510BD	HCC4510BE	HCC4510BF
HCC4510BK	HCF4510BE	HCF4510BF	HD14510B
HEF4510B	HEF4510BD	HEF4510BP	HEPC4031P
M4510BP	MC14510BAL	MC14510BCL	MC14510BCP
NTE4510B	SCL4510AC	SCL4510AD	SCL4510AE
SCL4510AF	SCL4510AH	SCL4510B	SCL4510BC
SCL4510BD	SCL4510BE	SCL4510BF	SCL4510BH
SK4510	SK4510B	TC4510BP	TCG4510B
TM4510B			

BLOCK-1122

1147-12	14511	4511	4511BDC
4511BDM	4511BPC	544-3001-143	760106
8000-00057-009	CD4511BCJ	CD4511BCN	CD4511BD
CD4511BE	CD4511BF	ECG4511B	F4511
F4511PC	HCC4511BD	HCC4511BF	HCC4511BK
HCF4511BE	HCF4511BF	HD14511B	HEF4511B
HEF4511BD	HEF4511BP	HEPC4041P	MC14511
MC14511B	MC14511B1	MC14511BAL	MC14511BCL
MC14511BCP	MC14511BP	MC14511L	MM14511BCN
MM14511CN	MM4511	NTE4511B	SCL4511AC
SCL4511AD	SCL4511AE	SCL4511AF	SCL4511AH
SCL4511BC	SCL4511BD	SCL4511BE	SCL4511BF
SCL4511BH	SK4511	SK4511B	TC4511BP
TCG4511B	TM4511B		

BLOCK-1123

34512PC	4512BDC	4512BDM	4512BPC
CD4512BD	CD4512BE	CD4512BF	ECG4512
ECG4512B	HD14512B	HEF4512BD	HEF4512BP
M4512BP	MC14512BAL	MC14512BCL	MC14512BCP
MC14512CP	NTE4512B	SCL4512BC	SCL4512BD
SCL4512BE	SK4512	SK4512B	TC4512BP
TC4512P	TCG4512B	TM4512B	TP4512
TP4512AN			

BLOCK-1124

ECG4513B	MC14513BCL	MC14513BCP	NTE4513B
SK4513B			

BLOCK-1125

443-871	4514BDC	4514BDM	4514BPC
CD4514BC	CD4514BD	CD4514BE	CD4514BF
CD4514BH	CD4514BK	CD4514BM	ECG4514B
HCC4514BD	HCF4514BD	HCF4514BE	HD14514B
HD14515B	HE-443-871	HEF4015BD	HEF4015BP
HEF4514BP	HEPC4054P	M4514BP	MC14514
MC14514B	MC14514BAL	MC14514BCL	MC14514BCP
MSM4514RS	NTE4514B	RH-IX0114PAZZ	SCL4514B
SCL4514BD	SCL4514BE	SCL4514BH	SK4514
SK4514B	TC4514BP	TC4514P	TCG4514B
TM4514B			

BLOCK-1126

443-707	4515BDC	4515BDM	4515BPC
CD4515BC	CD4515BD	CD4515BE	CD4515BF
CD4515BH	CD4515BK	CD4515BM	ECG4515B
HCC4515BD	HCF4515BD	HCF4515BE	HE-443-707
HEF4514BD	HEF4515BD	HEF4515BP	M4515BP
MC14515	MC14515BAL	MC14515BCL	MC14515BCP
NTE4515B	SCL4515B	SCL4515BD	SCL4515BE
SCL4515BH	SK4515	SK4515B	TCG4515B
TM4515B			

BLOCK-1127

443-708	4516BDC	4516BDM	4516BFC
4516BFM	4516BPC	CD4516BCJ	CD4516BCN
CD4516BD	CD4516BE	CD4516BF	CD4516BMJ
CD4516BMW	ECG4516B	HCC4516BD	HCC4516BF
HCC4516BK	HCF4516BE	HCF4516BF	HD14516B
HE-443-708	HEF4516B	HEF4516BD	HEF4516BP
M4516BP	MC14516BAL	MC14516BCL	MC14516BCP
MC14516CP	NTE4516B	SCL4516AC	SCL4516AD
SCL4516AE	SCL4516AF	SCL4516AH	SCL4516B
SCL4516BC	SCL4516BD	SCL4516BE	SCL4516BF
SCL4516BH	SK4516B	TC4516BP	TCG4516B
TM4516B			

BLOCK-1128

CD4517BE	CD4517BF	DM-110	ECG4517B
HD14517B	MC14517BAL	MC14517BCL	MC14517BCP
NTE4517B	SK4517B	TCG4517B	

BLOCK-1129

34518PC	443-737	4518	4518(IC)
4518BDC	4518BDM	4518BFC	4518BFM
4518BPC	4528BDC	CD4518B	CD4518BCJ
CD4518BCN	CD4518BD	CD4518BE	CD4518BF
CD4518BMJ	CD4518BMW	CM4518	ECG4518
ECG4518B	F4518	GE-4518	HCC4518BD
HCC4518BF	HCC4518BK	HCF4518BE	HCF4518BF
HD14518B	HD4518	HE-443-737	HEF4518BD
HEF4518BP	HEF4518P	HEPC4033P	M4518BP
MC14518	MC14518B	MC14518BAL	MC14518BCL
MC14518BCP	MC14518CP	MM4518	MSM4518
NTE4518B	RS4518	SCL4518	SCL4518AC
SCL4518AD	SCL4518AE	SCL4518AF	SCL4518AH
SCL4518B	SCL4518BC	SCL4518BD	SCL4518BE
SCL4518BF	SCL4518BH	SK4518	SK4518B
TC4518BP	TCG4518B	TM4518B	TP4518
TP4518AN	UPD4518		

BLOCK-1130

192062	34520PC	4520BDC	4520BDM
4520BFC	4520BFM	4520BPC	46-13491-3
8-759-245-20	8-759-985-20	905-262	905-265
CD4520B	CD4520BCJ	CD4520BCN	CD4520BD
CD4520BE	CD4520BEX	CD4520BF	CD4520BMJ
CD4520BMW	CM4520	ECG4520	ECG4520B
HCC4520BD	HCC4520BF	HCC4520BK	HCF4520BE
HCF4520BF	HD14520B	HEF4520BD	HEF4520BP
HEF4520P	M4520BP	MB84520B	MC14520BAL
MC14520BCL	MC14520BCP	MSM4520	NTE4520B
SCL4520AC	SCL4520AD	SCL4520AE	SCL4520AF
SCL4520AH	SCL4520BD	SCL4520BC	SCL4520BD
SCL4520BE	SCL4520BF	SCL4520BH	SK4520
SK4520B	TC4520BP	TCG4520B	TM4520B
TP4520AN	UPD4520C		

BLOCK-1131

CD4521BE	ECG4521B	MC14521BCL	MC14521BCP
NTE4521B	SK4521B	TCG4521B	

BLOCK-1132

CD4522BC	CD4522BE	CD4522BM	ECG4522
ECG4522B	HEF4522BD	HEF4522BP	HEPC4032P
MC14522BCL	MC14522BCP	NTE4522B	SCL4522BC
SCL4522BD	SCL4522BE	SCL4522BF	SCL4522BH
SK4522B	TC4522BP	TCG4522B	TM4522B
TP4522AN			

BLOCK-1133

8000-00047-003	CD4526BC	CD4526BCN	CD4526BM
ECG4526B	HD14526B	MC14526	MC14526BCL
MC14526BCP	NTE4526B	SCL4526B	SK4526B
TCG4526B			

BLOCK-1134

34527PC	CD4527BCJ	CD4527BCN	CD4527BD
CD4527BE	CD4527BF	CD4527BMJ	CD4527BMW
ECG4527	ECG4527B	HCC4527BD	HCC4527BF
HCC4527BK	HCF4527BE	HCF4527BF	HD14527B
MC14527BAL	MC14527BCL	MC14527BCP	MC14527CP
NTE4527B	SCL4527B	SCL4527BC	SCL4527BD
SCL4527BE	SCL4527BF	SCL4527BH	SK4527B
TC4527BP	TCG4527B	TM4527B	

BLOCK-1135

1240-3440	148737	152929	4528BDM
4528BPC	51X90458A90	51X90518A28	8-759-245-28
905-241	ECG4528B	EP84X241	F4528B
HEF4528B	HEF4528BD	HEF4528BP	M4528BP
MC14528BAL	MC14528BCL	MC14528BCP	MN4528B
NTE4528B	TC4528BP	TCG4528B	UPD4528B
UPD4528C			

BLOCK-1136

14529	8030529	AMX4585	CD4529BCN
ECG4529B	MC14529B	MC14529BCL	MC14529BCP
NTE4529B	SK4529B	TCG4529B	

BLOCK-1137

4531BDC	4531BDM	4531BPC	ECG4531B
HEF4531B	HEF4531BD	HEF4531BP	MC14531BCL
MC14531BCP	NTE4531B	SK4531B	TC4531BP
TCG4531B			

BLOCK-1138

4532BDC	4532BDM	4532BPC	CD4532BD
CD4532BE	CD4532BF	ECG4532B	HCC4532BF
HCF4532BE	HCF4532BF	HD14532B	MC14532BAL
MC14532BCL	MC14532BCP	NTE4532B	SK4532B
TCG4532B			

BLOCK-1139

CD4536BD	CD4536BE	CD4536BF	ECG4536B
HD14536B	HEPC4058P	MC14536BAL	MC14536BCP
NTE4536B	SK4536B		

BLOCK-1140

160554	8-759-045-38	CD4538BD	CD4538BE
CD4538BF	ECG4538B	HD14538B	HD14538BP
HEF4538B	I55D04538B	MC14538B	MC14538BAL
MC14538BCP	MC14538BP	NTE4538B	SK4538B
TC4538BP	UPD4538	UPD4538BC	

BLOCK-1141

34539PC	4539BDC	4539BDM	4539BPC
ECG4539	ECG4539B	HEF4539BD	HEF4539BP
M4539BP	MC14539BCL	MC14539BCP	MC14539CP
NTE4539B	SK4539B	TC4539BP	TC4539P
TCG4539B	TM4539B	UPD4539C	

BLOCK-1142

ECG4541B	HD14541B	MC14541BAL	MC14541BCL
MC14541BCP	NTE4541B	SCL4543BC	SK4541B
TCG4541B			

BLOCK-1143

4543BDC	4543BDM	4543BPC	CD4543BD
CD4543BE	CD4543BF	ECG4543B	HD14543B
MC14543BAL	MC14543BCL	MC14543BCP	NTE4543B
SCL4543BE	SK4543B	TCG4543B	

BLOCK-1144

ECG4547B	MC14547BCL	MC14547BCP	NTE4547B
SK4547B	TCG4547B		

BLOCK-1145

ECG4553B	MC14553BCL	MC14553BCP	NTE4553B
SK4553B			

BLOCK-1146

150419	34555PC	4555BDC	4555BDM
4555BPC	CD4555BD	CD4555BE	CD4555BF
ECG4555	ECG4555B	HCC4555BD	HCC4555BF
HCC4555BK	HCF4555BE	HCF4555BF	HD14555B
HEF4555BD	HEF4555BP	MC14555BAL	MC14555BCL
MC14555BCP	MC14555CP	NTE4555B	SCL4555B
SCL4555BC	SCL4555BD	SCL4555BE	SCL4555BF
SCL4555BH	SK4555B	TC4555BP	TCG4555B
TM4555B	UPD4555C		

BLOCK-1147

08SI-UPD4556BC	1270-4847	2119-103-2302	2119-201-3009
34556PC	443-738	4556BDC	4556BDM
4556BPC	CD4556BD	CD4556BE	CD4556BF
ECG4556	ECG4556B	HCC4556BD	HCC4556BF
HCC4556BK	HCF4556BE	HCF4556BF	HD14556B
HE-443-738	HEF4556BD	HEF4556BP	MC14556
MC14556BAL	MC14556BCL	MC14556BCP	MC14556CP
NTE4556B	SCL4556BC	SCL4556BD	SCL4556BE
SCL4556BF	SCL4556BH	SK4556B	TC4556BP
TCG4556B	TM4556B	UPD4556C	

BLOCK-1148

ECG4558B	HD14558B	MC14558BCL	MC14558BCP
NTE4558B	SK4558B	TCG4558B	

BLOCK-1149

ECG4562B	MC14562BCL	MC14562BCP	SK4562B

BLOCK-1150

CD4566	CD4566A	ECG4566B	MC14566BCL
MC14566BCP	NTE4566B	SK4566B	TCG4566B

BLOCK-1151

8000-00047-002	ECG4568B	HD14568B	HD14569B
MC14568	MC14568B	MC14568BCL	MC14568BCP
MC14568CP	SK4568B		

BLOCK-1152

ECG4569B	MC14569BCL	MC14569BCP	NTE4569B
SK4569B			

BLOCK-1153

4583BDC	4583BDM	4583BPC	ECG4583B
MC14583BCL	MC14583BCP	NTE4583B	SK4583B
TCG4583B			

BLOCK-1154

CD4585BD	CD4585BE	CD4585BF	ECG4585B
HEF4585BD	HEF4585BP	MC14585BAL	MC14585BCL
NTE4585B	SK4585B	TCG4585B	

BLOCK-1155

ECG4597B	MC14597BCL	MC14597BCP	NTE4597B
SK4597B			

BLOCK-1156

ECG4598B	MC14598BCL	MC14598BCP	NTE4598B
SK4598B			

BLOCK-1157

0073-0020	02-300574	07-28816-81	0ISS330000A
1001-9735	103-237	1039956300	1103-3669
11119016	1157-8689	1181-7194	1200-2382
1229-8725	1266-2053	1266-2953	1316-7911
1316-7929	1368-2638	1377-2777	142137
144923	145113	146138	146911
147464	148805	150365	158578
159974	167-006A	167-006B	177121
185898	186671	2002600107	2002600149
2002600198	2119-103-0603	2169-401-5808	21A119-045
23115822	23115878	23115922	23316411
2335991	264P24401	264P244010	264P244020
266P01002	35SI-UPC574J	35SZ-UPC574J	36003049
415100574J	4152056300	46-86465-3	46-86474-3
48-90445A97	4800155144	4835-209-87264	48S00155144
48S155144	48X90445A97	48X90456A34	51-90305A59
51X90305A59	530176-1	5302390086	5350611
5635-HZT33	56A113-1	6192140020	6192140620
67-10430-01	67-90430-01	8-759-157-40	8-759-157-41
9-901-371-01	903-10056	903-483	903-885
905-190	DGL5630	DHHZT33	DNMPC574J
DNUPC574J	ECG615	ECG615A	EP16X54
EW16X381	EW84X12	EW84X273	EW84X431
HZT-33	HZT-3301	HZT33	HZT33-01
HZT33-02	HZT33-02-TE	HZT33-02T	HZT33-04
HZT33-04T	HZT33-05	HZT33-10	HZT33-12
HZT33S1	I02190574J	I02990574J	I02I90574J
I031956310	I039956300	IX0037CE	KA33V
L5630	L5631	L5631-AA	MJC574J
MP0574J	MPC-574J	MPC574J	MX-4478
MZT33-01	NTE615	NTE615P	PC574J
RH-IX0037CEZZ	SI-UPC574J	SK9179	SK9976
SN16704C	TAA550-B	TVSHZT33	TVSHZT33S1
TVSMP0574J	TVSMPC574J	TVSUP574J	TVSUPC574J
U574	U574(IC)	U574K	UPC-574
UPC-574J	UPC574	UPC574-J	UPC5742
UPC574I	UPC574J	UPC574J(L)	UPC574J(M)
UPC574J(V)	UPC574J-G	UPC574J-KL	UPC574J-T
UPC574JA	UPC574JAG	UPC574JC	UPC574JK
UPC574JL	UPC574JM	UPC574JT	UPC574K

BLOCK-1157 (CONT.)

UPC574K02L	UPC574V	VHIUPC574JT-1	WEP615/615
X0249CE	X37C	Y15310158	ZTK33
ZTK33	ZTK33A		ZTK-33

BLOCK-1158

6502	6502A	6502AD	6502B
6502C	901435-01	C014377	C014795
C014806	C014806-03	CO14377	CO14806-12
ECG6502	IC-6502	MOS6502	MOS6502A
MPS6502	MPS6502A	NTE6502	P6502B
R6502-40	R6502C	R6502P	SY6502
SYC6502	SYD6502	SYP6502	TCG6502

BLOCK-1159

6507	C010745-03	C010745-12	CO-10745-03
ECG6507	NTE6507	R6507C	R6507P
SYC6507	SYD6507	SYP6507	TCG6507

BLOCK-1160

ECG6508	HM1-6508-5	HM3-6508-5	NMC6508J-5
NMC6508J-9	NMC6508N-5	NMC6508N-9	NTE6508
S6508	S6508-1	S6508A	S6508E
S6508P	TC5508P	TC5508P-1	TCG6508

BLOCK-1161

ECG65101	MN5101	MWS5101D	MWS5101E
NTE65101	SK9829	TCG65101	

BLOCK-1162

6532	901458-01	C0-10750	C010750-03
CO-10750-03	ECG6532	NTE6532	R6532P
SYC6532	SYD6532	SYP6532	TCG6532

BLOCK-1163

6664	ECG6664	M5K4164S-15	M5K4164S-20
MCM6664AL20	MCM6664AP	MCM6664AP15	MCM6664AP20

BLOCK-1164

443-827	6800	ECG6800	HE-443-827
HEPC6800P	MC6800CL	MC6800CP	MC6800P
NTE6800	S6800	SK1901	TCG6800

BLOCK-1165

6802	ECG6802	HEPC6802P	MC6802P
NTE6802	S6802	SK9797	TCG6802

BLOCK-1166

ECG6809	HD6809	MBM2732A25	MC6809L
MC6809P	MC6809S	NTE6809	SK9796

BLOCK-1167

AXX3051	ECG6809E	HD6809EP	HD68A09EP
MC6809E	MC6809EL	MC6809EP	MC6809ES
MX5560	NTE6809E	SK10456	

BLOCK-1168

612192-1	ECG6810	EF6810C	EF6810P
F6810CP	F6810CS	F6810DM	F6810P
F6810S	HEPC6810P	MCM6810CL	MCM6810CP
MCM6810P	NTE6810	S6810P	SK6810
TCG6810			

BLOCK-1169

443-843	6821	AMX4578	ECG6821
F6821CP	F6821DM	F6821P	F6821S
HD6821P	HE-443-843	HEPC6821P	MBL6821N
MC6820	MC6821	MC6821CP	MC6821P
NTE6821	S6821P	SK1900	TCG6821

BLOCK-1170

6850	ECG6850	F6850CP	F6850CS
F6850DL	F6850DM	F6850P	F6850S
HEPC6850P	IC-6850	MC6850CP	MC6850P
NTE6850	S6850	S6850P	SK9702
TCG6850			

BLOCK-1171

ECG6860	HEPC6860P	MC6860L	MC6860P
NTE6860	SK9795		

BLOCK-1172

443-840	ECG6875	HE-443-840	MC6875
MC6875L	NTE6875	SK9794	TCG6875

BLOCK-1173

AMX4261	DS8T26AJ	DS8T26AN	ECG6880
IC-8T26	MB424	MC6880AL	MC6880AP
MC8T26AL	MC8T26AP	N8T26	N8T26A

BLOCK-1173 (CONT.)

N8T26AF	N8T26AN	NT826AB	NTE6880
SK9701	TCG6880	UA8T26	UA8T26D
UA8T26DC	UA8T26P	UA8T26PC	

BLOCK-1174

8T95	ECG6885	IC-8T95	MC6885L
MC6885P	MC8T95L	MC8T95P	N8T95B
N8T95F	N8T95N	N8T96B	NTE6885
SK9793	TCG6885		

BLOCK-1175

ECG6886	MC6886L	MC6886P	MC8T96L
MC8T96P	N8T96F	N8T96N	NTE6886
SK9792	TCG6886		

BLOCK-1176

ECG6887	MC6887L	MC6887P	MC8T97L
MC8T97P	N8T97	N8T97B	N8T97F
N8T97N	NTE6887	SK9700	TCG6887

BLOCK-1177

ECG6888	MC6888L	MC6888P	MC8T98L
MC8T98P	N8T98F	N8T98N	NTE6888
SK9791	TCG6888		

BLOCK-1178

6889	8T28	DS8T28J	DS8T28MJ
DS8T28N	ECG6889	IC-8T28	MC6889L
MC6889P	MC8T28L	MC8T28P	N8T28F
N8T28N	NTE6889	S8T28F	SK9809
TCG6889	UA8T28	UA8T28D	UA8T28DC
UA8T28DM	UA8T28P	UA8T28PC	

BLOCK-1179

221-104	CA3158G	ECG700	HA11231
NTE700	SK9028	TCG700	

BLOCK-1180

ECG7000	SK10469	TDA1905

BLOCK-1181

ECG7001	SK10470	TDA2593	TDA2593N

BLOCK-1182

06300323	1215-8671	1316-0963	221-291
32119-401-010	ECG7002	SK10471	TDA4600-2
TDA4601			

BLOCK-1183

1330-9208	AN5301K	AN5301NK	ECG7003
SK10472			

BLOCK-1184

409-083-0103	46-131772-3	ECG7004	LA7916
SK10473			

BLOCK-1185

ECG7005	HA13421	HA13421A	SK10474
TA7774P			

BLOCK-1186

ECG7006	L292	L292V	SK10475

BLOCK-1187

2914961	ECG7007	HA11510NT	SK10476

BLOCK-1188

06300514	1254-5869	1316-1039	2915192
32119-102-040	ECG7008	LA7629	SK10477

BLOCK-1189

409-117-9607	46-132303-3	AN5322NK	ECG7009
SK10478			

BLOCK-1190

221-105	221-105-01	6125710001	CA3154G
ECG701	NTE701	SC5858	SK9026
TCG701	ULT2260		

BLOCK-1191

06300763	0IT0868000A	11119399	1348-5032
200137	272P461010	33A14668	46-132931-3
46-133344-3	B0383100	B0383111	ECG7010
IX1190CE	NTE7010	RH-IX1190CEZZ	
RH-IX1712CEZZ	SK10479	TA8680AN	TA8680BN
TA8680BNFA-1	TA8680N	X1190CE	

BLOCK-1192

AN7161N	ECG7011	SK10480

BLOCK-1193

46-131577-3	67-33215-01	ECG7012	SK10481
TA7750P			

BLOCK-1194

1346-6347	4151014908	56A223-2	C1490HA
ECG7013	SK10482	UPC1490HA	

BLOCK-1195

ECG7014	EW84X977	IX0633CE	M51376BSP
M51376SP	RH-IX0633CEZZ	SK10483	TVSM51376SP

BLOCK-1196

06300488	06300621	1316-1003	1316-1005
2002400779	221-516	905-940	ECG7015
I06DA365SP	M51365P	M51365SP	SK10484

BLOCK-1197

221-346	CA3189E	ECG7016	LM3189N
SK10485			

BLOCK-1198

ECG7017	SK10486	TDA8175

BLOCK-1199

61250700001	6125070001	6126120001	ECG7018
N6125070001	TDA4505	TDA4505A	TDA4505B

BLOCK-1200

178697	409-135-5506	46-132649-3	ECG7019
MX-4969	NTE15042	SK10488	UPC1486
UPC1486C	VHIUPC1486C	VHIUPC1486C-1	

BLOCK-1201

221-106	221-106-01	CA3145	ECG702
NTE702	SK9025	TCG702	

BLOCK-1202

1231-8358	ECG7020	EW84X625	SK10489
UPC1520CA			

BLOCK-1203

ECG7021

BLOCK-1204

ECG7022

BLOCK-1205

ECG7023

BLOCK-1206

ECG7024

BLOCK-1207

ECG7025

BLOCK-1208

ECG7026

BLOCK-1209

ECG7027

BLOCK-1210

ECG7028

BLOCK-1211

ECG7029

BLOCK-1212

020-1114-007	020-1114-008	022-2844-002	036001
09-308004	09-308013	09-308019	1000-25
1014-25	1016-80	13-1-6	13-10-6
13-11-6	13-9-6	15-26587-1	19-020-079
1C91	1E703E	20B1M	221-31
221-32	2327302	46-5002-1	46-5002-4
46-5002-7	51-10302A01	51-10302A1	51S10302A01
55810-167	56A1	56A1-1	56C1
56C1-1	612020-1	8000-00004-305	8000-00012-041
880-101-00	9005	9006	96-5238-01
96-5238-02	995022	99S022	99SO22
A054-165	C-555A	C555A	CS5995
CXL703A	ECG703	ECG703A	FA6001D
FV5D770339	GEIC-12	HC1000109	HC10001090
HC1000111-0	HEPC6001	IC-5(ELCOM)	IC-7
IC-91	IC5	IC500	IC500(IR)

BLOCK-1212 (CONT.)

IC7	IC91	ICF-1	IP20-0174
LA703E	LM-703LN	LM703	LM703E
LM703L	LM703LH	LM703LN	LS703L
MC1311	NJ703N	NJM-703N	NJM703
NJM703N	NTE703	NTE703A	PA7703
PA7703E	PTC743	QA703E	QTVCM-44
RE-300-IC	RE300-IC	REN703A	RVILM703
SK3157	SL7059	SL7283	SL7308
SL7531	SL7593	SL8020	T0B
T1B	TCG703A	TM703A	TR09005
TR09006	TRO-9005	TRO-9006	TRO9005
TRO9006	TVCM-44	U5D7703312	U5D770331X
U5D7703393	U5D7703394	U5D770339X	U8B770339
U8B770339-825	U8B7703394	U8B770339X	UA-703E
UA703	UA703A	UA703C	UA703CT
UA703E	UA703HC	UA703L	UBB770339X
UPC-555A	UPC555A	USD7703394	UZ1-24
WEP500	WEP500/703A	ZEN601	

BLOCK-1213
ECG7030

BLOCK-1214
ECG7031

BLOCK-1215
ECG7032

BLOCK-1216
ECG7033

BLOCK-1217
ECG7034

BLOCK-1218
ECG7035

BLOCK-1219
ECG7036

BLOCK-1220
ECG7037

BLOCK-1221
ECG7038

BLOCK-1222

221-637-01	32119-110-050	4835 209 88003
4835-209-88003	ECG7039	LA7838

BLOCK-1223

09-308017	118361	119609	122199
221-Z9000	266P00101	266P00102	32-23555-1
32-23555-2	32-23555-3	32-23555-4	417-119
51-10276A01	51S10276A01	57C28	80053
80070	80071	80073	80074
80083	80090	80094	86X0024-001
86X0027-001	CA3013	CA3013S	CA3013T
CA3014	CA3014S	CA3014T	CXL704
ECG704	GE-205	GEIC-205	GEIC-84
HEP591	IC-8B	IC501	M5113
M5113T	MC1314G	NTE704	SK3023
T1A	TA7038M#	TCG704	TM704
TVCM-46	WEP501	WEP501/704	

BLOCK-1224
ECG7040

BLOCK-1225
ECG7041

BLOCK-1226
ECG7042	XRA6418N

BLOCK-1227
ECG7043

BLOCK-1228
AN5900	ECG7044

BLOCK-1229
ECG7045

BLOCK-1230
ECG7046

BLOCK-1231
ECG7047

BLOCK-1232
ECG7048

BLOCK-1233
ECG7049

BLOCK-1234
ECG7050

BLOCK-1235

4835-209-47005	ECG7051	TDA7052	TDA7052/N1
TDA7052N1	VHITDA7052/-1		

BLOCK-1236

ECG7052	TDA7056	VHITDA7056/-1

BLOCK-1237
ECG7053

BLOCK-1238

221-679	ECG7054	LA7670

BLOCK-1239
ECG7055

BLOCK-1240
ECG7056

BLOCK-1241

ECG7057	TA8703S	VHITA8703S/-1

BLOCK-1242
ECG7058

BLOCK-1243
ECG7059

BLOCK-1244

14-2007	14-2007-00	14-2007-02	221-36
442-10	C6089	ECG705	ECG705A
EX39	EX39-X	GEIC-3	HEP-C6089
HEPC6067P	IC-19#	IC-25	IC-502
IC-503	IC502	NTE705A	PTC708
RE302-IC	REN705A	SK3134	SL20721
TCG705A	TM705A	TVCM-1	U8F7737394
U8F7746394	UA737E	UA746E	ULN-2114W
WEP502	WEP502/705A		

BLOCK-1245

13-28-6	1FP70020H	21A101-001	21A101-001(IC)
221-Z9001	57A29-2	57C29-1	57C29-2
80114	80287	CA3041	CA3041P
CA3041T	ECG706	FH70401E	GE-706
GEIC-43	NTE706	R2434-1	SK3101
TCG706	TM706	TVCM-50	WEP2000
WEP2000/706			

BLOCK-1246

AN5160NK	AN5160NK-N	ECG7060

BLOCK-1247
ECG7061

BLOCK-1248
ECG7062

BLOCK-1249
ECG7063

BLOCK-1250
ECG7064

BLOCK-1251
ECG7065

BLOCK-1252

4835 209 87089	4835-209-87089	ECG7066	LA7222
LA7222-TV			

BLOCK-1253

1204-000506	ECG7067	LA7510

BLOCK-1254

B0376795	ECG7068	TA8200AH

BLOCK-1255
ECG7069

BLOCK-1256

14-2007-00B	14-2007-01	14-2007-03	142007-2
16605	1C-502	221-37	221-39
221-41	46-1352-3	746HC	C6067G
C6089(HEP)	CXL705A	ECG707	HEPC6067G
HEPC6089	IC-25(ELCOM)	IC-903	IC12
IC19	IC25(ELCOM)	IC503	IC503(IR)
IC903	LM746H	MC-1328G	MC1326
MC1328G	NTE707	PTC707	SL21441
TA-7053M	TA7053M	TCG707	TM707
TVCM-3	U5E7746394	UA737	UA746
UA746(METAL-CAN)	UA746HC	ULN-2114K	ULN2114K
ULN2114W	WEP503	WEP503/707	ZEN606

BLOCK-1257

ECG7070

BLOCK-1258

ECG7071

BLOCK-1259

ECG7072

BLOCK-1260

ECG7073	STR-S6301	STRS6301
STRS6301-LF953		

BLOCK-1261

ECG7074

BLOCK-1262

07B27	07B2B	07B3B	07B3M
07B3Z	116445	14-2008-01	15-34048-1
15-34408-1	19A11644P1	1C504	2111A
221-34	301-576-3	3135/709	442-28
51-10631A01	612007	612007-1	612007-2
612007-3	740-2002-111	7419	A780
C6062P	C6062P(HEP)	C6082P	C6082P(HEP)
CA2111	CA2111A	CA2111AE	CA2111AQ
CA211AE	CA211AQ	CXL708	ECG708
GEIC-10	GEL2111	GEL2111AL1	GEL2111F1
GEL211AL1	GEL211F1	HEP-C6062P	HEP-C6082P
HEPC6062P	IC-12	IC-504	IC504
LM-2111	LM-2111N	LM2111	LM2111A
LM2111M	LM2111N	LM2111N01	MC-1356P
MC1357	MC1357A	MC1357P	MC1357PQ
N5111	N5111A	NTE708	PTC701
RE-311-IC	RE311-IC	REN708	SK3135
SN76643	SN76643A	SN76643N	SN76653N
T2J	TCG708	TM708	TVCM-4
ULC3037	ULN-2111A	ULN-2111N	ULN-2136A
ULN2111	ULN2111A	ULN2111N	WEP504
WEP504/708	ZEN604		

BLOCK-1263

ECG7085	LA7836	LA7836-TV

BLOCK-1264

051-0022-00	0783Z	149019	15-34452
15-34452-1	1C505	221-Z9002	32980
7311325	7932980	88-B-7-258	88-B-7258
88B7258	916112	A455	CXL709
DM-11	DM-11A	DM-31	DM11
DM31	ECG709	GE-IC11	GEIC-11
GEL2113	GEL2113AL1	GEL2113F1	HEPC6082P
IC-505	IC505	IC505(IR)	LM2113N01
NTE709	PTC703	QTVCM-5	RE301-IC
REN709	SK3491	SN76642	SN76642N
SN76642P	TCG709	TM709	TVCM-5
TVSSN76642N	ULM2111A	ULN-2113A	ULN-2113N
ULN2113	ULN2113A	ULN2113N	WEP505
WEP505/709			

BLOCK-1265

002	126871	129871	133-002
1462434-1	221-Z9003	23119004	23119007
2432	2434	2711904	3633-1
3633-1(IC)	46-1347-3	46-1348-3	46-1356-3
51S10408A01	995042	99S042	B0351500
BO351500	CA3042	CA3042E	CXL710
ECG710	GEIC-89	IC-2(RCA)	IC506
ITTA75902P	LA1342	MC3303J	MC3303L
MC3303N	MC3403P	MC3403J	MC3403L
NTE710	R2432-1	SK3102	T1H
TA7051P	TA75902P	TCG710	TM710
TVCM-51	WEP506	WEP506/710	

BLOCK-1266

1LA7837	ECG7104	I03SD78370	LA7837
QSBLA0ZSY003			

BLOCK-1267

126604	13-27-6	1462445-1	1C6
21A101-002	221-Z9004	23119003	23119005
46-1340-3	46-1346-3	46-5002-6	CA3044
CA3044T	CA3044V	CA3044V1	CXL711
ECG711	GEIC-207	IC-6	IC-6(PHILCO)
IC6	NTE711	R2445-1	R24451
SK3070	TA-7050M	TA-705M	TA7050M
TCG711	TM711	TVCM-53	WEP2030
WEP2030/711			

BLOCK-1268

AN5302K	ECG7111

BLOCK-1269

0031015012	01-121365	02-091128	02-121365
02-165143	02-341125	02-341128	02-341358
02-343065	02-403065	02-537666	05B2B
05B2D	05B2M	06300009	06300069
06300120	07-28777-41	09-308098	09-308100
1001-5972	1001-9693	1010-9940	1025-2712
1025-4172	1025-4712	1045-2761	1061-5972
10655B13	111-4-2020-04600	111-4-2060-0400	
111-4-2060-04000	11119023	1149-0729	1149-6729
1165(GE)	1169-2076	1178	1181-7145
1199-7848	13-29-5	13-29-6	13-35059-1
13-350591	13-64-6	130751	144026
1462516	1462516-001	1462516-1	1463686-1
146896	147438	149252	15-31015-7
15-33201-1	15-33201-2	15-35059	15-35059-1
15-35059-2	15-45300-1	156477	167-001B
167-018A	167-019A	1C-142	1C15
1C26	1MC1358P	1S1699	1S1700
2002110206	2002110269	200X2110-269	2109-103-2006
2119-101-0500	2119-101-4306	2119-102-0309	2165
2169-103-2006	219-103-2006	21A120-008	221-0048
221-48	221-48-01	23119950	2360042
2360201	2360331	2360391	2360501(SIF)
2516	2516-1	26630109	266P30102
266P30103	266P30109	266P32301	266P32302
276-1759	30710	35059-2	3686-1
36SI-CA3065E	370010011	4-2060-04000	4-2060-04600
41SI-HA1128	41SI-LM3065N	4206004000	4206004600
4206104970	4206105470	4206105770	4206105870
46-13145-3	46-13340-3	46-1361-3	46-29-6
46-5002-15	46-5002-26	51-13753A11	51-90305A6
51-90305A61	51-90305A87	51-90305A91	51513753A09
51D70177A07	51S1048A01	51S10655B13	51S13753A07
51S13753A09	51S13753A11	51X90305A61	51X90458A51
5320200050	5329500060	5351351	56A3-1
56D3-1	612005-1	612005-2	612351-1
6123510001	61A030-6	61A0306	6644000100
6644004202	7149A	7221A	73C180475
73C180476-5	73C182186	73SI-MC1358P	7666
8-759-424-00	8-759-425-00	80710	86X0053-001
86X53-1	93ERH-1X1128	99S082-1	A1201-H-75
A1365	A13658E3	AM167001A	AN221
AN240	AN240D	AN240P	AN240PD
AN240PN	AN240S	B0316451	BO316415
C6063P	C6063P(HEP)	C6083P	C6083P(HEP)
CA1365	CA2065	CA3065	CA3065,RCA
CA3065D	CA3065E	CA3065F	CA3065F.C.
CA3065FC	CA3065N	CA3065PC	CA3065RCA
CH3065	CX-093	CX093D	ECG712
EP84X2	EP84X6	EP84X73	ES75X1
ES84X3	ES84X6	EW84X114	EW84X268
EW84X901	EX48	FL274	GE-IC2
GEIC-147	GEIC-148	GEIC-2	GL-3201
GL3201	HA11103	HA11107	HA1124
HA1124D	HA1124DS	HA1124S	HA1124Z
HA1125	HA1125D	HA1125S	HA1128
HA1128E	HA1141	HA1154	HEP-C6063P
HEP-C6083P	IC-15	IC-15(PHILCO)	IC-507
IC15	IC507	IC507(IR)	IMC1358P
ITA7176AP	ITT3065	IX0043CE	IX0243CE
J241222	KA2101	LA1363	LA1363W
LA1365	LA1365N	LAB1363	LCS1008P
LM3065	LM3065N	LM3065N01	LSC1008
LSC1008P	M5143	M5143P	M5144BP
M5144P	MA1125	MA1128	MA3065
MC1358	MC1358P	MC1358PQ	N5065A
NB-07014	NS142	NTC-21	NTE712
PTC726	R2516-1	RE-305-IC	RE305-IC
RE378-IC	REN712	RH-1X0038CEZZ	
RH-1X0043CEZZ	RH-IX0017TAZZ	RH-IX0043CE	
RH-IX0043CEZZ	SC5220P	SC9436P	SC9438P
SE5-0933	SI-CA3065E	SI-LM3065N	SI-MC1358P
SK3072	SL21654	SN16706N	SN7666
SN76664N	SN76665	SN76665N	SN76666
SN76666N	SN76666N-S	SN76666NS	SN7666N
TA5814	TA7071GL	TA7071P	TA7176AP
TA7176APFA	TA7176PFA-1	TCG712	TM712
TVCM-11	TVSJN76664V	TVSN76665	TVSSN76664N

BLOCK-1269 (CONT.)

TVSSN76665	TVSSN7666N	U6A7065354	U7F7065354
U9C7065354	UA3065	UA3065P	UA3065PC
ULN-2165A	ULN-2165N	ULN2165A	ULN2165N
WEP507	WEP507/712	X0043CE	X0243CE
Y15310157	Y15453001		

BLOCK-1270

09-308047	09-308048	1010-9973	1077
1137	14-2010-01	1C508	221-46
221-82	442-33	442-82	7313
73C180837-1	73C180837-2	73C180837-3	995081-1
99S096-1	AMD-746	CA3072	CA3072E
CXL713	ECG713	EP84X3	EP84X9
EX46-X	GEIC-5	GEL2114	HEPC6057P
IC-508	IC508	IC508(IR)	LM3072
LM3072N	LM746	MC1328	MC1329P
N5072A	NS746	NTE713	PTC705
RE-306-IC	RE306-IC	REN713	SC8709P
SK3077	SL20755	SN76246N	TA5912
TM713	TVCM-2	U6A7746394	U7F7746394
UA746(DIP)	UA746C	UA746DC	UA746PC
ULN2114A	WEP508	WEP508/713	ZEN603

BLOCK-1271

02-561410	09-308079	13-42-6	15-37702-1
1C509	2002501006	21A101-017	221-42
221-87	442-57	46-5002-12	56A4-1
56D4-1	612070-1	70B1C	70B1Z
99S094-1	AMD-780	AMD780	C6070P
C6070P(HEP)	CA3070	CA3070E	CA3070G
CXL714	ECG714	EX42-X	EX42X
FX274	GEIC-4	HEP-C6070P	HEPC6070P
IC-12(PHILCO)	IC-509	IC509	IC509(IR)
LM3070A	LM3070N	LM3070N01	MC1370
MC1370P	N5070B	N5070N	NTE714
PTC715	RE-307-IC	RE307-IC	REN714
SK3075	SL21122	SN76242N	TA5649
TA5649A	TCG714	TM714	TVCM-8
U6B7780394	U7F7780394	U9B7780394	UA780
UA780C	UA780DC	UA780PC	ULN-2124A
ULN2124	ULN2124A	ULN2124N	WEP509
WEP509/714	ZEN607		

BLOCK-1272

09-308045	09-308046	13-40-6	15-37703-1
1C13	1C510	21A101-016	221-43
274-69BIC	442-58	46-5002-13	56A5-1
56D5-1	612067	612069-1	69B1M
760048	99S095-1	A781	AMD-781
AMD781	C6071P	C6071P(HEP)	CA3017
CA3071	CA3071E	CXL715	ECG715
EX48-X	F274	FF274	GE-IC6
GEIC-6	HEP-C6071P	HEPC6058P	HEPC6071P
`IC-13	IC-13(PHILCO)	IC-510	IC510
IC510(IR)	LM3071A	LM3071N	LM3071N01
MC1371	MC1371P	MTC1370PQ	N5071A
NTE715	PTC719	RE308-IC	REN715
SC5219P	SK3076	SL21017	SN16726N
SN76243AN	SN76243N	TA-5702B	TA5702
TA5702B	TCG715	TVCM-9	U6A7781394
U7F7781394	U9A7781394	UA781	UA781C
UA781DC	UA781PC	ULN-2127A	ULN-2127N
ULN2127A	ULN2127N	WEP510	WEP510/715
ZEN608			

BLOCK-1273

202922-280	ECG716	GEIC-208	NTE716
TCG716	TM716	WEP2002	WEP2002/716

BLOCK-1274

46-5002-3	51S10432A01	EA1760	ECG717
GEIC-209	IC13	IC3	
IC3(PHILCO)@	PA234	T1M	TCG717
TM717	WEP2003/717		

BLOCK-1275

020-1114-006	09-004	13-26-6	13-36-6
181-000200	183013	1C511	21B1M
21B1Z	221-Z9005	45380	51-10422801
51-10422A01	51-10422A02	51-10437A01	515104221401
51S10422A01	51S10382A	51S10422A	51S10422A01
51S10437A01	612021-1	A6-5002-9	C6074P
C6074P(HEP)	CI-1004	CXL718	DM-14
DM14	ECG718	GE-IC8	GEIC-8
HC1000117-0	HEP-594	HEP-C6074P	HEP-C6094P
HEP594	HEPC6074P	HEPC6094P	IC-17(ELCOM)
IC-511	IC17	IC511	KLH5489
LM1304	LM1304N	LN1304N01	MC1304
MC1304P	MC1304PQ	MP1304P	MP1304PQ
NTE718	PTC709	RE-319-IC	RE319IC
REN718	SC5117	SC5177P	SC5199P

BLOCK-1275 (CONT.)

SC9314P	SK3159	SN76104N	T1J
T1N	TC4007BP	TCG718	TIE
TIJ	TIN	TM718	TR09004
TRO-9004	TRO9004	TVCM-6	U6A7732394
U7F7732394	UA732	UA732PC	ULN-2120A
ULN-2120N	ULN2120A	WEP511	WEP511/718

BLOCK-1276

15-34048	15-53201-1	15-53201-2	569-0886-031
7932367	905-55	A-U2569	BFW70C6065P
C6065P(HEP)	DM-24	DM-54	DM24
DM54	ECG719	GEIC-28	HEP-C6065P
HEPC6065P	PTC711	SC5172P	SC5172PC
SC5172PQ	SN76111	SN76111N	TM719
TVCM-12	ULN-2121A	ULN-2121N	ULN2121A
ULN2121N	WEP2004	WEP2004/719	

BLOCK-1277

001-0036	020-1114-009	051-0012-00	051-0012-11
09-011	1001-0036	15-34379-1	221-Z9006
2327421	2327422	276-1757	4-009
442-9	57A32-22	70622(RCA)	740-9016-105
740622	C6068P	C6068P(HEP)	C6095P
ECG720	FUN14LH026	GE-IC7	GEIC-7
HA11115W	HA1115SW	HA1115	HA1115(W)
HA1115W	HEP-595	HEP-C6068P	HEP-C6095P
HEP595	HEPC6095P	IC-18	IC18
IC512	IC512(IR)	LM1305	LM1305M
LM1305N	LM1305N01	LM3105	MC1305
MC1305P	MC1305P-C	MC1305PC	MC1305PQ
MC1330	NTE720	PC-20008	PC-20018
PTC713	RE-320-IC	RE320-IC	REN720
SC5118P	SK9014	SN76105N	TCG720
TM720	TR09011	TR0911	TR22873700104
TR2327422	TRO-9011	TVCM-7	U6A7729394
U6E7729394	U7F7729394	UA729	UA729CA
ULN-2122A	ULN-2122N	ULN2122A	ULN2122N
WEP512	WEP512/720		

BLOCK-1278

DM8214N	ECG7214	NTE7214

BLOCK-1279

221-90	221-Z9087	3075DC	
354(CHRYSLER)	3596354	3L4-9007-0	4082665
4082665-0001	4082665-0002	4082665-0003	51-10594A01
51-1059A01	51-10619A01	51S10594A01	88-9842F
88-9842R	88-9842RS	88-9842S	92BIC
92BLC	AE907@	C6101P	C6101P(HEP)
CA3075	CA3075E	CI-1003	ECG723
GEIC-15	HEP-C6101P	HEPC6101P	LM3075N01
LM3075N01A	MC-1375P	MC1375P	MC1375PQ
NTE723	PTC723	QTVCM-16	R2516
RE345-IC	REN723	SK3144	SN94018N
T2D	T2G	TCG723	TM723
TVCM-16	UA3075PC	ULN-2129A	ULN-2129N
ULN2129A	WEP2006	WEP2006/723	

BLOCK-1280

09-308002	09-308003	09-308007	0900296
1042-11	123-001	307-008-9-001	57A32-1
8000-00032-031	8000-0004-P090	94BLC	C92-0025-R0
CA3028	CA3028A	CA3028AF	CA3028AS
CA3028AT	CA3028B	CA3028BF	CA3028BS
CA3028T	CA3053	CA3053T	CI-1002
DDEY002001	ECG724	GEIC-86	LM3028AH
LM3028BH	LM3053H	M2002	NTE724
PL-307-047-9-002	PTC744	QTVCM-48	SK3525
TA-7045	TA-7045M	TA7045	TA7045M
TCG724	TM724	TVCM-48	WEP926
WEP926/724			

BLOCK-1281

266P71301	442-615	569-0320-813	612008-2
739	739DC	C6055L(HEP)	CXL725
ECG725	F4-005	GEIC-19	HE-442-615
HEP-C6055L	HEPC6055L	MC1303	MC1303L
MC1303P	NTE725	PTC765	QTVCM-45
RC4739DB	RC4739DP	SK3162	SN76131M
SN76131N	TCG725	TM725	TVCM-43
UA739DC	UA739PC	WEP514	WEP515/725

BLOCK-1282

004795	007-009-00	09-308010	221-Z9025
3596808	3L4-9006-1	3L4-9006-51	4082664-0001
43122874	544-2002-004	60213(1C)	80081
880-102-00	99S049	AE906	CA3011
CA3011T	CA3012	CA3012T	CXL726
ECG726	GEIC-81	GEIC-83	HC1000114-0
MA5113	NTE726	SK3129	TA-7027M
TA7027M	TCG726	TM726	TVCM-45
WEP927	WEP927/726		

BLOCK-1283

15-34005-1	15-34202-1	CA3048	CA3052
CA3052E	CXL727	E2495	ECG727
GEIC-210	LA3148	NTE727	SK3071
TCG727	TM727	TVCM-54	WEP2032
WEP2032/727			

BLOCK-1284

132314	1462559-1	1465188-1	221-Z9007
2559	2559-1	3066DC	86X0055-001
CA3066	CA3066AE	CA3066E	CXL728
ECG728	EP84X7	GEIC-22	NTE728
SK3073	TCG728	TVCM-32	UA3066DC
UA3066PC	ULN2266A	ULN2266N	WEP2008
WEP2008/728			

BLOCK-1285

132315	1462506-1	1462560-001	1462560-1
221-Z9008	2560	2560(J.C.PENNEY)	2560-1
3067DC	86X0056-001	86X0056-002	86X56-2
CA3067	CA3067AE	CA3067E	CXL729
ECG729	GEIC-23	LM3067N	NTE729
PTC776	RE346-IC	REN729	SK3074
TCG729	TM729	TVCM-33	UA3067DC
ULN-2267N	ULN2267	ULN2267A	ULN2267N
ULX-2267A	WEP2009	WEP2009/729	

BLOCK-1286

025-1	025B1C	132313	133600
1462554	1462554-1	1462554-2	1462554-3
1462554-4	15-43638-1	1C311	2025-1
2025-2	2025-3	221-Z9009	2554-1(RCA)
2554-2(RCA)	2554-3	2554-3(RCA)	2554-4
2554-4(RCA)	25B1C	25B1T	25B2C
25B2T	612025-1	612025-2	612025-3
CA3068	CXL730	ECG730	GE-730
H215	HA1133	IC311(ELCOM)	NTE730
SK3143	TCG730	TM730	TVCM-55
WEP2033	WEP2033/730		

BLOCK-1287

16778	21B1AH	221-45	221-45-01
51S10600A	56A46-1	612024-1	612024-3
A7205	C6085P(HEP)	CA3120E	CA3142
CA3142E	CP3120E	CXL731	ECG731
GEIC-13	HEP-C6085P	HEPC6085P	LM1845M
NTE731	PTC731	RE361-IC	REN731
SK3170	SK3204	TCG731	TM731
TVCM-15	ULN-2125A	ULN2125A	WEP2011
WEP2011/731			

BLOCK-1288

2135E	51-10534A01	51-10534A04	51-43684B01
ECG735	GEIC-212	SC5150P	TM735
TVCM-17			

BLOCK-1289

221-89	3596353	51-10672A01	7753
88-9779	88-9779F	CXL734	DM-9
ECG734	ECG736	GEIC-17	NTE736
RE-316-IC	RE316-IC	REN736	SK3163
SN76678	SN76678P	SN94017P	T2V
TCG736	TM736	TVCM-14	TVCM-20
UA753TC	ULN-2209M	ULN2209M	WEP2014
WEP2014/736			

BLOCK-1290

1S-0142	3L4-9007-1	3L4-9007-51	88-9574
AE-907	AE907	AE907-51	CA2136A
CA2136AE	DM-41	ECG737	GEIC-16
LM1841	LS-0142	MC1356P	NTE737
PTC737	RE-315-IC	RE315-IC	REN737
SK3375	SN76669N	TCG737	TM737
TVCM-18	UA2136P	UA2136PC	ULN2136A
WEP2015	WEP2015/737		

BLOCK-1291

1010-9932	10655A03	10655B03	13-56-6
1351	15-39075-1	15-43703-1	221-Z9010
229-1301-44	266P60502	4-2060-09200	4206009200
4206009700	44B1	44B1Z	46-13124-3
51-10655A03	51G10679A03	51M70177A03	51S10655B03
612044-1	66F1551	73C182764-1	73C182764-2
73C182764-4	A1368	C6075P	CA1398E
CXL738	ECG738	EP84X12	GEIC-29
HEP-C6075P	HEPC6075P	LA1368	LA1368BP
LA1368CW	M5190P	M519OP	MC1398
MC1398P	MP5190P	NTE738	PTC741
RE-313-IC	RE313-IC	REN738	SK3167
SN76298	SN76298N	TCG738	TM738
TVCM-27	ULN2298	ULN2298A	ULN2298P
ULX2298	WEP2035	WEP2035/738	Y15390751
Y15437031			

BLOCK-1292

13-57-6	13-67-6	1347	1348
221-Z9032	612072-1	72B1Z	73C182763-1
73C182763-2	C6072P	C6072P(HEP)	ECG739
GEIC-30	HEP-C6072P	IC-14	IC-14(PHILCO)
IC14	MC-1326	MC1324	MC1324P
MC1326P	MC1326PQ	NTE739	PTC739
RE-312-IC	RE312-IC	REN739	SK3235
TM739	TVCM-21	ULN-2224A	WEP2017
WEP2017/739			

BLOCK-1293

006-0000146	007-1695001	09-308022	10302-04
1065-4861	11216-1	138311	1741-0051
1820-0054	19A116180P1	21A120-016	221-Z9075
225A6946-P000	236-0005	2360741	2610786
276-1801	301-576-4	339300	339300-2
3520041-001	373401-1	398-13223-1	43200
435-21026-0A	443-1	4663001D907	51-10611A11
51320000	51320012	51S10611A11	5359031
55001	558875	569-0773-800	569-0882-600
611563	68A9025	733W00042	7400
7400-6A	7400-9A	7400/9N00	7400A
7400DC	7400PC	760011	78A200010P4
8000-00038-004	8000-00042-008	800024-001	9003151
9003151-03	901522-04	930347-3	9N00
9N00DC	9N00PC	A00	C3000P
C7400P	C7400P(HEP)	DDEY030001	DM7400
DM7400J	DM7400N	ECG7400	EP84X13
FD-1073-BF	FJH131	FJH131-7400	FLH101
GE-7400	HC10004110	HD2503	HD2503P
HD7400	HD7400P	HEPC3000L	HEPC3000P
HEPC7400P	HL18998	HL56420	IC-7400
IC-80(ELCOM)	IP20-0205	ITT7400N	J1000-7400
J4-1000	KS20967-L1	LB3000	M53200
M53200P	MB400	MB601	MC53200
MC7400	MC7400F	MC7400L	MC7400N
MC7400P	MIC7400J	MIC7400N	N-7400A
N7400A	N7400F	N7400N	N7401A
N7401F	ND-07001	NTE7400	PA7001/521
QTVCM-500	RE382-IC	REN7400	RS7400
SFC400E	SG7400N	SK7400	SL16793
SMC7400N	SN7400	SN7400A	SN7400J
SN7400N	SN7400N-10	T7400B1	T7400D1
T7400D2	TC7400	TCG7400	TD1401
TD1401P	TD3400A	TD3400AP	TD3400P
TL7400N	TVCM-500	UPB201C	UPB201D
UPB7400C	US7400A	US7400J	WEP7400
WEP7400/7400	XAA104	ZN7400E	ZN7400F

BLOCK-1294

1031-25	138312	1741-0085	1805
19-09973-0	19A116180P2	398-13226-1	43A223007
55032	559507	569-0882-601	7401-6A
7401-9A	7401DC	7401PC	801805
90-68	9004075-03	9N01DC	9N01PC
C3001P	C7401P	C7401P(HEP)	DM7401J
DM7401N	ECG7401	FJH231	FJH231-7401
FJH311	FJH311-7401	FLH201	GEIC-194
HD2509	HD2509P	HD7401P	HD740LP
HEPC3001L	HEPC3001P	HEPC7401P	IC-74(ELCOM)
IP20-0310	ITT7401N	M53201	M53201P
MB416	MC7401F	MC7401L	MC7401P
MIC7401J	MIC7401N	N7401	N7401N
N8875F	N8881F	N8881N	NTE7401
PA7001/526	SFC401E	SK7401	SL16794
SL53431	SN7401	SN7401J	SN7401N
T7401B1	T7401D1	T7401D2	TCG7401
TD3401A	TD3401AP	TL7401N	UPB215C
UPB215D	UPC215C	US7401A	US7401J
WEP7401	WEP7401/7401	ZN7401E	ZN7401F

BLOCK-1295

007-1696201	11207-1	138313	1741-0119
1820-0328	19A116180P3	2610783	276-1811
435-21027-0A	43A223009	443-46	51320001
55002	558876	569-0882-602	611564
68A9027	7012166	7402	7402-6A
7402-9A	7402DC	7402N	7402PC
800080-001	930347-11	9N02	9N02DC
9N02PC	C3002P	C7402P	C7402P(HEP)
DM7402J	DM7402N	ECG7402	F-7402PC
FD-1073-BG	FJH221	FJH221-7402	FLH191
GE-7402	HD2511	HD2511P	HD7402
HD7402P	HE-443-46	HEPC3002P	HEPC7402P
HL19004	IC-7402	IC-82(ELCOM)	ITT7402N
J1000-7402	J4-1002	LB3008	M53202
M53202P	MB417	MC-7402	MC7402F
MC7402L	MC7402P	MIC7402J	MIC7402N
N7402A	N7402F	N7402N	NTE7402
PA7001/525	RS7402	SFC402E	SG7402N
SK7402	SL16795	SMC7402N	SN7402
SN7402J	SN7402N	SN7402N-10	T7402B1

BLOCK-1295 (CONT.)

T7402D1	T7402D2	TCG7402	TD3402A
TD3402AP	TL7402N	UPB232C	UPB232D
UPB7402C	US7402A	US7402J	WEP7402
WEP7402/7402	ZN7402E	ZN7402F	

BLOCK-1296

003542	007-1699401	11274-1	1820-0864
225A6946-P003	236-0006	443-54	51320009
569-0882-603	7403-6A	7403-9A	7403DC
7403N	7403PC	9001570-02	9N03DC
9N03PC	A02(I.C.)	C7403P	DM7403J
DM7403N	ECG7403	EW84X979	FD-1073-BH
FJH291	FJH291-7403	FLH291	HD2528
HD2528P	HD7403	HD7403P	HE-443-54
HEPC7403P	HL19005	IC-7403	IC-83(ELCOM)
ITT7403N	M53203	M53203P	MC7403L
MC7403P	MIC7403J	MIC7403N	N7403A
N7403F	N7403N	NTE7403	SK7403
SN7403	SN7403J	SN7403N	T7403B1
T7403D1	T7403D2	TCG7403	TL7403N
TVSSN7403N	US7403A	US7403J	X420100030
ZN7403E	ZN7403J		

BLOCK-1297

007-1695301	015040/7	11202-1	138314
156-0148-00	1741-0143	1806	1820-0174
1820-0894	19A116180P20	221-Z9076	225A6946-P004
236-0007	276-1802	3520048-001	373404-1
398-13224-1	398-13632-1	43201	435-21028-0A
443-18	4663001A909	508590	51-10611A12
51320002	51S10611A12	55003	569-0882-604
611565	68A9028	733W00064	7404
7404-6A	7404-9A	7404A	7404DC
7404N	7404PC	8000-00028-042	8000-0028-042
800387-001	801806	9001551-02	9001551-03
930347-13	9N04	9N04DC	9N04PC
A03	AMX3655	C3004P	C7404P
DM7404N	EAQ00-02000	ECG7404	F-7404PC
F5404DM	FD-1073-BJ	FJH241	FLH211
GE-7404	HD2522	HD2522P	HD7404
HD7404P	HEPC3004L	HEPC3004P	HEPC7404P
HL19000	HL55862	HL56421	IC-7404
IC-84(ELCOM)	ITT7404N	J1000-7404	J4-1004
KS20967-L2	LB3006	M53204	M53204P
MB418	MC7404L	MC7404P	MIC7404J
MIC7404N	N-7404A	N7404A	N7404F
N7404N	NTE7404	PA7001/527	RH-IX0012PAZZ
RS7404	SK7404	SL16796	SN7404
SN7404J	SN7404N	SN7404N-10	T7404B1
TCG7404	TD3404A	TD3404AP	TL7404N
UPB235D	UPB7404C	US7404A	US7404J
ZN7404E			

BLOCK-1298

10302-06	17184600	1741-0176	19A116180-22
236-0008	398-13225-1	443-642	51320017
55004	569-0882-605	611064	611845
612185-1	7405	7405-6A	7405-9A
7405DC	7405PC	7406DC	930347-15
9N05DC	9N05PC	A04(I.C.)	C7405P
DM7405N	ECG7405	FJH251	FJH321
FLH271	HD2523	HD2523P	HD7405
HD7405P	HE-443-642	HEPC7405P	IC-85(ELCOM)
ID20-0308	IP20-0308	ITT7405N	M53205
M53205P	MC7405L	MC7405P	MIC7405J
MIC7405N	N7405A	N7405F	N7405N
NTE7405	PA7001/528	SK7405	SL16797
SN7405	SN7405A	SN7405J	SN7405N
T7405B1	TCG7405	TD3405A	TD3405AP
TL7405N	UPB236D	UPB7405C	US7405A
US7405J	ZN7405E		

BLOCK-1299

007-1696901	08210006	1607A80	178738
2361251	2420SN07406	373429-1	50254200
51X90458A96	55036	569-0882-606	612159-1
68A9032	7406	7406N	7406PC
800651-001	901522-06	9N06	9N06DC
9N06PC	AMX-3675	AMX-4591	AMX3675
C7406P	DM7406N	EAQ00-07500	EAQ00-07514
ECG7406	FLH481	GE-7406	HD7406
HD7406N	HD7406P	HE-443-967	HEPC7406P
I64D7406P0	IC-104(ELCOM)	ITT7406N	M53206P
MC7406L	MC7406P	N7406A	N7406F
N7406N	NTE7406	RH-IX0038PAZZ	RS7406(IC)
SK7406	SN7406	SN7406J	SN7406N
SN7406N-10	T7406B1	TCG7406	TL7406N
VHIHD7406//-1	X420100060	X440032060	

BLOCK-1300

2420SN07407	373721-1	55035	569-0882-607
7407	7407A	7407DC	7407N
7407PC	800806-001	901522-30	9N07DC
9N07PC	AMX3684	C7407P	DM7407N
EAQ00-15200	ECG7407	FLH491	HD7407
HD7407P	HE-443-1020	HEPC7407P	ITT7407N
M53207P	MC7407L	MC7407P	N7407A
N7407F	N7407N	NTE7407	ON7407N
POSN7407N	RH-IX0116CEZZ	SK7407	SN7407
SN7407J	SN7407N	T7407B1	TCG7407
TL7407N	X420100070		

BLOCK-1301

007-1699301	138315	1741-0200	1820-0870
21A120-017	2473-2109	276-1822	310254
374109-1	43202	435-21029-0A	443-45
51330005	5175460	569-0882-608	73W00124
7408	7408-6A	7408-9A	7408A
7408DC	7408N	7408PC	9003398-03
9003398-04	94152	9N08	9N08DC
9N08PC	C7408P	C7408P(HEP)	DM7408J
DM7408N	ECG7408	FD-1073-BM	FLH381
GE-7408	HD2550	HD2550P	HE-443-45
HEPC7408P	HL55763	IC-102(ELCOM)	IC-7408
IC-84N	ITT7408N	KS21282-L1	L-612099
M53208P	MC7408L	MC7408P	N-7408A
N7408A	N7408F	N7408N	NTE7408
RS7408(IC)	SFC408E	SK7408	SL14971
SL16798	SL17869	SMC7408N	SN7408
SN7408J	SN7408N	SN7408N-10	T7408B1
T7408D1	T7408D2	TCG7408	UPB234C
UPB234D	US7408A	US7408J	WEP7408
WEP7408/7408	ZN7408E	ZN7408J	

BLOCK-1302

11211-1	373423-1	443-89	55021
569-0882-609	7409DC	7409PC	9N09DC
9N09PC	C7409P	DM7409J	DM7409N
ECG7409	FLH391	HD2551	HD2551P
HE-443-89	HEPC7409P	ITT7409N	M53209N
MC7409L	MC7409P	N7409A	N7409F
N7409N	NTE7409	SFC409E	SK7409
SN7409	SN7409J	SN7409N	T7409B1
T7409D1	T7409D2	TCG7409	TD3409A
TD3409AP	TL7409N	TVSSN7409N	US7409A
US7409J	ZN7409E	ZN7409J	

BLOCK-1303

01-119185-01	01-119185-02	01-119185-03	15-3015130-1
276-1725	276-706	301-576-14	380(IC)
380A/C	380B/M	442-686	45394
46-5002-23	576-14	6123540001	83-11
ECG740	ECG740A	GEIC-31	HE-442-686
IC-23	IC-23(PHILCO)	IC-524(ELCOM)	IC-605(ELCOM)
IC23	IC23(PHILCO)	J4-1203	K4-598
LM380	LM380N	M380	MN380
MN380H	NTE740A	RE-321-IC	RE321-IC
REN740	RS380	S380	SK3328
TCA380	TCG740A	TM740A	TVCM-35
ULN-2280A	ULN-2280B	ULN-2280P	ULN2280A
ULN2280B	ULN2280N	ULN2280P	ULN2280Q
WEP2018	WEP2018/740A		

BLOCK-1304

006-0000147	007-1695901	1009-11	10302-03
11200-1	1741-0234	1820-0068	19A116180P4
225A6946-P010	3520042-001	373405-1	435-21030-0A
443-12	4663001A912	49A0005-000	51320003
55005	558877	569-0882-610	611566
68A9030	733W00065	7410	7410-6A
7410-9A	7410DC	7410N	7410PC
8000-00042-009	800023-001	8410	9003091-02
9003091-03	9N10	9N10DC	9N10PC
A05(I.C.)	C3010P	C7410P	DM7410J
DM7410N	ECG7410	FD-1073-BN	FJH121
FJH121-7410	FLH111	GE-7410	HD2507
HD2507P	HD7410	HD7410P	HE-443-12
HEPC3010L	HEPC3010P	HEPC7410P	HL19001
HL56899	IC-7410	IC-86(ELCOM)	ITT7410N
J1000-7410	J4-1010	LB3001	M53210
M53210P	MB401	MB602	MC7410F
MC7410L	MC7410P	MIC7410J	MIC7410N
N7410A	N7410F	N7410N	N8470F
NTE7410	PA7001/520	RLH111	RS7410(IC)
SG7410N	SK7410	SL16801	SMC7410N
SN7410	SN7410J	SN7410N	SN7410N-10
T7410B1	T7410D1	T7410D2	TCG7410
TD1402	TD1402P	TD3410A	TD3410AP
TD3410P	TL7410N	UPB202C	UPB202D
UPB7410C	US7410A	US7410J	ZN7410E
ZN7410F			

BLOCK-1305

006-0000162	1741-1018	2360561	569-0882-707
611872	611900	74107DC	74107PC
9002097-03	9003097-02	9N107DC	9N107PC
DM74107J	DM74107N	ECG74107	FJJ261
FJJ261-74107	FLJ271	HD2530	HD2530P
HD74107	HD74107P	HEPC74107P	ITT74107N
M53307	M53307P	MB410	MC74107P
MIC74107J	MIC74107N	N74107A	N74107F
N74107N	NTE74107	RH-IX0039PAZZ	SFC4107E
SK74107	SL18387	SN74107	SN74107N
T74107B1	TCG74107	TL74107N	UPB74107C
US74107A	US74107J	ZN74107E	

BLOCK-1306

569-0882-709	C74109P	ECG74109	HEPC74109P
ITT74109N	N74109B	N74109F	N74109N
NTE74109	SK74109	SN74109	SN74109J
SN74109N	TCG74109		

BLOCK-1307

007-1696701	007-1699201	138317	1741-1133
1820-0495	443-623	443-623-51644	4915705
5233-7100	569-0773-811	7411-6A	7411-9A
7411DC	7411PC	8411	9001346
9004898-04	9N11DC	9N11PC	C7411P
DM7411J	DM7411N	ECG7411	FLH581
HEPC7411P	HL55660	ITT7411N	N7411A
N7411F	N7411N	NTE7411	SK7411
SL16799	SL17887	TCG7411	US7411A
US7411J			

BLOCK-1308

ECG74110	NTE74110	SK74110	SN74110J
SN74110N	TCG74110		

BLOCK-1309

ECG74111	NTE74111	SK74111	SN74111
SN74111J	SN74111N	TCG74111	

BLOCK-1310

55022	569-0882-612	612091-1	7412DC
7412PC	8412	9N12DC	9N12PC
ECG7412	FLH501	HD7412	HD7412P
ITT7412N	N7412F	N7412N	NTE7412
SK7412	SN7412	SN7412J	SN7412N
SN7412W	TCG7412	TL7412N	ZN7412E
ZN7412J			

BLOCK-1311

11273-1	1741-1257	1820-0261	2610788
443-22	443-32	569-0882-721	72181
74121DC	74121N	74121PC	800491-001
C74121P	DM74121J	DM74121N	ECG74121
FJK101	FJK101-74121	FLK101	HD2543
HD2543P	HD74121	HD74121P	HE-443-22
HEPC74121P	HL19008	IC-106(ELCOM)	ITT74121N
J1000-74121	J4-1121	M53321	M53321P
MC74121P	MIC74121J	MIC74121N	N74121A
N74121F	N74121N	NTE74121	RH-IX0040PAZZ
SFC4121E	SK74121	SN74121	SN74121J
SN74121N	TCG74121	TD34121A	TD34121AP
TL74121N	US74121A	US74121J	ZN74121E

BLOCK-1312

443-23	569-0882-722	74122DC	74122PC
9N122DC	9N122PC	ECG74122	FLK111
HE-443-23	ITT74122N	M53322P	N74122A
N74122F	N74122N	N8T22N	NTE74122
SFC4122E	SK74122	SN74122	SN74122J
SN74122N	TCG74122	TL74122N	ZN74122E

BLOCK-1313

1462	221-Z9086	443-90	569-0882-723
74123	74123DC	74123PC	9N123
9N123DC	9N123PC	AMX3955	C74123P
DM74123	DM74123N	ECG74123	EP84X19
FLK121	GE-74123	HD2561	HD2561P
HD74123P	HE-443-90	HEPC74123P	IC-74123
ITT74123N	M53323P	MB440	N74123A
N74123B	N74123F	N74123N	NTE74123
RE388-IC	REN74123	RH-IX0041PAZZ	RS74123
SFC4123E	SK74123	SN74123	SN74123J
SN74123N	SN74123N-10	TCG74123	TL74123N
UPB74123C	X420101230	ZN74123E	

BLOCK-1314

74125DC	74125N	74125PC	C74125P
DM74125J	DM74125N	ECG74125	HD74125
HD74125P	HEPC74125P	M53325P	N74125F
N74125N	NTE74125	SK74125	SN74125
SN74125J	SN74125N	TCG74125	

BLOCK-1315

443-717	569-0882-726	74126DC	74126PC
C74126P	DM74126N	ECG74126	HD74126
HD74126P	HE-443-717	HEPC74126P	M53326P
N74126F	N74126N	NTE74126	SK74126
SN74126	SN74126J	SN74126N	TCG74126

BLOCK-1316

ECG74128	N74128F	NTE74128	SK74128
SN74128	SN74128J	SN74128N	TCG74128

BLOCK-1317

3531-021-000	443-44	443-44-2854	55027
569-0882-613	601-0100865	7413(IC)	7413DC
7413PC	8413	9N13	9N13DC
9N13PC	C7413P	DM7413J	DM7413N
ECG7413	FLH351	GE-7413	HD2545
HD2545P	HE-443-44	HEPC7413P	IC-103(ELCOM)
ITT7413N	M53213P	MIC7413J	MIC7413N
MIC77413J	N7413A	N7413F	N7413N
NTE7413	RS7413(IC)	SK7413	SN7413
SN7413J	SN7413N	SN7413N-10	TCG7413
TL7413N	UPB7413C		

BLOCK-1318

443-625	569-0882-732	74132DC	74132PC
9N132DC	9N132PC	C74132P	DM74132J
DM74132N	ECG74132	FLH601	HD74132
HD74132P	HE-443-625	HEPC74132P	M53332P
N74132B	N74132F	N74132N	NTE74132
RH-IX0042PAZZ	SK74132	SN74132	SN74132J
SN74132N	TCG74132		

BLOCK-1319

ECG74136	NTE74136	SK9833	SN74136
SN74136J	SN74136N	TCG74136	

BLOCK-1320

007-1696101	443-858	569-0882-614	7414(IC)
7414DC	7414PC	8414	9N14DC
9N14PC	C7414P	DM7414J	DM7414N
ECG7414	HD7414	HD7414P	HE-443-858
HEPC7414P	M53214P	N7414B	N7414F
N7414N	NTE7414	SK7414	SN7414
SN7414J	SN7414N	TCG7414	

BLOCK-1321

3531-023-000	443-17	569-0882-741	74141DC
74141PC	93141DC	93141PC	C74141P
DM74141J	DM74141N	ECG74141	FLL101
HD2558	HD2558P	HE-443-17	HEPC74141P
ITT74141N	N74141B	NTE74141	SFC4141E
SK74141	SN74141	SN74141J	SN74141N
TCG74141	TL74141N	UPB74141D	

BLOCK-1322

ECG74142	NTE74142	SK74142	SN74142
SN74142J	SN74142N		

BLOCK-1323

ECG74143	NTE74143	SK74143	SN74143
SN74143J	SN74143N		

BLOCK-1324

ECG74144	NTE74144	SK74144	SN74144
SN74144J	SN74144N	TCG74144	

BLOCK-1325

007-1696801	1542	1542(GE)	19-130-004
276-1828	443-87	569-0882-745	61SI-SN74145N
74145	74145DC	74145N	74145PC
8000-00028-044	86SI-74145N	93145DC	93145PC
C74145P	DM74145	DM74145N	ECG74145
FLL111T	GE-74145	HD2555	HD2555P
HE-443-87	HEPC74145P	ITT74145N	M53345P
MB443	MC74145P	MIC74145J	MIC74145N
N74145	N74145A	N74145B	N74145F
N74145N	NTE74145	PA7001/593	POSN74145N
RS74145	SK74145	SN74145	SN74145J
SN74145N	SN74145N-10	TCG74145	TL74145N
US74145A	WEP74145	WEP74145/74145	

BLOCK-1326

443-622	C74147P	DM74147J	DM74147N
ECG74147	HE-443-622	HEPC74147P	N74147F
N74147N	NTE74147	SK74147	SN74147
SN74147J	SN74147N	TCG74147	

BLOCK-1327

155708	160753	1741-1042	569-0882-750
733W00183	74150	74150DC	74150PC
93150DC	93150PC	C74150P	DM74150
DM74150N	DM74150	ECG74150	GE-74150
HD2548	HD2548P	HD74150	HD74150P
HEPC74150P	HL55861	ITT74150N	M53350P
MC74150P	MIC74150J	MIC74150N	N74150A
N74150B	N74150F	N74150N	NTE74150
RS74150	SK74150	SN74150	SN74150J
SN74150N	SN74150N-10	TCG74150	TL74150N
UPB2150D	UPB74150C		

BLOCK-1328

1741-1075	443-25	569-0882-751	68A9048
74151ADC	74151APC	74151DC	74151PC
93151DC	93151PC	C74151AP	DM74151
DM74151N	ECG74151	HD2549	HD2549P
HD74151A	HD74151AP	HE-443-25	HEPC74151AP
IC-74151	ITT74151N	M53351	M53351P
MB445	MC74415P	MIC74151J	MIC74151N
N74151A	N74151B	N74151F	N74151N
NTE74151	SK74151	SN74151A	SN74151AJ
SN74151AN	SN74151N	TCG74151	TL74151N
UPB2151D	UPB74151C		

BLOCK-1329

74152DC	74152PC	93152DC	93152PC
ECG74152	ITT74152N	MC74152P	N74152F
N74152N	NTE74152	SN74152A	SN74152N
SN74152S	TCG74152		

BLOCK-1330

160752	569-0882-753	68A9049	74153DC
74153PC	93153DC	93153PC	C74153P
DM74153	DM74153N	ECG74153	HD2564
HD2564P	HEPC74153P	HL55764	ITT74153N
L-612150	M53353P	MC74153P	N74153B
N74153F	N74153N	NTE74153	SK74153
SN74153	SN74153J	SN74153N	TCG74153
TL74153N	UPB74153C	US74153A	

BLOCK-1331

155709	569-0882-754	74154	74154DC
74154PC	9311PC	93154DC	93154PC
BS74154	C74154P	DM74154	DM74154J
DM74154N	ECG74154	GE-74154	HD2580
HD2580P	HE-443-623	HEPC74154P	ITT74154N
M53354P	MIC74154J	MIC74154N	N74154A
N74154F	N74154N	NTE74154	RS74154
SK74154	SN74154	SN74154J	SN74154N
SN74154N-10	TCG74154	TL74154N	UPB2154D
UPB74154C	US74154A	US74154AN	

BLOCK-1332

196374	569-0882-755	74155DC	74155PC
93155DC	93155PC	C74155P	DM74155N
ECG74155	HD74155	HD74155P	HEPC74155P
ITT74155N	M53355P	MC74155P	MIC74155J
MIC74155N	N74155B	N74155F	N74155N
NTE74155	SFC4155E	SK74155	SN74155
SN74155J	SN74155N	TCG74155	TL74155N
UPB74155C	ZN74155E		

BLOCK-1333

74156DC	74156PC	93156DC	93156PC
C74156P	DM74156N	ECG74156	HD74156
HD74156P	HEPC74156P	ITT74156N	M53356P
MC74156P	MIC74156J	MIC74156N	N74156B
N74156F	N74156N	NTE74156	SFC4156E
SK74156	SN74156	SN74156J	SN74156N
TCG74156	TL74156N	UPB74156	UPB74156C

BLOCK-1334

74157DC	74157PC	9322DM	9322PC
C74157P	DM74157N	ECG74157	HEPC74157P
N74157F	N74157N	NTE74157	SK74157
SN74157	SN74157J	SN74157N	TCG74157
UPB74157C			

BLOCK-1335

74158	ECG74158	N74158F	N74158N
NTE74158	SK74158	TCG74158	

BLOCK-1336

148742	1820-1172	443-73	4915702
51X90458A95	55024	569-0882-616	7416
7416DC	7416N	7416PC	800-0016
8000016	8416	9N16DC	9N16PC
C7416P	DM7416N	ECG7416	FLH481T
GE-7416	HD7416	HD7416P	HE-443-73

BLOCK-1336 (CONT.)

HEPC7416P	IC-105(ELCOM)	ITT7416N	KS10969-L5
L-612158	M53216P	MC7416L	MC7416P
MX-4313	N7416A	N7416F	N7416N
NTE7416	POSN7416N	SK7416	SN7416
SN7416J	SN7416N	T7416B1	TCG7416
TL7416N	TVSM53216P	X420100160	

BLOCK-1337

569-0882-760	74160DC	74160PC	9310PC
C74160AP	DM74160AN	DM74160N	ECG74160
FLJ401	HD74160	HD74160P	HEPC74160AP
ITT74160N	M53360P	MB451	N74160B
N74160F	N74160N	NTE74160	SK74160
SN74160	SN74160J	SN74160N	TCG74160

BLOCK-1338

569-0882-761	74161DC	74161PC	8000-00038-003
8000-00045-004	916138	9316/74161	DM74161AN
DM74161N	ECG74161	F9316PC	FLJ411
HD74161	HD74161P	HEPC74161AP	IP20-0209
ITT74161N	M53361P	MB450	MX3139
N74161B	N74161F	N74161N	NTE74161
RE386-IC	REN74161	SK74161	SL1622C
SL1626C	SN74161	SN74161J	SN74161N
TCG74161	TVCM-506	UPB2161D	UPB74161C
ZN74161E			

BLOCK-1339

74162PC	C74162AP	DDEY015001	DM74162
DM74162AN	DM74162N	ECG74162	FLJ421
HD74162	HD74162P	HEPC74162AP	ITT74162N
M53362P	N74162B	N74162F	N74162N
NTE74162	SK74162	SN74162	SN74162J
SN74162N	TCG74162	TL74162N	WEP74162
WEP74162/74162			

BLOCK-1340

74163DC	74163N	74163PC	C74163AP
DM74163AN	DM74163N	ECG74163	FLJ431
HD74163	HD74163P	HEPC74163AP	ITT74163N
M53363P	N74163B	N74163F	N74163N
NTE74163	SK74163	SN74163	SN74163J
SN74163N	TCG74163	TL74163N	ZN74163E

BLOCK-1341

1820-1064	569-0882-764	74164DC	74164PC
93164DC	93164DC	C74164P	DM74164N
DM8570N	ECG74164	FLJ441	HD74164
HD74164P	HEPC74164P	ITT74164N	L-612161
M53364P	MC74164AP	N74164A	N74164F
N74164N	NTE74164	SK74164	SN74164
SN74164J	SN74164N	TCG74164	TD3503A
TD3503AP	TL74164N	UPB74164C	ZN74164E

BLOCK-1342

293118	373413-1	569-0882-765	74165
74165DC	74165PC	93165DC	93165PC
C74165P	DM74165N	DM8590N	ECG74165
FLJ451	HEPC74165P	IC-74165	ITT74165N
M53365P	MC74165P	N74165B	N74165F
N74165N	NTE74165	SK74165	SN74165
SN74165J	SN74165N	TCG74165	TL74165N
ZN74165E			

BLOCK-1343

1348A12H01	1741-0291	443-72	569-0882-617
7417	7417DC	7417N	7417NA
7417PC	8000017	8417	901522-01
9N17DC	9N17PC	C7417P	DM7417N
ECG7417	F-7417PC	F7417PC	FLH491T
GE-7417	HD7417	HD7417P	HD7471P
HE-443-72	HEPC7417P	IC-7417	ITT7417N
M53217P	MC7417L	MC7417P	N7417A
N7417F	N7417N	NTE7417	POSN7417N
PTSN7417N	SK7417	SN7417	SN7417J
SN7417N	TCG7417	VHIHD7417//-1	X420100170

BLOCK-1344

51320006	587-033	74170DC	74170PC
93170DC	93170PC	C74170P	DM74170N
ECG74170	FLQ131	HD2540	HD2540P
HEPC74170P	M53370P	MB460	N74170B
N74170F	N74170N	NTE74170	SK74170
SN74170	SN74170J	SN74170N	UPB2170D
UPB74170D			

BLOCK-1345

C74173P	DM74173N	ECG74173	HEPC74173P
N74173N	NTE74173	SK74173	SN74173
SN74173J	SN74173N	TCG74173	

BLOCK-1346

569-0882-774	74174	74174DC	74174PC
93174DC	93174PC	C74174P	DM74174
DM74174N	ECG74174	FLJ531	HD74174
HD74174P	HEPC74174P	IC-74174	ITT74174N
M53374P	N74174	N74174F	N74174N
NTE74174	SK74174	SN74174	SN74174J
SN74174N	TCG74170	TCG74174	ZN74174E

BLOCK-1347

569-0882-775	74175	74175DC	74175PC
93175DC	93175PC	C74175P	DM74175N
ECG74175	FLJ541	HD74175	HD74175P
HEPC74175P	IC-74175	ITT74175N	M53375P
N74175B	N74175F	N74175N	NTE74175
SK74175	SN74175	SN74175J	SN74175N
TCG74175	UPB74175C		

BLOCK-1348

569-0882-776	74176DC	74176PC	8280
93176DC	93176PC	C74176P	DM74176J
DM74176N	DM8280J	DM8280N	ECG74176
HEPC74176P	MC74176P	N8280F	N8280N
NTE74176	SK74146	SK74176	SN74176
SN74176J	SN74176N	TCG74176	

BLOCK-1349

569-0882-777	74177	74177DC	74177PC
901522-03	93177DC	93177PC	C74177P
DM74177J	DM74177N	DM8281J	DM8281N
ECG74177	HD74177P	HEPC74177P	MC74177P
N8281F	N8281N	NTE74177	SK74177
SN74177J	SN74177N	TCG74177	

BLOCK-1350

74178DC	74178PC	93178DC	93178PC
ECG74178	N74178A	N8270F	N8270N
NTE74178	SK74178	SN74178	SN74178J
SN74178N	TCG74178		

BLOCK-1351

373708-1	569-0882-779	74179DC	74179PC
93179DC	93179PC	ECG74179	N74179B
N74179F	N8271N	NTE74179	SK74179
SN74179	SN74179J	SN74179N	TCG74179

BLOCK-1352

007-1697701	1741-1349	569-0882-780	611901
74180DC	74180PC	93180DC	93180PC
C74180P	DM74180J	DM74180N	ECG74180
FLH421	HD74180	HD74180P	HEPC74180P
ITT74180N	M53380	M53380P	MB447
MC74180P	MIC74180J	MIC74180N	N74180A
N74180F	N74180N	NTE74180	SFC4180E
SK74180	SN74180	SN74180J	SN74180N
T74180B1	TCG74180	TL74180N	UPB2180D
UPB74180C	US74180A	US74180J	

BLOCK-1353

569-0882-781	74181DC	74181PC	9341DC
9341PC	C74181P	DM74181J	DM74181N
ECG74181	FLH401	HD2547	HD2547P
HEPC74181P	ITT74181N	M53381P	MB458
MC74181P	N74181F	N74181N	NTE74181
SFC4181E	SK74181	SN74181	SN74181J
SN74181N	TCG74181	TL74181N	UPB2181P
UPB74181D			

BLOCK-1354

569-0882-782	74182DC	74182PC	9342DC
9942PC	C74182P	DM74182N	ECG74182
FLH411	HD2562	HD2562P	HEPC74182P
ITT74182N	M53382P	MB459	MC74182P
N74182B	N74182F	N74182N	NTE74182
SFC4182E	SK74182	SN74182	SN74182J
SN74182N	TCG74182	TL74182N	UPB2182D
UPB74182C			

BLOCK-1355

3531-022-000	569-0882-790	74190DC	74190PC
93190DC	93190PC	C74190P	DM74190
DM74190J	DM74190N	ECG74190	FLJ201
HD74190	HD74190P	HEPC74190P	ITT74190N
M53390P	MB457	N74190B	N74190F
N74190N	NTE74190	SK74190	SN74190
SN74190J	SN74190N	TCG74190	TL74190N

BLOCK-1356

569-0882-791	74191DC	74191PC	93191DC
93191PC	C74191P	DM74191J	DM74191N
ECG74191	FLJ211	HD74191	HD74191P
HEPC74191P	M53391P	MB456	N74191B
N74191F	N74191N	NTE74191	SK74191
SN74191J	SN74191N	TCG74191	TL74191N
ZN74191E			

BLOCK-1357

007-1698301	155710	160764	443-66
569-0882-792	611731	74192	74192DC
74192PC	9360DC	9360PC	C74192P
DM74192J	DM74192N	DM8560	DM8560J
DM8560N	ECG74192	FLJ241	GE-74192
HD2541	HD2541P	HE-443-66	HEPC74192P
HL56429	ITT74192N	M53392	M53392P
MC74192P	N74192A	N74192B	N74192F
N74192N	N74192P	NTE74192	RS74192
SFC4192E	SK74192	SN74192	SN74192J
SN74192N	SN74192N-10	TCG74192	TD34192A
TD34192AP	TL74192N	UPB2192D	UPB74192C
ZN74192E			

BLOCK-1358

007-1698401	11204-1	138320	43C216447
43C216447P1	443-162	443-612	569-0882-793
611730	7012142-03	733W00076	74193
74193DC	74193PC	8000-00042-011	800386-001
9366DC	9366PC	C74193P	DM74193
DM74193J	DM74193N	DM8551N	DM8563J
DM8563N	ECG74193	FLJ251	GE-74193
HD2542	HD2542P	HE-443-612	HEPC74193P
HL56430	ITT74193N	M53393	M53393P
MC74193P	N74193A	N74193B	N74193F
N74193N	NTE74193	RS74193	SFC4193E
SK74193	SN74193	SN74193J	SN74193N
SN74193N-10	T74193B1	TCG74193	TL74193N
UPB2193D	UPB74193C	ZN74193E	

BLOCK-1359

74195DC	74195N	74195PC	C74195P
DM74195N	ECG74195	FLJ561	HEPC74195P
ITT74195N	MC74195P	N74195B	N74195F
N74195N	NTE74195	SK74195	SN74195
SN74195J	SN74195N	TCG74195	UPB2195D
UPB74195C			

BLOCK-1360

443-628	569-0882-796	74196	74196DC
74196PC	93196DC	93196PC	C74196P
DM74196N	ECG74196	FLJ381	GE-74196
HD2572	HD2572P	HE-443-628	HEPC74196P
N74196A	N8290F	N8290N	NTE74196
RS74196	SK74196	SN74196	SN74196J
SN74196N	SN74196N-10	TCG74196	TL74196N

BLOCK-1361

569-0882-797	74197DC	74197PC	93197DC
93197PC	C74197P	DM74197N	ECG74197
FLJ391	HD2573	HD2573P	HEPC74197P
N8291F	NTE74197	SK74197	SN74197
SN74197J	SN74197N	TCG74197	TL74197N

BLOCK-1362

74198DC	74198PC	93198DC	93198PC
C74198P	DM74198N	ECG74198	FLJ311
HD74198	HD74198P	HEPC74198P	M53398P
MB455	N74198F	N74198N	NTE74198
SK74198	SN74198	SN74198J	SN74198N
TCG74198	TL74198N	UPB2198D	UPB74198D

BLOCK-1363

74199DC	74199PC	93199DC	93199PC
C74199P	DM74199N	ECG74199	FLJ321
HD74199	HD74199P	HEPC74199P	M53399P
N74199F	N74199N	NTE74199	SK74199
SN74199	SN74199J	SN74199N	TCG74199
TL74199N			

BLOCK-1364

221-Z9033	229-1301-40	612076-1	612076-2
612076-4	703639-1	703639-2	ECG742
EP84X36	GEIC-213	NTE742	RE-314-IC
RE314-IC	REN742	SK3453	TCG742
TM742	TVCM-28	ULN-2211P	ULN2211
ULN2211A	ULN2211B	WEP2020	WEP2020/742

BLOCK-1365

007-1695101	10302-02	11205-1	138318
1741-0325	1820-0069	19A116180P5	221-Z9077
225A6946-P020	373406-1	435-21033-0A	443-2
49A0006-000	51320004	55006	558878
569-0882-620	56A11-1	611567	68A9033
733W00043	7420(IC)	7420-6A	7420-9A
7420DC	7420PC	800020-001	885540026-3
9004076	9004076-03	9004076-04	930347-1
930347-10	930347-2	9N20	9N20DC
9N20PC	A06(I.C.)	C3020P	C7420P
DM7420J	DM7420N	ECG7420	EP84X11
FD-1073-BR	FJH111	FJH111-7420	FLH121
GE-7420	HD2504	HD2504P	HD7420
HD7420P	HEPC3020L	HEPC3020P	HEPC7420P
HL19003	HL56422	IC-7420	IC-87(ELCOM)
IC7420	ITT7420N	LB3002	M53220
M53220P	MB402	MB603	MC7420F
MC7420L	MC7420P	MIC7420J	MIC7420N
N7420A	N7420F	N7420N	NTE7420
PA7001/519	RS7420	SFC420E	SG7420N
SK7420	SL16800	SMC7420N	SN7420
SN7420J	SN7420N	SN7420N-10	T7420B1
T7420D1	T7420D2	TCG7420	TD1403
TD1403P	TD3420A	TD3420AP	TD3420P
TL7420N	UPB203C	UPB203D	UPB7420C
US7420A	US7420J	ZN7420E	ZN7420F

BLOCK-1366

138319	7421	7421DC	7421PC
9N21DC	9N21PC	ECG7421	ITT7421N
N7421A	N7421F	N7421N	NTE7421
SK9837	TCG7421		

BLOCK-1367

55023	7422DC	7422PC	ECG7422
FLH611	HEPC7422P	NTE7422	SK7422
SN7422	SN7422J	SN7422N	TCG7422

BLOCK-1368

74221N	ECG74221	N74221F	N74221N
NTE74221	SK74221	SN74221	SN74221J
SN74221N	TCG74221		

BLOCK-1369

146732	68A9034	7423DC	7423PC
9N23DC	9N23PC	C7423P	DM7423J
DM7423N	DM7423W	ECG7423	FLH511
HEPC7423P	N7423F	N7423N	NTE7423
SK7423	SN7423	SN7423J	SN7423N
TCG7423	TL7423N		

BLOCK-1370

ECG74249	NTE74249	SN74249	SN74249J
SN74249N	TCG74249		

BLOCK-1371

C74251P	DM74251N	DM8121N	ECG74251
HEPC74251P	NTE74251	SK74251	SN74251
SN74251J	SN74251N	TCG74251	

BLOCK-1372

443-65	569-0882-627	7427(IC)	7427DC
7427PC	9N27	9N27DC	9N27PC
C7427P	DM7427J	DM7427N	DM7427W
ECG7427	FLH621	GE-7427	HD7427
HD7427P	HEPC7427P	M53227P	N7427A
N7427F	N7427N	NTE7427	RS7427
SK7427	SN7427	SN7427J	SN7427N
SN7427N-10	TCG7427	US7427A	ZN7427E
ZN7427J			

BLOCK-1373

ECG7428	ITT7428N	MIC7428J	MIC7428N
N7428F	N7428N	NTE7428	SK7428
SN7428	SN7428J	SN7428N	T7428B1
T7428D1	T7428D2	TCG7428	ZN7428E
ZN7428J			

BLOCK-1374

74290PC	ECG74290	NTE74290	SK74290
SN74290	SN74290J	SN74290N	TCG74290

BLOCK-1375

74293PC	ECG74293	NTE74293	SK74293
SN74293	SN74293J	SN74293N	TCG74293

BLOCK-1376

1000-23	15-36446-1	221-Z9011	36446-2
3L4-9004-1	3L4-9004-3	3L4-9004-4	3L4-9004-51
3L4-9004-6	45390	51-10711A01	51-10711A02

BLOCK-1376 (CONT.)

760522-0008	905-00160	90500160	95286
AE-904-03	AE304	AE904	AE904-04
AE904-4	AE904-51	AE904-6	CA758E
DM-44	DM44	ECG743	GEIC-214
GEIC-32	LM1800	LM1800A	LM1800N
NTE743	PTC729	RE-317-IC	RE317-IC
REN743	SK3172	SN76116N	T3A
T3A-1	T3A-2	TCG743	TM743
TVCM-29	UA758	UA758B	UA758BA
UA758N	UA758PC	ULN-2244A	ULN2244
ULN2244A	ULN2244N	ULN2244R	ULX2244
WEP2021	WEP2021/743		

BLOCK-1377

10302-01	11208-1	1741-0416	1820-0070
3520047-001	373407-1	43A223015	443-3
4663001A915	51320011	55007	558879
569-0882-630	68A9035	7430-6A	7430-9A
7430DC	7430PC	800021-001	9003642-03
9N30DC	9N30PC	A08(I.C.)	C3030P
C7430P	DM7430J	DM7430N	ECG7430
F7430PC	FD-1073-BS	FJH101	FJH101-7430
FLH131	GE-7430	HD2508	HD2508P
HD7430	HD7430P	HE-443-3	HEPC3030L
HEPC3030P	HEPC7430P	HL19013	HL56423
IC-71(ELCOM)	IC-7430	ITT7430N	LB3003
M5310	M5310P	M53230	M53230P
MB403	MB604	MC7430F	MC7430L
MC7430P	MIC7430J	MIC7430N	N7430A
N7430F	N7430N	NTE7430	PA7001/518
SFC430E	SG7430N	SK7430	SL16802
SMC7430N	SN7430	SN7430J	SN7430N
T7430B1	T7430D1	T7430D2	TCG7430
TD1404	TD1404P	TD3430A	TD3430AP
TD3430P	TL7430N	UPB204C	UPB204D
UPB7430C	US7430A	US7430J	XAA105
ZN7430E	ZN7430F		

BLOCK-1378

138381	146742	4511424	569-0882-632
733W00126	7432(IC)	7432DC	7432PC
9N32	9N32DC	9N32PC	C7432P
DM7432J	DM7432N	ECG7432	FLH631
GE-7432	HD7432	HD7432P	HEPC7432P
ITT7432N	KS21282-L3	L-612107	N7432A
N7432F	N7432N	NTE7432	RS7432
SK7432	SN7432	SN7432J	SN7432N
SN7432N-10	TCG7432	US7432A	US7432J
ZN7432E	ZN7432J		

BLOCK-1379

569-0882-633	ECG7433	ITT7433N	N7433F
N7433N	NTE7433	SK7433	SN7433
SN7433J	SN7433N	T7433B1	T7433D1
T7433D2	TCG7433		

BLOCK-1380

C74365P	DM74365J	DM74365N	ECG74365
HEPC74365P	N74365AF	N74365AN	NTE74365
SK74365	SN74365A	SN74365AJ	SN74365AN
SN74365J	SN74365N	TCG74365	

BLOCK-1381

C74366P	DM74366J	DM74366N	ECG74366
HEPC74366P	N74366AF	N74366AN	N74367AF
NTE74366	SK74366	SN74366A	SN74366AJ
SN74366AN	SN74366J	SN74366N	TCG74366

BLOCK-1382

C74367P	DM74367J	DM74367N	ECG74367
HEPC74367P	N74367AN	NTE74367	SK74367
SN74367A	SN74367AJ	SN74367AN	SN74367J
SN74367N	TCG74367		

BLOCK-1383

C74368P	DM74368J	DM74368N	ECG74368
HEPC74368P	N74368AF	N74368AN	NTE74368
SK74368	SN74368A	SN74368AJ	SN74368AN
SN74368J	SN74368N	TCG74368	

BLOCK-1384

007-1696301	1741-0440	569-0882-637	68A9036
7437(IC)	7437DC	7437PC	9N37DC
9N37PC	C7437P	DM7437J	DM7437N
DM7437W	EAQ00-11300	ECG7437	FLH531
HD2552	HD2552P	HEPC7437P	ITT7437N
M53237P	MB435	MC7437F	MC7437L
MC7437P	N7437A	N7437F	N7437N
NTE7437	SFC437E	SK7437	SN7437
SN7437J	SN7437N	TCG7437	TL7437N
UPB237D	UPB7437C	ZN7437E	ZN7437J

BLOCK-1385

182510	443-77	569-0882-638	610002-038
68A9037	7438	7438DC	7438N
7438PC	9003234-04	9N38DC	AMX-3683
AMX3683	DM7438J	DM7438N	DM7438W
ECG7438	FLH541	HD2544	HD2544P
HD7438P	HE-443-77	HEPC7438P	HL56424
IC-7438	ITT7438N	M53238P	M5323BP
MB433	MB434	MC7438F	MC7438L
MC7438P	N7438A	N7438F	N7438N
NTE7438	ON7438N	SFC438E	SK7438
SL14972	SN7438	SN7438J	SN7438N
TCG7438	TL7438N	UPB238D	UPB7438C
US7438A	US7438J	X420100380	ZN7438E
ZN7438J			

BLOCK-1386

7439DC	7439PC	9N39DC	9N39PC
ECG7439	N7439A	N7439F	N7439N
NTE7439	SK7439	SN7439N	TCG7439

BLOCK-1387

74390	ECG74390	IC-74390	NTE74390
SK74390	SN74390	SN74390J	SN74390N
TCG74390			

BLOCK-1388

ECG74393	NTE74393	SK9835	SN74393
SN74393J	SN74393N	TCG74393	

BLOCK-1389

1-235-444-11	15-36995-1	221-107	720DC
88-9302	88-9302RS	88-9302S	889302
CA3123	CA3123E	DM-20	DM-32
DM32	ECG744	GEIC-215	GEIC-24
LM1820	LM1820A	LM1820N	LM3820N
NTE744	PTC734	RE362-IC	REN744
SK3171	SN76635N	TCG744	TM744
TVCM-19	UA720DC	ULN-2137A	ULN2137A
WEP2022	WEP2022/744		

BLOCK-1390

007-1695701	1-590-1261	11214-1	1741-0473
1820-1111	19A116180-7	19A116180P7	3520044-001
373408-1	40-065-19-016	463984-1	4663001A911
51320005	55034	558880	569-0882-640
611568	7440-6A	7440-9A	7440PC
800022-001	9001549-02	930347-5	9N40DC
9N40PC	A09	C3040P	C7440P
DM7440J	DM7440N	ECG7440	FD-1073-BU
FJH141	FJH141-7440	FLH141	HD2501
HD2501P	HD7440	HD7440P	HEPC3040L
HEPC3040P	HEPC7440P	HL19011	IC-88(ELCOM)
ITT7440N	LB3009	M53240	M53240P
MB404	MB605	MC7440F	MC7440L
MC7440P	MIC7440J	MIC7440N	N7440A
N7440F	N7440N	N8455F	NTE7440
PA7001/522	SFC440E	SG7440N	SK7440
SL16803	SMC7440N	SN7440	SN7440J
SN7440N	T7440B1	T7440D1	T7440D2
TCG7440	TD1405	TD1405P	TD3440A
TD3440AP	TD3440P	TL7440N	UPB205C
UPB205D	UPB7440C	US7440A	US7440J
XAA106	ZN7440E	ZN7440F	

BLOCK-1391

007-1697801	1741-1190	443-35	569-0882-641
7441	7441-6A	7441-9A	7441DC
7441PC	9315DC	9315PC	C3041P
C7441AP	DM7441AJ	DM7441AN	DM7441N
ECG7441	FJL101	GE-7441	HEPC3041L
HEPC3401L	HEPC7441AP	HEPC7441P	M53241
M53241P	MC7441AL	MC7441AP	MIC7441AJ
MIC7441AN	N7441B	N7441F	NTE7441
RS7441	SK7441	SN7441N-10	T7441AB1
T7441AD1	T7441AD2	TCG7441	TD3441A
TD3441AP	UPB218C	UPB218D	US7441A
USN-7441B	ZN7441AE		

BLOCK-1392

1741-1224	1807	1820-0214	43A223029P1
443-53	55008	569-0882-642	611071
7442ADC	7442APC	7442DC	7442PC
800385-001	801807	9352DC	9352PC
B00	DM7442J	DM7442N	ECG7442
FJH261	FJH261-7442	FLH281	HD2536
HD2536P	HD7442A	HD7442AP	HE-443-53
HEPC7442P	HL19009	IC-90(ELCOM)	ITT7442N
M53242	M53242P	M5362	M5362P
MB442	MC7442L	MC7442P	MIC7442J
MIC7442N	N7442A	N7442BA	N7442BF
N7442F	N7442N	NTE7442	SFC442E

BLOCK-1392 (CONT.)

SK7442	SN7442A	SN7442AJ	SN7442AN
SN7442N	T7442B1	T7442D1	T7442D2
TCG7442	TD3442A	TD3442AP	TL7442N
UPB227D	UPB7442C	US7442A	ZN7442E
ZN7442J	ZNM442E		

BLOCK-1393

158624	7443	7443ADC	7443APC
7443DC	7443PC	9353DC	9353PC
ECG7443	FLH361	HD2537	HD2537P
HD7443A	HD7443AP	ITT7443N	M5323P
M53243	M53243P	MC7443L	MC7443P
MIC7443J	MIC7443N	N7443A	N7443F
N7443N	NTE7443	SK7443	SN7443A
SN7443AJ	SN7443AN	SN7443N	T7443B1
T7443D1	T7443D2	TCG7443	TL7443N
US7443A			

BLOCK-1394

158625	158626	7444	7444ADC
7444APC	7444DC	7444PC	9354DC
9354PC	ECG7444	FLH371	HD252538
HD2538P	HD7444A	HD7444AP	ITT7444N
M53244	M53244P	MC7444L	MC7444P
MIC7444J	MIC7444N	N7444B	N7444F
N7444N	NTE7444	SK7444	SN7444A
SN7444AJ	SN7444AN	SN7444N	T7444B1
T7444D1	T7444D2	TCG7444	TL7444N
US7444A			

BLOCK-1395

56A13-1	612188-1	7445	7445DC
7445PC	9345DC	9345PC	C7445P
DM7445J	DM7445N	DM7445W	ECG7445
FLL111	HD2531	HD2531P	HEPC7445P
ITT7445N	M53245P	MC7445L	MC7445P
MIC7445J	MIC7445N	N7445B	N7445F
N7445N	NTE7445	POSN7445N	SK7445
SN7445	SN7445J	SN7445N	TCG7445
TL7445N	UPB7445C	US7445A	

BLOCK-1396

7446	7446ADC	7446APC	7446PC
9317BDC	9317BDM	9317BPC	9317CDC
9317CDM	9357A	9357ADC	9357APC
C7446AP	DM7446AJ	DM7446AN	ECG7446
FLL121	FLL121U	HEPC7446AP	ITT7446AN
MC7446L	MC7446P	N7446AF	N7446AN
N7446B	NTE7446	SK7446	SN7446A
SN7446AJ	SN7446AN	TCG7446	TL7446AN
US7446A			

BLOCK-1397

276-1805	443-36	55009	569-0882-647
612100-1	7447	7447ADC	7447APC
7447BDC	7447BPC	7447DC	7447PC
9357B	9357BDC	9357BPC	C7447AP
DM7447AJ	DM7447AN	DM7447N	ECG7447
FLL121T	FLL121V	GE-7447	HD2532
HD2532P	HE-443-36	HEPC7447AP	HEPC7447P
IC-101(ELCOM)	ITT7447AN	J1000-7447	J4-1047
M53247	M53247P	MC7447L	MC7447P
N7447A	N7447AF	N7447B	N7447F
NTE7447	RS7447	SK7447	SN7447A
SN7447AJ	SN7447AN	SN7447J	SN7447N
SN7447N-10	TCG7447	TD3447A	TD3447AP
TL7447AN	TVCM-503	UPB2047D	UPB7447C
US7447A			

BLOCK-1398

7448	7448DC	7448PC	9307PC
9358	9358DC	9358PC	C7448P
DM7448J	DM7448N	DM7448W	ECG7448
FLH551	GE-7448	HEPC7448P	ITT7448N
M53248	M53248A	M53248P	MC7448L
MC7448P	N7448A	N7448B	N7448F
N7448N	NTE7448	RS7448	SK7448
SN7448	SN7448J	SN7448N	SN7448N-10
TCG7448	TL7448N	US7448A	

BLOCK-1399

ECG74490	NTE74490	SN74490	SN74490J
SN74490N	TCG74490		

BLOCK-1400

221-Z9088	ECG745	GEIC-216	MC1306P
NTE745	SK3276	TCG745	TM745
WEP2036	WEP2036/745		

BLOCK-1401

007-1698901	19A116180-8	19A116180P8	225A6946-P050
443-15	477-0415-002	51-10611A15	51S10611A15
55010	559509	569-0882-650	611569
68A9038	7450-6A	7450-9A	7450DC
7450PC	800026-001	9003096-02	9003096-03
9N50DC	9N50PC	C3050P	C7450P
DM7450J	DM7450N	ECG7450	FJH151
FJH151-7450	FLH151	HD2506	HD2506P
HD7450	HD7450P	HE-443-15	HEPC3050L
HEPC3050P	HEPC7450P	IC-75(ELCOM)	ITT7450N
LB3004	M53250	M53250P	M5352
M5352P	MB405	MB606	MC7450F
MC7450L	MC7450P	MIC7450J	MIC7450N
N7450A	N7450F	N7450N	N8840F
NTE7450	SFC450E	SG7450N	SK7450
SL16804	SN7450	SN7450J	SN7450N
T7450B1	T7450D1	T7450D2	TCG7450
TD1406	TD1406P	TD3450A	TD3450AP
TD3450P	TL7450N	UPB206C	UPB206D
UPB7450C	US7450A	US7450J	ZN7450E
ZN7450F			

BLOCK-1402

10302-05	1741-0564	1820-0063	3520045-001
373715-1	435-21034-0A	51320016	569-0882-651
7451	7451-6A	7451-9A	7451DC
7451PC	930347-12	9N51	9N51DC
9N51PC	A12(I.C.)	C7451P	DM7451J
DM7451N	ECG7451	FD-1073-BW	FJH161
FJH161-7451	FLH161	GE-7451	HD2505
HD2505P	HD7451	HD7451P	HEPC7451P
IC-91(ELCOM)	ITT7451N	MC7451F	MC7451L
MC7451P	MIC7451J	MIC7451N	N7451A
N7451F	N7451N	NTE7451	PA7001/523
RS7451	SFC451E	SG7451N	SK7451
SMC7451N	SN7451	SN7451J	SN7451N
SN7451N-10	T7451B1	T7451D1	T7451D2
TCG7451	TD1419	TD1419P	TD3451A
TD3451AP	TD3451P	TL7451N	UPB207C
UPB207D	UPB7451C	US7451A	US7451J
ZN7451E	ZN7451F		

BLOCK-1403

1741-0598	373714-1	477-0417-002	559613
569-0882-653	611570	7453-6A	7453-9A
7453DC	7453PC	800025-001	9004896-04
9N53DC	9N53PC	C7453P	DM7453J
DM7453N	ECG7453	FJH171	FJH171-7453
FLH171	HD2512	HD2512P	HD7453
HD7453P	HEPC7453P	IC-92(ELCOM)	ITT7453N
M53253	M53253P	MB411	MC7453F
MC7453L	MC7453P	MIC7453J	MIC7453N
N7453A	N7453F	N7453N	NTE7453
PA7001/524	SFC453E	SG7453N	SK7453
SL16805	SN7453	SN7453J	SN7453N
T7453B1	T7453D1	T7453D2	TCG7453
TL7453N	UPB208D	UPB7453C	US7453A
US7453J	ZN7453E	ZN7453F	

BLOCK-1404

11233-2	1741-0622	19A116180-11	19A116180P11
373714-2	55029	569-0882-654	7454-6A
7454-9A	7454DC	7454PC	9004360-03
930347-6	9N54DC	9N54PC	C7454P
DM7454J	DM7454N	ECG7454	FJH181
FJH181-7454	FLH181	HD2514	HD2514P
HD7454	HD7454P	HEPC7454P	IC-93(ELCOM)
ITT7454N	MC7454F	MC7454L	MC7454P
MIC7454J	MIC7454N	N7454A	N7454F
N7454N	NTE7454	PA7001/539	SFC454E
SG7454N	SK7454	SN7454	SN7454J
SN7454N	T7454B1	T7454D1	T7454D2
TCG7454	TL7454N	UPB209D	UPB7454C
US7454A	US7454J	ZN7454E	ZN7454F

BLOCK-1405

02-217660	051-0021-00	09A02	1C17
221-Z9028	276-1758	2906-005	442-18
51-03009A02	76600P	8000-00053-004	8000-00058-006
8000-0058-006	9010	916111	C6059P
C6059P(HEP)	ECG746	GEIC-217	HEP-C6059P
HEPC6059P	IC-17	IC-17(PHILCO)	IC-34
MC1350	MC1350P	MX3336	NTE746
PTC1101	PTC1623	QTVCM-59	SK3234
SN76008	SN76600	SN766008	SN76600P
TCG746	TR09010	TVCM-59	WEP2024
WEP2024/746			

BLOCK-1406

1741-0689	374110-1	398-8972-1	559510
569-0882-660	68A9040	7460	7460-6A
7460-9A	7460DC	7460PC	9N60DC

BLOCK-1406 (CONT.)

9N60PC	C7460P	DM7460J	DM7460N
ECG7460	FJY101	FJY101-7460	HD2502
HD2502P	HD7460	HD7460P	HEPC7460P
ITT7460N	LB3005	M5304	M5304P
M53260	M53260P	MB406	MB607
MC7460F	MC7460L	MC7460P	MIC7460J
MIC7460N	N7460A	N7460F	N7460N
NTE7460	PA7001/533	SFC460E	SG7460N
SK7460	SN7460	SN7460J	SN7460N
T7460B1	T7460D1	T7460D2	TCG7460
TD1407	TD1407P	TD3460A	TD3460AP
TD3460P	TL7460N	UPB210C	UPB210D
UPB7460C	US7460A	US7460J	ZN7460E
ZN7460F			

BLOCK-1407

15-39060-1	15-43251-1	15-43251-2	167-014A
1C18	1C202	21A101-015	21A120-002
221-Z9029	229-1301-42	23119014	266P10201
266P10202	AM167014A	C1330P	C6079P
CXL747	ECG747	GEIC-218	HEP-C6079P
IC-18(PHILCO)	ITT1330	M5169	M5169P
M5176P	MC1330A	MC1330A1P	MC1330P
MC133OP	NTE747	RE-310-IC	RE310-IC
REN747	SC9430P	SK3279	TCG747
TDA1330P	TM747	TVCM60	WEP2037
WEP2037/747	Y15432511	Y15432512	

BLOCK-1408

11206-1	19-05589	19A116180-P13	373411-1
7470-6A	7470-9A	7470DC	7470PC
9N70DC	9N70PC	C7470P	DM7470J
DM7470N	ECG7470	FJJ101	FLJ101
HD2539	HD2539P	HEPC7470P	ITT7470N
M53270	M53270P	M5375	M5375P
MC7470L	MC7470P	MIC7470J	MIC7470N
N7470A	N7470F	N7470N	NTE7470
PA7001/530	SK7470	SN7470	SN7470J
SN7470N	TCG7470	TL7470N	UPB211D
US7470A	US7470J	ZN7470E	ZN7470F

BLOCK-1409

006-0000151	075-046270	1121-1	1820-0304
3520050-001	373424-1	443-4	51320008
569-0882-672	56A14-1	669A471H01	7472
7472-6A	7472-9A	7472DC	7472PC
930347-4	9N72DC	9N72PC	C7472P
DM7472J	DM7472N	ECG7472	FJJ111
FJJ111-7472	FLJ111	HD2529	HD2529P
HD7472	HD7472P	HE-443-4	HEPC7472P
IC-94(ELCOM)	ITT7472N	M53272	M53272P
M5372	M5372P	MB407	MB609
MC7472F	MC7472L	MC7472P	MIC7472J
MIC7472N	N7472A	N7472F	N7472N
NTE7472	PA7001/532	SFC472E	SK7472
SN7472	SN7472J	SN7472N	T7472B1
T7472D1	T7472D2	TCG7472	TD1408
TD1408P	TD3472A	TD3472AP	TL7472N
UPB211C	UPB212D	US7472A	US7472J
ZN7472E	ZN7472F		

BLOCK-1410

1030-25	138403	1820-0075	19A116180P15
2002010072	236-0009	276-1803	3520043-001
43205	435-23006-0A	43A223025	443-5
477-0412-004	49A0002-000	558881	569-0882-673
7473	7473-6A	7473-9A	7473DC
7473F	7473PC	9004093-03	930347-7
9N73	9N73DC	9N73PC	B0075660
C3073P	C7473P	C7473P(HEP)	DM7373N
DM7473J	DM7473N	EA33X8385	EA33X8386
ECG7473	FJJ121	FJJ121-7473	FLJ121
GE-7473	HD2515	HD2515P	HD7473AP
HD7473P	HE-443-5	HEPC3073L	HEPC3073P
HEPC7473P	HL19002	IC-95(ELCOM)	ITT7473N
M53273	M53273P	M58472P	M7641
MC7473F	MC7473L	MC7473P	MIC7473J
MIC7473N	N-7473A	N7473	N7473A
N7473F	N7473N	NTE7473	PA7001/531
RE383-IC	RE385-IC	REN7473	RS7473
SFC473E	SK7473	SL16806	SL17242
SMC7473N	SN7473	SN7473J	SN7473N
SN7473N-10	T7473B1	T7473D1	T7473D2
TCG7473	TD1409	TD1409P	TD3473A
TD3473AP	TL7473N	UPB212C	UPB225C
UPB225D	UPB7473C	US7473A	US7473J
WEP7473	WEP7473/7473	XAA107	ZN7473E
ZN7473F			

BLOCK-1411

007-1699801	075-045037	11213-1	169403
1820-0077	19A116180P16	276-1818	2868536-1

REPLACEMENTS

BLOCK-1411 (CONT.)

3520046-001	373409-1	435-23007-0A	43A223026P1
443-6	4663001A905	49A0012-000	55011
558882	569-0882-674	611571	612092-1
733W00067	742725	7474	7474-6A
7474-9A	7474/9N74	7474DC	7474F
7474N	7474PC	8000-00038-007	800400-001
8474	881916	90-39	90-67
9003152	9003152-01	9N74	9N74DC
9N74PC	A15	C7474P	C7474P(HEP)
DM7474J	DM7474M	DM7474N	ECG7474
F7474PC	FJJ131	FJJ131-7474	FLJ141
GE-7474	GEIC-193	HD2510	HD2510P
HD7474	HD7474P	HE-443-6	HEPC7474P
HL18999	HL56425	IC-7474	IC-97(ELCOM)
IP20-0206	IP20-0316	M53274	M53274P
M5374	M5374P	MB420	MC7474P
MIC7474J	MIC7474N	N7474A	N7474F
N7474N	N8828F	NTE7474	PA7001/529
QTVCM-502	RE384-IC	REN7474	RS7474
SFC474E	SK7474	SL16807	SMC7474N
SN-7474	SN7474	SN7474J	SN7474M
SN7474N	SN7474N-10	T7474B1	T7474D1
T7474D2	TCG7474	TD3474A	TD3474AP
TD3474P	TL7474N	TVCM-502	UPB213C
UPB214C	UPB214D	UPB7474C	UPC213C
US7474A	US7474J	WEP7474	WEP7474/7474
YEAM53274P	ZN7474E	ZN7474F	

BLOCK-1412

1741-0747	1820-0301	36188000	373713-1
40-065-19-027	443-13	49A0000	51-10611A16
51S10611A16	569-0882-675	611065	68A9041
7011203-02	7011203-03	7475	7475-6A
7475-9A	7475DC	7475PC	800382-001
8475	9375DC	9375PC	B01
C3075P	C7475P	DM7475J	DM7475N
DM7485N	ECG7475	FJJ181	FJJ181-7475
FLJ151	GE-7475	HD2517	HD2517P
HD7475	HD7475P	HE-443-13	HEPC3075L
HEPC3075P	HEPC7475P	HL19012	IC-7475
IC-96(ELCOM)	IC7475	ITT7475N	J4-1075
M53275	M53275P	MC7475L	MC7475P
MIC7475J	MIC7475N	N7475A	N7475B
N7475F	N7475N	NTE7475	RS7475
SFC475E	SK7475	SM63	SM73(I.C.)
SMC7475N	SN7475	SN7475J	SN7475N
SN7475N-10	T7475B1	TCG7475	TD3475A
TD3475AP	TL7475N	UPB217C	UPB217D
US7475A	US7475J	ZN7475E	

BLOCK-1413

1348A14H01	276-1813	373414-1	43A223028
443-16	55012	569-0882-676	611870
68A9042	72185	733W00069	7476
7476-6A	7476-9A	7476DC	7476N
7476PC	760015	8476	9004300-03
9N76	9N76DC	9N76PC	B02
C7476P	C7476P(HEP)	DM7476J	DM7476N
ECG7476	F7476PC	FJJ191	FJJ191-7476
FLJ131	GE-7476	HD2516	HD2516P
HE-443-16	HEPC7476P	HL19010	IC-7476
IC-99(ELCOM)	ITT7476N	J1000-7476	J4-1076
M53276	M53276P	MC7476L	MC7476P
MIC7476J	MIC7476N	N7476A	N7476B
N7476F	N7476N	NTE7476	
RH-1X0005PAZZ	RH-IX0005PAZZ	RS7476	SFC476E
SK7476	SL16808	SMC7476N	SN7476
SN7476B	SN7476J	SN7476N	SN7476N-10
T7476B1	T7476D1	T7476D2	TCG7476
TL7476N	UPB224C	UPB224D	UPB7476C
US7476A	US7476J	WEP7476	WEP7476/7476
XAA108	ZN7476E		

BLOCK-1414

569-0882-680	7480DC	7480PC	9380DC
9380PC	ECG7480	FJH191	FLH221
ITT7480N	LB3060	M53280	M53280P
MB408	MC7480L	MC7480P	N7480A
N7480F	N7480N	N8268F	NTE7480
SK7480	SN7480	SN7480J	SN7480N
TCG7480	TD3480A	TD3480AP	TL7480N
UPB2080D	UPB7480C	US7480A	USN-7480A

BLOCK-1415

7483ADC	7483APC	DM7483J	ECG7481
FLQ111	MC4004	MIC7481J	MIC7481N
NTE7481	SK7481	SN7481A	SN7481AN
SN7483AJ	T7481B1	TCG7481	TL7481N

BLOCK-1416

2666293	373716-1	569-0882-682	7012132
7482-6A	7482-9A	7482DC	7482PC
9382DC	9382PC	ECG7482	FJH201
FLH231	HD2513	HD2513P	IC-500(ELCOM)
ITT7482N	MIC7482J	MIC7482N	NTE7482
SN7482	SN7482J	SN7482N	TCG7482
TD3482P	TL7482N	US7482A	USN-7482A

BLOCK-1417

11209-1	1741-0770	2666294-1	373412-1
43A223033P1	7012133-02	7483DC	7483PC
800383-001	9383DC	9383PC	C7483P
DM7483N	ECG7483	FJH211	FLH241
HD2535	HD2535P	HEPC7483P	HL56842
ITT7483N	KS20969-L2	M53283	M53283P
MC7483L	MC7483P	MIC7483J	MIC7483N
N7483B	N7483F	N7483N	NTE7483
SK7483	SL18386	SN7483A	SN7483AN
T7483B1	TCG7483	TD3483P	TL7483N
UPB230D	US7483A	USN-7483B	

BLOCK-1418

007-1696001	569-0882-685	7485	7485DC
7485PC	9385DC	9385PC	C7485P
ECG7485	FLH431	GE-7485	HD7485
HD7485P	HEPC7485P	HL56426	M53285P
MB448	N7485A	N7485B	N7485F
N7485N	NTE7485	RS7485	SFC485E
SK7485	SN7485	SN7485J	SN7485N
SN7485N-10	TCG7485	TL7485N	UPB2085D
UPB7485C			

BLOCK-1419

1741-0804	19A116180-18	19A116180P18	339486
373410-1	40-065-19-029	435-21035-0A	443-698
51320018	569-0882-686	611066	611844
733W00068	7486	7486DC	7486PC
9001349-02	9001349-03	9N86	9N86DC
9N86PC	C7486P	D7486J	DM7486J
DM7486N	ECG7486	FD-1073-CA	FLH341
GE-7486	HD2526	HD2526P	HD7486
HD7486P	HEPC7486P	HL19014	ITT7486N
KS20967-L3	M53286	M53286P	MB449
MC7486F	MC7486L	MC7486P	MIC7486J
MIC7486N	N7486A	N7486F	N7486N
NTE7486	RS7486	SFC486E	SK7486
SN7486	SN7486J	SN7486N	SN7486N-10
T7486B1	T7486D1	T7486D2	TCG7486
TCG7486A	TL7486N	UPB2086D	UPB7486
UPB7486C	US7486A	US7486J	ZN7486E
ZN7486J			

BLOCK-1420

569-0882-689	7489DC	7489PC	DM7489N
ECG7489	FLQ101	M53289P	N7489B
NTE7489	SK7489	SN7489	SN7489N
TCG7489	TL7489N		

BLOCK-1421

10655A13	10655B02	1C8	221-Z9012
23119025	23119999	40306312	46-1366-3
46-5002-8	51-10655B02	51-10679A13	51D70177A02
51D70177B02	51G10679A13	51M70177B02	544-2003-002
56A9-1	56D9-1	62737654	70177A02
70177B02	C6060P	C6060P(HEP)	ECG748
ECG748A	GEIC-115	GEIC-183	GEIC-26
HEP-C6060P	HEP6060P	IC-26	IC-26(ELCOM)
IC-8	IC-8B#	IC8	IC8B
J241270	LM1351#	LM1351N	MC1351
MC1351N	MC1351P	NTE748	PTC745
RE331-IC	REN748	RH-1X0001TAZZ	RH-IX0001TAZZ
SC5245P	SK3236	SN76651N	T1F
TA7072P	TA7073AP	TA7073P	TCG748
TM748	TVCM-41	WEP2026	WEP2026/748

BLOCK-1422

03SI-MC1352P	09-308090	1042-7938	10655A05
10655C05	1138-9442	1151	1174
1278	1278(GE)	167-013A	18SI-SN76650N
2002120012	200X2120-012	21A120-001	221-Z9013
23119019	266P10101	266P10103	4-2061-05170
40306400	4206105170	46-1365-3	46-5002-28
51-90305A50	51-90305A90	51M70177A05	51S10655C05
72SI-TA7074P	8098	AM167013A	AN158
AN238S	AN7074P	C6076P	C6076P(HEP)
CA1352	CA1352E	CA3152E	CXL749
ECG749	EP84X8	GEIC-97	HEP-C6076P
HEPC6076P	IC-28	ITT1352	J241219
LK1352P	LKB52P	M1358P	M5183
M5183P	MC1352	MC1352P	MPC1355C
NTE749	PTC746	RE-304-IC	RE304-IC

BLOCK-1422 (CONT.)

REN749	RH-1X0004CEZZ	RH-IX00004CEZZ	RH-IX0004CE
RH-IX0004CEZZ	SC9431P	SI-MC1352P	SI-SN76650N
SI-TA7074P	SK3168	SN76650	SN76650N
SN76675	TA7074GL	TA7074P	TA7074PGL
TCG749	TM749	TVCM-42	TVSMPC1355C
UPC1355C	WEP2027	WEP2027/749	

BLOCK-1423

1-000-099-00	102005	16088	1808
1820-0055	19-130-005	19A116180-24	276-1808
373427-1	443-629	443-7	443-7-16088
558883	569-0882-690	56A15-1	611572
7012167-02	733W00039	7490	7490-6A
7490-9A	7490ADC	7490APC	7490DC
7490PC	760013	801808	8530
905-102	9390DC	9390PC	A17(I.C.)
C3800P	C7490AP	C7490P(HEP)	DDEY029001
DM7490	DM7490AN	DM7490J	DM7490N
ECG7490	FJJ141	FJJ141-7490	FLJ161
GE-7490	HD2519	HD2519P	HD7490A
HD7490AP	HE-443-629	HE-443-7	HEPC3800L
HEPC3800P	HEPC7490AP	HEPC7490P	HL19015
IC-98(ELCOM)	ITT7490N	J1000-7490	J4-1090
LB3150	M53290	M53290P	MC7490AP
MC7490F	MC7490L	MC7490P	MIC7490J
MIC7490N	N7490A	N7490F	N7490N
NTE7490	QTVCM-505	RS7490	SK7490
SMC7490N	SN7490A	SN7490AJ	SN7490AN
SN7490N	SN7490N-10	T7490B1	T7490D1
T7490D2	TCG7490	TD3490A	TD3490AP
TD3490P	TL7490N	TVCM-505	UPB219C
UPB219D	US7490A	US7490J	USN-7490A
WEP7490	WEP7490/7490	XAA109	ZN7490E
ZN7490F			

BLOCK-1424

19A116180-25	202920-150	7491ADC	7491APC
7491DC	7491PC	930347-9	9391DC
9391PC	C7491AP	DM7491AN	ECG7491
FLJ221	HD2524	HD2524P	HEPC7491AP
ITT7491AN	LB3175	M53291	M53291P
MB454	MC7491AL	MC7491AP	MIC7491AJ
MIC7491AN	N7491A	N7491AF	N7491AN
N7491F	NTE7491	SN7491A	SN7491AJ
SN7491AN	TCG7491	TD3491A	TD3491AP
TL7491N	UPB2091D	UPB7491C	US7491A
USN-7491A			

BLOCK-1425

1000100	1000100-000	19A116180-27	373712-1
436-10010-0A	443-34	569-0882-692	611573
7492	7492-6A	7492-9A	7492ADC
7492APC	7492DC	7492PC	9003445-03
9392DC	9392PC	C3801P	C7492AP
DM7492AN	DM7492J	DM7492N	ECG7492
FJJ152	FJJ251-7492	FLJ171	GE-7492
HD2521	HD2521P	HD7492A	HD7492AP
HE-443-34	HEPC3801L	HEPC3801P	HEPC7492AP
HEPC7492P	IC-100(ELCOM)	IC-7492	ITT7492N
J1000-7492	J4-1092	KS20969-L3	M53292
M53292P	MC7492F	MC7492L	MC7492P
MCT492P	MIC7492J	MIC7492N	N7492A
N7492F	N7492N	NTE7492	RS7492
SFC492E	SK7492	SL16809	SN74921-10
SN7492A	SN7492AJ	SN7492AN	SN7492N
SN7492N-10	T7492B1	T7492D1	T7492D2
TCG7492	TD3492A	TD3492AP	TD3492P
TL7492N	TM7492	UPB222C	UPB222D
US7492A	US7492J	USN-7492A	ZN7492AE
ZN7492AJ	ZN7492E	ZN7492F	

BLOCK-1426

1000101	1000101-000	11203-1	128C830H05
1741-0895	1820-0099	2002010061	221-Z9078
225A6946-P093	23119960	373718-1	43A223034P1
443-640	5359271	558885	569-0882-693
733W00133	7493	7493-6A	7493-9A
7493/9393	7493ADC	7493APC	7493DC
7493PC	8000-00028-043	8000-00038-006	9001345-02
9393	9393DC	9393PC	A18(I.C.)
C7493AP	CD73/187/72	DM7493AN	DM7493J
DM7493N	ECG7493A	EP84X18	FJJ211
FJJ211-7493	FLJ181	HD2520	HD2520P
HD7493A	HD7493AD	HD7493AP	HD7493P
HE-443-640	HEPC7493AP	HL19006	IC-81(ELCOM)
IP20-0207	IP20-0315	ITT7493N	M53293
M53293P	MC7493F	MC7493L	MC7493P
MIC7493J	MIC7493N	N7493A	N7493F
N7493N	NTE7493A	REN7493A	SK7493
SMC7493N	SN7493	SN7493A	SN7493AJ
SN7493AN	SN7493N	T7493B1	T7493D1
T7493D2	TCG7493A	TD3493A	TD3493AP

BLOCK-1426 (CONT.)

TD3493BP	TD3493P	TL7493N	TM7493A
TVCM-504	UPB223C	UPB223D	US7493A
US7493J	USN-7493A	WEP2279	
WEP2279/7493A	ZN7493AE	ZN7493AJ	ZN7493E
ZN7493F			

BLOCK-1427

7494DC	7494PC	9394DC	9394PC
ECG7494	FLJ231	HD2533	HD2533P
ITT7494N	MC7494L	MC7494P	MIC7494J
MIC7494N	N7494B	N7494F	N7494N
NTE7494	SK7494	SN7494	SN7494J
SN7494N	TCG7494	TL7494N	US7494A
ZN7494E			

BLOCK-1428

1741-0952	19A116180-29	225A6976-P095	40-065-19-030
43A223030	443-680	52335500	7495ADC
7495APC	7495DC	7495PC	9395DC
9395PC	A19(I.C.)	C7495P	DM7495N
ECG7495	FLJ191	HD2534	HD2534P
HE-443-680	HEPC7495P	HL56427	IC-79(ELCOM)
ITT7495AN	KS20969-L4	M53295	M53295P
M5395	M5395P	MB453	MC7495L
MC7495P	MIC7495J	MIC7495N	N7495A
N7495AF	N7495AN	N7495F	N7495N
NTE7495	SK7495	SN7495AJ	SN7495AN
TCG7495	TD3495A	TD3495AP	TL7495AN
UPB226C	UPB226D	US7495A	ZN7495AE

BLOCK-1429

11276-1	2610784	373428-1	436-10011-0A
43A223031	569-0882-696	7496DC	7496PC
9003420-03	9396DC	9396PC	C7496P
DM7496J	DM7496N	ECG7496	FLJ261
HD2546	HD2546P	HD7496	HD7496P
HEPC7496P	ITT7496N	M53296	M53296P
MB452	MC7496L	MC7496P	MIC7496J
N7496B	N7496F	N7496N	NTE7496
SK7496	SL16810	SN7496J	SN7496N
TCG7496	TL7496N	US7496A	US74H76J
ZN7496E			

BLOCK-1430

9377	DM7497J	ECG7497	FLJ331
NTE7497	SK7497	SN7497	SN7497J
SN7497N	TCG7497	TL7497N	

BLOCK-1431

09-308074	3102026	CM0SMM74C00N	ECG74C00
HD1-74C00	HD9-74C00	MM74C00	MM74C00N
NTE74C00	SK9806	TC7400BP	TCG74C00

BLOCK-1432

ECG74C02	HD1-74C02	HD9-74C02	MM74C02
MM74C02N	NTE74C02	SK74C02	TCG74C02

BLOCK-1433

3102027	74C04	905-233	ECG74C04
MM74C04	MM74C04J	MM74C04N	NTE74C04
SK74C04	TCG74C04		

BLOCK-1434

ECG74C08	HD1-74C08	HD9-74C08	MM74C08
MM74C08N	NTE74C08	SK9785	TCG74C08

BLOCK-1435

BZX74C10	ECG74C10	HD1-74C10	HD9-74C10
MM74C10	MM74C10N	NTE74C10	SK74C10
TCG74C10			

BLOCK-1436

ECG74C107	MM74C107N	NTE74C107	SK74C107
TCG74C107			

BLOCK-1437

ECG74C14	MM74C14J	MM74C14N	NTE74C14
SK9805	TCG74C14		

BLOCK-1438

ECG74C151	MM74C151J	MM74C151N	NTE74C151
SK74C151			

BLOCK-1439

ECG74C154	HD1-74C154	HD9-74C154	MM74C154J
MM74C154N	NTE74C154	SK74C154	TCG74C154

BLOCK-1440
ECG74C157	MM74C157J	MM74C157N	NTE74C157
SK74C157	TCG74C157		

BLOCK-1441
ECG74C161	HD1-74C161	HD9-74C161	MM74C161N
NTE74C161	SK74C161	TCG74C161	

BLOCK-1442
ECG74C164	MM74C164N	NTE74C164	SK74C164
TCG74C164			

BLOCK-1443
ECG74C173	MM74C173N	NTE74C173	SK74C173
TCG74C173			

BLOCK-1444
ECG74C174	MM74C174N	NTE74C174	SK74C174
TCG74C174			

BLOCK-1445
ECG74C175	MM74C175N	NTE74C175	SK74C175
TCG74C175			

BLOCK-1446
ECG74C192	HD1-74C192	HD9-74C192	MM74C192N
NTE74C192	SK74C192	TCG74C192	

BLOCK-1447
ECG74C193	HD1-74C193	HD9-74C193	MM74C193N
NTE74C193	SK74C193	TCG74C193	

BLOCK-1448
ECG74C20	HD1-74C20	HD9-74C20	MM74C20
MM74C20N	NTE74C20	SK74C20	TCG74C20

BLOCK-1449
443-785	ECG74C221	HD9-74C221	HE-443-785
MM74C221	MM74C221N	NTE74C221	SK74C221
TCG74C221			

BLOCK-1450
161392	ECG74C240	MM74C240N	NTE74C240
SK74C240	TCG74C240		

BLOCK-1451
ECG74C244	MM74C244N	NTE74C244	SK74C244
TCG74C244			

BLOCK-1452
ECG74C30	MM74C30N	NTE74C30	SK74C30
TCG74C30			

BLOCK-1453
ECG74C32	MM74C32N	NTE74C32	SK74C32
TCG74C32			

BLOCK-1454
74HC373N	ECG74C373	MM74C373N	NTE74C373
SK74C373			

BLOCK-1455
ECG74C374	MM74C374N	NTE74C374	SK74C374

BLOCK-1456
ECG74C48	HD1-74C48	HD9-74C48	MM74C48N
NTE74C48	SK74C48	TCG74C48	

BLOCK-1457
ECG74C73	HD1-74C73	HD9-74C73	MM74C73N
NTE74C73	SK74C73		

BLOCK-1458
ECG74C74	HD1-74C74	HD9-74C74	MM74C74
MM74C74N	NTE74C74	SK9783	TCG74C74

BLOCK-1459
ECG74C76	HD1-74C76	HD9-74C76	MM74C76N
NTE74C76	SK74C76	TC7476BP	TCG74C76
TCG74C85			

BLOCK-1460
ECG74C85	HD1-74C85	HD9-74C85	MM74C85
MM74C85N	NTE74C85	SK9782	

BLOCK-1461
ECG74C90	HD1-74C90	HD9-74C90	MM74C90
MM74C90N	NTE74C90	SK9781	TCG74C90

BLOCK-1462
ECG74C901	MM74C901N	NTE74C901	SK74C901
TCG74C901			

BLOCK-1463
ECG74C902	MM74C902N	NTE74C902	SK74C902
TCG74C902			

BLOCK-1464
ECG74C903	MM74C903N	NTE74C903	SK74C903
TCG74C903			

BLOCK-1465
ECG74C904	MM74C904N	NTE74C904	SK74C904
TCG74C904			

BLOCK-1466
ECG74C922	MM74C922N	NTE74C922	SK74C922

BLOCK-1467
ECG74C923	MM74C923N	NTE74C923	SK74C923

BLOCK-1468
ECG74C925	MM74C925N	NTE74C925	SK74C925
TCG74C925			

BLOCK-1469
443-709	ECG74C93	HD1-74C93	HD9-74C93
HE-443-709	MM74C93	MM74C93N	NTE74C93
SK74C93			

BLOCK-1470
ECG74C95	MM74C95N	NTE74C95	SK74C95
TCG74C95			

BLOCK-1471
11292312	1479-0240	1741-0069	339002
36188700	43A223006P1	43C216408P1	443-71
68A9026	74H00DC	74H00PC	9H00DC
9H00PC	DM74H00J	DM74H00N	ECG74H00
ITT74H00N	MC3000L	MC3000P	N74H00A
N74H00F	N74H00N	N8H80A	N8H80J
NTE74H00	S8H80J	SK74H00	SN74H00
SN74H00J	SN74H00N	T74H00B1	T74H00D1
T74H00D2	TCG74H00	US74H00A	US74H00J

BLOCK-1472
19-10476-0	2470-1724	43A223008	43C216409P1
569-0881-801	74H01DC	74H01PC	9H01DC
9H01PC	DM74H01J	DM74H01N	ECG74H01
ITT74H01N	MC3004L	MC3004P	N74H01A
N74H01F	N74H01N	NTE74H01	SN74H01
SN74H01J	SN74H01N	TCG74H01	US74H01A
US74H01J			

BLOCK-1473
1479-7971	1741-0150	43C216410P1	52335600
52335600HL	569-0881-804	717126-505	74H04DC
74H04PC	9H04DC	9H04PC	D34013890-002
DM74H04J	DM74H04N	ECG74H04	ITT74H04N
MC3008L	MC3008P	N74H04A	N74H04F
N74H04N	NTE74H04	S-2200-1135	SK74H04
SN74H04	SN74H04J	SN74H04N	TCG74H04
UPB74H04C	US74H04A	US74H04J	

BLOCK-1474
11369564	1741-0184	2470-1732	74H05DC
74H05PC	9H05DC	9H05PC	DM74H05
DM74H05J	DM74H05L	DM74H05N	ECG74H05
ITT74H05N	MC3009L	MC3009P	N74H05A
N74H05F	N74H05N	NTE74H05	SN74H05
SN74H05J	SN74H05N	TCG74H05	US74H05A
US74H05J			

BLOCK-1475
74H08PC	9H08DC	9H08PC	DM74H08J
DM74H08N	ECG74H08	MC3001L	MC3001P
N74H08A	N74H08F	N74H08N	NTE74H08
TCG74H08	US74H08A	US74H08J	

BLOCK-1476
1486-8780	1741-0242	196634	339003
43A223012	443-68	569-0881-810	68A9031
74H10DC	74H10PC	9H10DC	9H10PC
DM74H10J	DM74H10N	ECG74H10	HE-443-68
ITT74H10N	MC3005L	MC3005P	N74H10A
N74H10F	N74H10N	NTE74H10	SL14959
SN74H10	SN74H10J	SN74H10N	T74H10B1
T74H10D1	T74H10D2	TCG74H10	US74H10A
US74H10J			

BLOCK-1477

74H101DC	74H101PC	9H101DC	9H101PC
ECG74H101	N74H101A	N74H101F	N74H101N
NTE74H101	SN74H101	SN74H101J	SN74H101N
TCG74H101			

BLOCK-1478

443-43	74H102DC	74H102PC	9H102DC
9H102PC	ECG74H102	N74H102A	N74H102F
N74H102N	NTE74H102	SN74H102	SN74H102J
SN74H102N	TCG74H102		

BLOCK-1479

443-70	74H103DC	74H103PC	9H103DC
9H103PC	ECG74H103	HE-443-70	N74H103A
N74H103F	N74H103N	NTE74H103	SN74H103
SN74H103J	SN74H103N	TCG74H103	

BLOCK-1480

68A9047	74H106DC	74H106PC	9H106DC
9H106PC	ECG74H106	HL56431	N74H106B
N74H106F	N74H106N	NTE74H106	SN74H106
SN74H106J	SN74H106N	TCG74H106	

BLOCK-1481

569-0881-908	74H108DC	74H108PC	9H108DC
9H108PC	ECG74H108	N74H108A	N74H108F
N74H108N	NTE74H108	SN74H108	SN74H108J
SN74H108N	TCG74H108		

BLOCK-1482

1741-0275	443-58	569-0881-811	74H11DC
74H11PC	9H11DC	9H11PC	DM74H11J
DM74H11N	ECG74H11	ITT74H11N	MC3006L
MC3006P	N74H11A	N74H11F	N74H11N
NTE74H11	SL14960	SN74H11	SN74H11J
SN74H11N	T74H11B1	T74H11D1	T74H11D2
TCG74H11	UPB216C	UPB216D	US74H11A
US74H11J			

BLOCK-1483

1741-1323	55017	74H183DC	74H183PC
93H183DC	93H193PC	ECG74H183	FLH451
HD2563	HD2563P	NTE74H183	SN74H183
SN74H183J	SN74H183N	TCG74H183	TL74H183N

BLOCK-1484

1479-0265	1741-0333	182503	43C216411P1
52335700	569-0881-820	74H20DC	74H20PC
9H20DC	9H20PC	DM74H20J	DM74H20N
ECG74H20	ITT74H20N	MC3010L	MC3010P
N74H20A	N74H20F	N74H20N	NTE74H20
SN74H20	SN74H20J	SN74H20N	T74H20B1
T74H20D1	T74H20D2	TCG74H20	US74H20A
US74H20J			

BLOCK-1485

1479-0257	1741-1299	569-0881-821	74H21DC
74H21PC	829704-6	829704-7	885540031-2
9H21DC	9H21PC	DM74H21J	DM74H21N
ECG74H21	ITT74H21N	MB614	MC3011L
MC3011P	N74H21A	N74H21F	N74H21N
NTE74H21	SN74H21	SN74H21J	SN74H21N
T74H21B1	T74H21D1	T74H21D2	TCG74H21
UPB233C	UPB233D	US74H21A	US74H21J

BLOCK-1486

1741-0366	569-0881-822	74H22	74H22/94H22
74H22/9H22	74H22DC	74H22PC	8000-00038-005
9H22	9H22DC	9H22PC	DM74H22J
DM74H22N	ECG74H22	IP20-0208	MC3012L
MC3012P	N74H22A	N74H22F	N74H22N
NTE74H22	SN74H22	SN74H22J	SN74H22N
TCG74H22	TVCM-501	US74H22A	US74H22J

BLOCK-1487

172318	1741-0424	1820-1068	569-0881-830
74021-1	74H30DC	74H30PC	9H30DC
9H30PC	CD73/187/73	DM74H30J	DM74H30N
ECG74H30	ITT74H30N	MC3016L	MC3016P
N74H30A	N74H30F	N74H30N	N74H30W
NTE74H30	SN74H30	SN74H30J	SN74H30N
TCG74H30	US74H30A	US74H30J	

BLOCK-1488

11292313	1741-0481	3267390-01	339009
40-065-19-012	43A223017	43A223018	569-0881-840
74H40DC	74H40PC	9H40DC	9H40PC
A10	DM74H40J	DM74H40N	ECG74H40
ITT74H40N	MC3024L	MC3024P	N74H40A
N74H40F	N74H40N	NTE74H40	SN74H40

BLOCK-1488 (CONT.)

SN74H40J	SN74H40N	T74H40B1	T74H40D1
T74H40D2	TCG74H40	UPB74H40C	US74H40A
US74H40J			

BLOCK-1489

52335800	52335800HL	569-0881-850	74H50DC
74H50FC	74H50PC	9H50DC	9H50PC
DM74H50J	DM74H50N	ECG74H50	ITT74H50N
MC3020L	MC3020P	N74H50A	N74H50F
N74H50N	NTE74H50	SN74H50	SN74H50J
SN74H50N	T74H50B1	T74H50D1	T74H50D2
TCG74H50	US74H50A	US74H50J	

BLOCK-1490

11292314	1741-0572	43C216414P1	569-0881-851
74H51DC	74H51PC	9H51DC	9H51PC
C74107P	DM74H51J	DM74H51N	ECG74H51
ITT74H51N	MC3023L	MC3023P	N74H51A
N74H51F	N74H51N	NTE74H51	SN74H51
SN74H51J	SN74H51N	T74H51B1	T74H51D1
T74H51D2	TCG74H51	US74H51A	US74H51J

BLOCK-1491

1741-1380	569-0881-852	74H52DC	74H52PC
9H52DC	9H52PC	DM74H52J	DM74H52N
ECG74H52	MC3031L	MC3031P	N74H52A
N74H52F	N74H52N	NTE74H52	SN74H52
SN74H52J	SN74H52N	T74H52B1	T74H52D1
T74H52D2	TCG74H52	US74H52A	US74H52J

BLOCK-1492

1741-0606	569-0881-853	74H53DC	74H53PC
9H53DC	9H53PC	DM74H53J	DM74H53N
ECG74H53	ITT74H53N	MB613	MC3032L
MC3032P	N74H53A	N74H53F	N74H53N
NTE74H53	SN74H53	SN74H53J	SN74H53N
SN74LS03	T74H53B1	T74H53D1	T74H53D2
TCG74H53	US74H53A	US74H53J	

BLOCK-1493

11292315	1741-0630	74H54DC	74H54PC
9H54DC	9H54PC	DM74H54J	DM74H54N
ECG74H54	ITT74H54N	MC3033L	MC3033P
N74H54A	N74H54F	N74H54N	NTE74H54
SN74H54	SN74H54DC	SN74H54J	SN74H54N
T74H54B1	T74H54D1	T74H54D2	TCG74H54
US74H54A	US74H54J		

BLOCK-1494

1741-0663	569-0881-855	611878	74H55DC
74H55PC	9H55DC	9H55PC	DM74H55J
DM74H55N	ECG74H55	MC3034L	MC3034P
N74H55A	N74H55F	N74H55N	NTE74H55
SN74H55	SN74H55J	SN74H55N	TCG74H55
US74H55A	US74H55J		

BLOCK-1495

1741-0697	569-0881-860	74H60DC	74H60PC
9H60DC	9H60PC	DM74H60J	DM74H60N
ECG74H60	ITT74H60N	MC3030L	MC3030P
N74H60A	N74H60F	N74H60N	NTE74H60
SN74H60	SN74H60J	SN74H60N	TCG74H60
US74H60A	US74H60J		

BLOCK-1496

74H61DC	74H61PC	9H61DC	9H61PC
DM74H61J	DM74H61N	ECG74H61	MC3019L
MC3019P	N74H61A	N74H61F	N74H61N
NTE74H61	SN74H61	SN74H61J	SN74H61N
T74H61B1	T74H61D1	T74H61D2	TCG74H61
US74H61A	US74H61J		

BLOCK-1497

74H62DC	74H62PC	9H62DC	9H62PC
DM74H62J	DM74H62N	ECG74H62	MC3018L
MC3018P	N74H62A	N74H62F	N74H62N
NTE74H62	SN74H62	SN74H62J	SN74H62N
T74H62B1	T74H62D1	T74H62D2	TCG74H62
US74H62A	US74H62J		

BLOCK-1498

74H71DC	74H71PC	9H71DC	9H71PC
DM74H71J	DM74H71N	ECG74H71	MC3054L
MC3054P	N74H71A	N74H71F	N74H71N
NTE74H71	SN74H71	SN74H71J	SN74H71N
T74H71B1	T74H71D2	TCG74H71	US74H71A
US74H71J			

BLOCK-1499

74H72DC	74H72PC	9H72DC	9H72PC
DM74H72J	DM74H72N	ECG74H72	ITT74H72N
MC3055L	MC3055P	N74H72A	N74H72F
N74H72N	NTE74H72	SN74H72	SN74H72J
SN74H72N	T74H72B1	T74H72D1	T74H72D2
TCG74H72	US74H72A	US74H72J	

BLOCK-1500

74H73DC	74H73PC	9H73DC	9H73PC
DM74H73J	DM74H73N	ECG74H73	ITT74H73N
MC303P	MC3063L	N74H73A	N74H73F
N74H73N	NTE74H73	SN74H73	SN74H73J
SN74H73N	TCG74H73	US74H73A	US74H73J

BLOCK-1501

1741-0721	1820-0512	52335900	52335900HL
569-0881-874	74H74DC	74H74PC	9H74DC
9H74PC	DM74H74J	DM74H74N	ECG74H74
ITT74H74N	MC74H74AF	MC74H74AL	MC74H74P
N74H74A	N74H74F	N74H74N	N8848F
NTE74H74	SN74H74	SN74H74J	SN74H74N
TCG74H74	US74H74A	US74H74J	

BLOCK-1502

74H76DC	74H76PC	9H76DC	9H76PC
B03	DM74H76J	DM74H76N	ECG74H76
ITT74H76N	N74H76B	N74H76F	N74H76N
NTE74H76	SN74H76	SN74H76J	SN74H76N
TCG74H76	US74H76A		

BLOCK-1503

74H78DC	74H78PC	9H78DC	9H78PC
DM74H78J	DM74H78N	ECG74H78	MB618
NTE74H78	SN74H78	SN74H78J	SN74H78N
TCG74H78	US74H78A	US74H78J	

BLOCK-1504

443-59	ECG74H86	MC3021L	MC3021P

BLOCK-1505

74H87DC	74H87PC	93H87DC	93H87PC
ECG74H87	FLH441	MC74H87F	MC74H87L
MC74H87P	NTE74H87	SN74H87	SN74H87J
SN74H87N	TCG74H87	TL74H87N	

BLOCK-1506

ECG74HC00	MC74HC00N	MLC74HC00A	SK7C00

BLOCK-1507

74HC02N	ECG74HC02	I55D0HC020	MC74HC02N
MLC74HC02A	SK7C02	SN74HC02N	TC74HC02
TC74HC02P			

BLOCK-1508

74HC04AP	ECG74HC04	I55D4HC040	MC74HC04F
MLC74HC04A	SK7C04	SN74HC04ANS	TC74HC04
TC74HC04P			

BLOCK-1509

ECG74HC08	MC74HC08	MLC74HC08A	SK7C08

BLOCK-1510

ECG74HC10	MC74HC10N	SK7C10

BLOCK-1511

ECG74HC109	MC74HC109N	MLC74HC109	SK7C109

BLOCK-1512

ECG74HC11	MLC74HC11	SK7C11

BLOCK-1513

8-759-206-28	ECG74HC123	SK7C123	TC74HC123A
TC74HC123F	TC74HC123F-TP1		

BLOCK-1514

ECG74HC125	SK7C125

BLOCK-1515

ECG74HC126	SK7C126

BLOCK-1516

ECG74HC132	MC74HC132N	SK7C132

BLOCK-1517

ECG74HC138	MC74HC138N	SK7C138

BLOCK-1518

ECG74HC139	MC74HC139N	SK7C139

BLOCK-1519

ECG74HC14	MLC74HC14A	SK7C14

BLOCK-1520

ECG74HC151	MC74HC151N	SK7C151

BLOCK-1521

ECG74HC153	SK7C153

BLOCK-1522

ECG74HC154	SK7C154

BLOCK-1523

ECG74HC161	MC74HC161	SK7C161

BLOCK-1524

ECG74HC163	MC74HC163N	SK7C163

BLOCK-1525

ECG74HC164	MC74HC164N	SK7C164

BLOCK-1526

ECG74HC165	MC74HC165N	SK7C165

BLOCK-1527

ECG74HC173	MC74HC173N	SK7C173

BLOCK-1528

ECG74HC174	MC74HC174N	SK7C174

BLOCK-1529

ECG74HC175	MC74HC175N	SK7C175

BLOCK-1530

74HC240N	ECG74HC240	MC74HC240N	MM74HC240N
SK7C240	SN74HC240N		

BLOCK-1531

ECG74HC244	MC74HC244N

BLOCK-1532

ECG74HC257	MC74HC257N

BLOCK-1533

ECG74HC259

BLOCK-1534

ECG74HC273	MC74HC273N

BLOCK-1535

ECG74HC299	SK7C299

BLOCK-1536

ECG74HC32	MLC74HC32A	SK7C32	SN74HC32N

BLOCK-1537

ECG74HC373	MC74HC373N	SK7C373

BLOCK-1538

ECG74HC374	MC74HC374N	SK7C374

BLOCK-1539

ECG74HC377	SK7C377

BLOCK-1540

ECG74HC390	MC74HC390N	SK7C390

BLOCK-1541

ECG74HC393	MC74HC393N	SK7C393

BLOCK-1542

ECG74HC40105	SK7C40105

BLOCK-1543

ECG74HC4020	MC74HC4020N	SK7C4020

BLOCK-1544

ECG74HC4040	MC74HC4040N	SK7C4040

BLOCK-1545

263P701010	4835-209-17005	ECG74HC4053	MC74HC4053N
SK7C4053			

BLOCK-1546

ECG74HC4060	MC74HC4060N	SK7C4060

BLOCK-1547
ECG74HC4067	SK7C4067		

BLOCK-1548
ECG74HC573	SK7C573		

BLOCK-1549
ECG74HC574	SK7C574		

BLOCK-1550
4835-209-17004	ECG74HC86	MC74HC86N	MLC74HC86
SK7C86	SN74HC86N		

BLOCK-1551
ECG74HCT00	SK7CT00		

BLOCK-1552
74HCT04	74HCT04N	8026004	CD74HCT04E
ECG74HCT04	SK7CT04		

BLOCK-1553
74HCT08	74HCT08N	8026008	CD74HCT08E
ECG74HCT08	SK7CT08		

BLOCK-1554
74HCT158N	CD74HCT138E	ECG74HCT138	SK7CT138
SK7CT158			

BLOCK-1555
74HCT14	74HCT14N	8026014	CD74HCT14E
ECG74HCT14	SK7CT14		

BLOCK-1556
ECG74HCT161	SK7CT161		

BLOCK-1557
ECG74HCT163	SK7CT163		

BLOCK-1558
ECG74HCT174	SK7CT174		

BLOCK-1559
ECG74HCT240	SK7CT240		

BLOCK-1560
74HCT244	74HCT244N	8026244	CD74HCT244E
ECG74HCT244	SK7CT244		

BLOCK-1561
ECG74HCT273	SK7CT273		

BLOCK-1562
74HCT32	74HCT32N	8026032	CD74HCT32E
ECG74HCT32	SK7CT32		

BLOCK-1563
74HCT373N	8026373	CD74HCT373E	ECG74HCT373
HCT373	SK7CT373		

BLOCK-1564
ECG74HCT374			

BLOCK-1565
ECG74HCT573	SK7CT573		

BLOCK-1566
ECG74HCT574	SK7CT574		

BLOCK-1567
DM74L93F	DM74L93J	DM74L93N	ECG74L93
NTE74L93	SN74L93J	SN74L93N	TCG74L93

BLOCK-1568
08210001	1313-5561	2109-301-0707	307-152-9-007
3102006	32109-301-070	443-728	612199-1
612261-1	74LS00	74LS00CH	74LS00DC
74LS00J	74LS00N	74LS00NA	74LS00PC
74LS00W	901521-01	AMX3550	C74LS00P
DM74LS00N	EA33X8605	EAQ00-12100	ECG74LS00
GD74LS00	HD74LS00	HD74LS00P	HE-443-728
HEPC74LS00P	LS00	M74LS00P	MB74LS00
MB74LS00M	N74LS00	N74LS00F	N74LS00N
NTE74LS00	P61LS00**H	PCMB74LS00M	SK74LS00
SN74LS00	SN74LS00J	SN74LS00JD	SN74LS00JDS
SN74LS00JS	SN74LS00N	SN74LS00ND	SN74LS00NDS
SN74LS00NS	SN74LS00W	T74LS00B1	T74LS00D1
TCG74LS00	TISM74LS00N	VHIHD74LS00P/	X420300000
X420500000			

BLOCK-1569
DM74LS01N	ECG74LS01	HD74LS01	HD74LS01P
MB74LS01	MB74LS01M	N74LS01F	N74LS01N
ND74LS01F	NTE74LS01	SK74LS01	SN74LS01
SN74LS01J	SN74LS01JD	SN74LS01JDS	SN74LS01JS
SN74LS01N	SN74LS01ND	SN74LS01NDS	SN74LS01NS
T74LS01B1N	TCG74LS01		

BLOCK-1570
08210072	08220072	3102007	443-779
54LS02	54LS02DM	74LS02	74LS02CH
74LS02DC	74LS02J	74LS02N	74LS02PC
74LS02W	901521-21	AMX-3551	AMX3551
C74LS02P	DM74LS02N	DN74LS02	ECG74LS02
HD74LS02	HD74LS02P	HE-443-779	HEPC74LS02P
IC-74LS02	LS02	M74LS02P	MB74LS02
MB74LS02M	N74LS02F	N74LS02N	NTE74LS02
PCMB74LS02M	SK74LS02	SN74LS02	SN74LS02J
SN74LS02JD	SN74LS02JDS	SN74LS02JS	SN74LS02N
SN74LS02NDS	SN74LS02NS	SN74LS02W	T74LS02B1
T74LS02D1	TCG74LS02	UPD23128EC-088	X420300020
X420500020			

BLOCK-1571
443-745	74L03	74LS03	74LS03CH
74LS03DC	74LS03J	74LS03P	74LS03PC
74LS03W	C74LS03P	DM74LS03N	ECG74LS03
EW84X1007	HD74LS03	HD74LS03P	HE-443-745
HEPC74LS03P	IC-74L03	M74LS03P	MB74LS03
MB74LS03M	N74LS03F	N74LS03N	NTE74LS03
ON74LS03N	RH-IX0072PAZZ	SK74LS03	SN74L03
SN74LS03J	SN74LS03JD	SN74LS03JDS	SN74LS03JS
SN74LS03N	SN74LS03ND	SN74LS03NDS	SN74LS03NS
SN74LS03W	T74LS03B1	T74LS03D1	TCG74LS03

BLOCK-1572
06300248	08210002	2360951	3102008
443-755	74LS04	74LS04DC	74LS04NA
74LS04P	74LS04PC	8-759-900-04	8409-08004
901521-02	AMX3552	C74LS04P	DM74LS04N
EAQ00-1200	EAQ00-12200	ECG74LS04	F-74LS04PC
GD74LS04	HD74LS04	HD74LS04P	HE-443-755
HEPC74LS04P	IC-74LS04	M74LS04P	MB74LS04
MB74LS04M	MOS8713	N74LS04F	N74LS04N
NTE74LS04	PCMB74LS04M	PO74LS04	SK74LS04
SN74LS04	SN74LS04J	SN74LS04JD	SN74LS04JDS
SN74LS04JS	SN74LS04N	SN74LS04ND	SN74LS04NDS
SN74LS04NS	T74LS04B1	TCG74LS04	X420300040
X420500040			

BLOCK-1573
08210017	3102009	443-818	74LS05
74LS05DC	74LS05PC	76-005-003	AMX3553
C74LS05P	DM74LS05N	ECG74LS05	HD74LS05
HD74LS05P	HE-443-818	HEPC74LS05P	LS05
M74LS05P	MB74LS05	MB74LS05M	N74LS05F
N74LS05N	NTE74LS05	PCMB74LS05M	SK74LS05
SN74LS05	SN74LS05J	SN74LS05JD	SN74LS05JDS
SN74LS05JS	SN74LS05N	SN74LS05ND	SN74LS05NDS
SN74LS05NS	TCG74LS05	X420300050	

BLOCK-1574
08210032	178740	2362051	443-780
74LS08	74LS08CH	74LS08DC	74LS08J
74LS08N	74LS08NA	74LS08PC	74LS08W
901521-03	AMX3698	C017097	C74LS08P
DM74LS08N	DN74LS08	EAQ00-12300	ECG74LS08
HD74LS08	HD74LS08P	HE-443-780	HEPC74LS08P
M74LS08P	MB74LS08	MB74LS08M	MOS7712
MOS8712	N74LS08F	N74LS08N	NTE74LS08
ON74LS08N	SK74LS08	SN74LS08	SN74LS08J
SN74LS08JD	SN74LS08JDS	SN74LS08JS	SN74LS08N
SN74LS08ND	SN74LS08NDS	SN74LS08NS	SN74LS08W
T74LS08B1	T74LS08D1	TCG74LS08	VHISN74LS08-1
X420300080	X420500080		

BLOCK-1575
443-816	74LS09	74LS09PC	8-759-900-09
DM74LS09N	DN74LS09	ECG74LS09	HD74LS09
HD74LS09P	HE-443-816	M74LS09P	MB74LS09
MB74LS09M	N74LS09F	N74LS09N	NTE74LS09
SK1924	SN74LS09	SN74LS09J	SN74LS09JD
SN74LS09JDS	SN74LS09JS	SN74LS09N	SN74LS09ND
SN74LS09NDS	SN74LS09NS	T74LS09B1	T74LS09D1
TCG74LS09	TVSSN74LS09N		

BLOCK-1576
443-797	74LS10	74LS10CH	74LS10DC
74LS10J	74LS10N	74LS10PC	74LS10W
8-759-900-10	AMX3898	C74LS10P	CO14341
DM74LS10N	EAQ00-12400	ECG74LS10	HD74LS10
HD74LS10P	HE-443-797	HEPC74LS10P	M74LS10P

BLOCK-1576 (CONT.)

MB74LS10	MB74LS10M	N74LS10F	N74LS10N
NTE74LS10	PCMB74LS10M	SK74LS10	SN74LS10
SN74LS10J	SN74LS10JD	SN74LS10JDS	SN74LS10JS
SN74LS10N	SN74LS10ND	SN74LS10NDS	SN74LS10NS
SN74LS10W	T74LS10B1	T74LS10D1	TCG74LS10
X420300100	X420500100		

BLOCK-1577

74LS107CH	74LS107DC	74LS107J	74LS107PC
74LS107W	AN74LS107AN	ECG74LS107	HD74LS107
HD74LS107P	M74LS107AP	M74LS107P	MB74LS107
MB74LS107A	MB74LS107AM	MB74LS107M	N74LS107F
N74LS107N	NTE74LS107	SK74LS107	SN74LS107A
SN74LS107AJ	SN74LS107AJD	SN74LS107AJDS	SN74LS107AJS
SN74LS107AN	SN74LS107AND	SN74LS107ANDS	SN74LS107ANS
SN74LS107AW	SN74LS107N	TCG74LS107	

BLOCK-1578

74LS109	74LS109CH	74LS109DC	74LS109J
74LS109PC	74LS109W	C74LS109P	DM74LS109AN
ECG74LS109A	HD74LS109A	HD74LS109AP	HEPC74LS109P
M74LS109AP	M74LS109P	MB74LS109A	MB74LS109AM
N74LS109AF	N74LS109AN	N74LS109F	N74LS109N
NTE74LS109A	SK74LS109	SN74LS109A	SN74LS109AJ
SN74LS109AJD	SN74LS109AJDS	SN74LS109AJS	SN74LS109AN
SN74LS109AND	SN74LS109ANDS	SN74LS109ANS	SN74LS109AW
SN74LS109N	T74LS109B1	T74LS109D1	TCG74LS109A

BLOCK-1579

08210058	3102010	443-864	610002-011
74LS11	74LS11CH	74LS11DC	74LS11J
74LS11N	74LS11PC	74LS11W	AMX3554
C74LS11P	DM74LS11N	ECG74LS11	HD74LS11
HD74LS11P	HE-443-864	HEPC74LS11P	M74LS11P
MB74LS11	MB74LS11M	N74LS11F	N74LS11N
NTE74LS11	SK74LS11	SN74LS11	SN74LS11J
SN74LS11JD	SN74LS11JDS	SN74LS11JS	SN74LS11N
SN74LS11ND	SN74LS11NDS	SN74LS11NS	SN74LS11W
T74LS11B1	T74LS11D1	TCG74LS11	X420500110

BLOCK-1580

443-948	74LS112	74LS112DC	74LS112PC
DM74LS112AN	ECG74LS112A	HD74LS112	HD74LS112P
M74LS112AP	MB74LS112A	MB74LS112AM	N74LS112F
N74LS112N	NTE74LS112A	SN74LS112A	SN74LS112AJ
SN74LS112AJD	SN74LS112AJDS	SN74LS112AJS	SN74LS112AN
SN74LS112AND	SN74LS112ANDS	SN74LS112ANS	T74LS112B1
T74LS112D1	TCG74LS112A	X420501120	

BLOCK-1581

74LS114PC	ECG74LS114	HD74LS114	HD74LS114P
M74LS114AP	MB74LS114A	MB74LS114AM	N74LS114F
N74LS114N	NTE74LS114	SK74LS114	SN74LS114A
SN74LS114AJ	SN74LS114AJD	SN74LS114AJDS	SN74LS114AJS
SN74LS114AN	SN74LS114AND	SN74LS114ANDS	SN74LS114ANS
T74LS114B1	T74LS114D1	TCG74LS114	

BLOCK-1582

74LS12	DM74LS12N	ECG74LS12	HD74LS12
HD74LS12P	IC-74LS12	M74LS12P	MB74LS12
MB74LS12M	N74LS12F	N74LS12N	NTE74LS12
SK74LS12	SN74LS12	SN74LS12J	SN74LS12JD
SN74LS12JDS	SN74LS12JS	SN74LS12N	SN74LS12ND
SN74LS12NDS	SN74LS12NS	TCG74LS12	

BLOCK-1583

ECG74LS122	HD74LS122	HD74LS122P	M74LS122P
MB74LS122	MB74LS122M	NTE74LS122	SK74LS122
SN74LS122	SN74LS122J	SN74LS122JD	SN74LS122JDS
SN74LS122JS	SN74LS122N	SN74LS122ND	
SN74LS122NDS	SN74LS122NS		

BLOCK-1584

08210059	443-942	46-13455-3	610002-123
7416NA	74LS123	74LS123CH	74LS123J
74LS123N	74LS123W	8-759-901-23	AMX3803
EAQ00-12900	EAQ00-12914	ECG74LS123	HD74LS123
HD74LS123P	HE-443-942	M74LS123P	MB74LS123
MB74LS123M	N74LS123AF	N74LS123AN	NTE74LS123
POSN74LS123N	SK74LS123	SK74LS123P	SN74LS123
SN74LS123J	SN74LS123JD	SN74LS123JDS	SN74LS123JS
SN74LS123N	SN74LS123ND	SN74LS123NDS	SN74LS123NS
SN74LS123W	TCG74LS123	X420301230	

BLOCK-1585

443-811	74LS125	74LS125A	74LS125ADC
74LS125AN	74LS125APC	901521-20	DM74LS125AN
DN74LS125P	ECG74LS125A	GD74LS125A	HD74LS125AP
HE-443-811	I61DLS1250	LS125	M74LS125AP

BLOCK-1585 (CONT.)

MB74LS125A	MB74LS125AM	N74LS125F	N74LS125N
NTE74LS125	SK74LS125A	SN74LS125A	SN74LS125AJ
SN74LS125AJD	SN74LS125AJDS	SN74LS125AJS	SN74LS125AN
SN74LS125AND	SN74LS125ANDS	SN74LS125ANS	TCG74LS125A

BLOCK-1586

443-919	74LS126	74LS126DC	74LS126PC
DM74LS126AJ	DM74LS126AN	DM74LS126AW	ECG74LS126
HE-443-919	M74LS126AP	MB74LS126A	MB74LS126AM
N74LS126F	N74LS126N	NTE74LS126A	SN74LS126A
SN74LS126AJ	SN74LS126AJD	SN74LS126AJDS	SN74LS126AJS
SN74LS126AN	SN74LS126AND	SN74LS126ANDS	SN74LS126ANS
SN74LS126J	SN74LS126N	TCG74LS126A	

BLOCK-1587

443-729	74LS13DC	74LS13N	74LS13PC
C74LS13P	DM74LS13J	DM74LS13N	DM74LS13W
ECG74LS13	HD74LS13	HD74LS13P	HE-443-729
HEPC74LS13P	M74LS13P	MB74LS13	MB74LS13M
N74LS13F	N74LS13N	NTE74LS13	SK74LS13
SN74LS13	SN74LS13J	SN74LS13JD	SN74LS13JDS
SN74LS13JS	SN74LS13N	SN74LS13ND	SN74LS13NDS
SN74LS13NS	T74LS13B1	TCG74LS13	

BLOCK-1588

3102018	443-792	6126760001	74LS132
74LS132DC	74LS132N	74LS132PC	AMX3561
DM74LS132J	DM74LS132N	DM74LS132W	DN74LS132
ECG74LS132	HE-443-792	M74LS132P	MB74LS132
MB74LS132M	N74LS132F	N74LS132N	NTE74LS132
S4LS132N	SN74LS132	SN74LS132J	SN74LS132JD
SN74LS132JDS	SN74LS132JS	SN74LS132N	SN74LS132ND
SN74LS132NDS	SN74LS132NS	TCG74LS132	

BLOCK-1589

74LS133DC	74LS133PC	ECG74LS133	M74LS133P
NTE74LS133	SK74LS133	SN74LS133J	SN74LS133JD
SN74LS133JDS	SN74LS133JS	SN74LS133N	SN74LS133ND
SN74LS133NDS	SN74LS133NS		

BLOCK-1590

178743	74LS136CH	74LS136DC	74LS136J
74LS136PC	74LS136W	C74LS136P	ECG74LS136
HD74LS136	HD74LS136P	HEPC74LS136P	M74LS136P
MB74LS136	MB74LS136M	N74LS136F	N74LS136N
NTE74LS136	SK74LS136	SN74LS136	SN74LS136J
SN74LS136JD	SN74LS136JDS	SN74LS136JS	SN74LS136N
SN74LS136ND	SN74LS136NDS	SN74LS136NS	SN74LS136W
T74LS136B1	T74LS136D1	TCG74LS136	VHID74LS136P
VHIHD74LS136P			

BLOCK-1591

08210029	150440	443-877	74LS138
74LS138DC	74LS138N	74LS138P	74LS138PC
8-759-901-38	AMX4583	C061428	C74LS138P
CO14344	DM74LS138N	EAQ00-18700	ECG74LS138
GD74LS138	HD74LS138	HD74LS138P	HE-443-877
HEPC74LS138P	IC-74LS138	INS82LS05N	M74LS138BP
M74LS138P	MB74LS138	MB74LS138M	N74LS138F
N74LS138N	NTE74LS138	ON74S138N	
POSN74LS138N	SK74LS138	SN74LS138	SN74LS138J
SN74LS138JD	SN74LS138JDS	SN74LS138JS	SN74LS138N
SN74LS138ND	SN74LS138NDS	SN74LS138NS	SN74LS138X
T74LS138B1	TCG74LS138		

BLOCK-1592

08210046	443-822	74LS139	74LS139DC
74LS139N	74LS139PC	901521-18	AMX3800
DM74LS139N	EAQ00-21500	ECG74LS139	HD74LS139
HD74LS139P	HE-443-822	IC-74LS139	LS139
M74LS139P	MB74LS139	MB74LS139M	MX-4314
N74LS139F	N74LS139N	NTE74LS139	P61LS139*H
SK74LS139	SN74LS139	SN74LS139AN	SN74LS139J
SN74LS139JD	SN74LS139JDS	SN74LS139JS	SN74LS139N
SN74LS139ND	SN74LS139NDS	SN74LS139NS	SN74LS139X
TCG74LS139	X420301390		

BLOCK-1593

08210023	443-872	610002-014	74LS14
74LS14DC	74LS14N	74LS14PC	74LS14PF
901251-30	901521-30	AMX-3716	AMX3716
C061850	C74LS14P	DM74LS14J	DM74LS14N
DM74LS14W	DN74LS14P	EA000-17200	EAQ00-17200
ECG74LS14	HD74LS14	HD74LS14P	HE-443-872
HEPC74LS14P	I61D4LS140	LS14	M74LS14P
MB74LS14	MB74LS14M	N74LS14F	N74LS14N
NTE74LS14	P61LS14**H	POSN74LS14N	SK74LS14
SN74LS14	SN74LS14J	SN74LS14JD	SN74LS14JDS
SN74LS14JS	SN74LS14N	SN74LS14ND	SN74LS14NDS
SN74LS14NS	T74LS14B1	T74LS14BI	TCG74LS14
VHIHD74LS14-1	X420300140		

BLOCK-1594

443-889	74LS145	AMX4659	ECG74LS145
HD74LS145	HD74LS145P	HE-443-889	M74LS145P
MB74LS145	MB74LS145M	N74LS145F	N74LS145N
NTE74LS145	ON74LS145N	SK74LS145	SN74LS145
SN74LS145J	SN74LS145JD	SN74LS145JDS	SN74LS145JS
SN74LS145N	SN74LS145ND	SN74LS145NDS	SN74LS145NS
TCG74LS145	X420301450		

BLOCK-1595

ECG74LS147	NTE74LS147	SN74LS147	SN74LS147J
SN74LS147JD	SN74LS147JDS	SN74LS147JS	SN74LS147N
SN74LS147ND	SN74LS147NDS	SN74LS147NS	

BLOCK-1596

443-912	612195-1	74LS148	ECG74LS148
HD74LS148	HD74LS148P	HE-443-912	M74LS148P
NTE74LS148	SK74LS148	SN74LS148	SN74LS148AJ
SN74LS148AJD	SN74LS148AJDS	SN74LS148AJS	SN74LS148AN
SN74LS148AND	SN74LS148ANDS	SN74LS148ANS	SN74LS148J
SN74LS148JD	SN74LS148JDS	SN74LS148JS	SN74LS148N
SN74LS148ND	SN74LS148NDS	SN74LS148NS	

BLOCK-1597

74LS15	74LS15PC	ECG74LS15	HD74LS15
HD74LS15P	LS15	M74LS15P	MB74LS15
MB74LS15M	N74LS15F	N74LS15N	NTE74LS15
SK74LS15	SN74LS15	SN74LS15J	SN74LS15JD
SN74LS15JDS	SN74LS15JS	SN74LS15N	SN74LS15ND
SN74LS15NDS	SN74LS15NS	T74LS15B1	T74LS15D1
TCG74LS15			

BLOCK-1598

08210071	197323	2420SNLS151	74LS151
74LS151DC	74LS151PC	AM25LS151PC	C74LS151P
DM74LS151N	ECG74LS151	HD74LS151	HD74LS151P
HEPC74LS151P	M74LS151P	MB74LS151	MB74LS151M
N74LS151F	N74LS151N	NTE74LS151	SK74LS151
SN74LS151	SN74LS151J	SN74LS151JD	SN74LS151JDS
SN74LS151JS	SN74LS151N	SN74LS151ND	
SN74LS151NDS	SN74LS151NS	SN74LS151X	TCG74LS151

BLOCK-1599

3102019	443-955	74LS153	74LS153DC
74LS153N	74LS153PC	AMX3562	C74LS153P
DM74LS153N	ECG74LS153	HD74LS153	HD74LS153P
HEPC74LS153P	M74LS153P	MB74LS153	MB74LS153M
N74LS153F	N74LS153N	NTE74LS153	SK74LS153
SN74LS153	SN74LS153J	SN74LS153JD	SN74LS153JDS
SN74LS153JS	SN74LS153N	SN74LS153ND	
SN74LS153NDS	SN74LS153NS	TCG74LS153	

BLOCK-1600

08210030	443-782	74LS155	74LS155DC
74LS155PC	C74LS155P	DM74LS155N	EAQ00-13000
EAQ00-22900	ECG74LS155	HD74LS155	HD74LS155P
HE-443-782	HEPC74LS155P	LS155	M74LS155P
MB74LS155	MB74LS155M	N74LS155F	N74LS155N
NTE74LS155	SK74LS155	SN74LS155	SN74LS155AN
SN74LS155J	SN74LS155JD	SN74LS155JDS	SN74LS155JS
SN74LS155N	SN74LS155ND	SN74LS155NDS	SN74LS155NS
T74LS155B1	TCG74LS155		

BLOCK-1601

3102028	409-006-0203	409-042-9000	410-051-5808
410-067-3300	46-132884-3	612230-1	74LS156
74LS156DC	74LS156N	74LS156PC	DM74LS156N
DN74LS156	EAQ00-24700	ECG74LS156	HD74LS156
HD74LS156P	HE-443-1036	M74LS156P	MB74LS156
MB74LS156M	N74LS156F	N74LS156N	NTE74LS156
SK74LS156	SN74LS156	SN74LS156J	SN74LS156JD
SN74LS156JDS	SN74LS156JS	SN74LS156N	SN74LS156ND
SN74LS156NDS	SN74LS156NS	TCG74LS156	X420301560

BLOCK-1602

08210039	3102020	443-799	74LS157
74LS157DC	74LS157N	74LS157PC	AMX3563
C74LS157P	ECG74LS157	GD74LS157	HD74LS157
HD74LS157P	HE-443-799	HEPC74LS157P	IC-74LS157
LS157N	M74LS157P	MB74LS157	MB74LS157M
N74LS157F	N74LS157N	NTE74LS157	
POSN74LS157N	SK74LS157	SN74LS157	SN74LS157J
SN74LS157JD	SN74LS157JDS	SN74LS157JS	SN74LS157N
SN74LS157ND	SN74LS157NDS	SN74LS157NS	T74LS157B1
TCG74LS157			

BLOCK-1603

74LS158P	74LS158DC	74LS158PC	C014345
C74LS158P	CO14345	ECG74LS158	GD74LS158
HD74LS158	HD74LS158P	HEPC74LS158P	M74LS158P
MB74LS158	MB74LS158M	N74LS158F	N74LS158N

BLOCK-1603 (CONT.)

NTE74LS158	SK1923	SN74LS158	SN74LS158J
SN74LS158JD	SN74LS158JDS	SN74LS158JS	SN74LS158N
SN74LS158ND	SN74LS158NDS	SN74LS158NS	TCG74LS158

BLOCK-1604

74LS160PC	DM74LS160AN	ECG74LS160A	HD74LS160
M74LS160AP	MB74LS160A	MB74LS160AM	N74LS160AF
N74LS160AN	NTE74LS160A	SN74LS160A	SN74LS160AJ
SN74LS160AJD	SN74LS160AJDS	SN74LS160AJS	SN74LS160AN
SN74LS160AND	SN74LS160ANDS	SN74LS160ANS	SN74LS160J
SN74LS160N	TCG74LS160A		

BLOCK-1605

443-757	74LS161	74LS161AN	74LS161CH
74LS161DC	74LS161J	74LS161PC	74LS161W
DM74LS161AN	ECG74LS161A	HD74LS161	HD74LS161P
HE-443-757	M74LS161AP	M74LS161P	MB74LS161A
MB74LS161AM	N74LS161AF	N74LS161AN	NTE74LS161A
POSN74LS161AN	SK74LS161	SN74LS161A	SN74LS161AJ
SN74LS161AJD	SN74LS161AJDS	SN74LS161AJS	SN74LS161AN
SN74LS161AND	SN74LS161ANDS	SN74LS161ANS	SN74LS161AW
SN74LS161J	SN74LS161N	T74LS161B1	TCG74LS161A
X420201610	X420301610		

BLOCK-1606

74LS162PC	DM74LS162AN	ECG74LS162A	HD74LS162
HD74LS162P	M74LS162AP	MB74LS162A	MB74LS162AM
N74LS162AF	N74LS162AN	NTE74LS162A	SN74LS162A
SN74LS162AJ	SN74LS162AJD	SN74LS162AJDS	SN74LS162AJS
SN74LS162AN	SN74LS162AND	SN74LS162ANDS	SN74LS162ANS
SN74LS162J	SN74LS162N	TCG74LS162A	

BLOCK-1607

443-934	74LS163	74LS163CH	74LS163DC
74LS163J	74LS163PC	74LS163W	74S163
DM74LS163AN	DM74LS163N	DM74S163N	EAQ00-20300
ECG74LS163A	HD74LS163	HD74LS163P	HE-443-934
IC-74LS163	M74LS163AP	M74LS163P	MB74LS163A
MB74LS163AM	N74LS163AF	N74LS163AN	NTE74LS163A
ON74LS163AN	SK74LS163	SN74LS163A	SN74LS163AJ
SN74LS163AJD	SN74LS163AJDS	SN74LS163AJS	SN74LS163AN
SN74LS163AND	SN74LS163ANDS	SN74LS163ANS	SN74LS163AW
SN74LS163J	SN74LS163N	SN74S163	SN74S163J
TCG74LS163A	TCG74S163	X420301630	

BLOCK-1608

74LS164	74LS164DC	74LS164FC	74LS164N
74LS164PC	9LS164PC	DM74LS164N	ECG74LS164
HD74LS164	HD74LS164P	HE-443-769	M74LS164P
N74LS164F	N74LS164N	NTE74LS164	SK74LS164
SN74LS164	SN74LS164J	SN74LS164JD	SN74LS164JDS
SN74LS164JS	SN74LS164N	SN74LS164ND	
SN74LS164NDS	SN74LS164NS	TCG74LS164	X420301640

BLOCK-1609

74LS165DC	74LS165FC	74LS165PC	DM74LS165N
ECG74LS165	IC-74LS165	M74LS165P	NTE74LS165
SN74LS165A	SN74LS165J	SN74LS165JD	SN74LS165JDS
SN74LS165JS	SN74LS165N	SN74LS165ND	
SN74LS165NDS	SN74LS165NS		

BLOCK-1610

3102021	443-892	74LS166	AMX3564
DM74LS166N	ECG74LS166	HD74LS166	HD74LS166P
HE-443-892	M74LS166AP	M74LS166P	NTE74LS166
POSN74LS166N	SN74LS166	SN74LS166A	SN74LS166AN
SN74LS166J	SN74LS166JD	SN74LS166JDS	SN74LS166JS
SN74LS166N	SN74LS166ND	SN74LS166NDS	SN74LS166NS

BLOCK-1611

74LS168DC	74LS168PC	ECG74LS168A	N74LS168AF
N74LS168AN	NTE74LS168A	SK74LS168A	SN74LS168N

BLOCK-1612

74LS169DC	74LS169PC	ECG74LS169A	N74LS169AF
N74LS169AN	NTE74LS169A	SK74LS169A	SN74LS169AJ
SN74LS169AN	SN74LS169B	SN74LS169J	SN74LS169N

BLOCK-1613

74LS170DC	74LS170FC	74LS170PC	C74LS170P
DM74LS170N	ECG74LS170	HEPC74LS170P	M74LS170P
N74LS170B	N74LS170F	N74LS170N	NTE74LS170
SN74LS170	SN74LS170J	SN74LS170JD	SN74LS170JDS
SN74LS170JS	SN74LS170N	SN74LS170ND	
SN74LS170NDS	SN74LS170NS	TCG74LS170	

BLOCK-1614

74LS173A	DM74LS173N	ECG74LS173	M74LS173AP
NTE74LS173A	SK1922	SN74LS173A	SN74LS173AJ
SN74LS173AJD	SN74LS173AJDS	SN74LS173AJS	SN74LS173AN
SN74LS173AND	SN74LS173ANDS	SN74LS173ANS	TCG74LS173A

BLOCK-1615

3102022	443-879	74LS174	74LS174N
74LS174PC	9LS174DC	9LS174DM	9LS174PC
AMX3565	C74LS174P	DM74LS174N	DN74LS174P
EAQ00-13100	ECG74LS174	HD74LS174	HD74LS174P
HE-443-879	HEPC74LS174P	I61DLS1740	LS174
M74LS174P	MB74LS174	MB74LS174M	N74LS174B
N74LS174F	N74LS174N	NTE74LS174	ON74LS174N
SK74LS174	SN74LS174	SN74LS174J	SN74LS174JD
SN74LS174JDS	SN74LS174JS	SN74LS174N	SN74LS174ND
SN74LS174NDS	SN74LS174NS	T74LS174B1	TCG74LS174

BLOCK-1616

13-117-6	3102023	443-752	74LS175
74LS175DC	74LS175N	74LS175PC	901521-34
9LS175DC	9LS175PC	AMX3566	C74LS175P
DM74LS175N	ECG74LS175	GD74LS175	HD74LS175
HD74LS175P	HE-443-752	HEPC74LS175P	M74LS175P
MB74LS175	MB74LS175M	N74LS175B	N74LS175F
N74LS175N	NTE74LS175	ON74LS175N	SK74LS175
SN74LS175	SN74LS175J	SN74LS175JD	SN74LS175JDS
SN74LS175JS	SN74LS175N	SN74LS175ND	
SN74LS175NDS	SN74LS175NS	T74LS175B1	TCG74LS175
X420301750			

BLOCK-1617

74LS181PC	C74LS181P	ECG74LS181	HD74LS181
HD74LS181P	HEPC74LS181P	MB74LS181	MB74LS181M
N74LS181F	N74LS181N	NTE74LS181	SN74LS181
SN74LS181J	SN74LS181JD	SN74LS181JDS	SN74LS181JS
SN74LS181N	SN74LS181ND	SN74LS181NDS	SN74LS181NS

BLOCK-1618

74LS190PC	C74LS190P	DM74LS190N	ECG74LS190
HD74LS190	HD74LS190P	HEPC74LS190P	M74LS190P
MB74LS190	MB74LS190M	N74LS190F	N74LS190N
NTE74LS190	SN74LS190	SN74LS190J	SN74LS190JD
SN74LS190JDS	SN74LS190JS	SN74LS190N	SN74LS190ND
SN74LS190NDS	SN74LS190NS	TCG74LS190	

BLOCK-1619

74LS191CH	74LS191DC	74LS191J	74LS191PC
74LS191W	C74LS191P	DM74LS191N	ECG74LS191
HD74LS191	HD74LS191P	HEPC74LS191P	M74LS191P
MB74LS191	MB74LS191M	N74LS191F	N74LS191N
NTE74LS191	SK74LS191	SN74LS191	SN74LS191J
SN74LS191JD	SN74LS191JDS	SN74LS191JS	SN74LS191N
SN74LS191ND	SN74LS191NDS	SN74LS191NS	SN74LS191W
TCG74LS191			

BLOCK-1620

443-817	74LS192	74LS192DC	74LS192PC
C74LS192P	DM74LS192N	ECG74LS192	HD74LS192
HD74LS192P	HE-443-817	HEPC74LS192P	M74LS192P
MB74LS192	MB74LS192M	N74LS192F	N74LS192N
NTE74LS192	SN74LS192	SN74LS192J	SN74LS192JD
SN74LS192JDS	SN74LS192JS	SN74LS192N	SN74LS192ND
SN74LS192NDS	SN74LS192NS	TCG74LS192	

BLOCK-1621

09A04	443-815	74LS193	74LS193CH
74LS193DC	74LS193J	74LS193N	74LS193PC
74LS193W	901521-26	C74LS193P	DM74LS193N
ECG74LS193	HD74LS193	HD74LS193P	HE-443-815
HEPC74LS193P	IC-74LS193	M74LS193P	MB74LS193
MB74LS193M	N74LS193F	N74LS193N	NTE74LS193
SK74LS193	SN74LS193	SN74LS193J	SN74LS193JD
SN74LS193JDS	SN74LS193JS	SN74LS193N	SN74LS193ND
SN74LS193NDS	SN74LS193NS	SN74LS193W	TCG74LS193

BLOCK-1622

74LS194	74LS194ADC	74LS194AFC	74LS194APC
74LSP14APC	9LS194DC	9LS194DM	9LS194PC
DM74LS194AN	ECG74LS194A	HD74LS194A	HD74LS194AP
M74LS194AP	N74LS194AF	N74LS194AN	NTE74LS194
NTE74LS194A	SK74LS194	SK74LS194A	SN74LS194A
SN74LS194AJ	SN74LS194AJD	SN74LS194AJDS	SN74LS194AJS
SN74LS194AN	SN74LS194AND	SN74LS194ANDS	SN74LS194ANS
TCG74LS194			

BLOCK-1623

443-718	74LS195ADC	74LS195AFC	74LS195APC
93L00DC	93L00DM	9LS195PC	DM74LS195AN
ECG74LS195A	HD74LS195A	HD74LS195AP	HE-443-718
M74LS195AP	N74LS195AF	N74LS195AN	NTE74LS195A
SK74LS195	SN74LS195	SN74LS195A	SN74LS195AJ
SN74LS195AJD	SN74LS195AJDS	SN74LS195AJS	SN74LS195AN
SN74LS195AND	SN74LS195ANDS	SN74LS195ANS	SN74LS195N
TCG74LS195A	X420301950		

BLOCK-1624

307-152-9-005	443-801	74LS196	74LS196PC
C74LS196P	DM74LS196	DM74LS196N	ECG74LS196
FC74LS196N	HE-443-801	HEPC74LS196P	M74LS196P
N74LS196F	N74LS196N	NTE74LS196	SK74LS196
SN74LS196	SN74LS196J	SN74LS196JD	SN74LS196JDS
SN74LS196JS	SN74LS196N	SN74LS196ND	
SN74LS196NDS	SN74LS196NS	TCG74LS196	

BLOCK-1625

74LS197	74LS197DC	74LS197PC	901521-54
C74LS197P	DM74LS197N	ECG74LS197	HEPC74LS197P
M74LS197P	N74LS197F	N74LS197N	N8293F
N8293N	NTE74LS197	SK1921	SN74LS197
SN74LS197J	SN74LS197JD	SN74LS197JDS	SN74LS197JS
SN74LS197N	SN74LS197ND	SN74LS197NDS	SN74LS197NS
TCG74LS197			

BLOCK-1626

3102011	443-798	74LS20	74LS20CH
74LS20DC	74LS20J	74LS20N	74LS20PC
74LS20W	AMX3555	C74LS20P	DM74LS20N
ECG74LS20	HD74LS20	HD74LS20P	HE-443-798
HEPC74LS20P	IC-74LS20	LS20	M74LS20P
MB74LS20	MB74LS20M	N74LS20F	N74LS20N
NTE74LS20	SK74LS20	SN74LS20	SN74LS20J
SN74LS20JD	SN74LS20JDS	SN74LS20JS	SN74LS20N
SN74LS20ND	SN74LS20NDS	SN74LS20NS	SN74LS20W
T74LS20B1	T74LS20D1	TCG74LS20	X420300200

BLOCK-1627

443-961	74LS21	74LS21DC	74LS21PC
ECG74LS21	GD74LS21	HD74LS21	HD74LS21P
M74LS21P	MB74LS21	MB74LS21M	N74LS21F
N74LS21N	NTE74LS21	SK1920	SN74LS21
SN74LS21J	SN74LS21JD	SN74LS21JDS	SN74LS21JS
SN74LS21N	SN74LS21ND	SN74LS21NDS	SN74LS21NS
T74LS21B1	T74LS21D1	TCG74LS21	

BLOCK-1628

08210063	74LS22	74LS22PC	ECG74LS22
HD74LS22	HD74LS22P	M74LS22P	MB74LS22
MB74LS22M	N74LS22F	N74LS22N	NTE74LS22
SK74LS22	SN74LS22	SN74LS22J	SN74LS22JD
SN74LS22JDS	SN74LS22JS	SN74LS22N	SN74LS22ND
SN74LS22NDS	SN74LS22NS	T74LS22B1	T74LS22D1
TCG74LS22			

BLOCK-1629

06300656	178744	74221	74LS221
74LS221CH	74LS221J	74LS221W	AMX3810
ECG74LS221	HD74LS221	HD74LS221P	M74LS221P
MB74LS221	MB74LS221M	N74LS221F	N74LS221N
NTE74LS221	POSN74LS221N	S4LS221N	SK74LS221
SN74LS221	SN74LS221J	SN74LS221JD	SN74LS221JDS
SN74LS221JS	SN74LS221N	SN74LS221ND	
SN74LS221NDS	SN74LS221NS	SN74LS221W	TCG74LS221
TVSS4LS221N	VHIHD74LS221/	X420302210	

BLOCK-1630

443-754	74LS240	74LS240DC	74LS240N
74LS240PC	AMX4225	DM74LS240N	DN74LS240P
ECG74LS240	GD74LS240	HD74LS240	HD74LS240P
HE-443-754	HP74LS240P	I61DLS2400	LS240
M74LS240P	MB74LS240	MB74LS240M	N74LS240F
N74LS240N	NTE74LS240	ON74LS240N	SK74LS240
SN74LS240	SN74LS240J	SN74LS240JD	SN74LS240JDS
SN74LS240JS	SN74LS240N	SN74LS240ND	
SN74LS240NDS	SN74LS240NS	TCG74LS240	

BLOCK-1631

443-824	74LS241	74LS241DC	74LS241N
74LS241PC	DM74LS241N	ECG74LS241	HD74LS241
HD74LS241P	HE-443-824	M74LS241P	MB74LS241
N74LS241F	N74LS241N	NTE74LS241	SK74LS241
SN74LS241	SN74LS241J	SN74LS241JD	SN74LS241JDS
SN74LS241JS	SN74LS241N	SN74LS241ND	
SN74LS241NDS	SN74LS241NS	TCG74LS241	

BLOCK-1632

443-884	74LS242	74LS242DC	74LS242PC
DM74LS242N	ECG74LS242	HE-443-884	M74LS242P
MB74LS242	MB74LS242M	N74LS242F	N74LS242N
NTE74LS242	SK74LS242	SN74LS242	SN74LS242J
SN74LS242JD	SN74LS242JDS	SN74LS242JS	SN74LS242N
SN74LS242ND	SN74LS242NDS	SN74LS242NS	TCG74LS242

BLOCK-1633

443-839	74LS243DC	74LS243N	74LS243PC
DM74LS243N	ECG74LS243	HE-443-839	M74LS243P
MB74LS243	MB74LS243M	N74LS243F	N74LS243N
NTE74LS243	SK74LS243	SN74LS243	SN74LS243J
SN74LS243JD	SN74LS243JDS	SN74LS243JS	SN74LS243N
SN74LS243ND	SN74LS243NDS	SN74LS243NS	SSN74LS243
TCG74LS243			

BLOCK-1634

08210069	74LS244	74LS244DC	74LS244N
74LS244PC	8020244	AMX3864	CO14313
DM74LS244N	DN74LS244	EAQ00-19400	ECG74LS244
GD74LS244	HD74LS244	HD74LS244P	HE-443-791
I61DLS2440	LS244	M74LS244P	MB74LS244
MB74LS244M	N74LS244F	N74LS244N	NTE74LS244
ON74LS244N	PBM74LS244P	SK74LS244	SN74LS244
SN74LS244J	SN74LS244JD	SN74LS244JDS	SN74LS244JS
SN74LS244N	SN74LS244ND	SN74LS244NDS	SN74LS244NS
T74LS244B1	TCG74LS244	X420302440	

BLOCK-1635

443-885	74LS245	74LS245DC	74LS245N
74LS245NA	74LS245PC	802-0245	8020245
AMX4470	DM74LS245N	DN74LS245P	EAQ00-22100
ECG74LS245	GD74LS245	HD74LS245WP	HE-443-885
I61DLS2450	LS245	M74LS245P	MB74LS245
MB74LS245M	N74LS245F	N74LS245N	NTE74LS245
ON74LS245N	POSN74LS245N	SK74LS245	SN74LS245
SN74LS245J	SN74LS245JD	SN74LS245JDS	SN74LS245JS
SN74LS245N	SN74LS245ND	SN74LS245NDS	SN74LS245NS
TCG74LS245	X420302450		

BLOCK-1636

443-711	74LS247DC	74LS247PC	74LS248DC
DM74LS247N	ECG74LS247	HD74LS247	HD74LS247P
HE-443-711	M74LS247P	MB74LS247	MB74LS247M
NTE74LS247	SK74LS247	SN74LS247	SN74LS247J
SN74LS247JD	SN74LS247JDS	SN74LS247JS	SN74LS247N
SN74LS247ND	SN74LS247NDS	SN74LS247NS	SW74LS247
TCG74LS247			

BLOCK-1637

74LS248PC	ECG74LS248	HD74LS248	HD74LS248P
M74LS248P	MB74LS248	MB74LS248M	NTE74LS248
SK74LS248	SN74LS248	SN74LS248J	SN74LS248JD
SN74LS248JDS	SN74LS248JS	SN74LS248N	SN74LS248ND
SN74LS248NDS	SN74LS248NS	TCG74LS248	

BLOCK-1638

74LS249DC	74LS249PC	ECG74LS249	HD74LS249
HD74LS249P	MB74LS249	MB74LS249M	MH746
NTE74LS249	SK74LS249	SN74LS249	SN74LS249J
SN74LS249JD	SN74LS249JDS	SN74LS249JS	SN74LS249N
SN74LS249ND	SN74LS249NDS	SN74LS249NS	TCG74LS249

BLOCK-1639

74LS251DC	74LS251PC	C74LS251P	DM74LS251N
ECG74LS251	HD74LS251	HD74LS251P	HEPC74LS251P
M74LS251P	MB74LS251	MB74LS251M	N74LS251AF
N74LS251AN	NTE74LS251	SK1919	SN74LS251
SN74LS251J	SN74LS251JD	SN74LS251JDS	SN74LS251JS
SN74LS251N	SN74LS251ND	SN74LS251NDS	SN74LS251NS
SN74LS251X	TCG74LS251		

BLOCK-1640

74LS253DC	74LS253PC	C74LS253P	DM74LS253N
ECG74LS253	HD74LS253	HD74LS253P	HEPC74LS253P
M74LS253P	MB74LS253	MB74LS253M	N74LS253F
N74LS253N	NTE74LS253	SK74LS253	SN74LS253
SN74LS253J	SN74LS253JD	SN74LS253JDS	SN74LS253JS
SN74LS253N	SN74LS253ND	SN74LS253NDS	SN74LS253NS
TCG74LS253			

BLOCK-1641

08210040	443-802	74LS257	74LS257A
74LS257AN	74LS257APC	74LS257DC	74LS257N
74LS257PC	901521-57	9334DC	9334DM
C74LS257P	DM74LS257BN	EAQ00-23600	ECG74LS257
HD74LS257	HD74LS257P	HE-443-1037	HE-443-802
HEPC74LS257P	M74LS257AP	MB74LS257	MB74LS257M
N74LS257AF	N74LS257AN	NTE74LS257	SK74LS257
SN74LS257	SN74LS257AJ	SN74LS257AJD	
SN74LS257AJDS	SN74LS257AJS	SN74LS257AN	
SN74LS257AND	SN74LS257ANDS	SN74LS257ANS	SN74LS257BN
SN74LS257J	SN74LS257N	SN74LS257X	T74LS257B1
TCG74LS257			

BLOCK-1642

74LS258	74LS258APC	74LS258DC	74LS258N
74LS258PC	901521-58	C74LS258P	DM74LS258BN
ECG74LS258	HD74LS258	HD74LS258P	HEPC74LS258P
M74LS258AP	MB74LS258	MB74LS258M	N74LS258AF
N74LS258AN	NTE74LS258	SK74LS258	SN74LS258
SN74LS258AJ	SN74LS258AJD	SN74LS258AJDS	SN74LS258AJS
SN74LS258AN	SN74LS258AND	SN74LS258ANDS	SN74LS258ANS
SN74LS258J	SN74LS258N	SN74LS258X	TCG74LS258

BLOCK-1643

443-804	74LS259	74LS259DC	74LS259PC
9334PC	C74LS259P	DM74LS259N	ECG74LS259
HD74LS259	HD74LS259P	HE-443-804	HEPC74LS259P
M74LS259P	N74LS259F	N74LS259N	N9334N
NTE74LS259	SK74LS259	SN74LS259	SN74LS259J
SN74LS259JD	SN74LS259JDS	SN74LS259JS	SN74LS259N
SN74LS259ND	SN74LS259NDS	SN74LS259NS	SN74LS259W
TCG74LS259			

BLOCK-1644

23119742	610002-026	74LS26	74LS26DC
74LS26PC	76-005-005	DM74LS26N	ECG74LS26
HD74LS26	HD74LS26P	MB74LS26	MB74LS26M
N74LS26F	N74LS26N	NTE74LS26	SK74LS26
SN74LS26	SN74LS26J	SN74LS26JD	SN74LS26JDS
SN74LS26JS	SN74LS26N	SN74LS26ND	SN74LS26NS
T74LS26B1	T74LS26D1	TCG74LS26	

BLOCK-1645

74LS260DC	74LS260PC	802-0260	ECG74LS260
N74LS260F	N74LS260N	NTE74LS260	SK74LS260
SN74LS260J	SN74LS260JD	SN74LS260JDS	SN74LS260JS
SN74LS260N	SN74LS260ND	SN74LS260NDS	SN74LS260NS

BLOCK-1646

197376	443-719	74LS266	74LS266CH
74LS266DC	74LS266J	74LS266PC	74LS266W
802-0266	9386DC	9386DM	9386PC
AMX4660	ECG74LS266	HD74LS266	HD74LS266P
HE-443-719	M74LS266P	MB74LS266	MB74LS266M
N74LS266F	N74LS266N	NTE74LS266	SK74LS266
SN74LS266	SN74LS266J	SN74LS266JD	SN74LS266JDS
SN74LS266JS	SN74LS266N	SN74LS266ND	
SN74LS266NDS	SN74LS266NS	SN74LS266W	TCG74LS266

BLOCK-1647

443-800	74LS27	74LS27CH	74LS27DC
74LS27J	74LS27PC	74LS27W	802-0027
901521-22	AMX4658	C74LS27P	DM74LS27N
ECG74LS27	HD74LS27	HD74LS27P	HE-443-800
HEPC74LS27P	M74LS27P	MB74LS27	MB74LS27M
N74LS27F	N74LS27N	NTE74LS27	SK74LS27
SN74LS27J	SN74LS27JD	SN74LS27JDS	SN74LS27N
SN74LS27ND	SN74LS27NDS	SN74LS27NS	SN74LS27W
T74LS27B1	T74LS27BI	T74LS27D1	TCG74LS27
X420300270			

BLOCK-1648

148684	443-805	74LS273	74LS273DC
74LS273N	74LS273PC	AMX4227	ECG74LS273
HD74LS273P	HE-443-805	M74LS273P	MB74LS273M
MX-4295	N74LS273F	N74LS273N	NTE74LS273
SK74LS273	SN74LS273	SN74LS273J	SN74LS273JD
SN74LS273JDS	SN74LS273JS	SN74LS273N	SN74LS273ND
SN74LS273NDS	SN74LS273NS	TCG74LS273	X420302730

BLOCK-1649

197381	443-854	74LS279DC	74LS279PC
C74LS279P	DM74LS279N	ECG74LS279	HD74LS279
HD74LS279P	HE-443-854	HEPC74LS279P	M74LS279P
N74LS279F	N74LS279N	NTE74LS279	SK74LS279
SN74LS279	SN74LS279J	SN74LS279JD	SN74LS279JDS
SN74LS279JS	SN74LS279N	SN74LS279ND	
SN74LS279NDS	SN74LS279NS	T74LS279B1	TCG74LS279

BLOCK-1650

74LS28DC	74LS28PC	ECG74LS28	MB74LS28
MB74LS28M	N74LS28F	N74LS28N	NTE74LS28
SK74LS28	SN74LS28	SN74LS28J	SN74LS28JD
SN74LS28JDS	SN74LS28JS	SN74LS28N	SN74LS28ND
SN74LS28NDS	SN74LS28NS	T74LS28B1	

BLOCK-1651

74LS280	ECG74LS280	HD74LS280	HD74LS280P
HE-443-1001	M74LS280P	MB74LS280	MB74LS280M
NTE74LS280	SK74LS280	SN74LS280	SN74LS280J
SN74LS280JD	SN74LS280JDS	SN74LS280JS	SN74LS280N
SN74LS280ND	SN74LS280NDS	SN74LS280NS	SN74LS280W
TCG74LS280	TCG74S280		

BLOCK-1652

443-855	74LS283	74LS283DC	74LS283PC
DM74LS283N	ECG74LS283	HD74LS283	HD74LS283P
HE-443-855	M74LS283P	MB74LS283	MB74LS283M
N74LS283F	N74LS283N	NTE74LS283	SN74LS283
SN74LS283J	SN74LS283JD	SN74LS283JDS	SN74LS283JS
SN74LS283N	SN74LS283ND	SN74LS283NDS	SN74LS283NS
TCG74LS283			

BLOCK-1653

443-731	74LS290	74LS290DC	74LS290PC
DM74LS290N	ECG74LS290	HD74LS290	HD74LS290P
HE-443-731	M74LS290P	N74LS290F	N74LS290N
NTE74LS290	SK74LS290	SN74LS290	SN74LS290J
SN74LS290JD	SN74LS290JDS	SN74LS290JS	SN74LS290N
SN74LS290ND	SN74LS290NDS	SN74LS290NS	TCG74LS290

BLOCK-1654

443-733	74LS293	74LS293DC	74LS293PC
DM74LS293N	ECG74LS293	HD74LS293	HD74LS293P
HE-443-733	M74LS293P	N74LS293F	N74LS293N
NTE74LS293	SK74LS293	SN74LS293	SN74LS293J
SN74LS293JD	SN74LS293JDS	SN74LS293JS	SN74LS293N
SN74LS293ND	SN74LS293NDS	SN74LS293NS	TCG74LS293

BLOCK-1655

74LS295ADC	74LS295AFC	74LS295APC	ECG74LS295A
HD74LS295B	HD74LS295BP	M74LS295AP	M74LS295BP
N74LS295BF	N74LS295BN	NTE74LS295	NTE74LS295A
SN74LS295AJ	SN74LS295AJD	SN74LS295AJDS	SN74LS295AJS
SN74LS295AN	SN74LS295AND	SN74LS295ANDS	SN74LS295ANS
SN74LS295B	SN74LS295BJ	SN74LS295BN	

BLOCK-1656

74LS298DC	74LS298PC	C74LS298P	ECG74LS298
HD74LS298	HD74LS298P	HEPC74LS298P	M74LS298P
N74LS298F	N74LS298N	NTE74LS298	SK74LS298
SN74LS298	SN74LS298J	SN74LS298JD	SN74LS298JDS
SN74LS298JS	SN74LS298N	SN74LS298ND	
SN74LS298NDS	SN74LS298NS	TCG74LS298	

BLOCK-1657

74LS299DC	74LS299FC	ECG74LS299	HD74LS299
HD74LS299P	M74LS299P	N74LS299F	N74LS299N
NTE74LS299	SN74LS299	SN74LS299J	SN74LS299JD
SN74LS299JDS	SN74LS299JS	SN74LS299N	SN74LS299ND
SN74LS299NDS	SN74LS299NS		

BLOCK-1658

3102013	443-732	74LS30	74LS30CH
74LS30DC	74LS30J	74LS30PC	74LS30W
AMX3556	C74LS30P	ECG74LS30	HD74LS30
HD74LS30P	HE-443-732	HEPC74LS30P	M74LS30P
MB74LS30	MB74LS30M	N74LS30F	N74LS30N
NTE74LS30	PBM74LS30P	SK74LS30	SN74LS30
SN74LS30J	SN74LS30JD	SN74LS30JDS	SN74LS30JS
SN74LS30N	SN74LS30ND	SN74LS30NDS	SN74LS30NS
SN74LS30W	T74LS30B1	T74LS30D1	TCG74LS30
X420300300			

BLOCK-1659

08210053	2362301	276-1915	3102014
443-875	612194-1	74LS32	74LS32CH
74LS32DC	74LS32J	74LS32L	74LS32N
74LS32P	74LS32PC	74LS32W	AMX3557
C74LS32P	DM74LS32N	DN74LS32	EAQ00-17600
ECG74LS32	GD74LS32	HD74LS32	HD74LS32P
HE-443-875	HEPC74LS32P	LS32	LS32N
M74LS32P	MB74LS32	MB74LS32M	N74LS32F
N74LS32N	NTE74LS32	P61LS32**H	PCMB74LS32M
SK74LS32	SN74LS32	SN74LS32J	SN74LS32JD
SN74LS32JDS	SN74LS32JS	SN74LS32N	SN74LS32ND
SN74LS32NDS	SN74LS32NS	SN74LS32W	T74LS32B1
T74LS32BI	T74LS32D1	TCG74LS32	VHINS74LS32-1
VHISN74LS32-1	X420300320	X420500320	

BLOCK-1660

74LS33	74LS33PC	EAQ00-24000	ECG74LS33
MB74LS33	MB74LS33M	N74LS33F	N74LS33N
NTE74LS33	SK74LS33	SN74LS33	SN74LS33J
SN74LS33N	SN74LS33NS		

BLOCK-1661

ECG74LS348	NTE74LS348	SK74LS348	SN74LS348
SN74LS348AJ	SN74LS348AJD	SN74LS348AJDS	SN74LS348AJS
SN74LS348AN	SN74LS348AND	SN74LS348ANDS	SN74LS348ANS
SN74LS348J	SN74LS348JD	SN74LS348JDS	SN74LS348JS
SN74LS348N	SN74LS348ND	SN74LS348NDS	SN74LS348NS
TCG74LS348			

BLOCK-1662

74LS352	74LS352DC	74LS352PC	DM74LS352N
DM74LS352W	ECG74LS352	M74LS352P	MB74LS352
MB74LS352M	N74LS352N	NTE74LS352	SK74LS352
SN74LS352	SN74LS352J	SN74LS352JD	SN74LS352JDS
SN74LS352JS	SN74LS352N	SN74LS352ND	
SN74LS352NDS	SN74LS352NS	T74LS352B1	TCG74LS352

BLOCK-1663

74LS353DC	74LS353PC	DM74LS352J	DM74LS353J
DM74LS353N	DM74LS353W	ECG74LS353	M74LS353P
MB74LS353	MB74LS353M	N74LS353N	NTE74LS353
SK74LS353	SN74LS353	SN74LS353J	SN74LS353JD
SN74LS353JDS	SN74LS353JS	SN74LS353N	SN74LS353ND
SN74LS353NDS	SN74LS353NS	T74LS353B1	TCG74LS353

BLOCK-1664

ECG74LS363	N74LS363F	N74LS363N	NTE74LS363
TCG74LS363			

BLOCK-1665

ECG74LS364	N74LS364F	N74LS364N	NTE74LS364
SK74LS364	TCG74LS364		

BLOCK-1666

08210068	612197-1	74LS365	74LS365A
74LS365ADC	74LS365APC	DM74LS365AJ	DM74LS365AN
DM74LS365AW	EAQ00-19000	ECG74LS365A	HD74LS365AP
M74LS365AP	MB74LS365A	MB74LS365AM	N74LS365AF
N74LS365AN	NTE74LS365A	SN74LS365A	SN74LS365AJ
SN74LS365AJD	SN74LS365AJDS	SN74LS365AJS	SN74LS365AN
SN74LS365AND	SN74LS365ANDS	SN74LS365ANS	SN74LS365J
SN74LS365N	TCG74LS365A		

BLOCK-1667

74LS366ADC	74LS366APC	DM74LS366AJ	DM74LS366AN
DM74LS366AW	ECG74LS366A	M74LS366AP	MB74LS366A
MB74LS366AM	N74LS366AN	NTE74LS366A	SN74LS366A
SN74LS366AJ	SN74LS366AJD	SN74LS366AJDS	SN74LS366AJS
SN74LS366AN	SN74LS366AND	SN74LS366ANDS	SN74LS366ANS
SN74LS366J	SN74LS366N	TCG74LS366A	

BLOCK-1668

276-1835	3102024	443-857	74LS367
74LS367A	74LS367ADC	74LS367APC	76-005-013
AMX3567	C74LS367P	DM74LS367AJ	DM74LS367AN
DM74LS367AW	DM74LS367N	ECG74LS367	HD74LS367AP
HE-443-857	HEPC74LS367P	M74LS367AP	MB74LS367A
MB74LS367AM	N74LS367AF	N74LS367AN	NTE74LS367
SK74LS367	SN74LS367	SN74LS367A	SN74LS367AJ
SN74LS367AJD	SN74LS367AJDS	SN74LS367AJS	SN74LS367AN
SN74LS367AN-X	SN74LS367AND	SN74LS367ANDS	SN74LS367ANS
SN74LS367J	SN74LS367N	TCG74LS367	

BLOCK-1669

3102025	74LS368	74LS368A	74LS368ADC
74LS368AN	74LS368APC	AMX3568	C74LS368P
DM74LS368AJ	DM74LS368AN	DM74LS368AW	ECG74LS368
HE-443-1024	HEPC74LS368P	M74LS368AP	MB74LS368A
MB74LS368AM	N74LS368AF	N74LS368AN	NTE74LS368
PO74LS368A	POSN74LS368AN	SN74LS368A	SN74LS368AJ
SN74LS368AJD	SN74LS368AJDS	SN74LS368AJS	SN74LS368AN
SN74LS368AND	SN74LS368ANDS	SN74LS368ANS	SN74LS368J
SN74LS368N	TCG74LS368		

BLOCK-1670

74LS37	74LS37DC	74LS37FC	74LS37PC
ECG74LS37	HD74LS37	HD74LS37P	M74LS37P
MB74LS37	MB74LS37M	MD74LS37M	N74LS37F
N74LS37N	NTE74LS37	SK74LS37	SN74LS37
SN74LS37J	SN74LS37JD	SN74LS37JDS	SN74LS37JS
SN74LS37N	SN74LS37ND	SN74LS37NDS	SN74LS37NS
TCG74LS37	X420500370		

BLOCK-1671

08210038	443-837	74LS373	74LS373DC
74LS373N	74LS373PC	8020373	901521-29
AM74LS373J	AM74LS373N	AM74LS373X	DM74LS373N
DN74LS373	EA000-22600	EAQ00-22600	EAQ00-22614
ECG74LS373	GD74LS373	HD74LS373P	HE-443-837
HE-443-867	I61DLS3730	LS373	M74LS373P
MB74LS373	MB74LS373P	N74LS373F	N74LS373N
NTE74LS373	ON74LS373N	P61LS373*H	
POSN74LS373N	SK74LS373	SN74LS373	SN74LS373J
SN74LS373JD	SN74LS373JDS	SN74LS373JS	SN74LS373N
SN74LS373ND	SN74LS373NDS	SN74LS373NS	T74LS373B1
TCG74LS373	X420303730		

BLOCK-1672

08210052	443-863	74LS374	74LS374DC
74LS374N	74LS374P	74LS374PC	C4LS374AP
DM74LS374N	EAQ00-19300	ECG74LS374	HD74LS374P
HE-443-863	LS374	M74LS374P	MB74LS374
N74LS374F	N74LS374N	NTE74LS374	ON74LS374N
P61LS374*H	S4LS374N	SK1918	SN74LS374
SN74LS374J	SN74LS374JD	SN74LS374JDS	SN74LS374JS
SN74LS374N	SN74LS374ND	SN74LS374NDS	SN74LS374NS
T74LS374B1	TCG74LS374	TVSS4LS374N	X420303740

BLOCK-1673

74LS377DC	74LS377PC	ECG74LS377	M74LS377P
N74LS377F	N74LS377N	NTE74LS377	SK74LS377
SN74LS377	SN74LS377J	SN74LS377JD	SN74LS377JDS
SN74LS377JS	SN74LS377N	SN74LS377ND	
SN74LS377NDS	SN74LS377NS	TCG74LS377	

BLOCK-1674

74LS378	74LS378DC	74LS378FC	74LS378PC
EAQ00-30300	ECG74LS378	N74LS378F	N74LS378N
NTE74LS378	SK74LS378	SN74LS378	SN74LS378J
SN74LS378JD	SN74LS378JDS	SN74LS378JS	SN74LS378N
SN74LS378ND	SN74LS378NDS	SN74LS378NS	TCG74LS378

BLOCK-1675

74LS379DC	74LS379FC	74LS379PC	ECG74LS379
N74LS379F	N74LS379N	NTE74LS379	SK74LS379
SN74LS379	SN74LS379J	SN74LS379JD	SN74LS379JDS
SN74LS379JS	SN74LS379N	SN74LS379ND	
SN74LS379NDS	SN74LS379NS	TCG74LS379	

BLOCK-1676

74LS38	74LS38-1	74LS38CH	74LS38DC
74LS38FC	74LS38J	74LS38N	74LS38PC
74LS38W	802-0038	AMX4328	DM74LS38N
ECG74LS38	HD74LS38	HD74LS38P	HE-443-1034
M74LS38P	MB74LS38	MB74LS38M	N74LS38F
N74LS38N	NTE74LS38	SK74LS38	SN74LS38
SN74LS38J	SN74LS38JD	SN74LS38JDS	SN74LS38JS
SN74LS38N	SN74LS38ND	SN74LS38NDS	SN74LS38NS
TCG74LS38			

BLOCK-1677

ECG74LS386	HD74LS386	HD74LS386P	M74LS386P
MB74LS386	MB74LS386M	N74LS386F	N74LS386N
NTE74LS386	SK74LS386	SN74LS386	SN74LS386J
SN74LS386JD	SN74LS386JDS	SN74LS386JS	SN74LS386N
SN74LS386ND	SN74LS386NDS	SN74LS386NS	

BLOCK-1678

443-893	443-921	74LS390	74LS390DC
74LS390PC	DM74LS390N	ECG74LS390	F-74LS390PC
HD74LS390	HD74LS390P	HE-443-921	IC-74LS390
M74LS390P	N74LS390F	N74LS390N	NTE74LS390
SN74LS390	SN74LS390J	SN74LS390JD	SN74LS390JDS
SN74LS390JS	SN74LS390N	SN74LS390ND	
SN74LS390NDS	SN74LS390NS		

BLOCK-1679

08210064	443-973	74LS393	74LS393DC
74LS393N	74LS393NA	74LS393PC	AMX3706
DM74LS393N	ECG74LS393	HD74LS393	HD74LS393P
HE-443-973	M74LS393P	N74LS393F	N74LS393N
NTE74LS393	PBM74LS393P	SK74LS393	SN74LS393
SN74LS393J	SN74LS393JD	SN74LS393JDS	SN74LS393JS
SN74LS393N	SN74LS393ND	SN74LS393NDS	SN74LS393NS
T74LS393B1	TCG74LS393	X420303930	

BLOCK-1680

74LS395A	74LS395DC	74LS395FC	74LS395PC
ECG74LS395A	M74LS395AP	M74LS395P	N74LS395AF
N74LS395AN	NTE74LS395A	SN74LS395A	SN74LS395AN
SN74LS395J	SN74LS395JD	SN74LS395JDS	SN74LS395JS
SN74LS395N	SN74LS395ND	SN74LS395NDS	SN74LS395NS
TCG74LS395A			

BLOCK-1681

ECG74LS398	N74LS398F	N74LS398N	NTE74LS398
SK74LS398	SN74LS398J	SN74LS398JD	SN74LS398JDS
SN74LS398JS	SN74LS398N	SN74LS398ND	
SN74LS398NDS	SN74LS398NS	T74LS398B1	

BLOCK-1682

ECG74LS399	N74LS399F	N74LS399N	SK74LS399
SN74LS399	SN74LS399J	SN74LS399JD	SN74LS399JDS
SN74LS399JS	SN74LS399N	SN74LS399ND	
SN74LS399NDS	SN74LS399NS	T74LS399B1	

BLOCK-1683

74LS40DC	74LS40PC	DM74LS40N	ECG74LS40
HD74LS40	HD74LS40P	LS40	M74LS40P
MB74LS40	MB74LS40M	N74LS40F	N74LS40N
NTE74LS40	SK74LS40	SN74LS40	SN74LS40J
SN74LS40JD	SN74LS40JDS	SN74LS40JS	SN74LS40N
SN74LS40ND	SN74LS40NDS	SN74LS40NS	T74LS40B1
T74LS40D1	TCG74LS40	X420500400	

BLOCK-1684

08210004	30901080	443-807	74LS42
74LS42CH	74LS42DC	74LS42J	74LS42PC
74LS42W	901521-17	C74LS42P	DM74LS42N
ECG74LS42	HD74LS42	HD74LS42P	HE-443-807
HEPC74LS42P	M74LS42P	MB74LS42	MB74LS42M
N74LS42F	N74LS42N	NTE74LS42	SK74LS42
SN74LS42	SN74LS42J	SN74LS42JD	SN74LS42JDS
SN74LS42JS	SN74LS42N	SN74LS42ND	SN74LS42NDS
SN74LS42NS	SN74LS42W	TCG74LS42	

BLOCK-1685

74LS47DC	74LS47PC	DM74LS47N	ECG74LS47
HD74LS47	HD74LS47P	M74LS47P	MB74LS47
MB74LS47M	NTE74LS47	SK74LS47	SN74LS47
SN74LS47J	SN74LS47JD	SN74LS47JDS	SN74LS47JS
SN74LS47N	SN74LS47ND	SN74LS47NDS	SN74LS47NS
TCG74LS47			

BLOCK-1686

74LS48DC	74LS48PC	DM74LS48N	ECG74LS48
HD74LS48	HD74LS48P	M74LS48P	MB74LS48
MB74LS48M	NTE74LS48	SK74LS48	SN74LS48J
SN74LS48JD	SN74LS48JDS	SN74LS48JS	SN74LS48N
SN74LS48ND	SN74LS48NDS	SN74LS48NS	TCG74LS48

BLOCK-1687

74LS49DC	74LS49PC	ECG74LS49	HD74LS49
HD74LS49P	MB74LS49	MB74LS49M	NTE74LS49
SK74LS49	SN74LS49	SN74LS49J	SN74LS49JD
SN74LS49JDS	SN74LS49JS	SN74LS49N	SN74LS49ND
SN74LS49NDS	SN74LS49NS	TCG74LS49	

BLOCK-1688

74LS490	74LS490PC	ECG74LS490	HD74LS490
HD74LS490P	M74LS490P	N74LS490F	N74LS490N
NTE74LS490	SK74LS490	SN74LS490	SN74LS490J
SN74LS490JD	SN74LS490JDS	SN74LS490JS	SN74LS490N
SN74LS490ND	SN74LS490NDS	SN74LS490NS	TCG74LS490

BLOCK-1689

443-951	74LS51	74LS51CH	74LS51DC
74LS51J	74LS51PC	74LS51W	C060474
C74LS51P	DM74LS51	DM74LS51N	ECG74LS51
HD74LS51	HD74LS51P	HEPC74LS51P	LS51N
M74LS51P	MB74LS51	MB74LS51M	N74LS51F
N74LS51N	NTE74LS51	SK74LS51	SN74LS51
SN74LS51J	SN74LS51JD	SN74LS51JDS	SN74LS51JS
SN74LS51N	SN74LS51ND	SN74LS51NDS	SN74LS51NS
SN74LS51W	T74LS51B1	T74LS51D1	TCG74LS51
X420300510			

BLOCK-1690

74LS54DC	74LS54PC	C74LS54P	ECG74LS54
HD74LS54	HD74LS54P	HEPC74LS54P	MB74LS54
MB74LS54M	MB75LS54	MB75LS54M	N74LS54F
N74LS54N	NTE74LS54	SK74LS54	SN74LS54
SN74LS54J	SN74LS54JD	SN74LS54JDS	SN74LS54JS
SN74LS54N	SN74LS54ND	SN74LS54NDS	SN74LS54NS
T74LS54B1	T74LS54D1	TCG74LS54	X420300540

BLOCK-1691

74LS540DC	74LS540PC	ECG74LS540	NTE74LS540
SN74LS540	SN74LS540J	SN74LS540JD	SN74LS540JDS
SN74LS540JS	SN74LS540N	SN74LS540ND	
SN74LS540NDS	SN74LS540NS	TCG74LS540	

BLOCK-1692

74LS541	74LS541DC	74LS541N	74LS541PC
ECG74LS541	HE-443-1058	NTE74LS541	SN74LS541
SN74LS541J	SN74LS541JD	SN74LS541JDS	SN74LS541JS
SN74LS541N	SN74LS541ND	SN74LS541NDS	SN74LS541NS
TCG74LS541	X420305410		

BLOCK-1693

74LS55PC	C74LS55P	DM74LS55N	ECG74LS55
HD74LS55	HD74LS55P	HEPC74LS55P	MB74LS55
MB74LS55M	N74LS55F	N74LS55N	NTE74LS55
SK74LS55	SN74LS55	SN74LS55J	SN74LS55JD
SN74LS55JDS	SN74LS55JS	SN74LS55N	SN74LS55ND
SN74LS55NDS	SN74LS55NS	T74LS55B1	T74LS55B1N
T74LS55D1	TCG74LS55		

BLOCK-1694

443-999	74LS624	ECG74LS624	HE-443-999
NTE74LS624	SK74LS624	SN74LS624	SN74LS624J
SN74LS624N	TCG74LS624		

BLOCK-1695

ECG74LS625	NTE74LS625	SK74LS625	SN74LS625
SN74LS625J	TCG74LS625		

BLOCK-1696

ECG74LS626	NTE74LS626	SK74LS626	SN74LS625N
SN74LS626	SN74LS626J	SN74LS626N	TCG74LS626

BLOCK-1697

ECG74LS627	NTE74LS627	SK74LS627	SN74LS627
SN74LS627J	SN74LS627N	TCG74LS627	

BLOCK-1698

74LS629N	901521-68	AMX4663	ECG74LS629
NTE74LS629	SK74LS629	SN74LS124J	SN74LS124N
SN74LS629	SN74LS629J	SN74LS629N	TCG74LS629

BLOCK-1699

ECG74LS640	M74LS640-1P	MB74LS640	MB74LS640M
N74LS640-1F	N74LS640-1N	N74LS640F	N74LS640N
NTE74LS640	SN74LS640	SN74LS640J	SN74LS640JD
SN74LS640JDS	SN74LS640JS	SN74LS640N	SN74LS640ND
SN74LS640NDS	SN74LS640NS	TCG74LS640	

BLOCK-1700

ECG74LS641	M74LS641-1P	MB74LS641	MB74LS641M
N74LS641-1F	N74LS641-1N	N74LS641F	N74LS641N
NTE74LS641	SK74LS641	SN74LS641	SN74LS641J
SN74LS641JD	SN74LS641JDS	SN74LS641JS	SN74LS641N
SN74LS641ND	SN74LS641NDS	SN74LS641NS	TCG74LS641

BLOCK-1701

ECG74LS642	M74LS642-1P	MB74LS642	MB74LS642M
N74LS642-1N	N74LS642F	N74LS642N	NTE74LS642
SK74LS642	SN74LS642	SN74LS642J	SN74LS642JD
SN74LS642JDS	SN74LS642JS	SN74LS642N	SN74LS642ND
SN74LS642NDS	SN74LS642NS	TCG74LS642	

BLOCK-1702

ECG74LS643	M74LS643-1P	MB74LS643	MB74LS643M
NTE74LS643	SN74LS643	SN74LS643J	SN74LS643JD
SN74LS643JDS	SN74LS643JS	SN74LS643N	SN74LS643ND
SN74LS643NDS	SN74LS643NS	TCG74LS643	

BLOCK-1703

08210050	74LS645	ECG74LS645	M74LS645-1P
MB74LS645	MB74LS645M	N74LS645-1F	N74LS645F
N74LS645N	NTE74LS645	SK74LS645	SN74LS645
SN74LS645J	SN74LS645JD	SN74LS645JDS	SN74LS645JS
SN74LS645N	SN74LS645ND	SN74LS645NDS	SN74LS645NS
TCG74LS645			

BLOCK-1704

74LS670	74LS670DC	74LS670FC	74LS670PC
C74LS670P	DM74LS670J	DM74LS670N	DM74LS670W
ECG74LS670	HE-443-1173	HEPC74LS670P	M74LS670P
N74LS670B	N74LS670F	N74LS670N	NTE74LS670
ON74LS670N	SN74LS670	SN74LS670J	SN74LS670JD
SN74LS670JDS	SN74LS670JS	SN74LS670N	SN74LS670ND
SN74LS670NDS	SN74LS670NS	T74LS670B1	TCG74LS670

BLOCK-1705

443-828	612308-1	74LS73	74LS73CH
74LS73J	74LS73W	ECG74LS73	HD74LS73
HD74LS73P	HE-443-828	M74LS73AP	M74LS73P
MB74LS73A	MB74LS73AM	N74LS73F	N74LS73N
NTE74LS73	SK74LS73	SK74LS73A	SN74LS73A
SN74LS73AJ	SN74LS73AJD	SN74LS73AJDS	SN74LS73AJS
SN74LS73AN	SN74LS73AND	SN74LS73ANDS	SN74LS73ANS
SN74LS73AW	SN74LS73J	SN74LS73N	TCG74LS73
X420300730			

BLOCK-1706

08210033	3102015	443-730	46-13593-3
569-0886-800	610002-074	612200-1	74LS74
74LS74A	74LS74AN	74LS74APC	74LS74CH
74LS74J	74LS74PC	74LS74W	76-005-007
8-759-900-74	901521-06	905-344	AMX3558
DM74LS74AN	EAQ00-12700	ECG74LS74A	GD74LS74A
HD74LS74	HD74LS74A	HD74LS74AP	HE-443-730
HEPC74LS74P	LS74	M74LS74AP	M74LS74P
MB74LS74A	MB74LS74AM	MX-3808	N74LS74AF
N74LS74AN	N74LS74F	N74LS74N	NTE74LS74A
P61LS74**H	PBM74LS74AP	SK74LS74	SK74LS74A

BLOCK-1706 (CONT.)

SN74LS74	SN74LS74A	SN74LS74AJ	SN74LS74AJD
SN74LS74AJDS	SN74LS74AJS	SN74LS74AN	SN74LS74AND
SN74LS74ANDS	SN74LS74ANS	SN74LS74AW	SN74LS74N
T74LS74B1	T74LS74D1	TCG74LS74A	X420300740
X420500740			

BLOCK-1707

443-781	74LS75	74LS75CH	74LS75J
74LS75W	DM74LS75N	ECG74LS75	HD74LS75
HD74LS75P	HE-443-781	IC-74LS75	M74LS75P
N74LS75F	N74LS75N	NTE74LS75	SK74LS75
SN74LS75	SN74LS75J	SN74LS75JD	SN74LS75JDS
SN74LS75JS	SN74LS75N	SN74LS75ND	SN74LS75NDS
SN74LS75NS	SN74LS75W	TCG74LS75	X420300750

BLOCK-1708

443-829	74LS76	74LS76DC	74LS76PC
ECG74LS76A	HD74LS76	HD74LS76P	HD74LSS76P
HE-443-829	M74LS76AP	MB74LS76A	MB74LS76AM
MB74LS76P	N74LS76F	N74LS76N	NTE74LS76A
SK74LS76A	SN74LS76A	SN74LS76AJ	SN74LS76AJD
SN74LS76AJDS	SN74LS76AJS	SN74LS76AN	SN74LS76AND
SN74LS76ANDS	SN74LS76ANS	TCG74LS76A	

BLOCK-1709

74LS77	ECG74LS77	HD74LS77	HD74LS77P
NTE74LS77	SK74LS77	SN74LS77J	SN74LS77JD
SN74LS77JDS	SN74LS77JS	SN74LS77N	SN74LS77ND
SN74LS77NDS	SN74LS77NS		

BLOCK-1710

ECG74LS78	HD74LS78	HD74LS78P	MB74LS78
MB74LS78AM	N74LS78F	N74LS78N	NTE74LS78
SK74LS78	SN74LS78A	SN74LS78AJ	SN74LS78AJD
SN74LS78AJDS	SN74LS78AJS	SN74LS78AN	SN74LS78AND
SN74LS78ANDS	SN74LS78ANS	TCG74LS78	

BLOCK-1711

74LS83ACH	74LS83ADC	74LS83AJ	74LS83APC
74LS83AW	DM74LS83AN	ECG74LS83A	HD74LS83A
HD74LS83AP	M74LS83AP	M74LS83P	MB74LS83A
MB74LS83AM	N74LS83AF	N74LS83AN	N74LS83F
NTE74LS83A	SK74LS83A	SK74LS83A	SN74LS83A
SN74LS83AJ	SN74LS83AJD	SN74LS83AJDS	SN74LS83AJS
SN74LS83AN	SN74LS83AND	SN74LS83ANDS	SN74LS83ANS
SN74LS83AW	SN74LS83N	T74LS83B1	T74LS83D1
TCG74LS83A			

BLOCK-1712

443-920	74LS85	74LS85CH	74LS85DC
74LS85J	74LS85PC	74LS85W	DM74LS85N
ECG74LS85	HD74LS85	HD74LS85P	HE-443-920
M74LS85P	MB74LS85	MB74LS85M	N74LS85F
N74LS85N	NTE74LS85	SK74LS85	SN74LS85J
SN74LS85JD	SN74LS85JDS	SN74LS85JS	SN74LS85N
SN74LS85ND	SN74LS85NDS	SN74LS85NS	SN74LS85W
TCG74LS85			

BLOCK-1713

08210070	443-891	46-131126-3	610002-086
74LS86	74LS86CH	74LS86DC	74LS86J
74LS86N	74LS86PC	74LS86W	8-759-900-86
901521-32	AMX3701	C74LS86P	DM74LS86N
ECG74LS86	F-74LS86PC	HD74LS86	HD74LS86P
HE-443-891	HEPC74LS86P	IC-74LS86	LS86
M74LS86BP	M74LS86P	MB74LS86	MB74LS86M
MB74LS86P	N74LS86F	N74LS86N	NTE74LS86
PBM74LS86P	Q7350	Q7402	SK74LS86
SN74LS86	SN74LS86AN	SN74LS86J	SN74LS86JD
SN74LS86JDS	SN74LS86JS	SN74LS86N	SN74LS86ND
SN74LS86NDS	SN74LS86NS	SN74LS86W	T74LS86B1
T74LS86D1	TCG74LS86	X420300860	

BLOCK-1714

443-813	74LS90	74LS90DC	74LS90PC
AMX3804	C74LS90P	ECG74LS90	FC74LS90
HD74LS90	HD74LS90P	HE-443-813	HEPC74LS90P
M74LS90P	N74LS90F	N74LS90N	NTE74LS90
SK74LS90	SN74LS90	SN74LS90J	SN74LS90JD
SN74LS90JDS	SN74LS90JS	SN74LS90N	SN74LS90ND
SN74LS90NDS	SN74LS90NS	TCG74LS90	

BLOCK-1715

74LS91	ECG74LS91	HD74LS91	HD74LS91P
M74LS91P	NTE74LS91	SN74LS91	SN74LS91J
SN74LS91JD	SN74LS91JDS	SN74LS91JS	SN74LS91N
SN74LS91ND	SN74LS91NDS	SN74LS91NS	

BLOCK-1716

3102016	74LS92	74LS92PC	DM74LS92N
ECG74LS92	HD74LS92	HD74LS92P	M74LS92P
N74LS92F	N74LS92N	NTE74LS92	SK74LS92
SN74LS92	SN74LS92J	SN74LS92JD	SN74LS92JDS
SN74LS92JS	SN74LS92N	SN74LS92ND	SN74LS92NDS
SN74LS92NS	TCG74LS92		

BLOCK-1717

08210022	307-152-9-009	3102017	612291-1
74LS93	74LS93CH	74LS93DC	74LS93J
74LS93N	74LS93PC	74LS93W	8-759-900-93
AMX3560	C74LS93P	DM74LS93N	ECG74LS93
FC74LS93	HD74LS93	HD74LS93P	HE-443-838
HEPC74LS93P	M74LS93P	N74LS93F	N74LS93N
NTE74LS93	SK74LS93	SN74LS93	SN74LS93J
SN74LS93JD	SN74LS93JDS	SN74LS93N	SN74LS93ND
SN74LS93NDS	SN74LS93NS	SN74LS93W	TCG74LS93
X420300930			

BLOCK-1718

ECG74LS95

BLOCK-1719

443-26	50254600	5076204	569-0883-200
717136-1	74S00	74S00DC	74S00N
74S00PC	9S00DC	9S00PC	DM74S00J
DM74S00N	ECG74S00	HD74S00	HD74S00P
HE-443-26	HL-55661	MB74S00	N74S00A
N74S00F	N74S00N	NTE74S00	ON74S00N
SK74S00	SN74500N	SN74S00	SN74S00J
SN74S00N	TCG74S00		

BLOCK-1720

443-896	74S02	74S02PC	ECG74S02
HE-443-896	N74S02F	N74S02N	NTE74S02
SK1917	SN74S02	SN74S02J	SN74S02N
TCG74S02	X420200020		

BLOCK-1721

569-0883-203	74S03DC	74S03PC	9S03DC
9S03PC	DM74S03J	DM74S03N	ECG74S03
HD74S03	HD74S03P	N74S03A	N74S03F
N74S03N	NTE74S03	SN74S03	SN74S03J
SN74S03N	TCG74S03		

BLOCK-1722

15109200	178737	443-897	569-0883-204
74S04	74S04ADC	74S04APC	74S04DC
74S04PC	9S04DC	9S04PC	AMX-4945
DM74S04J	DM74S04N	ECG74S04	GD74S04
HD74S04	HD74S04P	HE-443-897	N74S04A
N74S04F	N74S04N	NTE74S04	SK74S04
SN74S04	SN74S04J	SN74S04N	TCG74S04
VHIHD74S04/-1	VHISN74S04/-1	X420200040	

BLOCK-1723

569-0883-205	74S05	74S05ADC	74S05APC
74S05DC	74S05PC	9S05DC	9S05PC
DM74S05J	DM74S05N	ECG74S05	HD74S05
HD74S05P	N74S05A	N74S05F	N74S05N
NTE74S05	SN74S05	SN74S05J	SN74S05N
TCG74S05			

BLOCK-1724

443-976	74S08	74S08DC	74S08PC
ECG74S08	GD74S08	N74S08F	N74S08N
NTE74S08	SN74S08	SN74S08J	SN74S08N
TCG74S08			

BLOCK-1725

74S09DC	74S09PC	ECG74S09	N74S09F
N74S09N	NTE74S09	SN74S09	SN74S09J
SN74S09N	TCG74S09		

BLOCK-1726

50254700	569-0883-210	74S10	74S10DC
74S10PC	9S10DC	9S10PC	DM74S10J
DM74S10N	ECG74S10	HD74S10	HD74S10P
HL56320	N74S10A	N74S10F	N74S10N
NTE74S10	ON74S10N	SN74S10	SN74S10J
SN74S10N	TCG74S10		

BLOCK-1727

569-0883-211	74S11	74S11DC	74S11PC
9S11DC	9S11PC	DM74S11J	DM74S11N
ECG74S11	HD74S11	HD74S11P	HD74S11S
MX-6406	N74S11A	N74S11F	N74S11N
NTE74S11	SK74S11	SN74S11	SN74S11J
SN74S11N	TCG74S11	VHISN74S11/-1	

BLOCK-1728

443-42	569-0883-312	74S112	74S112A
74S112DC	74S112N	74S112PC	9S112DC
9S112PC	DM74S112N	ECG74S112	HD74S112
HD74S112P	N74S112B	N74S112F	N74S112N
NTE74S112	SK74S112	SN74LS112N	SN74S112
SN74S112J	SN74S112N	TCG74S112	

BLOCK-1729

50255000	74S113DC	74S113PC	9S113DC
9S113PC	DM74S113N	ECG74S113	HD74S113
HD74S113P	MC3062P	N74S113A	N74S113F
N74S113J	NTE74S113	SK9704	SN74S113
SN74S113J	SN74S113N	TCG74S113	

BLOCK-1730

ECG74S124	NTE74S124	SN74S124	SN74S124J
SN74S124N	TCG74S124		

BLOCK-1731

148716	74S133DC	74S133PC	9S133DC
9S133PC	ECG74S133	HD74S133	HD74S133P
N74S133B	N74S133F	N74S133N	N74S133W
NTE74S133	SN74S133	SN74S133J	SN74S133N
TCG74S133			

BLOCK-1732

74S134DC	74S134PC	9S134DC	9S134PC
ECG74S134	HD74S134	HD74S134P	N74S134B
N74S134F	N74S134N	N74S134W	NTE74S134
SN74S134	SN74S134J	SN74S134N	TCG74S134

BLOCK-1733

443-981	569-0883-338	74S138	74S138DC
74S138PC	ECG74S138	F-74S138PC	F74S138
HL55663	IC-74S138	N74S138B	N74S138F
N74S138N	NTE74S138	SN74S138	SN74S138J
SN74S138N	SN74S138X	TCG74S138	

BLOCK-1734

569-0883-340	74S140DC	74S140PC	9S140DC
9S140PC	DM74S140N	ECG74S140	HD74S140
HD74S140P	N74S140A	N74S140F	N74S140N
NTE74S140	SN74S140	SN74S140J	SN74S140N
TCG74S140			

BLOCK-1735

569-0883-215	74S15DC	74S15PC	9S15DC
9S15PC	DM74S15N	ECG74S15	HD74S15
HD74S15P	N74S15A	N74S15F	N74S15N
NTE74S15	SN74S15	SN74S15J	SN74S15N
TCG74S15			

BLOCK-1736

74S151	74S151DC	74S151PC	DM74S151N
ECG74S151	HD74S151	HD74S151P	N74S151B
N74S151F	NTE74S151	SN74S151	SN74S151J
SN74S151N	SN74S151X	TCG74S151	ZN74154E

BLOCK-1737

443-935	569-0883-353	74S153	74S153DC
74S153PC	93S153DC	DM74S153N	ECG74S153
HE-443-935	HL55723	N74S153F	N74S153J
N74S153N	NTE74S153	POSN74S153N	SK1915
SN74S153	SN74S153J	SN74S153N	SN74S153X
TCG74S153			

BLOCK-1738

443-964	569-0883-357	74S157	74S157DC
74S157PC	84S157PC	93S157DC	93S157PC
DM74S157N	ECG74S157	N74S157B	N74S157F
N74S157N	NTE74S157	SK1914	SN74S157
SN74S157J	SN74S157N	TCG74S157	

BLOCK-1739

569-0883-358	74S158DC	74S158PC	93S158DC
93S158PC	ECG74S158	N74S158B	N74S158F
N74S158N	NTE74S158	SN74S158	SN74S158J
SN74S158N	TCG74S158		

BLOCK-1740

569-0883-374	74S174	74S174N	74S174NA
74S174PC	93S174DC	93S174PC	DM74S174N
ECG74S174	HD74S174	HD74S174P	HE-443-1053
N74S174B	N74S174F	N74S174N	NTE74S174
SK74S174	SN74S174	SN74S174J	SN74S174N
TCG74S174			

BLOCK-1741

443-983	74S175	74S175DC	74S175N
74S175PC	93S175DC	93S175PC	DM74S175N
ECG74S175	HD74S175P	HE-443-983	HE-443-985
N74S175B	N74S175F	N74S175N	NTE74S175
SN74S175	SN74S175J	SN74S175N	TCG74S175

BLOCK-1742

74S181DC	74S181PC	DM74S181N	ECG74S181
HD74S181	HD74S181P	N74S151N	N74S181F
N74S181J	N74S181N	N74S181W	NTE74S181
SN74S181	SN74S181J	SN74S181N	TCG74S181

BLOCK-1743

AM27S18DC	DM74S188J	DM74S188N	ECG74S188
HM1-7602-5	HM3-7602-5	MB7056C	SN74S188
TBP18SA030	TBP18SA030J	TBP18SA030N	

BLOCK-1744

2360751	74S194DC	74S194FC	74S194PC
93S194DC	93S194PC	AM74S194N	DM74S194N
ECG74S194	HD74194	HD74194P	N74S194B
N74S194F	N74S194J	N74S194N	N74S194W
NTE74S194	SN74S194	SN74S194J	SN74S194N
TCG74S194			

BLOCK-1745

50254900	5076205BZ	569-0883-220	717136-15
74S20DC	74S20PC	9S20DC	9S20PC
DM74S20J	DM74S20N	ECG74S20	HD74S20
HD74S20P	N74S20A	N74S20F	N74S20N
NTE74S20	ON74S20N	SK74S20	SN74S20
SN74S20J	SN74S20N	TCG74S20	

BLOCK-1746

569-0883-222	74S22DC	74S22PC	9S22DC
9S22PC	DM74S22J	DM74S22N	ECG74S22
HD74S22	HD74S22P	N74S22A	N74S22F
N74S22N	NTE74S22	SK74S22	SN74S22
SN74S22J	SN74S22N	TCG74S22	

BLOCK-1747

74S251DC	DM74S251N	ECG74S251	HD74S251
HD74S251P	N74S251B	N74S251F	N74S251N
NTE74S251	SN74S251	SN74S251J	SN74S251N
SNM74S251J	TCG74S251		

BLOCK-1748

74S258DC	74S258PC	DM74S258N	ECG74S258
N74S258B	N74S258F	N74S258J	N74S258N
NTE74S258	SN74S258J	SN74S258N	TCG74S258

BLOCK-1749

24S10	AM27S21DC	DM74S287J	DM74S287N
ECG74S287	HE-444-228	HM1-7611-5	HM3-7611-5
HM3-7611A5	MB7052Z	N82S129N	NTE74S287
SN74S287	TBP24S10	TBP14S10J	TBP14S10N
TBP24S10N			

BLOCK-1750

74S288	82S123	AM27S19DC	DM74S288J
DM74S288N	ECG74S288	HM1-7603-5	HM3-7603-5
MB7051C	N82S123N	NTE74S288	SN74S288
TBP18S030	TBP18S030J	TBP18S030N	Y440800101

BLOCK-1751

74S30DC	74S30PC	ECG74S30	NTE74S30
SN74S30	SN74S30J	SN74S30N	TCG74S30

BLOCK-1752

6300-1N	AM27S20DC	DM74S387J	DM74S387N
ECG74S387	HM1-7610-5	HM3-7610-5	HM3-7610A5
MB7057Z	NTE74S387	SN74S387	TBP14SA10
TBP14SA10J	TBP14SA10N		

BLOCK-1753

443-982	569-0883-240	74S240DC	74S40
74S40PC	9S40DC	9S40PC	DM74S40N
ECG74S40	HD74S40	HD74S40P	N74S40A
N74S40F	N74S40N	NTE74S40	SN74S40
SN74S40J	SN74S40N	TCG74S40	

BLOCK-1754

29651ADC	29651DC	AM27S185DC	DM87S185J
DM87S185N	ECG74S454	HM1-7685-5	HM3-7685-5
HN25085	HN25085S	MB7128EZ	MB7128HZ
MCM7685D	SN74S454	TBP24S81	TBP24S81J
TBP24S81N			

BLOCK-1755

29621ADC	29621DC	AM27S29DC	DM74S472AJ
DM74S472AN	DM74S472N	ECG74S472	HM1-7649-5
HM3-7649-5	MB7124EZ	NTE74S472	SN74S472
TBP18S42	TBP18S42J	TBP18S42N	

BLOCK-1756

AM27S28DC	DM74S473AJ	DM74S473AN	DM74S473J
DM74S473N	ECG74S473	HM1-7648-5	HM3-7648-5
MB7123EZ	SN74S473	TBP18SA42	TBP18SA42J
TBP18SA42N			

BLOCK-1757

93448DC	93448DM	93448PC	AM27S31DC
DM74S474AJ	DM74S474AN	DM74S474J	DM74S474N
ECG74S474	HM1-7641-5	HM3-7641-5	HM3-7641A5
MCM7641D	NTE74S474	SN74S474	TBP18S46
TBP18S46J	TBP18S46N		

BLOCK-1758

93438DM	93438PC	AM27S30DC	DM74S475AJ
DM74S475AN	DM74S475J	DM74S475N	ECG74S475
HM1-7640-2	HM1-7640-5	HM3-7640-5	HM3-7640A5
NTE74S475	SN74S475	TBP18SA46	TBP18SA46J
TBP18SA46N			

BLOCK-1759

29631ADC	29631DC	93451DC	93451DM
93451PC	AM27S181DC	D3628A	D3628A-1
ECG74S478	HM1-7681-5	HM3-7681-5	HN25089
HN25089S	MB7132EC	MB7132HC	MCM7681
MCM7681D	MCM7681P	SK9703	SN74S478
TBP28S86J	TBP28S86N		

BLOCK-1760

93450DC	93450DM	93450PC	AM27S180DC
ECG74S479	HM1-7680-5	HM3-7680-5	HN25088
HN25088S	MB7131EC	MB7131HC	SN74S479
TBP28SA86J	TBP28SA86N		

BLOCK-1761

74S51	74S51DC	74S51NA	74S51PC
DM74S51N	ECG74S51	N74S51F	N74S51N
N74S551F	NTE74S51	SN74S51	SN74S51J
SN74S51N	TCG74S51		

BLOCK-1762

AM27S12DC	DM74S570J	DM74S570N	ECG74S570
HM1-7620-5	HM3-7620-5	HM3-7620A5	MB7058Z
NTE74S570			

BLOCK-1763

29611ADC	29611DC	AM27S13DC	DM74S571J
DM74S571N	ECG74S571	HM1-7621-5	HM3-7621-5
HM3-7621A5	MB7053Z	NTE74S571	

BLOCK-1764

93452DC	93452DM	93452PC	93453DM
AM27S32DC	DM74S572AJ	DM74S572AN	DM74S572J
DM74S572N	ECG74S572	HM1-7642-5	HM1-7642A-5
HM3-7642-5	HN25044	MB7121EC	NTE74S572
SN74S477	TBP24SA41	TBP24SA41J	TBP24SA41N

BLOCK-1765

93453PC	AM27S33DC	DM74S573AJ	DM74S573AN
DM74S573J	DM74S573N	ECG74S573	HM1-7643-5
HM1-7643A-5	HM3-7643-5	HM3-7643A-5	HN25045
MB7122EC	MCM7643D	N82S137BN	NTE74S573
SN74S476	TBP24S41	TBP24S41J	TBP24S41N

BLOCK-1766

569-0883-264	74S64	74S64DC	74S64PC
9S64DC	9S64PC	DM74S64N	ECG74S64
HD74S64	HD74S64P	N74S64A	N74S64F
N74S64N	NTE74S64	SN74S64	SN74S64J
SN74S64N	TCG74S64		

BLOCK-1767

443-899	569-0883-265	74S65	74S65DC
74S65PC	9S65DC	9S65PC	DM74S65N
ECG74S65	HD74S65	HD74S65P	HE-443-899
N74S65A	N74S65F	N74S65N	NTE74S65
SN74S65	SN74S65J	SN74S65N	TCG74S65

BLOCK-1768

443-900	569-0883-274	74S74	74S74DC
74S74PC	9S74DC	9S74PC	C74LS74P
DM74S74N	ECG74S74	GD74S74	H31100130
HD74S74	HD74S74P	HE-443-900	N74S74A
N74S74F	N74S74N	NTE74S74	SK74S74
SN74S74	SN74S74J	SN74S74N	TCG74S74

BLOCK-1769

443-915	74S86	74S86DC	74S86PC
DM74S86N	ECG74S86	HD74S86	HD74S86P
HE-443-915	N74S86A	N74S86F	N74S86N
N74S86W	NTE74S86	SK74S86	SN74S86
SN74S86J	SN74S86N	TCG74S86	

BLOCK-1770

1C204	221-Z9089	C6061P	ECG750
GEIC-219	HEP-C6061P	HEP6061P	HEPC6061P
MC1355P	NTE750	SK3280	TCG750
TM750	TR09007		

BLOCK-1771

08210047	2420SN75188	276-2520	443-794
75188	75188N	AMX3867	DS1488J
DS1488N	EAQ00-30000	ECG75188	GD75188
HD75188	HD75188P	HE-443-794	I6PDC14880
MC1488	MC1488F	MC1488L	MC1488N
MC1488P	MC1488PD	NTE75188	SG1488J
SK5188	SN75188	SN75188J	SN75188N
TCG75188	UA1488	UA1488D	UA1488DC
UA1488P	UA1488PC	UA4188DC	X440751880
XR1488N	XR1488P		

BLOCK-1772

08210048	2420SN75189	276-2521	443-795
75189	75189A	75189AN	AMX3868
DS1489AJ	DS1489AN	DS1489J	DS1489N
EAQ00-30200	ECG75189	GD75189A	HD75189
HD75189P	HE-443-795	I6PDC14890	IC-1489
MC1489	MC1489/SN75189N	MC1489A	MC1489AP
MC1489F	MC1489L	MC1489N	MC1489P
NTE75189	SG1489J	SK5189	SN75189
SN75189A	SN75189AN	SN75189N	TCG75189
UA1489	UA1489APC	UA1489D	UA1489DC
UA1489P	UA1489PC	UA1489PCQR	X440751890
XR1489AN			

BLOCK-1773

137161	657161	7161	
7161(WURLITZER)	7874(WURLITZER)	C6003	ECG752
HEPC6003	HEPC6003P	IC-21(ELCOM)	IC21(ELCOM)
MC3360P	MFC4000	MFC4000A	MFC4000B
TM752	WE4110	WEP928	

BLOCK-1774

301-680-1	51-10600A	51-10600A01	51S10600A01
680-1(IC)	C6010	ECG753	HEPC6010
HEPC6010P	MFC4010	MFC4010A	T2E
TM753			

BLOCK-1775

569-0885-850	75450APC	75450BDC	75450BN
75450BPC	DS75450N	ECG75450B	HD75450A
NTE75450B	SK9828	SN75450B	SN75450BN
TCG75450B			

BLOCK-1776

75451ATC	75451BN	75451BRC	75451BT
75451BTC	DS75451N	ECG75451B	HD75451A
HD75451AP	NTE75451B	SK9807	SN75451B
SN75451BP	TCG75451B	UDN-3611M	

BLOCK-1777

3106002	443-74	569-0885-854	75452
75452ATC	75452BN	75452BRC	75452BT
75452BTC	AMX3573	AMX4321	DS75452N
EAQ00-05000	ECG75452B	HD75452	HD75452P
HE-443-74	MC75452P	MC75452P1	NTE75452B
SK9834	SN75452B	SN75452BP	SN75452P
TCG75452B	UDN-3612M		

BLOCK-1778

569-0885-853	75453ATC	75453BN	75453BRC
75453BT	75453BTC	DS75453N	ECG75453B
HD75453	HD75453P	NTE75453B	SK9827
SN75453B	SN75453BP	TCG75453B	UDN-3613M

BLOCK-1779

75454BN	75454BT	DS75454N	ECG75454B
HD75454	HD75454P	NTE75454B	SK75454B
SN75454BN	TCG75454B	UDN-3614M	

BLOCK-1780

75491PC	DS75491N	ECG75491B	MC75491P
NTE75491B	SK9786	SN75491AN	SN75491N
TCG75491B			

BLOCK-1781

75492PC	9664PC	DS75492N	ECG75492B
NTE75492B	SN75492N	TCG75492B	

BLOCK-1782

DS75493N	ECG75493	NTE75493	SK75493
SN75493N	TCG75493		

BLOCK-1783

DS75494N	ECG75494	NTE75494	SK9825
SN75494N	TCG75494		

BLOCK-1784

ECG75497	SN75497N

BLOCK-1785

ECG75498	NTE75498	SN75498N

BLOCK-1786

221-Z9090	51-10534A03	51-43639802	51-43639B02
51-43684B02	51-43684B0P	51C43684B02	51C436884B01
51S10534A01	51S10534A03	A7M42394B39	A7M42894B39
C6013	ECG755	HEPC6013	IC-290(ELCOM)
IC-435(ELCOM)	IC-442(ELCOM)	IC290(ELCOM)	M304/4050
MC4050	MFC4050	MFC4052	REN755
SC5221P	SCF40501	SFC4050	SFC4052
T1T			

BLOCK-1787

134509	1464295-1	4295-1	4295-1(RCA)
4295-2	C6014	ECG756	ECG757
HEPC6014	IC-433(ELCOM)	IC-443(ELCOM)	MFC4060
MFC4060A	MFC4062	MFC4062A	R4295-1
R4295-1(RCA)	R4295-2	TM756	TM757

BLOCK-1788

ECG758	IC-444(ELCOM)	MFC4063	MFC4063A
TM758			

BLOCK-1789

ECG759	IC-445(ELCOM)	MFC4064	MFC4064A
TM759			

BLOCK-1790

181-000100	221-Z9091	51-10611A09	51S10611A09
770339	9008	C6001	CXL760
ECG760	IC-298(ELCOM)	IC298(ELCOM)	IC501(ELCOM)
MFC6010	NTE760	SN76246A	TM760
TR09008	TR09009		

BLOCK-1791

013-005005	013-005005-6	013-005007	013-005007-6
15942-1	5000	5005	5005(THOMAS)
5007	5007(THOMAS)	514-048	551A
7874	991-019542-1	C6011	ECG761
GE-761	HEPC6011	IC-286(ELCOM)	IC286(ELCOM)
MCF6021	MFC6020	TM761	WEP930

BLOCK-1792

134340	1464295-2	15-34401-1	162451
4295-2(RCA)	ECG763	IC-446(ELCOM)	MFC6032
MFC6032A	TM763		

BLOCK-1793

ECG764	ECG765	IC-447(ELCOM)	IC-448(ELCOM)
MFC6033	MFC6033A	MFC6034	MFC6034A
TM764	TM765		

BLOCK-1794

013-005009-6	13-50-6	177A07	221-76
221-76-01	5009	5009(THOMAS)	51-10542A01
51-70177A07	51M70177A07	70177A07	72133-1
72133-2	86-5009-2	A072133-001	ECG767
HEPC6015	IC-244(ELCOM)	MC3340P1	MFC6050
SFC6050	SN16890	TM767	

BLOCK-1795

ECG768	IC-449(ELCOM)	MFC6060	TM768

BLOCK-1796

ECG770	IC-450(ELCOM)	MFC6080	TM770

BLOCK-1797

142251(IC)	51-10636A	51-10636A01	51-84320A09
51R84320A09	51R8432A09	51S10636A	51S10636A01
C6016	ECG772	ECG772A	ECG796
HEPC6016	HEPC6016C	IC-451(ELCOM)	M2009
MFC8020	MFC8021	MFC8021A	T2L
TM772A	TM796		

BLOCK-1798

679-1	C6017	ECG773	HEPC6017
IC-217(ELCOM)	IC217(ELOCM)	MFC8030	TM773

BLOCK-1799

ECG775

BLOCK-1800

ECG776	IC-453(ELCOM)	MC3370P	MFC8070
TM776			

BLOCK-1801

02-004558	02-455804	10650A01	1231-0132
1231-0140	1458	148235	148724
148728	15-34502-1	15-34502-2	15-34502-3
161743	162037	178741	191868
2002800112	2102400203	2102400429	221-240
221-Z9034	22114594	2362601	2362605
276-038	30900361	30900740	30900741
4558C	4558D	4558DD	4558DV
46-131073-3	46-13310-3	46-133325-3	46-13345-3
46-13404-3	51-10650A01	51X90458A83	51X90458A84
569-0540-358	569-0773-610	612113-1	61249-1
612494-1	6124940001	6458D	6552
6572	8-759-145-58	8-759-945-58	8-795-145-58
905-255	921-1188	AMX4661	AN1458
AN6352	AN6552	AN6552(F)	AN6553
AN6558	AN6572	B0350000	B0350500
B0350602	BA4558	BA4558DX	C4558C
C6102P	C6102P(HEP)	CA1458	CA1458E
CA1458F	CA1458G	CA1458S	CA1558E
DM-40	EAS00-06100	ECG778	ECG778A
EP84X69	EW84X192	EW84X483	EW84X488
EW84X788	EW84X865	GEIC-220	GL4558
HA17458GS	HA17458PS	HA17558	HE-442-21
HEP-C6102P	HEPC6102P	IGLA6458DF	IJNJM4558DM
LA6458D	LA6458DF	LM1458	LM1458M
LM1458N	M5218P	M5223P	M5R4558P
MC1458	MC1458CP1	MC1458G	MC1458N
MC1458P	MC1458P1	MC1458V	MC1558G
MC4558ACP1	MC4558CP	MC4558CP1	MC4558CU
ML1458S	MPC1458C	NIM4558D	NJM4558C
NJM4558D	NJM4558D-1	NJM4558DD	NJM4558DD
NJM4558DM	NJM4558DMA	NJM4558DV	NJM4558DX
NJM4558K	NJM4558MDA	NJM4559D	NTE778A
PC4558C	PTC736	Q7219	RC1458DE
RC1458DN	RC1458NB	RC2041M	RC2043NB
RC4558	RC4558DE	RC4558DQ	RC4558JG
RC4558NB	RC4558P	RC4559DE	RC4559NB
RM4558	RS1458	RV4558DE	RV4558NB
RV4559DE	RV4559NB	SFC2458DC	SG1458M
SK3465	T2R	TA75458P	TA75557P
TA75558P	TA75559P	TA75559P(FA-1)	TA75559PFA-1
TBB1458B	TCG778A	TDA1458D	TL4558P
TM778	TM778A	TVCM-61	UA1458CP
UA1458CTC	UA1458P	UA1458TC	UA4558T
UPC1458	UPC1458C	UPC251C	UPC4558
UPC4558C	UPC4558D	VHIIR94558/-1	XR1458CP
VHINJM4558D-1	WEP2053	WEP2053/778	
XR4558CP			

BLOCK-1802

1330-9240	156932	180217	5369181
BA15218N	ECG778S	M5218L	NTE778S
SK10139	TVSM5218L		

BLOCK-1803

BA4558F	ECG778SM	MC4558CD	NJM4558M
NJM4558M-TE2	NTE778SM	SK10061	

BLOCK-1804

14-2011-01	14-201I-02	1C-19	1C288
442-74	46-5002-19	46-5002-27	ECG779
ECG779-1	ECG779A	GEIC-221	HE-442-74
IC-19	IC-19(PHILCO)	IC-27(PHILCO)	IC-288(ELCOM)
IC-32	IC288(ELCOM)	MC1344	MC1344P
MC1345	MC1345P	MC1345PQ	SK3240
TM779-1	TVCM-62	ZEN610	

BLOCK-1805

13-30-6	130130	137245	14636723
1463677-1	1463677-2	1463677-3	1473677-3
221-Z9014	3064TC	442-5	46-5002-16
CA3064	CXL780	ECG780	GEIC-20
GEIC-222	HEP-C6069G	HEPC6069G	IC-16
IC-16(PHILCO)	LM3064M	MC1364G	NTE780
RE364-IC	REN780	SK3141	TCG780
TM780	TVCM-31	UA3064HC	UA3064TC
ULN2264	ULN2264K	WEP2055	WEP2055/780

BLOCK-1806

4082802-0001	4082802-0002	51-10637A01	5757
CA3076	CA3076T	CXL781	ECG781
GEIC-223	NTE781	PTC1956	SK3169
SN76676HC	T2M	TCG781	TM781
TVCM-56			

BLOCK-1807

1C21	46-5002-21	46-5002-5	ECG782
GEIC-224	IC-5	IC-5(PHILCO)	IC21
TM782	WEP2057/782		

BLOCK-1808

02-561201	1010-9965	1079	1097
1154	13-73-6	15SI-CA3064E	221-Z9036
2N5466	40SI-MC1364P	56A20-1	56D20-1
5D20-1	612061-1	61B1C	73C180843
73C180843-1	73C180843-2	73C180843-3	73C180843-4
99S097-1	C6100P	C6100P(HEP)	CA3064E
CXL783	ECG783	EP84X4	EP84X5
GEIC-21	GEIC-225	H221	HEP-C6100P
HEPC6100P	ITT3064	LM3064N	LM3064N-01
MC-1364	MC1364	MC1364P	MC1364PQ
NTE783	PTC730	RE-303-IC	RE303-IC
REN783	SI-CA3064E	SI-MC1364P	SK3215
SN76564	SN76565N	T2062	TCG783
TM783	TVCM-30	UA3064PC	ULN-2264A
ULN2264A	WEP2058	WEP2058/783	

BLOCK-1809

125C3	125C3RB	221-Z9139	CA3020
CA3020A	CA3020T	ECG784	GEIC-236
ME32865-00001-B	NA1302	NTE784	SK3524
TCG784	TM784	TVCM-47	WEP2059
WEP2059/784			

BLOCK-1810

127166	1473528-1	1473528-2	4020
CA3035	CA3035T	CA3035V	CA3035V1
ECG785	GEIC-226	HE-442-4	SK3254
TCG785	TM785	TVCM-49	WEP2039
WEP2039/785			

BLOCK-1811

1C319	3L4-9002-1	AE900	CA3043
CA3043T	CXL786	ECG786	GEIC-227
NTE786	SK3140	TCG786	TM786
TVCM-52	WEP2061	WEP2061/786	

BLOCK-1812

09A07	136145	1464437-2	1464437-3
3130-3193-512	4437-3	93BLC	CA-3088E
CA3088E	CXL787	DM-88	DM88
ECG787	GEIC-228	NTE787	SK3146
TCG787	TM787	TVCM-57	

BLOCK-1813

07B2Z	136146	1464438-2	1464438-3
15-36994-1	18410-148	2068-022-108	2077-2
21192010402	221-108	221-Z9037	266P32801
29810-179	3596810	38510-169	4154011370
4438-3	45810-167	51-10658A01	51-10658A02
51-10658A03	612077-2	65-11137-23	668-009A
810	88-20372	905-120	95287
CA-3089E	CA3089E	CA3089F	CA3090
CA3289E	CXL788	DM-51	ECG788
G09-018-A	GEIC-229	HA1137	HA1137(W)
HA1137P	HA1137W	LA1230	LM3089N
NTE788	PTC767	PTC787	RE332-IC
REN788	SK3147	SK3829	SN76689
SN76689N	T2T	T2T-1	T2T-2
TA5628	TCG788	TDA1200	TM788
TR09018	TRLA1230	TVCM-39	UA3089E
ULN2289	ULN2289(A)	ULN2289A	ULX-2289A
WEP2040	WEP2040/788		

BLOCK-1814

136147	138380	138380(IC)	1464439-2
1464460-2	1464460-3	146460-2	221-Z9015
2799-1	2799-2	3596809	3L4-9008-01
3L4-9008-51	4082799-0001	4082799-0002	4082799-1
4082799-2	4439-1	4439-1(RCA)	4439-2
4439-2(RCA)	4460-2	4460-3	51-10638A01
51-10678A01	51S10638A01	5932	80829
88-9841R	AE908	AE908-51	CA3090
CA3090AB	CA3090AQ	CA3090E	CA3090Q
EA33X8373	ECG789	IC320(ELCOM)	NTE789
R4439-1	R4439-1(RCA)	R4439-2	R4439-2(RCA)
SK3078	T2N	T2N-1	T2N1
TCG789	TM789	TVCM-58	WEP2064
WEP2064/789			

BLOCK-1815

1264	13-41-6	14-2010-03	15-37704-1
221-62	229-1301-41	29B1B	29B1Z
46-5002-11	561103	56A6-1	56D6-1
612029-1	612029-3	66F1271	AMD746
C6057P	C6057P(HEP)	CXL790	ECG790
EX62-X	GEIC-18	GEIC-230	GEL3072F1
HEP-C6057P	LM746N	MC1328P	MC1328PQ
NTE790	PTC728	RE-323-IC	RE323-IC
REN790	SC8723P	SK3454	SN16727N
SN764246N	TCG790	TM790	TVCM-22
U9A7746394	ULM2114A	ULN-2114A	ULN-2114N
ULN-2228A	ULN2114	ULN2114N	ULN2228A
ULN2264AN	WEP2065	WEP2065/790	

BLOCK-1816

02-561171	09-308099	15-39209-1	2002501300
221-69	221-69-01	221-78	39209-1
51S1059401	56A17-1	56C17-1	56D17-1
57D17-1	CA3021E	CA3121	CA3121E
CA3121E-G	CA3121EG	CA3121G	ECG791
GEIC-231	GEIC-33	NTE791	PTC733
RE324-IC	REN791	SK3149	SK3209
SN16840C	SN16859	SN16869A	SN94174
TCG791	TM791	TVCM-34	ULN-2269A
ULN2269A	WEP2066	WEP2066/791	

BLOCK-1817

15-37700-1	221-Z9038	66F125-1	ECG792
GE-792	Y15377001		

BLOCK-1818

15-37701-1	15-37701-2	221-Z9039	56A24-1
66F136-1	DM8890N	ECG793	GE-793
NTE793	TM793	ULN-2298A	WEP2281
WEP2281/793			

BLOCK-1819

15-14504-1	15-37534-1	15-37534-2	15-39600-1
15-39600-2	15-41545-1	221-Z9031	229-1301-39
ECG794	GE-794	MC1393AP	MC1393P
SK3974	Y15415451		

BLOCK-1820

15-39061-1	15-41856-1	221-Z9040	229-1301-43
C6099P	ECG795	GEIC-232	HEP-C6099P
HEPC6099P	MC1349P	NTE795	SK3237
TM795	WEP2041	WEP2041/795	Y15418561

BLOCK-1821

1063-5019	138699	141259	141279
1465158-1	1465158-2	15-3015129-1	221-Z9041
443-909	51-13753-A19	51-13753A19	5158
5158-1	5320600070	612334-1	6123340001
86X0082-001	86X82-1	CA3126	CA3126EM
CA3126EM1	CA3126EMI	CA3126FM1	CA3126Q
CA3162E	ECG797	GEIC-233	HE-443-909
LM1329	LM1829	NTE797	PTC769
RE333-IC	REN797	SK3158	SK3211
TCG797	TM797	TVSCA3126	UA787
UA787PC	ULN2262	ULN3262A	WEP2043
WEP2043/797			

BLOCK-1822

10655A01	10655B01	1C-289	51D70177A01
51M70177A01	51S10655A01	51S10655B01	70177A01
C6066P	CA3125	CA3125E	ECG798
GEIC-234	HEP-C6066P	HEPC6066P	HEPC6072P
LC-289(ELCOM)	NTE798	SC5204P	SK3216
TCG768	TCG798	WEP2044	WEP2044/798
WEP2046	ZEN609		

BLOCK-1823

15-34906-1	159588	1C-30	21M506
221-Z9093	86X0064-001	EA33X8364	ECG799
GEIC-34	HC1000417	HEPC6126P	MC1312P
MXC-1312	MXC1312A	NTE799	SK3238
TCG799	TM799	UA1312PC	XC-1312A
XC1312			

BLOCK-1824

221-131	ECG800	LM373N	LM375N

BLOCK-1825

003515	051-0050-00	051-0050-01	1-235-971-12
15-40183-1	2002400617	2057A32-33	2068-018-104
21192021706	221-91	221-91-01	221-Z9027
26810-157	280-0002	29810-180	38510-170
415401560	415401160	4157013102	45810-168
51-40464P01	51-42908J01	5147013102	56-4834
5652-HA1156	57A32-32	612075-1	612075-3

BLOCK-1825 (CONT.)

6120750003	61K001-10	668-033A	741519
87-0234	88-20404	905-141	916113
C6096P	C6096P(HEP)	CA1310	CA1310E
DM-36	EA33X8375	EA33X8511	ECG801
ETI-22	G09-028-A	G09-034-A	GEIC-155
GEIC-35	HA1150	HA1156	HA1156(W)
HA1156-6C	HA1156H	HA1156W	HA1156WP
HC10004010	HC10006170	HEP-C6096P	HEPC6096P
KB4409	LM-1310N	LM1310	LM1310N
LM1310N01	LM131ON	LN1310	M5153L
M5153P	MC-1310P	MC1310	MC1310A
MC1310P	MX-3243	NTE801	PC-20045
PTC735	QTVCM-73	RE334-IC	REN801
SK3160	SN76115	SN76115N	TCG801
TM801	TR09019	TR2367171	TVCM-73
ULN-2210A	ULN2110(A)	ULN2210	ULX2210
ULX2210D	WEP2075	WEP2075/801	

BLOCK-1826

221-Z9094	ECG802	GEIC-235	HEPC6127P
MC1314P	NTE802	SK3277	TCG802
TM802	UA1314PC		

BLOCK-1827

221-Z9095	ECG803	GEIC-237	HEPC6128P
MC1315P	NTE803	SK3278	TCG803
TM803	UA1315PC		

BLOCK-1828

13-100000	13-1000000	13-1000001	13-59-6
13-61-6	183044	183405	1C-602
221-Z9016	4090605-1	4150003770	45393
45395	612045-1	612144-1	69-3116
7204A	889304	905-27	905-28
DN6851H1	DN6851HI	ECG732	ECG733
ECG804	GEIC-27	HEPC6090	IC-603(ELCOM)
LM-377N	LM377N10	LM378N	M51366P
NTE804	PA277	PA306	PTC732
PTC738	PTC742	RE-322-IC	RE322-IC
REN804	SK3455	SK9068	SN76177ND
TCG804	TM804	TVCM-23	TVCM-25
TVCM-26	TVCM-65	UA705	ULM2274P
ULN-2274P	ULN-2275A	ULN-2275P	ULN-2277P
ULN-2278P	ULN2274B	ULN2274P	ULN2275
ULN2275B	ULN2275P	ULN2276B	ULN2276P
ULN2276Q	ULN2277	ULN2277B	ULN2277P
ULN2278	ULN2278B	ULN2278P	ULN2278Q
WEP2046/804			

BLOCK-1829

DM-35	ECG806	GEIC-239	NTE806
SK3734	TM806		

BLOCK-1830

221-Z9140	ECG807	GEIC-240	NTE807
RG602	RG602T	SK3451	TCG807
TM807	ULN-2212A	ULN-2212P	ULN2212B

BLOCK-1831

15-392007-1	15-39207-1	66F175-1	66F1751
ECG808	TM808		

BLOCK-1832

443-762	8080A	C8080A	ECG8080A
HE-443-762	NTE8080A	TCG8080A	TMS8080AN
UPD8080AFC			

BLOCK-1833

15-39208-1	66F176-1	66F1761	ECG809
NTE809	RG604	RG604T	SK7727
TM809			

BLOCK-1834

ECG8092

BLOCK-1835

ECG80C95	NTE80C95	TCG80C95

BLOCK-1836

15-3015740-1	ECG80C96	NTE80C96	TCG80C96

BLOCK-1837

443-720	80C97	ECG80C97	HE-443-720
MC14503BCL	NTE80C97	TCG80C97	

BLOCK-1838

221-77	800-619	ECG810	ECG810A
GEIC-241	NTE810A	SK7730	SN16869A#
TM810			

REPLACEMENTS

BLOCK-1839

ECG812	GE-812	GEIC-242	PC-20015
PL20015	SN76007	SN76007N	TCG812
TM812			

BLOCK-1840

DM8123N	ECG8123	NTE8123	TCG8123

BLOCK-1841

15-36647-1	ECG813	QSI-5022	TM813

BLOCK-1842

C6073P	ECG814	ECG814A	GEIC-243
MC1316P	TM814		

BLOCK-1843

02-561251	221-141	221-Z9096	56A23-1
56A25-1	612082-1	612082-2	612082-3
86X0084-001	CA1391E	ECG815	EP84X37
GEIC-244	LM1391N	MC1391P	MC1394
NTE815	SK3255	SN76591P	TCG815
TM815	UA1391TC	ULN2291M	

BLOCK-1844

1C20	221-Z9097	590	C6091
C6091G	ECG816	HEP590	HEPC6091
HEPC6091G	MC1550	MC1550G	MC155DG
SK3242	TM816	WEP931	

BLOCK-1845

301-679-1	ECG817	MC-1550P	MC1550P
SC5282P	TM817		

BLOCK-1846

007-0150-00	221-96	221-96C	CA3135E
CA3135G	ECG818	GE-818	NTE818
SC5896	SK3207	T2370	TA6896
TCG818	TM818	ULN2261	ULN2261A
WEP2285	WEP2285/818		

BLOCK-1847

15-43098-1	221-Z9153	CA3221	CA3221G
ECG819	NTE819	SK3928	TCG819
TM819			

BLOCK-1848

15-43312-1	15-43312-2	221-Z9167	612331-2
6123310002	CA3159	CA3159G	ECG820
NTE820	SK3927	TCG820	TM820

BLOCK-1849

15-41627-1	15-41627-2	15-41627-3	221-Z9154
612347-1	6123470002	CA3172	CA3172E
CA3172G	ECG821	NTE821	SK3882
TCG821	TM821	WEP821/821	Y15416272

BLOCK-1850

15-41764-1	15-41764-2	221-Z9155	612338-3
6123380002	6123380003	CA3156	CA3156E
CA3156G	ECG822	GE-822	NTE822
SK3919	TCG822	TM822	WEP822/822

BLOCK-1851

16103860	276-1731	386	442-612
8-759-903-86	8050386	EA33X8548	ECG823
JRC386D	LM386	LM3862M	LM386A
LM386LM	LM386N	LM386N-1	LM386N-3
NJM386D	NTE823	SK9210	TCG823
TM823	X440073860		

BLOCK-1852

1099-3624	276-1737	3L4-9015-1	AE915
ECG824	LM387	LM387AN	LM387N
NE542N	NTE824	PC-20046	SK9013
TCG824	TM824		

BLOCK-1853

ECG825	LM390N	NTE825	SK9126
TCG825	TM825		

BLOCK-1854

ECG8250	N8250F	N8250N	

BLOCK-1855

8255A	8255A-5	8255AC-5	AM8255A-5PC
D8255A-5	D8255AC-2	D8255AC-5	EA000-44600
EA001-02200	ECG8255	HE-443-1021	I8255
IP8255A-5	JB8255AP-5	M5L8255AP-5	NTE8255
P8255A-5	TCG8255	UPD8255AC-5	X400082550

BLOCK-1856

ECG826	NTE826	SK7813	TM826

BLOCK-1857

ECG8266	N8266F	N8266N	NTE8266

BLOCK-1858

ECG828	LM388N	LM388N-1	NTE828
SK9127	TCG828		

BLOCK-1859

022-2844-702	06300311	2275-1	39-047-1
44T-300-101	612275-1	6122750001	619550-1
72132-1	760105	9550-1	C6009P
ECG766	ECG766A	ECG829	HEPC6009
HEPC6009P	IC-243(ELCOM)	MC-3340P	MC3340
MC3340P	MC3340PA	MFC6040	NTE829
SK3891	TCG829		

BLOCK-1860

145803	221-Z9168	56A55-1	56A551
ECG830	SK9030	TM830	

BLOCK-1861

9301DC	9301PC	DM8301J	DM8301N
DM9301J	DM9301N	ECG8301	N8252F
N8252N	NTE8301	TCG8301	

BLOCK-1862

9308DC	9308PC	ECG8308	N74116N
N9308F	N9308N	NTE8308	S9308F
SN74116	SN74116N	TCG8308	

BLOCK-1863

9309DC	9309PC	ECG8309	N9309F
N9309N	NTE8309	TCG8309	

BLOCK-1864

10655B14	ECG831		

BLOCK-1865

9314DC	9314PC	ECG8314	N9314F
N9314N	NTE8314	S9314F	TCG8314

BLOCK-1866

9316DC	9316DM	9316PC	DM8316N
ECG8316	NTE8316	TCG8316	

BLOCK-1867

9318DC	9318PC	DM8318N	DM9318J
ECG8318	NTE8318	SK9831	TCG8318

BLOCK-1868

276-1721	442-688	567	ECG832
HE-442-688	IR3N05	NE567N	NTE832
SIIR3N05XXXXX	SK9089	TCG832	TM832

BLOCK-1869

9321DC	9321PC	ECG8321	NTE8321
TCG8321			

BLOCK-1870

9328DC	9328PC	ECG8328	NTE8328
TCG8328			

BLOCK-1871

ECG832SM	LM567CM	NTE832SM	SK10062

BLOCK-1872

207484	276-1743	7555	ECG833
MC1422P1	SK7751	TM833	

BLOCK-1873

08220092	1154-9938	1200-2416	148740
150418	152364	153456	2119-101-3204
2119-201-1300	221-121	2365411	2420SLM339N
276-1712	295-95-1	339	39-077-0
4835-209-87261	4835-209-87263	5190458A07	51X90458A07
51X90458A93	51X90458A94	668-045A	8-759-133-90
8050339	905-288	AMX4200	AN1339
AN6912	AN6912N	BA10339	CA139AE
CA139F	CA139G	CA239	CA239A
CA239AE	CA239AG	CA239E	CA239G
CA339	CA339A	CA339AE	CA339AG
CA339E	CA339G	DM-87	DM87
EAS00-05800	ECG834	EW84X156	EW84X372
HA17339	HA17901G	HA17901P	LA6339
LM239	LM239A	LM239AJ	LM239AN
LM239J	LM239N	LM2901	LM2901J

BLOCK-1873 (CONT.)

LM2901N	LM3302	LM3302J	LM3302N
LM339	LM339A	LM339AJ	LM339AN
LM339DP	LM339J	LM339N	MB4204
MB4204C	MB4204M	MC3302	MC3302P
MLM139	MLM139AL	MLM139L	MLM239AL
MLM239L	MLM2901P	MLM339(P)	MLM339AL
MLM339L	MLM339P	NJM2901N	NTE834
P61XX0027	PC339C	RC3302DB	SK3569
TA75339	TA75339P	TCG834	TM834
TVSUPC339C	UA2901PC	UA339ADC	UPC177C
UPC339C			

BLOCK-1874

2901	2901(SM)	905-1372	ECG834SM
LM2901D	LM339D	NJM2901M	NTE834SM
SK10063	UPC339G2		

BLOCK-1875

221-Z9169	612120-1	ECG835	MC1399P
NTE835	SK9281		

BLOCK-1876

221-86	CA1394E	CA1394G	ECG836
MC1394P	NTE836	SK3206	SN76594
TCG836	ULN2294M		

BLOCK-1877

9368DC	9368PC	ECG8368	NTE8368
SK4368	TCG8368		

BLOCK-1878

544-0003-043	ECG837

BLOCK-1879

9370DC	9370PC	ECG8370	NTE8370
TCG8370			

BLOCK-1880

9374DC	9374PC	ECG8374	NTE8374
TCG8374			

BLOCK-1881

442-635	C6010P	ECG838	HE-442-635
IC-213(ELCOM)	IC213(ELCOM)	MC3310P	

BLOCK-1882

ECG839	HEPC6016P	MC3320P	SK7614

BLOCK-1883

ECG840	RH-IX0025PAZZ	SAA1025

BLOCK-1884

146150	CA3192E	ECG841	NTE841
SK9202	TCG841		

BLOCK-1885

146149	221-Z9170	CA3191E	ECG842
NTE842	SK7686		

BLOCK-1886

221-97	221-97-01	CA3136E	CA3136G
ECG843	NTE843	SK3208	TA6699B
TCG843			

BLOCK-1887

1145-2794	1166-3829	1166-3879	146858
149254	152051	153685	221-Z9171
ECG844	NTE844	SK9381	

BLOCK-1888

221-140	ECG845	NTE845	SK9382

BLOCK-1889

149037	ECG846	LM1889	LM1889N
NTE846	SK9178	TCG846	

BLOCK-1890

ECG847

BLOCK-1891

15-45693-1	612352-1	6123520001	ECG848
NTE848			

BLOCK-1892

221-103	221-103A	221-193	221-193A
CA3157G	ECG849	NTE849	SK9027
TCG849	ULN2296		

BLOCK-1893

1159-2342	149870	ECG850	NTE850
SK9383			

BLOCK-1894

1099-6395	143696	612294-1	CA3163G
ECG851	NTE851	SK9206	TCG851

BLOCK-1895

1138-0789	13-3017119-2	15-3017119-1	15-3017119-2
221-Z9172	6123630001	6123630005	ECG852
EP84X82	NTE852	SK9248	TCG852

BLOCK-1896

ECG8520	NTE8520	TCG8520

BLOCK-1897

905-209	905-243	ECG853	IR3N06
MC3357P	NTE853	SK7645	TCG853

BLOCK-1898

56A49-1	56A49-2	56C49-1	56D49-1
56D49-2	ECG854	NTE854	SK7737

BLOCK-1899

DM8542N	ECG8542	NTE8542	TCG8542

BLOCK-1900

DM8546N	ECG8546	NTE8546	TCG8546

BLOCK-1901

AMX4574	ECG855	MC1372	MC1372P
NTE855	SK7616	TCG855	

BLOCK-1902

ECG8552

BLOCK-1903

DM8553N	ECG8553

BLOCK-1904

DM8554N	ECG8554

BLOCK-1905

DM8555N	ECG8555

BLOCK-1906

DM8556N	ECG8556	NTE8556	TCG8556

BLOCK-1907

ECG856	MC1373P	NTE856	SK7708
TCG856			

BLOCK-1908

612189-1	612189-2	6123590001	ECG857M
LF351N	LF355	NTE857M	SK9800
TCG857M	TL071ACP	TL071CP	TL081
TL081ACP	TL081CP	UA771ARC	UA771ATC
UA771BRC	UA771BTC	UA771LRC	UA771LTC
UA771RC	UA771TC		

BLOCK-1909

ECG857SM	NTE857SM	SK9799	TL081CFP

BLOCK-1910

276-1715	353(IC)	ECG858M	HA5062-5
LF353	LF353N	MC34002P	NJM072D
NTE858M	SK7641	TL072CP	TL082CP
UA772ARC	UA772ATC	UA772BRC	UA772BTC
UA772RC	UA772TC	XR082CP	

BLOCK-1911

ECG858SM	NJM082M	NTE858SM	SK10064
TL082CD			

BLOCK-1912

06300761	205919	276-1714	4835-209-17032
6124800001	6125390001	ECG859	HA5084-5
LF347	LF347N	MC34004P	NTE859
SK4826	TCG859	TL074CN	TL084
TL084CDP	TL084CN	UA774LDC	UA774LPC
XR084CP			

BLOCK-1913

ECG859SM	NTE859SM	SK10065	TL084CD

BLOCK-1914

ECG860	MC3359P	NTE860	SK7731
TCG860	ULN3859A		

BLOCK-1915

ECG861	LF13331N	NTE861	SK4843

BLOCK-1916

DM8613N	ECG8613	NTE8613

BLOCK-1917

ECG741	ECG862	LM384N	NTE862
SK3456	SK9822	TCG741	TM741
ULN-2281B	ULN2281B		

BLOCK-1918

221-132	CA3177G	ECG863	NTE863
SK9207			

BLOCK-1919

276-2334	ECG864	ICL8038	NTE684
NTE864	SK9851	XR-8038CN	XR-8038CP
XR-8038N	XR-8038P	XR8038CN	XR8038CP
XR8038P			

BLOCK-1920

6123360001	ECG865

BLOCK-1921

6123370003	6123370005	ECG866

BLOCK-1922

ECG867	NTE867	SK9386

BLOCK-1923

1287-7908	1287-8989	149253	153875
154515	176222	177857	ECG868
NTE868	SK7692		

BLOCK-1924

ECG869	LM359J	LM359N	NTE869
SK10066			

BLOCK-1925

ECG870	LM13600N	SK9992

BLOCK-1926

CA3100S	CA3100T	CA3100TY	ECG871
NTE871	SK3646		

BLOCK-1927

ECG872	SN76514

BLOCK-1928

221-179-01	221-179-01-E-01	221-179-02	221-179-03
221-179-1	ECG873	NTE873	SK7718

BLOCK-1929

221-175	221-175-01	ECG874	NTE874
SK7717			

BLOCK-1930

221-190-01	221-190-01A	221-190-E-01	ECG875
NTE875	SK7725		

BLOCK-1931

276-1705	3909	ECG876	LM3909
LM3909N	NTE876	SK10067	

BLOCK-1932

ECG877	MC3458G	NTE877	SK9257

BLOCK-1933

221-190	ECG878	LM1823	LM1823N
NTE878	SK4814		

BLOCK-1934

ECG879	MC1377P	SK10514

BLOCK-1935

221-261	221-261-01	221-261B03	ECG880

BLOCK-1936

DM8853N	ECG8853

BLOCK-1937

ECG887M	NTE887M	SK9862	TL061ACP
TL061CP			

BLOCK-1938

ECG888M	MC1776CP1	NTE888M	SK9863
UA776TC			

BLOCK-1939

ECG889M	NTE889M	SK9864	TL062AC
TL062ACP	TL062C	UPC4062	

BLOCK-1940

0N101880	569-0690-300	CA3000	CA3000T
ECG900	GEIC-245	NTE900	SK3547
TCG900	TM900		

BLOCK-1941

0N088598	CA3001	CA3001T	ECG901
GEIC-180	NTE901	SK3549	TCG901
TM901			

BLOCK-1942

442-640	CA3080	CA3080A	CA3080S
ECG902	HE-442-640	NTE902	SK9200
TCG902			

BLOCK-1943

0N101881	1081-4739	156-0017	156-0017-00
1700-5182	221G3114	221G3114-1	3034725-1
3400-2412-3	34002412	34002412-3	3412986-1
8508331RB	CA3010	CA3010A	CA3010AT
CA3010T	CA3015	CA3015A	CA3015AT
CA3015T	ECG903	GEIC-288	HA1301
IC-211(ELCOM)	IC41(ELCOM)	NTE903	SK3540
TCG903	TM903		

BLOCK-1944

12951-1	129821	133-0021-0	1330021-0
1330021-1	1463681-1	1477-3436	1820-0352
20-0352	221-Z9141	3412951	3681-1
A066-125	CA3018	CA3018A	CA3018T
D-TEX-8	ECG904	EZT308	GEIC-289
IC2(ECS)	IC29(ELCOM)	KD2114	LA3018
NTE904	SK342	SK3542	SP180CI
TA7047	TA7047M	TCG904	TM904
UA3018AHM	UA3018H		

BLOCK-1945

1901-0557	198801-1	38927-00000	45299
CA3019	CA3019T	ECG905	GEIC-290
NTE905	SK3546	TA7057M	TCG905
TM905	UA3019HC		

BLOCK-1946

CA3026	CA3026T	CA3049T	ECG906
GEIC-246	NTE906	SK3548	TCG906
TM906	UA3026HC		

BLOCK-1947

CA3039	CA3039T	CA3545	ECG907
GEIC-247	NTE907	SK3545	TCG907
TM907	UA3039HC		

BLOCK-1948

221-Z9142	CA3029	CA3029A	CA3029E
CA3030	CA3030A	CA3030AE	CA3030E
CA3037	CA3037A	CA3037AE	CA3037E
CA3038	CA3038A	CA3038AE	CA3038E
CA3539	ECG908	GEIC-248	NTE908
SK3539	TCG908	TM908	

BLOCK-1949

126-40	198400	198410-1	221-Z9017
709CE	C6103P	CA741S	ECG909
GEIC-185	GEIC-249	HEP-C6103P	HEPC6103P
LM709CH	M51709T	MC1709CG	MC1709CP1
MC1709G	ML709CT	NTE909	RC709T
RL709T	SFC2709C	SK9177	SN72709T
TAA521	TAA522	TCG909	TM909
UA7094HC	UA709CL	UA709CT	UA709HC
UA709HM	UA709T	WEP2054	WEP2054/909

BLOCK-1950

10795-9	131312	158066	1627-2064
1820-0122	19A115913-10	2068510-0704	3007573-00
326853	423-800236	43A168135-9	443-27
48-1050-SL12326	56552	569-0320-814	586-425
65600800	6900K95-002	733W00030	762200-14
77027-1	77027-2	862200-16	9093-1-6A
9093-9-6A	9093-9-7A	9093DC	9093PC
94825000-05	C1053P	CD2315E	DM9093N
DTML9093	ECG9093	HD2211	HD2211P
HEPC1053P	IC-455(ELCOM)	IC33(ELCOM)	ITT9093-5

BLOCK-1950 (CONT.)

ITT9093N	LB2031	M5953	M5953P
MC853L	MC853P	MIC9093-5D	MIC9093-5P
NTE9093	PD9093-59	PE9093-59	SN158093J
SN158093N	SW705-2M	SW705-2P	TCG9093
TD1073	TD1073P		

BLOCK-1951

179-46445-10	1820-0341	1895989-1	50210-7
51310007	51717000	65611000	9094-1-6A
9094-9-6A	9094-9-7A	9094DC	9094PC
DM9094N	ECG9094	ITT9094-5	ITT9094N
LB2131	M5956	M5956P	MC856L
MC856P	MIC9094-5D	MIC9094-5P	NTE9094
PD9094-59	PE9094-59	SK9789	SN158094J
SN158094N	SW708-2M	SW708-2P	TCG9094
TCG9904			

BLOCK-1952

40-065-19-014	423-800223	51717100	52000-055
65611100	9097-1-6A	9097-9-6A	9097-9-7A
9097DC	9097PC	DM9097N	ECG9097
ITT9097-5	ITT9097N	LB2132	M5955P
MC855L	MC855P	MIC9097-5D	MIC9097-5P
NTE9097	PD9097-59	PE9097-59	SK9788
SN158097J	SN158097N	SW709-2M	SW709-2P
TCG9097			

BLOCK-1953

10795-10	151548	2068510-0705	
27A10446-101-11	326852	48-1050-SL12327	52000-052
569-0544-328	586-154	65600900	6900K95-001
9099-1-6A	9099-9-6A	9099-9-7A	9099DC
9099PC	C1052P	CD2318E	DM9099N
DTML9099	ECG9099	HD2210	HD2210P
HEPC1052P	IC-234(ELCOM)	ITT9099-5	ITT9099N
LB2032	MC852L	MC852P	MIC9099-5D
MIC9099-5P	NTE9099	PD9099-59	PE9099-59
SL16209	SL16522	SN158099J	SN158099N
SW706-2M	SW706-2P	TCG9099	TD1074
TD1074P			

BLOCK-1954

15-37833-2	442-2	569-0964-001	ECG909D
FU6A7709393	GEIC-250	LM709CN	MC1709CL
MC1709CP2	NTE909D	RC709D	RC709DC
SK3590	SN72709	SN72709J	TAA521A
TCG909D	TM909D	UA709A	UA709C
UA709CJ	UA709CN	UA709DC	UA709N
UA709PC	WEP2056	WEP2056/909D	

BLOCK-1955

128C213H01	221-Z9143	710CE	ECG910
GEIC-251	LM710CH	MC1710CG	NTE910
RC710T	SFC2710C	SG710CT	SK9175
TCG910	TCG910D	TM910	UA710
UA710CL	UA710CT	UA710HC	UA710T
UPC71A	WEP2060	WEP2060/910	

BLOCK-1956

006-0004779	133P100	143062	351-7577-010
68995-1	68A7652P1	7012130	717399-22
9109DC	ECG9109		

BLOCK-1957

561-0884-006	ECG910D	GEIC-252	LM710CN
MC1710CL	MC1710CP	MC1710P	NTE910D
RC710DC	SG710CN	SK9176	TM910D
UA710A	UA710CA	UA710CJ	UA710CN
UA710DC	UA710DM	UA710PC	

BLOCK-1958

ECG911	GEIC-253	LM711CH	MC1711CG
NTE911	QRT-272	RT-271	RT-272
SFC2711C	SG711CT	SK9070	SN72558L
TCG911	TM911	UA711A	UA711CK
UA711CL	UA711HC	UA711HM	UA711K

BLOCK-1959

1627-1413	68A7652P2	72045600	9110DC
ECG9110			

BLOCK-1960

9111DC	ECG9111	MC1816L	MC1816P

BLOCK-1961

006-0005545	007-1681301	1478-6982	148991
569-0320-815	7012131	717399-73	9112DC
ECG9112	MC1820P		

BLOCK-1962

ECG911D	GEIC-254	LM711CN	MC1711CL
MC1711CP	MC1711L	NTE911D	SG711CN
SK9071	TCG911D	UA711CA	UA711CJ
UA711CN	UA711DC	UA711DM	UA711PC

BLOCK-1963

007-1671101	020-1114-016	116623	150371
180010-001	221-Z9018	2626-1	4082626-0001
544-2002-008	569-0690-301	C010174-02	CA0345
CA3045	CA3045E	CA3045F	CA3045L
CA3046	CA3046E	CA3086	CA3086E
CO-10174-02	ECG912	GEIC-172	LA3086N
LM3045D	LM3045J	LM3045N	LM3046N
LM3086N	LM3086W	M2003	M2016
MC3346P	MC3386P	NTE912	SK3543
TCG912	TM912	TVCM-66	UA3045DC
UA3045PC	UA3046DC	UA3046PC	UA3086DC
UA3086DM	UA3086PC	ULN-2046A	ULN-2086A
ULN2046A	ULN2086A	ULS-2045H	ULS2045H
WEP2104	WEP2104/912		

BLOCK-1964

CA3047E	ECG913	GEIC-255	NTE913
TCG913	TM913		

BLOCK-1965

1485-9896	3007604-00	9135PC	ECG9135
NTE9135	SN15838J	SN15838N	TCG9135

BLOCK-1966

CA3056AE	CA3058	CA3058E	CA3059
CA3059E	CA3079	CA3079E	ECG914
GEIC-256	NTE914	SK3541	TCG914
TM914			

BLOCK-1967

143041(IC)	156-0151-00	157564	179-46447-21
1820-0476	2902798-2	50254400	715HC
82M432B2	9101600	ECG915	FU5F7715393
G390503S1	IC35(ELCOM)	NTE915	SK9090
SL21384	SL21577	TCG915	TM915
UA715(METAL-CAN)			

BLOCK-1968

36186200	52000-057	586-415	6900K91-009
9001567	9001567-02	9157PC	ECG9157
HD2213	HD2213&	HEPC1057P	IC-419(ELCOM)
MC857L	MC857P	NTE9157	SN15857J
SN15857N	SW957-2M	SW957-2P	TCG9157

BLOCK-1969

1820-0863	193022	301-576-2	380049
52000-058	569-0880-458	586-442	9158PC
C1057P	C1058P	DM958N	ECG9158
HD2214	HD2214#	HEPC1058P	IC-420(ELCOM)
MC-858	MC858	MC858L	MC858P
NTE9158	SCD-1062013	SN15858J	SN15858N
SW-10549	SW958-2M	SW958-2P	TCG9158
WEP9158	WEP9158/9158		

BLOCK-1970

569-0690-681	CA3081	CA3081E	ECG916
GEIC-257	M2032	NTE916	SG3081N
SK3550	TCG916	TM916	ULN-2081A
ULN2081A			

BLOCK-1971

266P75201	612054-2	CA3054	CA3054E
ECG917	GEIC-258	M5109	M5109P
NTE917	SK3544	TCG917	TM917
TVCM-67	UA3054DC		

BLOCK-1972

ECG918	LM318H	NTE918	SK9144
TCG918	ULN-2158D	ULN-2172D	ULN-2174D
ULN-2178D	ULS-2158D	ULS-2172D	ULS-2174D
ULS-2176D			

BLOCK-1973

ECG918M	LM318J-8	LM318JG	LM318N
LM318P	MC1741SCP1	NTE918M	SK9145
TCG918M	ULN-2158M	ULN-2172M	ULN-2174M
ULN-2176D	ULN-2176M	ULN-2178M	ULS-2158M
ULS-2172M	ULS-2174M	ULS-2176M	ULS-2178D
ULS-2178M			

BLOCK-1974

ECG918SM	LM318	LM318M	NTE918SM
SK10069			

BLOCK-1975

ECG919	LM319H	LM319K	NTE919
SK9217	TCG919		

BLOCK-1976

AMLM319N	ECG919D	LM319F	LM319J
LM319N	NTE919D	SK9218	TCG919D

BLOCK-1977

ECG920

BLOCK-1978

144022	51-90305A31	DN819	ECG9200
NTE9200	SK10447	TCG9200	

BLOCK-1979

ECG921	MC1468L	NTE921	SK7615

BLOCK-1980

901523-04	CA211	CA311	CA311S
CA311T	ECG922	LM211H	LM311
LM311H	LM311T	MLM111AG	MLM211AG
NTE922	SK3567	SK3567A	TCG922
UA311HC			

BLOCK-1981

200420	442-75	8-759-131-11	905-202
AMLM111D	AMLM311D	AMX4327	CA211E
CA211G	CA311E	CA311G	ECG922M
HE-442-75	LM311D	LM311F	LM311N
LM311NDS	LM311P	MLM311P1	MSM311EL
NTE922M	SK3668	SN5211P	SN72311P
TCG922M	UA311TC	UPC271C	UPC311C
X440223110			

BLOCK-1982

ECG922SM	NTE922SM	SK10070	UPC311G
UPC311G2			

BLOCK-1983

221-Z9019	276-009	723BE	723CE
AM723HM	AMU5R7723312	CA3085A	CA723
CA723CT	CA723T	ECG923	GEIC-259
GEVR-115	GEVR-116	HEPC6104L	LAS723
LAS723B	LM317K	LM723CG	LM723CH
LM723H	MC1723CG	MC1723CG	MIC723-1
ML723CT	ML723T	N5723T	NTE923
RC723T	RM723T	RS723	S5723T
SG723CT	SG723T	SK3164	SN52723L
SN72723L	TBA281	TCG923	TDB0723
TM923	UA723	UA723CL	UA723CT
UA723HC	UA723L	ULN-2723K	ULS-2723K
WEP2110	WEP2110/923		

BLOCK-1984

149255	221-Z9020	276-1740	3100001
569-0320-831	569-0540-623	723	723C
723CJ	AMX3548	CA723CE	CA723E
EA33X8469	ECG923D	GEIC-260	HA17723G
IC-20	IC-20(PHILCO)	L123CB	LM723CN
MC1723CL	MC1723CP	ML723CM	ML723CP
NE550A	NE550F	NE550N	NTE923D
RC723D	SA723CN	SG723CN	SK3165
SN52723N	SN72723J	SN72723N	TCG923D
TDB0723A	TM923D	UA723CA	UA723CJ
UA723CN	UA723DC	UA723DM	UA723HM
UA723ML	UA723PC	ULN-2723A	ULS-2723A
WEP2331	WEP2331/923D		

BLOCK-1985

442-25	ECG924	HE-442-25	LM302H
LM310H	NTE924	SG310T	SK9128
TCG924			

BLOCK-1986

442-648	ECG924M	HE-442-648	LM310N
NTE924M	SG310M	SK9129	TCG924M

BLOCK-1987

09391000A	ECG925	GEIC-261	LM725CH
NTE925	PM725CJ	SK9174	TCG925
TM925	UA725HC	UPC154A	

BLOCK-1988

558CP	ECG926	NE558	NE558CP
NE558N	NTE926	SK7721	TCG926

BLOCK-1989

ECG927	LM733CH	MC1733CG	MC1733G
NE592H	NE592K	NTE927	SK9017
TCG927	UA733C	UA733CH	UA733HC

BLOCK-1990

76-100-001	901523-08	AMX3688	AMX4326
ECG927D	LM733CN	MC1733CL	MC1733CP
NE592	NE592A	NE592D	NE592F
NE592N	NE592N14	NTE927D	SG733CJ
SG733CT	SK7617	TCG927D	UA733CF
UA733CJ	UA733CN	UA733DC	UA733PC
X440057330			

BLOCK-1991

ECG927SM	NE592D14	NTE927SM	SK10071

BLOCK-1992

CA158AS	CA158AT	CA158S	CA158T
CA258S	CA258T	CA358AS	CA358AT
CA358S	CA358T	ECG928	LM258AH
LM258AN	LM258AT	LM258H	LM258T
LM2904L	LM358AH	LM358H	LM358L
LM358NB	LM358T	NE532AT	NE532T
NTE928	SK3691	TCG928	UA798HM

BLOCK-1993

06300565	1027-4231	1175-7911	1175-8000
1217-9024	152921	153568	154581
160832	195606	23319802	23904303
23904391	2904D	3130-3167-909	442-728
46-13279-3	46-13297-3	4835-209-87079	4835-209-87262
5190507A39	5190528A01	5190528A01	51X90266A45
51X90507A39	51X90507A40	6122850001	6122850003
6562	8-759-135-80	8050358	905-1238
905-658	AN1358	AN6562	B0347500
BA10358	BA728	C061702	C358C
CA2904E	ECG928M	EW84X232	EW84X233
GL358	HA17094PS	HA17358	HA17904
HA17904PS	IR9358	LA6358	LM258JG
LM258N	LM258P	LM2904J	LM2904JG
LM2904N	LM2904P	LM358	LM358AN
LM358AT	LM358J	LM358JG	LM358N
LM358P	MX-5679	NE532AN	NE532N
NJM2904	NJM2904D	NTE928M	QC0069
SK3692	TA75558P	TA75558S	TA75558S-1
TCG928M	UPC358	UPC358C	UPC358P
VHIIR9358//-1	VHITA75558S-1	X440029040	

BLOCK-1994

2003321	A6358S	AN6561	ECG928S
EW84X224	I03S063580	LA6358S	LM358S
NTE928S	SK10072		

BLOCK-1995

483520987112	483520987113	8-759-701-01	8-759-981-69
AN1358S	ECG928SM	LM2904H	LM358D
LM358M	NJM2904M	NTE928SM	SK10073

BLOCK-1996

905-238	CA3083	ECG929	NTE929
SK3695	ULN2083A		

BLOCK-1997

442-623	442-715	CA3130	CA3130A
CA3130AE	CA3130AS	CA3130AT	CA3130B
CA3130S	CA3130T	DM-90	DM90
ECG930	HE-442-623	HE-442-715	IC-3130
NTE930	SK3568	SK3696	TCG930

BLOCK-1998

301AL	301CL	ECG9301

BLOCK-1999

302AL	302CL	ECG9302

BLOCK-2000

303AL	303CL	ECG9303

BLOCK-2001

304AL	304CL	ECG9304

BLOCK-2002

306AL	306CL	ECG9306

BLOCK-2003

307AL	307CL	ECG9307

BLOCK-2004

442-702	78HV05CDA	CJSE071	ECG931
HE-442-702	LM223K	LM323K	MC7805ACK
MC7805AK	MC7805BK	MC7805CK	MC7805K
MC78T05CK	MC78T05K	NTE931	SA7805CDA
SA78HV05CDA	SFC2109RM	SFC2209R	SFC2309R
SG323K	SG340-05K	SH323SC	SK9067
TCG931			

BLOCK-2005

311AL	311CL	ECG9311

BLOCK-2006

312AJ	312AL	312CJ	312CL
ECG9312			

BLOCK-2007

442-651	78H05-KC	78H05KC	78HO5KC
ECG932	HE-442-651	NTE932	SK9340
TCG932	UA78H05KC		

BLOCK-2008

321AJ	321AL	321CJ	321CL
ECG9321			

BLOCK-2009

322AJ	322AL	322CJ	322CL
ECG9322			

BLOCK-2010

323AJ	323AL	323CJ	323CL
ECG9323			

BLOCK-2011

324AL	324CL	ECG9324

BLOCK-2012

325AJ	325AL	325CJ	325CL
ECG9325			

BLOCK-2013

326AJ	326AL	326CJ	326CL
ECG9326			

BLOCK-2014

7812CDA	78HV12CDA	CJSE067	ECG933
MC7812ACK	MC7812AK	MC7812BK	MC7812CK
MC7812K	MC78T12K	NTE933	SA7812CDA
SA78HV12CDA	SFC2812RC	SG340-12K	SG7812CK
SK9341	TCG933	TDB7812	UA78H12KC

BLOCK-2015

331CL	ECG9331

BLOCK-2016

332AJ	332AL	ECG9332

BLOCK-2017

333AJ	333AL	333CJ	333CL
ECG9333			

BLOCK-2018

334AL	334CL	ECG9334

BLOCK-2019

335AJ	335AL	335CJ	335CL
ECG9335			

BLOCK-2020

ECG934	NTE934	SK9342	TCG934

BLOCK-2021

342AJ	342AL	342CJ	342CL
ECG9342			

BLOCK-2022

343AJ	343AL	343CJ	343CL
ECG9343			

BLOCK-2023

347AL	ECG9347

BLOCK-2024

ECG935	LM338K	NTE935	SK9344
TCG935	UA78HGKC		

BLOCK-2025

ECG936	NTE936	TCG936

BLOCK-2026

ECG9361

BLOCK-2027

ECG9362

BLOCK-2028

363AL	363CL	ECG9363

BLOCK-2029

ECG9367

BLOCK-2030

ECG9368

BLOCK-2031

ECG937	LF355AH	LF355H	LF356AH
LF356H	LF357AH	LF357H	LM307H
NTE937	PM355AJ	PM355J	PM356AJ
PM356J	PM357AJ	PM357J	SK9146
TCG937			

BLOCK-2032

370AJ	370AL	370CJ	370CL
ECG9370	NTE9370	TCG9370	

BLOCK-2033

371AJ	371AL	371CJ	371CL
ECG9371			

BLOCK-2034

372AL	372CL	ECG9372

BLOCK-2035

ECG9375

BLOCK-2036

905-239	ECG937M	LF355BN	LF355J
LF355N	LF356	LF356BN	LF356J
LF356N	LF357BN	LF357J	LF357N
MA356	MC34001P	NTE937M	SK9147
TCG937M			

BLOCK-2037

AMLM308A	ECG938	LM308AH	LM308AJ-8
LM308AT	LM308H	LM308T	NTE938
OP-05CT	OP-05ET	OP-07CT	OP-07DT
OP-07ET	OP-27ET	OP-27FT	OP-27GT
OP-37ET	OP-37FT	OP-37GT	OP-47FT
OP-47GT	RC714CH	RC714EH	RC714LH
SG308AT	SG308T	SK9166	TCG938
TCG938M	UA308AHC	UA308HC	

BLOCK-2038

ECG9380

BLOCK-2039

ECG9381

BLOCK-2040

ECG9382

BLOCK-2041

ECG9383	FZL111

BLOCK-2042

442-66	ECG938M	HE-442-66	LM308AN
LM308D	LM308J-8	LM308N	NTE938M
OP-05CDE	OP-05CNB	OP-05EDE	OP-05ENB
OP-07CDE	OP-07CNB	OP-07DDE	OP-07DNB
OP-07EDE	OP-07ENB	OP-27EDE	OP-27ENB
OP-27FDE	OP-27FNB	OP-27GDE	OP-27GNB
OP-37EDE	OP-37ENB	OP-37FDE	OP-37FNB
OP-37GDE	OP-37GNB	OP-47GNB	PM308AP
RC714CDE	RC714EDE	RC714LDE	SG308AM
SG308AY	SG308M	SG308Y	SK9167
UA308TC			

BLOCK-2043

ECG939	NTE939	TCG939

BLOCK-2044

ECG9390

BLOCK-2045

ECG9391

BLOCK-2046

ECG9392

REPLACEMENTS

BLOCK-2047
ECG9393

BLOCK-2048
ECG9394

BLOCK-2049

93L08DC	93L08DM	93L08PC	ECG93L08
NTE93L08	TCG93L08		

BLOCK-2050

93L16DC	93L16PC	ECG93L16	NTE93L16
TCG93L16			

BLOCK-2051

6H740AC	ECG940	GEIC-262	NE536T
NTE940	SK9171	TCG940	TM940
UA740CT	UA740EHC		

BLOCK-2052

51-90305A30	DN811	ECG9400	

BLOCK-2053

DN850	ECG9401	SK10448	

BLOCK-2054

07-28778-78	144855	51-90305A46	51X90305A46
61C002-3	DN838	ECG9402	NTE9402
SK9062	TCG9402		

BLOCK-2055

07-28816-83	144967	152920	51-90433A12
51X90433A12	51X90518A35	DN852	DN852P
ECG9403	NTE9403	SK9063	TCG9403

BLOCK-2056

183532	188660-01	193207	221-40
221-Z9144	276-010	32119-201-360	351-029-020
51750	569-0320-809	AD741	AD741C
AD741CH	AD741H	AD741KH	AM741HC
C6052M	CA3056	CA741CT	CA741T
ECG941	GEIC-263	HA1304	HC1000217
HC10002170	HEPC6052G	LM741AH	LM741C
LM741CH	LM741EH	MC1731G	MC1741
MC1741-5C	MC1741CG	MC1741G	MC1741NCG
MC1741SCG	MIC741-5C	ML741CT	NTE941
PM741CJ	RC741H	RC741T	RM741TE
RS741	SC51750	SC5175G	SFC2741C
SG741CT	SG741SCT	SK3514	SK3553
SN52741	SN52741L	SN52741P	SN74741L
TA7504M	TBA221	TBA222	TCG941
TM941	UA741	UA741CL	UA741CT
UA741HC	UA741ML	UC4741	UC4741C
ULN-2151D	ULN-2156D	ULN-2159D	ULN-2171D
ULN-2173D	ULN-2177D	ULN-2741D	ULN2741D
ULS-2151D	ULS-2156D	ULS-2159D	ULS-2171D
ULS-2173D	ULS-2177D	ULS-2741D	ULS2741D
UPC151A	WEP2114	WEP2114/941	

BLOCK-2057

221-Z9145	561-0884-041	AD741CN	ECG941D
GEIC-264	HC1000217-0	MC1741C2P	MC1741CL
MC1741CP2	NTE941D	PM741CY	RC741D
RC741DC	SG741CN	SN72741J	TBA2214
TCG941D	TM941D	UA741C	UA741CA
UA741CJ	UA741CN	UA741PC	ULN-2159H
ULN-2177H	ULS-2159H	ULS-2177H	WEP2063
WEP2063/941D			

BLOCK-2058

2002800034	221-93	266P62103	276-007
442-22	44A417779-001	46-13178-3	561-0884-043
6120480004	6190000200	6570	741
741C	AD741KN	AMX4258	AN1741
AN6570	B0345450	C6052P	C6052P(HEP)
CA3741CE	CA3741E	CA741	CA741C
CA741CE	CA741CG	CA741CS	CA741E
CA741G	EA33X8479	EAS00-09000	ECG941M
EW84X785	GEIC-265	HA17741	HA17741G
HA17741GS	HA17741PS	HE-442-22	HEPC6052P
J4-1215	K4-590	LM207N	LM741
LM741CJ	LM741CN	LM741EN	LM741N
MA741	MC1741CP	MC1741CP1	MC1741NCP1
MC1741P1	ML741CS	MM741	N5741V
NTE941M	RC741DE	RC741DN	RC741N
RC741NB	RV741NB	SG741CM	SK3552
SN72741P	T3B	TA7504	TA7504P
TA7504S	TCG941M	TM941M	UA741A
UA741CJG	UA741CP	UA741CV	UA741DC
UA741HM	UA741TC	ULN-2151M	ULN-2156M
ULN-2159M	ULN-2171M	ULN-2173M	ULN-2177M

BLOCK-2058 (CONT.)

ULN-2709CM	ULN-2709M	ULS-2151M	ULS-2156M
ULS-2159M	ULS-2171M	ULS-2173M	ULS-2177M
UPC151C	UPC741C	WEP933	WEP933/941M

BLOCK-2059

ECG941S	NTE941S	SK9323

BLOCK-2060

ECG941SM	LM741CM	MC1741CD	NTE941SM
SK10074	UA741CD		

BLOCK-2061

ECG942	LM381A	LM381AN	LM381N
NTE942	SK3924	TCG942	

BLOCK-2062

ECG943	NTE943	TCG943

BLOCK-2063

06300105	06300394	08200091	1175-7937
1175-8018	1191-8547	1231-8333	1231-8366
151056	181183	191296	191869
2002800133	2002800146	2002800178	2365452
266P15401	266P154010	2903D	46-13495-3
4835-209-87081	51X90266A15	58SI-LM393N	58SI-LM393P
59SI-LM393N	6914	79SI-UPC393C	8-759-103-93
8-759-987-16	8175-7937	905-1085	905-659
905-922	AN1393	AN6914	BA6993
C393C	C393C(IC)	CA3290AE	CA3290E
EAS00-12900	ECG943M	EW84X196	HA17393
I03S063930	I07D069930	I52D003930	IR9393
LA6393D	LM393JG	LM393N	LM393NB
LM393P	M5233P	NJM2901	NJM2901D
NJM2903D	NTE943M	PC393C	RC2403NB
SK9721	SK9993	TA75393	TA75393P
TCG943M	TDB0193DP	UPC373C	UPC393
UPC393C	X440072920		

BLOCK-2064

ECG943SM	LM393M	NTE943SM	SK10075

BLOCK-2065

ECG944	LM4250CH	MC3476G	NTE944
SK10076	TCG944		

BLOCK-2066

ECG944M	LM4250CN	MC3476P1	NTE944M
SK10077	TCG944M		

BLOCK-2067

ECG945	UA749DHC

BLOCK-2068

ECG946	GEIC-266	HEPC6049R	MC1461R
MC1469R	NTE946	TCG946	TM946

BLOCK-2069

14207	221-Z9147	AM747HC	AM747HM
CA747CT	CA747T	ECG947	GEIC-267
GEIC-268	LM747AH	LM747CH	LM747EH
LM747H	MC1747CG	MC1747G	ML747CT
NTE947	RC747T	SG747CT	SK3526
SN52747L	SN72747	SN72747L	TBB0747
TCG947	TM947	UA747C	UA747CK
UA747CL	UA747HC	UA747HM	UA747K
UA747ML	ULN-2157K	ULS-2157K	UPC251A
WEP2117	WEP2117/947		

BLOCK-2070

221-Z9148	561-0884-047	CA747CE	CA747CF
CA747CG	CA747E	CA747F	CA747G
ECG947D	HA17747G	HA17747P	LM747AJ
LM747CJ	LM747CN	LM747EJ	LM747J
MC1747CL	MC1747CP2	MC1747L	ML747CP
NTE947D	RC747DB	RC747DC	SK3556
SN72747J	SN72747JA	SN72747N	TBB0747A
TCG947D	TM947D	UA747CA	UA747CF
UA747CJ	UA747CN	UA747DC	UA747N
UA747PC	ULN-2157A	ULN-2157H	ULN-2747A
ULS-2157A	ULS-2157H		

BLOCK-2071

221-Z9173	AMLM348N	ECG948	HA1-4741-5
HA3-4741-5	HA4741-5	HEPC6129P	LM348
LM348J	LM348N	MC4741CL	MC4741CP
MC4741L	NTE948	RC4156DB	RC4156DC
RV4156DC	SK9173	TCG948	UA348DC
UA348PC			

BLOCK-2072

ECG948SM	LM348D	NTE948SM	SK10079

BLOCK-2073

138681	1464846-1	1464846-2	4084117-0001
4084117-0002	749DHC	88A752271	CXL949
ECG949	F4846-1	GEIC-25	MC1458CG
NTE949	SK3166	TCG949	TM949
UA749	UA749D	WEP2118	WEP2118/949

BLOCK-2074

08200045	221-167-05	442-644	78L12
78L12A	78L12ACZ	8-759-178-12	8-759-982-26
AN78L12	AN78L12-Y	ECG950	GEVR-109
HE-442-644	HEP-C6133P	HEPC6133P	LM78L12
LM78L12ACZ	MC78L12ACG	MC78L12ACP	MC78L12CG
MC78L12CP	NJM78L12A	NTE950	RC78L12A
SG7812T	SK9169	TA78L012AP	TCG950
UA78L12AC	UA78L12ACLP	UA78L12AWC	UA78L12C
UA78L12CLP	UPC78L12		

BLOCK-2075

AN78L15	ECG951	HEPC6134P	MC78L15ACG
MC78L15ACP	MC78L15CG	MC78L15CP	NTE951
SG7815T	SK9170	TCG951	UA78L15AC
UA78L15ACLP	UA78L15C	UA78L15CLP	UPC78L15

BLOCK-2076

ECG952	NTE952	SK7752	TCG952

BLOCK-2077

ECG953	NTE953	SK7806	TCG953
UA78GU1C	UA78MGU1C		

BLOCK-2078

ECG954	NTE954	SK7807	TCG954
UA79MGU1C			

BLOCK-2079

08210008	155712	221-217	221-Z9042
2420SNNE555	276-1723	51X90518A69	544-2009-555
555	555D	569-0540-355	569-0773-600
76-120-001	901523-01	C6130P	C6131P
CA555CE	CA555CG	CA555E	CA555G
EA33X8468	EAQ00-11100	ECG955M	GEIC-269
HA17555	HA17555PS	HEPC6130P	HEPC6131P
IC-520(ELCOM)	IC-555	J1000-NE555	J4-1555
JA-1555	LM555	LM555C	LM555CJ
LM555CN	LM555CN#1	LM555CN#2	M51841P
M51848P	MC1455D	MC1455P1	MCC555A
MCC555B	MX5741	NE555	NE555JG
NE555N	NE555P	NE555V	NJM555D
NTE955M	P61XX0002	RC555	RC555DE
RC555NB	RC555T	RS555	RV555NB
SE555V	SG555CM	SG555M	SK3564
SN72555JP	SN72555P	TCG955M	TDB0555B
TDB0555DP	TL1555P	TM955M	UA555
UA555IC	UA555TC	WEP2119	WEP2119/955M

BLOCK-2080

276-1718	ECG955MC	ICM7555IPA	NTE955MC
SK10449	TLC555	TLC555CP	TS555CN
XRL555CN			

BLOCK-2081

8-759-618-48	905-256	ECG955S	M51848L
NTE955S	SK9397		

BLOCK-2082

ECG955SM	ICM7555CBA	LMC555CM	NTE955SM
SK10450	TS555CD		

BLOCK-2083

1288-0183	164602	276-1778	317T
442-704	ECG956	LM317KC	LM317T
NTE956	SK9215	TCG956	UA317UC
UPC317H			

BLOCK-2084

442-705	ECG957	LM337MT	LM337T
LM377KC	NTE957	SK9216	TCG957

BLOCK-2085

07-28815-101	144921	4835-209-87275	51-90433A08
51-90433A27	51-90433A30	7818CU	78M18A
AN7818F	ECG958	HA17818P	HA178M18P
HEPC6115P	LM340-18KC	LM340-18U	LM340T-18
MC7818ACT	MC7818BT	MC7818CT	MC78M18CT
NJM7818FA	NJM78M18A	NTE958	SG7818ACP
SG7818CP	SK3699	TCG958	TDD1618S

BLOCK-2085 (CONT.)

UA7818C	UA7818CKC	UA7818CKG	UA7818CU
UA7818UC	UA7818UV	UPC7818HF	UPC78M18
UPC78M18H			

BLOCK-2086

ECG959	HEPC6124P	MC7819CT	MC7918CT
NTE959	SG7918ACP	SK9283	TCG959
UA7918C	UA7918CKC	UA7918CKG	UA7918UC

BLOCK-2087

0061-247-90001	06300039	06300089	06300144
06300342	08200043	1035-7564	1147-08
14305	148745	149017	154782
1643	191297	191299	194305
2003251	221-166	221-166-0	221-166-00
221-213	221-213-00	221-Z9043	2420T78005P
266P922010	266P922020	266P934060	276-1770
3100004	409 172 1509	409 241 5407	409 320 5700
442-54	46-131793-3	46-131801-3	46-132850-3
4835 209 87067	4835 209 87486	4835-209-87067	4835-209-87259
4835-209-87274	51X90480A07	51X90480A08	51X90518A07
612103-3	612107-4	6123600001	612448
6124480001	612479-1	6124790001	6126710001
6192140080	6192149480	6640000410	7805
7805A	7805C	7805CT	7805H
78HV05CU	78M05	78M05A	78M05C
78M05HF	8-759-013-06	8-759-171-05	8-759-604-29
8-759-701-56	8-759-924-12	8-759-982-31	901527-02
95KUCB0027AZ	AN7805	AN7805F	AN7805LB
AN78M05	AN78M05F	AN78M05LB	B0372540
BA178M05T	C014348	CO-14348	DM-94
EAS00-00700	ECG960	EP15X90(IC)	EW84X453
GEIC-190	GEVR-101	GEVR-102	GL7805
HA17805	HA17805H	HA17805P	HA178M05
HA178M05P	HE-442-54	HEPC6110P	I03A98M050
I03B98M050	IGL78M05	IP7805	IP7805A
IX0638CE	KIA7805P	L7805CV	L78M05
L78M05-A	L78M05-LU	L78M05-RA	L78M05ABV
L78M05C-V	L78M05CV	L78M05P	LM340
LM340-5KC	LM340-5U	LM340AT5.0	LM340T
LM340T-5	LM340T-5.0	LM340T-5.0R	LM340T5
LM340T5.0	LM340U5	LM341-5	LM341P-5.0
LM341P5.0	LM342P5.0	LM7805CT	LM7805CV
M5F7805	MA7805	MC7805	MC7805ACT
MC7805BT	MC7805C	MC7805CP	MC7805CT
MC7805CTDS	MC7805UC	MC78M05CT	MPC14305
N6124790001	NJM7805A	NJM7805FA	NJM78M05
NJM78M05A	NJM78M05FA	NTE960	PC17805H
PC7805H	PC7805HF	RC78M05FA	RH-IX0014PAZZ
SA7805CU	SA78HV05CU	SFC2805EC	SK3591
TA78005AP	TA78005P	TCG960	TDB7805T
TDD1605S	TM960	UA7508A	UA7805
UA7805A	UA7805C	UA7805CKC	UA7805CU
UA7805UC	UA7805UV	UA78M05C	UA78M05CKC
UA78M05UC	UC7805UC	UPC143-05	UPC14305H
UPC7805	UPC7805H	UPC78M05	UPC78M05AHF
UPC78M05H	UPD7805	VHIKA7805PI-1	X440063050
X440140050			

BLOCK-2088

9600DC	9600DM	9600PC	ECG9600

BLOCK-2089

443-67	443-722	9601DC	9601PC
DM8601J	DM8601N	DM9601J	DM9601N
ECG9601	HE-443-722	HEPC3803P	MC8601P
MC9601L	N8T22A	NTE9601	TCG9601

BLOCK-2090

901510-01	9602	9602DC	9602DM
9602PC	DM8602N	ECG9602	HE-443-1112
HEPC3807P	N9602B	N9602N	NTE9602
S9602F	TCG9602		

BLOCK-2091

276-1773	442-630	7905	7905CU
7905UC	79M05	79M05C	79M05CT
8190005	AMX4260	AN7905T	AN79M05
C6118P	ECG961	GE-961	GEVR-104
GEVR-105	GM320MP-5.2	HE-442-630	HEPC6118P
HEPC6136P	IP7905	LM320MP-5.2	LM320T-5.0
LM320T-5.2	LM320T5.0	LM7905CT	MC7905
MC7905.2CT	MC7905ACT	MC7905C	MC7905CT
MC79L05CP	MH745	NJM79M05A	NTE961
SK3671	SN72905	TCG961	TDB2905SP
TM961	UA7905	UA7905C	UA7905CKC
UA7905KC	UA7905UC	UA79M05AUC	UA79M05C
UA79M05CKC	X440069050		

BLOCK-2092

9615DC	9615DM	9615PC	ECG9615
SN75115J	SN75115N		

BLOCK-2093

103B98N060	148956	149598	221-Z9150
51X90458A98	51X90480A46	51X90507A24	7806CU
AN7806	C6111P	ECG962	GE-962
GL7806	HA17806P	HA178M06P	HEPC6111P
I03A98M060	I03B98M060	IGL78M06	L78M06
L78M06SA	L78N06	L78N06-SE	LM340-6KC
LM340-6U	LM340T-6	LM340T-6.0	LM340T-6.0R
LM340T6.0	LM340U6	LM341-6	LM341P-6.0
MC7806ACT	MC7806BT	MC7806CT	MC78M06CT
NTE962	SFC2806EC	SK3669	TCG962
TDB7806T	TDD1606S	TM962	UA7806C
UA7806CKC	UA7806CU	UA7806KC	UA7806UC
UA7806UV	UA78M06C	UA78M06CKC	UA78M06UC

BLOCK-2094

7906C	7906CU	AN79M06	C6120P
ECG963	HEPC6120P	LM320T-6.0	LM320T6.0
MC7906CT	NJM7906A	NJM79M06A	NTE963
SK3672	SN72906	TCG963	TM963
UA7906C	UA7906CKC	UA7906U	UA7906UC
UA79M06AUC	UA79M06C	UA79M06CKC	

BLOCK-2095

10-010#	10-010(IC)	10-010IC	14308
221-166-02	221-213-02	442-691	6124790003
7808C	7808CU	78M08	8-759-170-08
AN78M08	C6112P	DDEY028001	DM-33
DM-91	DM-92	DM91	ECG964
GEIC-191	GEIC191	GEVR-107	GEVR-108
HA17808P	HA178M08	HA178M08P	HE-442-691
HEPC6112P	L7808CY	LM320T8.0	LM340-8KC
LM340-8U	LM340T-8.0	LM340T8.0	LM340U8
LM341P-8.0	LM341P8.0	MC7808BT	MC7808C
MC7808CP	MC7808CT	MC78M08CT	MCT7808CT
MX-3452	NTE964	SFC2808EC	SK3630
TCG964	TDB7808T	TDD1608S	TM964
UA7808	UA7808C	UA7808CKC	UA7808CU
UA7808UC	UA7808UC	UA7808UV	UA78M08C
UA78M08CKC	UA78M08CKD	UA78M08UC	UPC14308
UPC14308H	UPC143C08	UPC78M08H	WEP2289
WEP2289/964			

BLOCK-2096

7908CU	ECG965	HEPC6121P	MC7908CT
NTE965	SG7908ACP	SG7908CP	SK9168
TCG965	UA7908C	UA7908CKC	UA7908KC
UA7908UC	UA79M08AUC	UA79M08C	

BLOCK-2097

06300219	0IKE780120A	1172-0042	1254-1694
1298-3409	1298-3888	1299-7862	1330-9224
1359-0161	144072-2	15-40140-2	186773
192285	194285	1LM340T12	2119-102-7605
2119-901-1109	221-101-01	221-166-04	221-166-4
221-213-04	221-213-4	221-296-04	2361461
2380426	266P93402	266P934020	266P934040
276-1771	3119-901-110	32117-901-110	32119-102-760
32119-901-110	33A17059	35684-107-534	
3H82-00020-000	442-663	442-674	46-131262-3
4835-209-87283	4835-209-87827	5190528A29	5190538A22
51X90480A09	51X90480A66	5653-TA78012A	6124790004
78012AP	7812	7812(IC)	7812A
7812CT	7812CU	7812P	78HV12CU
78M12	78M12KC	8-759-013-09	8-759-929-62
8-759-982-13	901527-01	905-242	AMX4577
AN7812	AN7812F	AN78M12	AN78M12(LB)
AN78M12-(LB)	AN78M12-LB	AN78M12LB	B0373230
C6113P	EAS00-13500	ECG966	EP84X217
EW84X747	FS-7812	FS7812	GEVR-110
GEVR-111	GL7812	HA17812	HA17812P
HA178M12P	HE-442-663	HE-442-674	HEPC6113P
I02A978120	I03A98M120	I03B98M120	ILM340T12
IP7812	KA7812	KIA78012AP	
KIA78012AP(KEC)	KIA7812P	L7812CV	L78M12
L78M12-LU	L78M12-RA	L78M12-SA	L78M12CV
L78M12RL	L78N12	L78N12-RA	LM340-12KC
LM340-12U	LM340T-12	LM340T-12R	LM340T12
LM340U12	LM341-12	LM341P-12	LM341P12
LM342P12	LM7812	LM7812CT	MC7812
MC7812AC	MC7812ACT	MC7812BT	MC7812C
MC7812CT	MC78M12	MC78M12CT	MC78T12CT
MPC7812H	NJM7812A	NJM78M12	NJM78M12A
NTE966	PC7812HF	QC0042	RC7812FA
SA7812CU	SA78HV12CU	SFC2812EC	SG7812
SK3592	TA78012A	TA78012AP	TCG966
TDA1412	TDB7812T	TDD1612S	TM966
TVSL78M12RL	TVSUC78M12H2	TVSUPC78M-12H2	
TVSUPC78M12H2	UA7812C	UA7812CKC	UA7812CU

BLOCK-2097 (CONT.)

UA7812UC	UA7812UV	UA78M12C	UA78M12CKC
UA78M12CKD	UA78M12UC	UC78M12H2	UC7933
UGH7812	UPC14312	UPC14312H	UPC7812
UPC7812H	UPC7812HA	UPC7812HF	UPC78M12
UPC78M12H	UPC78M12H2	WEP966L/966	X440058120

BLOCK-2098

569-0541-000	ECG9660	H104D1	H104D2
H104D6	H204B1	MC660P	NTE9660
TCG9660	TD2008P		

BLOCK-2099

C0905P	ECG9661	H124D1	H124D2
H124D6	H224B1	HEPC0905P	IC-413(ELCOM)
MC661P	MPB123D	NTE9661	TCG9661
TD2001P			

BLOCK-2100

C0900P	ECG9662	HEPC0900P	IC-220(ELCOM)
MC662P	MPB125D	TD2011P	

BLOCK-2101

569-0541-006	C0901P	C0912P	ECG9663
HEPC0901P	HEPC0912P	IC-225(ELCOM)	IC-409(ELCOM)
MC663P	NTE9663	TCG9663	TD2005P

BLOCK-2102

C0906P	ECG9664	HEPC0906P	IC-414(ELCOM)
MC664P	NTE9664	TCG9664	TD2004P

BLOCK-2103

C0910P	ECG9665	HEPC0910P	IC-221(ELCOM)
MC665P			

BLOCK-2104

ECG9666	IC-222(ELCOM)	MC666P	NTE9666
TCG9666			

BLOCK-2105

569-0541-014	C0911P	ECG9667	HEPC0911P
IC-223(ELCOM)	MC667P	TD2015P	

BLOCK-2106

569-0541-018	C0902P	ECG9668	H122D1
H122D2	H122D6	H222B1	HEPC0902P
IC-410(ELCOM)	MC668P	MPB121D	NTE9668
TCG9668	TD2003P		

BLOCK-2107

C0903P	ECG9669	HEPC0903P	IC-441(ELCOM)
MC669P	NTE9669	TCG9669	TD2006P

BLOCK-2108

276-1774	442-664	442-675	7912
7912C	79M12	79M12C	79M12CKC
AMX4188	C6122P	EAS00-09700	ECG967
GEVR-113	GEVR-114	HE-442-664	HE-442-675
HEPC6122P	IDUPC7912H	IP7912	L7912CV
LM320MP-12	LM320T-12	LM320T12	LM7912CT
MC7912ACT	MC7912CT	NJM79M12A	NTE967
SK3673	SN72912	TCG967	TDB2912SP
TM967	UA7912	UA7912C	UA7912CKC
UA7912U	UA7912UC	UA79M12AUC	UA79M12C
UA79M12CKC	UA79M12CKD	UPC7912H	X440079120

BLOCK-2109

C0904P	ECG9670	HEPC0904P	IC-412(ELCOM)
MC670P	NTE9670	TCG9670	TD2002P

BLOCK-2110

ECG9671	H103D1	H103D2	H103D6
MC671P	NTE9671	TCG9671	TD2009P

BLOCK-2111

569-0541-024	C0909P	ECG9672	H102D1
H102D2	H102D6	H202B1	H203B1
HEPC0909P	IC-224(ELCOM)	MC672	MC672P
NTE9672	TCG9672	TD2010P	

BLOCK-2112

ECG9673	H105D1	H105D2	H105D6
H205B1	MC673P	TCG9673	

BLOCK-2113

ECG9674	MC674P		

BLOCK-2114

569-0541-030	ECG9675	MC675P	NTE9675
TCG9675			

BLOCK-2115

569-0541-031	ECG9676	MC676P	NTE9676
TCG9676			

BLOCK-2116

ECG9677	H119D1	H119D6	MC677P
TD2013P			

BLOCK-2117

ECG9678	H115D1	H115D6	MC678P
MPB124D	NTE9678	TCG9678	

BLOCK-2118

569-0541-036	ECG9679	MC679P	NTE9679
TCG9679			

BLOCK-2119

0IGS781200A	0IGS781500A	149038	276-1772
442-63	46-13311-3	544-2003-005	612479-5
6124790005	7815	7815(IC)	7815CU
AN7815	AN78M15	AN78M15(LB)	AN78M15LB
C6114P	ECG968	EW84X526	GL7815
HA17815P	HA178M15P	HE-442-63	HEPC6114P
IP7815	L7815CV	LM340-15U	LM340T-15
LM340T-15R	LM340T15	LM340U15	LM341-15
LM341P-15	LM341P15	LM342P15	MC7815
MC7815ACT	MC7815BT	MC7815CP	MC7815CT
MC7815P	MC78M15CT	ML7815P	NTE968
SFC2815EC	SK3593	TCG968	TDA1415
TDB7815T	TDD1615S	TM968	UA7815
UA7815C	UA7815CKC	UA7815CU	UA7815UC
UA78M15C	UA78M15CKC	UA78M15CKD	UA78M15UC
UPC14315D	UPC14315H	UPC78M15H	

BLOCK-2120

569-0541-037	ECG9680	H118D1	H118D6
MC680P	NTE9680	TCG9680	TD2012P

BLOCK-2121

569-0541-038	C0908P	ECG9681	H112D1
H112D6	HEPC0908P	MC681P	NTE9681
TCG9681			

BLOCK-2122

C0907P	ECG9682	HEPC0907P	IC-415(ELCOM)
MC682P	NTE9682	TCG9682	

BLOCK-2123

ECG9683	MC683P

BLOCK-2124

569-0541-041	ECG9684	MC684P

BLOCK-2125

569-0541-042	ECG9685	MC685P

BLOCK-2126

569-0541-043	ECG9686	MC686P

BLOCK-2127

ECG9689	MC689P	NTE9689	TCG9689

BLOCK-2128

442-613	7915C	C6123P	ECG969
HE-442-613	HEPC6123P	IP7915	LM320T-15
LM320T15	MC7915ACT	MC7915CP	MC7915CT
MC7918CK	NTE969	SK3674	SN72915
TCG969	TDB2915SP	TM969	UA7915C
UA7915CKC	UA7915U	UA7915UC	UA79M15AUC
UA79M15C	UA79M15CKC	UA79M15CKD	UPC7815H

BLOCK-2129

ECG9690	MC690P

BLOCK-2130

ECG9691	MC691P	NTE9691	TCG9691

BLOCK-2131

ECG9696	MC696P

BLOCK-2132

443-727	96L02	96L02DC	96L02DM
96L02PC	AM26L02PC	ECG96L02	HE-443-727
NTE96L02	TCG96L02		

BLOCK-2133

96LS02DC	96LS02DM	96LS02PC	ECG96LS02
HE-443-1040	NTE96LS02	TCG96LS02	

BLOCK-2134

96S02DC	96S02PC	ECG96S02	NTE96S02
TCG96S02			

BLOCK-2135

276-1777	317K	ECG970	LM350K
NTE970	SK9339	TCG970	

BLOCK-2136

ECG971	HEPC6125P	LM320MP-24	LM320T-24
LM320T24	MC7924CT	NTE971	SK3675
TCG971	TM971	UA7924C	UA7924CKC
UA7924U	UA7924UC	UA79M24AUC	UA79M24C
UA79M24CKC	UA79M24CKD		

BLOCK-2137

442-60	56A21-1	612103-4	612107-2
7824CU	C6105	C6116P	ECG972
GE-972	HA17824P	HA178M24P	HE-442-60
HEPC6105P	HEPC6116P	LM340-24KC	LM340-24U
LM340T-24	LM340T-24R	LM340T24	LM340U24
LM341-24	LM341P-24	MC7824	MC7824ACT
MC7824BT	MC7824CT	MC78M24CP	MC78M24CT
NTE972	SFC2824EC	SK3670	SN72924
TCG972	TDB7824T	TDD1624S	TM972
UA7824C	UA7824CKC	UA7824CU	UA7824UC
UA7824UV	UA78M24C	UA78M24CKC	UA78M24CKD
UA78M24UC			

BLOCK-2138

2056-05	221-Z9098	307-047-9-001	307-112-9-005
442-96	C6050G	DDEY001001	ECG973
GE-973	HEPC6050G	LM1496	LM1496H
LM1596H	LM1946H	MC1496	MC1496G
MC1496K	MC1946A	N5596K	NTE973
PL-307-047-9-001	SG1496	SG1496T	SK3233
TCG973	TM973	TVCM69	UA796
UA796HC	WEP2067	WEP2067/973	

BLOCK-2139

1496	2000-009	DDEY020001	ECG973D
LM1496J	LM1496N	MC1496A	MC1496L
MC1496N	MC1496P	MX-3369	NTE973D
SK3892	TCG973D	TM973D	UA796DC
UA796PC			

BLOCK-2140

11C44DC	443-62	569-0542-744	8000-00038-002
906128-01	DDEY019001	ECG974	IP20-0210
IP20-2010	MC4040P	MC4044	MC4044CP
MC4044D	MC4044L	MC4044P	NTE974
SK3965	TCG974	TM974	

BLOCK-2141

442-39	569-0964-002	AM748DC	AM748DM
C6107P	CA301AE	CA301AG	CA301E
CA301G	CA748CE	CA748CG	CA748CJ
CA748CN	CA748E	CA748G	CA748J
CA748N	CM301AN	DDEY026001	ECG975
GEIC-304	HE-442-39	HEPC6107P	LM201AN
LM301ADE	LM301AN	LM301AP	LM301AT
LM301AV	LM748CN7	LM748N	ML748CS
MLM101AG	MLM301AU	NTE975	SG301AM
SK3641	SK3644	SN52301	SN52748J
SN52748N	SN52748P	SN72301	SN72301AP
SN72748J	SN72748JG	SN72748N	SN72748P
TBB0748B	TCG975	TM975	UA301AT
UA301ATC	UA748CJ	UA748CJG	UA748CN
UA748CP	UA748DC	UA748DM	UA748MJ
UA748MJG	UA748N	UA748TC	UA748V
UPC157C	UPC301AC	UPC301AN	

BLOCK-2142

ECG975SM	LM301AD	NTE975SM	SK10455
UA748CD			

BLOCK-2143

612048-4	AD301AN	AMLM207D	CA307
CA307E	CA307G	ECG976	LM307DE
LM307F	LM307J	LM307N	LM307P
LS307B	ML307S	MLM307P1	MLM307U
NTE976	SK3596	SN72307	SN72307JA
SN72307JP	SN72307N	SN72307P	TCG976
TM976	UA307T	UA307TC	

BLOCK-2144

0061-969-90130	02-781050	08200069	159355
163794	178745	2000-004	2000-008
2049-03	2061-42	2065-52	221-167-01
221-167-01A	221-Z9044	266P92304	307-095-9-003
307-112-9-002	307-112-9-003	307-113-9-003	30901050
3403	4835-209-87252	4835-209-87821	5364601
5652-TA78L005	6123999001	6192140081	742726
78005AP	78L05	78L05-AN	78L05-AV
78L05A	78L05ACP	78L05AV	78L05AVP
78L05J	78L05V	8-759-108-05	8-759-150-61
8-759-178-05	8-759-982-21	8052805	905-219
905-267	916150	AN78L05	AN78L05-Y
AN78N05	AN78N05LB	C6132P	DDEY042001
DDEY046001	DDEY088001	DDEY093001	DM-104
ECG977	F78105AV	F78L05AC	F78L05AV
F78L05AWC	GEVR-100	HE-442-627	HEP-C6132P
HEPC6132P	HEPC6142P	I02A98L050	I02J98L050
I0QT98L050	K0201E	KIA78L005AP	KIA78L05AP
KIA78L05BP	L78N05	LM7805ACZ	LM78L05A
LM78L05ACH	LM78L05ACI	LM78L05ACZ	M5278L-05
M5278L05	MC78L05	MC78L05,CP	MC78L05ACG
MC78L05ACP	MC78L05ACPRP	MC78L05C	MC78L05CG
MC78L05CP	ML78L05A	MX-3198	NJM78L05
NJM78L05A	NJM78L05A(T3)	NJM78L05AV	NTE977
PC78L05	RC78L05A	RE387-IC	REN977
RH-IX0111CEZZ	RVIUPC78L05	SK3462	TA78L
TA78L005	TA78L005AP	TA78L005AP-Y	TA78L005P
TA78L05S	TCG977	TM977	TVSL78N05
UA78L05	UA78L05AC	UA78L05ACLP	UA78L05AWC
UA78L05C	UA78L05CLP	UA78L05S	UA78L05WC
UPC-781-05AN	UPC-78L-05AN	UPC14305	UPC78L05
UPC78L05A	UPC78L05J	UPC78L05J-T	UPC78L05J-TP
UPC78L05T	X440078051	X440078054	YEAMJM78L05
YEAMNJ58L05	YEAMNJM78L05		

BLOCK-2145

221-Z9152	276-1728	556	556D
901523-03	ECG978	LM556CN	LM55SCN
MC3456P	NE556	NE556A	NE556N
NTE978	RC556DB	RC556DC	RV556DB
SK3689	TCG978	TDB0556A	TM978
UA556DC	UA556DM	UA556PC	X440055560

BLOCK-2146

ECG978C	NTE978C	SK10454	TLC556
TLC556CP	TS556CN	XRL556CN	XRL556CP

BLOCK-2147

ECG978SM	LM556D	LM556M	NTE978SM
SK10453			

BLOCK-2148

ECG979

BLOCK-2149

CD4046AE	ECG980	MC14046BAL	NTE980
SK4046	SK4046B	TCG980	TM980

BLOCK-2150

1800DC	1800PC	376-0062	586-546
ECG9800	ITT1800-5	ITT1800N	MC1800P
NTE9800	SN151800J	SN151800N	TCG9800

BLOCK-2151

1801DC	1801PC	ECG9801	ITT1801N
MC1801P	NTE9801	SN151801J	SN151801N
TCG9801			

BLOCK-2152

1802DC	1802PC	586-528	ECG9802
MC1802P	NTE9802	SN151802J	SN151802N
TCG9802			

BLOCK-2153

1803DC	1803PC	ECG9803	MC1803P
NTE9803	SN151803J	SN151803N	TCG9803

BLOCK-2154

1804DC	1804PC	586-517	ECG9804
MC1804P	NTE9804	SN151804J	SN151804N
TCG9804			

BLOCK-2155

1805DC	1805PC	ECG9805	MC1805P
NTE9805	SN151805J	SN151805N	TCG9805

BLOCK-2156

1806DC	1806PC	586-331	ECG9806
ITT1806-5	ITT1806N	ITT1807-5	MC1806P
NTE9806	SN151806J	SN151806N	TCG9806

BLOCK-2157

1807DC	1807PC	236-0012	ECG9807
IC-237(ELCOM)	ITT1807N	MC1807P	NTE9807
SN151807J	SN151807N	TCG9807	

BLOCK-2158

1808DC	1808PC	586-547	ECG9808
ITT1808-5	ITT1808N	MC1808P	NTE9808
SN151808J	SN151808N	TCG9808	

BLOCK-2159

1809DC	1809PC	ECG9809	ITT1809-5
ITT1809N	MC1809P	NTE9809	SN151809J
SN151809N	TCG9809		

BLOCK-2160

13-0161	158040	221-Z9045	3130-0000-014
44T-300-100	5359281	78L08	78L08A
78L08AC	78L08AWC	78L82AC	AN78L08
DM-106	DM106	ECG981	GEVR-106
HEPC6144P	LM7808A-8	LM78L08ACH	LM78L08ACZ
LM78L08CH	MC78L08ACG	MC78L08ACP	MC78L08CG
MC78L08CP	NTE981	SK3724	TA78L008P
TCG981	TM981	UA78L	UA78L-8.2AWC
UA78L08	UA78L08AC	UA78L08ACLP	UA78L08AWC
UA78L08C	UA78L08CLP	UA78L08S	UA78L08WC
UA78L82	UA78L82AWC	UA78L82AWV	UA78L82W
UPC78L08			

BLOCK-2161

1810DC	586-780	ECG9810	ITT1810-5
ITT1810N	MC1810P	NTE9810	SN151810J
SN151810N	TCG9810		

BLOCK-2162

1811DC	1811PC	ECG9811	ITT1811-5
ITT1811N	MC1811P	NTE9811	SN151811J
SN151811N	TCG9811		

BLOCK-2163

1812DC	1812PC	569-0965-006	586-412
ECG9812	MC1812P	NTE9812	SN151812J
SN151812N	TCG9812		

BLOCK-2164

1813DC	1813PC	ECG9813	MC1813P
NTE9813	TCG9813		

BLOCK-2165

1814DC	1814PC	ECG9814	MC1814P
NTE9814	TCG9814		

BLOCK-2166

1C12	221-87-01	CA3170	CA3170E
ECG982	EX42	IC-312	NTE982
SE5-0930	SK3205	TM982	ULN2268A

BLOCK-2167

DM-50	ECG983	NTE983	SK3887
TM983	ULN2231A	ULX-2231A	

BLOCK-2168

1044-7035	141134	143808	1465316-1
221-Z9174	3153GM1	51-13753A47	51S13752A47
51S13753A47	51S23753A47	51S33753A47	5320500100
612273-1	CA3153G	CA3153GM1	CA3153GMI
ECG984	GE-984	NTE984	SK3185
SK9276	TA6472	TCG984	TM984
TVS3153GM1	TVSCA3153GM1	TVSCA3153GMI	ULN2297A

BLOCK-2169

142341	142903	146151	221-Z9175
CA3144E	CA3144G	CA3144Q	ECG985
NTE985	SK3214	TCG985	TM985

BLOCK-2170

1081-3558	142719	143822	146152
221-Z9176	51-13753A50	51S13753A50	51S13753A64
51S137A50	CA3151G	CA3151GM1	CA3151GMI
ECG986	EN11438	NTE986	SK3918
TCG986	TM986	TVSEN11438	WEP986/986

BLOCK-2171

149018	15-45186-1	163434	1754-3
207827	221-129	221-129A-01	221-188
276-1711	324(IC)	442-602	4H20980587
544-2020-002	6121630001	8-759-132-40	905-1227
AM224D	AM324D	AM324N	AN6564
BA10324	CA0324E	CA124E	CA224E
CA224G	CA324E	CA324G	ECG987

BLOCK-2171 (CONT.)

EP84X119	HA17902G	HA17902P	HE-442-602
I03D063240	IGLA6324	LA6324	LM224AD
LM224AF	LM224AJ	LM224AN	LM224D
LM224J	LM224N	LM2902	LM2902J
LM2902N	LM324	LM324A	LM324AD
LM324AF	LM324AJ	LM324AN	LM324D
LM324J	LM324N	M5224P	MC3403N
MC3403P	MLM224L	MLM224P	MLM324
MLM324L	MLM324P	MLM324P1	NJM2902N
NTE987	SG224J	SG224N	SG324N
SK3643	TCG987	TDB0124DP	TM987
UA224DM	UA2902	UA324DC	UA324PC
UA3303PC	UA3403DC	UA3403PC	UPC324C
UPC451C	VHIM5224P//-1	X440029020	XR3403CN
XR3403CP			

BLOCK-2172

905-1289	ECG987SM	LM2902D	NJM2902M
NTE987SM	SK10452	UPC324G2	

BLOCK-2173

02-781060	022-2844-501	022-2844-701	11-60
111825	309-033-0	39-033-0	39-033-2
78106C	78606C	78L06	78L06A
78L06AC	78L06C	78L62	78L62AC
78L62AWC	78L62WV	8000-00047-006	AN78L06
ECG988	F78L06	F78L062AC	F78L06AC
F78L06C	F78L62AC	F78L62AWC	F78L62WV
IP20-0220	IP20-0253	MC78L06AV	NJM78L06A
NTE988	SK3973	TCG988	TM988
UA78L062WV	UA78L06ACLP	UA78L06AS	UA78L06CLP
UA78L06S	UA78L6.2AHC	UA78L62AHC	UA78L62AWC
UA78L62AWV			

BLOCK-2174

221-171	276-1720	442-654	565
ECG989	HE-442-654	LM565CN	NE565A
NTE989	SK3595	TCG989	TM989
WEP989	WEP989/989		

BLOCK-2175

1099-3616	146164	ECG990	LM1877
LM1877N	LM1877N-10	LM1877N-3	LM1877N-9
LM377-N	LM377N	NTE990	SK9012
TCG990	TM990		

BLOCK-2176

10112563	2234494	336637-20	351-7011-020
3610005	84630	84630-1	84630-2
84630-3	900HC	99E16-1	D919695
ECG9900	IC-56(ELCOM)	MC700G	MC800G
SL02518	SL04217	US-0909D	

BLOCK-2177

161-011-0001	161-011-0002	19E19-2	351-7011-030
351-7011-040	903HC	D919698	ECG9903
G612994	MC703G	MC803G	

BLOCK-2178

19E18-1	398-8418-1	904HC	D919699
ECG9904	MC704G	MC804G	

BLOCK-2179

161-006-0001	161-006-0002	19E15-1	351-7011-050
351-7011-060	905HC	D919700	ECG9905
MC705G	MC805G		

BLOCK-2180

4914296	906HC	ECG9906	MC706G
MC806G			

BLOCK-2181

161-012-0002	351-7025-010	907HC	ECG9907
MC707G	MC807G		

BLOCK-2182

351-7015-010	351-7026-010	351-7206-080	351-7206-150
908HC	ECG9908	MC708G	MC808G
US-0908D			

BLOCK-2183

351-7015-020	351-7206-090	351-7206-160	559-1495-001
909HC	ECG9909	MC709G	MC809G

BLOCK-2184

3001-201	544-3001-201	7636	ECG991
NTE991	SC42502P	TCG991	TM991

BLOCK-2185

10541284	161-118-0001	179-46444-01	196639
351-7008-010	351-7011-070	351-7011-080	351-7015-030
351-7026-030	351-7206-100	351-7206-170	4-08018-667
559-1496-001	592-027	7528156-P3	7528158-P4
8505870-1	910HC	CD5328	ECG9910
IC-58(ELCOM)	LED21085	MC710G	MC810G
NTE9910	SL04194	SL16122	TCG9910
US-0910D			

BLOCK-2186

351-7011-090	351-7015-040	351-7206-110	351-7206-180
559-1492-001	592-028	911HC	C2007G
ECG9911	HEP581	HEPC2007G	IC-59(ELCOM)
MC711G	MC811G	MC911G	SL18699
US-0911D			

BLOCK-2187

0N049874	11253588	351-7015-050	351-7206-190
912HC	ECG9912	MC712G	MC812G
US-0912D			

BLOCK-2188

179-46444-05	351-7008-020	351-7015-060	351-7026-060
351-7206-200	559-1494-001	913HC	ECG9913
MC713G	MC813G	SL03667	US-0913D

BLOCK-2189

10176209	103508-2	156-0011-00	161-152-0101
161-152-0102	179-46444-02	2234496	351-3006
351-7011-100	351-7011-110	351-7011-120	351-7025-020
351-7121-020	3610003	4381P1	592-029-0
592-081-0	59B402787	84626	84626-1
84626-2	84626-3	84626-4	880-103-00
914HC	B77T0049	CD6039	D34002410-001
ECG9914	FUL914-28	HEP584	HEPC2006G
HEPC2010G	HEPC2012G	IC-61(ELCOM)	IC-62(ELCOM)
LED-15005	MC714G	MC814G	PC7327C
QB400428	SL01640	SL02519	SL02734
SL02779	SL03019	SL04218	SL04732
UL-914	US-0960D	V151(BENCO)	WEP9914
X-2408473			

BLOCK-2190

103508-3	400-1735	915HC	ECG9915
MC715G	MC815G		

BLOCK-2191

1147-10	276-1713	3100002	3900
442-71	612080-1	742727	CA3401E
CM3900	ECG992	EP84X20	HA17301G
HA17301P	HE-442-71	LM2900	LM2900N
LM3301N	LM3401N	LM3900	LM3900N
MC3301P	MC3401L	MC3401P	NTE992
SK3688	TCG992	TCG9926	TM992
UA3301P	UA3401PC	YEAMLM2900N	

BLOCK-2192

351-7011-150	351-7011-160	351-7015-070	351-7026-070
351-7206-140	351-7206-210	559-1493-001	84628
84628-1	84628-2	84628-3	84628-4
921HC	ECG9921	MC721G	MC821G
US-0921D			

BLOCK-2193

1001(EF-JOHNSON)	1800	544-3001-001	692-0016-00
801800	C2001P(HEP)	ECG9924	HEPC2001P
HEPC2502P	IC-55(ELCOM)	IC724	MC724P
MC824P	NTE9924	SC9963P	TCG9924
WEP9924	WEP9924/9924		

BLOCK-2194

2234497	400-1736	926HC	ECG9926
MC726G	MC826G	NTE9926	

BLOCK-2195

84353-2	927HC	ECG9927	MC727G
MC827G			

BLOCK-2196

ECG993	NTE993	TCG993

BLOCK-2197

103729	104830	10658276	131300
134195-001	1348A30H01	146643	1471-4380
17-12054-1	1895993-1	198409-13	19A115913-1
202914-010	221-Z9079	2501557-430	2656211-1
2899002-00	3007474-00	3007474-01	3172629-1
326830	349-113-011	401113-2	423-800234
43A168135-2	477-0377-001	48-1050-SL12318	51577500

BLOCK-2197 (CONT.)

52000-030	55977	56553	569-0544-300
569-0880-430	586-151	65600500	671A290H01
68A7349PD30	733W00024	77C800-005	810002-269
811790	860003-101	9003148	9003148-01
9003148-02	928512-101	930DC	930PC
932292-1C	998280-930	CD2300E	CD2300E/830
DM930N	DTML9930	ECC-01267	ECG9930
HD2204	HD2204P	HEPC1030P	HL53424
IC30(ELCOM)	ITT930-5	ITT930N	LB2002
M5930	M5930P	MC830L	MC830P
MC832N	MC9930	MIC930-5D	MIC930-5P
MIS-18101-3	NTE9930	PD9930-59	PE9930-59
SL04563	SL11877	SL16201	SL16516
SL16584	SL53424	SN15830J	SN15830N
SW930-2M	SW930-2P	TCG9930	TD1060
TD1060P			

BLOCK-2198

105412(5	10541285	84323	860003-111
931DC	931PC	93C	ECG9931
MC831L	MC831P	MIC931-5D	MIC931-5P
SL16585	SN15831J	SN15831N	

BLOCK-2199

103731	131301	134196-001	1348A32H01
151544	1527-5282	15405-4	17-12056-1
179-46445-01	1895994-1	198409-6	19A115913-3
202914-040	221-Z9080	236-0002	2500389-432
2500747	2808577	3007359-03	3007473-00
3007473-01	3177200	326832	331378
349-113-012	352-0023-001	40-065-19-006	401113-3
423-800175	423-800177	43A168135-1	477-0375-001
48-1050-SL12319	50210-2	51310000	51577600
52000-032	55982-1	569-0544-304	586-303
65600600	669A464H01	68A7349-D32	68A7349-PD32
6900K91-007	733W00025	7528046-P4	77C800-007
811794	860003-121	9003911	9003911-01
9003911-03	932DC	932PC	933044-2D
94333	C1032P	CD2306E	CD2306E/832
CPS15553-101	DM932N	DTML9932	ECC-01262
ECG9932	HD2201	HD2201P	HEPC1032P
IC-280(ELCOM)	ITT932-5	ITT932N	LB2003
M5932	M5932P	MC832L	MC832P
MIC932-5D	MIC932-5P	MIS-181C1-4	NTE9932
PD9932-59	PE9932-59	SL03914	SL04568
SL16206	SL16586	SL17284	SL53425
SN15832J	SN15832N	SW932-2M	SW932-2P
TCG9932	TD1062	TD1062P	

BLOCK-2200

104833	17-12057-1	179-46445-02	1895995-1
19A115913-4	2501337-433	3007572-00	3172626
326833	40-065-19-008	423-800-235	43A168135-3
48-1050-SL12320	51310001	51577700	52000-033
55980	569-0544-305	7011200-02	733W00026
89028-6	928517-101	933DC	933PC
94331	C1033P	CD2314E	CD2314E/833
DM933N	DTML9933	ECG9933	HD2202
HD2202P	HEPC1033P	IC-281(ELCOM)	ITT933-5
ITT933N	LB2005	M5933	M5933P
MC833L	MC833P	MIC933-5D	MIC933-5P
NTE9933	PD9933-59	PE9933-59	SL16517
SN15833J	SN15833N	SW933-2M	SW933-2P
TCG9933	TD1063	TD1063P	

BLOCK-2201

14500004-003	164981	1820-0869	
202914-010/250-070	2500390-435	400931	569-0544-312
935DC	935PC	C1035P	CD2312E
DM935N	DTML9935	ECG9935	HD2208
HD2208P	HEPC1035P	HL53426	ITT935-5
ITT935N	LB2007	M5935	M5935P
MC835L	MC835P	MC840L	MC840P
MIC935-5D	MIC935-5P	NTE9935	PD9935-59
PE9935-59	SL53426	SN15835J	SN15835N
SW935-2M	SW935-2P	TCG9935	TD1061
TD1061P			

BLOCK-2202

103743	104836	10658278	131303
1348A36H01	14500004-001D	14500004-001P	1471-4356
1471-9860	16271033	16271041	164982
1914062-1	198409-14	19A115913P14	
202914-010/250-100	2068510-0701	221-Z9081	2500390-436
2501557-436	2899003-00	3007477-00	3007477-01
3176135-1	326836	3520025-001	423-800202
43A168135-6	48-1050-SL12321	51310002	5165440
52000-036	56557	569-0320-818	569-0320-826
569-0544-308	586-152	68A7349PD36	6900K91-006
7012128	7012128-02	733W00027	860003-161
936DC	936PC	94825000-11	C1036P
CD2310E	CD2310E/836	DM936N	DTML9936

BLOCK-2202 (CONT.)

ECG9936	HD2206	HD2206P	HEPC1036P
IC31(ELCOM)	ITT936-5	ITT936N	LB2006
M5936	M5936P	MC836L	MC836P
MIC936-5D	MIC936-5P	NTE9936	PD9936-59
PE9936-59	R10254P936	SL04567	SL11878
SL16210	SL16518	SL16587	SL17289
SL53427	SN15836J	SN15836N	SW936-2M
SW936-2P	TCG9936	TD1072	TD1072P

BLOCK-2203

0404011-001	14500004-002	1479-0224	179-46445-03
2500390-437	2501557-437	3007359-00	3176135-2
40-065-19-004	52000-037	586-308	733W00048
937DC	937PC	94325	CD2311E/837
DM937N	ECG9937	HD2216	HD2216P
ITT937-5	ITT937N	LB2106	M5937
M5937P	MC837L	MC837P	MIC937-5D
MIC937-5P	NTE9937	PD9937-59	PE9937-59
SL03916	SL16208	SN15837J	SN15837N
SW-10548	SW937-2M	SW937-2P	TCG9937
VE79141			

BLOCK-2204

ECG994	LM566CH	LM566H	NE566T

BLOCK-2205

221-Z9082	349-113-025	505254	6900K93-002
941DC	941PC	ECG9941	ITT941-5
ITT941N	MC841P	MIC941-5D	MIC941-5P
SW941-2M	SW941-2P		

BLOCK-2206

104844	10658279	110472-003	131304
1348A44H01	14500005-001	1471-4398	147256
1820-0865	198409-5	202914-050	2500389-444
2652613	2656213	2656213-1	2656747
3172630	326844	40-065-19-005	43A168135-7
48-1050-SL12322	49A0010	49A0510	51310003
51577800	51624200	52000-044	55978-1
569-0544-318	569-0880-444	671A291H01	6900K91-002
7528159-P4	860003-99	89028-3	928560-1
928560-101	928571-1	94327	944DC
944PC	C1044P	CD2307E	CD2307E/844
CG24015A	DM944N	DTML9944	ECG9944
HD2209	HD2209P	HEPC1044P	HL53428
IC-276(ELCOM)	ITT944-5	ITT944N	LB2004
M5944	M5944P	MC1044P	MC844L
MC844P	MC944L	MIC944-5D	MIC944-5P
NTE9944	PD9944-59	PE9944-59	S-2041-8661
SL03915	SL16215	SL16811	SL53428
SN15844J	SN15844N	SW944-2M	SW944-2P
TCG9944	TD1064	TD1064P	VE79093

BLOCK-2207

103755	10658280	10795-5	131305
134197-001	1348A45H01	146644	1527-5308
17-12065-1	197666	202914-010/250-060	221-Z9083
2500391-445	2652614	2656214	2656214-1
2899004-00	3007472-00	326845	329814
349-113-013	349-113-022	401113-4	43A168135-8
477-0380-001	48-1050-SL12323	551-013-00	55975-1
56558	569-0880-445	586-187	671A292H01
68A7349PD45	77C813-002	860003-141	860003-151
89028-2	9137-C-1004	928515-101	945DC
945PC	998280-945	C1045P	CD2304E
CD2304E/845	DM945N	DTML9945	ECC-01263
ECG9945	HD2205	HD2205P	HEPC1045P
IC229(ELCOM)	ITT945-5	ITT945N	LB2030
M5945	M5945P	MC845L	MC845P
MIC945-5D	MIC945-5P	MIS-18101-5	NTE9945
PD9945-59	PE9945-59	R10254P945	R10254P945B
SL16211	SL16521	SL16588	SN15845J
SN15845N	SW945-2M	SW945-2P	TCG9945
TD1070	TD1070P	WS6945-2	

BLOCK-2208

09-308021	103705	104846	10658281
10795-6	131306	134198-001	1348A46H01
14500001-001	146641	1471-4364	15405-1
17-12058-1	1820-0861	1895991-1	198409-11
19A115913-19	202914-020	2068510-0702	221-Z9084
236-0003	2500390-446	2501557-446	2656212
2656212-1	2656748	2898431-2	2899000-00
3007476-00	3007476-01	317-2627-1	326846
349-113-014	3L4-9013-01	40-065-19-013	401113-1
423-800178	43A168135-4	477-0381-001	
48-1050-SL12324	5-113641	51310004	51577900
52000-046	551-010-00	55976-1	56519
569-0320-820	569-0320-822	569-0544-322	569-0880-446
586-153	65600700	671A293H01	68A7349-D46
68A7349PD46	6900K91-003	733W00028	7528048-P4
77C800-008	811791	89028-4	89028-4-1-4

BLOCK-2208 (CONT.)

9003149	9003149-02	928514-1	928514-101
946DC	946PC	C1046P	C1046P(HEP)
CD2302E	CD2302E/846	DM946N	DN1946
DTML9946	EA33X8399	ECC-01264	ECG9946
HD2203	HD2203P	HEPC1046P	HEPC1056P
HL53429	IC32(ELCOM)	ITT946-5	ITT946N
LB-2000	LB2000	M5946	M5946P
MC846L	MC846P	MIC946-5D	MIC946-5P
NTE9946	PD9946-59	PE9946-59	R10254P946
R10255P946B	SL11879	SL16204	SL16519
SL16589	SL53429	SN15846J	SN15846N
SW946-2M	SW946-2P	TCG9946	TD1065
TD1065P	WEP9946	WEP9946/9946	

BLOCK-2209

110240-005	179-46445-05	1820-0095	1820-0862
3172631	40-065-19-007	51310005	51578000
52000-048	7528153-P4	948DC	948PC
CD2305E	CD2305E/848	DM948N	DTML9948
ECG9948	IC-279(ELCOM)	ITT948-5	ITT948N
LB2130	M5948	M5948P	MC848L
MC848P	MIC948-5D	MIC948-5P	NTE9948
PD9948-59	PE9948-59	SL11880	SL16590
SN15848J	SN15848N	ST984539-006	SW948-2M
SW948-2P	TCG9948	TD1067	TD1067P

BLOCK-2210

0404012-001	14500001-002	15405-2	179-46445-06
198409-1	2500390-449	2501557-449	3007359-01
3172627-2	40-065-19-003	423-800194	52000-049
7011201-02	733W00049	949DC	949PC
CD2303E	CD2303E/849	DM949N	DTML9949
ECG9949	ITT949-5	ITT949N	LB2100
M5949	M5949P	MC849L	MC849P
MIC949-5D	MIC949-5P	NTE9949	PD9949-59
PE9949-59	SL03911	SL16216	SN15849J
SN15849N	SW949-2M	SW949-2P	TCG9949
TD1085	TD1085P	VE79092	

BLOCK-2211

610020-917	76-140-001	AMX4181	CS2917
ECG995	LM2917	LM2917-8	LM2917J
LM2917N	NTE995	SK9209	TCG995

BLOCK-2212

134254-001	1820-0087	3172638	349-113-024
569-0880-450	950DC	950PC	ECG9950
MC850L	MC850P	MIC950-5D	MIC950-5P
NTE9950	PD9950-59	PE9950-59	SN15850J
SN15850N	SW950-2M	SW950-2P	TCG9950

BLOCK-2213

10658282	128C212H01	17-12089-1	19A115913-2
19A115913P2	349-113-015	401182	55979-1
6900K93-001	7528160-P4	7528374P3	928510-1
928510-101	951DC	951PC	CPS-16676-1
ECC-01266	ECG9951	ITT951-5	ITT951N
LED21092	MC851L	MC851P	MIC951-5D
MIC951-5P	NTE9951	PD9951-59	PE9951-59
SL03917	SL04570	SL16203	SN15851J
SN15851N	SW951-2M	SW951-2P	TCG9951
X-2408784			

BLOCK-2214

ECG995M	LM2917N-8	LM2917N8	NTE995M
SK10451			

BLOCK-2215

CA3080E	ECG996	LM3080AN	NTE996
SK9201	TCG996		

BLOCK-2216

14500001-004	179-46445-08	19166123	2500390-461
2501557-461	3172629-2	40-065-19-001	52000-061
961DC	961PC	CD2301E	CD2301E/861
DM961N	DTML9961	ECG9961	ITT961-5
ITT961N	LB2102	M5961	M5961P
MC861L	MC861P	MIC961-5D	MIC961-5P
NTE9961	PD9961-59	PE9961-59	SL03912
SN15861J	SN15861N	SW961-2M	SW961-2P
TCG9961	TD1080	TD1080P	

BLOCK-2217

103717	104862	10795-8	110242-003
110242-004	131308	1348A62H01	146642
1471-4372	17-12064-1	1895992-1	198409-12
19A115913-20	202914-030	2068510-0703	221-Z9085
2295361	236-0017	2500390-462	2501557-462
2899001-00	3007475-00	3007475-01	3172632-1
326862	401113-5	423-800176	43A168135-10

BLOCK-2217 (CONT.)

477-0379-001	48-1050-SL12325	50210-8	51310006
51578100	52000-062	55987-1	56571
569-0320-821	569-0544-348	569-0880-462	586-155
669A492H01	68A7349-D62	68A7349PD62	6900K91-004
733W00029	811793	84324	862209-16
9003150	9003150-02	9003150-03	9137-C-1020
928533-101	962DC	962PC	998280-962
C1062P	CD2308E	CD2308E/862	DM962N
DTML9962	ECC-01265	ECG9962	HD2207
HD2207P	HD2215	HD2215P	HEPC1062P
IC34(ELCOM)	ITT962-5	ITT962N	LB2001
M5962	M5962P	MC862L	MC862P
MIC962-5D	MIC962-5P	NTE9962	PD9962-59
PE9962-59	R10256P962	R10256P962B	SL11881
SL16212	SL16520	SL16591	SN15862J
SN15862N	SN962-2M	SW962-2P	TCG9962
TD1066	TD1066P		

BLOCK-2218

14500001-003	179-46445-09	3007359-02	40-065-19-002
52000-063	733W00050	963DC	963PC
CD2309E	CD2309E/863	DM963N	DTML9963
ECG9963	ITT963-5	ITT963N	LB2101
M5963	M5963P	MC863L	MC863P
MIC963-5D	MIC963-5P	NTE9963	PD9963-59
PE9963-59	SL03913	SL16218	SN15863J
SN15863N	SW963-2M	SW963-2P	TCG9963
TD1086	TD1086P		

BLOCK-2219

ECG997	NJM58L05	NTE997	RC4136
RC4136DB	RC4136DC	RV4136DB	RV4136DC
SK9172	TCG997	UA4136	UA4136PC
XR4136CP	YEAMNJM58L05		

BLOCK-2220

4013373-0701	59B402788	84631	84631-1
84631-2	84631-3	84631-4	974HC
ECG9974	MC774G	MC874G	

BLOCK-2221

ECG9976	HEP572	HEPC2004P	MC776P
MC876P	SC9964P		

BLOCK-2222

ECG998	NTE998

BLOCK-2223

ECG9982

BLOCK-2224

692-0020-00	C2005P	ECG9989	HEP573
HEPC2005P	IC-65(ELCOM)	MC789	MC789P
MC889P	NTE9989	SC9965P	TCG9989
V181(BENCO)			

BLOCK-2225

ECG999

BLOCK-2226

ECG9990

BLOCK-2227

AN1431	ECG999M	SK10515	TL431ACP
TL431CP			

BLOCK-2228

ECG999SM	KA431CD	SK10516	TL431ACD
TL431CD			

ES&T Presents TV Troubleshooting & Repair
Electronic Servicing & Technology Magazine

ES&T Presents Computer Troubleshooting & Repair
Electronic Servicing & Technology

TV set servicing has never been easy. The service manager, service technician, and electronics hobbyist need timely, insightful information in order to locate the correct service literature, make a quick diagnosis, obtain the correct replacement components, complete the repair, and get the TV back to the owner.

ES&T Presents TV Troubleshooting & Repair presents information that will make it possible for technicians and electronics hobbyists to service TVs faster, more efficiently, and more economically, thus making it more likely that customers will choose not to discard their faulty products, but to have them restored to service by a trained, competent professional.

Originally published in *Electronic Servicing & Technology*, the chapters in this book are articles written by professional technicians, most of whom service TV sets every day.

ES&T is the nation's most popular magazine for professionals who service consumer electronics equipment. PROMPT® Publications, a rising star in the technical publishing business, is combining its publishing expertise with the experience and knowledge of *ES&T's* best writers to produce a new line of troubleshooting and repair books for the electronics market. Compiled from articles and prefaced by the editor in chief, Nils Conrad Persson, these books provide valuable, hands-on information for anyone interested in electronics and product repair.

Computer Troubleshooting & Repair is the second book in the series and features information on repairing Macintosh computers, a CD-ROM primer, and a color monitor. Also included are hard drive troubleshooting and repair tips, computer diagnostic software, networking basics, preventative maintenance for computers, upgrading, and much more.

Video Technology
226 pages • Paperback • 6 x 9"
ISBN: 0-7906-1086-8 • Sams: 61086
$18.95 ($25.95 Canada) • August 1996

Computer Technology
288 pages • Paperback • 6 x 9"
ISBN: 0-7906-1087-6 • Sams: 61087
$18.95 ($26.50 Canada) • February 1997

CALL 1-800-428-7267 TODAY FOR THE NAME OF YOUR NEAREST PROMPT PUBLICATIONS DISTRIBUTOR

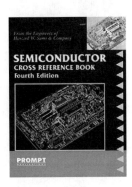

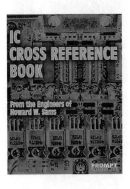

Semiconductor Cross Reference Book Fourth Edition
Howard W. Sams & Company

IC Cross Reference Book Second Edition
Howard W. Sams & Company

This newly revised and updated reference book is the most comprehensive guide to replacement data available for engineers, technicians, and those who work with semiconductors. With more than 490,000 part numbers, type numbers, and other identifying numbers listed, technicians will have no problem locating the replacement or substitution information needed.

There is not another book on the market that can rival the breadth and reliability of information available in the fourth edition of the *Semiconductor Cross Reference Book*.

The engineering staff of Howard W. Sams & Company assembled the *IC Cross Reference Book* to help readers find replacements or substitutions for more than 35,000 ICs and modules. It is an easy-to-use cross reference guide and includes part numbers for the United States, Europe, and the Far East.

This reference book was compiled from manufacturers' data and from the analysis of consumer electronics devices for PHOTOFACT® service data, which has been relied upon since 1946 by service technicians worldwide.

Professional Reference
688 pages • Paperback • 8-1/2 x 11"
ISBN: 0-7906-1080-9 • Sams: 61080
$24.95 ($33.95 Canada) • August 1996

Professional Reference
192 pages • Paperback • 8-1/2 x 11"
ISBN: 0-7906-1096-5 • Sams: 61096
$19.95 ($26.99 Canada) • November 1996

CALL 1-800-428-7267 TODAY FOR THE NAME OF YOUR NEAREST PROMPT PUBLICATIONS DISTRIBUTOR

The Component Identifier and Source Book

Victor Meeldijk

Tube Substitution Handbook

William Smith & Barry Buchanan

Because interface designs are often reverse engineered using component data or block diagrams that list only part numbers, technicians are often forced to search for replacement parts armed only with manufacturer logos and part numbers.

This source book was written to assist technicians and system designers in identifying components from prefixes and logos, as well as find sources for various types of microcircuits and other components. There is not another book on the market that lists as many manufacturers of such diverse electronic components.

The most accurate, up-to-date guide available, the *Tube Substitution Handbook* is useful to antique radio buffs, old car enthusiasts, and collectors of vintage ham radio equipment. In addition, marine operators, microwave repair technicians, and TV and radio technicians will find the *Handbook* to be an invaluable reference tool.

The *Tube Substitution Handbook* is divided into three sections, each preceded by specific instructions. These sections are vacuum tubes, picture tubes, and tube basing diagrams.

Professional Reference
384 pages • Paperback • 8-1/2 x 11"
ISBN: 0-7906-1088-4 • Sams: 61088
$24.95 ($33.95 Canada) • November 1996

Professional Reference
149 pages • Paperback • 6 x 9"
ISBN: 0-7906-1036-1 • Sams: 61036
$16.95 ($22.99 Canada) • March 1995

CALL 1-800-428-7267 TODAY FOR THE NAME OF YOUR NEAREST PROMPT PUBLICATIONS DISTRIBUTOR

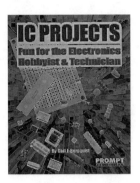

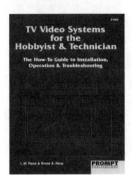

IC Projects: Fun for the Electronics
Hobbyist & Technician
Carl Bergquist

TV Video
Systems
L.W. Pena & Brent A. Pena

IC Projects was written for electronics hobbyists and technicians who are interested in projects for integrated circuits — projects that you will be able to breadboard quickly and test easily. Designed to be fun and user-friendly, all of the projects in this book employ integrated circuits and transistors.

IC projects presents the experienced electronics enthusiast with instructions on how to construct such interesting projects as a LED VU meter, infrared circuit, digital clock, digital stopwatch, digital thermometer, electronic "bug" detector, sound effects generator, laser pointer, and much more!

Knowing which video programming source to choose, and knowing what to do with it once you have it, can seem overwhelming. Covering standard hard-wired cable, large-dish satellite systems, and DSS, *TV Video Systems* explains the different systems, how they are installed, their advantages and disadvantages, and how to troubleshoot problems. This book presents easy-to-understand information and illustrations covering installation instructions, home options, apartment options, detecting and repairing problems, and more. The in-depth chapters guide you through your TV video project to a successful conclusion.

Electronic Projects
256 pages • Paperback • 6 x 9"
ISBN: 0-7906-1116-3 • Sams: 61116
$21.95 • May 1997

Video Technology
124 pages • Paperback • 6 x 9"
ISBN: 0-7906-1082-5 • Sams: 61082
$14.95 ($20.95 Canada) • June 1996

Surface-Mount Technology for PC Boards
James K. Hollomon, Jr.

Digital Electronics
Stephen Kamichik

The race to adopt surface-mount technology, or SMT as it is known, has been described as the latest revolution in electronics. This book is intended for the working engineer or manager, the student or the interested layman, who would like to learn to deal effectively with the many trade-offs required to produce high manufacturing yields, low test costs, and manufacturable designs using SMT. The valuable information presented in *Surface-Mount Technology for PC Boards* includes the benefits and limitations of SMT, SMT and FPT components, manufacturing methods, reliability and quality assurance, and practical applications.

Although the field of digital electronics emerged years ago, there has never been a definitive guide to its theories, principles, and practices — until now. *Digital Electronics* is written as a textbook for a first course in digital electronics, but its applications are varied.

Useful as a guide for independent study, the book also serves as a review for practicing technicians and engineers. And because *Digital Electronics* does not assume prior knowledge of the field, the hobbyist can gain insight about digital electronics.

Some of the topics covered include analog circuits, logic gates, flip-flops, and counters. In addition, a problem set appears at the end of each chapter to test the reader's understanding and comprehension of the materials presented. Detailed instructions are provided so that the readers can build the circuits described in this book to verify their operation.

Professional Reference
510 pages • Paperback • 7 x 10"
ISBN: 0-7906-1060-4 • Sams: 61060
$26.95 ($36.95 Canada) • July 1995

Electronic Theory
150 pages • Paperback • 7-3/8 x 9-1/4"
ISBN: 0-7906-1075-2 • Sams: 61075
$16.95 ($22.99 Canada) • February 1996

CALL 1-800-428-7267 TODAY FOR THE NAME OF YOUR NEAREST PROMPT PUBLICATIONS DISTRIBUTOR

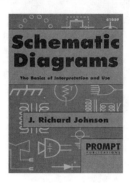

The Microcontroller
Beginner's Handbook
Lawrence A. Duarte

Schematic Diagrams
J. Richard Johnson

Microcontrollers are found everywhere — microwaves, coffee makers, telephones, cars, toys, TVs, washers and dryers. This book will bring information to the reader on how to understand, repair, or design a device incorporating a microcontroller. *The Microcontroller Beginner's Handbook* examines many important elements of microcontroller use, including such industrial considerations as price vs. performance and firmware. A wide variety of third-party development tools is also covered, both hardware and software, with emphasis placed on new project design. This book not only teaches readers with a basic knowledge of electronics how to design microcontroller projects, it greatly enhances the reader's ability to repair such devices. Lawrence A. Duarte is an electrical engineer for Display Devices, Inc. In this capacity, and as a consultant for other companies in the Denver area, he designs microcontroller applications.

Step by step, *Schematic Diagrams* shows the reader how to recognize schematic symbols and determine their uses and functions in diagrams. Readers will also learn how to design, maintain, and repair electronic equipment as this book takes them logically through the fundamentals of schematic diagrams. Subjects covered include component symbols and diagram formation, functional sequence and block diagrams, power supplies, audio system diagrams, interpreting television receiver diagrams, and computer diagrams. *Schematic Diagrams* is an invaluable instructional tool for students and hobbyists, and an excellent guide for technicians.

Electronic Theory
240 pages • Paperback • 7-3/8 x 9-1/4"
ISBN: 0-7906-1083-3 • Sams: 61083
$18.95 ($25.95 Canada) • July 1996

Electronic Theory
196 pages • Paperback • 6 x 9"
ISBN: 0-7906-1059-0 • Sams: 61059
$16.95 ($22.99 Canada) • October 1994

CALL 1-800-428-7267 TODAY FOR THE NAME OF
YOUR NEAREST PROMPT PUBLICATIONS DISTRIBUTOR

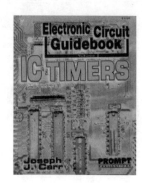

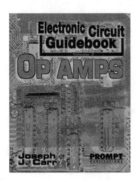

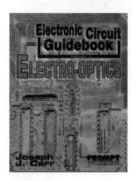

Electronic Circuit Guidebook
Volume 3: Op Amps
Joseph J. Carr

The operational amplifier is the most commonly used linear IC amplifier in the world. The range of applications for the op amp is truly awesome – it has become a mainstay of audio, communications, TV, broadcasting, instrumentation, control, and measurement circuits. Third in a series covering electronic instrumentation and circuitry, *Electronic Circuit Guidebook, Volume 3: Op Amps* is design to give you some insight into how practical linear IC amplifiers work in actual real-life circuits. Because of their widespread popularity, operational amplifiers figure heavily in this book, though other types of amplifiers are not overlooked. This book allows you to design and configure your own circuits, and is intended to be a practical workshop aid. Some of the topics covered in detail include linear IC amplifiers, ideal operational amplifiers, instrumentation amplifiers, isolation amplifiers, active analog filter circuits, waveform generators, and many more.

Electronics Technology
273 pages • Paperback • 7-3/8 x 9-1/4"
ISBN: 0-7906-1131-7 • Sams: 61131
$24.95 • August 1997

Electronic Circuit Guidebook
Volume 4: Electro-Optics
Joseph J. Carr

Electronic Circuit Guidebook, Volume 4: Electro-Optics is mostly about E-O sensors — those electronic transducers that convert light waves into a proportional voltage, current, or resistance. The coverage of the sensors is wide enough to allow you to understand the physics behind the theory of operation of the device, and also the circuits used to make these sensors into useful devices. This book examines the photoelectric effect, photoconductivity, photovoltaics, and PN junction photodiodes and phototransistors. Also examined is the operation of lenses, mirrors, prisms, and other optical elements keyed to light physics.

Electronic Circuit Guidebook, Volume 4: Electro-Optics is intended to teach the physics and operation of E-O devices, then proceed to circuits and methods for actual application of the devices in real situations.

Electronics Technology
416 pages • Paperback • 7-3/8 x 9-1/4"
ISBN: 0-7906-1132-5 • Sams: 61132
$29.95 • October 1997

CALL 1-800-428-7267 TODAY FOR THE NAME OF
YOUR NEAREST PROMPT PUBLICATIONS DISTRIBUTOR

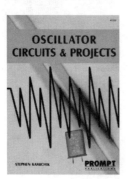

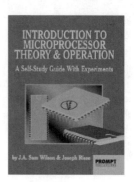

Oscillator Circuits & Projects
Stephen Kamichik

Introduction to Microprocessor Theory & Operation
J.A. Sam Wilson & Joseph Risse

Oscillator circuits usually are not taught at technology and engineering schools and universities. Electronics is a vast field; therefore, some areas of expertise, including oscillators, must be acquired in the field. *Oscillator Circuits & Projects* helps to make this process easier by presenting the information you need to master oscillator circuitry.

Oscillator Circuit & Projects was written as a textbook and project book for individuals who need to know more about oscillator circuits. Students, technicians, and electronics hobbyists can build and enjoy the informative and entertaining projects described at the end of this book. Compete information about oscillator circuits is presented in an easy-to-follow manner with many illustrations to help guide you through the theory stage to the hands-on stage.

This book takes readers into the heart of computerized equipment and reveals how microprocessors work. By covering digital circuits in addition to microprocessors and providing self-tests and experiments, *Introduction to Microprocessor Theory & Operation* makes it easy to learn microprocessor systems. The text is fully illustrated with circuits, specifications, and pinouts to guide beginners through the ins-and-outs of microprocessors, as well as provide experienced technicians with a valuable reference and refresher tool.

Electronic Projects
256 pages • Paperback • 7-3/8 x 9-1/4"
ISBN: 0-7906-1111-2 • Sams: 61111
$19.95 • April 1997

Electronic Theory
211 pages • Paperback • 6 x 9"
ISBN: 0-7906-1064-7 • Sams: 61064
$16.95 ($22.99 Canada) • February 1995

CALL 1-800-428-7267 TODAY FOR THE NAME OF YOUR NEAREST PROMPT PUBLICATIONS DISTRIBUTOR

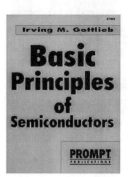

Semiconductor Essentials
Stephen Kamichik

Basic Principles of Semiconductors
Irving M. Gottlieb

Readers will gain hands-on knowledge of semiconductor diodes and transistors with help from the information in this book. *Semiconductor Essentials* is a first course in electronics at the technical and engineering levels. Each chapter is a lesson in electronics, with problems included to test understanding of the material presented. This generously illustrated manual is a useful instructional tool for the student and hobbyist, as well as a practical review for professional technicians and engineers. The comprehensive coverage includes semiconductor chemistry, rectifier diodes, zener diodes, transistor biasing, and more.

Despite their ever-growing prominence in the electronics industry, semiconductors are still plagued by a stigma which defines them merely as poor conductors. This narrow-sighted view fails to take into account the fact that semiconductors are truly unique alloys whose conductivity is enhanced tenfold by the addition of even the smallest amount of light, voltage, heat, or certain other substances. *Basic Principles of Semiconductors* explores the world of semiconductors, beginning with an introduction to atomic physics before moving onto the structure, theory, applications, and future of these still-evolving alloys. Such a theme makes this book useful to a wide spectrum of practitioners, from the hobbyist and student, right up to the technician and the professional electrician.

Electronic Theory
112 pages • Paperback • 6 x 9"
ISBN: 0-7906-1071-X • Sams: 61071
$16.95 ($22.99 Canada) • September 1995

Electronic Theory
158 pages • Paperback • 6 x 9"
ISBN: 0-7906-1066-3 • Sams: 61066
$14.95 ($20.95 Canada) • April 1995

CALL 1-800-428-7267 TODAY FOR THE NAME OF
YOUR NEAREST PROMPT PUBLICATIONS DISTRIBUTOR

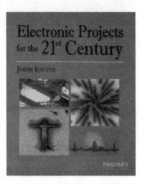

Basic Digital Electronics
Alvis J. Evans

Explains digital system functions and how digital circuits are used to build them! Digital — what does it mean? Why is it that electronic systems are being designed using digital electronic circuits? Find the answer to these questions and more, as you learn the difference between analog and digital systems, the functions required to design digital systems, the circuits used to make decisions, code conversions, data selections, adding and subtracting, interfacing and storage, and the circuits that keep all operations in time and under control.

Learn about logic circuits, flip-flops, registers, multivibrators, counters, 3-state bus drivers, bidirectional line drivers and receivers, and more using easy-to-read, easy-to-understand explanations coupled with detailed illustrations.

Electronic Projects for the 21st Century
John Iovine

If you are an electronics hobbyist with an interest in science, or are fascinated by the technologies of the future, you'll find *Electronic Projects for the 21st Century* a welcome addition to your electronics library. It's filled with nearly two dozen fun and useful electronics projects designed to let you use and experiment with the latest innovations in science and technology — innovations that will carry you and other electronics enthusiasts well into the 21st century!

Electronic Projects for the 21st Century contains the expert, hands-on guidance and detailed instructions you need to perform experiments that involve genetics, lasers, holography, Kirlian photography, and more. Among the projects are a lie detector, an ELF monitor, air pollution monitor, pinhole camera, laser power supply for holography, synthetic fuel, and an expansion cloud chamber.

Electronic Theory
192 pages • Paperback • 8-1/2 x 11"
ISBN: 0-7906-1118-X • Sams: 61118
$19.95 • April 1997

Electronic Projects
256 pages • Paperback • 7-3/8 x 9-1/4"
ISBN: 0-7906-1103-1 • Sams: 61103
$19.95 • June 1997

CALL 1-800-428-7267 TODAY FOR THE NAME OF YOUR NEAREST PROMPT PUBLICATIONS DISTRIBUTOR

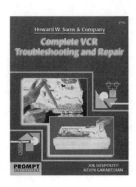

Howard W. Sams Complete VCR Troubleshooting & Repair
Joe Desposito & Kevin Garabedian

Complete VCR Troubleshooting and Repair contains sound VCR troubleshooting procedures beginning with an examination of the external parts of the VCR, then narrowing the view to gears, springs, pulleys, belts, and other mechanical parts. This book also shows how to troubleshoot tuner/demodulator circuits, audio and video circuits, display controls, servo systems, video heads, TV/VCR combination models, and more.

This book also contains nine VCR case studies, each focusing on a particular model of VCR with a specific problem. The case studies guide you through the repair from start to finish, using written instruction, helpful photographs, and Howard W. Sams' own *VCRfacts®* schematics.

Howard W. Sams Computer Monitor Troubleshooting & Repair
Joe Desposito & Kevin Garabedian

Computer Monitor Troubleshooting & Repair makes it easier for any technician, hobbyist or computer owner to successfully repair dysfunctional monitors. Learn the basics of computer monitors with chapters on tools and test equipment, monitor types, special procedures, how to find a problem and how to repair faults in the CRT. Other chapters show how to troubleshoot circuits such as power supply, high voltage, vertical, sync and video.

This book also contains six case studies which focus on a specific model of computer monitor. Using carefully written instructions and helpful photographs, the case studies guide you through the repair of a particular problem from start to finish. The problems addressed include a completely dead monitor, dysfunctional horizontal width control, bad resistors, dim display and more.

Video Technology
184 pages • Paperback • 8-1/2 x 11"
ISBN: 0-7906-1102-3 • Sams: 61102
$29.95 • March 1997

Troubleshooting & Repair
308 pages • Paperback • 8-1/2 x 11"
ISBN: 0-7906-1100-7 • Sams: 61100
$29.95 • July 1997

CALL 1-800-428-7267 TODAY FOR THE NAME OF YOUR NEAREST PROMPT PUBLICATIONS DISTRIBUTOR

Home Security Projects
Robert Gaffigan

Security Systems for Your Home & Automobile
by Gordon McComb

Home Security Projects presents the reader with many projects about home security, safety and nuisance elimination that can easily be built in the reader's own home for less than it would cost to buy these items ready-made. Readers will be able to construct devices that will allow them to protect family members and electrical appliances from mishaps and accidents in the home, and protect their homes and belongings from theft and vandalism.

This book shows the reader how to construct the many useful projects, including a portable CO detector, trailer hitch alignment device, antenna saver, pool alarm, dog bark inhibitor, and an early warning alarm system. These projects are relatively easy to make and the intent of *Home Security Projects* is to provide enough information to allow you to customize them.

Security Systems is about making homes safer places to live and protecting cars from vandals and thieves. It is not only a buyer's guide to help readers select the right kind of alarm system for their home and auto, it also shows them how to install the various components. Learning to design, install, and use alarm systems saves a great deal of money, but it also allows people to learn the ins and outs of the system so that it can be used more effectively. This book is divided into eight chapters, including home security basics, warning devices, sensors, control units, remote paging automotive systems, and case histories.

Gordon McComb has written over 35 books and 1,000 magazine articles which have appeared in such publications as *Omni* and *PC World*. In addition, he is the coauthor of PROMPT® Publication's *Speakers for Your Home and Auto*.

Projects
256 pages • Paperback • 6 x 9"
ISBN: 0-7906-1113-9 • Sams: 61113
$19.95 • September 1997

Projects
130 pages • Paperback • 6 x 9"
ISBN: 0-7906-1054-X • Sams: 61054
$16.95 • July 1994